ISNM
International Series of Numerical Mathematics
Vol. 148

Domain Decomposition Methods in Optimal Control of Partial Differential Equations

John E. Lagnese
Günter Leugering

Springer Basel AG

Authors:

John E. Lagnese
Department of Mathematics
Georgetown University
Washington, DC 20057
USA
e-mail: jack@lagnese.us

Günter Leugering
Angewandte Mathematik II
Universität Erlangen-Nürnberg
91058 Erlangen
Germany
e-mail: leugering@am.uni-erlangen.de

2000 Mathematics Subject Classification 49J20, 90C20, 49K20, 49M27, 65K10, 65M55, 34K10, 35L10, 35J25, 37N35

A CIP catalogue record for this book is available from the Library of Congress, Washington D.C., USA

Bibliographic information published by Die Deutsche Bibliothek
Die Deutsche Bibliothek lists this publication in the Deutsche Nationalbibliografie; detailed bibliographic data is available in the Internet at <http://dnb.ddb.de>.

ISBN 978-3-0348-9610-8 ISBN 978-3-0348-7885-2 (eBook)
DOI 10.1007/978-3-0348-7885-2

The cover pictures show a set of three membranes coupled along a flexible joint. The first picture shows the initial configuration while the second picture provides a snapshot of the corresponding vibrations after some seconds. The second picture amply demonstrates the transmission of vibrations across the joint. Transverse motion is thereby converted to in-plane motion and vice versa. The results and methods in this monograph concentrate on the decomposition in space and time of such and much more complex dynamics subject to boundary and distributed controls.

Originally published by Birkhäuser Verlag in 2004
Softcover reprint of the hardcover 1st edition 2004

Printed on acid-free paper produced of chlorine-free pulp. TCF ∞

ISBN 978-3-0348-9610-8 www.birkhauser-science.com

Contents

Preface

This monograph considers problems of optimal control for partial differential equations of elliptic and, more importantly, of hyperbolic types on networked domains. The main goal is to describe, develop and analyze iterative space and time domain decompositions of such problems on the infinite-dimensional level.

While domain decomposition methods have a long history dating back well over one hundred years, it is only during the last decade that they have become a major tool in numerical analysis of partial differential equations. A keyword in this context is parallelism. This development is perhaps best illustrated by the fact that we just encountered the 15th annual conference precisely on this topic. Without attempting to provide a complete list of introductory references let us just mention the monograph by Quarteroni and Valli [91] as a general up-to-date reference on domain decomposition methods for partial differential equations.

The emphasis of this monograph is to put domain decomposition methods in the context of so-called virtual optimal control problems and, more importantly, to treat optimal control problems for partial differential equations and their decompositions by an all-at-once approach. This means that we are mainly interested in decomposition techniques which can be interpreted as virtual optimal control problems and which, together with the real control problem coming from an underlying application, lead to a sequence of individual optimal control problems on the subdomains that are iteratively decoupled across the interfaces. This task can be accomplished in a very general way that applies to elliptic, parabolic, and hyperbolic or Petrowski-type systems.

While this is already a formidable task, we also develop a general theory for systems of partial differential equations on multi-linked or networked domains; see the introductory chapter for extensive motivation. It is obvious that from a physical point of view, any domain decomposition of a multi-link system should respect the structural elements as individuals coupled at multiple interfaces. Thus it is the decoupling of multiple interface conditions or, in other words, a nonoverlapping domain decomposition procedure, that one is seeking. The engineering terminology for this approach is 'substructuring'.

As the overall requirement of decomposing, optimizing and simulating systems described by partial differential equations on networked domains is beyond the scope of any in-depth treatment in a single book, in this monograph we have concentrated on the infinite-dimensional level. However, the philosophy outlined above, and made more transparent in the introductory chapter, recommends this selection. Indeed, once the decomposition into local optimal control problems is established and the convergence of the corresponding scheme guaranteed on the infinite-dimensional level, the implementation in parallel of numerical techniques for standard optimal control problems on standard domains is a separate and independent task. The infinite-dimensional algorithmic procedures can thus be seen as a general framework which is completely independent of any discretization scheme. We therefore develop not only the convergence of those schemes in an infinite-dimensional framework but also prove a posteriori error estimates with respect to the matching conditions along multiple interfaces. After local discretization, and after incorporating discrete a posteriori error estimates with respect to finite-dimensional approximations of the local equations into the former ones, the basis is laid for a parallel and adaptive numerical treatment of optimal control problems involving partial differential equations on networked domains. Moreover, the iterative domain decomposition techniques may be used as preconditioners of global-local type.

In order to stay within a common theoretical framework, in this work we consider optimal control problems for elliptic and hyperbolic problems only. Elliptic problems, apart from their natural physical significance, are also more accessible at the theoretical level and are used for introductory expositions. Moreover, after a semi-discretization in time, hyperbolic systems may be viewed as sequences of elliptic problems. We further limit our considerations to linear problems, since a nonlinear theory of domain decompositions of optimal control problems is not available on a general level at this time. In addition, for the sake of clarity and convenience, we consider mainly control problems without control or state constraints. Let us remark, however, that at the expense of some additional technical complexities, one may develop domain decomposition algorithms for all of the examples considered in this book when control constraints are present. Moreover, in Chapter 8 we exhibit a domain decomposition procedure for a state constrained optimal control problem arising from a problem of exact controllability.

Another selection we have made is to consider, in the case of hyperbolic systems, only final value optimal control problems, that is, problems in which the goal is optimal steering of a trajectory from a given initial state to a neighborhood of a specified target state at a specified time. This choice is motivated not only by important physical and engineering applications, but also by the fact that domain decomposition methods for such problems are technically more challenging and far less studied than, for example, domain decomposition methods for trajectory tracking problems, which have been considered extensively in the literature. For exactly the same reasons, we generally restrict our attention to boundary controls as opposed to distributed, or locally distributed, controls; the latter are easily handled by the same theoretical framework that we develop for boundary control problems, and with fewer technical complexities.

Although, as stated, we do not concern ourselves with the numerical implementation of the algorithms that we develop at the infinite-dimensional level, some selected numerical tests are displayed in order to illustrate the efficacy of the methods.

Outline of the Book

In Chapter 1 we give a somewhat detailed introduction to the field of domain decomposition, the state of the art and the scope of this monograph.

In Chapter 2 we give an introduction to some basic ideas typically used in the field of domain decomposition. The material presented is highly selective and is not intended to give a survey of all general concepts of domain decomposition. Rather we have selected those methods and ideas which admit an interpretation as optimal control problems in a wide sense.

We will deal almost exclusively with problems on an infinite-dimensional level. The first part of Chapter 2 is devoted to dynamical problems on one-dimensional domains. We discuss the wave equation and the diffusion-advection equation in an unbounded domain as exemplary models, and describe the most basic decomposition method going back to Schwarz [95] for elliptic problems with overlapping domains. Even on this simple level, the analysis clearly reveals basic features (see [27] and also [24], [25], [26]). For instance, for the wave equation one may develop optimal interface conditions such that the iterative domain decomposition procedure terminates after a few steps. Those interface conditions correspond to transparent boundary conditions which admit complete transmission of data across the interfaces, with no reflections taking place. Typically, however, this desirable property cannot be realized via local-in-time boundary operators across the interface. Similarly, optimal but nonlocal interface operators may be constructed for the diffusion-advection equation such that the iterative domain decomposition procedure terminates after two steps. Also discussed are finite termination and convergence properties for overlapping and nonoverlapping decompositions of bounded one-dimensional domains. It turns out that for the one-dimensional wave equation with transparent boundary conditions at the outer boundary, it is possible to achieve finite termination with local operators. As is to be expected, such is not the case for diffusion-advection equations. A consequence of this discussion is that a general concept of how to construct domain decomposition procedures for nonoverlapping domains begins to appear.

In Section 2.3 some basic concepts for elliptic domain decomposition methods are reviewed. The reader is referred to [91] for an excellent textbook on this material, also with respect to numerical realizations. The classical Dirichlet-Dirichlet method and a Neumann-Neumann method are considered. In both cases the desired transmission condition across the interface can be rewritten as a Steklov-Poincaré type equation. These variants of the Steklov-Poincaré equation may also be interpreted as a controllability constraints. This point of view is not typical in the domain decomposition literature but has been promoted by J.-L. Lions and by the authors of this monograph. Once this interpretation is accepted, many methods which have been developed in the area of optimal control of partial differential equations can be applied.

Subsections 2.3.3–2.3.7 provide a detailed analysis of the most basic algorithm given by P.-L. Lions. The approach that we follow there will be repeated frequently in this book: we first describe the method, then give an easy to access, two domain proof of convergence for the unrelaxed iterations. We then generalize to multi-domain splittings and present interpretations as standard iterative schemes, such as a Richardson iteration, and give a posteriori error estimates. A more explicit treatment of multi-link serial decompositions is then given and the iteration discussed in the context of preconditioned iterations.

Chapter 3 is devoted to partial differential equations on graphs. The concept of linear differential operators of a general vector-valued Sturm-Liouville type on graphs is introduced, after which domain decomposition procedures, which are motivated by the methods and ideas of Chapter 2, are derived. Convergence of the iterations, which are again seen to be canonical Richardson-type iterations, is then established. The last section of Chapter 3 is devoted to hyperbolic problems on graphs, corresponding domain decomposition methods, and their convergence properties.

While Chapters 2 and 3 focus on domain decomposition methods without 'real' controls, Chapter 4 initiates the discussion of domain decomposition methods for optimal control problems for partial differential equations. In order to fix the ideas, the relatively simple, but nonetheless fundamental, elliptic case is treated. First distributed controls, and then boundary controls, are considered. It is shown that the distributed case can be handled by direct application of the methods developed for elliptic problems without control. However, this simple connection ceases in the case of boundary controls. Nevertheless, the basic methods can be extended to this situation as well. After a discussion of the convergence of the iterates we provide, for the first time, a posteriori error estimates for the procedure of domain decomposition of an optimal control problem. The derivation is strongly motivated by the work of Otto and Lube [90].

Chapter 5 is devoted to domain decomposition of optimal control problems for partial differential equations on graphs. Once again we start with the basic elliptic case and discuss distributed as well as boundary control problems. Domain decomposition procedures are formulated and convergence of the corresponding iterative schemes is established. We then proceed to develop and analyze a domain decomposition procedure for optimal final value control problems for linear hyperbolic equations on graphs.

Chapters 6 through 9 are devoted to the study of domain decomposition of optimal control problems for (primarily) hyperbolic partial differential equations in space dimension greater than one. This enterprise starts with Chapter 6, which deals with the domain decomposition of optimal final value control problems for general wave equations with absorbing (dissipative) boundary conditions, this choice being motivated by the spatial domain decompositions in Chapter 2. First of all the model problem is introduced and well-posedness questions are considered in detail. In particular, time and space regularity results are developed. The section on well-posedness is followed by a description of a general time domain decomposition procedure which is reminiscent of those devised by J.-L. Lions et al. [75]. The method is also motivated by the spatial domain decomposition method of

Chapters 2 to 5. The time domain decomposition procedure is formulated, and its convergence is proved. Then pointwise-in-time and uniform a posteriori error estimates are developed. Following that analysis, decomposition of the spatial domain is considered. Once again, convergence of the iterative procedure, and a posteriori error estimates, are provided. The last section of Chapter 6 concentrates then on the possible combinations of time and space domain decomposition procedures.

Chapter 7 is devoted to domain decomposition of optimal final value control problems for the dynamic Maxwell system with a dissipative boundary condition known as the Silver-Müller condition, which is in fact the first-order local approximation to a transparent boundary condition. The flow of the presentation is the same as in Chapter 6 and results analogous to those established there are obtained.

In Chapter 8 we reconsider general wave equations, but this time with a conservative boundary condition. The lack of dissipativity in the boundary condition causes a loss of regularity so that, in particular, when considering optimal final value control the deviation of the final state from the target state must be penalized in a weaker norm than previously. This leads to an optimality system that differs from that occurring when dissipativity is present. While this has little effect on the analysis of our time domain decomposition algorithm, it necessitates significant changes in the spatial domain decomposition algorithm and its analysis. Once again, appropriate convergence results, and a posteriori error estimates, are established.

In addition, in Chapter 8 is presented a spatial domain decomposition procedure for the optimality system associated with the problem of *exact controllability*, that is, the problem of minimum norm control of a trajectory from a given initial state to a prescribed target state. It is shown, in particular, that an appropriate domain decomposition for this optimality system is obtained in the limit, as the penalty parameter goes to infinity, of the domain decomposition algorithm associated with optimal final value control to the target state, discussed earlier in the chapter.

Chapter 9 is concerned with domain decomposition of elliptic equations (without control), and hyperbolic equations (with control), on 2-dimensional polygonal networks in n-dimensional space. The discussion parallels that of Chapters 3 and 5 dealing with such systems on graphs, but with an unavoidable increase in technical complexity.

Acknowledgment. The authors gratefully acknowledge the advice of K.-H. Hoffmann, the editor of this series, and the technical support of its publisher, Birkhäuser-Verlag. We also wish to express our gratitude to Mrs. Garhammer for her professional typing of a significant part of the material, and to Dr. R. Hundhammer, who performed many of the numerical simulations displayed in this book. Research of the first author was partially funded by the National Science Foundation through grant DMS-9972034. The second Author was partially funded by the Deutsche Forschungsgemeinschaft through grant DFG-Le 595/9-1(2).

Chapter 1

Introduction

In this monograph we consider problems of optimal control for partial differential equations of elliptic and, more importantly, of hyperbolic type on networked domains. The main goal is to describe, develop and analyze iterative space and time domain decompositions of such problems on the infinite-dimensional level.

In order to better understand this enterprise, we first of all describe some typical examples of engineering applications which genuinely lead to a mathematical modeling involving partial differential equations of various types on complex domains which, in turn, can often be directly identified as, or more generally, can be interpreted as networks.

We may consider flexible structures consisting of strings, cables, shells, beams and plates that are coupled at mechanical joints. These structures can be frames or skeletons, such as in buildings, bridges, off shore structures etc., or they can be mechanisms which admit not only local but also global motions. Most of mechanical structures are mixtures of the two types as in flexible robots, space stations and satellites, antenna systems, aircraft, cars, ships, etc.

Typically one is faced with requirements such as stability and operability. In the first case one has to react by some controls, which may be physically realized in various ways (Piezo-elements, magneto-restrictive devices, semi-active elements and many others) to measurements taken by various sensors in order to remove unwanted perturbations from the system. This task is usually achieved according to some feedback policy. In the second case, one seeks optimal performance, e.g., tracking of a desired system behavior. This requirement is typically dealt with by an open-loop strategy. In a real environment, in particular under real-time requirements, open-loop and closed-loop control strategies are mixed. The unwanted behavior is viewed as a perturbation of the entire system and may be periodic or of a general global nature, for instance the periodic change in the temperature of the ISS-station while orbiting the earth, or wind loads, fluid structure interactions, etc. in other contexts. On the other hand, the behavior may be local in space and/or time, such as occurs in a docking maneuver or from an impact by some space material. In both cases the impacts will ultimately spread over the entire structure and may cause it to enter an unoperational state.

There is a long history of problems of this type where a dramatic failure was a result of not properly modeling the system. As a consequence, systems may fail to properly react to controls because of lack of resolution. Optimal control problems for such flexible structures have been discussed in the engineering literature for more than 30 years. But is only during the last 15 years that the fully continuum mechanical structural elements, represented mathematically as partial differential equations, have been taken into account on a systematic basis. This development came along with the tendency to reduce the weight of the structural constitutives of such complex systems. Thinner and lighter structures exhibit many more and particularly large modes of oscillations. Thus PDE modeling and optimal control of PDE modeled connected systems appeared as a major challenge of the 1990's. See the monograph of the authors together with G. Schmidt [61] for an account of the state of affairs until 1994.

Once the modeling of flexible multi-link structures is understood, many other applications which are extremely important in civil engineering, ecology and medicine can be seen to immediately fall into the same general framework, namely, partial differential equations on networked domains. In order to illustrate the scope of this theory, let us mention in particular the problems of flood management and the control of sewer or irrigation systems. These are systems of hundreds or thousands of open channels or closed pipes that constitute a network carrying the flow of water. The flow is modeled by the shallow-water equations with source terms due to friction. The equations are systems of hyperbolic conservation laws with nonlinear driving forces, that are sometimes subject to diffusion. The controls consist of valves, gates, weirs and pumps located at various nodes of the network. The control objective is to operate those instruments according to physical, legal and economic constraints such that certain flow requirements are met in specified links. In flood management, for example, one has to avoid overflows in cities and purification plants; in irrigational systems one has to supply a specified amount of water, etc. A very similar problem occurs in the management of gas pipeline networks, traffic flow problems on road networks, neuronal networks, the blood vessel system and root dynamics, just to mention a few more challenging applications. Problems of this type have been considered in the engineering literature for a long time. However, as resources become more and more expensive, and as the open global economy allows for an increasing number of different providers operating on the same networks, the dynamics also on the local level becomes more and more important. As a result, applied mathematics is left with a grand challenge: to make a difference.

The point to be made with respect to this book is that most of these applications are driven by hyperbolic equations on network-like structures. Indeed, when operating a multi-link flexible mechatronic system, or a water or gas network near its actual working conditions, linear hyperbolic equations on networked domains arise from linearization around trajectories. Such systems, in turn, are the objects discussed in this book.

For ongoing research on nonlinear hyperbolic networks see Leugering and Schmidt [71], Coron et al. [13], Gugat and Leugering [33], Gugat, Leugering and Schmidt [34] and Gugat [31]. These papers are concerned with water flow in networks and corresponding optimal control problems. The work by Gugat, Herty, Klar and Leugering [32] is devoted to control problems concerning traffic flow on road networks.

Once the mathematical analysis had been made available, the demand for numerical simulation of such uncontrolled and controlled systems became more dominant. Obviously flexible structures, general infrastructures or complex biological systems with several hundred or even thousand structural elements cannot be numerically simulated on a very fine spatial and time-wise scale as the dimensions easily approach any technical limitation. Therefore model hierarchies and model reduction and, in particular, decomposition methods have been the main focus of research in recent years. In particular, real-time requirements for the controls strongly enhanced this development. Consequently, domain decomposition methods, going back to the work of Schwarz [95] at the end of the nineteenth century, have been reintroduced and further developed. We are currently approaching the 15th annual conference on *Domain decomposition methods for partial differential equations*, and the vast number of individual contributions precludes any attempt to give proper credit. We just mention the recent monograph by Quarteroni and Valli [91] as a general up-to-date reference on domain decomposition methods for partial differential equations.

It is obvious from the physical point of view that such decompositions should respect the structural elements as individuals coupled at multiple joints. Thus it is the decoupling of multiple joint conditions or, in other words, a nonoverlapping domain decomposition procedure one is seeking. The engineering terminology for this approach is 'substructuring'.

Even though decomposing a complex structure, by whatever means, into simple, more standard pieces is itself a formidable task, the challenge increases when it comes to the problem of decomposing optimal control problems for such structures. The ultimate goal is to break the globally defined optimal control problem into a collection of smaller, standard, optimal control problems on substructures for which standard software is available.

Typically in continuous control systems, optimal controls are determined via optimality conditions which, in turn, can be expressed as coupled forward and backward running equations, still defined on the entire structure. As parallelism is the governing criterion, a decoupling of both the forward and the backward running problems separately does not lead to a complete and parallelizable decoupling. Therefore, a decoupling of the entire optimality system appears to be necessary. However, if one decouples an optimality system, the result is not necessarily an optimality system by itself. In other words, such a decoupling does not necessarily lead to decoupled optimal control problems on the substructures.

Therefore it is natural to investigate methods that achieve these simultaneous requirements, that is, the substructuring of the system and decoupling of the optimization problem. The design and analysis of such methods is the goal of this monograph. The challenge as a whole is prohibitive with respect to a comprehensive presentation in the given space and time frame. Moreover, the theory is still far from being complete.

Much is known for elliptic problems on simple domains which admit a natural decomposition. Benamou and Desprès [10] and Benamou [6], [7], [8], [9] where probably the first to consider this kind of procedure. They employed a simple but powerful nonoverlapping adaption, due to P.-L. Lions [82], of the classical Schwarz overlapping domain decomposition procedure. In the same proceedings, Glowinski and Le Tallec [29] observed that Lions' algorithm could be interpreted as a saddle-point iteration of the Uzawa-type corresponding to an augmented Lagrangian formulation of the problem. By this interpretation a domain decomposition procedure, without any underlying optimal control requirement, falls naturally into the framework of optimization. Starting from an elliptic optimal control problem Benamou and Desprès (loc. cit.) observed that the corresponding optimality system could be reformulated as a complex "elliptic" system with the real part of the solution being the direct solution and the imaginary part corresponding to the adjoint solution. Once this observation was made, P.-L. Lions' method for domain decomposition could be applied. Even though this observation holds only in the very special case of distributed controls, it has triggered much of the research in the directions outlined above.

However, the methods which evolved from this starting point do not apply to multiple nodes or junctions between more than two subdomains. Moreover, in the dynamic case the results do not cover the important final value control problem which genuinely appears for relaxed controllability constraints.

The extension of nonoverlapping domain decomposition methods to systems of partial differential equations of hyperbolic type took place only in recent years. See papers of the authors of this monograph and their collaborators [64], [66], [65], [67], [68], [69], [57], [52], [53], [54], [55], [58], [59], [60].

Parallel to this development, J.-L. Lions and O. Pironneau [76], [77], [79], [81], [80] and J.-L. Lions [1] (see also [14], [42], [78]) reconsidered the problem of overlapping and nonoverlapping domain decomposition from the general point of view of optimization. They realized that the interpretation of the P.-L. Lions' method as the result of an optimization procedure could be generalized to other domain decomposition techniques such that a general concept of virtual controls, that is, controls that are responsible for the realization of transmission conditions between the subdomains, could be formulated. For Robin type boundary conditions as transmission operators across the interfaces, which are typical of the P.-L. Lions-type procedures, this observation had been made already by Benamou [8] and, for networks, by Leugering [68]. The concept of virtual controls, in turn, triggered a more recent development in the field based on the observation that general domain decomposition problems, nonoverlapping as well as overlapping ones, can be formulated as optimal control problems for partial differential equations. See also Gunzburger et al. [36], [38], [19], [37], and Gunzburger and Heinkenschloss [35] for a penalty method. Therefore, methods that have been developed with respect to optimal control of partial differential equations can now be applied to decomposition problems. On the other hand, as soon as domain decomposition procedures are interpreted as virtual optimal control problems, they are genuinely amenable to optimal control methods for partial differential equations with real controls, that is, with controls entering via the physical model. This is because, as stated above, the decomposition procedure and the optimization involving the physical process can be handled simultaneously. The controls that undergo optimization during the entire process have just a different interpretation. In this sense we can speak of an all-at-once approach.

The more classical procedures would either tend to first decompose the system and then optimize, or to first optimize on the global system and then decompose. In both cases the decomposition would naturally take place on the discrete level.

Decomposition of domains thus adds yet another aspect to the more classical controversy that is still actively discussed, namely, whether one should first discretize and then optimize or the other way round. The current tendency, even though in particular cases one strategy can be seen to be superior to the other, is to use an all-at-once approach in which discretization, decomposition and optimization are intrinsically coupled. There is currently little theoretical basis for the numerical success that can be observed in various applications. We hope that this book will be helpful in remedying that situation.

While on the discrete level we are faced with an abundance of methods and their corresponding analyses of convergence, the infinite-dimensional case is far less understood. See however Quarteroni and Valli [91] and the references therein. When it comes to the analysis of domain decomposition methods for optimal control of partial differential equations the literature becomes very sparse. See the papers [16], [7], [10], [65], [66], [64], [67], [53], [54], [55], [69], [68], [57], [78], [79], [81], [1], [14] and the references therein.

As domain decomposition procedures of nonoverlapping type originate from a decomposition of transmission conditions across the interfaces between adjacent subdomains, the matching of such transmission conditions can be expressed as certain boundary operator equalities involving the so-called Steklov-Poincaré operators. Virtual controls are then used to achieve exact matching. This is equivalent to the classical notion of exact controllability. On the discrete level the corresponding Schur-complement equation is analyzed. Whenever exact matching is relaxed, the virtual control problem turns into a problem of approximate controllability or into a quadratic optimal control problem. This approach again finds its analog on the discrete level in so-called inexact methods. After solving subproblems on the local level of the subdomains, a global interface problem is established and solved either by direct solvers or in an iterative way. While for a long time iterative methods have been favored over noniterative ones, the recent work of Hackbusch [40] on elliptic problems provides new alternatives. However, the all-at-once approach which we consider is genuinely iterative in nature, having natural discrete analogues.

In this monograph we consider such methods on the infinite-dimensional level. Because of space limitations, we have elected to not include convergence results on the discrete level and refer instead to forthcoming publications. Obviously, the infinite-dimensional analysis provides the basis for the finite-dimensional analysis, as the proper space setup for the latter is governed by the infinite-dimensional problem.

In order to make the focus even narrower, we consider mainly (1) linear, second-order, problems of elliptic or hyperbolic types on 1-dimensional and 2-dimensional networks; (2) transmission problems for second-order hyperbolic equations in spaces of any dimension; and (3) transmission problems for the time-dependent Maxwell system in space dimension 3. In general, we do not discuss in this work systems of order greater than 2, for example, fourth-order problems that occur in modeling mechanical structures involving elements such as elastic beams and plates. An exception, however, is found in Section 6.3.4 where a time domain decomposition procedure is discussed for the dynamic Kirchhoff plate equation. See also Leugering [65] and Leugering and Rathmann [70] for results on domain decomposition of optimal control problems for Euler-Bernoulli and Rayleigh beams. Methods similar to whose discussed in this book are also available for parabolic and diffusion-advection problems on networked domains, however with different convergence properties. See Fischer and Leugering [22] for domain decomposition results in problems of optimal control for diffusion-advection and transport dominant equations in the context of root dynamics.

As methods of domain decomposition of optimal control problems for partial differential equations are intrinsically motivated by the necessity of simulations, where parallelism and also adaptivity play key roles, we also focus on *a posteriori error estimates* with respect to the mismatch of transmission conditions. Indeed, one may interpret the partition into subdomains as a coarse grain discretization of the problem by introducing interfaces or by interpreting the natural nodes in a network as coarse grid points. On the other hand we have the local problems on the subdomains, or on the edges of the underlying graphs. If those local problems are discretized by some methods, their corresponding a posteriori error estimates can then be combined with those obtained on the coarse interfacial level. The resulting estimates would then govern the convergence and, in particular, any adaptive regime. This two-level paradigm (see also [91]) can be used to enhance convergence properties, because any iteration on the interfacial level will provide global information on the entire domain, or the entire network, respectively. This global information is then used to obtain the local solutions in parallel. Therefore the methods considered here can also be seen as preconditioning methods of a global versus local nature. The latter interpretation, in turn, triggered the most recent development in this area, namely time domain decompositions. See J.-L. Lions, Maday and Turnici [75], and Maday, Turnici [84] for first-order in time equations, and the authors in [57] for hyperbolic problems.

In this work, a posteriori error estimates are developed only on the continuous level although, at least for the time domain decomposition procedure discussed in Chapter 6, similar estimates hold also at the discrete level; see [58]. Let us also mention Otto and Lube [90], [83], [4], where a posteriori error estimates are obtained in the elliptic case, however without (real) control.

The book is organized in a way that readers with a background in functional analysis and partial differential equations will have self-contained access to the theory presented. For those individuals, Chapters 2 through 5 provide the basic ideas, fundamental methods and basic convergence proofs. We have chosen a descriptive style in the sense that the models, methods and the corresponding ideas of proofs are successively developed. The corresponding sections will typically end with a summary of results in the form of a Theorem. This presentation is chosen in order to make it easy to follow the flux of ideas.

The second part of the book, even though strongly motivated by the material of the earlier chapters, is mainly directed to specialists and is itself largely self-contained. After reading the first part of the book, a nonspecialist should also be motivated and able to follow the material in Chapters 6 through 9. These latter chapters are organized in a somewhat different style, where results on well-posedness, convergence and a posteriori error estimates are formalized as theorems with subsequent proofs, in order to have precise statements of results and complete proofs that are readily accessible to researchers.

Chapter 2

Background Material on Domain Decomposition

2.1 Introduction

In this chapter we present some basic concepts of domain decomposition for partial differential equations. We do not attempt to provide a general introduction to that area, which has undergone a dramatic development in recent years. Rather we want to focus on those methods that we view as natural candidates to be extended to optimal control problems.

We begin the discussion in Section 2.2 with simple and standard one-dimensional dynamic problems, the wave equation and the diffusion-advection-reaction equation on an unbounded interval; see Section 2.2.1. We consider the basic form of domain decomposition methods, namely, a Schwarz-type iteration on overlapping domains. The simple arguments proving finite-time and super-linear convergence of these methods, for the wave equation and the diffusion equation, respectively, are provided. It turns out that, upon vanishing of the overlaps, convergence is lost. This is indicative of an absence of transmission information across the interface. In order to restore convergence, dynamic Robin-type conditions are introduced. The transmission conditions can be optimized for unbounded subdomains by the introduction of transparent interface conditions.

In Section 2.2.2 we consider bounded intervals. It is seen that finite termination is lost and, instead of transparent conditions which are then given by nonlocal operators, one has to find local approximations which, in turn, typically lead to Robin-type transmission conditions. In Section 2.2.3 we briefly discuss a semi-discretization method for wave equations that reduces the dynamic problem to a sequence of static elliptic problems. An optimized domain decomposition is used for this type of semi-discretization.

Elliptic problems in higher dimensions are considered in Section 2.3. The reader interested in such problems should consult the book of Quarteroni and Valli [91] as a general reference for problems without control. We first consider the classical methods which are based on taking either the continuity conditions or the first-order transmission conditions at the interfaces as an exact equation, while the complementary condition is then relaxed by iteration. The resulting methods can be reinterpreted as iterative solves of so-called Steklov-Poincaré equations. On the discrete level one then deals with Schur complements. Moreover, the iterations involved can also be interpreted as damped Richardson iterations.

Next, we reinterpret the domain decomposition problem as a problem of exact controllability. Indeed, taking for example the common Dirichlet traces across an interface as a control, we may view the remaining transmission condition in the Neumann traces as a controllability constraint, i.e., we want to achieve exact tracking by choosing controls with, say, least norm. We may then consider relaxing the exact controllability constraints by an approximate matching using optimal controls via regularization. This may also be seen as an inexact method. Note, however, that the controls entering this problem are artificial in that they do not correspond to any physical controls. Hence the name *virtual*, or *artificial*, controls. Various types of virtual control problems are considered. More precisely, we focus our interest on those virtual control problems where the corresponding optimality condition decomposes into local optimality conditions for problems on the subdomains. Such methods are fundamental to understanding the rest of the material in this book. Indeed, it will be seen in subsequent chapters how the methods developed in this chapter can be generalized to space and/or time domain decomposition procedures for many other optimal control problems, such as elliptic or hyperbolic problems on one-dimensional graphs, on two-dimensional polygonal networks, and to hyperbolic problems on higher-dimensional domains.

2.2 Domain Decomposition Methods for One-Dimensional Problems

2.2.1 Unbounded Domains

This section is intended serve as a first motivation for dynamic domain decomposition methods. The presentation is highly selective and does not attempt to account for the vast literature in this area. See the monograph of Quarteroni and Valli [91] for an excellent general reference on domain decomposition methods for partial differential equations. We begin our discussion with two very simple one-dimensional (1-d) partial differential equations, namely the wave equation and the diffusion advection reaction equation. These two models are representative of many problems in applied mathematics.

Define

$$\begin{aligned} L_1(u) &:= u_{tt} - c^2 u_{xx} &&= f \qquad \text{in } \mathbf{R} \times [0,\infty) \\ L_2(u) &:= u_t - \nu u_{xx} + a u_x + bu &&= f \qquad \text{in } \mathbf{R} \times [0,\infty). \end{aligned}$$

To fix ideas, we consider the simplest problem possible, namely, an unbounded interval where we do not have to account for boundary conditions. Our goal is to decompose the real line into half-lines at points $x = 0$ and $x = L$ such that the interval $[0, L]$ is a region of overlap, i.e., $\mathbf{R} = \Omega_1 \cup \Omega_2 := (-\infty, L] \cup [0, \infty)$, with $\Omega_1 \cap \Omega_2 = [0, L]$. We consider the partial differential equations on each half-line and exchange information on the overlapping interval, or on its boundary.

The following simple overlapping domain decomposition algorithm is a variant of the classical alternating additive Schwarz algorithm and has been reconsidered in the form below by Gander, Halpern and Nataf [27] (see also [26]):

Algorithm 2.2.1.1 *Given initial data, right-hand sides f and boundary data w^k, v^k on the iteration level k, compute v^{k+1} and w^{k+1} according to:*

$$\begin{aligned} L_i(v^{k+1}) &= f && \text{in} \quad (-\infty, L] \times (0,T) \\ v^{k+1} &= w^k && \text{on} \quad x = L,\ t \in (0,T) \\ L_i(w^{k+1}) &= f && \text{in} \quad [0,\infty) \times (0,T) \\ w^{k+1} &= v^k && \text{on} \quad x = 0,\ t \in (0,T). \end{aligned}$$

We note that the style in which the algorithms are represented in this monograph are somewhat conceptual. It is remarkable that, in the case of the wave equation on the real line, this algorithm terminates in finite 'time':

Theorem 2.2.1.1 *For $i = 1$, Algorithm 2.2.1.1 terminates for $k > \frac{Tc}{2L}$.*

Proof. As the proof is very simple we reproduce it here. Since we are interested in the error, we let $f = 0$ and the initial data be equal to zero. We apply the Laplace transform $\mathcal{L}$ to $L_1(u) = 0$ and obtain:

$$\begin{aligned} \hat{v}^{k+1}(x,s) &= \hat{w}^k(L,s) \exp\left(\frac{s(x-L)}{c}\right) \\ \hat{w}^{k+1}(x,s) &= \hat{v}^k(0,s) \exp\left(-\frac{sx}{c}\right) \end{aligned}$$

Thus

$$\hat{v}^{k+1}(0,s) = \hat{v}^{k-1}(0,s) \underbrace{\exp\left(-\frac{2sL}{c}\right)}_{=:\rho},$$

and hence

$$\hat{v}^{2k}(0,s) = \rho^k \hat{v}^0(0,s), \quad \hat{w}^{2k}(L,s) = \rho^k \hat{w}^0(L,s).$$

But

$$\exp\left(-\frac{2kLs}{c}\right) = \mathcal{L}(\delta_{t-\frac{2kL}{c}})$$

implying

$$v^{2k}(0,t) = v^0\left(0, t - \frac{2kL}{c}\right).$$

As seen from the proof, the finite speed of propagation, peculiar to the wave equation, is responsible for finite-time-termination. Indeed the 'time' for finite termination is the time the errors need for to travel across the overlapping domain and back until they leave the overlapping slab at the top. This property nicely shows the transmission mechanism in the case of wave equations. Accordingly, for the diffusion-advection problem one obtains ([27]):

Theorem 2.2.1.2 *For $i = 2$, Algorithm 2.2.1.1 converges superlinearly.*

Proof. Again, the proof is simple and instructive. Therefore we sketch the argument given in [27]. We apply the Laplace transform to $L_2(u) = 0$ and obtain by similar arguments as above

$$\hat{v}^{2k} = \rho^k(s)\hat{v}^0(0,s),$$

where $\rho(s) := \exp(-\frac{\sqrt{a^2+4\nu(s+b)}}{\nu}L)$, and similarly for $\hat{w}^{2k}(L,s)$. This time, however, the inverse transform gives

$$v^{2k}(0,t) = \int_0^t K_x(\frac{2kL}{\sqrt{\nu}}, t-\tau)\exp(-(\frac{a^2}{4\nu}+b)(t-\tau))v^0(0,\tau)d\tau$$

where

$$K(x,t) = \frac{x}{2\sqrt{\pi}t^{3/2}}\exp(-\frac{x^2}{4t}).$$

Obviously this implies the error-estimate:

$$\|v^{2k}\|_{L^\infty(0,T)} \leq \max(1, \exp(b-\frac{a^2}{4\nu})T)\mathrm{erfc}(\frac{kL}{\sqrt{\nu T}})\|v^0\|_{L^\infty(0,T)}.$$

A similar analysis holds for w^{2k}. Thus we obtain super-linear convergence. Note that the convergence rate depends on the size of the overlap.

Remark 2.2.1.1 For $i = 1, 2$, if $L \to 0$ convergence is lost.

In order to restore finite termination or even convergence for vanishing overlaps, we need to enhance the exchange of information across the interface. The first and simplest idea to approach this problem is to introduce generalized Robin-type conditions instead of Dirichlet (Neumann) conditions. This can also be done for the overlapping case as follows.

Algorithm 2.2.1.2 *Given initial data, right-hand sides f and boundary data w^k, v^k on the iterative level k, compute v^{k+1},w^{k+1} according to:*

$$\begin{aligned}
L_i(v^{k+1}) &= f && \text{in} \quad (-\infty, L] \times (0,T)\\
v_x^{k+1} + \Lambda_v(v^{k+1}) &= w_x^k + \Lambda_v(w^k) && \text{on} \quad x = L, (0,T)\\
L_i(w^{k+1}) &= f && \text{in} \quad [0,\infty) \times (0,T)\\
w_x^{k+1} + \Lambda_w(w^{k+1}) &= v_x^k + \Lambda_w(v^k) && \text{on} \quad x = 0, (0,T)
\end{aligned}$$

Here Λ_v and Λ_w are either constants or local-in-time operators such as time derivatives or convolution operators.

Remark 2.2.1.2

- For $L = 0$ and $\Lambda_v(u) = \lambda u$, $\lambda \in \mathbf{R}$ this is a nonoverlapping Robin-type-Algorithm. See the next section.
- For $\Lambda_{v,w}(u) := \lambda \star u$ (convolution in time), see Gander, Halpern and Nataf [27].

The following results on finite termination ([27]) for transmission operators $\Lambda_{v,w}(u) = \lambda_{v,w} \star u$ reflect the fact that the corresponding transmission conditions realize exact transparent boundary conditions on each half-line:

Theorem 2.2.1.3 *For* $i = 1$, $\Lambda_v = \frac{1}{c}\partial_t$, $\Lambda_w = -\frac{1}{c}\partial_t$ *Algorithm* 2.2.1.2 *terminates in 2 steps.*

Proof. One applies the Laplace transform and obtains after some elementary calculations

$$\hat{v}^{k+1}(0,s) = \frac{(\frac{s}{c} + \lambda_w(s))(-\frac{s}{c} + \lambda_v(s))}{(\frac{s}{c} + \lambda_v(s))(-\frac{s}{c} + \lambda_w(s))} \exp(-2\frac{s}{c}L)\hat{v}^{k-1}(0,s)$$

A similar expression is derived for $\hat{w}^{k+1}$. The obvious choice

$$\lambda_w(s) = -\lambda_v(s) = -\frac{s}{c}$$

leads to two-step-termination.

Remark 2.2.1.3 It is important to note that the finite-termination property of Theorem 2.2.1.3 does not depend on the size of the overlap. Note, however, that due to the unboundedness of the domain, the 'optimal' transmission operators are local operators in time.

As to be expected, the finite-termination property will not hold in general for diffusion-advection-reaction equations, unless we are able to implement non-local boundary operators.

Remark 2.2.1.4

i) For $i = 2$ and $\Lambda_{v,w}$ according to

$$\hat{\lambda}_{v,w}(s) = -\frac{a}{2\nu} \pm \frac{1}{2\nu}\sqrt{a^2 + 4\nu(s+b)},$$

Algorithm 2.2.1.2 *terminates after* 2 *steps.*

ii) For $i = 1, 2$ finite termination is independent of L.

iii) For $i = 2$, the *optimal* interface-conditions are *non-local.*

iv) We may approximate the non-local transmission conditions by local operators:

$$\Lambda_{v,w} = \pm\frac{-a \pm \sqrt{a^2 + 4\nu b}}{2\nu\sqrt{a^2 + 4\nu b}}\partial_t$$

2.2.2 Bounded Domains

We now consider a bounded interval $[-L, L]$ with interface at $x = 0$. For the sake of simplicity, we consider a nonoverlapping version of the domain decomposition Algorithm 2.2.1.2. While in the 1-d situation it is always possible to consider the entire real axis by applying transparent boundary conditions, the question as to whether finite termination is achievable in the case of more general boundary conditions, or what kind of approximations are useful in this context, is still not completely understood. By using arguments similar to those of [27] we can state the following.

Remark 2.2.2.1

i) For $i = 1$ and $\Lambda_{v,w}$ chosen according to $\hat{\lambda}_{v,w}(s) = \pm\frac{s}{c}\coth(\frac{s}{c}L)$, Algorithm 2.2.1.2 *terminates after* 2 *steps.*

ii) The 'optimal' conditions are non-local (even for $i = 1$).

iii) The local approximations (for s small):

$$\hat{\lambda}_{v,w}(s) := \pm(\frac{1}{L} + \frac{L}{3c^2}s^2 + O(s^4))$$

yield, upon transformation back into the time-domain,

$$w_x(0,t) + \frac{1}{L}w(0,t) + \frac{L}{3c^2}w_{tt}(0,t)$$

Note that this adds a *mass* at the interface.

To complete the picture we ask: what happens if one works with a transparent boundary condition at one end (equivalent to a semi-infinite domain)? By using arguments similar to those above, it turns out that in this case one can derive 'optimal' transmission conditions which are local; once again, one chooses impedance transmission conditions. For a numerical realization of this method see Figure 2.1, where two strings are connected at $x = 0$ and two initial displacements on each string cause interacting waves. The Figure 2.2 clearly reveals that essentially no reflection occurs at the interface. The mismatch at the interface is shown in Figure 2.3. See also Hundhammer [46]. We will use the idea of optimizing transmission conditions later in this book for higher-dimensional problems.

2.2.3 Semi-discretization

As we have seen, the convergence or termination analysis of Algorithm 2.2.1.2 is done in the frequency domain. Indeed, we are dealing with elliptic problems parameterized by the Laplace variable. Now, after time discretization of our equations we once again obtain with 'elliptic' problems. Hence it appears natural to consider applying the nonoverlapping version of Algorithm 2.2.1.2 to that situation.

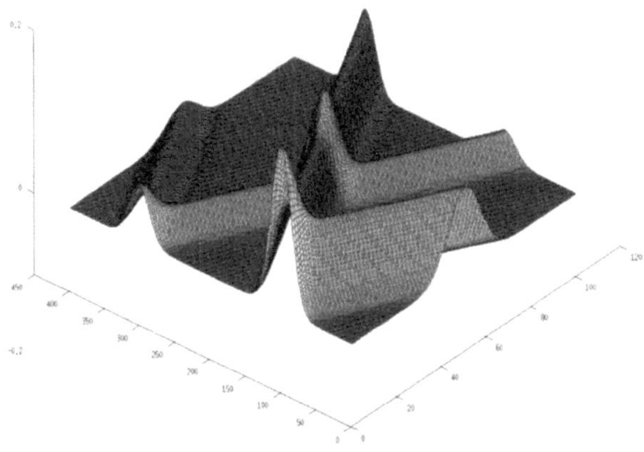

Figure 2.1: Interacting waves: serial connection

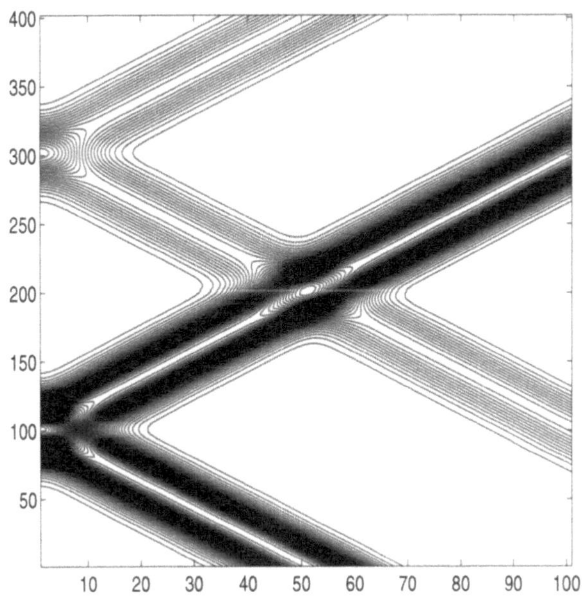

Figure 2.2: Interacting waves: serial connection

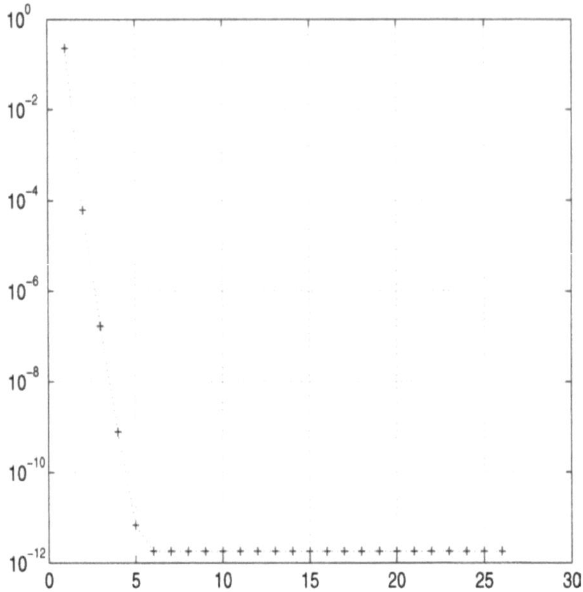

Figure 2.3: Iteration history: dynamic problem

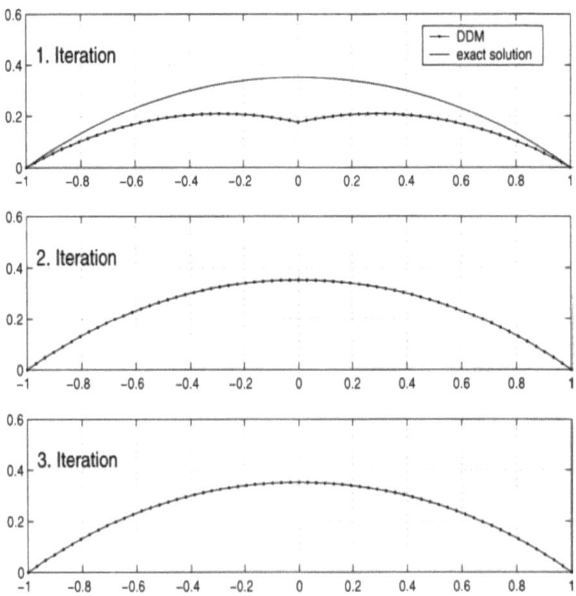

Figure 2.4: Two-step convergence in the static case: numerical/exact solution

We use Newmark's scheme as in Raviart-Thomas [92] for the wave equation:

$$\begin{aligned}
w_i^m &\doteq w_i(\cdot, m\delta t),\\
z_i^m &\doteq w_{i,t}(\cdot, m\delta t),\\
w_i^{m+1} &= w_i^m + \delta t z_i^m + \delta t^2 c_i^2 (\beta w_{i,xx}^{m+1} + (\frac{1}{2} - \beta) w_{i,xx}^m),\\
z_i^{m+1} &= z_i^m + \delta t c_i^2 (\gamma w_{i,xx}^{m+1} + (1-\gamma) w_{i,xx}^m).
\end{aligned}$$

(With $\beta = \frac{1}{4}$, $\gamma = \frac{1}{2}$ there is conservation of energy.) This time-stepping procedure leads to the elliptic problem (followed by an update of the velocity z_i^{m+1})

$$\begin{aligned}
w_i^{m+1} &- \delta t^2 \beta c_i^2 w_{i,xx}^{m+1} \\
&= w_i^m + \delta t z_i^m + \delta t^2 c_i^2 (\frac{1}{2} - \beta) w_{i,xx}^m =: f_i^m
\end{aligned} \tag{2.2.3.1}$$

We have to add the continuity and transmission conditions, say, at the interface $x = 0$:

$$w_1^{m+1}(0) = w_2^{m+1}(0), \quad w_{1,x}^{m+1}(0) = w_{2,x}^{m+1}(0) \tag{2.2.3.2}$$

Thus, for each time step we have to solve an elliptic problem on a serially connected domain. At each time step with step size δt and with

$$\Lambda_{1,2} := \pm \frac{1}{\delta t^2 \beta c} \coth(\frac{L}{\delta t^2 \beta c}),$$

we apply Algorithm 2.2.1.2 to (2.2.3.1), (2.2.3.2). The behavior of the algorithm is displayed numerically in Figure2.4.

As we have seen, Robin-type iterations are motivated from the point of view of optimal transmission of information across the interface. In particular, classical Robin conditions typically correspond to zero-order or first-order approximations of such optimal conditions. In general, therefore, one cannot expect finite termination.

2.3 Nonoverlapping Domain Decomposition Methods for Elliptic Problems

2.3.1 Review of Basic Methods

After the preliminary discussion of domain decomposition algorithms for one-dimensional dynamic problems presented in the previous section, we embark on partial differential equations on higher-dimensional domains. In order to fix ideas, we begin our discussion with a standard elliptic problem. However, as the main purpose of this monograph is to deal with nonoverlapping domain decompositions for optimal control problems, we do not pursue the discussion of overlapping domain decompositions beyond this point.

Here we consider the following model problem

$$\begin{aligned} -\Delta u &= f \text{ in } \Omega \\ u &= 0 \text{ on } \partial\Omega, \end{aligned} \tag{2.3.1.1}$$

where $\Omega \subset \mathbb{R}^d$ is a bounded domain with piecewise Lipschitz boundary $\partial\Omega$. Let $V = H_0^1(\Omega)$, $H = L^2(\Omega)$, $D_{\mathcal{A}} = H^2(\Omega) \cap V$ and $f \in V^*$ (the dual of V). (See, e.g., Grisvard [30] for standard notions and properties of Sobolev spaces.) Problem (2.3.1.1) is a special case of a class of problems that can be expressed in variational form as follows. Introduce the bilinear form

$$a(u,v) := \sum_{k,\ell} \int_\Omega a_{k\ell} \frac{\partial u}{\partial x_k} \frac{\partial u}{\partial x_\ell}\, dx + \int_\Omega c\, v\, u\, dx \tag{2.3.1.2}$$

where c, $a_{k\ell} \in L^\infty(\Omega)$, $c(x) \geq c_0 \geq 0$ a.e. in Ω, $a_{k\ell}$ is piecewise C^1 such that $a_{k\ell} = a_{\ell k}$ a.e. in Ω and satisfies

$$\sum_{k\ell} a_{k\ell}(x)\, \xi_k\, \xi_\ell \geq \alpha \sum_k |\xi_k|^2 \text{ a.e.}, \tag{2.3.1.3}$$

with $\alpha > 0$. Thus,

$$a(u,u) \geq \alpha\, \|u\|_V^2, \qquad \forall u \in V. \tag{2.3.1.4}$$

The elliptic operator $\mathcal{A}$ corresponding to the bilinear form is given by

$$\mathcal{A}\, u := -\sum_{k,\ell} \frac{\partial}{\partial x_k} (a_{k\ell} \frac{\partial}{\partial x_\ell}) + c\, u. \tag{2.3.1.5}$$

We may pose the elliptic boundary value problem

$$\begin{aligned} \mathcal{A}\, u &= f \text{ in } \Omega \\ u &= 0 \text{ on } \partial\Omega. \end{aligned} \tag{2.3.1.6}$$

For $a_{k_\ell} = \delta_{k_\ell} := \begin{cases} 1 & k = \ell \\ 0 & \text{else} \end{cases}$, $c = 0$, (2.3.1.6) reduces to (2.3.1.1) and (2.3.1.2) reduces to the standard form

$$a(u,v) = \int_\Omega \nabla\, u\, \nabla\, v\, dx. \tag{2.3.1.7}$$

Problem (2.3.1.6) admits the variational formulation

$$a(u,v) = \langle f, v \rangle_V \;\forall\, v \in V, \tag{2.3.1.8}$$

where $\langle f, v \rangle_V$ denotes the duality pairing between elements $f \in V^*$ and $v \in V$. By applying the Lax-Milgram Lemma to (2.3.1.8) we obtain, for any given $f \in V^*$, a unique solution $u \in V$ to (2.3.1.8), which is a weak (mild) solution of (2.3.1.6). A strong solution $u \in D(A)$ is obtained for $f \in H$.

We now decompose the domain Ω into disjoint subdomains Ω_i, $i = 1, \dots, I$, $I \geq 2$. To this end we introduce the notation

$$\begin{aligned} \bar{\Omega} &= \bigcup_{i=1}^{I} \bar{\Omega}_i, \quad \Omega_i \cap \Omega_j = \emptyset \\ \Gamma_{ij} &:= \partial\Omega_i \cap \partial\Omega_j = \Gamma_{ji}, \quad \Gamma_i := \partial\Omega_i \cap \partial\Omega \\ \gamma_i &= \bigcup_{j \in \mathcal{I}_i} \Gamma_{ij}, \quad \Gamma := \bigcup_{1 \leq i \neq j \leq I} \Gamma_{ij}, \\ \mathcal{I} &= \{1, \dots, I\}, \quad \mathcal{I}_i := \{j \in \mathcal{I} : \Gamma_{ij} \neq \emptyset\}. \end{aligned} \tag{2.3.1.9}$$

With this notation, Γ denotes to the total transmission boundary, γ_i denotes the transmission boundary for Ω_i, $\mathcal{I}_i$ is the set of indices of adjacent domains for Ω_i. Let us introduce the following spaces associated with the splitting (2.3.1.9):

$$\begin{aligned} V_i &= \{u \in H^1(\Omega_i) : u = 0 \text{ on } \Gamma_i\} \\ U_{ij} &= L^2(\Gamma_{ij}) = U_{ji} = L^2(\Gamma_{ji}). \end{aligned} \tag{2.3.1.10}$$

Then we may consider the operators

$$\begin{aligned} r_i &: V \to V_i \\ r_i\, u &:= u|_{\Omega_i}, \end{aligned} \tag{2.3.1.11}$$

$$\begin{aligned} r_{ji} &: V_i \to L^2(\Gamma_{ij}) \\ r_{ji}\, u_i &:= u_i|_{\Gamma_{ij}}. \end{aligned} \tag{2.3.1.12}$$

Obviously, r_i is linear and bounded, i.e.,

$$r_i \in \mathcal{L}(V, V_i). \tag{2.3.1.13}$$

Further

$$r_{ji} \in \mathcal{L}(V_i, W_{ji}) \tag{2.3.1.14}$$

where $W_{ij} = H_{00}^{1/2}(\Gamma_{ij}) := \{v|_{\Gamma_{ij}} : v \in V_i\} \hookrightarrow U_{ij}$. A fortiori,

$$r_{ji} \in \mathcal{L}(V_i, U_{ji}). \tag{2.3.1.15}$$

We introduce the forms on Ω_i

$$a_i(u_i, v_i) = \int_{\Omega_i} \Big[\sum_{k,\ell} a_{k\ell} \frac{\partial u_i}{\partial x_k} \frac{\partial u_i}{\partial x_\ell} + c\, u_i\, v_i\Big]\, dx, \tag{2.3.1.16}$$

with $a_{k\ell}|_{\Omega_i} \in C^1(\Omega_i)$. For regions Ω_i with $\text{meas}(\Gamma_i) > 0$ we have for some $\alpha_i > 0$

$$a_i(u_i, u_i) \geq \alpha_i\, \|u_i\|_{V_i}^2. \tag{2.3.1.17}$$

For interior domains Ω_i with $\Gamma_i = \emptyset$, (2.3.1.17) will hold if $c(x) > 0$ on a set in Ω_i of positive measure. Otherwise we assume (2.3.1.17). It turns out, however, that one can dispense with this local coercivity condition.

Now, to (2.3.1.2) with (2.3.1.4) there corresponds a self-adjoint positive definite operator $\mathcal{A}$ in H such that (2.3.1.1) or (2.3.1.6) correspond to the operator equation

$$\mathcal{A}u = f, \tag{2.3.1.18}$$

as long as $f \in H$, $u \in D(\mathcal{A}) = D_{\mathcal{A}}$ or $u \in V$ if $f \in V^*$. In the latter case we consider the unique unitary extension of $\mathcal{A}$ from $D(\mathcal{A})$ in H to V. Thus, $\mathcal{A}$ may be regarded as the Riesz isomorphism from V onto V^*.

Accordingly, to $a_i(\cdot,\cdot)$, given by (2.3.1.16), there corresponds a self-adjoint positive definite operator $\mathcal{A}_i$. Moreover

$$\begin{aligned} \langle \mathcal{A}u, v\rangle_V &= a(u,v) \\ &= \sum_{i=1}^{I} a_i(u_i, v_i) \\ &= \sum_{i=1}^{I} \langle \mathcal{A}_i\, u_i, v_i\rangle_{V_i}, \end{aligned} \tag{2.3.1.19}$$

where now $\mathcal{A}_i : V_i \to V_i^*$ is the canonical Riesz isomorphism. The splitting (2.3.1.9) thus corresponds to the operator splitting (2.3.1.19).

With this basic notation at hand we may now review various classical domain decomposition techniques. The material to follow is, however, selective, because of our special interests outlined in the introduction. We refer the reader to [91] for a general treatment.

For the sake of simplicity we go back to (2.3.1.1) and consider two subdomains only. Problem (2.3.1.1) is equivalent to

$$\begin{aligned} -\Delta\, u_i &= f_i \ \text{ in } \ \Omega_i \\ u_i &= 0 \ \text{ on } \ \Gamma_i \\ u_i &= u_j \ \text{ on } \ \Gamma_{ij} = \Gamma \\ \frac{\partial u_i}{\partial \nu_i} + \frac{\partial u_j}{\partial \nu_j} &= 0 \ \text{ on } \ \Gamma_{ij} = \Gamma, \end{aligned} \tag{2.3.1.20}$$

$i = 1, 2$. In (2.3.1.20) the third and fourth equation constitute so-called transmission conditions. The decomposition of those transmission conditions is at the heart of nonoverlapping domain decomposition methods (DDM).

One may either directly apply an iterative procedure to (2.3.1.20) by some alternating step between the problems on Ω_i or formulate the transmission conditions $(2.3.1.20)_3$, $(2.3.1.20)_4$ in an operator format as follows. We first concentrate on methods that start with $(2.3.1.20)_3$. The following procedure parallels the one in [91, Chapter 1.2]. We include the presentation for the sake of self-consistency. We enforce $(2.3.1.20)_3$ by setting $r_{ji}\, u_i = u_i|_\Gamma = \lambda = u_j|_\Gamma = r_{ij}\, u_j$:

$$\begin{aligned} -\Delta\, u_i &= f_i \ \text{ in } \ \Omega_i \\ u_i &= 0 \ \text{ on } \ \Gamma_i \\ u_i &= \lambda \ \text{ on } \ \Gamma, \end{aligned} \tag{2.3.1.21}$$

$i = 1, 2$. Problems (2.3.1.21) can be solved in parallel yielding the unique solutions u_i. Thus $(2.3.1.20)_4$, the other transmission condition, may be seen as an equality constraint as follows. To this end we decompose $u_i = w_i + z_i$ such that

$$\begin{aligned} -\Delta\, w_i &= 0 \ \text{ in } \ \Omega_i \\ w_i &= 0 \ \text{ on } \ \Gamma_i \\ w_i &= \lambda \ \text{ on } \ \Gamma \end{aligned} \tag{2.3.1.22}$$

and

$$\begin{aligned} -\Delta\, z_i &= f_i \ \text{ in } \ \Omega_i \\ z_i &= 0 \ \text{ on } \ \Gamma_i \\ z_i &= 0 \ \text{ on } \ \Gamma. \end{aligned} \tag{2.3.1.23}$$

The solution w_i is given by the Dirichlet map, the harmonic extension of λ into the domain Ω_i:

$$\begin{aligned} &D_i : H^s(\Gamma_{ij}) = H^s(\Gamma) \to H^{s+\frac{1}{2}}(\Omega_i) \\ &w_i =: D_i\, \lambda. \end{aligned} \tag{2.3.1.24}$$

In fact, in our situation we may extend (2.3.1.24) as a continuous operator in

$$\begin{aligned} &D_i : H_{00}^{\frac{1}{2}}(\Gamma) := W \to V_i \\ &w_i =: D_i\, \lambda. \end{aligned} \tag{2.3.1.25}$$

In order to express $(2.3.1.20)_4$ in operator notation, we need to consider Neumann traces

$$\begin{aligned} &T_i : H^{s+2}(\Omega_i) \to H^{s+\frac{1}{2}}(\Gamma_{ij}) = H^{s+\frac{1}{2}}(\Gamma) \\ &T_i\, u_i = \frac{\partial u_i}{\partial \nu_i}\Big|_\Gamma \end{aligned} \tag{2.3.1.26}$$

for $s \geq 0$. According to (2.3.1.24), (2.3.1.26) we may write

$$\frac{\partial w_i}{\partial \nu_i} = \frac{\partial}{\partial \nu_i} D_i\, \lambda = T_i\, D_i\, \lambda \ \text{ for } \ \lambda \in H^1(\Gamma). \tag{2.3.1.27}$$

This operator can be extended as a continuous map in $L^2(\Gamma)$. See [91] for the properties of the Steklov-Poincaré operator. It can also be extended to a map between $H_{00}^{\frac{1}{2}}(\Gamma)$ and its dual. The solution z_i of (2.3.1.23) is

$$z_i = \mathcal{A}_i^{-1}\, f \tag{2.3.1.28}$$

and hence,

$$\frac{\partial z_i}{\partial \nu_i} = \frac{\partial}{\partial \nu_i} \mathcal{A}_i^{-1} f = T_i \mathcal{A}_i^{-1}. \tag{2.3.1.29}$$

With (2.3.1.27) and (2.3.1.29) the transmission condition $(2.3.1.20)_4$ reads

$$\frac{\partial u_i}{\partial \nu_i} + \frac{\partial u_j}{\partial \nu_j} = (\sum_i T_i D_i) \lambda + \sum_i T_i \mathcal{A}_i^{-1} f_i = 0. \tag{2.3.1.30}$$

We denote by

$$\mathcal{P}_i := T_i D_i : H^1(\Gamma) \to L^2(\Gamma), \tag{2.3.1.31}$$

the Steklov-Poincaré interface operator for Ω_i and by

$$\begin{aligned} &\mathcal{P} := \mathcal{P}_1 + \mathcal{P}_2 \\ &\mathcal{P} : H^1(\Gamma) \to L^2(\Gamma) \end{aligned} \tag{2.3.1.32}$$

the total Steklov-Poincaré operator . Let

$$\begin{aligned} g_i &:= -T_i \mathcal{A}_i^{-1} f_i \\ g &:= g_1 + g_2. \end{aligned} \tag{2.3.1.33}$$

Then (2.3.1.30) reads

$$\begin{aligned} \mathcal{P}_1 \lambda + \mathcal{P}_2 \lambda = g_1 + g_2 \\ \mathcal{P} \lambda = g. \end{aligned} \tag{2.3.1.34}$$

Thus, solving (2.3.1.20) is equivalent to solving the Steklov-Poincaré equation (2.3.1.34) on Γ.

The problem (2.3.1.22) admits a variational formulation

$$\begin{cases} a_i(D_i \lambda, v_i) = 0 \quad \forall\, v_i \in V_i^0 := H_0^1(\Omega_i) \\ r_{ji} D_i \lambda = \lambda \end{cases} \tag{2.3.1.35}$$

where $r_{ji} D_i \lambda = \lambda$ is the operator notation for $tr|_\Gamma D_i \lambda = \lambda$ on Γ (see (2.3.1.12)–(2.3.1.15)). The problem (2.3.1.23) has the variational representation

$$a_i(\mathcal{A}_i^{-1} f, v_i) = (f, v_i),\ \forall\, v_i \in V_i^0. \tag{2.3.1.36}$$

In order to express the dependence of z_i on f_i in a more compact way, we may introduce

$$\begin{aligned} &\mathcal{G}_i : V_i^* \to V_i^0 \\ &\mathcal{G}_i f := \mathcal{A}_i^{-1} f. \end{aligned} \tag{2.3.1.37}$$

Let R_i be any extension operator from Γ into V_i (with dense range in V_i)

$$\begin{aligned} &\mathcal{R} \mu := \begin{cases} \mathcal{R}_1 \mu \text{ in } \Omega_1 \\ \mathcal{R}_2 \mu \text{ in } \Omega_2 \end{cases} \\ &\mathcal{R} : W \to V. \end{aligned} \tag{2.3.1.38}$$

We have $r_{ji}\,\mathcal{R}_i\,\mu = \mu$, for all i, j and hence

$$r_{ji}\,\mathcal{R}_i\,\mu = r_{ij}\,\mathcal{R}_j\,\mu. \tag{2.3.1.39}$$

Now, consider

$$a(u, \mathcal{R}\,\mu) = a_1(u_1, \mathcal{R}_1\,\mu) + a_2(u_2, \mathcal{R}_2\,\mu), \tag{2.3.1.40}$$

where u_i solves

$$a_i(u_i, v_i) = (f_i, v_i) \;\; \forall\, v_i \,\in\, V_i^0 \tag{2.3.1.41}$$

and $(2.3.1.20)_3$ holds. We have

$$\begin{aligned} a_i\,(u_i, \mathcal{R}_i\,\mu) &= \int_\Gamma \frac{\partial u_i}{\partial \nu_i}\,\mu\,d\gamma - \int_{\Omega_i} \Delta\,u_i\,\mathcal{R}_i\,\mu\,dx \\ &= \int_\Gamma \frac{\partial u_i}{\partial \nu_i}\,\mu\,d\gamma + \int_{\Omega_i} f_i\,\mathcal{R}_i\,\mu\,dx. \end{aligned} \tag{2.3.1.42}$$

Therefore, (2.3.1.40) amounts to

$$a(u, \mathcal{R}\mu) = \int_\Gamma (\frac{\partial u_i}{\partial \nu_i} + \frac{\partial u_j}{\partial \nu_j})\,\mu d\gamma + \int_\Omega f\,\mathcal{R}\,\mu\,dx. \tag{2.3.1.43}$$

Thus, (2.3.1.20) is equivalent to

$$\begin{aligned} &a_i(u_i, v_i) = (f_i, v_i) \;\; \forall\, v_i \in V_i^0 \\ &r_{21}\,u_1 = r_{12}\,u_2 \\ &a_1(u_1, \mathcal{R}_{1\mu}) + a_2(u_2, \mathcal{R}_{2\mu}) = (f_1, \mathcal{R}_1\mu) + (f_2, \mathcal{R}_2\mu), \;\; \forall\,\mu \in W, \end{aligned} \tag{2.3.1.44}$$

which may be combined to

$$a(u, \mathcal{R}\mu) = (f, \mathcal{R}\mu), \;\; \forall\,\mu \in W. \tag{2.3.1.45}$$

On the other hand, for $v \in V$, $v_i = r_i v \in V_i$, $tr|_\Gamma\, v =: \mu \in W$, we have $r_i\,v - \mathcal{R}_i\,\mu \in V_i^0$, since $r_{ji}(r_i\,v - \mathcal{R}_i\mu) = tr|_\Gamma v - \mu = 0$ on Γ. Thus, from (2.3.1.44)

$$a_i(u_i, r_i v - \mathcal{R}_i\mu) = (f_i, r_i v - \mathcal{R}_i\mu)$$

and

$$\begin{aligned} a(u, v) &= \sum_{i=1}^{2} [a_i(u_i, r_i v - \mathcal{R}_i\mu) + a_i(u_i, \mathcal{R}_i\mu)] \\ &= \sum_{i=1}^{2} [(f_i, r_i v - \mathcal{R}_i\mu) + (f_i, \mathcal{R}_i\mu)] \\ &= \sum_{i=1}^{2} (f_i, r_i v) = (f, v), \end{aligned}$$

which implies (2.3.1.8). The reverse inclusion has been already shown.

It is now obvious that the variational form of the Steklov-Poincaré equation (2.3.1.34) can be written as

$$\begin{aligned}
\langle \mathcal{P}\,\lambda,\,\mu\rangle &= \sum_{i=1}^{2}\langle \mathcal{P}_i\,\lambda,\mu\rangle \\
&= \sum_{i=1}^{2}\int_{\Gamma}\frac{\partial}{\partial\nu_i}\,D_i\,\lambda\,\mu\,d\gamma \\
&= \sum_{i=1}^{2}\int_{\Omega_i}\nabla D_i\,\lambda\,\nabla\,\mathcal{R}_i\,\mu\,dx \\
&= \sum_{i=1}^{2} a_i(D_i\,\lambda,\mathcal{R}_i\,\mu),\quad \forall\,\mu\in W.
\end{aligned} \tag{2.3.1.46}$$

Moreover,

$$\begin{aligned}
\langle g,\mu\rangle &= \sum_{i=1}^{2}\langle g_i,\mu\rangle \\
&= -\sum_{i=1}^{2}\langle\frac{\partial}{\partial\nu_i}\,\mathcal{G}_i\,f,\mu\rangle \\
&= \sum_{i=1}^{2}\int_{\Omega_i}(f_i\,\mathcal{R}_i\,\mu-\nabla\,\mathcal{G}_i\,f,\,\nabla\mathcal{R}_i\,\mu)\,dx \\
&= \sum_{i=1}^{2}[(f,\mathcal{R}_i\,\mu)-a_i(\mathcal{G}_i\,f,\mathcal{R}_i\,\mu)]\;\;\forall\,\mu\in W.
\end{aligned} \tag{2.3.1.47}$$

Thus, according to (2.3.1.46) and (2.3.1.47) the equation (2.3.1.34) reads in variational form as

$$\langle\mathcal{P}\,\lambda,\mu\rangle=\langle g,\mu\rangle,\quad\forall\,\mu\in W \tag{2.3.1.48}$$

which is, in turn, equivalent to $(2.3.1.44)_3$. Thus, any iterative procedure for the solution of the Steklov-Poincaré equation (2.3.1.34) results in such a procedure for (2.3.1.44), and vice versa. This is one of the basic aspects for iterative nonoverlapping domain decomposition methods. The most standard such decompositions are the Dirichlet-Neumann and the Neumann-Neumann method. Let us display their variational forms based on the decomposition of (2.3.1.44) (see [91]).

Algorithm 2.3.1.1 *Given λ^k at iteration level k:*

1) *Solve for $u_1^{k+1}\in V_1$:*

$$\begin{aligned}
a_1(u_1^{k+1},v_1)&=(f_1,v_1)\;\;\forall\,v_1\in V_1^0\\
u_1^{k+1}&=\lambda^k\;\;\textit{on}\;\;\Gamma
\end{aligned}$$

2) *Solve for* $u_2^{k+1} \in V_2$*:*

$$
\begin{aligned}
a_2(u_2^{k+1}, v_2) &= (f_2, v_2) \;\; \forall \, v_2 \in V_2^0 \\
a_2(u_2^{k+1}, \mathcal{R}_2 \, \mu) &= (f_2, \mathcal{R}_2 \, \mu) + (f_1, \mathcal{R}_1 \, \mu) \\
&\quad - a_1(u_1^{k+1}, \mathcal{R}_1 \, \mu), \;\; \forall \, \mu \in W
\end{aligned}
$$

3) *update* l^k*:* $\lambda^{k+1} = \theta \, r_{12} \, u_2^{k+1} + (1-\theta) \, \lambda^k, \; \theta > 0.$

Algorithm 2.3.1.1 has an operator format as follows:

$$\lambda^{k+1} = \lambda^k + \theta \, \mathcal{P}_2^{-1} \, (-\mathcal{P} \, \lambda^k + g). \tag{2.3.1.49}$$

Moreover, it has a sequential character, as the subproblem on Ω_2 has to be solved before the step can be completed.

The so-called Neumann-Neumann method resolves this difficulty with respect to parallelism at the expense of introducing an additional Neumann-solve with the residual of the transmission condition $(2.3.1.20)_4$ on the right-hand side of the Neumann condition. As we will not use this method, we just display its operator form, which is

$$\lambda^{k+1} = \lambda^k + \theta(\delta_1 \, \mathcal{P}_1^{-1} + \delta_2 \, \mathcal{P}_2^{-1}) \, (g - \mathcal{P} \, \lambda^k). \tag{2.3.1.50}$$

Both iterations (2.3.1.49) and (2.3.1.50) can be interpreted as preconditioned damped Richardson iterations in the Hilbert space W. See [91] for a detailed discussion of the methods and their discretizations and extensions.

We start the procedure with equation $(2.3.1.20)_4$. Thus, we consider

$$
\begin{aligned}
-\Delta \, u_i &= f_i \;\; \text{in} \;\; \Omega_i \\
u_i &= 0 \;\; \text{on} \;\; \Gamma_i \\
\frac{\partial u_i}{\partial \nu_i} &= (-1)^i \, \lambda \;\; \text{on} \;\; \Gamma.
\end{aligned} \tag{2.3.1.51}
$$

Problem (2.3.1.51) splits into $u_i = w_i + z_i$ according to

$$
\begin{aligned}
-\Delta \, w_i &= 0 \;\; \text{in} \;\; \Omega_i \\
w_i &= 0 \;\; \text{on} \;\; \Gamma_i \\
\frac{\partial w_i}{\partial \nu_i} &= (-1)^i \, \lambda \;\; \text{on} \;\; \Gamma
\end{aligned} \tag{2.3.1.52}
$$

and

$$
\begin{aligned}
-\Delta \, z_i &= f_i \;\; \text{in} \;\; \Omega_i \\
z_i &= 0 \;\; \text{on} \;\; \Gamma_i \\
\frac{\partial z_i}{\partial \nu_i} &= 0 \;\; \text{on} \;\; \Gamma.
\end{aligned} \tag{2.3.1.53}
$$

Let the Neumann-map be defined by

$$\begin{aligned} \mathcal{N}_i &: L^2(\Gamma) \to H^{3/2}_\Gamma(\Omega) \\ w_i &:= (-1)^i \mathcal{N}_i \lambda. \end{aligned} \tag{2.3.1.54}$$

Problem (2.3.1.53) admits a unique solution $z_i \in V_i$, $z_i = \mathcal{G}_i f$ where, now, $\mathcal{G}_i$ is the solution of the homogeneous mixed boundary value problem (2.3.1.53). The continuity condition $(2.3.1.20)_3$ reads

$$r_{21}(-\mathcal{N}_1 \lambda + z_1) = r_{12}(\mathcal{N}_2 \lambda + z_2),$$

which is equivalent to

$$(r_{12}\mathcal{N}_2 + r_{21}\mathcal{N}_1)\lambda = r_{21}\mathcal{G}_1 f_1 - r_{12}\mathcal{G}_2 f_2 =: g. \tag{2.3.1.55}$$

We may then define the Neumann-Dirichlet-type Steklov-Poincaré operators

$$\begin{aligned} \mathcal{P}_i &:= r_{ji}\mathcal{N}_i \\ \mathcal{P} &:= \mathcal{P}_1 + \mathcal{P}_2, \end{aligned} \tag{2.3.1.56}$$

so that, in this context, we arrive at the Steklov-Poincaré equation

$$\mathcal{P}\lambda = g. \tag{2.3.1.57}$$

The variational form of (2.3.1.57) is given by

$$a_i(w_i, v_i) = \langle (-1)^i B_i \lambda, v_i \rangle_{V_i}, \quad \forall v_i \in V_i, \tag{2.3.1.58}$$

where $B_i \lambda$ is the extension of $\lambda \in L^2(\Gamma)$ into V_i^*, such that $\langle B_i\lambda, v_i\rangle_{V_i} = \langle \lambda, r_{ji} v_i\rangle_\Gamma$. Thus $B_i = r_{ji}^*$. As

$$a_i(z_i, v_i) = (f_i, v_i), \quad \forall v_i \in V_i, \tag{2.3.1.59}$$

we may write the variational form of (2.3.1.51) as

$$a_i(u_i, v_i) = (f_i, v_i) + b_i(\lambda_i, v_i), \quad \forall v_i \in V_i, \tag{2.3.1.60}$$

$i = 1, 2$, where $b_i(\lambda_i, v_i) := \langle B_i\lambda, v_i\rangle_{V_i}$ and where $\lambda_i + \lambda_j = 0$ (or $\lambda_i = (-1)^i \lambda$, $i = 1, 2$).

In order to solve (2.3.1.20) we need to require in addition to (2.3.1.60)

$$r_{ji} u_i = r_{ij} u_j \;\; \forall\, i, j. \tag{2.3.1.61}$$

Equations (2.3.1.60) and (2.3.1.61) give again rise to another family of iterative domain decompositions which can then be recast into the format of damped preconditioned Richardson iterations. One can also combine the approaches (2.3.1.44) and (2.3.1.60), (2.3.1.61). See also [91, Chapter 1.3].

2.3.2 Virtual Controls

We now come to a major point that is fundamental to understanding the philosophy of this monograph. Either of the transmission conditions $(2.3.1.20)_3$ or (2.3.1.30) and (2.3.1.61) can be interpreted as exact controllability constraints wherein λ is viewed as a 'control' function in the control space W or $L^2(\Gamma)$, respectively. As it is much more convenient to work in the control space $L^2(\Gamma)$, we take the second approach (2.3.1.60). Rather than factoring the kernel of the trace operators as above we are now looking for solutions λ of (2.3.1.61) with minimal norm; see J.-L. Lions and O. Pironneau [77]. We may envision various ways to approach the problem (2.3.1.60), (2.3.1.61), such as the formulation

$$\inf \frac{1}{2} \sum_{i=1}^{2} |\lambda_i|^2_{L^2(\Gamma)} \tag{2.3.2.1}$$

such that

$$a_i(u_i, v_i) = (f_i, v_i) + b_i(\lambda_i, v_i), \quad \forall\, v_i \in V_i \tag{2.3.2.2}$$

$$r_{ji}\, u_i = r_{ij}\, u_j, \quad \forall\, i, j \tag{2.3.2.3}$$

$$\lambda_i + \lambda_j = 0, \quad \forall\, i, j. \tag{2.3.2.4}$$

Problem (2.3.2.1), (2.3.2.2) is an optimal control problem with equality state and control constraints (2.3.2.3) and (2.3.2.4), respectively. The optimal controls which realize (2.3.2.1)–(2.3.2.4) are artificial or *virtual* controls, the objective of the control problem being the continuity of traces along Γ.

In the general case of a multi-domain splitting (2.3.1.9) we define

$$\lambda_i = (\lambda_{ij})_{j\in\mathcal{I}_i} \tag{2.3.2.5}$$

and, recalling (2.3.1.10), we denote by

$$U_i = \bigcup_{j\in\mathcal{I}_i} U_{ij}, \qquad U = \bigcup_{i=1}^{I} U_i. \tag{2.3.2.6}$$

The requirement (2.3.2.4) then reads as

$$\lambda_{ij} + \lambda_{ji} = 0, \quad \forall\, j \in \mathcal{I}_i, \tag{2.3.2.7}$$

$i = 1, \ldots, I$. Thus in the case of multi-domain splittings the state and control constrained optimal control problem is given by

$$\begin{aligned} &\inf\{J(\lambda_i) := \sum_{i=1}^{I} \sum_{j\in\mathcal{I}_i} |\lambda_{ij}|^2_{U_{ij}}\} \\ &a_i(u_i, v_i) = (f_i, v_i) + b_i\,(\lambda_i, v_i), \quad \forall\, v_i \in V_i, \\ &r_{ji}\, u_i = r_{ij}\, u_j, \quad j \in \mathcal{I}_i, \\ &\lambda_{ij} + \lambda_{ji} = 0, \quad j \in \mathcal{I}_i, \quad i = 1, \ldots, I. \end{aligned} \tag{2.3.2.8}$$

We add that $b_i(\lambda_i, v_i)$ is now replaced with

$$\begin{aligned} b_i(\lambda_i, v_i) &= \sum_{j\in\mathcal{I}_i} (\lambda_{ij}, r_{ji}\, v_i)_{U_{ij}} \\ &= \big\langle \sum_{j\in\mathcal{I}_i} r_{ji}^* \,\lambda_{ji}, v_i\big\rangle_{V_i}. \end{aligned} \tag{2.3.2.9}$$

Thus

$$B_i\,\lambda_i := \sum_{j\in\mathcal{I}_i} r_{ji}^*\,\lambda_{ij}. \tag{2.3.2.10}$$

Obviously, the problem (2.3.2.8) is coupled in i via the adjacency structure of the decomposition. The main goal is to derive iterative procedures in order to decouple the constraints $(2.3.2.8)_3$ and $(2.3.2.8)_4$.

We first consider a two domain decomposition and eliminate the constraint $\lambda_i + \lambda_j$ by setting $\lambda_1 = -\lambda_2 = \lambda$. Then (2.3.2.8) reads

$$\begin{gathered} \inf_{\lambda} \frac{1}{2}\,|\lambda|^2_{L^2(\Gamma)=:U} \\ a_i(u_i, v_i) = (f_i, v_i) + (-1)^{i+1}(\lambda_i,\, r_{ji}\, v_i)\ \forall\, v_i \in V_i,\ i = 1,2 \\ r_{21}\,u_1 = r_{12}\,u_2. \end{gathered} \tag{2.3.2.11}$$

We introduce a Lagrangian relaxation of $(2.3.2.11)_3$ via a Lagrange-Multiplier $q \in L^2(\Gamma)$. Thus, we consider the Lagrangian

$$\mathcal{L}(\lambda, q) := \frac{1}{2}\,|\lambda|^2_U + (q, r_{21}\,u_1 - r_{12}\,u_2)_U \tag{2.3.2.12}$$

and then the saddle-point problem

$$\inf_{\lambda}\,\sup_{q}\,\mathcal{L}(\lambda, q) = -\inf_{q}\,\inf_{\lambda}\,\mathcal{L}(\lambda, q)$$

subject to $(2.3.2.11)_2$. We define the dual functional $\tilde{J}(q)$ as

$$\tilde{J}(q) := -\inf_{\lambda}\,\mathcal{L}(\lambda, q). \tag{2.3.2.13}$$

As the problem

$$\inf_{\lambda}\,\{\mathcal{L}(\lambda, q) =: J(q; \lambda)\}\ \text{ subject to } (2.3.2.11)_2 \tag{2.3.2.14}$$

admits a unique optimal solution λ^{opt}, the original problem reduces to the unconstrained minimization problem

$$\inf_{q}\,\{-J(q; \lambda^{opt})\}. \tag{2.3.2.15}$$

We proceed to solve (2.3.2.14). To this end we take variations in $J(q;\lambda)$ with respect to λ and write down the optimality condition

$$0 = \delta J(q;\lambda)(\tilde{\lambda}) = (\lambda, \tilde{\lambda})_U + (q, r_{21}\,\tilde{u}_1 - r_{12}\,\tilde{u}_2)_U, \tag{2.3.2.16}$$

where $\tilde{u}_i$ solves $(2.3.2.11)_2$ with $f_i = 0$, $\lambda = \tilde{\lambda}$. We then introduce the adjoint states p_i corresponding to

$$a_i^*(p_i, \hat{p}_i) := a_i(\hat{p}_i, p_i) = (-1)^{i+1}\,(q, r_{ji}\,\hat{p}_i), \quad \forall\, \hat{p}_i \in V_i. \tag{2.3.2.17}$$

Thus

$$\begin{aligned} a_i^*(p_i, \tilde{u}_i) &= (-1)^{i+1}\,(q, r_{ji}\,\tilde{u}_i) \\ &= a_i(\tilde{u}_i, p_i) = (-1)^{i+1}\,(\tilde{\lambda}, r_{ji}\,p_i), \end{aligned}$$

and hence the optimality condition (2.3.2.16) gives

$$\lambda + r_{21}\,p_1 - r_{12}\,p_2 = 0. \tag{2.3.2.18}$$

We insert the unique optimal λ given by (2.3.2.18) into (2.3.2.15) to obtain

$$\begin{aligned} -J(q;\lambda^{opt}) &= -\frac{1}{2}\,|r_{21}\,p_1 - r_{12}\,p_1|^2 - (q, r_{21}\,u_1 - r_{12}\,u_2) \\ &= \frac{1}{2}\,|r_{21}\,p_1 - r_{12}\,p_2|^2 - (q, r_{21}\,w_1 - r_{12}\,w_2) \\ &\quad - (q, r_{21}\,z_1 - r_{12}\,z_2) - |r_{21}\,p_1 - r_{12}\,p_2|^2, \end{aligned}$$

where w_i solves $(2.3.2.11)_2$ with $\lambda = 0$ and z_i solves $(2.3.2.11)_2$ with $f_i = 0$ (thus $v_i = w_i + z_i$). But according to (2.3.2.17)

$$\begin{aligned} a_i(z_i, p_i) &= (-1)^{i+1}\,(q, r_{ji}\,p_i) \\ &= (-1)^{i+1}\,(\lambda, r_{ji}\,p_i) \\ &= (-1)^{i+1}\,(r_{21}\,p_1 - r_{12}\,p_2, r_{ji}\,p_i) \end{aligned}$$

or

$$-J(q;\lambda^{opt}) = \frac{1}{2}\,|r_{21}\,p_1 - r_{12}\,p_2|^2 - (q, r_{21}\,w_1 - r_{12}\,w_2). \tag{2.3.2.19}$$

The directional derivative of $-J(q;\lambda^{opt})$ with respect to q is

$$\begin{aligned} \delta(-J(q;\lambda^{opt}))(\tilde{q}) =& (r_{21}\,p_1 - r_{12}\,p_2, r_{21}\,\tilde{p}_1 - r_{12}\,\tilde{p}_2) \\ &- (r_{21}\,w_1 - r_{12}\,w_2, \tilde{q}) \\ =& (r_{21}\,p_1 - r_{12}\,p_2, r_{21}\,\tilde{p}_1 - r_{12}\,\tilde{p}_2) \\ &- (r_{21}\,u_1 - r_{12}\,u_2, \tilde{q}) \\ &+ (r_{21}\,z_1 - r_{12}\,z_2, \tilde{q}) \\ =& -(r_{21}\,u_1 - r_{12}\,u_2, \tilde{q}), \end{aligned}$$

so the gradient of $-J(q;\lambda^{opt})$ thus is

$$\nabla(-J(q;\lambda^{opt})) = -(r_{21}\,u_1 - r_{12}\,u_2). \tag{2.3.2.20}$$

The gradient procedure for the Lagrange-multiplier then is

$$q^{k+1} = q^k + p(r_{21}\,u_1 - r_{12}\,u_2) \tag{2.3.2.21}$$

where

$$a_i^*(p_i,\hat{p}_i) = (-1)^{i+1}\,(q^k, r_{ji}\,\hat{p}_i), \quad \forall\, \hat{p}_i \in V_i, \tag{2.3.2.22}$$

$$a_i(u_i,\hat{u}_i) = (-1)^{i}\,(r_{21}\,p_1 - r_{12}\,p_2, r_{ji}\,\hat{u}_i), \quad \forall\, \hat{u}_i \in V_i. \tag{2.3.2.23}$$

Algorithm 2.3.2.1

0) *Given* q^k,
1) *solve* (2.3.2.22) *for* p_i^k,
2) *solve* (2.3.2.23) *for* u_i^k,
3) *update* q^k (2.3.2.21).
4) *Go to step* 1) *and continue until done.*

Remark 2.3.2.1

i) Steps 1) and 2) are forwardly decoupled. Thus the adjoint problems(2.3.2.22) can be solved in parallel first. Then the forward problems can be solved in parallel.
ii) However, (2.3.2.22) and (2.3.2.23) does not correspond to an optimality system of an optimal control problem on Ω_i, separately.

Thus, if we eliminate the constraint $\lambda_i + \lambda_j = 0$ in (2.3.2.8) we do obtain a parallelizable iterative domain decomposition procedure based on an optimization problem, in fact a virtual optimal control problem on Ω, but the local problems to solve on Ω_i do not correspond to local virtual optimal control problems on Ω_i.

In this monograph we are interested in multi-domain and heterogeneous problems. Therefore it is desirable to be able to directly apply existing optimization software developed for a broad variety of problems on standard domains. Hence, domain decomposition methods for optimal control problems that lead to local optimal control problems are preferred. It is obvious that certain extensions, variations and alternatives may also be considered.

For instance, instead of using a Lagrange-relaxation (2.3.2.12) one may introduce an augmented Lagrangian

$$\mathcal{L}_r(\lambda,q) := \mathcal{L}(\lambda,q) + \frac{r}{2}\,|r_{21}\,u_1 - r_{12}\,u_2|_U^2. \tag{2.3.2.24}$$

See Glowinski and Le Tallec [28] for a general discussion of such methods. Again, we can derive a saddle-point type iteration analogous to Algorithm 2.3.2.1. Similar remarks as above apply to its parallel features, e.g., There is no decoupling into separate optimization problems.

Yet another variant is to consider just a penalization of the state constraints $r_{21}\, u_1 - r_{12}\, u_2$:

$$J_r(\lambda) = \frac{1}{2}\, |\lambda|_U^2 + \frac{r}{2}\, |r_{21}\, u_1 - r_{12}|^2. \tag{2.3.2.25}$$

Gradient procedures associated with (2.3.2.25) have been investigated by Gunzburger and Lee [36] and Gunzburger and Heinkenschloss [35].

Penalty based iterative domain decomposition procedures are interesting in particular for nonlinear problems, where the existence and regularity of Lagrange multipliers is difficult to establish.

Consequently, we now consider the Lagrange relaxation of $(2.3.2.8)_3$ and $(2.3.2.8)_4$ followed by gradient procedure. This approach has been outlined in [76]. We obtain the following saddle point problem

$$\begin{aligned}
&\inf_{\lambda\in U}\ \sup_{\mu,q} [\{\frac{1}{2} \sum_i \sum_{j\in\mathcal{I}_i} |\lambda_{ij}|_{U_{ij}}^2 \\
&\quad + \sum_i \sum_{j\in\mathcal{I}_i} (q_{ji}, r_{ji}\, u_i - r_{ij}\, u_j)_{U_{ij}} \\
&\quad + \sum_i \sum_{j\in\mathcal{I}_i} (\mu_{ij}, \lambda_{ij} + \lambda_{ji})_{U_{ij}}\} =: \mathcal{L}(\lambda,\mu,q)] \quad \text{subject to} \\
&a_i(u_i, v_i) = (f_i, v_i) + b_i(\lambda_i, v_i), \quad \forall\, v_i \in V_i,
\end{aligned} \tag{2.3.2.26}$$

where $\mathcal{L}(\lambda,\mu,q)$ is the Lagrangian for (2.3.2.8). We have

$$\begin{aligned}
\mathcal{L}\,(\lambda,\mu,q) =& \frac{1}{2} \sum_i \sum_{j\in\mathcal{I}_i} |\lambda_{ij}|_{U_{ij}}^2 \\
&+ \sum_{i,j} (q_{ji} - q_{ij}, r_{ij}\, u_i)_{U_{ij}} \\
&+ \sum_{i,j} (\mu_{ij} + \mu_{ji}, \lambda_{ij})_{U_{ij}}.
\end{aligned} \tag{2.3.2.27}$$

Define $J(\mu,q;\lambda) := \mathcal{L}(\mu,q,\lambda)$. Upon taking variations in $J(\mu,q;\lambda)$ with respect to λ, the necessary optimality condition is

$$\begin{aligned}
\delta\, J\,(\mu,q;\lambda)(\tilde\lambda) =& \sum_{i,j} (\lambda_{ij}, \tilde\lambda_{ij})_{U_{ij}} \\
&+ \sum_{ij} (q_{ji} - q_{ij}, r_{ji}\, \tilde u_i)_{U_{ij}} \\
&+ \sum_{i,j} (\mu_{ij} + \mu_{ji}, \tilde\lambda_{ij})_{U_{ij}} = 0.
\end{aligned} \tag{2.3.2.28}$$

We introduce the following adjoint problem

$$a_i^*(p_i, \hat p_i) = \sum_{j\in\mathcal{I}_i} (q_{ji} - q_{ij}, r_{ji}\, \hat p_i)_{U_{ij}}, \quad \forall\, \hat p_i \in V_i, \tag{2.3.2.29}$$

where $a_i^*(p_i, \hat{p}_i) = a_i(\hat{p}_i, p_i)$. Thus,

$$a_i^*(p_i, \tilde{u}_i) = a_i(\tilde{u}_i, p_i) = \sum_{j\in\mathcal{I}_i} (\tilde{\lambda}_{ij}, r_{ji}\, p_i)_{U_{ij}}. \tag{2.3.2.30}$$

Then (2.3.2.29), (2.3.2.30) used in (2.3.2.28) give the optimality condition

$$\sum_{i,j} (\lambda_{ij} + r_{ji}\, p_i + \mu_{ij} + \mu_{ji}, \tilde{\lambda}_{ij})_{U_{ij}} = 0, \quad \forall\, \tilde{\lambda}_{ij} \in U_{ij}$$

or, taking variations in U_{ij}, $\forall i, j\, (i \in \{1, \dots, I\},\, j \in \mathcal{I}_i)$

$$\lambda_{ij} = -r_{ji}\, p_i - \mu_{ij} + \mu_{ji} \quad \forall\, i, j. \tag{2.3.2.31}$$

We are going to use (2.3.2.31) in order to express $J(q, \mu; \lambda^{opt})$ as a cost functional $\tilde{J}(\mu, q)$. Indeed,

$$\begin{aligned} -J(\mu, q; \lambda^{opt}) =& \frac{1}{2} \sum_{i,j} |r_{ji}\, p_i + \mu_{ij} + \mu_{ji}|^2_{U_{ij}} \\ & - \sum_{i,j} (q_{ji} - q_{ij}, r_{ji}\, w_i) - \sum_{i,j} (\mu_{ij} + \mu_{ji}, \lambda_{ij})_{U_{ij}} \\ & - \sum_{i,j} (q_{ji} - q_{ij}, r_{ji}\, z_i) - \sum_{i,j} |r_{ji}\, p_i + \mu_{ij} + \mu_{ji}|^2_{U_{ij}} \end{aligned} \tag{2.3.2.32}$$

where $u_i := w_i + z_i$ such that

$$a_i(w_i, v_i) = (f_i, v_i), \quad \forall\, v_i \in V_i, \tag{2.3.2.33}$$

$$a_i(z_i, v_i) = \sum_{j\in\mathcal{I}_i} (\lambda_{ij}, r_{ji}\, v_i)_{U_{ij}}, \quad \forall\, v_i \in V_i. \tag{2.3.2.34}$$

Equations (2.3.2.33), (2.3.2.34) and (2.3.2.30) imply

$$\begin{aligned} -J(\mu, q; \lambda^{opt}) =& \frac{1}{2} \sum_{i,j} |r_{ji}\, p_i + \mu_{ij} + \mu_{ji}|^2_{U_{ij}} \\ & - \sum_{i,j} (q_{ji} - q_{ij}, r_{ji}\, w_i)_{U_{ij}} - \sum_{i,j} (\mu_{ij} + \mu_{ji}, \lambda_{ij})_{U_{ij}} \\ & - \sum_{i} \{a_i(z_i, p_i) - \sum_{j\in\mathcal{I}_i} (\lambda_{ij}, r_{ji}\, p_i)_{U_{ij}}\} \\ & + \sum_{i,j} (\lambda_{ij} + \lambda_{ji}, \mu_{ij})_{U_{ij}} \\ =& \frac{1}{2} \sum_{ij} |r_{ji}\, p_i + \mu_{ij} + \mu_{ji}|^2_{U_{ij}} \\ & - \sum_{i,j} (q_{ji} - q_{ij}, r_{ij}\, w_i)_{U_{ij}} = \tilde{J}(\mu, q). \end{aligned} \tag{2.3.2.35}$$

We have to consider the dual problem

$$\begin{aligned}
&\inf_{\mu,q} \tilde{J}(\mu,q) \quad \text{subject to} \\
&a_i(u_i,\hat{u}_i) = (f_i,\hat{u}_i) - \sum_{j\in\mathcal{I}_i} (r_{ji}\,p_i + \mu_{ij} + \mu_{ji}, r_{ji}\,\hat{u}_i)_{U_{ij}}, \quad \forall\, \hat{u}_i \in V_i, \\
&a_i^*(p_i,\hat{p}_i) = \sum_{j\in\mathcal{I}_i} (q_{ji} - q_{ij}, r_{ji}\,\hat{p}_i), \quad \forall\, \hat{p}_i \in V_i.
\end{aligned} \tag{2.3.2.36}$$

The optimality conditions for the optimal control problem (2.3.2.36) are derived as follows:

$$\begin{aligned}
0 = \delta\,\tilde{J}\,(\mu,q)\,(\tilde{\mu},\tilde{q}) &= \sum_{i,j} (r_{ji}\,p_i + \mu_{ij} + \mu_{ji},\, r_{ji}\,\tilde{p}_i + \mu_{ij} + \mu_{ji})_{U_{ij}} \\
&\quad - \sum_{ij} (\tilde{q}_{ji} - \tilde{q}_{ij},\, r_{ji}\,w_i)_{U_{ij}} \\
&= \sum_{i,j} (r_{ji}\,p_i + \mu_{ij} + \mu_{ji},\, r_{ji}\,\tilde{p}_i)_{U_{ij}} \\
&\quad + \sum_{i,j} (r_{ji}\,p_i + r_{ij}\,p_j + 2(\mu_{ij} + \mu_{ij}),\, \tilde{\mu}_{ij})_{U_{ij}} \\
&\quad + \sum_i a_i(z_i,\tilde{p}_i) - \sum_{i,j} (\tilde{q}_{ji} - \tilde{q}_{ij},\, r_{ji}\,u_i)_{U_{ij}} \\
&= -\sum_{i,j} (\lambda_{ij} + \lambda_{ji},\, \tilde{\mu}_{ij})_{U_{ij}} - \sum_{i,j} (r_{ji}\,u_i - r_{ij}\,u_j,\, \tilde{q}_{ij})_{U_{ij}}.
\end{aligned} \tag{2.3.2.37}$$

Hence the gradient of $\tilde{J}(\mu,q)$ is

$$\nabla\,\tilde{J}(\mu,q) = -((r_{ji}\,u_i - r_{ij}\,u_j)_{i,j},\, (\lambda_{ij} + \lambda_{ji})_{i,j}). \tag{2.3.2.38}$$

Thus we arrive at the following gradient procedure for (2.3.2.36).

Algorithm 2.3.2.2

0) *Given* q^k, μ^k.
1) *Solve for* p_i^k:
   $a_i^*(p_i^k,\hat{p}_i) = \sum_{j\in\mathcal{I}_i} (q_{ji}^k - q_{ij}^k, r_{ji}\,\hat{p}_i)_{U_{ij}} \quad \forall\, \hat{p}_i \in V_i,$
2) *Solve for* u_i^k:
   $a_i(u_i^k,\hat{u}_i) = (f_i,\hat{u}_i) - \sum_{j\in\mathcal{I}_i} (r_{ji}\,p_i^k + \mu_{ij}^k + \mu_{ji}^k, r_{ji}\,\hat{u}_i)_{U_{ij}} \quad \forall\, \hat{u}_i \in V_i,$
3) *update* q^k, μ^k:
   $q_{ji}^{k+1} = q_{ji}^k + p(r_{ji}\,u_i^k - r_{ij}\,u_j^k)$
   $\mu_{ij}^{k+1} = \mu_{ij}^k + p\,(\lambda_{ij}^k + \lambda_{ji}^k),$
   with $\lambda_{ij}^k = r_{ji}\,p_i^k + \mu_{ij}^k + \mu_{ji}^k$.
4) *Go to step* 1) *and continue until done.*

This procedure has been proposed by J.-L. Lions and O. Pironneau [76] for overlapping domain decompositions, in case no constraint on λ_{ij} applies. They remark that the method can be extended also to the nonoverlapping case. The arguments above provide the details.

In retrospective, going back to the original saddle-point formulation (2.3.2.26) of (2.3.2.8) we realize that the cost, for given q^k, μ^k can be written as

$$\begin{aligned} J(\mu^k, q^k; \lambda) = &\sum_{i=1}^{I} \{ \sum_{j\in\mathcal{I}_i} \frac{1}{2} \sum_{j\in\mathcal{I}_i} |\lambda_{ij}|^2_{U_{ij}} \\ &\qquad + \sum_{j\in\mathcal{I}_i} (q^k_{ji} - q^k_{ij}, r_{ji}\, u_i)_{U_{ij}} \\ &\qquad + \sum_{j\in\mathcal{I}_i} (\lambda_{ij}, \mu^k_{ij} + \mu^k_{ji})_{U_{ij}} \} \\ := &\sum_i J_i(\mu^k,\, q^k,\, \lambda_i). \end{aligned} \tag{2.3.2.39}$$

But

$$\begin{aligned} 0 = \delta\, J_i(\mu^k, q^k; \lambda_i)(\tilde{\lambda}_i) &= \sum_{j\in\mathcal{I}_i} (\lambda_{ij}, \tilde{\lambda}_{ij})_{U_{ij}} \\ &\quad + \sum_{j\in\mathcal{I}_i} (q^k_{ji} - q^k_{ij}, r_{ji}\, \tilde{u}_i)_{U_{ij}} + \sum_{j\in\mathcal{I}_i} (\mu^k_{ij} + \mu^k_{ij},\, \tilde{\lambda}_{ij})_{U_{ij}} \\ &= \sum_{j\in\mathcal{I}_i} (\lambda_{ij},\, \tilde{\lambda}_{ij})_{U_{ij}} \\ &\quad + a_i^*(p_i, \tilde{u}_i) + \sum_{j\in\mathcal{I}_i} (\mu^k_{ij} + \mu^k_{ji}, \tilde{\lambda}_{ij})_{U_{ij}} \\ &= \sum_{j\in\mathcal{I}_i} (\lambda_{ij}, \tilde{\lambda}_{ij})_{U_{ij}} \\ &\quad + \sum_{j\in\mathcal{I}_i} (\tilde{\lambda}_{ij}, r_{ji}\, p_i) + \sum_{j\in\mathcal{I}_i} (\mu^k_{ij} + \mu^k_{ji},\, \tilde{\lambda}_{ij})_{U_{ij}}. \end{aligned}$$

if (2.3.1.60), (2.3.2.29) and (2.3.2.31) hold or, more compactly, if

$$\begin{aligned} &a_i(u_i, \hat{u}_i) + \sum_{j\in\mathcal{I}_i} (r_{ji}\, p_i, r_{ji}\, \hat{u}_i)_{U_{ij}} \\ &= (f_i, \hat{u}_i) - \sum_{j\in\mathcal{I}_i} (\mu^k_{ij} + \mu^k_{ji}, r_{ji}\, \hat{u}_i)_{U_{ij}}, \quad \forall\, \hat{u}_i \in V_i, \\ &a_i^*(p_i, \hat{p}_i) = \sum_{j\in\mathcal{I}_i} (q^k_{ji} - q^k_{ij}, r_{ji}\, \hat{p}_i)_{U_{ij}}, \quad \forall\, \hat{p}_i \in V_i. \end{aligned} \tag{2.3.2.40}$$

This shows that (2.3.2.40) is the optimality condition for the optimal control problem

$$\begin{aligned}&\inf_{\lambda_i} J_i(\mu^k, q^k; \lambda_i) \text{ subject to}\\ &a_i(u_i, \hat{u}_i) = (f_i, \hat{u}_i) + \sum_{j\in\mathcal{I}_i} (\lambda_{ij}, r_{ji}\,\hat{u}_i), \quad \forall\, \hat{u}_i \in V_i.\end{aligned} \tag{2.3.2.41}$$

Remark 2.3.2.2 The Lagrange relaxation (2.3.2.26) followed by the saddle-point iteration given in Algorithm 2.3.2.2 results in a decomposition of the global optimality system – which, in turn, is obtained from (2.3.2.40) without indices k and $\nabla\,\tilde{J}(\mu, q) = 0$ in (2.3.2.38) – into a local system (2.3.2.40) that is itself an optimality system of the virtual optimal control problem (2.3.2.41), defined on subdomain i. This is precisely the paradigm we intend to pursue throughout this monograph.

Remark 2.3.2.3 It is important to note that if we eliminate the control constraints $\lambda_{ij} + \lambda_{ji} = 0$ first and then optimize with respect to the remaining virtual controls (just one function on $\Gamma_{ij} = \Gamma_{ji}$ $\forall\, i, j$ such that Ω_i and Ω_j are adjacent), then the optimality conditions are coupled.

It appears to be very natural to stabilize and possibly accelerate Algorithm 2.3.2.2. The typical procedure is to consider an augmented Lagrangian instead of the original Lagrangian (2.3.2.26), where one introduces a penalty parameter $r > 0$ and considers

$$\begin{aligned}\mathcal{L}_r(\lambda, \mu, q) :=& \mathcal{L}(\lambda, \mu, q)\\ &+ \frac{r}{2}\{\sum_{i,j} |r_{ji}\,u_i - r_{ij}\,u_j|^2_{U_{ij}}\\ &+ \sum_{i,j} |\lambda_{ij} + \lambda_{ji}|^2_{U_{ij}}\}.\end{aligned} \tag{2.3.2.42}$$

It turns out, however, that the corresponding optimality systems would not decouple into a local penalized problem.

In order to derive an augmented Lagrangian method that leads to a decomposed set of optimization problems in each individual domain, we further relax the transmission condition $r_{ji}\,u_i = r_{ij}\,u_j$ by introducing yet another virtual control η

$$r_{ji}\,u_i = \eta_{ij} \quad \forall\, i, j,\ i \in \mathcal{I},\ j \in \mathcal{I}_i, \tag{2.3.2.43}$$

$\eta_{ij} = \eta_{ji}$. We may then consider the Lagrangian relaxations of (2.3.2.43). Let us first look at the two-domain case with $\lambda_{ij} = -\lambda_{ji} = \lambda$ on Γ, $\eta_{ij} = \eta_{ji} = \eta$,

$$\begin{aligned}&\inf_{\lambda,\eta} \sup_{q} \{\mathcal{L}_\beta(\lambda, \eta; q) = |\lambda|^2 + q_{21}(r_2, u_1 - \eta)\\ &+ q_{12}(r_{12}\,u_2 - \eta) + \frac{\beta}{2}\{|r_{21}\,u_1 - \eta|^2 + |r_{12}\,u_2 - \eta|\} \ s.t.\\ &a_1(u_1, v_1) = (f_1, v_1) + (\lambda, r_{21}\,u_1)_\Gamma, \quad \forall\, v_1 \in V_1\\ &a_2(u_2, v_2) = (f_2, v_2) - (\lambda, r_{12}\,v_2)_\Gamma, \quad \forall\, v_2 \in V_2.\end{aligned} \tag{2.3.2.44}$$

Equations $(2.3.2.44)_{2,3}$ imply

$$a_1(u_1, v_1) + a_2(u_2, v_2) = (f_1, v_1) + (f_2, v_2), \ \ \forall (v_1, v_2) \in V_1 \times V_2 \qquad (2.3.2.45)$$

which is, in turn, equivalent to

$$\inf_{u_1, u_2 \in V_1 \times V_2} \{\sum_{i=1}^{2} \frac{1}{2}\, a_i(u_i, u_i) - (f_i, u_i) =: J(u_1, u_2)\}. \qquad (2.3.2.46)$$

On the other hand, $(2.3.2.44)_{2,3}$ is equivalent to

$$\inf_{u_i \in V_i} \{\frac{1}{2}\, a_i(u_i, u_i) - (f_i, u_i) - (-1)^{i+1}(\lambda, r_{ji}\, u_i) =: J_i(u_i)\}. \qquad (2.3.2.47)$$

If in (2.3.2.46) we have $r_{ji}\, u_i = r_{ij}\, u_j$, then (2.3.2.46) is the original problem. In this case $J_1(u_1) + J_2(u_2) = J(u_1, u_2)$. Thus the PDE-constraints (2.3.2.44) are equivalent to yet another optimization problem, if the continuity of traces holds across Γ. Thus, we may rewrite problem (2.3.2.44) as the unconstrained problem

$$\begin{aligned} &\inf_{(u_i), \eta} \sup_{q} \{\mathcal{L}_\beta(u, \eta; q) =: \sum_{i=1}^{2} \frac{1}{2}\, a_i(u_i, u_i) - (f_i, u_i) \\ &+ \sum_{i,j} (q_{ji}, r_{ji}\, u_i - \eta) + \frac{\beta}{2} \sum_{i,j} |r_{ji}\, u_i - \eta|^2\}. \end{aligned} \qquad (2.3.2.48)$$

Now (2.3.2.48) is a saddle-point problem for which we may use various extensions of the classical Uzawa algorithm. See Glowinski and Le Tallec [28] for a very nice presentation of the general subject.

We use a fractional step method with respect to the gradient procedure for the dual problems as advocated in [28] (ALG3).

Algorithm 2.3.2.3

0) η^{k-1}, q^k *given*
1) *solve for* u_i^k, $i = 1, 2$:
   $\partial_{u_i} \mathcal{L}_\beta(u^k, \eta^{k-1}; q^k) = 0$
2) *update* q_{ji}^k, $i = 1, 2$:
   $q_{ji}^{k+\frac{1}{2}} = q_{ji}^k + \rho_k(r_{ji}\, u_i^k - \eta^{k-1})$
3) *solve for* η^k:
   $\partial_\eta \mathcal{L}_\beta(u^k, \eta^k; q^{k+\frac{1}{2}}) = 0$
4) *update* $q_{ji}^{k+\frac{1}{2}}$
   $q_{ji}^{k+1} = q_{ji}^{k+\frac{1}{2}} + \rho_k(r_{ji}\, u_i^k - \eta^k)$.

Now by 1), $\delta \mathcal{L}_\beta(u^k, \eta^{k-1}; q^k)(\tilde{u}) = 0 \Leftrightarrow$

$$0 = \sum_{i=1}^{2} a_i(u_i\, \tilde{u}_i) - (f_i, \tilde{u}_i) + \sum_{i,j} (q_{ji}^k, r_{ji}\, \tilde{u}_i) + \beta \sum_{i,j} (r_{ji}\, u_i^k - \eta^{k-1}, r_{ji}\, \tilde{u}_i)$$

or

$$a_i(u_i^k, \tilde{u}_i) + \beta(r_{ji}\, u_I^k, r_{ji}\, \tilde{u}_i) = (f_i, \tilde{u}_i) - (q_{ji}^k - \beta\, \eta^{k-1}, r_{ji}\, \tilde{u}_i), \qquad (2.3.2.49)$$

$i, j = 1, 2$. In fact for $k \to k+1$, (2.3.2.49) reads

$$a_i(u_i^{k+1}, \tilde{u}_i) + \beta\, (r_{ji}\, u_i^{k+1}, r_{ji}\, \tilde{u}_i) = (f_i, \tilde{u}_i) - (q_{ji}^{k+1} - \beta\, \eta^k, r_{ji}\, \tilde{u}_i). \qquad (2.3.2.50)$$

The third step gives:

$$\eta^k = \frac{1}{2\beta}\, (q_{ji}^{k+\frac{1}{2}} + q_{ij}^{k+\frac{1}{2}}) + \frac{1}{2}\, (r_{ji}\, u_i^k + r_{ij}\, u_j^k), \qquad (2.3.2.51)$$

or

$$q_{ji}^{k+1} + q_{ij}^{k+1} + (\beta - \rho_k)(r_{ji}\, u_i^k + r_{ij}\, u_j^k - 2\eta^k) = 0, \qquad (2.3.2.52)$$

after applying step 4). From (2.3.2.50) and (2.3.2.52) we obtain

$$a_i(u_i^{k+1}, \tilde{u}_i) + \beta(r_{ji}\, u_i^{k+1}, r_{ji}\, \tilde{u}_i) = (f_i, \tilde{u}_i) + (q_{ij}^{k+1} + q_{ji}^{k+1}) + (\beta - \rho_k)(r_{ji}\, u_i^k + r_{ij}\, u_j^k - 2\eta^k) + \beta(\eta^k, r_{ji}\, u_i), \qquad (2.3.2.53)$$

where we use variations satisfying $r_{ji}\, \tilde{u}_i = r_{ij}\, \tilde{u}_j$. According to steps 2) and 4), (2.3.2.52) gives

$$\begin{aligned} a_i(u_i^{k+1}, \tilde{u}_i) &+ \beta(r_{ji}\, u_i^{k+1}, r_{ji}\, \tilde{u}_i) \\ &= (f_i, \tilde{u}_i) + (q_{ij}^k + 2\rho_k\, r_{ij}\, u_j^k - \rho_k |\, r_{ji}\, \tilde{u}_i) \\ &\quad + (\beta - \rho_k)(r_{ij}\, u_j^k + r_{ji}\, u_i^k - \eta^k |, r_{ji}\, \tilde{u}_i) \\ &= (f_i, \tilde{u}_i) + (q_{ij}^k + 2\beta\, r_{ij}\, u_j^k - \beta\, \eta^{k-1}) \\ &\quad + (\beta - \rho_k)(\eta^{k+1} - \eta^k + r_{ji}\, u_i^k - r_{ij}\, u_j^k) \\ &= (f_i, \tilde{u}_i) + (f_j, \tilde{u}_j) - a_j(u_j^k, \tilde{u}_j) \\ &\quad + \beta(r_{ij}\, u_j^k, r_{ji}\, \tilde{u}_i) \\ &\quad + (\beta - \rho_k)(\eta^{k-1} - \eta^k + r_{ji}\, u_i^k, r_{ij}\, \tilde{u}_j). \end{aligned} \qquad (2.3.2.54)$$

In conclusion, Algorithm 2.3.2.3 results in the following iteration:

Algorithm 2.3.2.4

0) η^{k-1}, q^k *given,*

1) *solve for* u_i^k, $i = 1, 2$

$$a_i(u_i^k\, \hat{u}_i) + \beta(r_{ji}\, u_i^k, r_{ji}\, \hat{u}_i) =$$
$$(f_i, \hat{u}_i) - (q_{ji}^k - \beta\, \eta^{k-1}, r_{ji}\, \hat{u}_i) \;\; \forall\, \hat{u}_i \in V_i.$$

2) *Update:*

$$q_{ji}^{k+\frac{1}{2}} = q_{ji}^k + \rho_k(r_{ji}\, u_i^k - \eta^{k-1}).$$

3) *Compute* η^k *as*

$$\eta^k = \frac{1}{2\beta}(q_{ij}^{k+\frac{1}{2}} + q_{ji}^{k+\frac{1}{2}}) + \frac{1}{2}(r_{ji}\, u_i^k + r_{ij}\, u_j^k).$$

4) *Update:*

$$q_{ji}^{k+1} = q_{ji}^{k+\frac{1}{2}} + \rho_k(r_{ji}\, u_i^k - \eta^k).$$

Thus, Algorithm 2.3.2.4 can bee seen as a parallel iterative solution of

$$\begin{aligned} a_i(u_i, \hat{u}_i) &= (f_i, \hat{u}_i) + (c_i\, g_i^k, r_{ji}\, \hat{u}_i) \\ M_k\, g_i^{k+1} &= N_k\, g_i^k + d_i^k, \end{aligned} \tag{2.3.2.55}$$

where

$$g_i^k := (\eta^{k-1}, q_{ij}^k, q_{ij}^k)^T,\; c_1 = (\beta, -1, 0),\; c_2 = (\beta, 0, -1)$$

$$M_k := \begin{pmatrix} 2(\beta - \rho_k) & -1 & -1 \\ \rho_k & 1 & 0 \\ \rho_k & 0 & 1 \end{pmatrix},\; N_k = \begin{pmatrix} 0 & 0 & 0 \\ -\rho_k & 1 & 0 \\ -\rho_k & 0 & 1 \end{pmatrix} \tag{2.3.2.56}$$

$$d_i^k = (\beta - \rho_k)(r_{ji}\, u_i^k + r_{ij}\, u_j^k),\; 2\rho_k\, r_{ji}\, u_i^k,\; 2\rho_k\, r_{ij}\, u_j^k)^T.$$

Now

$$M_k^{-1} = \frac{1}{2\beta}\begin{pmatrix} 1 & 1 & 1 \\ -\rho_k & 2\beta - \rho_k & -\rho_k \\ -\rho_k & -\rho_k & 2\beta - \rho_k \end{pmatrix},$$

$$M_k^{-1}\, N_k = \frac{1}{2\beta}\begin{pmatrix} -2\rho_k & 1 & 1 \\ 2\rho_k(\rho_k - \beta) & 2\beta - \rho_k & -\rho_k \\ 2\rho_k(\rho_k - \beta) & -\rho_k & 2\beta - \rho_k \end{pmatrix},$$

$$\begin{aligned} M_k^{-1}\, d_k = \frac{\rho}{2\beta}\, (2[1 + \rho_k - \beta]\, &(r_{ij}\, u_j^k + r_{ij}\, u_j^k), \\ &(3\beta - \rho_k)\, r_{ji}\, u_i^k - (\beta + \rho_k)\, r_{ij}\, u_j^k, \\ &-(\beta + \rho_k)\, r_{ji}\, u_i^k + (3\beta - \rho_k,\, r_{ij}\, u_i^k)^T. \end{aligned}$$

The iteration matrix $M_k^{-1} N_k$ has the eigenvalues $\{0, 1, 1 - 2\,\frac{\rho_k}{\beta}\}$ with corresponding eigenvectors $(\frac{1}{\rho_k}, 1, 1)^T$, $(0, -1, 1)^T$, $(\frac{1}{\beta - \rho_k}, 1, 1)^T$ for $\beta \neq \rho_k$, and $\{0, -1, 1\}$, $(\frac{1}{\rho_k}, 1, 1)^T$, $(1, 0, 0)^T$, $(0, -1, 1)^T$ for $\beta = \rho_k$.

It therefore appears optimal to choose $\rho_k = \beta/2$. The limiting value is $\beta = \rho_k$. However, the convergence of Algorithm 2.3.2.4 cannot be analyzed from the spectrum and the eigenbasis of $M_k^{-1} N_k$ alone. Indeed, a convergence analysis for $\beta \neq \rho_k$ is not known in this context of domain decompositions. The case $\beta = \rho$ for all k will be precisely the one considered in this monograph. See Glowinski and Le Tallec [28] for a discussion of the convergence properties of this version of the saddle-point algorithm.

We return to Algorithm 2.3.2.4 and set $\beta = \rho$, $\forall\, k$. However, by (2.3.2.52) this implies

$$q_{ji}^{k+1} = -q_{ij}^{k+1}$$

and (2.3.2.53) implies

$$a_i(u_i^{k+1}, \tilde{u}_i) + \beta(r_{ji}\, u_i^{k+1}, r_{ji}\, \tilde{u}_i) = (f_i, \tilde{u}_i) + (q_{ij}^{k+1} + \beta\eta^k, r_{ji}\, \tilde{u}_i). \qquad (2.3.2.57)$$

In the strong form (2.3.2.57) leads to the boundary condition

$$\frac{\partial u_i^{k+1}}{\partial \nu_{A_i}} + \beta\, r_{ji}\, u_i^{k+1} = q_{ij}^{k+1} + \beta\, \eta^k$$

which, according to steps 2) and 4), reads as

$$\begin{aligned} \frac{\partial u_i^{k+1}}{\partial \nu_{A_i}} + \beta\, r_{ji}\, u_i^{k+1} &= q_{ij}^k + 2\beta\, r_{ij}\, u_j^k - \beta\, \eta^{k-1} \\ &= -\frac{\partial u_j^k}{\partial \nu_{A_j}} + \beta\, r_{ij}\, u_j^k. \end{aligned}$$

Thus, for $\beta = \rho$, $\forall\, k$ we can completely eliminate the Lagrangian multipliers and obtain an iteration based on the boundary conditions

$$\frac{\partial u_i^{k+1}}{\partial \nu_{A_i}} + \beta\, r_{ji}\, u_i^{k+1} = -\frac{\partial u_j^k}{\partial \nu_{A_j}} + \beta\, r_{ij}\, u_j^k \qquad (2.3.2.58)$$

for $i, j = 1, 2$. As in the previous, more general setting, consistency of this algorithm is easily assessed.

The decoupling of the original transmission conditions

$$\begin{aligned} r_{ji}\, u_i &= r_{ij}\, u_j \quad \text{on} \quad \Gamma \\ \frac{\partial u_i}{\partial \nu_{A_i}} + \frac{\partial u_j}{\partial \nu_{A_i}} &= 0 \quad \text{on} \quad \Gamma \end{aligned} \qquad (2.3.2.59)$$

according to (2.3.2.58) is due to P.-L. Lions [82].

2.3.3 The Basic Algorithm of P.-L. Lions

For easier reference, we now summarize the algorithm of P.-L.Lions [82] described at the end of the last section and show convergence of the algorithm. To fix ideas, we begin with the simple and standard Poisson problem. Let Ω be a bounded Lipschitz domain in $\mathbb{R}^2$ with piecewise Lipschitz boundary $\partial\Omega$ which admits the decomposition $\Omega = \Omega_1 \cup \Omega_2$, $\Omega_1 \cap \Omega_2 = \emptyset$, $\Omega_1 \cap \Omega_2 = \Gamma$, such that $\partial\Omega_i = \Gamma \cup (\partial\Omega_i \cap \partial\Omega)$, $i = 1, 2$; see Figure 2.5.

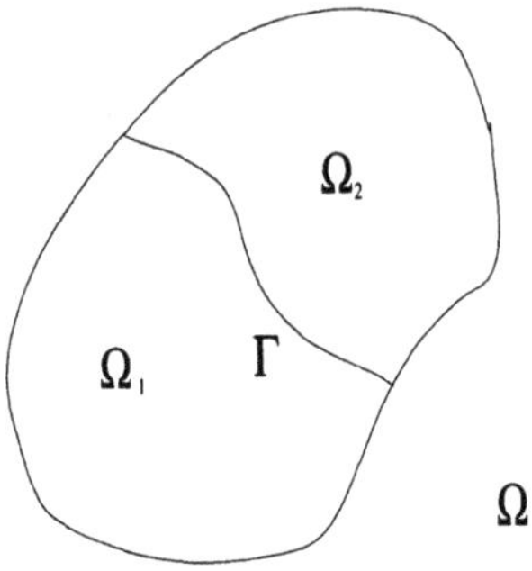

Figure 2.5: Nonoverlapping domain decomposition

In Ω we consider the problem

$$\begin{aligned} -\Delta u &= f \text{ in } \Omega \\ u &= 0 \text{ on } \partial\Omega. \end{aligned} \tag{2.3.3.1}$$

This problem is equivalent to solving the coupled problems

$$\begin{aligned} -\Delta u_i &= f_i \text{ in } \Omega_i \\ u_i &= 0 \text{ on } \partial\Omega_i \cap \partial\Omega \\ u_i &= u_j \text{ on } \Gamma \\ \frac{\partial}{\partial \nu_i} u_i + \frac{\partial}{\partial \nu_j} u_j &= 0 \text{ on } \Gamma,\ i, j = 1, 2, \end{aligned} \tag{2.3.3.2}$$

where ν_i, ν_j denotes the outward normal on Γ for Ω_i towards Ω_j, e.g., $\nu_i = -\nu_j$.

The idea pursued by P.-L. Lions [82] is very natural and simple. Namely, one reformulates the equality constraints on Γ in (2.3.3.2) as

$$\frac{\partial}{\partial \nu_i} u_i + \beta u_i = -\frac{\partial}{\partial \nu_j} u_j + \beta u_j,\ i, j = 1, 2 \tag{2.3.3.3}$$

and relaxes (2.3.3.3) by the way of an iteration in the spirit of the alternating direction or operator splitting paradigm as follows:

Algorithm 2.3.3.1 *Given* u_i^n, $\frac{\partial}{\partial \nu_i} u_i^n$ $i = 1,2$ *compute* u_i^{n+1}, $\frac{\partial}{\partial \nu_i} u_i^{n+1}$, $i = 1,2$ *according to*

$$-\Delta u_i^{n+1} = f_i \ \ in\ \ \Omega_i \tag{2.3.3.4}$$

$$u_i^{n+1} = 0 \ \ on\ \ \partial\Omega_i \cap \partial\Omega \tag{2.3.3.5}$$

$$\frac{\partial}{\partial \nu_i} u_i^{n+1} + \beta u_i^{n+1} = -\frac{\partial}{\partial \nu_j} u_j^n + \beta u_j^n \ \ on\ \Gamma,\ i,j = 1,2. \tag{2.3.3.6}$$

We first show that this algorithm is consistent with the original problem (2.3.3.1) and (2.3.3.2). Indeed, (2.3.3.3) is equivalent to the transmission conditions on Γ in (2.3.3.2) since (2.3.3.3) gives

$$\frac{\partial}{\partial \nu_i} u_i + \beta\, u_i = -\frac{\partial}{\partial \nu_j} u_j + \beta\, u_j$$

$$\frac{\partial}{\partial \nu_j} u_j + \beta\, u_j = -\frac{\partial}{\partial \nu_i} u_i + \beta\, u_i$$

which upon addition first yields

$$\frac{\partial}{\partial \nu_i} u_i + \frac{\partial}{\partial \nu_j} u_j = 0$$

and then $u_i = u_j$, provided $\beta \neq 0$ (we always take $\beta > 0$). Now for consistency, assume convergence in Algorithm 2.3.3.1. Then the iteration (2.3.3.6) reduces to (2.3.3.3), so that the limiting problem is precisely (2.3.3.1), or (2.3.3.2).

We proceed to prove convergence of Algorithm 2.3.3.1. To this end we introduce the spaces $H = L^2(\Omega)$, $V = H_0^1(\Omega)$ and $V_i = \{u \in H^1(\Omega_i) : u|_{\Omega_i \cap \partial\Omega} = 0\}$, $H_i = L^2(\Omega_i)$, $i = 1,2$, where standard notation from functional analysis is employed. Let us for simplicity assume

$$u \in H_0^1(\Omega) \cap H^2(\Omega) \tag{2.3.3.7}$$

such that (2.3.3.3) , (2.3.3.6) are well defined on $L^2(\Gamma)$. In fact $u \in H_0^1(\Omega) \cap H^{3/2}(\Omega)$ and $\frac{\partial}{\partial \nu} u \in L^2(\Gamma)$ would also be sufficient.

The local errors (that is, the differences $u_i^n - u_i$ between the local and global solutions) satisfy (2.3.3.4) with $f_i = 0$. Let $\lambda_{ij} \in L^2(\Gamma)$, $i,j = 1,2$, $i \neq j$, and let u_i solve

$$\begin{aligned} -\Delta u_i &= 0 \ \text{ in } \ \Omega_i, \\ u_i &= 0 \ \text{ on } \ \partial\Omega_i \cap \partial\Omega \\ \frac{\partial u_i}{\partial \nu_i} + \beta u_i &= \lambda_{ij} \ \text{ on } \ \Gamma \end{aligned} \tag{2.3.3.8}$$

Thus $u_i \in H^2(\Omega_i) \cap V_i$, and hence $-\frac{\partial u_j}{\partial \nu_j} + \beta\, u_j \in L^2(\Gamma)$, $i \neq j = 1,2$. Set $X = (\lambda_{12}, \lambda_{21})$ and define

$$(\mathcal{T}X)_i = (-\frac{\partial u_j}{\partial \nu_j} + \beta\, u_j)|_\Gamma, \quad i = 1,2, \tag{2.3.3.9}$$

$$\mathcal{T}X = (\mathcal{T}_1 X, \mathcal{T}_2 X). \tag{2.3.3.10}$$

Then

$$\mathcal{T} : \mathcal{X} \mapsto \mathcal{X} \tag{2.3.3.11}$$

where $\mathcal{X} := L^2(\Gamma) \times L^2(\Gamma)$ and, moreover, Algorithm 2.3.3.1 (with $f_i = 0$) is equivalent to the fixed point iteration in $\mathcal{X}$:

$$X^{n+1} = \mathcal{T}(X^n), \tag{2.3.3.12}$$

where $X^0 \in \mathcal{X}$ is arbitrary.

We proceed to show that $\mathcal{T}$ is non-expansive, i.e., $\|\mathcal{T}X\|_{\mathcal{X}} \leq \|X\|_{\mathcal{X}}$, where we set

$$\|X\|^2_{\mathcal{X}} := \frac{1}{\beta} \int_{\Gamma} (\,|\lambda_{12}|^2 + |\lambda_{21}|^2)\, d\gamma. \tag{2.3.3.13}$$

Now

$$\|X\|^2_{\mathcal{X}} = \frac{1}{\beta} \sum_{i=1}^{2} \int_{\Gamma} \{ |\frac{\partial u_i}{\partial \nu_i}|^2 + 2\beta (\frac{\partial u_i}{\partial \nu_i})\, u_i + \beta^2 |u_i|^2 \}\, d\gamma$$

and

$$\|\mathcal{T}X\|^2_{\mathcal{X}} = \frac{1}{\beta} \sum_{i=1}^{2} \int_{\Gamma} \{ |\frac{\partial}{\partial \nu_i} u_i|^2 - 2\beta (\frac{\partial}{\partial \nu_i} u_i)\, u_i + \beta^2 |u_i|^2 \}\, d\gamma$$

so that

$$\|\mathcal{T}X\|^2_{\mathcal{X}} - \|X\|^2_{\mathcal{X}} = -4 \sum_{i=1}^{2} \int_{\Gamma} (\frac{\partial}{\partial \nu_i}\, u_i)\, u_i\, d\gamma. \tag{2.3.3.14}$$

By integration by parts

$$\|\mathcal{T}\, X\|^2_{\mathcal{X}} - \|\mathcal{X}\|^2_{\mathcal{X}} = -4 \sum_{i=1}^{2} \int_{\Omega_i} |\nabla\, u_i|^2\, dx \tag{2.3.3.15}$$

which shows that $\mathcal{T}$ is indeed nonexpansive.

Convergence of Algorithm 2.3.3.1 is equivalent to showing that

$$\mathcal{T}\, X^n = X^{n+1} \to 0. \tag{2.3.3.16}$$

Before we embark on a proof of convergence, we emphasize that, from the point of view of numerical procedures based on Algorithm 2.3.3.1, one should also consider some (under-)relaxation, namely, instead of (2.3.3.6) we consider:

$$\begin{aligned} \frac{\partial}{\partial \nu_i}\, u_i^{n+1} + \beta\, u_i^{n+1} &= (1-\varepsilon)(-\frac{\partial}{\partial \nu_j}\, u_j^n + \beta\, u_j^u) \\ &+ \varepsilon(\frac{\partial}{\partial \nu_i}\, u_i^n + \beta\, u_i^n),\ i,j = 1,2. \end{aligned} \tag{2.3.3.17}$$

This amounts to the relaxed iteration

$$X^{n+1} = (1-\varepsilon)\,\mathcal{T}\,X^n + \varepsilon\,X^n \tag{2.3.3.18}$$

which, obviously, has the same fixed points. We proceed to show convergence of the iterations (2.3.3.18) for any $\varepsilon \in [0,1)$.

From (2.3.3.15) we have

$$\begin{aligned}
&\|X^{n+1}\|^2_{\mathcal{X}} \\
&= (1-\varepsilon)^2\,\|\mathcal{T}\,X^n\|^2_{\mathcal{X}} + \varepsilon^2\,\|X^n\|^2_{\mathcal{X}} + 2\varepsilon(1-\varepsilon)(X^n, \mathcal{T}\,X^n)_{\mathcal{X}} \\
&= (1-\varepsilon)^2(\|X^n\|^2_{\mathcal{X}} - 4\sum_{i=1}^{2}\int_{\Omega} |\nabla\, u_i^n|^2\,dx) + \varepsilon^2\|X^n\|^2_{\mathcal{X}} + 2\varepsilon(1-\varepsilon)(X^n, \mathcal{T}\,X^n) \\
&= ((1-\varepsilon)^2 + \varepsilon^2)\,\|X^n\|^2_{\mathcal{X}} - 4(1-\varepsilon)^2\sum_{j=1}^{2}\int_{\Omega_i} |\nabla u_i^n|^2\,dx + 2\varepsilon(1-\varepsilon)(X^n, \mathcal{T}\,X^n).
\end{aligned} \tag{2.3.3.19}$$

On the other hand

$$\begin{aligned}
\|X^n\|^2_{\mathcal{X}} &= \frac{1}{\beta}\sum_{i=1}^{2}\int_{\Gamma}\{|\frac{\partial}{\partial\nu_i}\,u_i^n|^2 + \beta^2\,|u_i^n|^2\}\,d\gamma \\
&\quad + 2\sum_{i=1}^{2}\int_{\Omega_i} |\nabla\, u_i^n|^2\,dx \\
&= E^n + 2\sum_{i=1}^{2}\int_{\Omega_i} |\nabla\, u_i^n|^2\,dx,
\end{aligned}$$

where

$$E^n := \frac{1}{\beta}\sum_{i=1}^{2}\int_{\Gamma}\{|\frac{\partial}{\partial\nu_i}\,u_i^n|^2 + \beta^2|u_i^n|^2\}\,d\gamma. \tag{2.3.3.20}$$

Thus, from (2.3.3.19)and (2.3.3.20) we obtain

$$\begin{aligned}
E^{n+1} =&((1-\varepsilon)^2 + \varepsilon^2)\,E^n - 2(1-2\varepsilon)\sum_{i=1}^{2}\int_{\Omega_i} |\nabla\, u_i^n|^2\,dx \\
&- 2\sum_{i=1}^{2}\int_{\Omega_i} |\nabla\, u_i^{n+1}|^2\,dx + 2\,\varepsilon(1-\varepsilon)(X^n, \mathcal{T}\,X^n).
\end{aligned} \tag{2.3.3.21}$$

Now

$$
\begin{aligned}
(X^n, \mathcal{T} X^n)_{\mathcal{X}} &= \sum_{i=1}^{2} \frac{1}{\beta} \int_{\Gamma} (\frac{\partial u_i^n}{\partial \nu_i} + \beta\, u_i^n)(-\frac{\partial u_j^n}{\partial \nu_j} + \beta\, u_j^n)\, d\gamma \\
&= \sum_{\substack{i=1\\ j\neq i}}^{1} \int_{\Gamma} \frac{1}{\beta}(-\frac{\partial u_i^n}{\partial \nu_i}\, \frac{\partial u_j^n}{\partial \nu_j} + \beta(\frac{\partial u_i^n}{\partial \nu_i}\, u_j - \frac{\partial u_j^n}{\partial \nu_j}\, u_i) + \beta^2\, u_i^n\, u_j^n)\, d\gamma \\
&= -\int_{\Gamma} (\frac{2}{\beta}\, \frac{\partial u_1^n}{\partial \nu_1}\, \frac{\partial u_2^n}{\partial \nu_2} + 2\beta\, u_1^n\, u_2^n) d\gamma \\
&\leq \frac{1}{\beta} \sum_{i=1}^{2} \int_{\Gamma} (|\frac{\partial u_i^n}{\partial \nu_i}|^2 + \beta^2 |u_i^n|^2)\, d\gamma = E^n.
\end{aligned}
\tag{2.3.3.22}
$$

Thus,

$$
\begin{aligned}
E^{n+1} \leq \underbrace{\left((1-\varepsilon)^2 + \varepsilon^2 + 2\varepsilon(1-\varepsilon)\right)}_{=1} E^n \\
- 2 \sum_{i=1}^{2} \int_{\Omega_i} ((1-2\varepsilon)\, |\nabla\, u_i^n|^2 + |\nabla\, u_i^{n+1}|^2)\, dx.
\end{aligned}
\tag{2.3.3.23}
$$

Define $c_1(\varepsilon) := 1 - 2\varepsilon$, $c_{n+1}(\varepsilon) = 1$, $c_\ell(\varepsilon) = 2(1-\varepsilon)$, $\ell = 2, \ldots, n$. We then have from (2.3.3.23)

$$
\begin{aligned}
E^{n+1} \leq & E^{n-1} - 2 \sum_{i=1}^{2} \int_{\Omega_i} ((1-2\varepsilon)\, |\nabla\, u_i^{n-1}|^2 + |\nabla\, u_i^n|^2)\, dx \\
& - 2 \sum_{i=1}^{2} \int_{\Omega_i} ((1-2\varepsilon)\, |\nabla\, u_j^n|^2 + |\nabla\, u_i^{n+1}|^2)\, dx \qquad (2.3.3.24) \\
= & E^{n-1} - 2 \sum_{i=1}^{2} \int_{\Omega_j} (c_1(\varepsilon)\, |\nabla\, u_i^{n-1}|^2 + c_2(\varepsilon)\, |\nabla\, u_i^n|^2 + c_{n+1}(\varepsilon)\, |\nabla\, u_i^{n+1}|^2)\, dx \\
= & E^1 - 2 \sum_{\ell=1}^{n+1} c_\ell(\varepsilon) \sum_{i=1}^{2} \int_{\Omega_i} |\nabla\, u_i^\ell|^2\, dx.
\end{aligned}
$$

Thus

$$
E^{n+1} + 2 \sum_{\ell=1}^{n+1} c_\ell(\varepsilon) \sum_{i=1}^{2} \int_{\Omega_i} |\nabla\, u_j^\ell|^2\, dx \leq E^1 \tag{2.3.3.25}
$$

and we conclude

$$\int_{\Omega_i} |\nabla u_i^n|^2 \, dx \to 0, \quad i = 1, 2, \quad n \to \infty \tag{2.3.3.26}$$

and

$$E^n \text{ is bounded.} \tag{2.3.3.27}$$

As the u_i^n satisfy Dirichlet boundary data on $\Omega_i \cap \partial\Omega$ with $\text{meas}(\Omega_i \cap \partial\Omega) > 0$ and $\partial\Omega_i = \Gamma_i \cup (\Omega_i \cap \partial\Omega)$, from

$$\|u_i^n\|_{H^1(\Omega_i)} \le C \, \|\nabla u_i^n\|_{L^2(\Omega_i)} \to 0 \;\; n \to \infty, \quad i = 1, 2 \tag{2.3.3.28}$$

we deduce that $u_i^n \to 0$ in $H^1(\Omega_i)$, $u_i^n \to 0$ in $L^2(\Gamma)$, $i = 1, 2$, $n \to \infty$. We also have by (2.3.3.27)

$$\int_{\Gamma} |\frac{\partial u_i^n}{\partial \nu_i}|^2 \, d\gamma < C \tag{2.3.3.29}$$

with a generic constant C.

For $\varphi_i \in V_i$ we have

$$\begin{aligned} 0 = -\int_{\Omega_i} \Delta u_i^{n+1} \, \varphi_i \, dx &= -\int_{\Gamma} \frac{\partial u_i^{n+1}}{\partial \nu_i} \, \varphi_i \, d\gamma + \int_{\Omega_i} \nabla u_i^{n+1} \, \nabla \varphi_i \, dx \\ &= \int_{\Gamma} (\beta \, u_i^{n+1} - \lambda_{ij}^n) \, \varphi_i \, d\gamma + \int_{\Omega_i} \nabla u_i^{n+1} \, \varphi_i \, dx. \end{aligned} \tag{2.3.3.30}$$

Thus by (2.3.3.26), (2.3.3.28), (2.3.3.30) and a compactness argument we conclude weak convergence of X^n to zero in $L^2(\Gamma)$. We summarize in the following Theorem.

Theorem 2.3.3.1 *Let $u \in H_0^1 \cap H^{\frac{3}{2}}(\Omega)$ be a solution (2.3.3.1) and $u_i^n \in V_i$, $i = 1, 2$ be the solution of (2.3.3.8) with*

$$\lambda_{ij}^{n-1} = (1 - \varepsilon)(-\frac{\partial}{\partial \nu_j} \, u_j^{n-1} + \beta \, u_j^{n-1}) + \varepsilon(\frac{\partial}{\partial \nu_i} \, u_i^{n-1} + \beta \, u_i^{n-1}) \tag{2.3.3.31}$$

and $\varepsilon \in [0, 1)$, $\lambda_{ij}^0 \in L^2(\Gamma)$ given. Then

$$\|u^n - u\|_{H^1} = (\sum_{i=1}^{2} \|u_i^n - u_i\|_{H^1(\Omega_i)}^2)^{\frac{1}{2}} \to 0 \text{ as } n \to \infty, \tag{2.3.3.32}$$

and the transmission conditions hold in the sense that, as $n \to \infty$,

$$\begin{aligned} &(u_1^n - u_2^n)|_\Gamma \to 0 \;\; \text{strongly in} \;\; L^2(\Gamma), \\ &\frac{\partial u_1^n}{\partial \nu_1} + \frac{\partial u_2^n}{\partial \nu_2} \to 0 \;\; \text{weakly in} \;\; L^2(\Gamma). \end{aligned}$$

From the point of view of numerical simulations, the fact that the update (2.3.3.6) involves a Neumann-derivative of u_j^n on Γ is not very convenient. We can reformulate (2.3.3.6) as

$$\frac{\partial u_i^{n+1}}{\partial \nu_i} + \beta\, u_i^{n+1} = g_i^{n+1} \quad g_i^{n+1} = 2\beta\, u_j^n - g_j^n. \tag{2.3.3.33}$$

But

$$g_i^{n+1} = 2\beta\, u_j^n - g_j^n = -\frac{\partial u_j^n}{\partial \nu_j} - \beta\, u_j^n + 2\beta\, u_j^n = -\frac{\partial u_j^n}{\partial \nu_j} + \beta\, u_j^n.$$

Thus (2.3.3.6) is equivalent to (2.3.3.33). This variant has been used by Deng [15], where a proof of convergence without a relaxation parameter can be found.

2.3.4 An Augmented Lagrangian Formulation

Let us repeat that it has been observed by Glowinski and Le Tallec [29] that Algorithm 2.3.3.1 could be retrieved from an augmented Lagrangian relaxation followed by an Uzawa-type saddle-point iteration. For the sake of clarity in this simple standard situation, we display the arguments in the strong form. More precisely, we relax the continuity condition

$$u_1 = u_2 \text{ on } \Gamma \tag{2.3.4.1}$$

by introducing the augmented Lagrangian

$$\begin{aligned} \mathcal{L}(u,\lambda,q) :=& \sum_i \int_{\Omega_i} (|\nabla u_i|^2 - u_i f_i)\, dx \\ &+ \sum_i \int_{\Gamma} \{\lambda_i(u_i - q) + \frac{\beta}{2}\, |u_i - q|^2\}\, d\gamma. \end{aligned} \tag{2.3.4.2}$$

with Lagrange multipliers λ_i $i = 1, 2$, and penalty parameter $\beta > 0$. Let us the following saddle-point iteration:

Algorithm 2.3.4.1

0) *Let* q^{n-1}, λ_i^n, $i = 1, 2$, *be given*

1) *Solve for* u_i^n, $i = 1, 2$*:*

$$\partial_{u_i}\mathcal{L}(u^n, q^{n-1}, \lambda^n) = 0$$

2) *update* λ_i^n; $i = 1, 2$*:*

$$\lambda_i^{n+\frac{1}{2}} = \lambda_i^n + \beta(u_i^n|_\Gamma - q^{n-1})$$

3) *solve for* q^n*:*

$$\partial_q\mathcal{L}(u^n, q^n, \lambda^{n+\frac{1}{2}}) = 0$$

4) *update* $\lambda_i^{n+\frac{1}{2}}$, $i = 1, 2$*:*

$$\lambda_i^{n+1} = \lambda_i^{n+\frac{1}{2}} + \beta(u_i^n|_\Gamma - q^n).$$

Indeed,

$$\begin{aligned}
0 =& \partial_{u_i} \mathcal{L}(u^n, q^{n+1}, \lambda^n)(\varphi_i) = \sum_i \int_{\Omega_i} (\nabla u_i^n \nabla \varphi_i - \varphi_i f_i)\, dx \\
&+ \sum_i \int_\Gamma [\lambda_i^n \varphi_i + \beta(u_i^n - q^{n-1})\, \varphi_i]\, d\gamma \\
=& \sum_i \int_\Gamma \frac{\partial u_i^n}{\partial \nu_i} \varphi_i \, d\gamma - \sum_i \int_{\Omega_i} (\Delta u_i^n + f_i)\, \varphi_i \, dx \\
&+ \sum_i \int_\Gamma (\lambda_i^n + \beta(u_i^n|_\Gamma - q^{n-1}))\, \varphi_i \, d\gamma, \ \forall \varphi \in V_i,
\end{aligned}$$

which implies, for $n+1$ instead of n,

$$\begin{aligned}
-\Delta u_i^{n+1} &= f_i \ \text{ on } \ \Omega_i \\
u_i^{n+1} &= 0 \ \text{ on } \ \bar{\Omega}_i \cap \partial\Omega \\
\frac{\partial u_i^{n+1}}{\partial \nu_i} + \beta\, u_i^{n+1} &= -\lambda_i^{n+1} + \beta\, q^n \ \text{ on } \ \Gamma.
\end{aligned} \tag{2.3.4.3}$$

The third step gives

$$0 = \sum_i \int_\Gamma (\lambda_i^{n+\frac{1}{2}} + \beta(u_i^n|_\Gamma - q^n))\, \xi \, d\gamma, \quad \forall \xi \in L^2(\Gamma). \tag{2.3.4.4}$$

The fourth step then says $\lambda_i^{n+1} = -\lambda_j^{n+1}$. Then (2.3.4.3) reads as

$$\begin{aligned}
\frac{\partial u_i^{n+1}}{\partial \nu_i} + \beta\, u_i^{n+1} &= \lambda_j^{n+1} + \beta\, q^n = \lambda_j^{n+\frac{1}{2}} + \beta\, u_j^n \\
&= \lambda_j^n + 2\beta\, u_j^n - \beta\, q^{n-1} = -(\frac{\partial u_j^n}{\partial \nu_j} + \beta\, u_j^n) + 2\beta\, u_j^n \\
&= -\frac{\partial u_j^n}{\partial \nu_j} + \beta\, u_j^n \text{ on } \Gamma
\end{aligned}$$

which is precisely (2.3.3.6) of Algorithm 2.3.3.1. Thus, Algorithm 2.3.4.1 and Algorithm 2.3.3.1 are equivalent. Algorithm 2.3.4.1 has been labeled *variant* 3 of the classical Uzawa Algorithm in Glowinski and Le Tallec [28].

The convergence proof in [29], however, is given in the finite-dimensional context. See also Quarteroni and Valli [91], where this algorithm is shown to be a variant of a more general one by Agoshkov. There it is also noted that convergence rates are still unknown. We will, however, provide numerical evidence that the convergence is linear.

Remark 2.3.4.1 Algorithm 2.3.3.1 and Algorithm 2.3.4.1 easily extend to the transmission problem

$$\begin{aligned} -a_i \, \Delta \, u_i &= f_i \ in \, \Omega \\ u_i &= 0 \ on \, \bar{\Omega}_i \cap \partial\Omega \\ u_1 &= u_2 \ on \, \Gamma \\ a_1 \frac{\partial u_1}{\partial \nu_1} + a_2 \frac{\partial u_2}{\partial \nu_2} &= 0 \ on \, \Gamma. \end{aligned} \tag{2.3.4.5}$$

2.3.5 General Elliptic Problems and More General Splittings

In this section we consider elliptic problems more general than the Poisson problem studied in Section 2.3.3 in order to fix ideas. Let $\mathcal{A}$ be the differential operator of second order in $\Omega \subset \mathbb{R}^d$ given by

$$\mathcal{A}u = -\sum_{k,\ell} \frac{\partial}{\partial x_k} (a_{k\ell}(x) \frac{\partial u}{\partial x_\ell}) + c(x)\, u, \tag{2.3.5.1}$$

with Ω as above, $c(\cdot) \in L^\infty(\Omega)$, $c \geq 0$ a.e, $a_{ij} \in L^\infty(\Omega)$, and piecewise C^1, $a_{k\ell} = a_{\ell k}$ a.e. such that $A := (a_{ij})$ satisfies

$$\sum_{k,\ell} a_{k\ell}(x)\, \psi_k\, \psi_\ell \geq \beta |\psi|^2, \ \forall \psi \in \mathbb{R}^d \text{ on } \Omega, \ \beta > 0. \tag{2.3.5.2}$$

Consider the model problem

$$\begin{aligned} \mathcal{A}u &= f \text{ in } \Omega \\ u &= 0 \text{ on } \partial\Omega \end{aligned} \tag{2.3.5.3}$$

We associate with $\mathcal{A}$ on Ω the bilinear form

$$a_\Omega(u,v) = \int_\Omega (\sum_{k,\ell} a_{k\ell}(x) \frac{\partial u}{\partial x_k} \frac{\partial u}{\partial x_\ell} + c(x)\, uv)\, dx \tag{2.3.5.4}$$

and denote the inner product in $L^2(\Omega)$ by $(\cdot,\cdot)_\Omega$. The canonical weak formulation of (2.3.5.3) then is

$$a_\Omega(u,v) = (f,v)_\Omega, \ \ \forall v \in H_0^1(\Omega). \tag{2.3.5.5}$$

By the Lax-Milgram-Lemma it is well known that (2.3.5.5) possesses a unique solution $u \in H_0^1(\Omega)$.

We decompose Ω into $I(\geq 2)$ arbitrary disjoint subdomains Ω_i, $i = 1, \ldots, I$, as in (2.3.1.9). For the sake of simplicity, we will assume that the subdomains are connected and the areas Γ_{ij} meet the boundary $\partial\Omega$ at right angles, in order to have convex subdomains. This assumption can be relaxed, however. We also assume the regularity

$$u|_{\Omega_i} \in H^{3/2}(\Omega_i) \tag{2.3.5.6}$$

and set

$$\frac{\partial u}{\partial \nu_{A_i}} := \sum_{k,\ell} (\alpha_{k\ell}(x) \frac{\partial u}{\partial x_\ell}) \nu_{ik}, \quad i = 1, \dots, I, \tag{2.3.5.7}$$

$\nu_i = (\nu_{i1}, \dots, \nu_{id})$ and $A_i := A|_i$. We generalize Algorithm 2.3.3.1 to the following algorithm.

Algorithm 2.3.5.1

1) *Given* u_i^n, $\frac{\partial u_i^n}{\partial \nu_{A_i}}$ *on* Γ_{ij}, $1 \le i \ne j \le I$, *with* u_i^n, $\frac{\partial u_i^n}{\partial \nu_i} \in L^2(\Gamma_{ij})$,
2) *compute* u_i^{n+1}, $i = 1, \dots, I$ *according to*

$$\begin{aligned} \mathcal{A}_i\, u_i^{n+1} &= f_i \quad in \ \Omega_i, \\ u_i^{n+1} &= 0 \quad on \ \Gamma_i \\ \frac{\partial u_i^{n+1}}{\partial \nu_{A_i}} + \beta_{ij}\, u_i^{n+1} &= -\frac{\partial u_j^n}{\partial \nu_{Aj}} + \beta_{ij}\, u_j^n := \lambda_{ij}^n \ on\, \Gamma_{ij} \end{aligned} \tag{2.3.5.8}$$

3) $n \to n+1$, *go to* 2).

Here $\mathcal{A}_i$ is the natural restriction of $\mathcal{A}$ to Ω_i. The weak formulation of (2.3.5.8) is obtained as follows:

$$\begin{aligned} (f_i, v_i)_{\Omega_i} &= \int_{\Omega_i} (-\sum_{k,\ell} \frac{\partial}{\partial x_k} (a_{k\ell}(x) \frac{\partial u_i^{n+1}}{\partial x_\ell}) + c(x)\, u_i^{n+1})\, v_i\, dx \\ &= -\int_{\partial\Omega_i} \sum_{k,\ell} (a_{k\ell}(x) \frac{\partial u_i^{n+1}}{\partial x_k}) \nu_{ik}\, v_i\, d\Gamma + a_{\Omega_i}(u_i^{n+1}, v_i) \\ &= -\int_{\Gamma_i} \frac{\partial u_i^{n+1}}{\partial \nu_{A_i}} v_i\, d\Gamma - \sum_{j \in \mathcal{I}_i} \int_{\Gamma_{ij}} \frac{\partial u_i^{n+1}}{\partial \nu_{A_i}} v_i\, d\Gamma + a_{\Omega_i}(u_i^{n+1}, v_i) \\ &= \sum_{j \in \mathcal{I}_i} [\beta_{ij} \int_{\Gamma_{ij}} u_i^{n+1}\, v_i\, d\Gamma - \int_{\Gamma_{ij}} \lambda_{ij}^n\, v_i\, d\Gamma] + a_{\Omega_i}(u_i^{n+1}, v_i) \end{aligned} \tag{2.3.5.9}$$

where we have used the notation of (2.3.1.9).

Let $H^1_{\Gamma_i}(\Omega_i)$ denote the subspace of $H^1(\Omega_i)$ of functions vanishing on Γ_i. Then (2.3.5.9) in condensed form is

$$\begin{aligned} a_{\Omega_i}(u_i^{n+1}, v_i) + \sum_{j \in \mathcal{I}_i} \beta_{ij} \int_{\Gamma_{ij}} u_i^{n+1}\, v_i\, d\Gamma \\ = \sum_{j \in \mathcal{I}_i} \int_{\Gamma_{ij}} \lambda_{ij}^n\, v_i\, d\Gamma + (f_i, v_i)_{\Omega_i}, \quad \forall v_i \in H^1_{\Gamma_i}(\Omega_i). \end{aligned} \tag{2.3.5.10}$$

In order to establish convergence of this scheme, we introduce the following space setting:

$$H := L^2(\Omega) = \prod_{i=1}^{I} L^2(\Omega_i) \tag{2.3.5.11}$$

$$\begin{aligned} V &:= \{(v_1, \dots, v_I) \,|v_i = v|_{\Omega_i},\ v \in H_0^1(\Omega)\} \\ &= \{(v_1, \dots, v_I)|\ v_i \in H^1_{\Gamma_i}(\Omega_i),\ v_i = v_j \text{ on } \Gamma_{ij}, \\ &\qquad \text{for } j \in \mathcal{I}_i\} \end{aligned} \tag{2.3.5.12}$$

$$W := \prod_{i=1}^{I} H^1_{\Gamma_i}(\Omega_i). \tag{2.3.5.13}$$

Let $u,\, u^n \in W$, $u_i = u|_{\Omega_i}$, $u_i^n = u^n|_{\Omega_i}$, $e^n = u^n - u$, $e_i^n = u_i^n - u_i = e^n|_{\Omega_i}$. Hence e^n denotes the error at the iteration step n if u is the solution to (2.3.5.3). Indeed, by linearity, e^n solves

$$\begin{aligned} \mathcal{A}_i\, e_i^{n+1} &= 0 \ \text{ in } \Omega_i \\ \frac{\partial e_i^{n+1}}{\partial \nu_{A_i}} + \beta_{ij}\, e_i^{n+1} &= -\frac{\partial e_j^n}{\partial \nu_{Aj}} + \beta_{ij}\, e_j^n = \lambda_{ij}^n \text{ on } \Gamma_{ij},\ j \in \mathcal{I}_i, \\ e_i^n &= 0 \ \text{ on } \Gamma_i \end{aligned} \tag{2.3.5.14}$$

or

$$a_{\Omega_i}(e_i^{n+1}, v_i) + \sum_{j\in\mathcal{I}_i} \beta_{ij}(e_j^{n+1}, v_i)_{\Gamma_i} = \sum_{j\in\mathcal{I}_i} (\lambda_{ij}^n, v_i)_{\Gamma_{ij}},\ \forall\, v_i \in H^1_{\Gamma_i}(\Omega_i), \tag{2.3.5.15}$$

where $(u_i, v_i)_{\Gamma_{ij}} = \int\limits_{\Gamma_{ij}} u_i\, v_i\, d\Gamma$.

We proceed as in Section 2.3.3. We introduce the space

$$\mathcal{X} := \prod_{i=1}^{I} L^2(\gamma_i) \tag{2.3.5.16}$$

with the norm given by

$$\|X\|^2_{\mathcal{X}} = \sum_{i=1}^{I} \sum_{j\in\mathcal{I}_i} \frac{1}{\beta_{ij}} \int\limits_{\Gamma_{ij}} |\lambda_{ij}|^2 d\Gamma \tag{2.3.5.17}$$

where $X = (\lambda_i,\ i = 1, \dots, I)$, $\lambda_i = (\lambda_{ij})_{j\in\mathcal{I}_i}$. We also introduce the operator

$$\begin{aligned} &\mathcal{T} : \mathcal{X} \to \mathcal{X}, \\ &(\mathcal{T}X)_{ij} := \Big(-\frac{\partial u_j}{\partial \nu_{Aj}} + \beta_{ij}\, u_j\Big)\Big|_{\Gamma_{ij}}, \\ &(\mathcal{T}X)_i = \{(\mathcal{T}X)_{ij} : j \in \mathcal{I}_i\}, \\ &\mathcal{T}X = \{(\mathcal{T}X)_i : i = 1, \dots, I\} \end{aligned} \tag{2.3.5.18}$$

Iteration (2.3.5.14) then is equivalent to

$$X^{n+1} = \mathcal{T}(X^n). \tag{2.3.5.19}$$

As in Section 2.3.3 we have

$$\begin{aligned}\|X\|^2_{\mathcal{X}} &= \sum_{i=1}^{I} \sum_{j\in\mathcal{I}_i} \frac{1}{\beta_{ij}} \int_{\Gamma_{ij}} |\frac{\partial u_i}{\partial \nu_{A_i}} + \beta_{ij}\, u_i|^2 \, d\Gamma \\ &= \sum_{i=1}^{I} \sum_{j\in\mathcal{I}_i} \int_{\Gamma_{ij}} (\frac{1}{\beta_{ij}} |\frac{\partial u_i}{\partial \nu_{A_i}}|^2 + 2 \frac{\partial u_i}{\partial \nu_{A_i}} u_i + \beta_{ij}|u_i|^2) \, d\Gamma\end{aligned}$$

and

$$\begin{aligned}\|\mathcal{T}X\|^2_{\mathcal{X}} &:= \sum_{i=1}^{I} \sum_{j\in\mathcal{I}_i} \frac{1}{\beta_{ij}} \int_{\Gamma_{ij}} |-\frac{\partial u_j}{\partial \nu_{Aj}} + \beta_{ij}\, u_j|^2 \, d\Gamma \\ &= \sum_{i=1}^{I} \sum_{j\in\mathcal{I}_i} \int_{\Gamma_{ij}} (\frac{1}{\beta_{ij}} |\frac{\partial u_j}{\partial \nu_{Aj}}|^2 - 2 \frac{\partial u_j}{\partial \nu_j} u_j + \beta_{ij}\, |u_j|^2) \, d\Gamma.\end{aligned}$$

As $\sum_{i=1}^{I} \sum_{j\in\mathcal{I}_i} = \sum_{j=1}^{I} \sum_{i\in\mathcal{I}_j}$ we have with $\beta_{ij} = \beta_{ji} > 0$

$$\|\mathcal{T}X\|^2_{\mathcal{X}} - \|X\|^2_{\mathcal{X}} = -4 \sum_{i=1}^{I} \sum_{j\in\mathcal{I}_i} \int_{\Gamma_{ij}} \frac{\partial u_i}{\partial \nu_{A_i}} u_i d\Gamma. \tag{2.3.5.20}$$

But with $e_i = u_i - u$,

$$a_{\Omega_i}(e_i, e_i) - \sum_{j\in\mathcal{I}_i} \int_{\Gamma_{ij}} \frac{\partial e_i}{\partial \nu_{A_i}} e_i d\Gamma = 0, \tag{2.3.5.21}$$

thus (2.3.5.20) applied to X corresponding to e is

$$\|\mathcal{T}X\|^2_{\mathcal{X}} - \|X\|^2_{\mathcal{X}} = -4 \sum_{i=1}^{I} a_{\Omega_i}(e_i, e_i) \tag{2.3.5.22}$$

which is analogous to (2.3.3.15).

We state this result in terms of the iteration (2.3.5.14).

Lemma 2.3.5.1 *Let e^n be given and $X^n = \{\frac{\partial e_i^n}{\partial \nu_{A_i}} + \beta_{ij}\, e_i^n : j \in \mathcal{I}_i, i = 1, \dots, I\}$. Then we have*

$$\|\mathcal{T}X^n\|^2_{\mathcal{X}} = \|X^n\|^2_{\mathcal{X}} - 4 \sum_{i=1}^{I} a_{\Omega_i}(e_i^n, e_i^n).$$

We concentrate on the (under-)relaxed iteration (2.3.3.18), i.e.

$$X^{n+1} = (1-\varepsilon)\, \mathcal{T} X^n + \varepsilon\, X^n,\ \varepsilon \in [0,1). \tag{2.3.5.23}$$

The calculations in (2.3.3.19) are reproduced to give

$$\begin{aligned}\|X^{n+1}\|^2_{\mathcal{X}} &= ((1-\varepsilon)^2+\varepsilon^2)\, \|X^n\|^2_{\mathcal{X}} \\ &\quad - 4(1-\varepsilon)^2 \sum_{i=1}^{I} a_{\Omega_i}(e_i^n, e_i^n) \ + 2\varepsilon(1-\varepsilon)(X^n, \mathcal{T} X^n).\end{aligned} \tag{2.3.5.24}$$

Also, with

$$E^n := \sum_{i=1}^{I} \sum_{j\in\mathcal{I}_i} \int_{\Gamma_{ij}} \{\frac{1}{\beta_{ij}} |\frac{\partial e_i^n}{\partial \nu_{A_i}}|^2 + \beta_{ij}\, |e_i^n|^2\}\, d\Gamma \tag{2.3.5.25}$$

we have

$$\|X^n\|^2_{\mathcal{X}} = E^n + 2 \sum_{i=1}^{I} a_{\Omega_i}\, (e_i^n, e_i^n) \tag{2.3.5.26}$$

In complete analogy to Section 2.3.3 we obtain

$$(X^n, \mathcal{T} X^n) \le E^n$$

and

$$E^{n+1} \le E^n - 2 \sum_{i=1}^{I} [(1-2\varepsilon)\, a_{\Omega_i}(e_i^n, e_i^n) + a_{\Omega_i}(e_i^{n+1}, e_i^{n+1})]. \tag{2.3.5.27}$$

Thus with the same coefficients $c(\varepsilon)$ as in (2.3.3.25) we obtain the crucial estimate

$$E^{n+1} + 2 \sum_{\ell=1}^{n+1} c_\ell(\varepsilon) \sum_{i=1}^{I} a_{\Omega_i}(e_i^n, e_i^n) \le E^1. \tag{2.3.5.28}$$

We, therefore, conclude

$$\sum_{i=1}^{I} a_{\Omega_i}(e_i^n, e_i^n) \to 0 \ \text{ as } \ n \to \infty \tag{2.3.5.29}$$

$$\|e_i^n\|_{H^1(\Omega_i)} \le C,\ \ \forall\, n \ge 0,\ i = 1, \dots, I. \tag{2.3.5.30}$$

From this point on we may argue as in [15]: If $c(x) \ge c_0 > 0$ a.e., we obtain

$$\|e^n\|_{H^1} := (\sum_{i=1} \|e_i^n\|^2_{H^1(\Omega_i)})^{\frac{1}{2}} \to 0,\ \ \text{as } n \to \infty. \tag{2.3.5.31}$$

However, in the general case $c(x) \geq 0$ a.e. we obtain only

$$\|\nabla e_i^n\|_{L^2(\Omega_i)} \to 0 \quad \text{as} \quad n \to \infty, \tag{2.3.5.32}$$

$i = 1, \ldots, I$. In order to achieve convergence in this case, we have to proceed from the boundary $\partial\Omega$ of Ω where Dirichlet conditions hold. To this end we introduce the set G_1 of domains adjacent to the boundary, i.e.,

$$G_1 := \{\cup\Omega_i, \ \text{meas}(\Gamma_i) > 0\}. \tag{2.3.5.33}$$

We proceed into the 'interior' regions as follows

$$G_{r+1} := \{\cup\Omega_i, \ \text{meas}\ (\partial\Omega_i \cap \bar{G}_r) > 0, \ \partial\Omega_i \cap G_\ell = \emptyset, \ \forall\, \ell \leq r\}. \tag{2.3.5.34}$$

Now, for all i with $\Omega_i \subset G_1$ we have

$$e_i^n = 0 \quad \text{on} \quad \Gamma_i. \tag{2.3.5.35}$$

Hence, from (2.3.5.32), (2.3.5.35) and the Poincaré inequality we infer

$$\|e_i^n\|_{H^1(\Omega_i} \leq C\,\|\nabla e_i^n\|_{L^2(\Omega_i} \to 0, \quad \text{as} \ \ n \to \infty, \ \forall\, i : \Omega_i \subset G_1. \tag{2.3.5.36}$$

This, again, implies for those subdomains

$$\|e_i^n\|_{L^2(\Gamma_{ij})} \to 0 \quad \text{as} \quad n \to \infty, \ j \in \mathcal{I}_i. \tag{2.3.5.37}$$

From the weak representation (2.3.5.15) we deduce that

$$\|\lambda_{ij}^n\|_{H^{-\frac{1}{2}}(\Gamma_{ij})} \to 0, \ n \to \infty, \ j \in \mathcal{I}_i, \ i : \Omega_i \subset G_1. \tag{2.3.5.38}$$

We are now in position to penetrate further into the layer of interior subdomains $\Omega_i \subset G_2$. In particular,

$$\|\lambda_{ij}^n\|_{H^{-\frac{1}{2}}(\Gamma_{ij})} \to 0, \ n \to \infty, \ j \in \mathcal{I}_i, \ \Omega_j \subset G_1, \tag{2.3.5.39}$$

i.e., for those subdomains adjacent to G_1. We may use as test functions

$$v_i := \begin{cases} e_i^n & \text{on} \quad \Gamma_{ij}, \ j \in \mathcal{I}_i, \ \Omega_j \subset G_1 \\ 0 & \text{else} \end{cases} \tag{2.3.5.40}$$

together with (2.3.5.15) and (2.3.5.29), (2.3.5.39) in order to conclude

$$\|e_i^n\|_{L^2(\Gamma_{ij})} \to 0, \ n \to \infty, \ j \in \mathcal{I}_i, \ \Omega_j \in G_1. \tag{2.3.5.41}$$

As for each $\Omega_i \subset G_2$ there is at least one j such that $\Omega_j \subset G_1$ with $\text{meas}(\Gamma_{ij}) > 0$, we conclude

$$\|e_i^n\|_{H^1(\Omega_i)} \leq G(\|\nabla e_i^n\|_{L^2(\Omega_i)} + \sum_{\substack{j:\Omega_j \subset G_1 \\ j \in \mathcal{I}_i}} \|e_i^n\|_{L^2(\Gamma_{ij})}) \to 0,$$

$$n \to \infty, \ \Omega_i \subset G_2. \tag{2.3.5.42}$$

By induction on r one obtains (2.3.5.31) also in the case $\alpha_0(x) \geq 0$ a.e. Thus, we have proved the following result.

Theorem 2.3.5.1 *Let $u \in H_0^1 \cap H^{\frac{3}{2}}(\Omega)$ with $\frac{\partial u}{\partial \nu_{A_i}}|_{\Gamma_{ij}} \in L^2(\Gamma_{ij})$, $1 \leq i \neq j \leq I$, be the solution of* (2.3.5.3), *and $u_i^n \in H^1_{\Gamma_i}(\Omega)$, $i = 1, \dots, I$ the solution of*

$$\begin{aligned} \mathcal{A}_i\, u_i^n &= f_i \ \ in\ \ \Omega_i \\ u_i^n &= 0 \ \ on\ \ \Gamma_i \\ \frac{\partial u_n^i}{\partial \nu_{A_i}} + \beta_{ij}\, u_i^n &= (1-\varepsilon)(-\frac{\partial u_j^{n-1}}{\partial \nu_{Aj}} + \beta_{ij}\, u_j^{n+1}) \\ &+\varepsilon(\frac{\partial u_i^{n-1}}{\partial \nu_{A_i}} + \beta_{ij}\, u_i^{n-1}) \end{aligned} \tag{2.3.5.43}$$

with $\varepsilon \in [0,1)$. If

$$\frac{\partial u_i^0}{\partial \nu_{A_i}} + \beta_{ij}\, u_i^0 \in L^2(\Gamma_{ij}),\ 1 \leq i \neq j \leq m \tag{2.3.5.44}$$

then

$$\|u^n - u\|_{H^1} = (\sum_{i=1}^{I} \|u_i^n - u_i\|^2_{H^1(\Omega_i)})^{\frac{1}{2}} \to 0,\ u \to \infty. \tag{2.3.5.45}$$

2.3.6 An a Posteriori Error Estimate

The convergence results established so far guarantee convergence. However, for a numerical implementation, and in particular for adaptivity with respect to the splitting, we need reasonable stopping criteria, first on the infinite-dimensional level. Moreover, the question how the parameters of the algorithms have to be chosen in order to obtain fast convergence is not answered. Otto and Lube [90], [83] have obtained 'a posteriori' estimates for the algorithm of this section at least for the two-domain case. We are going to reproduce their arguments for the sake of self-consistency. We adopt the notation of Section 2.3.5.

Consider the problem

$$a_\Omega(u,v) = L(v),\ \forall\, v \in H_0^1(\Omega) = V, \tag{2.3.6.1}$$

which is equivalent to

$$\sum_{i=1}^{2} a_{\Omega_i}(u_i, v_i) = \sum_{i=1}^{2} L_i(v_i),\ v_i \in H^1_{\Gamma_i}(\Omega_i),\ i = 1,2. \tag{2.3.6.2}$$

The domain decomposition method of Section 2.3.6 gives at iteration level $n+1$:

$$a_{\Omega_i}(u_i^{n+1}, v_i) + \int_\Gamma \beta_i\, u_i^{n+1}\, v_i\, d\Gamma = \int_\Gamma \lambda_j^n\, v_i\, d\Gamma + L_i(v_i), \tag{2.3.6.3}$$

$$\lambda_i^{n+1} = (\beta_i + \beta_j)\, u_j^{n+1} - \lambda_j^n \ \text{ on } \Gamma. \tag{2.3.6.4}$$

In fact we may extend, if necessary, the $L^2(\Gamma)$ pairing to the duality pairing between $W := H_{00}^{\frac{1}{2}}(\Gamma) := \{v|_\Gamma : v \in H_0^1(\Omega)\} \subset H^{\frac{1}{2}}(\Gamma)$ and $H_{00}^{\frac{1}{2}}(\Gamma)^* = H^{-\frac{1}{2}}(\Gamma)$, and denote this by $\langle \cdot, \cdot \rangle_\Gamma$. Thus, in general, (2.3.6.3) and (2.3.6.4) can be rewritten as

$$\begin{aligned} & a_{\Omega_i}(u_i^{n+1}, v_i) + \langle \beta_i \, u_i^{n+1}, v_i \rangle_\Gamma \\ & = L_i(v_i) + \langle \lambda_j^n, v_i \rangle_\Gamma \quad \text{on} \quad V_i := H^1_{\Gamma_i}(\Omega_j), \end{aligned} \tag{2.3.6.5}$$

$$\langle \lambda_i^{n+1}, \varphi \rangle_\Gamma = \langle (\beta_i + \beta_j) \, u_j^{n+1}, \varphi \rangle_\Gamma - \langle \lambda_j^n, \varphi \rangle_\Gamma, \ \forall \, \varphi \in W. \tag{2.3.6.6}$$

It is well known that the trace operator

$$tr_i : V_i \to L^2(\Gamma) \tag{2.3.6.7}$$

is continuous and that there is a continuous extension of tr_i^{-1}

$$tr_i^{-1} : W \to V_i. \tag{2.3.6.8}$$

Moreover, W is continuously embedded into $L^2(\Gamma)$, see, e.g., [90]. We may therefore express elements $v_i \in V_i$ by $v_i = tr_i^{-1} \, \varphi + v_i^0$ where $\varphi \in W$ and $v_i^0 \in V_i \cap H_0^1(\Omega_i)$. Hence, (2.3.6.5) reads as

$$a_{\Omega_i}(u_i^{n+1}, v_i) + \langle \beta_i \, u_i^{n+1}, \varphi \rangle_\Gamma = L_i(v_i) + \langle \lambda_j^n, \varphi \rangle_\Gamma. \tag{2.3.6.9}$$

Moreover, v_i^0 can be extended by zero across Γ to an element of V. Thus, if we consider the errors

$$e_i^n = u_i^n - u_i, \ \eta_i^n = \lambda_i^n - \lambda_i, \tag{2.3.6.10}$$

where λ_i^n satisfies (2.3.6.6) (for n) and λ_i is given by

$$\langle \lambda_i, \varphi \rangle_\Gamma = a_{\Omega_i}(u_i, tr_i^{-1} \, \varphi) + \langle \beta_i, \, u_i, \varphi \rangle_\Gamma - L_i(tr_i^{-1} \, \varphi), \ \ \forall \varphi \in W, \tag{2.3.6.11}$$

then the error satisfies

$$a_{\Omega_i}(e_i^{n+1}, v_i) + \langle \beta_i \, e_i^{n+1}, v_i \rangle_\Gamma = \langle \eta_i^n, v_i \rangle_\Gamma, \ \forall \, v_i \in V_i, \tag{2.3.6.12}$$

$$\langle \eta_i^{n+1}, \varphi \rangle_\Gamma = \langle (\beta_i + \beta_j) \, e_i^{n+1}, \varphi \rangle_\Gamma - \langle \eta_j^n, \varphi \rangle, \ \forall \, \varphi \in W. \tag{2.3.6.13}$$

Suppose that $\beta_i \in L^\infty(\Omega)$. Then

$$\langle \eta_i^{n-1} - \beta_i \, e_i^n, \varphi \rangle = a_{\Omega_i}(e_i^n, tr_i^{-1} \varphi) \le C_i \, \|tr_i^{-1}\| \, \|e_i^n\|_{V_i} \, \|\varphi\|_W \tag{2.3.6.14}$$

with $C_i = \max\{\|A\|_{L^\infty}, \, \|c\|_{L^\infty}\}$ as in Section 2.3.5, and

$$\begin{aligned} \langle \beta_i \, \varphi, e_i^{n+1} \rangle_\Gamma &= \int_\Gamma \beta_i \, \varphi \, e_i^{n+1} \, d\gamma \\ &\le \|\beta_i\|_{L^\infty(\Gamma)} \, \|\varphi\|_{L^2(\Gamma)} \, \|tr_i \, e_i^{n+1}\|_{L^2(\Gamma)} \\ &\le C_W \, \|tr_i\| \, \|\beta_i\|_{L^\infty(\Gamma)} \, \|\varphi\|_W \, \|e_i^{n+1}\|_{V_i}. \end{aligned} \tag{2.3.6.15}$$

Let us now consider $a_{\Omega_1}(e_1^{n+1}, v_1) + a_{\Omega_2}(e_2^n, v_2)$. We obtain, with (2.3.6.13)

$$\begin{aligned} a_{\Omega_1}(e_1^{n+1}, v_1) + a_{\Omega_2}(e_2^n, v_2) &= \langle \eta_2^n - \beta_1 e_1^{n+1}, v_1\rangle_\Gamma + \langle \eta_1^{n-1} - \beta_2 e_2^n, v_2\rangle_\Gamma \\ &= \langle \beta_1(e_2^n - e_1^{n+1}), v_1\rangle_\Gamma + \langle \eta_2^n - \beta_1 e_2^n, v_1\rangle_\Gamma \\ &+ \langle \eta_1^{n-1} - \beta_2 e_2^n, v_2 - v_1\rangle_\Gamma + \langle \eta_1^{n-1} - \beta_2 e_2^n, v_1\rangle_\Gamma \\ &= \langle \beta_1(u_2^n - u_1^{n+1}), v_1\rangle_\Gamma + \langle \eta_1^{n-1} - \beta_2 e_2^n, v_2 - v_1\rangle_\Gamma \end{aligned} \tag{2.3.6.16}$$

Thus upon setting $v_1 = e_1^{n+1}$, $v_2 = e_2^n$ we obtain using (2.3.6.14)

$$\begin{aligned} &a_{\Omega_1}(e_1^{n+1}, e_1^{n+1}) + a_{\Omega_2}(e_2^n, e_2^n) \\ &\quad \le \langle \beta_1(u_2^n - u_1^{n+1}), e_1^{n+1}\rangle_\Gamma + C_2\|tr_2^{-1}\|\,\|e_2^n\|_{V_2}\,\|u_2^n - u_1^{n+1}\| \\ &\le C_W\|tr_1\|\,\|\beta_1\|_{L^\infty(\Omega)}\,\|e_1^{n+1}\|_{V_1} + C_2\|tr_2^{-1}\|\,\|e_2^n\|_{V_2})\,\|u_2^n - u_1^{n+1}\|_{H_{00}^{\frac12}(\Gamma)} \end{aligned} \tag{2.3.6.17}$$

$$\begin{aligned} &(\|e_1^{n+1}\|_{V_1} + \|e_2^n\|_{V_2})\cdot\|u_2^n - u_1^{n+1}\|_{H_{00}^{\frac12}(\Gamma)} \\ &\quad \le \max\{C_W\|tr_1\|,\ \|\beta_1\|_{L^\infty(\Omega)}\},\ \max\{\|A\|_{L^\infty(\Omega)},\ \|C\|_{L^\infty(\Omega)}\}\,\|tr_2\| \\ &\qquad\qquad \cdot(\|e_1^{n+1}\|_{V_1} + \|e_2^n\|_{V_2})\,\|u_2^n - u_1^{n+1}\|_{H_{00}^{\frac12}(\Gamma)}. \end{aligned} \tag{2.3.6.18}$$

On the other hand, we have $a_{\Omega_i}(v,v) \ge m_1\,\|v\|_i^2$, so that (2.3.6.17), (2.3.6.18) become

$$\begin{aligned} &\min\{m_1, m_2\}\,(\|e_1^{n+1}\|_{V_1}^2 + \|e_2^n\|_{V_2}^2) \\ &\quad \le C(\|\beta_1\|_{L^\infty},\ \|a\|_{L^\infty},\ \|c\|_{L^\infty})\,(\|e_1^{n+1}\|_{V_1} + \|e_2^n\|_{V_2}) \\ &\qquad \cdot\|u_2^n - u_1^{n+1}\|_{H_{00}^{\frac12}(\Gamma)} \end{aligned} \tag{2.3.6.19}$$

and hence

$$\|e_1^{n+1}\|_{V_1} + \|e_2^n\|_{V_2} \le C(\beta_1, A, c)\,\|u_2^n - u_1^{n+1}\|_{H_{00}^{\frac12}(\Gamma)} \tag{2.3.6.20}$$

with

$$\begin{aligned} C(\beta_1, A, c) := &\frac{2C(\|\beta_1\|_{L^\infty}, \|A\|_{L^\infty}, \|c\|_{L^\infty})}{\min\{m_1, m_2\}} \\ &= \min\{m_1, m_2\}^{-1}\ \max\{C_W\|tr_1\|\,\|\beta_1\|_{L^\infty}, \\ &\qquad \max\{\|A\|_{L^\infty}, \|c\|_{L^\infty}\}\cdot\|tr_2^{-1}\|\}. \end{aligned} \tag{2.3.6.21}$$

Theorem 2.3.6.1 *Let u be a solution of* (2.3.6.1) *and u_i^{n+1} be solutions of* (2.3.6.3), (2.3.6.4) *or* (2.3.6.5), (2.3.6.6), *for $i = 1, 2$. Let the errors e_i^n, η_i^n, $i = 1, 2$, be given as in* (2.3.6.10). *Then we have the a posteriori estimate*

$$\|e_1^{n+1}\|_{V_1} + \|e_2^n\|_{V_2} \le C(\beta_1, a, c)\|u_2^n - u_1^{n+1}\|_{H_{00}^{\frac{1}{2}}(\Gamma)}. \tag{2.3.6.22}$$

Remark 2.3.6.1

i) Note that the error estimate is not symmetric with respect to the domains. But this asymmetry can easily be removed by starting at domain 2 and then combining the estimates.
ii) The estimate (2.3.6.22) allows one to optimize the error bound $C(\beta, a, c)$ with respect to the transmission coefficients β_i appearing in the Robin data. This is a major advantage as the convergence results do not give any hint in this direction.

2.3.7 Interpretation as a Damped Richardson iteration

In order to fix ideas we consider the simplest problem as in (2.3.3.5), (2.3.3.6) and (2.3.3.7), i.e.,

$$\begin{aligned} -\Delta u &= f \quad \text{in} \quad \Omega \\ u &= 0 \quad \text{on} \quad \partial\Omega \end{aligned} \tag{2.3.7.1}$$

and

$$\begin{aligned} -\Delta u_i &= f_i \quad \text{in} \quad \Omega \\ u_i &= 0 \quad \text{on} \quad \partial\Omega \cap \partial\Omega_i \\ \frac{\partial}{\partial\nu_i} u_i + \beta u_i &= -\frac{\partial}{\partial\nu_j} u_j + \beta u_j \text{ on } \Gamma \end{aligned} \tag{2.3.7.2}$$

$i = 1, 2$.

We decompose u_i as follows. Let $G_i(f_i) =: z_i$ solve

$$\begin{aligned} -\Delta z_i &= f_i \quad \text{in} \quad \Omega \\ z_i &= 0 \quad \text{on} \quad \partial\Omega \cap \partial\Omega_i \\ \frac{\partial}{\partial\nu_i} z_i + \beta z_i &= 0 \text{ on } \Gamma, \quad i = 1, 2. \end{aligned} \tag{2.3.7.3}$$

Moreover, let $w_i := R_i \lambda_{ij}$ solve

$$\begin{aligned} -\Delta w_i &= 0 \quad \text{in} \quad \Omega \\ w_i &= 0 \quad \text{on} \quad \partial\Omega \cap \partial\Omega_i \\ \frac{\partial}{\partial\nu_i} w_i + \beta w_i &= \lambda_{ij} \text{ on } \Gamma, \quad i = 1, 2, \ j \ne i. \end{aligned} \tag{2.3.7.4}$$

Finally, define $S_i(\lambda_{ij}) = w_i|_\Gamma$. The operator S_i is an example of a Poincaré-Steklov-type operator, and $u_i = w_i + z_i$ solves (2.3.7.2), iff $\lambda_{ij} = -\frac{\partial}{\partial\nu_j} u_j + \beta u_j$, $i \ne j$.

Let us express (2.3.7.2) in terms of the operators G_i, R_i, S_i as follows:

$$
\begin{aligned}
-\frac{\partial}{\partial \nu_j} u_j + \beta u_j &= -\frac{\partial}{\partial \nu_j} R_j(\lambda_{ji}) \\
-\frac{\partial}{\partial \nu_j} G_j(f_j) + \beta(S_j(\lambda_{ji}) + G_j(f_j)|_\Gamma) &= -(\frac{\partial}{\partial \nu_j} R_j(\lambda_{ji}) + \beta S_j(\lambda_{ji})) \\
-(\frac{\partial}{\partial \nu_j} G_j(f_j) + \beta G_j(f_j)|_\Gamma) + 2\beta(S_j(\lambda_{ji}) + G_j(f_j)|_\Gamma) & \\
&= -\lambda_{ji} + 2\beta(S_j(\lambda_{ji}) + G_j(f_j)|_\Gamma).
\end{aligned}
\tag{2.3.7.5}
$$

On the other hand, (2.3.7.5) equals λ_{ij}. Thus

$$
\lambda_{ij} + \lambda_{ji} - 2\beta S_j(\lambda_{ji}) = 2\beta G_j(f_j)_{\Gamma_{ij}} \tag{2.3.7.6}
$$

$i, j = 1, 2,\ i \neq j$.

Equations (2.3.7.6) can be interpreted as a Poincaré-Steklov-type equation to be solved on the common interface. This interpretation of the nonoverlapping domain decomposition of P.-L. Lions seems to not have been recognized in the literature. For easier reference, let us denote

$$
\lambda := \lambda_{ij}, \quad \mu = \lambda_{ji}.
$$

Thus, (2.3.7.6) reads as

$$
\begin{aligned}
\lambda &= 2\beta S_2 \mu - \mu + 2\beta G_2 f_2|_\Gamma \\
\mu &= 2\beta S_1 \lambda - \lambda + 2\beta G_1 f_1|_\Gamma .
\end{aligned}
\tag{2.3.7.7}
$$

Yet another way of writing this is as follows

$$
\begin{pmatrix} I & -(2\beta S_2 - I) \\ -(2\beta S_1 - I) & I \end{pmatrix} \begin{pmatrix} \lambda \\ \mu \end{pmatrix} = \begin{pmatrix} 2\beta G_2 f_2|_\Gamma \\ 2\beta G_1 f_1|_\Gamma \end{pmatrix}. \tag{2.3.7.8}
$$

The damped Richardson iteration of (2.3.7.8) is given as follows. First set

$$
\mathcal{A} := \begin{pmatrix} I & -(2\beta S_2 - I) \\ -(2\beta S_1 - I) & I \end{pmatrix}
$$

$$
x := \begin{pmatrix} \lambda \\ \mu \end{pmatrix}, \quad b = \begin{pmatrix} 2\beta G_2 f_2|_\Gamma \\ 2\beta G_1 f_1|_\Gamma \end{pmatrix}.
$$

Then (2.3.7.8) is the same as

$$
\mathcal{A} x = b
$$

which is solved iteratively as follows for some $\theta \in \mathbb{R}$:

$$
x^{n+1} = x^n - \theta(\mathcal{A} x^n - b). \tag{2.3.7.9}
$$

This Richardson iteration is rewritten as

$$\begin{aligned}\lambda^{n+1} &= \lambda^n - \theta(\lambda^n - (2\beta\, S_2\, \mu^n - \mu^n) - 2\beta\, G_2\, f_2|_\Gamma)\\ \mu^{n+1} &= \mu^n - \theta(\mu^n - (2\beta\, S_1\, \lambda^n - \lambda^n) - 2\beta\, G_1\, f_1|_\Gamma)\end{aligned}$$

or

$$\begin{aligned}\lambda^{n+1} &= (1-\theta)\,\lambda^n + \theta(2\beta\, S_2\, \mu^n - \mu^n + 2\beta\, G_2\, f_2|_\Gamma)\\ \mu^{n+1} &= (1-\theta)\,\mu^n + \theta(2\beta\, S_1\, \lambda^n - \lambda^n + 2\beta\, G_1\, f_1|_\Gamma).\end{aligned} \tag{2.3.7.10}$$

Going back to the variable u_i we can express iteration (2.3.7.10) as

$$\begin{aligned}\lambda^{n+1} &= (1-\theta)\,\lambda^n + \theta(2\beta\, u_2^n|_\Gamma - \mu^n)\\ \mu^{n+1} &= (1-\theta)\,\mu^n + \theta(2\beta\, u_1^n|_\Gamma - \lambda^n).\end{aligned} \tag{2.3.7.11}$$

Converting θ into $1-\varepsilon$ converts (2.3.7.11) into

$$\begin{aligned}\lambda^{n+1} &= \varepsilon\,\lambda^n + (1-\varepsilon)(2\beta\, u_2^n|_\Gamma - \mu^n)\\ \mu^{n+1} &= \varepsilon\,\mu^n + (1-\varepsilon)(2\beta\, u_1^n|_\Gamma - \lambda^n).\end{aligned} \tag{2.3.7.12}$$

Thus it is seen that the damped Richardson iteration (2.3.7.9), (or (2.3.7.10)) for (2.3.7.8) with damping parameter θ is precisely the underrelaxed fixed point iteration considered in this chapter, where we set

$$\mathcal{T}(\lambda,\mu) = (2\beta\, u_2|_\Gamma - \mu,\ 2\beta\, u_1|_\Gamma - \lambda) \tag{2.3.7.13}$$

and recall that $g_1 = \lambda$, $g_2 = \mu$.

Obviously the problem (2.3.7.8) is not always symmetric, which makes it difficult to apply convergence theory for the Richardson iteration, in particular in the given case of a Hilbert space setting in infinite dimensions. If, however, the subregions are identical, then the problem is symmetric and the theory of J. Bialy [11] applies; for parameters $\theta \in (0, \frac{2}{\|\mathcal{A}\|})$ we have strong convergence of the g_i's in $L^2(\Gamma)$. This implies convergence for the fixed point iteration (2.3.7.12) with relaxation parameter $\varepsilon = 1-\theta$. As the norm of $\mathcal{A}$ can in principle be small, depending on the choice of β, θ may be large. Therefore, ε may even be negative and still there is convergence (for $\mathcal{A}$ symmetric).

A Schur Operator Formulation

According to (2.3.7.3), (2.3.7.4) the solution u_i of (2.3.7.2) can be uniquely expressed as

$$u_i = w_i + z_i \tag{2.3.7.14}$$

where $w_i = r_i\,\lambda_{ij}$, $z_i = G_i(f_i) := A_i^{-1} f_i$. Furthermore, $S_i(\lambda_{ij}) := R_i\,\lambda_{ij}|_\Gamma =: tr_i\, R_i\,\lambda_{ij}$. We have $w_i \in H^2(\Omega_i)$, and with the extension of $A_i = -\Delta$ on Ω_i to $A_i : V_i \to V_i^*$ we can write

$$A_i\, u_i = A_i\, R_i\, \lambda_{ij} + f_i \quad i = 1,2. \tag{2.3.7.15}$$

Moreover, according to (2.3.7.7)

$$\begin{aligned}\lambda_{12} &= 2\beta\, S_2\,\lambda_{21} - \lambda_{21} + 2\beta\, tr_2\, z_2\\ \lambda_{21} &= 2\beta\, S_1\,\lambda_{12} - \lambda_{12} + 2\beta\, tr_1\, z_1.\end{aligned} \tag{2.3.7.16}$$

We can rewrite (2.3.7.15), (2.3.7.16) in terms of the variables z_i, λ_{ij} as follows

$$\begin{aligned}&-\Delta\, z_1 - \Delta\, R_1\,\lambda_{12} = f_1\\ &-\Delta\, z_2 - \Delta\, R_2\,\lambda_{21} = f_2\\ &-2\beta\, tr_2\, z_2 + \lambda_{12} + \lambda_{21} - 2\beta\, tr_2\, R_2\,\lambda_{21} = 0\\ &-2\beta\, tr_1\, z_1 + \lambda_{12} + \lambda_{21} - 2\beta\, tr_1\, R_1\,\lambda_{12} = 0\end{aligned} \tag{2.3.7.17}$$

or in operator matrix format

$$\begin{pmatrix} -\Delta & 0 & -\Delta R_1 & 0\\ 0 & -\Delta & 0 & -\Delta R_2\\ 0 & -2\beta tr_2 & I & I - 2\beta tr_2 r_2\\ -2\beta tr_1 & 0 & I - 2\beta tr_1 R_1 & I \end{pmatrix} \begin{pmatrix} z_1\\ z_2\\ \lambda_{12}\\ \lambda_{21}\end{pmatrix} = \begin{pmatrix} f_1\\ f_2\\ 0\\ 0\end{pmatrix} \tag{2.3.7.18}$$

The Schur complement of this block-operator-matrix equation is precisely (2.3.7.16). Equation (2.3.7.18) can easily extended into a weak operator-matrix equation

$$\begin{pmatrix} A_1 & 0 & B_{11} & 0\\ 0 & A_2 & 0 & B_{22}\\ 0 & C_{12} & I & I - 2\beta S_2\\ C_{21} & 0 & I - 2\beta S_1 & I \end{pmatrix} \begin{pmatrix} z_1\\ z_2\\ \lambda_{12}\\ \lambda_{21}\end{pmatrix} = \begin{pmatrix} h_1\\ h_2\\ h_3\\ h_4\end{pmatrix} \tag{2.3.7.19}$$

It is evident that for more general splittings the following pattern for the operator matrix in (2.3.7.19) evolves; for example, for p domains we have

$$\left(\begin{array}{ccc|ccc} A_1 & & 0 & B_{11} & \dots & B_{1q}\\ 0 & A_2 & 0 & B_{21} & \dots & B_{2q}\\ \vdots & & & \vdots & & \\ 0 & & A_p & B_{p1} & \dots & B_{pq}\\ \hline C_{11} & & C_p & \Sigma_1 & & 0\\ C_{21} & & C_{2p} & 0 & \Sigma_2 & 0\\ C_{q1} & & C_{qp} & 0 & & \Sigma_q \end{array}\right) \tag{2.3.7.20}$$

where the matrices (B_{ij}), (C_{jk}) have a sparsity pattern according to the adjacency structure of the subdomains. In particular, the number q equals the sum of all cardinalities of adjacent subdomains

$$q = \sum_{i=1}^{p} |\mathcal{I}_i|$$

where $\mathcal{I}_i = \{j \in \{1, \dots, p\}\backslash\{i\}|\, \Gamma_{i_j} \neq \phi\}$.

The structure of the operator (2.3.7.20) remains after discretization. We note that Robin-type decomposition discussed here differ from the more standard splittings in the structure of the Schur operator. This can be seen from Quarteroni and Valli [91]. For the sake of self-consistency, we reconsider the standard problem and decompose

$$\begin{aligned} -\Delta u_i &= f_i \text{ in } \Omega_i \\ u_i &= 0 \text{ on } \partial\Omega_i \cap \partial\Omega \\ u_1 &= u_2 \text{ on } \Gamma \\ \frac{\partial u_1}{\partial \nu_1} &+ \frac{\partial u_2}{\partial \nu_2} \text{ on } \Gamma \end{aligned} \tag{2.3.7.21}$$

as follows. Introduce y_i by

$$\begin{aligned} -\Delta y_i &= f_i \text{ in } \Omega_i \\ y_i &= 0 \text{ on } \partial\Omega_i \cap \partial\Omega \\ y_i &= \lambda \text{ on } \Gamma. \end{aligned} \tag{2.3.7.22}$$

Then y_i can be decomposed into

$$y_i = z_i + w_i, \tag{2.3.7.23}$$

where

$$\begin{aligned} -\Delta z_i &= f_i \text{ in } \Omega_i \\ z_i &= 0 \text{ on } \partial\Omega_i \cap \partial\Omega \\ z_i &= 0 \text{ on } \Gamma \end{aligned} \tag{2.3.7.24}$$

and

$$\begin{aligned} -\Delta w_i &= 0 \text{ in } \Omega_i \\ w_i &= 0 \text{ on } \partial\Omega_i \cap \partial\Omega \\ w_i &= \lambda \text{ on } \Gamma, \end{aligned} \tag{2.3.7.25}$$

see Quarteroni and Valli [91] Section 1.1.

Now with $w_i = R_i\,\lambda$ we obtain

$$\begin{aligned} &\frac{\partial u_1}{\partial \nu_1} + \frac{\partial u_2}{\partial \nu_2} = 0 \Longleftrightarrow \\ &\quad \frac{\partial}{\partial \nu_1} R_1\,\lambda + \frac{\partial}{\partial \nu_2} R_2\,\lambda = -\frac{\partial}{\partial \nu_1} z_1 - \frac{\partial}{\partial \nu_2} z_2. \end{aligned}$$

In other words, (2.3.7.21) is solved if and only if, with $S_i := \frac{\partial}{\partial \nu_i} R_i$, $\mathcal{S} = \mathcal{S}_1 + \mathcal{S}_2$, and $\mathcal{X} := -\frac{\partial}{\partial \nu_1} A_1^{-1} f_1 - \frac{\partial}{\partial \nu} A_2^{-1} f_2$, we have the Steklov-Poincaré interface equation

$$\mathcal{S}\,\lambda = \mathcal{X}. \tag{2.3.7.26}$$

The counterpart of (2.3.7.18) can now be expressed as

$$\begin{aligned} & -\Delta z_i - \Delta R_i \lambda = f_i, \quad i = 1,2 \\ & -tr_1 z_1 - tr_2 z_2 + (\mathcal{S}_1 + \mathcal{S}_2)\lambda = 0 \end{aligned} \tag{2.3.7.27}$$

or, in block format

$$\begin{pmatrix} -\Delta & 0 & -\Delta R_1 \\ 0 & -\Delta & -\Delta r_2 \\ -tr_1 & -tr_2 & \mathcal{S}_1 + \mathcal{S}_2 \end{pmatrix} \begin{pmatrix} z_1 \\ z_2 \\ \lambda \end{pmatrix} = \begin{pmatrix} f_1 \\ f_2 \\ 0 \end{pmatrix} \tag{2.3.7.28}$$

The extension to more general splittings can again be put into the formal pattern

$$\begin{pmatrix} A_1 & 0 & \dots & 0 & B_{11} & \dots & B_{19} \\ 0 & A_2 & \dots & 0 & B_{21} & & B_{29} \\ \vdots & & \ddots & & \vdots & & \vdots \\ 0 & & & A_p & B_{p1} & & B_{pq} \\ C_{11} & \dots & \dots & C_{1p} & \Sigma_{11} & \dots & \Sigma_{1q} \\ C_{21} & \dots & \dots & C_{2p} & \Sigma_{21} & \dots & \Sigma_{2q} \\ C_{q1} & & & C_{qp} & \Sigma_{q1} & \dots & \Sigma_{qq} \end{pmatrix} \tag{2.3.7.29}$$

where the matrices (B_{ij}), (C_{ij}) have a somewhat condensed sparsity pattern when compared with the corresponding matrices of (2.3.7.20) and (Σ_{ij}) is dense. However, the q of (2.3.7.28) is half of the q in (2.3.7.20).

Typically, (2.3.7.29) is expressed compactly as

$$\left(\begin{array}{ccc|c} A_1 & & & A_{1\Sigma} \\ & \ddots & & \vdots \\ & & A_p & A_{p\Sigma} \\ \hline A_{\Sigma,1} & \dots & A_{\Sigma,p} & A_{\Sigma\Sigma} \end{array}\right) \tag{2.3.7.30}$$

with the obvious definitions $(B_{i1} \dots B_{iq}) =: A_{i,\Sigma}$, $i = 1, \dots, p$, $(C_{1i}^T, \dots, C_{qi}^T)^T =: A_{\Sigma,i}$, $A_{\Sigma\Sigma} := \Sigma$. After discretization, the structure remains as in (2.3.7.30). The matrix representation (2.3.7.30) is the paradigm for most of the known domain-decomposition techniques, whereas the special sparsity structure of the Robin method resulting in (2.3.7.20) does not seem to have been noticed in the literature.

It is important to note that equations with matrices according to (2.3.7.30) have been shown to possess the structure of the hierarchical matrices, or H-matrices, introduced by W. Hackbusch [18]; see also [40] for the particular situation of (2.3.7.30). Corresponding equations, after discretization, can be very effectively solved in order $O(n)$, where n is the order of the matrix, by a direct solve.

As a very simple example to illustrate these remarks consider the problem

$$\begin{aligned} -u'' &= f \quad \text{in} \quad (-L, L) \\ u(-1) &= u(1) = 0. \end{aligned} \tag{2.3.7.31}$$

The subproblems to be solved are, for example,

$$\begin{aligned} &-u_1'' = f_1 \quad \text{in} \quad (-L,0)\\ &u_1(-L) = 0\\ &u_1'(0) + \beta\, u_1(0) = \lambda \end{aligned} \tag{2.3.7.32}$$

$$\begin{aligned} &-u_2'' = f_2 \quad \text{in} \quad (0,L)\\ &u_2(L) = 0\\ &-u_2'(0) + \beta\, u_2(0) = \mu. \end{aligned} \tag{2.3.7.33}$$

In discussing the properties of $\mathcal{A}$ (see (2.3.7.8)) we are interested in the solutions w_i to (2.3.7.32), (2.3.7.33) with $f_1 = f_2 = 0$. Thus

$$\begin{aligned} w_1(x) &= \frac{1}{1+\beta L}\,(x+L)\cdot\lambda\\ w_2(x) &= \frac{1}{1+\beta L}\,(-x+L)\cdot\mu. \end{aligned} \tag{2.3.7.34}$$

But $S_1\,\lambda = w_1(0) = \frac{\lambda}{1+\beta L}$, $S_2\,\mu = w_2(0) = \frac{\mu}{1+\beta L}$ and, therefore, $S_1 = S_2 = \frac{1}{1+\beta L}$. In this case $\mathcal{A}$ is given by

$$\mathcal{A} = \begin{pmatrix} 1 & -\frac{\beta L-1}{\beta L+1}\\ -\frac{\beta L-1}{\beta L+1} & 1 \end{pmatrix}$$

and, as λ, μ are just scalars, the problem is 2×2. The eigenvalues of $\mathcal{A}$ are

$$w_1 = \frac{2}{\beta L+1}, \quad w_2 = \frac{2\beta L}{\beta L+1}$$

hence, $w_i \in [0,2]$, $\forall\beta\in[0,\infty)$. The minimum with respect to β of the maximum eigenvalue occurs for $\beta = 1/L$, where $w_1 = w_2 = 1$. Thus $\theta\in(0,2)$ and hence $\varepsilon\in(-1,1)$. Obviously, for $\beta = 1/L$ we have $\mathcal{A} = I$ and the Richardson iteration matrix $I-\mathcal{A} = 0$ for $\theta = 1$. Thus, after step 2 we are done. The case $\theta = 1$ relates to $\varepsilon = 0$, that is to the unrelaxed iteration. In addition, $\theta = 1$ is also the optimal choice according to the standard theorems; see Hackbusch [39]. The case $\theta \neq 1$, $\varepsilon\in(0,1)$ leads to a convergence rate $\max\{|1-\theta\,\frac{2}{1+\beta}|,\ |1-\theta\,\frac{2\beta}{1+\beta}|\}$. Note also that the optimal choice of $\beta > 0$ depends on the geometry.

This trivial exercise shows that, under certain circumstances, the iteration is easy to analyze. Note, however, that even in the two-dimensional completely symmetric case, the iteration takes place in a trace space which is infinite-dimensional. Hence, a spectral analysis of the Poincaré-Steklov-operators is needed to proceed further. The more common situation in this book will be one where there is no symmetry with respect to the subdomains. Thus the matrix operator in (2.3.7.8) will generically not be symmetric. While to our knowledge the nonsymmetric case is handled only in the finite-dimensional setting in the framework of the Richardson-iteration, we can resort to theorems of Schaefer [94] and Opial [89]. See also Fortin and Glowinski [23].

To apply these theorems, we need the notion of asymptotic regular operators in a Banach space. In fact, one can work in a closed convex subset $\mathcal{C}$. Let $\mathcal{X}$ be a Banach space, $\mathcal{C} \subseteq \mathcal{X}$ be a closed convex subset. A nonexpansive map $\mathcal{T} : \mathcal{C} \to \mathcal{C}$ is said to be *asymptotically regular* if, for any $x \in \mathcal{C}$, the sequence $\{\mathcal{T}^{n+1} x - \mathcal{T}^n x\} = \{(\mathcal{T} - I)\mathcal{T}^n x\}$ tends to zero as $n \to \infty$. We also introduce the family of mappings

$$\mathcal{T}_\varepsilon := \varepsilon I + (1 - \varepsilon)\mathcal{T}, \quad \varepsilon \in [0, 1). \tag{2.3.7.35}$$

Now suppose that $\mathcal{X}$ is uniformly convex and has a weakly continuous duality map. Under these assumptions we can state the following three results.

Proposition 2.3.7.1 (Schaefer [94]) *If $\mathcal{T}$ has at least one fixed point in $\mathcal{C}$, then the mapping $\mathcal{T}_\varepsilon$ is asymptotically regular for all $\varepsilon \in (0, 1)$.*

Proposition 2.3.7.2 (Opial [89]) *If the asymptotically regular map $\mathcal{T}$ has a fixed point $x \in \mathcal{C}$ then, for all $x \in \mathcal{C}$, $\{\mathcal{T}^n x\}$ is weakly convergent to a fixed point of $\mathcal{T}$.*

Proposition 2.3.7.3 (Opial [89]) *Let T be nonexpansive with at least one fixed point. Then for each $\varepsilon \in (0, 1)$ and $x \in C$ the sequence $\{\mathcal{T}_\varepsilon^n x\}$ is weakly convergent to a fixed point.*

As the iteration with $\mathcal{T}_\varepsilon$ corresponds exactly with the underrelaxed iteration (2.3.7.12) and, in turn, with the damped Richardson iteration for $\theta = (1 - \varepsilon)$, we infer weak convergence of the iterates to the only fixed point $x = 0$. Thus, generically we will expect convergence in the relaxed case $\varepsilon \in (0, 1)$. However, proving convergence of the solutions of the partial differential equations involved remains a difficult problem. This is particularly true for time-dependent problems which are at the heart of this monograph.

2.3.8 A Serial One-Dimensional Problem

The Steklov-Poincaré Equation

In order to further explore the structure of the iteration and the Schur operator equation, we consider a serial 1-d situation. To this end, we decompose the interval $(0, \ell)$ into $N + 1$ subintervals

$$0 = \ell_0 < \ell_1 < \cdots < \ell_N < \ell_{N+1} = \ell,$$

and we set $L_i := (\ell_i, \ell_{i+1})$, $i = 0, \ldots, N$. Let us assume that at $0 = \ell_0$ we have a Dirichlet condition, whereas at $\ell_{N+1} = \ell$ a Neumann condition is imposed. Let $u_i = u|_{L_i}$ be the restriction of u to L_i. Thus,

$$u_0(\ell_0) = 0, \quad u'_N(\ell_{N+1}) = 0. \tag{2.3.8.1}$$

The transmission conditions at the interfaces, represented by the break-points ℓ_i, $i = 1, \ldots, N$, are

$$u_i(\ell_{i+1}) = u_{i+1}(\ell_{i+1}), \quad i = 0, \ldots, N-1, \tag{2.3.8.2}$$

$$u_i'(\ell_{i+1}) = u_{i+1}'(\ell_{i+1}), \quad i = 0, \ldots, N-1. \tag{2.3.8.3}$$

Obviously, the standard Robin-Robin-type domain decomposition procedure leads one to consider the relaxation of (2.3.8.2), (2.3.8.3) as usual:

$$\begin{aligned} u_i'(\ell_{i+1}) + \beta\, u_i(\ell_{i+1}) &= \lambda_{i,i+1} \\ &= u_{i+1}'(\ell_{i+1}) + \beta\, u_{i+1}(\ell_{i+1}), \quad i = 0, \ldots, N \end{aligned} \tag{2.3.8.4}$$

and

$$\begin{aligned} -u_i'(\ell_i) + \beta\, u_i(\ell_i) &= \lambda_{i,i-1} \\ &= -u_{i-1}'(\ell_i) + \beta\, u_{i-1}(\ell_i), \quad i = 1, \ldots, N. \end{aligned} \tag{2.3.8.5}$$

On the bounding intervals $i = 0$, $i = N$ we have to solve

$$\begin{aligned} &-u_0'' + c_0\, u_0 = f_0 \quad \text{in} \quad L_0 \\ &u_0(0) = 0 \\ &u_0'(\ell_1) + \beta\, u_0(\ell_1) = \lambda_{0,1} \end{aligned} \tag{2.3.8.6}$$

and

$$\begin{aligned} &-u_N'' + c_N\, u_N = f_N \quad \text{in} \quad L_N \\ &u_N'(\ell_{N+1}) = 0 \\ &-u_N'(\ell_N) + \beta\, u_N(\ell_N) = \lambda_{N,N-1} \end{aligned} \tag{2.3.8.7}$$

respectively, whereas in L_i $i = 2, \ldots, N-1$, we have

$$\begin{aligned} &-u_i'' + c_i\, u_i = f_i \quad \text{in} \quad L_i \\ &-u_i'(\ell_i) + \beta\, u_i(\ell_i) = \lambda_{i,i-1} \\ &u_i'(\ell_{i+1}) + \beta\, u_i(\ell_{i+1}) = \lambda_{i,i+1}. \end{aligned} \tag{2.3.8.8}$$

Problems (2.3.8.6), (2.3.8.7), (2.3.8.8) are to be decomposed according to

$$u_i = z_i + w_i \quad i = 0, \ldots, N \tag{2.3.8.9}$$

where z_i solves (2.3.8.6), (2.3.8.7), (2.3.8.8) with $\lambda_{0,1} = \lambda_{i,i-1} = \lambda_{i,i+1} = \lambda_{N,N-1} = 0$ and w_i satisfies (2.3.8.6), (2.3.8.7), (2.3.8.8) with $f_i = 0$. Define

$$z_i =: G_i\, f_i, \quad i = 0, \ldots, N \tag{2.3.8.10}$$

$$\lambda_i = \begin{pmatrix} \lambda_{i,i-1} \\ \lambda_{i,i+1} \end{pmatrix}, \quad i = 1, \ldots, N-1, \tag{2.3.8.11}$$

$$\mathcal{S}_i(\lambda_i) = \begin{pmatrix} w_i(\ell_i) \\ w_i(\ell_{i+1}) \end{pmatrix}, \quad i = 1, \ldots, N-1, \tag{2.3.8.12}$$

$$\begin{aligned} S_{i,i+1}(\lambda_i) &= w_i(\ell_{i+1}) \\ S_{i,i-1}(\lambda_i) &= w_i(\ell_i), \quad i = 1,\dots,N-1, \end{aligned} \tag{2.3.8.13}$$

$$R_i\,\lambda_i := w_i, \quad i = 1,\dots,N-1. \tag{2.3.8.14}$$

Going back to (2.3.8.4), (2.3.8.5) we have

$$\begin{aligned} \lambda_{i,i+1} =& w'_{i+1}(\ell_{i+1}) + \beta\, w_{i+1}(\ell_{i+1}) \\ &+ z'_{i+1}(\ell_{i+1}) + \beta\, z_{i+1}(\ell_{i+1}) \\ =& (R_{i+1}(\lambda_{i+1}))'(\ell_{i+1}) + \beta\, S_{i+1,i}(\lambda_{i+1}) \\ &+ G_{i+1}(f_{i+1})'(\ell_{i+1}) + \beta\, G_{i+1}(f_{i+1})(\ell_{i+1}) \\ =& -\lambda_{i+1,i} + 2\beta\, S_{i+1,i}(\lambda_{i+1}) \\ &+ 2\beta\, G_{i+1}(f_{i+1})(\ell_{i+1}) \end{aligned} \tag{2.3.8.15}$$

and

$$\begin{aligned} \lambda_{i,i-1} =& -w'_{i+1}(\ell_i) + \beta\, w_{i-1}(\ell_i) \\ &- z'_{i-1}(\ell_i) + \beta\, z_{i-1}(\ell_i) \\ =& -\lambda_{i-1,i} + 2\beta\, S_{i-1,i}(\lambda_{i-1}) \\ &+ 2\beta\, G_{i-1}(f_{i-1})(\ell_i). \end{aligned} \tag{2.3.8.16}$$

From (2.3.8.15), (2.3.8.16) we obtain

$$\begin{aligned} \begin{pmatrix} \lambda_{i,i-1} \\ \lambda_{i,i+1} \end{pmatrix} =& -\begin{pmatrix} 1 & 0 \\ 0 & 0 \end{pmatrix} \begin{pmatrix} \lambda_{i+1,i} \\ \lambda_{i+1,i+2} \end{pmatrix} + 2\beta \begin{pmatrix} S_{i-1,i}(\lambda_{i-1}) \\ 0 \end{pmatrix} \\ & -\begin{pmatrix} 0 & 0 \\ 0 & 1 \end{pmatrix} \begin{pmatrix} \lambda_{i-1,i-2} \\ \lambda_{i-1,i} \end{pmatrix} + 2\beta \begin{pmatrix} 0 \\ S_{i+1,i}(\lambda_{i+1}) \end{pmatrix} \\ & + 2\beta \begin{pmatrix} G_{i-1}(f_{i-1})(\ell_i) \\ G_{i+1}(f_{i+1})(\ell_{i+1}) \end{pmatrix}. \end{aligned} \tag{2.3.8.17}$$

Now using the elementary matrices $a^{ij}_{\ell k} = \delta_{i,\ell}\delta_{j,k}$, $E_{ij} = (a^{ij}_{\ell k})_{\ell,k}$, we write (2.3.8.17) as

$$\begin{aligned} \lambda_i =& -E_{11}\,\lambda_{i+1} + 2\beta\, E_{12}\, S_{i-1}(\lambda_{i-1}) \\ &- E_{22}\,\lambda_{i-1} + 2\beta\, E_{21}\, S_{i+1}(\lambda_{i+1}) \\ &+ 2\beta(E_{12}\, G_{i-1}(f_{i-1})(\ell_i) + E_{21}\, G_{i+1}(f_{i+1})(\ell_{i+1})). \end{aligned} \tag{2.3.8.18}$$

Let us define

$$\begin{aligned} h_i &:= 2\beta(E_{12}\, G_{i-1}(f_{i-1})(\ell_i) + E_{21}\, G_{i+1}(f_{i+1})(\ell_{i+1})) \\ \Gamma_{i,i+1} &:= 2\beta(E_{21}\, S_{i+1} - E_{11}) \\ \Gamma_{i,i-1} &:= 2\beta(E_{12}\, S_{i-1} - E_{22}). \end{aligned} \tag{2.3.8.19}$$

Then (2.3.8.18) can be written as

$$-\Gamma_{i,i-1}\,\lambda_{i-1} + \lambda_i - \Gamma_{i,i+1}\,\lambda_{i+1} = h_i, \tag{2.3.8.20}$$

$i = 2, \ldots, N-2$. Accordingly, for $i = 1$, $i = N-1$ we have

$$\begin{aligned} \lambda_1 - \Gamma_{1,2}\,\lambda_2 &= h_1 \\ -\Gamma_{N,N-1}\,\lambda_{N-1} + \lambda_N &= h_N, \end{aligned} \tag{2.3.8.21}$$

where $\lambda_1 := \lambda_{1,2}$, $\lambda_N := \lambda_{N,N-1}$. Note that (2.3.8.21) are scalar equations.

Therefore, the "Steklov-Poincaré" equation associated with this serial problem, i.e., (2.3.8.20), (2.3.8.21), can be written as

$$\begin{pmatrix} I & -\Gamma_{12} & & & & \\ -\Gamma_{21} & I & -\Gamma_{22} & & & \\ & & \cdots & & & \\ & & & \Gamma_{N-1,N-2} & I & -\Gamma_{N-1,N} \\ & & & & -\Gamma_{N,N-1} & I \end{pmatrix} \times \begin{pmatrix} \lambda_1 \\ \lambda_2 \\ \vdots \\ \lambda_{N-1} \\ \lambda_N \end{pmatrix} = \begin{pmatrix} h_1 \\ h_2 \\ \vdots \\ h_{N-1} \\ h_N \end{pmatrix}. \tag{2.3.8.22}$$

We may write (2.3.8.22) compactly as

$$[\mathcal{I} - \mathcal{G}]\,\Lambda = H. \tag{2.3.8.23}$$

A Preconditioning Method

It is common practice in the numerical implementation of domain decomposition methods to discretize the Steklov-Poincaré equation which, in the context of a serial 1-d problem, is (2.3.8.22) or (2.3.8.23), thereby obtaining the Schur complement equation to which is applied a preconditioning step. The general philosophy is discussed in Quarteroni and Valli [91, Chapter 3]. The strategy can be most easily explained in the context of the very simple 1-d serial problem which, in the case of constant coefficients, can be solved analytically. The reason for the choice of this model problem, besides its exemplary character, is that the same procedure can be applied to time domain decomposition problems discussed in Chapters 6–8 of this book. For the latter decomposition problems, there will be a strong relation to the very recent preconditioning methods of Maday, Lions and Turinici [75], [84]. As discretizations are not the focus of this book, however, we confine our attention to this 1-d problem.

The idea behind the two-level approach discussed in [91] is to use global information that can be computed fast and efficiently to accelerate the iteration of the local problems. As in our situation of a Robin-Robin-Jabobi-type iteration (see, e.g., [91, Chapter 1.3, iteration 3 and 4] for the Gauss-Seidel version and [91, Remark 1.3.1]), the Steklov-Poincaré operator is not necessarily symmetric and elliptic but rather an operator-valued matrix and the two-level approach is not identical to the one explained in [91].

In order to obtain global information on the λ_i's fast, we consider a coarse grid discrete problem as follows. Let $U_i \sim u(\ell_i)$ be the approximation to the solution of

$$\begin{aligned} -u'' + u &= f \ \text{ in } \ (0,\ell) \\ u(0) = 0, \ & u'(\ell) = 0 \end{aligned} \tag{2.3.8.24}$$

evaluated at $x = \ell_i$, the breakpoints, as follows:

$$\begin{aligned} &-\frac{u_{i+1} - 2u_i + u_{i-1}}{\Delta \ell_i^2} + c\, u_i = f_i, \ i = 1,\dots,N \\ &u_0 = 0, \\ &(1 + c\Delta \ell_{N+1}^2)\, u_{N+1} - u_{N-1} = f_{N+1}. \end{aligned} \tag{2.3.8.25}$$

This is a second order in $\Delta \ell_i$ approximation to (2.3.8.24). We proceed to retrieve information on λ_i's from the u_i's by simply approximating (2.3.8.4), (2.3.8.5), even to first order only.

$$\begin{aligned} \frac{u_{i+2} - u_{i+1}}{\Delta \ell_{i+1}} + \beta\, u_{i+1} &=: \lambda_{i,i+1} \\ -\frac{u_{i+1} - u_i}{\Delta \ell_i} + \beta\, u_i &=: \lambda_{i,i-1}. \end{aligned} \tag{2.3.8.26}$$

With λ_i given by (2.3.8.26), $i = 1,\dots,N-1$, we have a first data set in order to start the iterative solve of (2.3.8.23):

$$\Lambda^{n+1} = \Lambda^n - \Theta\left((\mathcal{I} - \mathcal{G})\,\Lambda^n - H\right). \tag{2.3.8.27}$$

In order to proceed further, we need a numerical solution based on the data Λ. To this end we transform the local equations (2.3.8.6), (2.3.8.7), (2.3.8.8) into first-order two-point boundary value systems

$$\begin{aligned} &\begin{pmatrix} u_0 \\ u_0' \end{pmatrix}' = \begin{pmatrix} 0 & 1 \\ c & 0 \end{pmatrix} \begin{pmatrix} u_0 \\ u_0' \end{pmatrix} + \begin{pmatrix} 0 \\ -f_0 \end{pmatrix} \ \text{ in } L_0 \\ &(1,0) \begin{pmatrix} u_0 \\ u_0' \end{pmatrix} (0) = 0 \\ &(\beta,1) \begin{pmatrix} u_0 \\ u_0' \end{pmatrix} (\ell_1) = \lambda_{0,1}, \end{aligned} \tag{2.3.8.28}$$

$$\begin{aligned} &\begin{pmatrix} u_i \\ u_i' \end{pmatrix}' = \begin{pmatrix} 0 & 1 \\ c & 0 \end{pmatrix} \begin{pmatrix} u_i \\ u_i' \end{pmatrix} + \begin{pmatrix} 0 \\ -f_i \end{pmatrix} \ \text{ in } L_i \\ &(\beta,1) \begin{pmatrix} u_i \\ u_i' \end{pmatrix} (\ell_{i+1}) = \lambda_{i,i+1} \\ &(\beta,-1) \begin{pmatrix} u_i \\ u_i' \end{pmatrix} (\ell_i) = \lambda_{i,i-1}, \end{aligned} \tag{2.3.8.29}$$

$$\begin{pmatrix} u_N \\ u_N' \end{pmatrix}' = \begin{pmatrix} 0 & 1 \\ c & 0 \end{pmatrix} \begin{pmatrix} u_N \\ u_N' \end{pmatrix} + \begin{pmatrix} 0 \\ -f_N \end{pmatrix} \quad \text{in } L_N$$

$$(0,1) \begin{pmatrix} u_N \\ u_N' \end{pmatrix} (\ell_{N+1}) = 0 \tag{2.3.8.30}$$

$$(\beta,-1) \begin{pmatrix} u_N \\ u_N' \end{pmatrix} (\ell_N) = \lambda_{N,N-1},$$

with the matching condition

$$\begin{aligned} \lambda_{i,i+1} &= (\beta,1) \begin{pmatrix} u_{i+1} \\ u_{i+1} \end{pmatrix} (\ell_{i+1}) \\ \lambda_{ii-1} &= (\beta,1) \begin{pmatrix} u_{i-1} \\ u_{i-1} \end{pmatrix} (\ell_i). \end{aligned} \tag{2.3.8.31}$$

We may use the methods of Asher, Matheij and Russell [3] in order to solve (2.3.8.28)–(2.3.8.30). In particular, we may use a single shooting step on each interval L_i, $i = 0, \dots, N$. Note that this can be done in parallel. Even though in the 1-d context of a decomposition in space, this observation may not seem relevant, it turns out that in optimal control of hyperbolic problems discussed in Chapters 6 to 9, the analogous time domain decomposition will benefit from parallelism. In any case, we will obtain a numerical solution of (2.3.8.28)–(2.3.8.30), a discrete analog $\mathcal{G}_\Delta$, H_Δ of $\mathcal{G}$ and H given by (2.3.8.23), (2.3.8.22), respectively, such that

$$(\mathcal{I} - \mathcal{G}_\Delta)\,\Lambda = H_\Delta. \tag{2.3.8.32}$$

The idea is to precondition the damped Richardson iteration (2.3.8.23) by rewriting it as

$$\Lambda^{n+1} - \mathcal{G}_\Delta\,\Lambda^{n+1} = -\mathcal{G}_\Delta\,\Lambda^n + \Lambda^n - \Theta\left((\mathcal{I} - \mathcal{G})\,\Lambda^n - H\right)$$

or as

$$\Lambda^{n+1} = \Lambda^n - \Theta\,(\mathcal{I} - \mathcal{G}_\Delta)^{-1}\left((\mathcal{I} - \mathcal{G})\,\Lambda^n - H\right). \tag{2.3.8.33}$$

The interpretation of (2.3.8.33) is very close to the one given by Maday et al. [75], [84]. Indeed, replacing the right-hand side of (2.3.8.33) by the residual

$$res^n = (\mathcal{I} - \mathcal{G})\,\Lambda^n - H, \tag{2.3.8.34}$$

we propagate the error on the coarse grid (using a one-step shooting or a two-point discretization of (2.3.8.28)–(2.3.8.30)), do an updating step (2.3.8.33) and perform a local solve on a fine grid (or even an exact solve in this setting of a 1-d problem) in order to obtain the new residuals. By its nature, the process is completely parallel.

Chapter 3

Partial Differential Equations on Graphs

3.1 Introduction

In this chapter we focus on one-dimensional partial differential equations on graphs. Partial differential equations on graphs, or on higher-dimensional 'networked' domains, have frequently been considered in the literature, beginning with the work of G. Lumber. We dispense with displaying a list of papers on this topic, but rather refer the reader to a recent proceedings volume [2] for an account of current research in this area. For higher-dimensional problems and, more importantly, for optimal control problems for partial differential equations on graphs or networked domains, see [61].

Applications of the theory of partial differential equations on multi-link structures are many. For instance, we encounter problems on graphs or, more generally, on networked domains in structural control, civil engineering, infrastructures, mechatronics, biology, medicine, traffic flow, electrical and communication networks, fluid flow through a network of pipes or canals. See the introductory chapter for more details.

The study of partial differential equations on graphs or networked domains is not just the investigation of well-known mathematical objects on some special domains. Graphs and networked domains, in our terminology, are not manifolds. Thus, we encounter partial differential equations locally on single links, substructures or subdomains, which themselves are indeed manifolds, plus certain transmission conditions at multiple nodes, interfaces, etc., where more than of these substructures meet. Such conditions can be called continuity and Kirchhoff conditions, as in the theory of electrical networks.

Thus, the local equations plus the transmission conditions at multiple interfaces constitute a new mathematical entity that is globally defined on the entire structure. A full account of this subject is far beyond the scope of this book. In very recent applications one considers so-called heterogeneous, multi-physical systems on complex structures. This terminology expresses the fact that one considers a various types of local behavior, necessarily described by different partial differential equations on different parts of the structure. One may, for example, model the flow of water in channels by a one-dimensional shallow water equation, while flooded areas, reservoirs and river sections with higher curvature are taken into account by two- or even three-dimensional models. This applies to the modeling of blood vessels and heart dynamics, as well as to mechatronical structures, where one-dimensional elements such as strings, cables and beams are connected to plates and shells. See [61] for some of these models. Moreover, the coupled dynamics at multiple nodes or interfaces may be of first or second order in time. Static problems may be coupled with dynamical ones, and so forth.

In this book we will concentrate on elliptic and hyperbolic problems on graphs and networked domains. In particular, in this chapter we want to introduce the reader to the general philosophy of partial differential equations on networked domains by examining one-dimensional problems on graphs. In order to allow for more generality in this context, we describe a general class of second order vector-valued Sturm-Liouville problems on graphs. We describe the operators, then discuss elliptic problems and give a first introduction to how a domain decomposition into equations on single links, the edges of the graph, can be designed.

Domain decomposition of problems on graphs is motivated by the analysis of Chapter 2. More precisely, in the elliptic case we consider Robin-type decompositions, as discussed in Section 2.3 of Chapter 2, while in the hyperbolic case we appeal to ideas that were introduced in Section 2.2.2 of Chapter 2. We further introduce an analog of the Steklov-Poincaré operator on graphs, which in the context of wave equations can be interpreted as a scattering operator at multiple nodes. Indeed, it turns out that the domain decomposition of wave equations on graphs is essentially a Schwarz-type nonoverlapping domain decomposition applied to the Riemann invariants. Obviously, such a natural decomposition has the potential to be extended to higher dimensions, as will be seen in Chapters 6 to 9. We also provide an interpretation as a Richardson iteration in order to embed the problem into more standard notions. We dispense with a posteriori estimates in this chapter, as they will be obtained for the more complex problems with controls in subsequent chapters.

3.2 Partial Differential Operators on Graphs

Consider a simple graph $G = (V, G)$ in $\mathbb{R}^d$ with vertices $V = \{v_J : J \in \mathcal{J}\}$ and edges $E = \{e_i : i \in \mathcal{I}\}$, where we index vertices and edges with capital and lower case letters, respectively, and where $\mathcal{J}$ and $\mathcal{I}$ denote the index sets for vertices and edges, respectively. For a given vertex v_J we consider the set of edges that are incident at v_J. The corresponding set of edge-indices is denoted by

$$\mathcal{I}_J := \{i \in \mathcal{I} : e_i \text{ is incident at } v_J\}.$$

The cardinality of $\mathcal{I}_J$ is the edge degree at v_J, that is, $d_J := |\mathcal{I}_J|$. To each edge incident at v_J we assign a positive or negative sign according to whether the edge

e_i ends at v_J or starts at v_J, respectively. That is $\varepsilon_{iJ} = +1$, if e_i ends at v_J and $e_{iJ} = -1$, if e_i starts at v_J. This incidence structure turns G into a directed graph. We are going to consider functions

$$r : G \to \mathbb{R}^p = \prod_{i \in \mathcal{I}} \mathbb{R}^{p_i} \tag{3.2.1.1}$$

where p_i denotes the number of degrees of freedom of r_i. Indeed, we localize the function r to the edges

$$r_i := r|_{e_i} \quad i \in \mathcal{I}. \tag{3.2.1.2}$$

The functions r_i act along e_i as follows:

$$r_i : [0, \ell_i] \to \mathbb{R}^{p_i},\ i \in \mathcal{I}, \tag{3.2.1.3}$$

where the edge e_i is parameterized by $x \in [0, \ell_i]$ and ℓ_i denotes the length of the edge e_i. Indeed, it is possible to consider the edges as smooth arcs parameterized by the arclength parameter; see [98]. We will also consider such functions depending on time:

$$r_i : [0, \ell_i] \times [0, T] \to \mathbb{R}^{p_i},\ i \in \mathcal{I}. \tag{3.2.1.4}$$

The possibility of having different p_i's on different edges relates to the fact that the functions r_i may have a different physical meanings on these edges. For instance, one edge might represent an elastic string and another a Timoshenko beam; see [61] for details.

An edge e_i incident at the vertex v_J will either start or end at v_J. This can be represented by the notation x_{iJ}, where

$$x_{iJ} = \begin{cases} 0 & \text{if} \quad \varepsilon_{iJ} = -1 \\ \ell_i & \text{if} \quad \varepsilon_{iJ} = 1 \end{cases}. \tag{3.2.1.5}$$

Later on, $r_i(x_{iJ})$ and $r_i(x_{iJ}, t)$ will be replaced by the more descriptive notation $r_i(v_J)$ and $r_i(v_J, t)$, respectively.

We will consider one-dimensional partial differential equations on the graph G. Therefore, it will be necessary to specify boundary and compatibility conditions as well as transmission conditions at the vertices V of G. To this end, we subdivide the set of vertices (nodes) as follows:

$$V = V_S \cup V_M, \tag{3.2.1.6}$$

where V_S denotes the set of simple nodes such that $d_J = 1$, and V_M signifies the set of multiple nodes where $d_J > 1$. The set of simple nodes V_S is in turn divided as

$$V_S = V_D \cup V_N, \tag{3.2.1.7}$$

where the suffixes D and N are chosen to indicate nodes at which Dirichlet conditions and Neumann conditions (or Robin conditions), respectively, are imposed. Correspondingly, the index set $\mathcal{J}$ is divided as

$$\mathcal{J} = \mathcal{J}_M \cup \mathcal{J}_D \cup \mathcal{J}_N$$

where

$$\mathcal{J}_M = \{J \in \mathcal{J} : v_J \in V_M\},\ \mathcal{J}_D = \{J \in \mathcal{J} : v_J \in V_D\},\ \mathcal{J}_N = \{J \in \mathcal{J} : v_J \in V_N\}.$$

We will consider differential operators L_i on the edges of the following form:

$$L_i\, r_i = -[K_i(r_i' + R_i\, r_i)]' + R_i^T\, K_i(r_i' + R_i\, r_i) + S_i\, r_i, \tag{3.2.1.8}$$

where

$$\begin{gathered} K_i,\ R_i,\ S_i \in C(0, \ell_i; \mathbb{R}^{p_i \times p_i}), \\ K_i \text{ is symmetric and uniformly positive definite} \\ S_i \text{ is either symmetric and positive definite or the zero matrix.} \end{gathered} \tag{3.2.1.9}$$

A simple calculation shows that the differential operator L_i is symmetric. In fact

$$\begin{aligned} \int_0^{\ell_i} L_i\, r_i \cdot \phi_i\, dx &= \int_0^{\ell_i} \{-[K_i(R_i' + R_i r_i)]' + R_i^T K_i(r_i' + R_i r_i) + S_i r_i\} \cdot \phi_i\, dx \\ &= -K_i(r_i' + R_i r_i) \cdot \phi_i|_{x=0}^{x=\ell_i} + \int_0^{\ell_i} [(K_i(r_i' + R_i r_i)) \cdot (\phi_i' + R_i \phi_i) + S_i r_i \cdot \phi_i]\, dx \\ &= [K_i(r_i' + R_i r_i)] \cdot \phi_i|_0^{\ell_i} + r_i \cdot [K_i(\phi_i' + R_i \phi_i)]|_0^{\ell_i} \\ &\qquad + \int_0^{\ell_i} r_i \cdot \{[-K_i(\phi_i' + R_i \phi_i)]' + R_i^T K_i(\phi_i' + R_i \phi_i) + S_i\, r_i\, \phi_i\}\, dx \\ &= B(r_i, \phi_i)|_0^{\ell_i} + \int_0^{\ell_i} r_i \cdot L_i\, \phi_i\, dx, \end{aligned} \tag{3.2.1.10}$$

where

$$B(r_i, \phi_i) = [K_i(r_i' + R_i r_i)] \cdot \phi_i + r_i \cdot [K_i(\phi_i' + R_i \phi_i)].$$

Thus for $\phi_i \in C_0^\infty(0, \ell_i)^{p_i}$ we have

$$(L_i\, r_i, \phi_i)_{H_i} = (r_i, L_i\, \phi_i)_{H_i}, \tag{3.2.1.11}$$

where

$$\begin{gathered} H_i := L^2(0, \ell_i), \\ (r_i, \phi_i)_{H_i} := \sum_{j=1}^{p_i} \int_0^{\ell_i} r_{ij}\, \phi_{ij}\, dx = \int_0^{\ell_j} (r_i, \phi_i)_{\mathbb{R}^{p_i}}\, dx = \int_0^{\ell_i} r_i \cdot \phi_i\, dx \end{gathered} \tag{3.2.1.12}$$

is the canonical Hilbert space of square integrable $\mathbb{R}^{p_i}$-valued distributions. The definition (3.2.1.8) of L_i is motivated by applications to multi-link structures such as networks of elastic strings, or Timoshenko beams, or combinations thereof; see [61, Chapter II]. Suffice it to say that L_i is a generalization of a classical Sturm-Liouville operator to vector-valued functions, since obviously the choice $R_i = 0$ reduces L_i to the classical case. Eventually one might wish to consider also nonsymmetric differential expressions obtained by adding a "transport"-form to L_i, for example,

$$\mathcal{P}_i r_i = L_i r_i + P_i r_i'. \tag{3.2.1.13}$$

In order to describe boundary and transmission conditions at simple and multiple nodes, we introduce matrices

$$C_{iJ} : \mathbb{R}^{p_i} \to \mathbb{R}^{q_J},\ i \in \mathcal{I}_J,\ J \in \mathcal{J} \tag{3.2.1.14}$$

of full rank q_J with

$$\begin{aligned} q_J &= p_i \ \text{ if } \ d_J = 1 \\ q_J &\leq p_i \ \text{ if } \ d_J > 1. \end{aligned} \tag{3.2.1.15}$$

We impose continuity conditions across multiple nodes of the form

$$C_{iJ}\, r_i(x_{iJ}) = C_{jJ}\, r_j(x_{jJ}),\ \forall\, i,j \in \mathcal{I}_J,\ J \in \mathcal{J}_M. \tag{3.2.1.16}$$

This means that

$$C_{iJ}\, r_i(x_{iJ}) = C_{iJ}\, r_i(v_J) = \lambda_J,\ \forall\, i \in \mathcal{I}_J,\ J \in \mathcal{J}_M. \tag{3.2.1.17}$$

Obviously, if $C_{iJ} = I \in \mathbb{R}^{p_i \times p_i}$, then all variables are continuous across the node v_J. As C_{jJ} has rank q_J, there is a right inverse C_{iJ}^+ such that $C_{iJ}\, C_{iJ}^+ = I$. Let Π_{iJ} denote the orthoprojector onto the kernel of C_{iJ}, and $\Pi_{iJ}^\perp$ the corresponding orthoprojector onto $(\ker C_{iJ})^\perp = \operatorname{Rg} C_{iJ}^T$, that is,

$$C_{iJ}^+\, C_{iJ} = \Pi_{iJ}^\perp, \quad C_{iJ}\, C_{iJ}^+ = I. \tag{3.2.1.18}$$

At simple nodes Dirichlet-type conditions may be imposed, for example. For the sake of simplicity, we restrict ourselves to "full" Dirichlet conditions

$$r_i(x_{iJ}) = 0,\ \forall\, i \in \mathcal{I}_J. \tag{3.2.1.19}$$

Notice that $|\mathcal{I}_J| = 1$ for $J \in \mathcal{J}_D$.
Let us introduce the space

$$\mathcal{H}^s(0,\ell_i) := \prod_{k=1}^{p_i} H^s(0,\ell_i) \tag{3.2.1.20}$$

with the standard product norm, where $H^s(0,\ell_i)$ is the standard Sobolev space of order s on scalar-valued functions defined on $(0,\ell_i)$. We also introduce the Hilbert spaces $(\mathcal{H}, \|\cdot\|)$, $(\mathcal{V}, \|\cdot\|_{\mathcal{V}})$ as follows. Let $\phi \in G \to \mathbb{R}^p$ such that $\phi_i : [0, \ell_i] \to \mathbb{R}^{p_i}$. Set

$$\mathcal{H} := \{\phi \in G \to \mathbb{R}^p : \phi_i \in \mathcal{H}^0(0, \ell_i),\ i \in \mathcal{I}\},$$
$$\|\phi\|^2 := \sum_{i\in\mathcal{I}} \int_0^{\ell_j} M_i\, \phi_i \cdot \phi_i\, dx, \tag{3.2.1.21}$$

where $M_i \in \mathbb{R}^{p_i \times p_i}$ is a symmetric and uniformly positive definite matrix with smooth entries, and $p = \max_{1\le i\le n} p_i$. Further set

$$\begin{aligned}\mathcal{V} := \{\phi \in \mathcal{H} : \phi_i \in \mathcal{H}^1(0, \ell_i), \phi_i(v_J) = 0,\ J \in \mathcal{J}_D,\ i \in \mathcal{I}_J,\\ C_{iJ}\, \phi_i(v_J) = C_{kJ}\, \phi_k(v_J),\ \forall\, i, k \in \mathcal{I}_J,\ J \in \mathcal{J}_M\},\\ \|\phi\|_{\mathcal{V}} := a(\phi, \phi)\end{aligned} \tag{3.2.1.22}$$

where

$$a(r, \phi) := \sum_{i\in\mathcal{I}} \int_0^{\ell_j} [K_i(r_i' + R_i) \cdot (\phi_i' + R_i\, \phi_i) + S_i\, r_i \cdot \phi_i]\, dx,\ \forall\, r,\, \phi \in \mathcal{V}. \tag{3.2.1.23}$$

By standard reasoning, we see that $\mathcal{V}$ is densely and compactly embedded in $\mathcal{H}$. The space $\mathcal{H}$ is identified with its dual space and dual space of $\mathcal{V}$ is denoted by $\mathcal{V}^*$. Thus we have the dense and continuous embedding $\mathcal{V} \hookrightarrow \mathcal{H} \hookrightarrow \mathcal{V}^*$.

3.3 Elliptic Problems on Graphs

We consider the following problem on the graph G

$$a(r, \phi) = \langle f, \phi\rangle, \quad \forall \phi \in \mathcal{V} \tag{3.3.1.1}$$

where $a(\cdot, \cdot)$ is given by (3.2.1.23) and $\langle f, \phi\rangle_{\mathcal{V}}$ denotes the duality pairing between elements $f \in \mathcal{V}^*$ and $\phi \in \mathcal{V}$. It is assumed that either

i) for each $i \in \mathcal{I}$ either S_i is uniformly positive definite on $[0, \ell_i]$ or, when $S_i = 0$, at least one the nodes $x = 0$, $x = \ell_i$ is a Dirichlet node; or (3.3.1.2)

ii) $V_D \neq \emptyset$ and for each vertex in V there is a path in G to a vertex in V_D along which the corresponding matrices C_{ik} are regular.

It has been shown in [61, Lemma 3.1] that, under assumption (3.3.1.2),

$$\|\phi\|_{\mathcal{V}} = [a(\phi, \phi)]^{1/2} \tag{3.3.1.3}$$

is a norm on $\mathcal{V}$ equivalent to that induced on $\mathcal{V}$ by the standard Sobolev norm

$$\|r\|_{\mathcal{H}^1(G)} := \left\{ \sum_{i\in\mathcal{I}} \int_0^{\ell_i} [|r_i|^2 + |r_i'|^2]\, dx \right\}^{1/2},$$

where $\mathcal{H}^1(G) := \prod_{i\in\mathcal{I}} \mathcal{H}^1(0,\ell_j)$. Therefore, we may and do equip $\mathcal{V}$ with the norm given in (3.3.1.3).

We now define local spaces on the edges of the graph whose norms are consistent with those defined globally. Set

$$\mathcal{H}_i := \mathcal{H}^0(0,\ell_i), \quad \|\phi\|^2_{\mathcal{H}_i} := \int_0^{\ell_i} M_i\,\phi\cdot\phi\, dx \tag{3.3.1.4}$$

where M_i is symmetric and positive definite. We further define the local forms $a_i(\cdot,\cdot)$ by

$$a_i(r_i,\phi_i) := \int_0^{\ell_i} [K_i(r_i' + R_i r_i)\cdot(\phi_i' + R_i,\phi_i)$$

$$+ S_i\, r_i\cdot\phi_i]\, dx, \quad \forall\, r_i,\ \phi_i\, \mathcal{H}^1(0,\ell_i), \tag{3.3.1.5}$$

so that for $r,\phi\in\mathcal{V}$ we have

$$a(r,\phi) = \sum_{i\in\mathcal{I}} a_i(r_i,\phi_i). \tag{3.3.1.6}$$

However, $a_i(\phi_i,\phi_i)$ defines a norm on $\mathcal{V}_i := \mathcal{H}^1(0,\ell_i)$ only when (3.3.1.2) i) is fulfilled. Denote by $\mathcal{V}_i^*$ the dual of $\mathcal{V}_i$ with respect to the pivot space $\mathcal{H}_i$. Then $\mathcal{V}$ is a closed subspace of $\prod_{i\in\mathcal{I}} \mathcal{V}_i$. Let

$$\mathcal{A} : \mathcal{V} \to \mathcal{V}^*, \qquad \mathcal{A}_i : \mathcal{V}_i \to \mathcal{V}_i^* \tag{3.3.1.7}$$

be the corresponding Riesz isomorphisms. Then

$$\mathcal{A} = \{\mathcal{A}_i\phi_i\}_{i\in\mathcal{I}}, \quad \forall\,\phi = \{\phi_i\}_{i\in\mathcal{I}} \in \mathcal{V}, \tag{3.3.1.8}$$

and the relation to the forms can be expressed as

$$\langle \mathcal{A}\, r,\phi\rangle_{\mathcal{V}} = a(r,\phi), \quad \forall\, r,\phi\in\mathcal{V}, \tag{3.3.1.9}$$

$$\langle \mathcal{A}_i\, r_i,\phi_i\rangle_{\mathcal{V}_i} = a_i(r_i,\phi_i), \quad \forall\, r_i,\phi_i\in\mathcal{V}_i. \tag{3.3.1.10}$$

On the other hand, taking functions $r,\ \phi \in \prod_{i\in\mathcal{I}} \mathcal{H}^2(0,\ell_i)$ we obtain from (3.2.1.10)

$$a(r,\phi) = \sum_{i\in\mathcal{I}} \int_0^{\ell_i} [K_i(r_i' + R_i\,r_i)\cdot(\phi_i' + R_i\,\phi_i) + S_i\,r_i\cdot\phi_i]\,dx$$
$$= \sum_{i\in\mathcal{I}} \int_0^{\ell_i} L_i\,r_i\cdot\phi_i\,dx + \sum_{i\in\mathcal{I}} [B_i(r_i,\phi_i)]_0^{\ell_i}.$$

Now

$$\begin{aligned}
\sum_{i\in\mathcal{I}} B_i(r_i,\phi_i)|_0^{\ell_i} &= \sum_{i\in\mathcal{I}} [K_i(\ell)(r_i' + R_i\,r_i)](\ell_i)\cdot\phi_i(\ell_i) \\
&\quad - K_i(0)(r_i' + R_i\,r_i)(0)\cdot\phi_i'(0) \\
&= \sum_{J\in\mathcal{J}} \sum_{i\in\mathcal{I}_J} \varepsilon_{iJ}\,K_i(v_J)(r_i'(v_J) + R_i(v_J)\,r_i(v_J))\cdot\phi_i(v_J) \\
&= \sum_{J\in\mathcal{J}\setminus\mathcal{J}_D} \sum_{i\in\mathcal{I}_J} \varepsilon_{iJ}[K_i(r_i' + R_i\,r_i)](v_J)(\Pi_{iJ} + \Pi_{iJ}^{\perp})\,\phi_i(v_J) \\
&= \sum_{J\in\mathcal{J}\setminus\mathcal{J}_D} \sum_{i\in\mathcal{I}_J} \varepsilon_{iJ}\,\Pi_{iJ}\,[K_i(r_i' + R_i\,r_i)](v_J)\,\Pi_{iJ}\,\phi_i(v_J) \\
&\quad + \sum_{J\in\mathcal{J}\setminus\mathcal{J}_D} \sum_{i\in\mathcal{I}_J} \varepsilon_{iJ}(C_{iJ}^{+})^{\top}\,K_i(r_i' + R_i\,r_i)(v_J)\,C_{iJ}\,\phi_i(v_J).
\end{aligned}$$

That is, if we impose the boundary and transmission conditions :

$$r_i(v_J) = 0,\ i\in\mathcal{I}_J,\ J\in\mathcal{J}_D, \tag{3.3.1.11}$$

$$C_{iJ}\,r_i(v_J) = C_{kJ}\,r_k(v_J),\ \forall\, i,k\in\mathcal{I}_J,\ J\in\mathcal{J}_M, \tag{3.3.1.12}$$

$$\sum_{i\in\mathcal{I}_J} \varepsilon_{iJ}(C_{iJ}^{+})^{T}[K_i(r_i' + R_i\,r_i)](v_J) = 0,\ \forall\, J\in\mathcal{J}\setminus\mathcal{J}_D, \tag{3.3.1.13}$$

$$\Pi_{iJ}[K_i(r_i' + R_i\,r_i)](v_J) = 0,\ \forall\, i\in\mathcal{I}_J,\ J\in\mathcal{J}\setminus\mathcal{J}_D, \tag{3.3.1.14}$$

then we obtain

$$a(r,\phi) = \sum_{i\in\mathcal{I}} \int_0^{\ell_i} L_i\,r_i\cdot\phi_i\,dx = \sum_{i\in\mathcal{I}} (M_i^{-1}\,L_i\,r_i,\ \phi_i)_{\mathcal{H}_i}. \tag{3.3.1.15}$$

If we define an operator $\mathcal{A}$ in $\mathcal{H}$ by

$$\begin{aligned}
\mathcal{A}r &= (M_i^{-1}\,L_i\,r_i)_{i\in\mathcal{I}} \\
\mathcal{D}(\mathcal{A}) &= \{r\in\prod_{i\in\mathcal{I}} \mathcal{H}^2(0,\ell_i) : r \text{ satisfies (3.3.1.11)–(3.3.1.14)}\},
\end{aligned} \tag{3.3.1.16}$$

then $\mathcal{A}$ is self-adjoint and positive definite, provided assumption (3.3.1.2) holds, and its unique extension to $\mathcal{V}$, denoted by the same letter $\mathcal{A}$, is precisely the Riesz-isomorphism between $\mathcal{V}$ and $\mathcal{V}^*$ when $\mathcal{V}$ is equipped with the norm $\|\phi\|_{\mathcal{V}}^2 = a(\phi, \phi)$. It is clear that as an operator acting in $\mathcal{H}$, $\mathcal{A}$ has a compact resolvent and, consequently, $\mathcal{A}$ has a discrete spectrum whereas, as an operator acting between $\mathcal{V}$ and $\mathcal{V}^*$, $\mathcal{A}$ is unitary.

3.3.2 Domain Decomposition

In order to decompose the transmission conditions (3.3.1.11)–(3.3.1.14), we introduce the following 'scattering' operator $\mathcal{S}_J$ defined at the vertex v_J, $J \in \mathcal{J}_M$, acting on $z = (z_{i_1}, \ldots, z_{i_{d_J}}) \in \mathbb{R}^{q_J}$, $q_J := \sum_{i \in \mathcal{I}_J} q_i$:

$$(\mathcal{S}_J(z))_i := \frac{2}{d_J} \sum_{j \in \mathcal{I}_J} (z_j) - z_i, \;\; i \in \mathcal{I}_J. \tag{3.3.2.1}$$

Obviously, $\mathcal{S}_J^2 = I$ on $\mathbb{R}^{q_J}$ and $\mathcal{S}_J 1 = 1$, $1 := (1, \ldots, 1)$. We apply this operator to the displacements $C_{iJ}\, r_i(v_J)$ and the strains $\varepsilon_{iJ}(C^+)^T K_i(r_i' + R_i r_i)(v_J)$. Our candidate for domain decomposition then is obtained from

$$\begin{aligned}
&\varepsilon_{iJ}(C_{iJ}^+)^T\, K_i(r_i' + R_i r_i)(v_J) + \beta_J\, C_{iJ}\, r_i(v_J) \\
&= \beta_J\, \mathcal{S}_J(C_{*J} r_*(v_J))_i - \mathcal{S}_J(\varepsilon_{*J}(C_{*J}^+)^T\, K_*(r_*' + R_* r_*)(v_J))_i \\
&= \mathcal{S}_J[\beta_J\, C_{*J} r_*(v_J) - \varepsilon_{*J}(C_{*J}^+)^T\, K_*(r_*' + R_* r_*)(v_J)] \\
&=: g_{iJ},\; \forall i \in \mathcal{I}_J,\; J \in \mathcal{J}_M
\end{aligned} \tag{3.3.2.2}$$

where $g_* := (g_i)_{i \in \mathcal{I}_J}$ denotes the vector g. As $\mathcal{S}_J 1 = 1$ and hence $\langle 1, \mathcal{S}_J(z)\rangle_{\mathbb{R}^{q_J}} = \langle 1, z\rangle_{\mathbb{R}}$, summation of (3.3.2.2) over $\mathcal{I}_J$ leads to

$$\begin{aligned}
&\sum_{i \in \mathcal{I}_J} \varepsilon_{iJ}(C_{iJ}^+)^T K_i(r_i' + R_i r_i)(v_J) + \beta_J \sum_{i \in \mathcal{I}_J} C_{iJ}\, r_i(v_J) \\
&\qquad = \beta_J \sum_{i \in \mathcal{I}_J} C_{iJ}\, r_i(v_J) - \sum_{i \in \mathcal{I}_J} \varepsilon_{iJ}(C_{iJ}^+)^T\, K_i(r_i' + R_i r_i)(v_J),
\end{aligned}$$

which shows that

$$\sum_{i \in \mathcal{I}_J} \varepsilon_{iJ}(C_{iJ}^+)^T\, K_i(r_i' + R_i r_i)(v_J) = 0.$$

For z such that $\sum_{i \in \mathcal{I}_j} z_i = 0$ we have $\mathcal{S}_J\, z = -z$. Hence, (3.3.2.2) reduces to

$$C_{iJ}\, r_i(v_J) = \mathcal{S}_J(C_{*J}\, r_*(v_J)),$$

which implies $C_{*J}\, r_*(v_J) = \alpha\, 1$, or, in other words,

$$C_{iJ}\, r_i(v_J) = C_{jJ}\, r_j(v_J), \quad i, j \in \mathcal{I}_J.$$

Setting $q^J := \varepsilon_{*J}(C_{*J}^+)^T K_*(r_*' + R_* r_*)(v_J)$, $r^J := C_{*J}\, r_*(v_J)$ and using $g^J := (g_{*J})$, from (3.3.2.2) we obtain

$$g^J + \mathcal{S}_J\, g^J = 2\,\beta_J\, \mathcal{S}_J\, r^J. \tag{3.3.2.3}$$

On the other hand,

$$q^J + \beta_J\, r^J = \mathcal{S}_J(\beta_J\, r^J - q^J).$$

Thus, we may decouple the dynamic node condition as follows

$$\begin{aligned}(q^J)^{n+1} + \beta_J (r^J)^{n+1} &= \mathcal{S}_J(\beta_J (r^J)^n - (q^J)^n)\\ &= \mathcal{S}_J(2\beta_J (r^J)^n - (g^J)^n) =: (g^J)^{n+1}.\end{aligned} \tag{3.3.2.4}$$

We have, in fact, two choices corresponding to the first and second equality sign in (3.3.2.4). For the sake of easier reference we restate (3.3.2.4) as two statements:

$$\begin{aligned}(q^J)^{n+1} + \beta_J (r^J)^{n+1} &= (\lambda^J)^n,\\ (\lambda^J)^n &= \mathcal{S}_J(-(q^J)^n + \beta_J (r^J)^n)\end{aligned} \tag{3.3.2.5}$$

or, equivalently

$$\begin{aligned}(q^J)^{n+1} + \beta_J (r^J)^{n+1} &= (g^J)^{n+1},\\ (g^J)^{n+1} &= \mathcal{S}_J(2\beta_J (r^J)^n - (g^J)^n).\end{aligned} \tag{3.3.2.6}$$

Set $(g^J)_i = g_{iJ}$ and $(\lambda^J)_i = \lambda_{iJ}$, respectively. As before, one may introduce a relaxation parameter $\varepsilon \in [0,1)$ and restate (3.3.2.6) as follows

$$\begin{aligned}(q^J)^{n+1} + \beta_J (r^J)^{n+1} = (1-\varepsilon)\, \mathcal{S}_J(-(q^J))^n + \beta_J (r^J)^n)\\ +\varepsilon((q^J)^n + \beta_J (r^J)^n) =: (g^{\varepsilon,J})^{n+1}.\end{aligned} \tag{3.3.2.7}$$

Then the equivalent condition in terms of g^J is

$$\begin{aligned}(q^J)^{n+1} + \beta_j (r^J)^{n+1} &= (g^J)^{n+1},\\ (g^J)^{n+1} + (1-\varepsilon)\, \mathcal{S}_J (g^J)^n - \varepsilon (g^J)^n &= 2(1-\varepsilon)\, \beta_J\, \mathcal{S}_J (r^J).\end{aligned} \tag{3.3.2.8}$$

In order to analyze the convergence properties of the above iteration, let us first display the corresponding local problems that are to be solved on each individual edge.

$$-[K_i(r_2' + R_i r_i]' + R_i^T K_i(r_i' + R_i r_i) + \mathcal{S}_i r_i = F_i \ \text{ on } \ (0,\ell_i), \tag{3.3.2.9}$$

$$r_i(v_J) = 0,\ J \in \mathcal{J}_D,\ i \in \mathcal{I}_J, \tag{3.3.2.10}$$

$$\varepsilon_{iJ}(C_{iJ}^+)^T\, K_i(r_i' + R_i r_i)(v_J) = f_J,\ J \in \mathcal{J}_N,\ i \in \mathcal{I}_J, \tag{3.3.2.11}$$

$$\varepsilon_{iJ}\, \Pi_{iJ}\, K_i(r_i' + R_i r_i)(v_J) = m_J,\ J \in \mathcal{J}_N,\ i \in \mathcal{I}_J, \tag{3.3.2.12}$$

$$\varepsilon_{iJ}(C_{iJ}^+)^T\, K_i(r_i' + R_i r_i)(v_J) + \beta_J\, C_{iJ}\, r_i(v_J) = \lambda_{iJ},\ J \in \mathcal{J}_M,\ i \in \mathcal{I}_J, \tag{3.3.2.13}$$

$$\varepsilon_{iJ}\, \Pi_{iJ}\, K_i(r_i' + R_i r_i)(v_J) = 0,\ J \in \mathcal{J}_M,\ i \in \mathcal{I}_J. \tag{3.3.2.14}$$

In particular, if e_i is an edge incident to a Dirichlet vertex v_J, $J \in \mathcal{J}_D$, then, unless the connected graph is trivial, the edge e_i is also incident at a multiple vertex (node) v_J, $J \in \mathcal{J}_M$. In that case we consider problems (3.3.2.9), (3.3.2.10), (3.3.2.13) and (3.3.2.14). Similarly, an edge incident at a Neumann node will correspond to (3.3.2.9), (3.3.2.11)–(3.3.2.14), while such an edge connecting multiple nodes will correspond to (3.3.2.9), (3.3.2.13) and (3.3.2.14).

Remark 3.3.2.1 It should be pointed out that the global problem on the graph can be interpreted as a discrete dynamical system (in the iteration index) on the vertices of that graph, the state of which serve as boundary inputs for the local continuous systems on the edges.

We proceed to describe the variational formulation of these local problems. Multiply (3.3.2.9) by φ_i and integrate over $[0, \ell_i]$. With the notation of the first section we have by (3.3.2.9)

$$\int_0^{\ell_i} F_i\,\varphi_i\,dx = \int_0^{\ell_i} L_i\varphi_i\,dx = \int_0^{\ell_i} [K_i(r_i' + R_i r_i)\cdot(\varphi_i' + R_i\varphi_i) + \mathcal{S}_i r_i \varphi_i]\,dx$$
$$+ \mathcal{B}_i(r_i,\varphi_i)|_0^{\ell_i} = a_i(r_i,\varphi_i) + \mathcal{B}_i(r_i,\varphi_i)|_0^{\ell_i}. \quad (3.3.2.15)$$

In particular, for an edge connecting a 'controlled' Neumann node v_N with a multiple node v_J we have

$$\int_0^{\ell_i} F_i\,\varphi_i\,dx = a_i(r_i,\varphi_i) + m_N\cdot\Pi_{iN}\,\varphi_i(v_N) - f_N\cdot C_{iN}\,\varphi_i(v_N)$$
$$- (\lambda_{iJ}^n - \beta_J C_{iJ}\,r_i(v_J))\cdot C_{iJ}\,\varphi_i(v_J). \quad (3.3.2.16)$$

For simplicity we set the 'moment controls' m_N to zero, set $(F_i,\varphi_i) := \int_0^{\ell_i} F_i\cdot\varphi_i\,dx$ and define

$$b_{iN}(f_N,\varphi_i) := f_N\cdot C_{iN}\,\varphi_i(v_N),$$
$$a_{iJ}(r_i,\varphi_i) := \beta_J\,C_{iJ}\,r_i(v_J)\cdot C_{iJ}\,\varphi_i(v_J).$$

We may then rewrite (3.3.2.16) as

$$a_i(r_i,\varphi_i) + a_{iJ}(r_i,\varphi_i) = (F_i,\varphi_i)$$
$$+ b_{iN}(f,\varphi_i) + b_{iJ}(\lambda_{iJ}^n,\varphi_i),\ \forall\,\varphi_i \in \mathcal{H}^1(0,\ell_i). \quad (3.3.2.17)$$

Obviously, the form

$$\tilde{a}_i(r_i,\varphi_i) := a_i(r_i,\varphi_i) + a_{iJ}(r_i,\varphi_i) \quad (3.3.2.18)$$

is positive definite, as long as assumption (3.3.1.2) holds. Moreover, for given $f_N \in \mathbb{R}^{q_i}$, $b_{iN}(f_N,\cdot)$ is a bounded functional and so is $b_{iJ}(\lambda^n_{iJ},\cdot)$ for given data λ^n_{iJ}. Thus, taking $F_i \in L^2(0,\ell_i)$ or in $(\mathcal{H}^1(0,\ell_i))^*$ the right-hand side of (3.3.2.17), rewritten as $\ell_i(\varphi_i)$, is a bounded linear form on $\mathcal{V}_i$ and, hence, problem (3.3.2.17) is a classical variational problem

$$\tilde{a}_i(r_i,\varphi_i) = \ell_i(\varphi_i), \quad \forall \varphi_i \in \mathcal{H}^1(0,\ell_i), \tag{3.3.2.19}$$

which, by the Lax-Milgram Theorem, admits a unique solution $r_i \in \mathcal{H}^1(0,\ell_i)$.

We may write down problem (3.3.2.17) together with the update at the common interface v_J, $J \in V_M$ as follows:

$$\begin{aligned} a_i(r_i^{n+1},\varphi_i) + a_{iJ}(r_i^{n+1},\varphi_i) &= (F_i,\varphi_i) \\ &+ b_{iN}(f_N,\varphi_i) + b_{iJ}(\lambda^{\varepsilon,n}_{iJ},\varphi_i), \quad \forall \varphi_i \in \mathcal{H}^1_0, \end{aligned} \tag{3.3.2.20}$$

$$\lambda^{\varepsilon,n}_{iJ} := (1-\varepsilon)\,\mathcal{S}_J(-(q^J)^n + \beta_J(r^J)^n) + \varepsilon((q^J)^n + \beta_J(r^J)^n), \tag{3.3.2.21}$$

where again

$$\begin{aligned} (q^J)^n_i &= q^n_{iJ} = \varepsilon_{iJ}(C^+_{iJ})^T\, K_i((r_i^n)' + R_i r_i^n)(v_J), \\ (r^J)^n_i &= r^n_{iJ} = C_{iJ}\, r_i^n(v_J), \end{aligned} \tag{3.3.2.22}$$

$(q^J)^n = (q^J)^n_{i\in\mathcal{I}_J}$, $(r^J)^n = (r_i^J)^n_{i\in\mathcal{I}_J}$. An alternative to (3.3.2.20), (3.3.2.21) is

$$\begin{aligned} a_i(r_i^{n+1},\varphi_i) + a_{iJ}(r_i^{n+1},\varphi_i) &= (F_i,\varphi_i) \\ &+ b_{iN}(f_N,\varphi_i) + b_{iJ}((g^{\varepsilon,J})^{n+1}_i,\varphi_i), \; \forall \varphi_i \in \mathcal{H}^2(0,\ell_j) \end{aligned} \tag{3.3.2.23}$$

$$(g^J)^{n+1} + (1-\varepsilon)\,\mathcal{S}_J(g^J)^n - \varepsilon(g^J)^n = 2(1-\varepsilon)\,\beta_J\,\mathcal{S}_J(r^J)^n. \tag{3.3.2.24}$$

3.3.3 Convergence

We introduce the space $\mathcal{X} = \prod\limits_{J\in\mathcal{J}_M} \prod\limits_{i\in\mathcal{I}_J} \mathbb{R}^{p_i}$ with the norm

$$\|g\|^2_{\mathcal{X}} := \sum_{J\in\mathcal{J}_M} \sum_{i\in\mathcal{I}_J} \frac{1}{\beta_J}\, |g_{iJ}|^2, \tag{3.3.3.1}$$

where

$$g = ((g^J)_{J\in\mathcal{J}_M}), \tag{3.3.3.2}$$

and define an operator $\mathcal{T} : \mathcal{X} \mapsto \mathcal{X}$ by

$$\begin{aligned} (\mathcal{T}g)_{iJ} &:= 2\beta_J\,\mathcal{S}_J(r^J)_i - \mathcal{S}_J(g^J)_i \\ (\mathcal{T}g)_J &:= \{(\mathcal{T}g)_{iJ}, i \in \mathcal{I}_J\} \\ \mathcal{T}g &:= \{(\mathcal{T}g)_J, J \in \mathcal{J}_M\}. \end{aligned} \tag{3.3.3.3}$$

We have

$$\|g\|_{\mathcal{X}}^2 = \sum_{J\in\mathcal{J}_M} \sum_{i\in\mathcal{I}_J} \frac{1}{\beta_J} |g_{iJ}|^2, \tag{3.3.3.4}$$

$$\|\mathcal{T}g\|_{\mathcal{X}}^2 = \sum_{J\in J_M} \sum_{i\in\mathcal{I}_J} \frac{1}{\beta_J} |2\beta_J\, \mathcal{S}_J(r^J)_i - \mathcal{S}_J(g^J)_i|^2. \tag{3.3.3.5}$$

Clearly

$$\sum_{i\in\mathcal{I}_J} (\mathcal{S}_J r^J)_i^2 = \sum_{i\in\mathcal{I}_J} r_{iJ}^2 \tag{3.3.3.6}$$

and

$$\sum_{i\in\mathcal{I}_J} (\mathcal{S}_J r^J)_i (\mathcal{S}_J g^J)_i = \sum_{i\in\mathcal{I}_J} r_{iJ}\, g_{iJ}. \tag{3.3.3.7}$$

Hence

$$\begin{aligned} \|\mathcal{T}g\|_{\mathcal{X}}^2 &= \sum_{J\in\mathcal{J}_M} \sum_{i\in\mathcal{I}_J} \frac{1}{\beta_J} \{4\beta_J^2 (\mathcal{S}_J(r^J)_j)^2 \\ &\qquad + (\mathcal{S}_J(g^J)_i)^2 - 4\beta_J\, \mathcal{S}_J(r^J) i\, \mathcal{S}_J(g^J)_j\} \\ &= \sum_{J\in\mathcal{J}_M} \frac{1}{\beta_J} \sum_{i\in\mathcal{I}_J} \{4\beta_j^2\, r_{iJ}^2 + g_{iJ}^2 - 4\beta_J\, r_{iJ}\, g_{iJ}\} \\ &= \|g\|_{\mathcal{X}}^2 - 4 \sum_{J\in\mathcal{J}_M} \sum_{i\in\mathcal{I}_J} (g_{iJ} - \beta_J\, r_{iJ})\, r_{iJ}. \end{aligned} \tag{3.3.3.8}$$

From (3.3.2.22) and (3.3.2.17) we then obtain

$$\begin{aligned} &\sum_{J\in\mathcal{J}_M} \sum_{i\in\mathcal{I}_J} (g_{iJ} - \beta_J C_{iJ} r_i(v_J))\, C_{iJ} r_i(v_J) \\ &\quad = \sum_{J\in\mathcal{J}_M} \sum_{i\in\mathcal{I}_J} b_{iJ}(g_{iJ}, r_i) - a_{iJ}(r_i, r_i) \\ &\quad = \sum_{i\in\mathcal{I}} \{a_i(r_i, r_i) - (F_i, r_i)\}, \end{aligned} \tag{3.3.3.9}$$

As we will deal with errors, without loss of generality we may set $F_i \equiv 0$. Thus from (3.3.3.9) and (3.3.3.8) we obtain

$$\|\mathcal{T}g\|_{\mathcal{X}}^2 = \|g\|_{\mathcal{X}}^2 - 4 \sum_{i\in\mathcal{I}} a_i(r_i, r_i) \tag{3.3.3.10}$$

which is completely analogous to (2.3.5.22). One may now consider the relaxed procedure which consists of the iteration

$$g^{n+1} = (1-\varepsilon)\, \mathcal{T}(g^n) + \varepsilon\, g^n, \tag{3.3.3.11}$$

where, as usual, we consider the errors $r_i^{n+1} - r_i =: e_i^{n+1}$ between the uncoupled local and the coupled global solutions after restriction to edge i.

In order to establish convergence of the iteration (3.3.3.11) we proceed in much the same manner as in Section 2.3, beginning with the calculation

$$\|g^{n+1}\|_{\mathcal{X}}^2 = ((1-\varepsilon)^2 + \varepsilon^2)\,\|g^n\|_{\mathcal{X}}^2 - 4(1-\varepsilon)^2 \sum_{i\in\mathcal{I}} a_i(e_i^n, e_i^n) + 2\varepsilon(1-\varepsilon)(g^n, \mathcal{T} g^n). \quad (3.3.3.12)$$

Now

$$\begin{aligned}(g^n, \mathcal{T} g^n) &= \sum_{J\in\mathcal{J}_M}\sum_{i\in\mathcal{I}_J} \frac{1}{\beta_J}\, g_{iJ}^n\, \mathcal{T}(g^n)_{iJ} \\ &= \sum_{J\in\mathcal{J}_M}\sum_{i\in\mathcal{I}_J} \frac{1}{\beta_J}\, g_{iJ}^n (2\beta_J\, \mathcal{S}_J(e^J)_i^n - \mathcal{S}_J(g^J)_i^n.\end{aligned} \quad (3.3.3.13)$$

From (3.3.3.13) and (3.3.2.24) we infer

$$2\varepsilon(1-\varepsilon)(g^n, \mathcal{T}\, g^n) = 2\varepsilon \sum_{J\in\mathcal{J}_M}\sum_{i\in\mathcal{I}_J} \frac{1}{\beta_J}\, g_{iJ}^n\, g_{iJ}^{n+1} - 2\varepsilon^2 \sum_{J\in\mathcal{J}_M}\sum_{i\in\mathcal{I}_J} \frac{1}{\beta_J}\, (g_{iJ}^n)^2$$

which, together with (3.3.3.12), implies

$$\|g^{n+1}\|_{\mathcal{X}}^2 = (1-2\varepsilon)\,\|g^n\|_{\mathcal{X}}^2 + 2\varepsilon \sum_{J\in\mathcal{J}_M}\sum_{i\in\mathcal{I}_J} \frac{1}{\beta_J}\, g_{iJ}^n\, g_{iJ}^{n+1} - 4(1-\varepsilon)^2 \sum_{i\in\mathcal{I}} a_i(e_i^n, e_i^n). \quad (3.3.3.14)$$

For $\varepsilon \in [0,1]$ we obtain from (3.3.3.14)

$$\|g^{n+1}\|_{\mathcal{X}}^2 \le \|g^n\|_{\mathcal{X}}^2 - 4(1-\varepsilon) \sum_{i\in\mathcal{I}} a_i(e_i^n, e_i^n). \quad (3.3.3.15)$$

Iterating (3.3.3.15) down to $n = 0$ we conclude

$$g^n \text{ is bounded} \quad (3.3.3.16)$$

$$a_i(e_i^n, e_i^n) \to 0 \quad n \to \infty, \;\; i \in \mathcal{I}. \quad (3.3.3.17)$$

Moreover, according to (3.3.3.16) and (3.3.2.23), (3.3.2.24) (these can be rewritten in terms of (3.3.2.20) and (3.3.2.21)) with f_N, F_i equal to zero, we infer that e_i is bounded in $\mathcal{H}^1(0,\ell_i)$. Thus, for some numbers $\hat{e}_i$, $\hat{w}_i$ we have

$$\begin{aligned} e_i^n &\to \hat{e}_i \quad \text{weakly in } \mathcal{H}^1(0,\ell_i), \\ C_{iJ} e_i^n(v_J) &\to \hat{e}_i(v_J), \quad (C_{iJ}^+)^T K_i (e_i^n)' \to \hat{w}_i, \end{aligned} \quad (3.3.3.18)$$

on a subsequence n_k. Now, in the case $\mathcal{S}_i$ is positive definite for $i \in \mathcal{I}$ we conclude directly from (3.3.3.17) and the local equation that

$$\begin{aligned} e_i^n &\overset{n\to\infty}{\longrightarrow} 0 \text{ strongly in } \mathcal{H}^1(0,\ell_i) \\ C_{iJ}e_i^n(v_J) &\overset{n\to\infty}{\longrightarrow} 0, \ \forall J \in \mathcal{J}_M,\ i \in \mathcal{I}_J \\ (C_{iJ}^+)^T K_i(e_i^n)' &\overset{n\to\infty}{\longrightarrow} 0, \ \forall J \in \mathcal{J}_M,\ i \in \mathcal{I}_J. \end{aligned} \tag{3.3.3.19}$$

In fact, in contrast to (3.3.3.18) which holds possibly only on a subsequence, the convergence in (3.3.3.19) holds on the entire sequence. If, however, some $\mathcal{S}_i$ are only semidefinite, we may proceed as follows. If we have a tree with rigid joints, i.e., $C_{iJ} = I$, $\forall J \in V_M$, $i \in \mathcal{I}_J$, and all but possibly one simple node obeying Dirichlet (or Robin) conditions, we may proceed from the simple Dirichlet or Robin nodes, where we have uniqueness in the limit and where consequently $e_i^n \overset{n\to\infty}{\longrightarrow} 0$ strongly in $\mathcal{H}^1(0,\ell_i)$, to the set of penultimate nodes, where, in the limit, zero boundary conditions hold. We proceed through the entire tree and obtain

$$e_i^n \overset{n\to\infty}{\longrightarrow} 0 \text{ strongly in } \mathcal{H}^1(0,\ell_i),\ \forall i. \tag{3.3.3.20}$$

Note that the arguments above are true for $\epsilon \in [0,1)$. We will state the results for $\epsilon = 0$ in the theorem below, separately.

If now $\epsilon \in (0,1)$, even without the special assumptions on the graph above, according to Opial's Lemma 2.3.7.3 and the uniqueness assumption (3.3.1.2) we have

$$g^n \to 0, \quad \mathcal{T}g^n \to 0, \quad n \to \infty \tag{3.3.3.21}$$

on the entire sequence n.

But then, according to (3.3.2.23), (3.3.2.24) and (3.3.3.18), we infer from (3.3.3.21) that the convergence in (3.3.3.18) take place on the entire sequence as well. Moreover, the traces $C_{iJ}e_i^n(v_J)$ tend to zero. By assumption (3.3.1.2) at least one of the bounding vertices for edge i carries a matrix C_{iJ} with zero kernel. Therefore, we may conclude that the convergence of the errors e_i^n to zero takes place strongly in $\mathcal{H}(0,\ell_i)$. We collect these results in the following theorem.

Theorem 3.3.3.1 *Let G be a graph, and let (3.3.1.2) be satisfied.*

i) *Let $\epsilon = 0$. Then*
   a) $a_i(e_i^n, e_i^n) \longrightarrow 0$, *as* $n \to \infty$.
   b) *If all $\mathcal{S}_i$ are positive definite, then the errors e_i^n converges to zero strongly in $\mathcal{H}^1(0,\ell_i)$.*
   c) *If some or all $\mathcal{S}_i$ are only semidefinite, the errors e_i^n converge to zero strongly in $\mathcal{H}^1(0,\ell_i)$ provided the graph G is a tree with all but possibly one of the simple nodes satisfying Dirichlet or Robin conditions.*

ii) *Let $\epsilon \in (0,1)$. Then the errors e_i^n converge to zero strongly in $\mathcal{H}^1(0,\ell_i)$.*

Remark 3.3.3.1 It should be pointed out that the iteration operator for the one-dimensional static problem on a graph acts between finite-dimensional trace spaces, which allows a relatively easy convergence analysis. This case, however, is not typical for the control problems considered in this monograph. Consequently, we dispense with a more detailed analysis here. Suffice it to say that, already at this point, we observe that networks whose geometry admits their exact controllability (see in particular [61]) are also well designed for domain decomposition with an unrelaxed iteration, while relaxed iterations require fewer geometric presumptions.

3.3.4 Interpretation as a Richardson Iteration

For the sake of simplicity, in this section we focus on the simple scalar elliptic equation on an graph G

$$\begin{aligned} -u_i'' &= f_i \quad \text{in} \quad (0,\ell_i),\\ u_i(v_J) &= 0, \quad i\in\mathcal{I}_J,\ J\in\mathcal{J}_0,\\ \varepsilon_{iJ}\,u_i'(v_J) &= 0, \quad i\in\mathcal{I}_J,\ J\in\mathcal{J}_N,\\ u_i(v_J) &= u_j(v_J), \quad i,j\in\mathcal{I}_J,\ J\in\mathcal{J}_M\\ \sum_{i\in\mathcal{I}_J}\varepsilon_{iJ}u_i'(v_J) &= 0, \quad J\in\mathcal{J}_M. \end{aligned} \tag{3.3.4.1}$$

In this case the matrices are as follows: $K_i = 1$, $R_i = 0$, $S_i = 0$, $C_{iJ} = 1$, $\Pi_{iJ} = 0$, $M_i = 1$. The standard decomposition of $(3.3.4.1)_3$–$(3.3.4.1)_5$ is based on the iterative decoupling of

$$\begin{aligned} &\varepsilon_{iJ}\,u_i'(v_J) + \beta\,u_i(v_J) = \lambda_{iJ}\\ &= \beta\,\Big(\frac{2}{d_J}\sum_{j\in\mathcal{I}_J}u_j(v_J) - u_i(v_J)\Big) - \Big(\frac{2}{d_J}\sum_{j\in\mathcal{I}_J}\varepsilon_{jJ}\,u_J'(v_J) - \varepsilon_{iJ}\,u_{iJ}'(v_J)\Big), \end{aligned} \tag{3.3.4.2}$$

$\forall\, i\in\mathcal{I}_J$, $J\in\mathcal{J}_M$. As in Section 2.3.7 we introduce z_i and w_i as follows. To $i\in\mathcal{I}$ there correspond two nodes, say $v_{J_i}, v_{M_i}\in M$, such that edge i is bounded by nodes v_{J_i}, v_{M_i}. Accordingly, to each $i\in\mathcal{I}$, there are two parameters $\lambda_{iJ}, \lambda_{iM}$, if edge i is incident at two multiple nodes. If edge i is incident at a simple node only one parameter, say λ_{iJ}, $J\in\mathcal{J}_M$, is assigned to i. Let z_i solve

$$\begin{aligned} -z_i'' &= f_i \quad \text{in} \quad (0,\ell_i),\\ z_i(v_J) &= 0, \quad i\in\mathcal{I}_J,\ J\in\mathcal{J}_D,\\ \varepsilon_{iJ}\,z_i'(v_J) &= 0, \quad i\in\mathcal{I}_J,\ J\in\mathcal{J}_N,\\ \varepsilon_{iJ}\,z_i'(v_J) + \beta\,z_i(v_J) &= 0, \quad i\in\mathcal{I}_J,\ J\in\mathcal{J}_M. \end{aligned} \tag{3.3.4.3}$$

We denote these solutions by

$$G_i\,f_i =: z_i. \tag{3.3.4.4}$$

Let now w_i be the solutions to the systems

$$\begin{aligned} -w_i'' &= 0 \quad \text{in} \quad (0,\ell_i), \\ w_i(v_J) &= 0, \quad i \in \mathcal{I}_J,\ J \in \mathcal{J}_0, \\ \varepsilon_{i_J}\, w_i'(v_J) &= 0, \quad i \in \mathcal{I}_J,\ J \in \mathcal{J}_0, \\ \varepsilon_{i_J}\, w_i'(v_J) + \beta\, w_i(v_J) &= \lambda_{iJ}, \end{aligned} \tag{3.3.4.5}$$

$$\begin{aligned} \lambda_{iJ} = {} & \beta\,(\frac{2}{d_J} \sum_{j\in\mathcal{I}_J} w_j(v_J) - w_i(v_J)) \\ & - (\frac{2}{d_J} \sum_{j\in\mathcal{I}_J} \varepsilon_{j_J}\, w_j'(v_J) - \varepsilon_{iJ}\, w_i'(v_J)) \\ & + \beta\,(\frac{2}{d_J} \sum_{j\in\mathcal{I}_J} z_j(v_J) - z_i(v_J)) \\ & - (\frac{2}{d_J} \sum_{j\in\mathcal{I}_J} \varepsilon_{j_J}\, z_j'(v_J) - \varepsilon_{iJ}\, z_i'(v_J)) \end{aligned} \tag{3.3.4.6}$$

for $i \in \mathcal{I}_J$, $J \in \mathcal{J}_M$.

Define $R_i\begin{pmatrix} \lambda_{iJ} \\ \lambda_{iM} \end{pmatrix} := w_i$ or, in short, $R_i(\lambda_i) := w_{iJ}$ with the understanding that $\lambda_i = \begin{pmatrix} \lambda_{iJ} \\ \lambda_{iM} \end{pmatrix}$. Let us further set $\mathcal{S}_i(\lambda_i) := \begin{pmatrix} w_i(v_J) \\ w_i(v_M) \end{pmatrix}$, $\mathcal{S}_{iJ}(\lambda_i) = w_i(v_J)$. Then (3.3.4.6) takes the form

$$\begin{aligned} \lambda_{iJ} = {} & \beta\,(\frac{2}{d_J} \sum_{j\in\mathcal{I}_J} \mathcal{S}_{jJ}(\lambda_j) - \mathcal{S}_{iJ}(\lambda_i)) \\ & - (\frac{2}{d_J} \sum_{j\in\mathcal{I}_J} \varepsilon_{j_J}\, R_j'(\lambda_j)(v_J) - \varepsilon_{iJ}\, R_i'(\lambda_i)(v_J)) \\ & + \beta(\frac{2}{d_J} \sum_{j\in\mathcal{I}_J} G_j(f_j)(v_J) - G_i(f_i)(v_J)) \\ & - (\frac{2}{d_J} \sum_{j\in\mathcal{I}_J} \varepsilon_{j_J}\, G_j'(f_j)(v_J) - \varepsilon_{iJ}\, G_{iJ}'(f_i)(v_J)) \\ = {} & 2\beta\,(\frac{2}{d_J} \sum_{j\in\mathcal{I}_J} \mathcal{S}_{jJ}(\lambda_j) - \mathcal{S}_{iJ}(\lambda_i)) \\ & - (\frac{2}{d_J} \sum_{j\in\mathcal{I}_J} \lambda_{jJ} - \lambda_{iJ}) \\ & + 2\beta\,(\frac{2}{d_J} \sum_{j\in\mathcal{I}_J} G_j(f_j)(v_J) - G_i(f_i)(v_J)). \end{aligned} \tag{3.3.4.7}$$

Thus

$$\frac{2}{d_J}\sum_{j\in\mathcal{I}_J}\lambda_{iJ} = 2\beta\,(\frac{2}{d_J}\sum_{j\in\mathcal{I}_J}\mathcal{S}_{jJ}(\lambda_j)-\mathcal{S}_{iJ}(\lambda_j)) \\ +\ 2\beta\,(\frac{2}{d_J}\sum_{j\in\mathcal{I}_J}G_j(f_j)(v_J)-G_i(f_i)(v_J)) \quad (3.3.4.8)$$

or, in a another format,

$$\lambda_{iJ} = \frac{2}{d_J}\sum_{j\in\mathcal{I}_J}(2\beta\,\mathcal{S}_{jJ}(\lambda_j)-\lambda_{jJ}) -\ (2\beta\,\mathcal{S}_{iJ}(\lambda_i)-\lambda_{iJ})) \\ +\ 2\beta\,(\frac{2}{d_J}\sum_{j\in\mathcal{I}_J}G_j(f_i)(v_J)-G_i(f_i)(v_J).$$

As $S_{iJ}(\lambda_i)$ is linear, we may split $S_{iJ}(\lambda_i)=\tilde{S}_{iJ}(\lambda_{iJ})+\hat{S}_{iJ}(\lambda_{iJ})$ by slight abuse of notation. Then the trace of w_i at v_J is obtained from $\tilde{w}_i(v_J):=\tilde{S}_{iJ}(\lambda_{iJ})$, $\hat{w}_i(v_J)=\hat{S}_{iJ}(\lambda_{iM})$ such that

$$\begin{aligned} \tilde{w}_i'' &= 0 \quad \text{in } (0,\ell_i),\\ \varepsilon_{iJ}\,\tilde{w}_i'(v_J)+\beta\,\tilde{w}_i(v_J) &= \lambda_{iJ},\\ \varepsilon_{iM}\,\tilde{w}_i'(v_M)+\beta\,\tilde{w}_i(v_M) &= 0, \end{aligned} \quad (3.3.4.9)$$

and

$$\begin{aligned} \hat{w}_i'' &= 0 \quad \text{in } (0,\ell_i),\\ \varepsilon_{iJ}\,\hat{w}_i'(v_J)+\beta\,\hat{w}_i(v_J) &= 0,\\ \varepsilon_{iM}\,\hat{w}_i'(v_M)+\beta\,\hat{w}_i(v_M) &= \lambda_{iM}, \end{aligned} \quad (3.3.4.10)$$

so that $w_i=\tilde{w}_i+\hat{w}_i$. Accordingly, $\tilde{w}_i(v_M)=\tilde{S}_{iM}(\lambda_{iJ})$, $\hat{w}_i(v_M)=\hat{S}_{iM}(\lambda_{iM})$. With this notation in mind we have from (3.3.4.8)

$$\frac{2}{d_J}\sum_{j\in\mathcal{I}_J}\lambda_{iJ} = 2\beta\,(\frac{2}{d_J}\sum_{j\in\mathcal{I}_J}(\tilde{S}_{jJ}(\lambda_{jJ})+\hat{S}_{jJ}(\lambda_{jM_J})) \\ -\,(\tilde{S}_{iJ}(\lambda_{iJ})+\hat{S}_{iJ}(\lambda_{iM}))+2\beta\,(\frac{2}{d_J}\sum_{j\in\mathcal{I}_J}G_j(f_j)(v_J)-G_i(f_i)(v_J)),$$

or

$$\lambda_{iJ} = \frac{2}{d_J}\sum_{j\in\mathcal{I}_J}[2\beta(\tilde{S}_{jJ}(\lambda_{jJ}+\hat{S}_{jJ}(\lambda_{jM_j}))-\lambda_{jJ})] \\ -\ [2\beta(\tilde{S}_{iJ}\,(\lambda_{iJ})+\hat{S}_{iJ}(\lambda_{iM}))-\lambda_{iJ}]) \\ +\,2\beta(\frac{2}{d_J}\sum_{j\in\mathcal{I}_J}G_j(f_j)(v_J)-G_i(f_i)(v_J)). \quad (3.3.4.11)$$

By using the 'scattering' operator $\mathcal{S}_J$, (3.3.4.11) may be written as

$$\lambda_{iJ} = \mathcal{S}_J\left(2\beta(\tilde{S}_{*J}(\lambda_{*J}) + \hat{S}_{*J}(\lambda_{*M})) - \lambda_{*J}\right)_i + 2\beta\,\mathcal{S}_J(G_*(f_*)(v_J))_i. \quad (3.3.4.12)$$

Now, (3.3.4.12) is an equation in the nodal values $(\lambda_{iJ})_{\substack{J\in\mathcal{J}_M\\ i\in\mathcal{I}_J}} \in \mathbb{R}^{(\mathcal{J}_M)d_J}$. Let us explicitly solve for $w_i = \tilde{w}_i + \hat{w}_i$.

$$\tilde{w}_i(x) = (\varepsilon_{iJ} - \varepsilon_{iM} + \beta(x_{iJ} + x_{iM}))^{-1}(x - \frac{1}{\beta}\varepsilon_{iM} - x_{iM})\,\lambda_{iJ} \quad (3.3.4.13)$$

$$\hat{w}_i(x) = (\varepsilon_{iM} - \varepsilon_{iJ} + \beta(x_{iM} - x_{iJ}))^{-1}(x - \frac{1}{\beta}\varepsilon_{iJ} - x_{iJ})\,\lambda_{iM} \quad (3.3.4.14)$$

for $M, J \in \mathcal{J}_M$, that is, for an interior edge $i \widehat{=} MJ$. If, however, $J \in \mathcal{J}_M$ and $M \in \mathcal{J}_D$ we have

$$w_i(x) = \frac{\lambda_{iJ}}{\varepsilon_{iJ} + \beta(x_{iJ} - x_{iM})}(x - x_{iM}) \quad (3.3.4.15)$$

or, if $M \in \mathcal{J}_N$,

$$w_i(x) = \frac{\lambda_{iJ}}{\beta}. \quad (3.3.4.16)$$

Thus, for edges i connecting a multiple node v_J to a simple node v_M we have

$$\mathcal{S}_{iJ}(\lambda_{iJ}) = \frac{\lambda_{iJ}(x_{iJ} - x_{iM})}{\varepsilon_{iJ} + \beta(x_{iJ} - x_{iM})}, \quad M \in \mathcal{J}_D \quad (3.3.4.17)$$

$$\mathcal{S}_{iJ}(\lambda_{iJ}) = \frac{1}{\beta}\lambda_{iJ}, \quad M \in \mathcal{J}_N. \quad (3.3.4.18)$$

Let $\mathcal{S}_{iJ}(\lambda_{iJ}) = \tilde{s}_{iJ}(\lambda_{iJ})$, $\tilde{s}_{iJ} := 0$ for a v_J connecting to a simple node. For v_J connecting to a multiple node we have

$$\begin{aligned}\tilde{w}_i(x_{iJ}) &= \tilde{w}_i(v_J) = \tilde{S}_{iJ}(\lambda_{iJ})\\ &= \frac{1}{\beta}(\varepsilon_{iJ} - \varepsilon_{iM} + \beta(x_{iJ} - x_{iM}))^{-1}(\beta(x_{iJ} - x_{iM}) - \varepsilon_{iM})\,\lambda_{iJ}\end{aligned} \quad (3.3.4.19)$$

$$\begin{aligned}\tilde{w}_i(x_{iJ}) &= \tilde{w}_i(v_J) = \tilde{S}_{iJ}(\lambda_{iM})\\ &= -(\varepsilon_{iM} - \varepsilon_{iJ} + \beta(x_{iM} - x_{iJ}))^{-1}\,\frac{1}{\beta}\,\varepsilon_{iJ}\,\lambda_{iM}.\end{aligned} \quad (3.3.4.20)$$

Thus

$$\begin{aligned}\tilde{S}_{iJ}(\lambda_{iJ}) &= \frac{1}{\beta}(1 - \frac{\varepsilon_{iJ}}{\varepsilon_{iJ} - \varepsilon_{iM} + \beta(x_{iJ} - x_{iM})}) =: \tilde{s}_{iJ}\,\lambda_{iJ}\\ \hat{S}_{iJ}(\lambda_{iM}) &= \frac{1}{\beta}(\frac{\varepsilon_{iJ}}{\varepsilon_{iJ} - \varepsilon_{iM} + \beta(x_{iJ} - x_{iM}}) =: \hat{s}_{iJ}\,\lambda_{iM}.\end{aligned} \quad (3.3.4.21)$$

Therefore (3.3.4.9) is equivalent to

$$\lambda_{iJ} = \frac{2}{d_J} \sum_{j \in \mathcal{I}_J} (2\beta(\tilde{s}_{jJ}\,\lambda_{jJ} + \hat{s}_{jJ}\,\lambda_{jM}) - \lambda_{jJ}) \\ - (2\beta\,(\tilde{s}_{iJ}\,\lambda_{iJ} + \hat{s}_{iJ}\,\lambda_{iM}) - \lambda_{iJ}) \\ + 2\beta\,(\frac{2}{d_J} \sum_{j \in \mathcal{I}_J} G_j(f_j)(v_J) - G_i(f_i)(v_J)). \quad (3.3.4.22)$$

Equation (3.3.4.22) holds for a given edge i incident at the nodes v_J, v_M, where λ_{jM_j} refers to nodal values at v_{M_j} such that $j \in \mathcal{I}_J$, so that v_J and v_{M_j} bound the edge j.

The single equation (3.3.4.22) involves all nodes adjacent to v_J. Therefore we may alternatively express (3.3.4.22) as a purely nodal system by setting

$$\begin{aligned} \lambda_{iJ} &=: \lambda_{MJ} \quad \text{if} \quad i \in \mathcal{I}_J \cap \mathcal{I}_M \\ \lambda_{iM} &=: \lambda_{JM} \quad \text{if} \quad i \in \mathcal{I}_J \cap \mathcal{I}_M \\ \lambda_J &:= (\lambda_{JM}, \lambda_{JQ}, \ldots, \lambda_{JP})^T \in \mathbb{R}^{d_J} \end{aligned}$$

such that $JM, JQ, \ldots, JP \in E$.

As an example, we consider a star configuration with just one multiple node, v_J and, say, three branches emanating from this node. For the sake of simplicity, let v_J be located at $x_{iJ} = 0$, $i = 1, 2, 3$. Thus, $x_{iM} = \ell_i$, $MJ \in E$. Accordingly, $\varepsilon_{iJ} = -1$, $\varepsilon_{iM} = 1$. We consider, as a first case, the situation where all 3 simple nodes satisfy Dirichlet conditions. Then, according to (3.3.4.17),

$$S_{iJ}(\lambda_{iJ}) = \tilde{s}_{iJ} \cdot \lambda_{iJ} = \frac{\ell_i\,\lambda_{iJ}}{1 + \beta\,\ell_i}. \quad (3.3.4.23)$$

By (3.3.4.9)

$$\lambda_{iJ} = \frac{2}{3} \sum_{j=1}^{3} (2\beta\,\tilde{s}_{jJ} \cdot \lambda_{jJ} - \lambda_{iJ}) - (2\beta\,\tilde{s}_{iJ}\,\lambda_{iJ} - \lambda_{iJ})) \\ + 2\beta\,(\frac{2}{3} \sum_{j=1}^{3} G_j(f_j)(0) - G_i(f_i)(0)) \quad (3.3.4.24)$$

or, with $\lambda_{1_J} = \lambda_1$, $\lambda_{2_J} = \lambda_2$, $\lambda_{3_J} = \lambda_3$, $\tilde{s}_{iJ} = \tilde{s}_i$

$$\lambda_1 = \frac{2}{3}\,(2\beta\,\tilde{s}_2\,\lambda_2 + 2\beta\,\tilde{s}_3\,\lambda_3 - \lambda_2 - \lambda_3) - \frac{1}{3}\,(2\beta\,\tilde{s}_1\,\lambda_1 - \lambda_1) + \tilde{b}_1 \quad (3.3.4.25)$$

where $\tilde{b}_1$ is the last term on the right-hand side of (3.3.4.24). Equation (3.3.4.25) can be rewritten as

$$(1 + \beta\,\tilde{s}_1)\,\lambda_1 = (2\beta\,\tilde{s}_2 - 1)\,\lambda_2 + (2\beta\,\tilde{s}_3 - 1)\,\lambda_3 + \hat{b}_1 \quad (3.3.4.26)$$

where $\hat{b}_1 = \frac{3}{2}\tilde{b}_1$. By symmetry, we obtain

$$\begin{aligned} (1+\beta\,\tilde{s}_1)\,\lambda_1 - (2\beta\,\tilde{s}_2 - 1)\,\lambda_2 - (2\beta\,\tilde{s}_3 - 1)\,\lambda_3 &= \hat{b}_1 \\ -(2\beta\,\tilde{s}_1 - 1)\,\lambda_1 + (1+\beta\,\tilde{s}_2)\,\lambda_2 - (2\beta\,\tilde{s}_3 - 1)\,\lambda_3 &= \hat{b}_2 \\ -(2\beta\,\tilde{s}_1 - 1)\,\lambda_1 - (2\beta\,\tilde{s}_2 - 1) + (1+\beta\,\tilde{s}_3)\,\lambda_3 &= \hat{b}_2. \end{aligned} \tag{3.3.4.27}$$

According to (3.3.4.23), $2\beta\,\tilde{s}_i - 1 = \frac{2\beta\ell_i}{1+\beta\ell_i} - 1 = \frac{\beta\ell_i - 1}{\beta\ell_i+1}$ and $1+\beta\,\tilde{s}_i = \frac{2\beta\ell_i+1}{1+\beta\ell_i}$. Thus (3.3.4.27) is equivalent to

$$\begin{pmatrix} 1 & -\frac{\beta\ell_2-1}{2\beta\ell_1+1} & -\frac{\beta\ell_3-1}{2\beta\ell_1+1} \\ -\frac{\beta\ell_1-1}{2\beta\ell_2+1} & 1 & -\frac{\beta\ell_3-1}{2\beta\ell_2+1} \\ -\frac{\beta\ell_1-1}{2\beta\ell_3+1} & -\frac{\beta\ell_2-1}{2\beta\ell_3+1} & 1 \end{pmatrix} \begin{pmatrix} \lambda_1 \\ \lambda_2 \\ \lambda_3 \end{pmatrix} = \begin{pmatrix} b_1 \\ b_2 \\ b_3 \end{pmatrix} \tag{3.3.4.28}$$

where $b_2 = \hat{b}_i(1+\beta\,\tilde{s}_i)$, $i = 1,2,3$. Equation (3.3.4.28) may be written in the format $\mathcal{A}\,\lambda = b$. This is a Schur-complement equation.

The corresponding damped Richardson iteration, again equivalent to the domain decomposition with relaxation ε, is then

$$\lambda^{n+1} = \lambda^n - \theta(\mathcal{A}\,\lambda^n - b). \tag{3.3.4.29}$$

Unless $\ell_i = \ell$ for all i, the problem is nonsymmetric. In case $\ell_i = \ell$ we see that a single parameter $\alpha = \frac{\beta\ell-1}{2\beta\ell+1}$ is involved in (3.3.4.28). In that case

$$\mathcal{A} = I - \alpha \begin{pmatrix} 0 & 1 & 1 \\ 1 & 0 & 1 \\ 1 & 1 & 0 \end{pmatrix} = I - \alpha\, B$$

The damped Richardson iteration matrix is then $R_\theta = (1-\theta)\,I + \alpha\,\theta\,B$. In this case, however, we may select $\theta = 1$, which corresponds to $\varepsilon = 0$ in the relaxed standard fixed point iteration. Then $R = \alpha\,B$. Moreover, the choice $\beta = \frac{1}{\ell}$ leads to $B = 0$. Thus after the second step the iteration terminates in the solution.

3.4 Hyperbolic Problems on Graphs

3.4.1 The Model

With the notation of Section 3.2, we now consider the following second-order in time equations on a graph.

$$M_i\,\ddot{r}_i = [K_i(r_i' + R_i\,r_i)]' - R_i^T\,K_i(r_i' + R_i\,r_i) - S_i\,r_i + f_i \quad \text{in } Q_i := (0,\ell_i)\times(0,T), \tag{3.4.1.1}$$

$$r_i(v_J,t) = 0,\ J\in\mathcal{J}_D,\ i\in\mathcal{I}_J,\ t\in(0,T), \tag{3.4.1.2}$$

$$C_{iJ}\, r_i(v_J,t) = C_{jJ}\, r_j(v_J,t),\ J\in\mathcal{J}_M,\ i,j\in\mathcal{I}_J,\ i\neq j,\ t\in(0,T), \tag{3.4.1.3}$$

$$\varepsilon_{iJ}\, K_i(r_i' + R_i\, r_i)(v_J,t) + \alpha_i\, \dot r_i(v_J,t) = g_J(t),$$
$$J\in\mathcal{J}_N,\ i\in\mathcal{I}_J,\ t\in(0,T), \tag{3.4.1.4}$$

$$\sum_{i\in\mathcal{I}_J} \varepsilon_{iJ}(C_{iJ}^{+})^T\, [K_i(r_i' + R_i\, r_i)](v_J,t) = 0,\ J\in\mathcal{J}_M,\ t\in(0,T). \tag{3.4.1.5}$$

$$\varepsilon_{iJ}\, \Pi_{iJ}\, K_i(r_i' +\ R_i\, r_i)(v_J,t) = 0,\ J\in\mathcal{J}_M, i\in\mathcal{I}_J,\ t\in(0,t), \tag{3.4.1.6}$$

$$r_i(\cdot,0) = r_{i0},\ \dot r_i(\cdot,0) = r_{i1}\ \text{ in }\ (0,\ell_i), \tag{3.4.1.7}$$

$i\in\mathcal{I}$. Note, that a superimposed dot represents differentiation with respect to the time t. Obviously (3.4.1.1) is, locally on an edge e_i, a hyperbolic system of equations, representing wave equations, Timoshenko or Bresse beam equations, depending on the choices of the matrices K_i, R_i, S_i.

Within the operator notation of Section 3.2, (3.4.1.1) can be rewritten as

$$M_i\, \ddot r_i + L_i\, r_i = f_i\ \text{ in }\ Q_i. \tag{3.4.1.8}$$

If we consider the homogeneous problem $g_J\equiv 0$ and if (3.4.1.4) holds with $\alpha_i = 0$, then (3.4.1.1)–(3.4.1.7) can be written as

$$\begin{aligned} \ddot r + Ar &= \hat f \\ r(0) = r_0, &\quad \dot r(0) = r_1, \end{aligned} \tag{3.4.1.9}$$

where $\hat f = M^{-1}f$, $f = \{f_i\}_{i\in\mathcal{I}}$ and M is a block diagonal matrix with diagonal elements M_i. Problem (3.4.1.9) is a standard second-order in time equation in a Hilbert space.

When $\alpha_i\neq 0$ and $g_J(0)\not\equiv 0$ we proceed as follows. Consider the extension operator

$$\mathcal{N}:\ \prod_{J\in\mathcal{J}_N} \mathbb{R}^{p_J}\mapsto\mathcal{H} \tag{3.4.1.10}$$

defined by setting $h = \mathcal{N}\, g$ where h satisfies

$$\begin{aligned} L_i\, h_i &= 0\ \text{ in }\ (0,\ell_i), \\ h_i(v_J) &= 0,\ \ J\in\mathcal{J}_D,\ i\in\mathcal{I}_J, \\ \varepsilon_{iJ}\, K_i(h_i' + R_i\, h_i)(v_J) &= g_J,\ \ J\in\mathcal{J}_N,\ i\in\mathcal{I}_J, \end{aligned} \tag{3.4.1.11}$$

$$\begin{aligned} &C_{iJ}\, h_i(v_J) = C_{jJ}\, h_j(v_J),\ \ J\in\mathcal{J}_M, i,j\in\mathcal{I}_J, \\ &\sum_{i\in\mathcal{I}_J} \varepsilon_{iJ}(C_{iJ}^{+})^T\, K_i(h_i' + R_i\, h_i)(v_J) = 0, \\ &\prod_{iJ} K_i(h_i' + R_i\, h_i)(v_J) = 0,\ \ J\in\mathcal{J}_M,\ i\in\mathcal{I}_J. \end{aligned} \tag{3.4.1.12}$$

We may then consider the operator

$$\mathcal{B} : \prod_{J\in\mathcal{J}_N} \mathbb{R}^{p_J} \mapsto \mathcal{V}^*, \tag{3.4.1.13}$$

$$\mathcal{B}\, g := -\mathcal{A}\mathcal{N}\, g. \tag{3.4.1.14}$$

Then

$$\mathcal{B}^* : \mathcal{V} \mapsto \prod_{J\in\mathcal{J}_N} \mathbb{R}^{p_J} \tag{3.4.1.15}$$

acts as follows:

$$\langle w, \mathcal{B}\, g\rangle_{V,V^*} = \langle w, \mathcal{A}^{-1}\,\mathcal{B}\, g\rangle_V = -\langle \mathcal{A}\, w, \mathcal{N}\, g\rangle_{V^*,V} = -(\mathcal{N}^*\,\mathcal{A}\, w, g).$$

For $w \in \{w \in \mathcal{V}|\ L_i\, w_i \in H_i,\ i \in \mathcal{I}\}$ we have

$$\begin{aligned}\langle \mathcal{A}\, w, \mathcal{N}\, g\rangle &= \sum_{i\in\mathcal{I}} \int\limits_0^{\ell_i} \{-[K_i(w_i' + R_i\, w_i)]' + R_i^T\, K_i(w_i' + R_i\, w_i) + S_i\, w_i\}\, (\mathcal{N}\, g)_i\, dx\\ &= -\sum_{i\in\mathcal{I}} B(w_i, (\mathcal{N} g)_i)|_0^{\ell_i},\end{aligned}$$

according to (3.4.1.11) and (3.4.1.10). However, if $w \in \mathcal{D}(\mathcal{A})$ we have

$$\begin{aligned}&\sum_{i\in\mathcal{I}} B_i(w_i, (Ng)_i)|_0^{\ell_i}\\ &\quad= \sum_{J\in\mathcal{J}\backslash\mathcal{J}_D}\sum_{i\in\mathcal{I}_J} \varepsilon_{iJ}\, \Pi_{iJ}\, [K_i(w_i' + R_i\, w_i)](v_J)\cdot \pi_{iJ}\, (Ng)_i(v_J)\\ &\qquad+ \sum_{J\in\mathcal{J}\backslash\mathcal{J}_D}\sum_{i\in\mathcal{I}_J} \varepsilon_{iJ}(C^+_{\hat{i}_J})^T\, K_i(w_i' + R_i\, w_i)(v_J)\cdot C^+_{iJ}(Ng)_i(v_J)\\ &\qquad+ \sum_{J\in\mathcal{J}\backslash\mathcal{J}_D}\sum_{i\in\mathcal{I}_J} \Pi_{iJ}\, w_i(v_J)\, \varepsilon_{iJ}\, \Pi_{iJ}\, K_i(g_i' + R_i\, g_i)(v_J)\\ &\qquad+ \sum_{J\in\mathcal{J}\backslash\mathcal{J}_D} C_{\hat{i}_J}\, w_i(v_J) \sum_{i\in\mathcal{I}_J} \varepsilon_{iJ}(C^+_{iJ})^T\, K_i(g_i' + R\, g_i)(v_J)\\ &\quad= \sum_{J\in\mathcal{J}_N} w_{\hat{i}}\,(v_J)\, g_J\ \ \forall\, g \in \prod_{J\in\mathcal{J}_N} \mathbb{R}^{p_J},\end{aligned} \tag{3.4.1.16}$$

where $\hat{\imath}$ is the single index in $\mathcal{I}_J$, $j \in \mathcal{J}_N$. Thus,

$$(\mathcal{B}^* w, g) = \sum_{J\in\mathcal{J}_N} w_{\hat{i}}(v_J)\, g_J,\ \ i \in \mathcal{I}_J,\ \forall\, g_J \in \prod_{j\in\mathcal{J}_N} \mathbb{R}^{P_J}. \tag{3.4.1.17}$$

In other words

$$(\mathcal{B}^* w)_i = w_i(v_J),\ J \in \mathcal{J}_N,\ i \in \mathcal{I}_J. \tag{3.4.1.18}$$

Thus, in case $\alpha_i \neq 0$ and $g_J \not\equiv 0$ on $\mathcal{J}_N$ we can restate (3.4.1.1)–(3.4.1.7) as

$$\begin{aligned} \ddot{r} + \mathcal{A}r + \alpha\,\mathcal{B}\mathcal{B}^*\,\dot{r} &= \mathcal{B}g + f \\ r(0) = r_0,\ \dot{r}(0) &= r_1, \end{aligned} \tag{3.4.1.19}$$

where we replaced f_i with $M_i^{-1}\,f_i$. Indeed, without loss of generality we may set $M_i = \mathrm{id},\ \ \forall i \in \mathcal{I}$.

We introduce the total energy

$$E(t) := \frac{1}{2}\{(\dot{r}(t), \dot{r}(t))_{\mathcal{H}} + a(r(t), r(t))\} \tag{3.4.1.20}$$

with (3.3.1.5), (3.3.1.6) and (3.3.1.3) for the scalar product in $\mathcal{H}$ and for $a(\cdot,\cdot)$. From (3.4.1.20) we obtain after differentiation with respect to time

$$\begin{aligned} \dot{E}(t) &= (\ddot{r}(t), \dot{r}(t))_{\mathcal{H}} + a(\dot{r}(t), r(t)) \\ &= (-\mathcal{A}r(t) - \alpha\mathcal{B}\mathcal{B}^*\,\dot{r}(t) + \mathcal{B}g(t) + f(t), \dot{r}(t)) + a(\dot{r}(t), r(t)) \\ &= (-\mathcal{A}r(t), \dot{r}(t))_{\mathcal{H}} - \alpha\langle\mathcal{B}^*\,\dot{r}(t), \mathcal{B}^*\,\dot{r}(t)\rangle \\ &\quad + (f(t), \dot{r}(t))_{\mathcal{H}} + (g(t), \mathcal{B}^*, \dot{r}(t)) + a(\dot{r}(t), r(t)) \\ &= \sum_{J\in\mathcal{J}_N} (-g_J(t) - \alpha_i\,\ddot{r}_i(v_J,t))\,\dot{r}_i(v_J,t) + \sum_{i\in\mathcal{I}} \int_0^{\ell_i} f_i(t)\cdot\dot{r}_i(t)\,dx. \end{aligned} \tag{3.4.1.21}$$

But

$$\begin{aligned} |g_J(t)|^2 &= |\varepsilon_{iJ}\,K_i(r_i' + R_i\,r_i)(v_J,t) + \alpha_i\,\dot{r}_i(v_J,t)|^2 \\ &= |K_i(r_i' + R_i\,r_i)(v_J,t)|^2 + K_i^2\,|\dot{r}_i(v_J,t)|^2 \\ &\quad + 2\alpha\,\varepsilon_{iJ}\,K_i(r_i' + R_i\,r_i)(v_J,t)\cdot\dot{r}_i(v_J,t) \end{aligned}$$

so that

$$\begin{aligned} \varepsilon_{iJ}\,K_i\,(r_i' + R_i\,r_i)(v_J,t)\cdot\dot{r}_i(v_J,t) = \frac{1}{2\alpha_i}\{&|g_J(t)|^2 - \alpha_i^2\,|\dot{r}_i(v_J,t)|^2 \\ &- |K_i|(r_i' + R_i\,r_i)(v_J,t)|^2\}. \end{aligned} \tag{3.4.1.22}$$

Use of (3.4.1.22) in (3.4.1.21) allows us to write

$$\begin{aligned} \dot{E}(t) = (f(t), \dot{r}(t)) + \sum_{J\in\mathcal{J}_N} \frac{1}{2\alpha_i}\{&|g_J(t)|^2 - \alpha_i^2\,|\dot{r}_i(v_J,t)|^2 \\ &- |K_i(r_i' + R_i\,r_i)(v_J,t)|^2\}. \end{aligned} \tag{3.4.1.23}$$

If we integrate (3.4.1.23) from 0 to T we obtain

$$E(T) + \frac{1}{2}\int_0^T \sum_{J\in\mathcal{J}_N} \frac{1}{\alpha_i}\left[\alpha_i^2\,|\dot r_i(v_J,t)|^2 + |K_i(r_i' + R_i\,r_i)(v_J,t)|^2\right] dt$$
$$= E(0) + \int_0^T (f(t),\dot r(t))\,dt + \frac{1}{2}\int_0^T \sum_{J\in\mathcal{J}_N} \frac{1}{\alpha_i}|g_J(t)|^2\,dt \qquad (3.4.1.24)$$

as long as $\alpha_i > 0$.

The point of the energy identity (3.4.1.24) is that it ensures extra L^2-trace regularity for the velocity at v_J. If $\alpha_i = 0$ for some or for all i, then we still have trace the regularity $K_i(r_i' + R_i\,r_i)(v_J,t) \in L^2(0,T)$. Following typical a priori estimates starting from (3.4.1.24) for $\alpha_i > 0$ or from (3.4.1.21) in case $\alpha_i = 0$, we obtain the following existence, uniqueness and regularity result:

Theorem 3.4.1.1 *Let* $f_i \in L^1(0,T;L^2(0,\ell_i))$ $i \in \mathcal{I}$, $g_J \in L^2(0,T)$, $J \in \mathcal{J}_N$, $(r_0,r_1) \in \mathcal{V}\times\mathcal{H}$. *Then problem* (3.4.1.1)–(3.4.1.7) *has a unique solution* $r = (r_i)_{i\in\mathcal{I}}$ *such that*

$$(r,\dot r) \in C(0,T;\mathcal{V}\times\mathcal{H}), \qquad (3.4.1.25)$$

$$K_i(r_i' + R_i\,r_i)(v_J,\cdot) \in L^2(0,T),\quad J \in \mathcal{J}_N, \qquad (3.4.1.26)$$

$$\dot r_i(v_J,\cdot) \in L^2(0,T),\ J \in \mathcal{J}_N \ \ \text{if}\ \ \alpha_j > 0, \qquad (3.4.1.27)$$

where, in (3.4.1.26) *and* (3.4.1.27), i *is the single element in* $\mathcal{I}_J$, $J \in \mathcal{J}_N \subset \mathcal{J}_S$.

Remark 3.4.1.1 Note that for $K_i = c_i^2\,\mathrm{id}$, $R_i = 0$, $\mathcal{S}_i = 0$, $M_i = \mathrm{id}$, system (3.4.1.1)–(3.4.1.7) reduces to the standard wave problem on the graph $G = (V,E)$. In this case the boundary condition (3.4.1.4) reduces to

$$\varepsilon_{iJ}\,c_i^2\,r_i'(v_J,t) + \alpha_i\,\dot r_i(v_J,t) = g_J(t). \qquad (3.4.1.28)$$

If now $\alpha_i > 0$ matches the impedance of the half-infinite string, in other words, if $\alpha_i = c_i$, then for $g_J \equiv 0$ (3.4.1.28) describes transparent boundary conditions. See Gander and Halpern and Nataf [26] and Halpern [41] for more details. In the general case, transparent or even absorbing boundary conditions are typically nonlocal in time and, hence, cannot be described by (3.4.1.4). Nevertheless, even in the general case the boundary condition (3.4.1.4), for suitable choices of $\alpha_i > 0$, is a first-order approximation of transparent boundary conditions. This fact is precisely the reason we investigate nonoverlapping domain decompositions for the wave equation with velocity traces appearing at the transmission interface.

3.4.2 The Domain Decomposition Procedure

We consider the problem on edge i:

$$\begin{aligned} M_i \ddot{r}_i = {} & [K_i(r_i' + R_i r_i)]' \\ & - R_i^T K_i(r_i' + R_i r_i) - S_i r_i + f_i, \text{ in } Q_i,\ i \in \mathcal{I}, \end{aligned} \tag{3.4.2.1}$$

$$r_i(v_J, t) = 0,\ J \in \mathcal{J}_D,\ i \in \mathcal{I}_J,\ t \in (0,T), \tag{3.4.2.2}$$

$$\begin{aligned} \varepsilon_{iJ} K_i(r_i' + R_i r_i)(v_J,t) + \alpha_i \dot{r}_i(v_J,t) = g_J(t), \\ J \in \mathcal{J}_N,\ i \in \mathcal{I}_J,\ t \in (0,T) \end{aligned} \tag{3.4.2.3}$$

$$\begin{aligned} \varepsilon_{iJ} \Pi_{iJ} K_i(r_i' + R_i r_i)(v_J,t) = 0, \\ J \in \mathcal{J}_M,\ i \in \mathcal{I}_J,\ t \in (0,T), \end{aligned} \tag{3.4.2.4}$$

$$\begin{aligned} \varepsilon_{iJ}(C_{iJ}^+)^T K_i(r_i' + R_i r_i)(v_J,t) + \beta_J C_{iJ} \dot{r}_i(v_J,t) = \lambda_{iJ}(t), \\ J \in \mathcal{J}_M,\ i \in \mathcal{I}_J,\ t \in (0,T), \end{aligned} \tag{3.4.2.5}$$

$$r_i(\cdot,0) = r_{i_0},\ \dot{r}_i(\cdot,0) = r_{i_1} \text{ in } (0,\ell_i),\ i \in \mathcal{I}. \tag{3.4.2.6}$$

Obviously, if the edge e_i connects a Dirichlet node with an interior node, then (3.4.2.2) and (3.4.2.4), (3.4.2.5) apply in addition to (3.4.2.1) and (3.4.2.6). If e_i connects a Neumann node and an interior node, (3.4.2.3) holds instead of (3.4.2.2), the other conditions remaining. Finally, if e_i connects two interior nodes then (3.4.2.4) and (3.4.2.5) apply on both ends. Of course, the two-point graph with Dirichlet-Neumann conditions at both ends is also included. Clearly (3.4.2.1)–(3.4.2.6) is a special case of (3.4.1.1)–(3.4.1.7). Therefore, Theorem 3.4.1.1 applies to the local problem (3.4.2.1)–(3.4.2.6).

We introduce an iterative domain decomposition as follows. Consider $\lambda_{iJ}^n \in L^2(0,T)$, where n is now an iteration counter. We solve the system (3.4.2.1)–(3.4.2.6) with the data λ_{iJ}^n for $J \in \mathcal{J}_M$, and g_J, f_i, r_{i_0}, r_{i_1} as above. We denote the corresponding unique solution by $r_i^{n+1}(x,t)$. Let us then introduce the transmission condition

$$\begin{aligned} \lambda_{iJ}^n(t) := {} & \beta_J \Big(\frac{2}{d_J} \sum_{j \in \mathcal{I}_J} C_{jJ} \dot{r}_j^n(v_J,t) - C_{iJ} \dot{r}_i^n(v_J,t)\Big) \\ & - \Big(\frac{2}{d_J} \sum_{j \in \mathcal{I}_J} \varepsilon_{jJ}(C_{jJ}^+)^T K_j(r_j^{n'} + R_j r_j^n)(v_J,t) \\ & - \varepsilon_{iJ}(C_{iJ}^+)^T K_i(r_i^{n'} + R_i r_i^n)(v_J,t)\Big). \end{aligned} \tag{3.4.2.7}$$

With $\lambda_{iJ}^0 \in L^2(0,T)$ given, (3.4.2.1)–(3.4.2.7), with r_i replaced with r_i^{n+1} on the left-hand sides of (3.4.2.1)–(3.4.2.6), defines an iteration in $n = 0,1,2,\ldots$. By standard calculations as in the preceding sections, we obtain, upon convergence as n tends to infinity, the limiting problem (3.4.1.1)–(3.4.1.7).

For purposes of analyzing convergence of the errors in the iteration process, we may suppose that the data g_J, f_i, r_{i_0}, r_{i_1} is zero. We introduce for each v_J, $J \in \mathcal{J}_M$, the vector $\lambda_J := (\lambda_{iJ})_{i\in\mathcal{I}_J}$ and collect those into

$$\Lambda = (\lambda_J)_{J\in\mathcal{J}_M}.$$

We may then write down (3.4.2.7) for all $J \in \mathcal{J}_M$ and $i \in \mathcal{I}_J$ and define the right-hand sides as a linear map in Λ as follows:

$$\begin{aligned}\Lambda &= (\lambda_J)_{J\in\mathcal{J}_M}\\ &= (\lambda_{iJ})_{\substack{J\in\mathcal{J}_M\\ i\in\mathcal{I}_J}}\end{aligned} \tag{3.4.2.8}$$

$$(\mathcal{T}\Lambda)_{iJ} = \text{r.h.s of (3.4.2.7)} \tag{3.4.2.9}$$

Then the iteration (3.4.2.7) can be rewritten after a shift in the index n as

$$\Lambda^{n+1} = \mathcal{T}\,\Lambda^n. \tag{3.4.2.10}$$

We note that Λ is a fixed point of $\mathcal{T}$ if and only if the solutions of the local problems collectively comprise the solution of the global problem with vanishing data. Thus the only fixed point of $\mathcal{T}$ is $\Lambda = 0$.

As usual, the iteration (3.4.2.10) can be underrelaxed by introducing a parameter $\varepsilon \in [0, 1)$ and replaced by

$$\Lambda^{n+1} = (1-\varepsilon)\,\mathcal{T}\,\Lambda^n + \varepsilon\,\Lambda^n. \tag{3.4.2.11}$$

Let us introduce the energy of the edge e_i by (3.4.1.20) with r replaced with r_i and $a(\cdot,\cdot)$ with $a_i(\cdot,\cdot)$. According to (3.4.1.24), with the changes just introduced we obtain, when g_J, f_i, r_{i_0}, $r_{i_1} = 0$,

$$\begin{aligned}\frac{1}{2\beta_J}\int_0^T |\lambda_{iJ}(t)|^2\,dt = E_i(t) &+ \frac{1}{2}\int_0^T \alpha_j\,|\dot r_i(v_N,t)|^2\,dt\\ &+ \frac{1}{2}\int_0^T \frac{1}{\beta_J}\{\beta_J^2\,|C_{iJ}\,\dot r_i(v_J,t)|^2\\ &+ |(C_{iJ}^+)^T\,K_i(r_i' + R_i\,r_i)(v_J,t)|^2\}\,dt.\end{aligned} \tag{3.4.2.12}$$

The energy relation (3.4.2.12) is for an edge e_i connecting the Neumann node v_N to an interior node v_J. The other cases follow in the same manner. By the definitions (3.4.2.8), (3.4.2.9) and (3.4.2.7), we have $(\mathcal{T}\,\Lambda) = \lambda_{iJ}$ $J \in \mathcal{J}_M$, $i \in \mathcal{I}_J$.

We need to analyze $|\lambda_{iJ}(t)|^2$. To this end we calculate

$$
\begin{aligned}
|\lambda_{iJ}|^2 = \beta_J^2 (& \frac{4}{d_J^2} \,|\sum_{j\in\mathcal{I}_J} C_{iJ}\,\dot r_j(v_J)|^2 + |C_{iJ}\,\dot r_i(v_J)|^2 \\
& - \frac{4}{d_J} (\sum_{j\in\mathcal{I}_J} C_{jJ}\,\dot r_j(v_J))\, C_{iJ}\,\dot r_i(v_J))) \\
& + \frac{4}{d_J^2} \,|(\sum_{j\in\mathcal{I}_J} \varepsilon_{jJ}(C_{jJ}^{+})^T K_j(r_j' + R_j\, r_j)(v_J))|^2 \\
& + |\varepsilon_{iJ}(C_{iJ}^{+})^T K_i(r_i' + R_i\, r_i)(v_J)|^2 \\
& - \frac{4}{d_J} (\sum_{j\in\mathcal{I}_J} \varepsilon_{jJ}(C_{jJ}^{+})^T K_j(r_j' + R_j\, r_j)(v_J)) \\
& \times (\varepsilon_{iJ}(C_{iJ}^{+})^T K_i(r_i' + R_i\, r_i)(v_J)) \\
& - 2\beta_J\, (\frac{2}{d_J} \sum_{j\in\mathcal{I}_j} C_{jJ}\,\dot r_j(v_J) - C_{iJ}\,\dot r_i(v_J)) \\
& \times (\frac{2}{d_J} \sum_{j\in\mathcal{I}_J} \varepsilon_{jJ}(C_{jJ}^{+})^T K_j(r_j' + R_j\, r_j)(v_J) \\
& - \varepsilon_{iJ}(C_{iJ}^{+})^T K_i(r_i' + R_i\, r_i)(v_J)),
\end{aligned}
\tag{3.4.2.13}
$$

where the dependence on the time variable has been suppressed for simplicity. We define

$$
E(T) := \sum_{i\in\mathcal{I}} E_i(T) \tag{3.4.2.14}
$$

$$
\mathcal{E} := \frac{1}{2} \sum_{J\in\mathcal{J}_M} \frac{1}{\beta_J} \sum_{i\in\mathcal{I}_J} \int_0^T \{|(C_{iJ}^{+})^T K_i(r_i' + R_i\, r_i)(v_J)|^2 + \beta_J^2\, |C_{iJ}\,\dot r_i(v_J)|^2\}\, dt. \tag{3.4.2.15}
$$

We now sum (3.4.2.13) over all $i \in \mathcal{I}_J$

$$
\begin{aligned}
\sum_{i\in\mathcal{I}_J} |\lambda_{iJ}|^2 = \beta_J^2 \sum_{i\in\mathcal{I}_J} |C_{iJ}\,\dot r_i(v_J)|^2 + \sum_{i\in\mathcal{I}_J} |(C_{iJ}^{+})^T K_i(r_i' + R_i\, r_i)(v_J)|^2 \\
- 2\beta_J \sum_{i\in\mathcal{I}_J} \varepsilon_{iJ}(C_{iJ}^{+})^T K_i(r_i' + R_i\, r_i)(v_J)\cdot C_{iJ}\,\dot r_i(v_J).
\end{aligned}
\tag{3.4.2.16}
$$

Now (3.4.2.14), (3.4.2.15) together with (3.4.2.12) imply

$$
E(T) + \sum_{J\in\mathcal{J}_N} \int_0^T \alpha_i\, |\dot r_i(v_J,t)|^2\, dt + \mathcal{E} = \sum_{J\in\mathcal{J}_M} \sum_{i\in\mathcal{I}_J} \int_0^T |\lambda_{iJ}(t)|^2\, dt. \tag{3.4.2.17}
$$

Moreover, (3.4.2.16) implies

$$
\begin{aligned}
\|\mathcal{T}\Lambda\|^2 &:= \sum_{J\in\mathcal{J}_M}\sum_{i\in\mathcal{I}_J}\frac{1}{2\beta_J}\int_0^T |\lambda_{iJ}(t)|^2\,dt \\
&= \frac{1}{2\beta_J}\sum_{j\in\mathcal{J}_M}\sum_{i\in\mathcal{I}_J}\int_0^T |(C_{iJ}^{+})^T K_i(r_i'+R_i r_i)(v_J,t)|^2 + |C_{iJ}\dot r_i(v_J,t)|^2\,dt \\
&\quad - \sum_{J\in\mathcal{J}_M}\sum_{i\in\mathcal{I}_J}\int_0^T \varepsilon_{iJ}\,(C_{iJ}^{+})^T K_i(r_i'+R_i r_i)(v_J,t)\cdot C_{iJ}\dot r_i(v_j,t)\,dt \\
&= \mathcal{E} - \sum_{J\in\mathcal{J}_M}\sum_{i\in\mathcal{I}_J}\int_0^T \varepsilon_{iJ}(C_{iJ}^{+})^T K_i(r_i'+R_i r_i)(v_J,t)\cdot C_{iJ}\dot r_i(v_j,t)\,dt \qquad (3.4.2.18)
\end{aligned}
$$

But, according to (3.4.2.5) we also have

$$
\begin{aligned}
\sum_{J\in\mathcal{J}_M}\sum_{i\in\mathcal{I}_J}\frac{1}{2\beta_J}\int_0^T |\lambda_{iJ}(t)|^2\,dt &= \|\Lambda\|^2 \\
&= \frac{1}{2\beta_J}\sum_{J\in\mathcal{J}_M}\sum_{i\in\mathcal{I}_J}\int_0^T \{|(C_{iJ}^{+})^T K_i(r_i'+R_i r_i)(v_j,t)|^2 + \beta_J^2\,|C_{iJ}\dot r_i(v_J,t)|^2 \\
&\qquad + 2\beta_J\,\varepsilon_{iJ}(C_{iJ}^{+})^T K_i(r_i'+R_i r_i)(v_J,t)\cdot C_{iJ}\dot r_i(v_J,t)\}\,dt \\
&= \mathcal{E} + \sum_{J\in\mathcal{J}_M}\sum_{i\in\mathcal{I}_J}\int_0^T \varepsilon_{iJ}(C_{iJ}^{+})^T K_i(r_i'+R_i r_i)(v_J,t)\cdot C_{iJ}\dot r_i(v_J,t)\,dt. \qquad (3.4.2.19)
\end{aligned}
$$

Combining (3.4.2.18) and (3.4.2.19) gives

$$
\|\mathcal{T}\Lambda\|^2 = \|\Lambda\|^2 - 2\mathcal{F}, \qquad (3.4.2.20)
$$

where

$$
\mathcal{F} := \sum_{J\in\mathcal{J}_M}\sum_{i\in\mathcal{I}_J}\int_0^T \varepsilon_{iJ}(C_{iJ}^{+})^T K_i(r_i'+R_i r_i)(v_J,t)\cdot C_{iJ}\dot r_i(v_J,t)\,dt \qquad (3.4.2.21)
$$

We proceed to show that $\mathcal{F}$ given by (3.4.2.21) is nonnegative. Then (3.4.2.20) shows that $\mathcal{T}$ is a nonexpansive map. To this end, we multiply (3.4.2.1) by $\dot r_i$, integrate with respect to x and t and then sum over $i \in \mathcal{I}$. The result is

$$
\begin{aligned}
0 = \sum_{i\in\mathcal{I}} \int_0^T \int_0^{\ell_i} \{M_i \ddot{r}_i - [K_i(r_i' + R_i\, r_i)]' + R_i^T K_i(r_i' + R_i\, r_i) + S_i\, r_i\}\, \dot{r}_i \, dx\, dt \\
= \sum_{i\in\mathcal{I}} \int_0^T \int_0^{\ell_i} \frac{1}{2}\frac{d}{dt}(M_i \dot{r}_i \cdot \dot{r}_i + S_i\, r_i \cdot r_i)\, dx\, dt \\
- \sum_{J\in\mathcal{J}\setminus J_D} \sum_{i\in\mathcal{I}_J} \int_0^T \varepsilon_{iJ}(C_{iJ}^+)^T K_i(r_i' + R_i\, r_i)(v_J)\cdot C_{iJ}\, \dot{r}_i(v_J)\, dt \\
+ \sum_{i\in\mathcal{I}_J} \int_0^T \int_0^{\ell_i} K_i(r_i' + R_i\, r_i)(\dot{r}_i' + R_i\, r_i)\, dx\, dt \\
= E(T) - E(0) - \mathcal{F} + \sum_{J\in\mathcal{J}_N} \int_0^T \alpha_i\, |\dot{r}_{\hat{i}}(v_J)|^2\, dt,
\end{aligned}
\tag{3.4.2.22}
$$

according to (3.4.2.14). But $E(0) = 0$. Thus

$$
\mathcal{F} = E(T) + \sum_{J\in\mathcal{J}_N} \int_0^T \alpha_{\hat{i}}\, |\dot{r}_{\hat{i}}(v_J)|^2\, dt, \tag{3.4.2.23}
$$

which together with (3.4.2.21) gives

$$
\|\mathcal{T}\Lambda\|^2 = \|\Lambda\|^2 - 2E(T) - 2\sum_{J\in\mathcal{J}_N} \int_0^T \alpha_{\hat{i}}\, |\dot{r}_{\hat{i}}(v_J)|^2\, dt. \tag{3.4.2.24}
$$

Equality (3.4.2.24) clearly shows that $\mathcal{T}$ is a nonexpansive map. Going back to (3.4.2.18), (3.4.2.21) we see that

$$
\|\mathcal{T}\Lambda\|^2 = \mathcal{E} - \mathcal{F}, \tag{3.4.2.25}
$$

whereas (3.4.2.19) yields in

$$
\|\Lambda\|^2 = \mathcal{E} + \mathcal{F}. \tag{3.4.2.26}
$$

We now iterate the map and obtain from (3.4.2.25), (3.4.2.26)

$$
\|\mathcal{T}\Lambda^n\|^2 = \mathcal{E}^n - \mathcal{F}^n \tag{3.4.2.27}
$$

$$
\|\Lambda^n\|^2 = \mathcal{E}^n + \mathcal{F}^n. \tag{3.4.2.28}
$$

Now, according to (3.4.2.11) and (3.4.2.27)

$$\begin{aligned}\mathcal{E}^{n+1} + \mathcal{F}^{n+1} =& \|\Lambda^{n+1}\|^2 = \|(1-\varepsilon)\,\mathcal{T}\Lambda^n + \varepsilon\,\Lambda^n\|^2 \\ =&(1-\varepsilon)^2\,\|\mathcal{T}\Lambda^n\|^2 + 2(1-\varepsilon)\,\varepsilon(\mathcal{T}\Lambda^n,\Lambda^n) + \varepsilon^2\,\|\Lambda^n\|^2 \\ \le&(1-\varepsilon)^2(\mathcal{E}^n - \mathcal{F}^n) + (1-\varepsilon)\,\varepsilon(\mathcal{E}^n - \mathcal{F}^n) \\ &+ (1-\varepsilon)\,\varepsilon(\mathcal{E}^n + \mathcal{F}^n) + \varepsilon^2(\mathcal{E}^n + \mathcal{F}^n)\end{aligned} \tag{3.4.2.29}$$

according to (3.4.2.27), (3.4.2.28). Thus,

$$\mathcal{E}^{n+1} + \mathcal{F}^{n+1} \le \mathcal{E}^n - (1-2\varepsilon)\,\mathcal{F}^n. \tag{3.4.2.30}$$

This crucial energy estimate can be iterated down to $n=1$ on the right-hand side of (3.4.2.30), resulting in

$$\mathcal{E}^{n+1} \le \mathcal{E}^1 + \sum_{k=1}^{n+1} c_k(\varepsilon)\,\mathcal{F}^k, \tag{3.4.2.31}$$

where $c_1(\varepsilon) = 1-2\varepsilon,\ c_k(\varepsilon) = 2(1-\varepsilon),\ k=2,\dots,n,\ c_{n+1}(\varepsilon)=1$. From (3.4.2.31) we conclude that

$$\mathcal{F}^n \to 0 \ \text{ as } \ n\to\infty,\ \forall\,\varepsilon\in[0,1) \tag{3.4.2.32}$$

$$\mathcal{E}^n \le C,\ \ \forall\, n\in\mathbb{N}. \tag{3.4.2.33}$$

We consider the two cases $\varepsilon=0$ and $\varepsilon\in(0,1)$ separately.

Case i) $\varepsilon = 0$.

In this case, from (3.4.2.32) and (3.4.2.23), (3.4.2.14) we deduce that

$$E_i^n(T) = \frac{1}{2}\,(r_i^n(T),\dot r_i^n(T)) + \frac{1}{2}\,a_i\,(r_i^n(T),r_i^n(T)) \to 0, \quad i\in\mathcal{I}, \tag{3.4.2.34}$$

and, for $\alpha_i>0$,

$$\|\dot r_i^n(v_J,\cdot)\|_{L^2(0,T)} \to 0, \quad i\in\mathcal{I}_J,\ J\in\mathcal{J}_N. \tag{3.4.2.35}$$

Moreover, by (3.4.2.33) and (3.4.2.15) we obtain

$$\begin{aligned}&\|(C_{iJ}^{+})^T\,K_i((r_i^n)' + R_i\,r_i^n)(v_J,\cdot)\|_{L^2(0,T)} \le C,\\ &\|\beta_J^2\,C_{iJ}\,\dot r_i^n(v_J,\cdot)\|_{L^2(0,T)} \le C,\end{aligned} \tag{3.4.2.36}$$

for all $J\in\mathcal{J}_M$, $i\in\mathcal{I}_J$. By (3.4.2.34), (3.4.2.36) and (3.4.2.12) we then obtain

$$\|\lambda_{iJ}^n\|_{L^2(0,T)} \le C, \quad J\in\mathcal{J}_M, \quad i\in\mathcal{I}_J. \tag{3.4.2.37}$$

Since $g_J\equiv 0$, $f_i\equiv 0$, we obtain in addition

$$E_i^n(\cdot) \le C \ \text{ in } \ L^\infty(0,T),\ \ i\in\mathcal{I}. \tag{3.4.2.38}$$

In other words

$$(r_i^n, \dot{r}_i^n)_{H_i \times V_i} \leq C \text{ in } L^\infty(0,T),\ i \in \mathcal{I} \tag{3.4.2.39}$$

which implies that, on a subsequence $n := n_k \to \infty$

$$(r_i^n, \dot{r}_i^n) \to (\tilde{r}_i, \dot{\tilde{r}}_i) \text{ weakly* in } L^\infty(0,T; H_i \times V_i),\ i \in \mathcal{I}. \tag{3.4.2.40}$$

$$\begin{cases} \varepsilon_{iJ}\,(C_{iJ}^+)^T\,K_i((r_i^n)' + R_i\,r_i^n)(v_J,\cdot) \to B_{iJ} \\ C_{iJ}\,r_i^n(v_J,\cdot) \to A_{iJ} \text{ weakly in } L^2(0,T),\ J \in \mathcal{J}_M,\ i \in \mathcal{I}_J. \end{cases} \tag{3.4.2.41}$$

Moreover $\tilde{r}_i$ satisfies

$$M_i\,\ddot{\tilde{r}}_i = [K_i(\tilde{r}_i' + R_i\,\tilde{r}_i)]' - R_i^T\,K_i(\tilde{r}_i' + R_i\,\tilde{r}_i) - S_i\,r_i \quad \text{in} \quad Q_i \tag{3.4.2.42}$$

$$\begin{aligned} &\varepsilon_{iJ}(C_{iJ}^+)^T\,K_i(\tilde{r}_i' + R_i\,\tilde{r}_i)(v_J,t) = B_{iJ}(t), \\ &C_{iJ}\,\dot{\tilde{r}}_i(v_J,t) = A_{iJ}(t),\ J \in \mathcal{J}_M,\ t \in (0,T) \end{aligned} \tag{3.4.2.43}$$

$$\begin{aligned} &\varepsilon_{jN}\,K_i(\tilde{r}_i' + R_i\,\tilde{r}_i)(v_N,t) = 0 \\ &\dot{\tilde{r}}_i(v_N,t) = 0, \quad v_N \in \mathcal{J}_N,\ t \in (0,T) \end{aligned} \tag{3.4.2.44}$$

$$\tilde{r}_i(\cdot,0) = \dot{\tilde{r}}_i(\cdot,0) = \tilde{r}_i(\cdot,T) = \dot{\tilde{r}}_i(\cdot,T) = 0. \tag{3.4.2.45}$$

Note that, according to (3.4.2.35), $\dot{r}_i^n(v_J,\cdot)$ converges on the entire sequence, strongly in $L^2(0,T)$.

Thus, for e_i connecting a Neumann node v_N to a multiple node v_J, (3.4.2.44) constitute Cauchy data at v_J on $(0,T)$, whereas (3.4.2.45) are zero initial and final data. Hence

$$\tilde{r}_i(x,t) = 0 \quad \text{in} \quad Q_i,\ i \in \mathcal{I}_J,\ J \in \mathcal{J}_N. \tag{3.4.2.46}$$

We now consider a special situation, where G is a tree with all simple node but one in $\mathcal{J}_N$ and the excluded one in $\mathcal{J}_D$. We proceed from the simple Neumann (ultimate) nodes to the penultimate nodes at the next layer. In other words: a penultimate node v_J is characterized as

$$\begin{aligned} &i \in \mathcal{I}_J \Rightarrow \exists\, d_J - 1 \text{ many } v_N,\ N \in \mathcal{J}_N \\ &\text{such that } i \in \mathcal{I}_N, \text{ while the remaining node, say } v_M, \text{ is in } J_M. \end{aligned} \tag{3.4.2.47}$$

The unique multiple node v_M is then located in the next layer.

We go back to the iteration. Let us look at (3.4.2.7). Consider $i \in \mathcal{I}_J \cap \mathcal{I}_N$. Then we may pass to the limit in (3.4.2.5), (3.4.2.7) to obtain

$$0 = \beta_J\,A_{jJ}(t) - B_{jJ}(t) \quad j \in \mathcal{I}_J \cap \mathcal{I}_M. \tag{3.4.2.48}$$

On the other hand, for $i \in \mathcal{I}_J \cap \mathcal{I}_M$, (3.4.2.5), (3.4.2.7) imply in the limit

$$\begin{aligned} &\varepsilon_{iJ}(C_{iJ}^+)^T\,K_i(\tilde{r}_i' + R_i\,\tilde{r}_i)(v_J,t) + \beta_J\,C_{iJ}\,\dot{\tilde{r}}_i(v_J,t) \\ &= \beta_J\,(\frac{2}{d_J} - 1)\,C_{iJ}\,\dot{\tilde{r}}_i(v_J,t) - (\frac{2}{d_J}\,\varepsilon_{iJ}(C_{iJ}^+)^T\,K_i(\tilde{r}_i' + R_i\,\tilde{r}_i)(v_J,t) \end{aligned} \tag{3.4.2.49}$$

or

$$0 = \beta_J\, A_{iJ}(t) + B_{iJ}(t), \quad i \in \mathcal{I}_J \cap \mathcal{I}_M. \tag{3.4.2.50}$$

Now, (3.4.2.49) and (3.4.2.50) imply

$$A_{iJ} = B_{iJ} \equiv 0 \tag{3.4.2.51}$$

and thus, by unique continuation, $\tilde{r}_i(x,t) = 0$ in $(0,\ell_i) \times (0,T)$ for edge e_i connecting the penultimate node v_J to $v_M \in \mathcal{J}_M$ in the next layer.

By this procedure we proceed from all simple Neumann nodes through the tree until we reach the single Dirichlet node. On that last edge we also conclude $\tilde{r}_i \equiv 0$. Thus, $\tilde{r}_i \equiv 0$ in Q_i, $\forall i \in \mathcal{I}$.

This shows that for $\varepsilon = 0$ (no relaxation) and for G a rooted tree with all leaves satisfying impedance type or Robin type conditions, we have weak* convergence of $(r_i^n, \dot{r}_i^n)$, the local errors, to zero in $L^\infty(0,T;V_i \times H_i)$. Moreover, the traces $\varepsilon_{iJ}(C_{iJ}^+)^T\, K_i(r_i^{n'} + r_i^n)(v_J,t)$, $C_{iJ}\,\dot{r}_i^n(v_J,t)$ at the interior nodes tend to zero in $L^\infty(0,T)$. We will collect these result in the statement of the next theorem below.

Case ii) $\varepsilon \in (0,1)$.

According to Opial's Lemma 2.3.7.3, with Λ^n given by (3.4.2.11) we have

$$\Lambda^n \to 0, \quad \mathcal{T}\,\Lambda^n \to 0 \text{ weakly in } \prod_{J\in\mathcal{J}_M}\prod_{i\in\mathcal{I}_J} L^2(0,T) =: \mathcal{X}. \tag{3.4.2.52}$$

Thus

$$\varepsilon_{iJ}(C_{iJ}^+)^T\, K_i((r_i^{n'}) + R_i\, r_i^n)(v_J,\cdot) + \beta_J\, C_{iJ}\,\dot{r}_i^n(v_J,\cdot) \to 0 \tag{3.4.2.53}$$

and

$$\begin{aligned}\beta_J\,(\frac{2}{d_J}\sum_{j\in\mathcal{I}_J} C_{iJ}\,\dot{r}_j^n(v_J,\cdot) - C_{iJ}\,\dot{r}_i^n(v_J,\cdot))&\\ -(\frac{2}{d_J}\sum_{j\in\mathcal{I}_J}\varepsilon_{jJ}(C_{jJ}^+)^T\, K_j((r_j^n)' + R_j\, r_j^n)(v_J,\cdot)&\\ -\varepsilon_{iJ}(C_{iJ}^+)^T\, K_i((r_i^n)' + R_i\, r_i^n)(v_J,\cdot)) \to 0&\end{aligned} \tag{3.4.2.54}$$

weakly in $L^2(0,T)$, as $n \to \infty$.

We proceed to show that the convergence take place in the strong topologies. To this end we sum (3.4.2.53), (3.4.2.54) over $i \in \mathcal{I}_J$ and obtain

$$\sum_{i\in\mathcal{I}_J}\varepsilon_{iJ}(C_{iJ}^+)^T\, K_i((r_i^n)' + R_i\, r_i^n)(v_J,\cdot) + \beta_J\sum_{i\in\mathcal{I}_J} C_{iJ}\,\dot{r}_i^n(v_J,\cdot) \to 0 \tag{3.4.2.55}$$

and

$$\beta_J\sum_{i\in\mathcal{I}_J} C_{iJ}\,\dot{r}_i^n(v_J,\cdot) - \sum_{i\in\mathcal{I}_J}\varepsilon_{iJ}(C_{iJ}^+)^T\, K_i((r_i^n)' + R_i\, r_i)(v_J,\cdot) \to 0 \tag{3.4.2.56}$$

weakly in $L^2(0,T)$. Now, (3.4.2.55) and (3.4.2.56) imply

$$\begin{aligned}\beta_J \sum_{i\in\mathcal{I}_J} C_{iJ}\,\dot{r}_i^n(v_J,\cdot) &\to 0,\\ \sum_{i\in\mathcal{I}_J} \varepsilon_{iJ}(C_{iJ}^+)^T K_i((r_i^n)' + R_i\, r_i)(v_J,\cdot) &\to 0,\end{aligned} \tag{3.4.2.57}$$

weakly in $L^2(0,T)$. By using (3.4.2.54) it is seen that (3.4.2.57) implies

$$\begin{aligned}C_{iJ}\,\dot{r}_i^n(v_J,\cdot) &\to 0,\\ \varepsilon_{iJ}\,(C_{iJ}^+)^T K_i((r_i^n)' + R_i\, r_i)(v_J,\cdot) &\to 0\end{aligned} \tag{3.4.2.58}$$

weakly in $L^2(0,T)$.

In order to obtain strong convergence in (3.4.2.57) and (3.4.2.58), we have to consider the convergence of the norms of the terms involved. Recall (3.4.2.11), that is

$$\Lambda^{n+1} = (1-\varepsilon)\,\mathcal{T}\,\Lambda^n + \varepsilon\,\Lambda^n =: \mathcal{T}_\varepsilon\,\Lambda^n. \tag{3.4.2.59}$$

Thus

$$\Lambda^{n+1} - \Lambda^n = (1-\varepsilon)\,(\mathcal{T}\,\Lambda^n - \Lambda^n)$$

and hence

$$\begin{aligned}\|\Lambda^{n+1} - \Lambda^n\| &= (1-\varepsilon)\,\|\mathcal{T}\,\Lambda^n - \Lambda^n\|\\ &= \|\mathcal{T}_\varepsilon^n\,\Lambda^1 - \mathcal{T}_\varepsilon^{n-1}\,\Lambda^1\| \to 0,\end{aligned} \tag{3.4.2.60}$$

as $n \to \infty$. On the other hand

$$\|\mathcal{T}\,\Lambda^n - \Lambda^n\|^2 = \|\mathcal{T}\,\Lambda^n\|^2 - 2(\mathcal{T}\,\Lambda^n, \Lambda^n) + \|\Lambda^n\|^2$$

which, according to (3.4.2.25)and (3.4.2.26), is equivalent to

$$\|\mathcal{T}\,\Lambda^n - \Lambda^n\|^2 = 2(\mathcal{E}^n - (\mathcal{T}\,\Lambda^n, \Lambda^n)). \tag{3.4.2.61}$$

This leaves us with the analysis of $(\mathcal{T}\,\Lambda, \Lambda)$. Now,

$$\begin{aligned}(\mathcal{T}\,\Lambda, \Lambda) = \sum_{J\in\mathcal{J}_M} \sum_{i\in\mathcal{I}_J} \frac{1}{2\beta_J} \int_0^T &[\varepsilon_{iJ}\,(C_{iJ}^+)^T K_i(r_i' + R_i\, r_i)(v_J) + \beta_J\, C_{iJ}\,\dot{r}_i(v_J)]\\ &\times [\beta_J\,(\frac{2}{d_J} \sum_{i\in\mathcal{I}_J} C_{jJ}\,\dot{r}_j(v_J) - C_{iJ}\,\dot{r}_i(v_J))\\ &-(\frac{2}{d_J} \sum_{j\in\mathcal{I}_J} \varepsilon_{iJ}(C_{iJ}^+)^T K_j(r_j' + R_j\, r_j)(v_J)\\ &\qquad - \varepsilon_{iJ}(C_{iJ}^+)^T K_i(r_i' + R_i\, r_i)(v_J))]\, dt\end{aligned}$$

A little calculation gives

$$(\mathcal{T}\,\Lambda,\Lambda) = \sum_{J\in\mathcal{J}_M} \frac{1}{2\beta_J} \int_0^T \{\sum_{i\in\mathcal{I}_J} |\varepsilon_{iJ}(C_{iJ}^+)^T K_i(r_i' + R_i\,r_i)(v_J)|^2$$
$$- \frac{2}{d_J} |\sum_{i\in\mathcal{I}_J} \varepsilon_{iJ}(C_{iJ}^+)^T K_i(r_i' + R_i\,r_i)(v_J)|^2$$
$$+ \beta_J^2 \frac{2}{d_J} |\sum_{i\in\mathcal{I}_J} C_{iJ}\,\dot r_i(v_J)|^2 - \beta_J^2 \sum_{i\in\mathcal{I}_J} |C_{iJ}\,\dot r_i(v_J)|^2\}\,dt. \qquad (3.4.2.62)$$

We recall the definition of $\mathcal{E}$ from (3.4.2.15), together with (3.4.2.62), and obtain

$$\mathcal{E} - (\mathcal{T}\,\Lambda,\Lambda) = \sum_{J\in\mathcal{J}_M} \frac{1}{2\beta_J} \int_0^T \{2\beta_J^2 \sum_{i\in\mathcal{I}_J} |C_{iJ}\,\dot r_i(v_J)|^2$$
$$+ \frac{2}{d_J} |\sum_{i\in\mathcal{I}_J} \varepsilon_{iJ}(C_{iJ}^+)^T K_i(r_i' + R_i\,r_i)(v_J)|^2$$
$$- \frac{2\beta_J^2}{d_J} |\sum_{i\in\mathcal{I}_J} C_{iJ}\,\dot r_i(v_J)|^2\}\,dt \qquad (3.4.2.63)$$
$$= \sum_{J\in\mathcal{J}_M} \int_0^T \{\beta_J\,[\sum_{i\in\mathcal{I}_J} |C_{iJ}\,\dot r_i(v_J)|^2 - \frac{1}{d_J} |\sum_{i\in\mathcal{I}_J} C_{iJ}\,\dot r_i(v_J)|^2$$
$$+ \frac{1}{\beta_J\,d_J} |\sum_{i\in\mathcal{I}_J} \varepsilon_{iJ}(C_{iJ}^+)^T K_i(r_i' + R_i\,r_i)(v_J)|^2\}\,dt.$$

But

$$\sum_{i\in\mathcal{I}_J} |a_i|^2 - \frac{1}{d_J} |\sum_{i\in\mathcal{I}_J} a_i|^2 = \frac{1}{2d_J} \sum_{i,j\in\mathcal{I}_J} |a_i - a_j|^2$$

which allows (3.4.2.63) to be written

$$\mathcal{E} - (\mathcal{T}\,\Lambda,\Lambda) = \sum_{J\in\mathcal{J}_M} \int_0^T \{\frac{\beta_J}{2d_J} \sum_{i,j\in\mathcal{I}_J} |C_{iJ}\,\dot r_i(v_J) - C_{jJ}\,\dot r_j(v_J)|^2$$
$$+ \frac{1}{\beta_J d_J} |\sum_{i\in\mathcal{I}_J} \varepsilon_{iJ}(C_{iJ}^+)^T K_i(r_i' + R_i\,r_i)(v_J)|^2\}\,dt. \qquad (3.4.2.64)$$

Equations (3.4.2.60), (3.4.2.61) and (3.4.2.64), with Λ replaced by Λ^n, imply that

$$\|\sum_{i\in\mathcal{I}_J} \varepsilon_{iJ}(C_{iJ}^+)^T K_i((r_i^n)' + R_i\,r_i^n)(v_J)\|_{L^2(0,T)} \to 0 \qquad (3.4.2.65)$$

and

$$\|C_{iJ}\,\dot{r}_i^n(v_J) - C_{jJ}\,\dot{r}_j^n(v_J)\|_{L^2(0,T)} \to 0, \tag{3.4.2.66}$$

for $i, j \in \mathcal{I}_J$. Thus the global transmission conditions are satisfied in the strong sense in $L^2(0,T)$ in the limit. According to (3.4.2.17) and (3.4.2.19) we also have

$$\begin{aligned} E^n(t) + \sum_{J \in \mathcal{J}_N} \int_0^t \alpha_{\hat{i}}\, |\dot{r}_{\hat{i}}^n(v_J)|^2\, dt \\ = \sum_{J \in \mathcal{J}_M} \sum_{i \in \mathcal{I}_J} \int_0^t \varepsilon_{iJ}\, (C_{iJ}^+)^T\, K_i((r_i^n)' + R_i\, r_i^n) \cdot C_{iJ}\, \dot{r}_i(v_J)\, dt \\ = \sum_{J \in \mathcal{J}_M} \frac{1}{d_J} \int_0^t \{ \sum_{i,j \in \mathcal{I}_J} \varepsilon_{iJ} (C_{iJ}^+)^T\, K_i((r_i^n)' + R_i\, r_i)(v_J) \\ \times (C_{iJ}\, \dot{r}_i^n(v_J) - C_{jJ}\, \dot{r}_j^n(v_J)) \\ + \sum_{i \in \mathcal{I}_J} \varepsilon_{iJ} (C_{iJ}^+)^t\, K_i((r_i^n)' + R_i\, r_i^n)(v_J) \cdot \sum_{i \in \mathcal{I}_J} C_{iJ}\, \dot{r}_i^n(v_J)\}\, dt. \end{aligned} \tag{3.4.2.67}$$

According to (3.4.2.36), (3.4.2.65) and (3.4.2.66) we infer from (3.4.2.67) that

$$\sup_{0 \le t \le T} E^n(t) \to 0, \quad n \to \infty. \tag{3.4.2.68}$$

Theorem 3.4.2.1

i) *Let $\epsilon = 0$ and let G be a rooted tree, with the root clamped and all leaves satisfying* (3.4.1.5). *Then for each $i \in \mathcal{I}$ the local solutions in the iterative procedure based on* (3.4.2.1)–(3.4.2.7) *converge weakly* in $L^\infty(0,T; V_i \times H_i)$ to the solution of the global problem* (3.4.1.1)–(3.4.1.7), *and there is convergence of the traces $\epsilon_{iJ}(C_{iJ}^+)^T K_i((r_i^n)' + R_i r_i^n)(v_J)$ and $C_{iJ}\dot{r}_i^n(v_J)$ in $L^\infty(0,T)$.*

ii) *Let $\epsilon \in (0,1)$. Then the iterative local solutions converges to the global solution in $C([0,T]; V_i \times H_i)$, $\forall\, i \in \mathcal{I}$.*

Chapter 4

Domain Decomposition for Elliptic Optimal Control Problems

4.1 Introduction

In this chapter we begin the study of domain decomposition of optimal control problems, which is the main focus of this monograph. This topic, in contrast to domain decomposition procedures as such, has attracted the attention of mathematicians only recently. We focus the discussion at the beginning on elliptic partial differential equations which, from a technical point of view, are the most accessible.

Control problems are ubiquitous. There is virtually no need for any motivation here. Suffice it to say that optimal control problems have been a major focus in pure and applied mathematics for many years. The necessity to describe the fine structure of processes, to account for local phenomena, has made control theory of partial differential equations, or control theory of distributed parameter systems, one of the major mathematical tools in applied mathematics. Typically, control problems involve constraints on the controls and also on the states evolving according to a partial differential equation. While simple box-type control constraints are more or less standard in the modern theory, state constraints are still a major area of current research. However, just the mere presence of controls, even without constraints, is a challenge from the point of view of numerical implementations.

Control problems add yet another difficulty to the treatment of partial differential equations, since, in the presence of controls, the goal is not only to show existence and possibly uniqueness of solutions under the influence of controls, but also to optimize some cost functional for admissible controls allowing a unique solution of the underlying control problem. From the point of view of numerical algorithms, typically an optimal solution is sought by the use of gradient steps in the unconstrained case, and by additional second-order information in the constrained case. Gradient information can be obtained on the discrete level by just discretizing and using finite differences, or by resorting to automatic differentiation tools. Gradient information can also be retrieved on the finite- as well as on the infinite-dimensional level by the method of adjoints. Adjoints, in turn, can be viewed as Lagrange multipliers for the state equation. In this context, the optimal control problem is viewed as an equality constrained optimization problem, the equation to be satisfied being the state equation. Then the unconstrained Lagrangian is checked for saddle-points and Uzawa-type algorithms can be devised, as in Chapter 2. One may also look into the first-order necessary and, in the convex case, sufficient optimality condition.

This optimality condition consists of a so-called forward equation, which is the original state equation, and a 'backwards' equation, which is the adjoint equation. These equations are coupled in a way dictated by the cost functional and the control inputs. For elliptic and hyperbolic equations, the adjoint, even though possibly requiring a different space state setting, is of the same general structure as the forward equation. Thus, considering optimal control problems, and pursuing the avenue of gradient type methods or directly exploiting the optimality system, leads to an additional equation of the same complexity as the forward state equation, the two equations being coupled at various locations and by various means.

It is then obvious that domain decompositions in the context of optimal control problems are highly desirable, just because of their potential in parallel computing. This is even more important for multi-link system or, more generally, for partial differential equations on graphs or networked domains, as discussed in the previous chapter and in Chapters 6 to 9.

In this book we pursue the idea of accomplishing such domain decompositions by an all-at-once approach in the following sense. As explained above, an optimal control problem can be seen as a constrained optimization problem in infinite dimensions. We may design methods based on the discretized optimization problem. This is what is called the direct approach. On the other hand, one may use the optimality condition, which is a coupled set of partial differential equations. In this method the optimal control problem is transported into a pure PDE problem. Accordingly, software for PDEs enters the focus of interest. This is typically referred to as the indirect approach to optimality. On this level, domain decomposition methods constitute a modern tool to decompose the problem into tractable pieces. One possible approach is to apply domain decomposition methods first to the forward equation for the computation of its solution, and then use that information in order to compute the solution of the backward equation, again by domain decomposition. Such a procedure would, however, not be completely parallel since there remains, at least, a forward coupling. This is true for gradient-type methods in general, where the gradient is computed via an adjoint state. If one insists on the optimality system, the forward and the backward equation are two-way coupled. In that case a domain decomposition of the entire optimality system is in order. The very natural question that arises here is whether the decoupled optimality system has itself any relation to an optimality system on the substructure. If this is the case, the global optimal control problems breaks down into iteratively coupled optimal control problems on the substructures. Such methods are of great importance, in particular for problems discussed in this book, namely, problems on networked domains, where possibly different equations govern the process on adjacent domains.

Thus performing optimization and domain decomposition together, as described above, will lead to algorithms that are completely parallel with respect to subdomains. Equipped with such algorithms one is able to resort to standard software developed for processes on standard domains. The description of those algorithms, proofs of their convergence and the development of a posteriori error estimates is central to this book.

To keep matters as simple as possible in the present chapter, we focus on elliptic problems and do not include a treatment of control and state constraints. Domain decomposition algorithms for such constrained optimal control problems will be the subject of future research, although we do provide such an algorithm for a state constrained optimal control problem for wave type equations in Chapter 8. The idea pursued in this chapter is very simple and attractive. It has been used by Benamou and Desprès [10] for a Helmholtz-type problem with a purely distributed control (see also [16]), i.e., a control acting on the entire domain rather than just on its boundary. They observed that the corresponding optimality system could be viewed as a complex Helmholtz equation. Therefore, a domain decomposition of the complex Helmholtz equation results in a domain decomposition of the original optimality system. We describe their idea, introduce the domain decomposition, prove convergence and provide a posteriori error estimates. It turns out, however, that this point of view cannot be carried over to boundary control problems. Nevertheless, once the domain decomposition method is established in the distributed case, one may use the method as a paradigm also in the case of boundary controls. Again, we consider a domain decomposition method, prove convergence and provide a posteriori error estimates.

4.2 Distributed Controls

In order to fix ideas let us begin with a standard elliptic optimal control problem which has been considered by Benamou [7] and Despres. [16]

$$\begin{cases} \min_{u\in U} \displaystyle\int_\Omega (|w(u) - w_d|^2 + \nu|u|^2)\,dx \\ \quad -\Delta w = f + u \quad \text{in}\,\Omega \\ \qquad w = 0 \quad \text{on}\,\partial\Omega, \end{cases} \tag{4.2.1.1}$$

where $U \subset L^2(\Omega)$ is closed and convex, $\Omega \subset \mathbf{R}^n$, $f \in L^2(\Omega)$.

There is a unique solution to (4.2.1.1). The optimality system is given by:

$$\begin{cases} -\Delta w = f + u \quad \text{in}\,\Omega \\ -\Delta p = w - w_d \quad \text{in}\,\Omega \\ w = 0 = p \quad \text{on}\,\partial\Omega \\ \displaystyle\int_\Omega (p + \nu u)(v - u)dx \geq 0, \quad \forall v \in U \end{cases}$$

The derivation of (4.2.1.2) follows standard procedures. In particular, if $U = L^2(\Omega)$ then u is given by the adjoint state p in the following typical form:

$$u = -\frac{1}{\nu}p$$

Thus, in this case the optimality system is given by

$$\begin{cases} -\Delta w = f - \dfrac{1}{\nu}p \quad \text{in}\,\Omega \\ -\Delta p = w - w_d \quad \text{in}\,\Omega \\ w = 0 = p \quad \text{on}\,\partial\Omega \end{cases} \tag{4.2.1.2}$$

Thus the optimality system is a coupled elliptic problem with the coupling occurring in two places.

In order to obtain a reasonable decomposition procedure of (4.2.1.2), Benamou and Despres developed the following embedding into a Helmholtz problem with complex coefficient $k \in \mathbb{C}$ as follows. Let z be the (complex-valued) solution of

$$\begin{cases} -\Delta z + kz = g \quad \text{in}\,\Omega \\ z = 0 \quad \text{on}\,\partial\Omega \\ z \in H_0^1(\Omega; \mathbf{C}) \end{cases} \tag{4.2.1.3}$$

and observe that for

$$z = w + \frac{\mathbf{i}}{\sqrt{\nu}}p,$$

$$k = -\frac{\mathbf{i}}{\sqrt{\nu}},$$

$$g = f - \frac{\mathbf{i}}{\sqrt{\nu}}w_d$$

(4.2.1.3) is equivalent to (4.2.1.2). Hence, a convergent domain decomposition of (4.2.1.3) will result in such an algorithm for (4.2.1.2). This observation leads one to suggest a Robin-type algorithm.

4.2.2 Domain Decomposition

Algorithm 4.2.2.1 (Benamou, Despres) *Given data $\lambda_{ij}^n, \mu_{ij}^n$ at level n, compute the solutions w_i^{n+1}, p_i^{n+1} of the local problems*

$$\begin{cases} -\Delta w_i^{n+1} + \frac{1}{\nu} p_i^{n+1} = f_i & in \quad \Omega_i \\ -\Delta p_i^{n+1} = w_i^{n+1} - w_{id} & in \quad \Omega_i \end{cases} \tag{4.2.2.1}$$

$$w_i^{n+1} = 0 = p_i^{n+1} \quad on \quad \partial\Omega \cap \partial\Omega_i \tag{4.2.2.2}$$

together with

$$\begin{cases} \frac{\partial}{\partial \nu_{ij}} w_i^{n+1} + \beta p_i^{n+1} = \lambda_{ij}^n \\ \frac{\partial}{\partial \nu_{ij}} p_i^{n+1} - \beta w_i^{n+1} = \mu_{ij}^n \end{cases}$$

where for $n \geq 1$

$$\begin{cases} \lambda_{ij}^n = -\frac{\partial}{\partial \nu_{ji}} w_j^n + \beta p_j^n \\ \mu_{ij}^n = -\frac{\partial}{\partial \nu_{ji}} p_j^n - \beta w_j^n \end{cases} \tag{4.2.2.3}$$

4.2.3 A Complex Helmholtz Problem and its Decomposition

Let us consider the complex model problem

$$\begin{cases} -\Delta z + kz = f & \text{in} \quad \Omega \\ \frac{\partial z}{\partial \nu} + \beta\, z = h & \text{on} \quad \Gamma_1 \\ z = 0 & \text{on} \quad \Gamma_0 \end{cases} \tag{4.2.3.1}$$

with $\partial\Omega = \Gamma_0 \cup \Gamma_1$, $k, \beta \in \mathbb{C}$, $f \in L^2(\Omega; \mathbb{C})$, $h \in L^2(\Gamma_{1j}, \mathbb{C})$. We are interested in solutions $z \in H^1(\Omega; \mathbb{C})$ of (4.2.3.1). We proceed as in Benamou ([7]). Consider the auxiliary problem

$$\begin{cases} -\Delta z + z = \tilde{f} & \text{in} \quad \Omega \\ \frac{\partial z}{\partial \nu} + \beta z = 0 & \text{on} \quad \Gamma_1 \\ z = 0 & \text{on} \quad \Gamma_0 \end{cases} \tag{4.2.3.2}$$

with $\tilde{f} \in L^2(\Omega; \mathbb{C})$.

We have

$$\begin{aligned} \int_\Omega (-\Delta z + z)\, \bar{z}\, dx &= -\int_{\Gamma_1} \frac{\partial}{\partial \nu} z\, \bar{z}\, d\gamma + \int_\Omega |\nabla z|^2 + |z|^2\, dx \\ &= \int_{\Gamma_1} \beta |z|^2\, d\gamma + \int_\Omega |\nabla z|^2 + |z|^2\, dx. \end{aligned}$$

Thus

$$\mathrm{Re}(-\Delta z + kz, z) \geq C \, \|z\|^2_{H^1(\Omega;\mathbb{C})}$$

if $\mathrm{Re}\,\beta \geq 0$. The problem (4.2.3.2) is therefore coercive in $H^1(\Omega;\mathbb{C})$, if $\mathrm{Re}\,\beta \geq 0$. We may then describe the original problem (4.2.3.1) as a perturbation of (4.2.3.2) as follows. Let $R : L^2(\Omega;\mathbb{C}) \times L^2(\Gamma_1;\mathbb{C}) \to H^1(\Omega;\mathbb{C})$ be defined by $R(\tilde{f}, h) =:$ z, where $z \in H^1(\Omega;\mathbb{C})$ is the unique weak solution of (4.2.3.2). Moreover, let $T : H^1(\Omega;\mathbb{C}) \to H^1(\Omega;\mathbb{C})$ be defined by $Tz = R((1-k)\,z, h)$. Then with $F = R(f,h)$ we see that z solves (4.2.3.1) weakly in $H^1(\Omega;\mathbb{C})$ if and only if z satisfies $(I-T)\,z = F$. This is because $(I-T)\,z = F \Leftrightarrow z = Tz + F = R((1-k)\,z + f, h)$. Now since Ω is bounded, T is a compact map in $H^1(\Omega;\mathbb{C})$. Upon applying the Fredholm alternative we are left with a uniqueness problem for (4.2.3.1). To this end it is sufficient to consider zero data $f = 0 = h$. By multiplying (4.2.3.1) by $\bar{z}$ and integrating over Ω we obtain

$$0 = \int_\Omega (-\Delta z + kz)\,\bar{z}\,dx = \int_{\Gamma_1} -\frac{\partial}{\partial\nu} z\,\bar{z}\,d\gamma + \int_\Omega k|z|^2\,dx$$
$$+ \int_\Omega |\nabla z|^2\,dx = \int_{\Gamma_1} \beta|z|^2\,d\gamma + \int_\Omega |\nabla z|^2\,dx + \int_\Omega k|z|^2\,dx.$$

This implies

$$\mathrm{Im}(\beta) \int_{\Gamma_1} |z|^2\,d\gamma + \mathrm{Im}(k) \int_\Omega |z|^2\,dx = 0.$$

If we assume that $\mathrm{Im}(\beta)$ and $\mathrm{Im}(k)$ have the same sign, then both terms are equal to zero. If, in addition, we assume that $\mathrm{Im}(\beta)$ and $\mathrm{Im}(k)$ do not vanish simultaneously, then either $\mathrm{Im}(k) \neq 0$, which implies $z \equiv 0$ in Ω, or $\mathrm{Im}(\beta) \neq 0$, which implies $z \equiv 0$ on Γ_1 and, in turn $\partial z/\partial\nu = 0 \equiv z$ on Γ_1 implies $z \equiv 0$ in Ω by unique continuation. Thus one obtains the following result.

Theorem 4.2.3.1 (Benamou) *Let $f \in L^2(\Omega;\mathbb{C})$, $h \in L^2(\Gamma_1;\mathbb{C})$ be given. Assume that $k, \beta \in \mathbb{C}$ satisfy*

$$\mathrm{Re}(\beta) \geq 0, \quad \mathrm{Im}(\beta)\,\mathrm{Im}(k) \geq 0;\ \mathrm{Im}(k) + \mathrm{Im}(\beta) \neq 0.$$

Then problem (4.2.3.1) *admits a unique weak solution z in $H^1(\Omega;\mathbb{C})$.*

With this result at hand we may now consider the following domain decomposition procedure which, for the sake of simplicity, we write down for two subdomains in an unrelaxed form. The general case may then be handled as in the preceding sections. Let $z_i^n \in H^1(\Omega;\mathbb{C})$ be the unique weak solution of

$$\begin{cases} -\Delta\, z_i^n + k\, z_i^n = f \quad \text{in} \quad \Omega_i \\ \dfrac{\partial z_i^n}{\partial \nu_i} + \beta\, z_i^n = g_i^n \quad \text{on} \quad \Gamma \\ z_i^n = 0 \quad \text{on} \quad \partial\Omega_i \cap \partial\Omega \end{cases} \tag{4.2.3.3}$$

with $f = L^2(\Omega;\mathbb{C})$ and $g_i^n \in L^2(\Gamma;\mathbb{C})$, $i = 1,2$. Update the inhomogeneity g_i^n by

$$g_i^{n+1} = 2\,\beta\, z_j^n - g_j^n, \quad i = 1,2, \quad j \neq i. \tag{4.2.3.4}$$

Together with (4.2.3.4), the transmission condition (4.2.3.3) at $n+1$ reads

$$\begin{aligned} \frac{\partial z_i^{n+1}}{\partial \nu_i} + \beta\, z_i^{n+1} &= g_i^{n+1} = 2\,\beta\, z_j^n - g_j^n \\ &= 2\,\beta\, z_j^n - \frac{\partial z_j^n}{\partial \nu_j} - \beta\, z_j^n \\ &= -\frac{\partial z_j^n}{\partial \nu_j} + \beta\, z_j^n, \quad i = 1,2, \quad j \neq i, \end{aligned}$$

which, upon convergence, reduces to the correct transmission conditions for the global problem across the interface Γ, as in the real case.

4.2.4 Convergence

In order to show convergence, we consider the errors

$$e_i^n := z_i^n - z_i \tag{4.2.4.1}$$

where z_i, $i = 1,2$, are the restrictions $z|_{\Omega_i}$ of the solution z to the global problem. Obviously, e_i^n satisfies (4.2.3.3) with $f \equiv 0$ and (4.2.3.4). Let $g = (g_1, g_2) \in L^2(\Gamma_i\mathbb{C})^2$ be given. The norm of g is defined by

$$\|g\|^2 = \sum_{i=1}^{2} \int_\Gamma |g_i|^2\, d\gamma. \tag{4.2.4.2}$$

We consider the map

$$\begin{aligned} \mathcal{T} : L^2(\Gamma;\mathbb{C})^2 &\to L^2(L;\mathbb{C})^2 \\ \mathcal{T} g &:= (2\beta\, e_2 - g_2,\ 2\beta\, e_1 - g_1) \end{aligned} \tag{4.2.4.3}$$

where e solves (4.2.3.3) with $f = 0$ and (4.2.3.4). We have

$$\begin{aligned} \|\mathcal{T} g\|^2 &= \sum_{i=1}^{2} \int_\Gamma |2\beta\, e_i - g_i|^2\, d\gamma \\ &= \sum_{i=1}^{2} \int_\Gamma [4|\beta\, e_i|^2 - 4\, Re\, \overline{\beta\, e_i} g_i + |g_i|^2]\, d\gamma \\ &= \|g\|^2 + 4 \sum_{i=1}^{2} \int_\Gamma Re\, (\overline{\beta\, e_i}\, (\beta\, e_i - g_i))\, d\gamma \\ &= \|g\|^2 - 4 \sum_{i=1}^{2} \int_\Gamma Re\, (\overline{\beta\, e_i}\, \frac{\partial}{\partial \nu_i} e_i)\, d\gamma. \end{aligned} \tag{4.2.4.4}$$

If we multiply (4.2.3.3) with $f \equiv 0$ by $\overline{\beta\, e_i}$ and integrate over Ω we obtain

$$0 = \int_{\Omega_i} (-\Delta\, e_i + k\, e_i)\, \bar{\beta}\, \bar{e}_i\, dx$$
$$= -\int_{\Gamma} \frac{\partial}{\partial \nu_i} e_i\, \bar{\beta}\, \bar{e}_i\, d\gamma + \bar{\beta} \int_{\Omega_i} |\nabla\, e_i|^2\, dx + k\, \bar{\beta} \int_{\Omega_i} |e_i|^2\, dx$$

or

$$\bar{\beta} \int_{\Gamma} \frac{\partial}{\partial \nu_i} e_i \bar{e}_i\, d\gamma = \bar{\beta} \int_{\Omega_i} |\nabla\, e_i|^2\, dx + k\, \bar{\beta} \int_{\Omega_i} |e_i|^2\, dx.$$

Therefore (4.2.4.4) gives

$$\|\mathcal{T} g\|^2 = \|g\|^2 - 4 \sum_{i=1}^{2} \{\mathrm{Re}(\bar{\beta}) \int_{\Omega_i} |\nabla\, e_i|^2\, dx + \mathrm{Re}(k\, \bar{\beta}) \int_{\Omega_i} |e_i|^2\, dx\}. \qquad (4.2.4.5)$$

It is clear from (4.2.4.5) that the operator $\mathcal{T}$ defined by (4.2.4.3) is nonexpansive if

$$\mathrm{Re}(\beta) \geq 0, \quad \mathrm{Re}(k\, \bar{\beta}) \geq 0. \qquad (4.2.4.6)$$

We define an iteration as follows. Given $g^0 \in L^2(\Gamma;\, \mathbb{C})$ set

$$g^{n+1} = \mathcal{T}(g^n), \quad n \geq 0. \qquad (4.2.4.7)$$

Then (4.2.4.7) is equivalent to the iterative process (4.2.3.3) with $f \equiv 0$ and z_i^n replaced with e_i^n, together with (4.2.3.4) (again with z_i^n replaced with e_i^n).

In order to prove convergence of (4.2.4.7) we assume in addition to (4.2.4.6)

$$\mathrm{Re}(\beta) + \mathrm{Re}(k\, \bar{\beta}) > 0. \qquad (4.2.4.8)$$

We have from (4.2.4.5) and (4.2.4.7)

$$\|g^{n+1}\|^2 = \|g^n\|^2 - 4 \sum_{i=1}^{2} \{\mathrm{Re}(\beta) \int_{\Omega_i} |\nabla\, e_i^n|^2\, dx + Re(k\bar{\beta}) \int_{\Omega_i} |e_i^n|^2\, dx\} \qquad (4.2.4.9)$$

We have three cases to consider

i) $\mathrm{Re}(\beta) > 0$ and $\mathrm{Re}(k\bar{\beta}) = 0$. In this case (4.2.4.9) implies

$$\nabla\, e_i^n \to 0 \quad \text{in} \quad L^2(\Omega_i; \mathbb{C}), \quad i = 1, 2$$

and, in turn, $e_i^n \to 0$ in $H^1(\Omega_j; \mathbb{C})$. But then also $e_i^n|_\Gamma \to 0$ in $L^2(\Gamma)$.

ii) $\mathrm{Re}(\beta) = 0$ and $\mathrm{Re}(k\bar{\beta}) > 0$.

In this case (4.2.4.9) implies only $e_i^n \to 0$ in $L^2(\Omega_i)$. However, upon multiplying (4.2.3.3) ($f = 0,\ z_i \leftrightarrow e_i,\ n \leftrightarrow n+1$) by $\overline{e_i^{n+1}}$ we obtain after integration

$$0 = \int_{\Omega_i} (-\Delta\, e_i^{n+1} + k\, e_i^{n+1})\, \overline{e_i^{n+1}}\, dx$$

$$= \int_{\Gamma} \frac{\partial}{\partial \nu} \overline{e_i^{n+1}}\, d\gamma + \int_{\Omega_i} |\nabla\, e_i^{n+1}|^2\, dx + k \int_{\Omega_i} |e_i^{n+1}|^2\, dx$$

which implies

$$\begin{aligned}\int_{\Omega_i} |\nabla\, e_i^{n+1}|^2\, dx &= -\mathrm{Re}(k) \int_{\Omega_i} |e_i^{n+1}|^2\, dx \\ &\quad + \mathrm{Re} \int_{\Gamma} (\beta\, e_i^{n+1} - g_i^{n+1})\, \overline{e_i^{n+1}}\, d\gamma \\ &= -\mathrm{Re}(k) \int_{\Omega_i} |e_i^{n+1}|^2\, dx - \mathrm{Re} \int_{\Gamma} g_i^{n+1}\, \overline{e_i^{n+1}}\, d\gamma\end{aligned} \tag{4.2.4.10}$$

where we have used (4.2.3.4) ($n \leftrightarrow n+1,\ z_j \leftrightarrow e_j$) and $Re\beta = 0$.

Since (g^n) is bounded in $L^2(\Gamma; \mathbb{C})$ according to (4.2.4.9), we infer from (4.2.3.4) (with $n \leftrightarrow n+1,\ z_j \leftrightarrow e_j$) that $(e_j^n)|_\Gamma$ is bounded in $L^2(\Gamma; \mathbb{C})$. By (4.2.4.10) we conclude that (e_i^n) is bounded in $H^1(\Omega_i; \mathbb{C})$. This implies that $e_i^n|_\Gamma \to 0$ strongly in $L^2(\Gamma; \mathbb{C})$. Going back to (4.2.4.10) we then conclude that $e_i^n \to 0$ in $H^1(\Omega_i; \mathbb{C})$.

iii) $\mathrm{Re}(\beta) > 0$ and $\mathrm{Re}(k\bar{\beta}) > 0$. This case is obvious.

We conclude that $e_i^n \to 0$ in $H^1(\Omega_i; \mathbb{C}),\ i = 1, 2$

Theorem 4.2.4.1 *Let g_i^0 $i = 1, 2$, be given in $L^2(\Gamma; \mathbb{C})$ and let $\beta, k \in \mathbb{C}$ satisfy*

$$\mathrm{Re}(\beta) \geq 0, \quad \mathrm{Re}(\bar{\beta} k) \geq 0, \quad \mathrm{Re}(\beta) + \mathrm{Re}(\bar{\beta} k) \geq 0.$$

Then as $n \to \infty$ the solutions z_i^n, $i = 1, 2$, of (4.2.3.3) and (4.2.3.4) converge strongly in $H^1(\Omega_i; \mathbb{C})$ to the restriction of z on Ω_i, where z is the solution of

$$\begin{cases} -\Delta\, z + k\, z = f & \text{in} \quad \Omega \\ z = 0 & \text{in} \quad \partial\Omega. \end{cases} \tag{4.2.4.11}$$

Remark 4.2.4.1

i) The method can be extended to an arbitrary number of subdomains.
ii) The method can be relaxed as in the preceding chapter.
iii) The proof is an adaption of the one given by Benamou[7] combined with the updating procedure given by Deng [15].
iv) The method also extends to more general second-order complex equations.

4.2.5 Methods for Elliptic Optimal Control Problems

We consider (4.2.4.11) with $k = \frac{-i}{\sqrt{\nu}}$, $\beta = \lambda + i\mu$, $z = w + \frac{i}{\sqrt{\nu}} p$, $g = f - \frac{i}{\sqrt{\gamma}} w_d$. Then

$$\begin{aligned} -\Delta z + k z &= -\Delta w + \frac{i}{\sqrt{\nu}}(-\Delta p - w) + \frac{1}{\nu} p \\ &= f - \frac{i}{\sqrt{\nu}} w_d = g, \end{aligned}$$

which is equal to

$$\begin{aligned} -\Delta w + \frac{1}{\nu} p = f, \quad -\Delta p = w - w_d \quad \text{in} \quad \Omega \\ w = 0, \quad p = 0 \quad \text{on} \quad \partial\Omega \end{aligned}$$

that is, to the original optimality system .

We go back to the domain decomposition procedure (4.2.3.3), (4.2.3.4) and replace the complex quantities according to our setting above. We obtain

$$\begin{cases} -\Delta w_i^n + \frac{1}{\nu} p_i^n = f, \quad -\Delta p_i^n = w_i^n - w_d \quad \text{in} \quad \Omega_i \\ \frac{\partial}{\partial \nu_j} w_i^n + \lambda w_i^n - \frac{\mu}{\sqrt{\nu}} p_i^n = g_i^n \quad \text{on} \quad \Gamma \\ \frac{\partial}{\partial \nu_i} p_i^n + \lambda p_i^n + \mu\sqrt{\nu}\, w_i^n = h_i^n \quad \text{on} \quad \Gamma \\ w_i = 0 = p_i \quad \text{on} \quad \partial\Omega \cap \partial\Omega, \end{cases} \tag{4.2.5.1}$$

together with the updates

$$\begin{cases} g_i^{n+1} = 2\,(\lambda w_j^n - \frac{\mu}{\sqrt{\nu}} p_j^n) - g_j^n \\ h_i^{n+1} = 2\,(\lambda p_j^n + \mu\sqrt{\nu}\, w_j^n) - h_j^n. \end{cases} \tag{4.2.5.2}$$

The iteration (4.2.5.1) and (4.2.5.2) contains, besides $\nu > 0$, the parameters λ, μ which have to satisfy

$$\lambda \geq 0,\ \mu \leq 0,\ \lambda - \mu > 0. \tag{4.2.5.3}$$

The canonical choice in the context under consideration is $\lambda = 0$, $\mu = -\delta \in \mathbb{R}$, $\nu = 1$ and results in the iteration

$$\begin{cases} -\Delta w_i^n + p_i^n = f \quad \text{in} \quad \Omega_i \\ -\Delta p_i^n = w_i^n - w_d \quad \text{in} \quad \Omega_i \\ w_i = p_i = 0 \quad \text{on} \quad \partial\Omega_i \cap \partial\Omega \end{cases} \tag{4.2.5.4}$$

$$\begin{cases} \frac{\partial w_i^n}{\partial \nu_i} + \delta\, p_i^n = g_i^n \\ \frac{\partial p_i^n}{\partial \nu_i} - \delta\, w_i^n = h_i^n \end{cases} \tag{4.2.5.5}$$

$$\begin{cases} g_i^{n+1} = 2\delta\, p_j^n - g_j^n \\ h_i^{n+1} = -2\delta\, w_j - h_j^n . \end{cases} \tag{4.2.5.6}$$

Of course (4.2.5.6) can equivalently be expressed as

$$\begin{cases} g_i^n = -\dfrac{\partial w_j^{n+1}}{\partial \nu_j} + \delta\, p_j^{n-1} \\ h_i^n = -\dfrac{\partial p_j^{n-1}}{\partial \nu_j} + \delta\, w_j^{n-1} \end{cases} \tag{4.2.5.7}$$

4.2.6 An A Posteriori Error Estimate

We consider (4.2.3.1) and its decomposed counterpart (4.2.3.3) and introduce the error

$$\tilde{z}_i^n := z_i^n - z_i \tag{4.2.6.1}$$

as usual. Then z_i^n satisfies

$$\begin{aligned} -\Delta\, \tilde{z}_i^n + k\, z_i^n &= 0 \ \text{ in } \ \Omega_i \\ \frac{\partial \tilde{z}_i^n}{\partial \nu_i} + \beta\, \tilde{z}_i^n &= \tilde{g}_i^n \ \text{ on } \ \Gamma \\ \tilde{z}_i^n &= 0 \ \text{ on } \ \partial\Omega_i \cap \partial\Omega \end{aligned} \tag{4.2.6.2}$$

$$\tilde{g}_i^{n+1} = 2\beta\, \tilde{z}_j^n - \tilde{g}_j^n \ \text{ on } \ \Gamma. \tag{4.2.6.3}$$

Then with

$$\begin{aligned} a_i(z_i, v_i) &:= \int_{\Omega_j} (\nabla\, z_i \cdot \overline{\nabla v_i} + k\, z_i \bar{v}_i)\, dx \\ \langle w_i, v_i \rangle_\Gamma &:= \int_\Gamma w_i\, \bar{v}_i\, d\gamma \end{aligned} \tag{4.2.6.4}$$

We obtain the variational form of (4.2.6.2)

$$a_i(\tilde{z}_i^n, v_i) = \langle \tilde{g}_i^n - \beta\, \tilde{z}_i^n, v_i \rangle_\Gamma, \ \ \forall\, v_i \in V_i, \tag{4.2.6.5}$$

together with the update (4.2.6.3), that is,

$$\begin{aligned} a_i(\tilde{z}_i^n, v_i) + \beta\, \langle \tilde{z}_i, v_i \rangle_\Gamma &= \langle \tilde{g}_i^n, v_i \rangle_\Gamma \ \ , \forall\, v_i \in V_i, \\ \langle\, \tilde{g}_i^{n+1}, \phi \rangle_\Gamma &= 2\beta \langle \tilde{z}_j^n - \tilde{g}_j^n, \phi \rangle_\Gamma \ \ , \forall\, \phi \in W, \end{aligned} \tag{4.2.6.6}$$

where, for simplicity, we have assumed that $\beta_i = \beta_j \equiv \beta \in \mathbb{C}$. Using the notation of Section 2.3.6, we obtain

$$\langle \tilde{g}_i^n - \beta\, \tilde{z}_i^n, \phi \rangle_\Gamma \leq C_i\, \| tr_i^{-1} \|\, \| \tilde{z}_i^n \|_{V_i}\, \|\phi\|_W \tag{4.2.6.7}$$

and

$$\beta\,\langle \tilde z_i^n, \phi\rangle_\Gamma \le |\beta|\,\tilde C_i\,\|\tilde z_i^n\|_{V_i}\,\|\phi\|_W. \tag{4.2.6.8}$$

Notice that

$$a_i(\tilde z_i^n, \tilde z_i^n) = -\int_\Gamma \beta\,|\tilde z_i^n|^2\,d\gamma + \int_\Gamma \tilde g_i^n\,\tilde z_i^n\,d\gamma \tag{4.2.6.9}$$

and thus, as $\mathrm{Re}(k)=0$,

$$\mathrm{Im}(k)\int_\Omega |\tilde z_k^n|^2\,dx + \mathrm{Im}(\beta)\int_\Gamma |\tilde z_i^n|^2\,d\gamma = \mathrm{Im}\int_\Gamma \tilde g_i^n\,\tilde z_i^n\,d\gamma, \tag{4.2.6.10}$$

$$\int_{\Omega_i} |\nabla\,\tilde z_i^n|^2 + \mathrm{Re}(\beta)\int_\Gamma |\tilde z_i^n|^2\,d\gamma = \mathrm{Re}\int_\Gamma \tilde g_i^n\,\tilde z_i^n\,d\gamma. \tag{4.2.6.11}$$

We have

$$\begin{aligned}
&Re\,a_1(\tilde z_1^{n+1}, v_1) + Re\,a_2(\tilde z_i^n, v_2)\\
&\quad= Re\,\langle \tilde g_1^{n+1} - \beta\,\tilde z_1^{n+1}, v_1\rangle_\Gamma\\
&\qquad+ Re\,\langle \tilde g_2^n - \beta\,\tilde z_2^n, v_2\rangle_\Gamma\\
&\quad= Re\,\langle \tilde g_1^{n+1} - \beta\,\tilde z_1^{n+1}, v_1 - v_2\rangle_\Gamma\\
&\qquad+ Re\,(\langle \tilde g_1^{n+1} - \beta\,\tilde z_1^{n+1}, v_2\rangle_\Gamma + \langle \tilde g_2^n - \beta\,\tilde z_2^n, v_2\rangle_\Gamma)\\
&\quad= Re\,\langle \tilde g_1^{n+1} - \beta\,\tilde z_1^{n+1}, v_1 - v_2\rangle_\Gamma\\
&\qquad+ Re\,\langle \tilde g_1^{n+1} + \tilde g_2^n - 2\beta\,\tilde z_2^n + \beta\,\tilde z_2^n - \beta\,\tilde z_1^{n+1}, v_2\rangle_\Gamma\\
&\quad= Re\,\langle \tilde g_1^{n+1} - \beta\,\tilde z_1^{n+1}, v_1 - v_2\rangle_\Gamma\\
&\qquad+ Re\,\langle \beta\,(\tilde z_2^n - \tilde z_1^{n+1}), v_2\rangle_\Gamma
\end{aligned} \tag{4.2.6.12}$$

according to (4.2.6.6). Set $v_1 := \tilde z_1^{n+1}$, $v_2 := \tilde z_2^n$ and insert these into (4.2.6.12) to obtain, according to (4.2.6.7) and (4.2.6.8),

$$\begin{aligned}
&\mathrm{Re}\,a_1(\tilde z_1^{n+1}, \tilde z_1^{n+1}) + \mathrm{Re}\,a_2(\tilde z_2^n, \tilde z_2^n)\\
&\le C_1\,\|tr_1^{-1}\|\,\|\tilde z_1^{n+1}\|_{V_1}\,\|z_1^{n+1} - z_2^n\|_\Gamma + |\beta|\,\tilde C_2\,\|\tilde z_2^n\|_{V_2}\,\|z_1^{n+1} - z_2^n\|_\Gamma.
\end{aligned} \tag{4.2.6.13}$$

Obviously, as $k \in \mathbb{C}\backslash\mathbb{R}$ and because of the boundary conditions on $\partial\Omega$ we have

$$\hat c_i\,\|z_i\|_{V_i}^2 \le \mathrm{Re}\,a_i(z_i, z_i) = \int_{\Omega_i} |\nabla\,z_i|^2\,dx \le \hat C_i\|z_i\|_{V_i}^2.$$

Thus, the estimate (4.2.6.13) gives

$$\|\tilde z_1^{n+1}\|_{V_1} + \|\tilde z_2^n\|_{V_2} \le C(\beta)\,\|z_1^{n+1} - z_2^n\|_\Gamma, \tag{4.2.6.14}$$

where now $C(\beta)$ is a generic constant depending on β. By changing the order of the domains we may write (4.2.6.14) in a more symmetric way:

$$\|\tilde{z}_1^{n+1}\|_{V_1} + \|\tilde{z}_1^n\|_{V_1} + \|\tilde{z}_2^{n+1}\|_{V_2} + \|\tilde{z}_2^n\|_{V_2} \\ \leq C(\beta)\{\|z_1^{n+1} - z_2^n\|_\Gamma + \|z_2^{n+1} - z_2^n\|_\Gamma) \quad (4.2.6.15)$$

By defining the total error as

$$e^n := \|\tilde{z}_1^n\|_{V_1} + \|\tilde{z}_2^n\|_{V_2} \quad (4.2.6.16)$$

and the mismatch along Γ as

$$\mathcal{E}_{ij}^{n,n+1} := \|z_i^{n+1} - z_j^n\|_\Gamma^2 \quad (4.2.6.17)$$

we obtain the a posteriori error estimate

$$e^{n,n+1} := e^n + e^{n+1} \leq C(\beta)\,(\mathcal{E}_{12}^{n,n+1} + \mathcal{E}_{21}^{n,n+1})^{\frac{1}{2}}. \quad (4.2.6.18)$$

Let us translate the estimate (4.2.6.18) into the notation of the original problem (4.2.1.2). Recall $z = w + \frac{i}{\sqrt{\nu}}\,p$, where w, p are real-valued functions. Then, of course,

$$\begin{aligned} \|\tilde{z}_i^n\|_{V_i}^2 &= \int_{\Omega_i} \nabla\,\tilde{z}_i^n\,\overline{\nabla\,\tilde{z}_i^n} + \tilde{z}_i^n\,\bar{\tilde{z}}_i^n\,dx \\ &= \int_{\Omega_i} |\nabla\,\tilde{w}_i^n|^2\,dx + \frac{1}{\nu}\int_{\Omega_i} |\nabla\,\tilde{p}_i^n|^2\,dx + \int_{\Omega_i} |\tilde{w}_i^n|^2\,dx + \frac{1}{\nu}\int_{\Omega_i} \|\tilde{p}_i^n\|^2\,dx \\ &= \|\tilde{w}_i^n\|_{V_i}^2 + \frac{1}{\nu}\,\|\tilde{p}_i^n\|_{V_i}^2. \end{aligned}$$

Thus $\|\tilde{z}_i^n\|_{V_i} \geq \frac{1}{\sqrt{2}}\,(\|\tilde{w}_i^n\|_{V_i} + \frac{1}{\sqrt{\nu}}\,\|\tilde{p}_i^n\|_{V_i})$. Hence with the redefinition of $\mathbf{e}_n$ as

$$e^n := \|\tilde{w}_1^n\|_{V_1} + \frac{1}{\sqrt{\nu}}\,\|\tilde{p}_1^n\|_{V_1} + \|\tilde{w}_2^n\|_{V_2} + \frac{1}{\sqrt{\nu}}\,\|\tilde{p}_2^n\|_{V_2} \quad (4.2.6.19)$$

and

$$\mathcal{E}_{ij}^{n,n+1} = \|w_i^{n+1} - w_j^n\|_\Gamma^2 + \frac{1}{\nu}\,\|p_i^{n+1} - p_j^n\|_\Gamma^2 \quad (4.2.6.20)$$

we obtain the following theorem.

Theorem 4.2.6.1 *Let the error measure be defined by* (4.2.6.19) *and the mismatch of the consecutive iterates along* Γ *be given by* (4.2.6.20). *Then, there is a constant* C *depending only on the geometry of* Ω_1, Ω_2 *and* $\beta > 0$ *along* Γ, *such that*

$$e^{n,n+1} \leq C\{\mathcal{E}_{12}^{n,n+1} + \mathcal{E}_{21}^{n,n+1}\}^{1/2}. \quad (4.2.6.21)$$

4.3 Boundary Controls

Let us consider a standard elliptic problem again, but now with a control acting through the boundary on Ω:

$$\begin{gathered}\inf_{f\in L^2(\Gamma_1)} \frac{1}{2}\int\limits_{\Gamma_1} f^2\, d\gamma + \frac{\kappa}{2}\|w-w_d\|^2 \\ \text{subject to} \\ -\Delta w = F \quad \text{in} \quad \Omega \\ w=0 \quad \text{on} \quad \Gamma_0, \quad \frac{\partial w}{\partial \nu} = f \quad \text{on} \quad \Gamma_1,\end{gathered} \tag{4.3.1.1}$$

where $\partial\Omega = \Gamma_0 \dot{\cup} \Gamma_1$, $\text{meas}(\Gamma_1) > 0$. The optimality condition for (4.3.1.1) is given by

$$f = -p \quad \text{on} \quad \Gamma_1, \tag{4.3.1.2}$$

where p solves the adjoint problem

$$\begin{gathered}-\Delta p = \kappa(w - w_d) \quad \text{in} \quad \Omega \\ p = 0 \quad \text{on} \quad \Gamma_0, \quad \frac{\partial p}{\partial \nu} = 0 \quad \text{on} \quad \Gamma_1.\end{gathered} \tag{4.3.1.3}$$

The complete optimality system is expressed as a system of coupled problems:

$$\begin{gathered}-\Delta w = F, \\ -\Delta p = \kappa(w - w_d) \quad \text{in} \quad \Omega, \\ w = 0 = p \quad \text{on} \quad \Gamma_0 \\ \frac{\partial w}{\partial \nu} + p = 0, \quad \frac{\partial p}{\partial \nu} = 0 \quad \text{on} \quad \Gamma_1.\end{gathered} \tag{4.3.1.4}$$

4.3.2 Domain Decomposition

As the coupling in (4.3.1.4) is both on the boundary, for the adjoint variable, and in the interior for the direct variable, a reformulation as a complex problem as in (4.2.3.1) is not possible. Still, the analysis in the last section leading to the decomposition method (4.2.5.1), (4.2.5.2) suggests the following procedure.

$$\begin{gathered}-\Delta w_i^n = F_i, \quad -\Delta p_i = \kappa(w_i^n - w_{di}) \quad \text{in} \quad \Omega_i \\ \frac{\partial}{\partial \nu_i} w_i^n + \lambda w_i^n + \mu p_i^n = g_i^n \quad \text{on} \quad \Gamma \\ \frac{\partial}{\partial \nu_i} p_i^n + \lambda p_i^n - \mu w_i^n = h_i^n \quad \text{on} \quad \Gamma \\ w_i^n = 0 = p_i^n \quad \text{on} \quad \partial\Omega_i \cap \Gamma_0 =: \Gamma_{0,i}, \\ \frac{\partial w^n}{\partial \nu_i} = -p_i^n, \ \frac{\partial p_i^n}{\partial \nu_i} = 0 \quad \text{on} \quad \partial\Omega_i \cap \Gamma_1 =: \Gamma_{1,i}\end{gathered} \tag{4.3.2.1}$$

together with the updates

$$
\begin{aligned}
g_i^{n+1} &= 2\,(\lambda\, w_j^n + \mu\, p_j^n) - g_j^n \\
h_i^{n+1} &= 2\,(\lambda\, p_j^n - \mu\, w_j^n) - h_j^n.
\end{aligned}
\tag{4.3.2.2}
$$

For simplicity we assume that $\text{meas}(\Gamma_{0,i}) > 0$ and $\text{meas}(\Gamma_{1,i}) > 0$. As always, (4.3.2.2) can be underrelaxed by the introduction of a parameter $\varepsilon \in [0,1)$ such that

$$
\begin{aligned}
g_i^{n+1} &= (1-\varepsilon)[2(\lambda\, w_j^n + \mu\, p_j^n) - g_j^n] + \varepsilon\, g_i^n \\
h_i^{n+1} &= (1-\varepsilon)[2(\lambda\, p_j^n - \mu\, w_j^n) - h_j^n] + \varepsilon h_j^n.
\end{aligned}
\tag{4.3.2.3}
$$

4.3.3 Convergence

For the sake of clarity we do not consider a relaxation in this section. Relaxation will become more essential for dynamic problems. We proceed much like in the last section. We introduce the local errors $\tilde{w}_i = w_i - w|_i$, $\tilde{p}_i = p_i - p|_i$, which solve (4.3.2.1), (4.3.2.2) with the superscripts removed and with $F = 0 = w_d$ and $(\tilde{g}_1, \tilde{g}_2, \tilde{h}_1, \tilde{h}_2) \in L^2(\Gamma)^4 =: \mathcal{X}$. We set

$$
\|X\|_{\mathcal{X}}^2 = \sum_{i=1}^{4} \int_{\Gamma} |X_i|^2 \, d\gamma. \tag{4.3.3.1}
$$

and introduce the map $\mathcal{T} : \mathcal{X} \mapsto \mathcal{X}$ defined by

$$
\begin{aligned}
\mathcal{T}(X) = (2(\lambda\, \tilde{w}_2 + \mu\, \tilde{p}_2) - \tilde{g}_2,\ &2(\lambda\, \tilde{w}_1 + \mu\, \tilde{p}_1) - \tilde{g}_1, \\
&2(\lambda\, \tilde{p}_2 - \mu\, \tilde{w}_2) - h_2,\ 2(\lambda\, \tilde{p}_1 - \mu\, \tilde{w}_1) - \tilde{h}_1).
\end{aligned}
\tag{4.3.3.2}
$$

A fixed point of $\mathcal{T}$ corresponds in a one-to-one way to a solution of (4.3.1.4) with $F = w_d = 0$. Hence $\mathcal{T}$ has zero as its only fixed point.

We first want to see under which conditions $\mathcal{T}$ is a nonexpansive map. We have

$$
\begin{aligned}
\|\mathcal{T}X\|_{\mathcal{X}}^2 &= \sum_{i=1}^{2} \int_{\Gamma} |2(\lambda\, \tilde{w}_i + \mu\, \tilde{p}_i) - \tilde{g}_i|^2 \, d\gamma + \sum_{i=1}^{2} \int_{\Gamma} |2(\lambda\, \tilde{p}_i - \mu\, \tilde{w}_i) - \tilde{h}_i|^2 \, d\gamma \\
&= \sum_{i=1}^{2} \int_{\Gamma} [4(\lambda\, \tilde{w}_i + \mu\, \tilde{p}_i)^2 - 4(\lambda\, \tilde{w}_i + \mu\, \tilde{p}_i)\, \tilde{g}_i + \tilde{g}_i^2] \, d\gamma \\
&\quad + \sum_{i=1}^{2} \int_{\Gamma} [4(\gamma\, \tilde{p}_i - \mu\, \tilde{w}_i)^2 - 4(\gamma\, \tilde{p}_i - \mu\, \tilde{w}_i)\, \tilde{h}_i + \tilde{h}_i^2] \, d\gamma.
\end{aligned}
$$

This may be written

$$\begin{aligned}
\|\mathcal{T}X\|_{\mathcal{X}}^2 =&\|X\|_{\mathcal{X}}^2 + \sum_{i=1}^{2} \int_{\Gamma} 4(\lambda\,\tilde{w}_i + \mu\,\tilde{p}_i)\{(\lambda\,\tilde{w}_i + \mu\,\tilde{p}_i) - \tilde{g}_i\}\,d\gamma \\
&+ \sum_{i=1}^{2} \int_{\Gamma} 4(\lambda\,\tilde{w}_i - \mu\,\tilde{p}_i)\{(\lambda\,\tilde{p}_i - \mu\,\tilde{w}_i) - \tilde{h}_i\}\,d\gamma \\
=&\|X\|_{\mathcal{X}}^2 - 4\sum_{i=1}^{2} \int_{\Gamma} (\lambda\,\tilde{w}_i + \mu\,\tilde{p}_i)\,\frac{\partial}{\partial\nu}\,\tilde{w}_i\,d\gamma \\
&- 4\sum_{i=1}^{2} \int_{\Gamma} (\lambda\,\tilde{p}_i - \mu\,\tilde{w}_i)\,\frac{\partial}{\partial\nu}\,\tilde{p}_i\,d\gamma.
\end{aligned} \tag{4.3.3.3}$$

We then multiply the homogeneous state equations in $(4.3.2.1)_1$ by $\tilde{w}_i$ and $\tilde{p}_i$ and integrate over Ω_i to obtain

$$0 = \int_{\Omega_i} \Delta\,\tilde{w}_i\,\tilde{w}_i\,dx = \int_{\Gamma} \frac{\partial\tilde{w}_i}{\partial\nu_i}\,\tilde{w}_i\,d\gamma - \int_{\Gamma_{1,i}} \tilde{p}_i\,\tilde{w}_i\,d\gamma - \int_{\Omega_i} \nabla\,\tilde{w}_i^2\,dx$$

$$0 = \int_{\Omega_i} (\Delta\,\tilde{p}_i + \kappa\,\tilde{w}_i)\,\tilde{p}_i = \int_{\Gamma} \frac{\partial\tilde{p}_i}{\partial\nu_i}\,\tilde{p}_i + \kappa \int_{\Omega_i} \tilde{w}_i\,\tilde{p}_i\,dx - \int_{\Omega_i} |\nabla\,\tilde{p}_i|^2\,dx$$

and

$$0 = \int_{\Omega_i} \Delta\,\tilde{w}_i\,\tilde{p}_i\,dx = \int_{\Gamma} \frac{\partial\tilde{w}_i}{\partial\nu_i}\,\tilde{p}_i\,d\gamma - \int_{\Gamma_{1,1}} \tilde{p}_i^2\,d\gamma - \int_{\Omega_i} \nabla\,\tilde{w}_i\,\nabla\,\tilde{p}_i\,dx$$

$$0 = \int_{\Omega_i} (\Delta\,\tilde{p}_i + \kappa\,\tilde{w}_i)\,\tilde{w}_i\,dx = \int_{\Gamma} \frac{\partial\tilde{p}_i}{\partial\nu_i}\,\tilde{w}_i\,d\gamma + \kappa \int_{\Omega_i} \tilde{w}_i^2\,dx - \int_{\Omega_i} \nabla\,\tilde{p}_i\,\nabla\,\tilde{w}_i\,dx.$$

Thus,

$$\begin{aligned}
\int_{\Gamma} (\lambda\tilde{w}_i + \mu\tilde{p}_i)\,\frac{\partial}{\partial\nu}\,\tilde{w}_i\,d\gamma = \lambda \int_{\Omega_i} |\nabla\,\tilde{w}_i|^2\,dx + \mu \int_{\Omega_i} \nabla\,\tilde{w}_i\,\nabla\tilde{p}_i\,dx \\
+ \lambda \int_{\Gamma_{1,i}} \tilde{p}_i\,\tilde{w}_i\,d\gamma + \mu \int_{\Gamma_{1,i}} \tilde{p}_i^2\,d\gamma
\end{aligned} \tag{4.3.3.4}$$

$$\begin{aligned}
\int_{\Gamma} (\lambda\tilde{p}_i - \mu\tilde{w}_i)\,\frac{\partial}{\partial\nu}\,\tilde{p}_i\,d\gamma = \lambda \int_{\Omega_i} |\nabla\,\tilde{p}_i|^2\,dx - \lambda\,\kappa \int_{\Omega} \tilde{w}_i\,\tilde{p}_i\,dx \\
+ \mu\kappa \int_{\Omega_i} \tilde{w}_i^2\,dx - \mu \int_{\Omega_i} \nabla\,\tilde{p}_i\,\nabla\,\tilde{w}_i\,dx.
\end{aligned} \tag{4.3.3.5}$$

If we insert (4.3.3.4) and (4.3.3.5) into (4.3.3.3) we obtain

$$\begin{aligned}\|\mathcal{T}X\|_{\mathcal{X}}^2 = \|X\|_{\mathcal{X}}^2 - 4\sum_{i=1}^{2}\lambda\int_{\Omega_i}\{|\nabla\tilde{w}_i|^2+|\nabla\tilde{p}_i|^2\}\,dx \\ -4\sum_{i=1}^{2}(\mu\,\kappa\int_{\Omega_i}\tilde{w}_i^2\,dx-\lambda\,\kappa\int_{\Omega_i}\tilde{w}_i\,\tilde{p}_i\,dx) \\ -4\sum_{i=1}^{2}(\lambda\int_{\Gamma_{1,i}}\tilde{p}_i\,\tilde{w}_i\,d\gamma+\mu\int_{\Gamma_{1,i}}\tilde{p}_i^2\,d\gamma)\end{aligned} \qquad (4.3.3.6)$$

We discuss three cases:

i) $\lambda = 0,\ \mu > 0$

ii) $\lambda > 0,\ \mu = 0$

iii) $\lambda > 0,\ \mu > 0$

i) In this case (4.3.3.6) reduces to

$$\|\mathcal{T}X\|_{\mathcal{X}}^2 = \|X\|_{\mathcal{X}}^2 - 4\sum_{i=1}^{2}\{\mu\,\kappa\int_{\Omega_i}\|\tilde{w}_i\|^2\,dx+\mu\int_{\Gamma_1}\tilde{p}_i^2\,d\gamma\}. \qquad (4.3.3.7)$$

Thus, the iteration of errors

$$X^{n+1} = \mathcal{T}(X^n)$$

satisfies

$$\|X^{n+1}\|_{\mathcal{X}}^2 = \|X^n\|_{\mathcal{X}}^2 - 4\sum_{i=1}^{2}\{\mu\,\kappa\int_{\Omega_i}\|\tilde{w}_i^n\|^2\,dx+\mu\int_{\Gamma_{1,i}}|\tilde{p}_i^n|^2\,d\gamma\} \qquad (4.3.3.8)$$

Iterating (4.3.3.8) to zero, we can conclude that

$$(X^n)_{n\in\mathbb{N}} \quad \text{is bounded} \qquad (4.3.3.9)$$

$$\begin{aligned}\tilde{w}_i^n &\to 0 \quad \text{strongly in} \quad L^2(\Omega_i) \\ \tilde{p}_i^n\big|_{\Gamma_{1,i}} &\to 0 \quad \text{strongly in} \quad L^2(\Gamma_{1,i}).\end{aligned} \qquad (4.3.3.10)$$

Hence $\partial\tilde{w}^n/\partial\nu_i \to 0$ strongly in $L^2(\Gamma_{1,i})$ and $\tilde{p}_i^n \to 0$ strongly in $L^2(\Omega_i)$, since the limit $\tilde{p}_i^n$ solves the Laplace equation with Cauchy data on $\Gamma_{1,i}$.

According to (4.3.3.9) and (4.3.2.2) and $\lambda = 0$, $\tilde{p}_i^n$, $\tilde{w}_i^n$ are bounded in $L^2(\Gamma_{1,i})$. But

$$\int_{\Omega_i} |\nabla \tilde{p}_i^n|^2\, dx = \kappa \int_{\Omega_i} \tilde{w}_i^n\, \tilde{p}_i^n\, dx + \mu \int_{\Gamma} \tilde{w}_i^n\, \tilde{p}_i^n\, d\gamma + \int_{\Gamma} \tilde{h}_i^n\, \tilde{p}_i^n\, d\gamma.$$

Hence, $\tilde{p}_i^n$ is bounded in $H^1(\Omega_i)$. This implies $\tilde{p}_i^n \to 0$ strongly in $L^2(\Gamma)$ and $\tilde{p}_i^n \to 0$ in $H^1(\Omega_i)$. Then, because

$$\int_{\Omega_i} |\nabla \tilde{w}_i^n|^2\, dx = \int_{\Gamma} (\tilde{g}_i^n - \mu\, \tilde{p}_i^n)\, \tilde{w}_i^n\, d\gamma - \int_{\Gamma_1} \tilde{p}_i^n\, \tilde{w}_i^n\, d\gamma,$$

we conclude that also $\tilde{w}_i^n$ is bounded in $H^1(\Omega_i)$ and hence $\tilde{w}_i^n \to 0$ in $H^1_{\Gamma_{i0}}(\Omega_i)$.

ii) In this case $\mu = 0$, $\lambda > 0$ and (4.3.3.6) reduces to

$$\begin{aligned} \|\mathcal{T}X\|_{\mathcal{X}}^2 = \|X\|_{\mathcal{X}}^2 - 4\,\lambda \sum_{i=1}^{2} \int_{\Omega_i} (|\nabla\, \tilde{w}_i|^2 + |\nabla\, \tilde{p}_i|^2)\, dx \\ -4\lambda \sum_{i=1}^{2} \{-\kappa \int_{\Omega_i} \tilde{w}_i\, \tilde{p}_i\, dx + \int_{\Gamma_i} \tilde{p}_i\, \tilde{w}_i\, d\gamma\}. \end{aligned} \tag{4.3.3.11}$$

The convergence in the case thus depends on the geometry and the choice of κ. There is no general convergence result that we can infer from this analysis.

iii) In the case $\lambda > 0$, $\mu > 0$ we write (4.3.3.6) as follows

$$\begin{aligned} \|\mathcal{T}X\|_{\mathcal{X}}^2 =& \|X\|_{\mathcal{X}}^2 - 4\,\lambda \sum_{i=1}^{2} \{ \int_{\Omega_i} [|\nabla\, \tilde{w}_i|^2 + |\nabla\, \tilde{p}_i|^2 \\ &+ \frac{\mu}{\lambda}\, \kappa\, \tilde{w}_i^2 - \kappa\, \tilde{w}_i\, \tilde{p}_i]\, dx + \int_{\Gamma_1} [\tilde{p}_i\, \tilde{w}_i + \frac{\mu}{\lambda}\, \tilde{p}_i^2]\, d\gamma\} \\ \le& \|X\|_{\mathcal{X}}^2 - 4\,\lambda \sum_{i=1}^{2} \{ \int_{\Omega_i} [|\nabla\, \tilde{w}_i|^2 + |\nabla\, \tilde{p}_i|^2 \\ &+ \frac{\mu}{\lambda}\, \kappa\, \tilde{w}_i^2 - \frac{\kappa}{2\varepsilon}\, \tilde{w}_i^2 - \frac{\kappa}{2}\, \varepsilon\, \tilde{p}_i^2]\, dx + \int_{\Gamma_1} [-\frac{1}{\delta}\, \tilde{p}_i^2 - \frac{\delta}{2}\, \tilde{w}_i^2 + \frac{\mu}{\lambda}\, \tilde{p}_i^2]\, d\gamma\}. \end{aligned} \tag{4.3.3.12}$$

Hence, for $\frac{\mu}{\lambda}$ sufficiently large we compensate the terms $\int_{\Omega_i} \frac{\kappa}{2}\, \varepsilon\, \tilde{w}_i^2\, dx$ and $\int_{\Gamma_1} \frac{\delta}{2}\, \tilde{w}_i^2\, d\gamma$ by the Poincaré inequality. We may, therefore, for $\frac{\mu}{\lambda}$ sufficiently large, conclude that

$$\tilde{w}_i^n \to 0 \quad \text{strongly in} \quad H^1(\Omega_i)$$
$$\tilde{p}_i^n \to 0 \quad \text{strongly in} \quad H^1(\Omega_i).$$

The condition on $\frac{\mu}{\lambda}$ can be made explicit in terms of the geometry of Ω, and the penalty term $\kappa > 0$.

Theorem 4.3.3.1 *For $\lambda = 0$ and $\mu > 0$, and $\lambda > 0$ and $\mu > 0$, where in the latter case μ/λ is sufficiently large, the errors of iterative scheme converge strongly to zero in the H^1-sense.*

4.3.4 An A Posteriori Error Estimate

We consider the errors $\tilde{w}_i$, $\tilde{p}_i$ as in the preceding section. They satisfy the optimality system (4.3.1.2) with $F_i = 0$, $w_{d_i} = 0$ and $\tilde{g}_i^n$, $\tilde{h}_i^n$ instead of g_i^n, h_i^n. For the sake of simplicity we consider the standard case $\lambda = 0$. Thus,

$$\begin{aligned}
&-\Delta \tilde{w}_i^n = 0, \quad -\Delta \tilde{p}_i^n = \kappa \tilde{w}_i \;\text{ in }\; \Omega_i \\
&\frac{\partial \tilde{w}_i^{n+1}}{\partial \nu_i} + \mu \tilde{p}_i^{n+1} = \tilde{g}_i^n \;\text{ on }\; \Gamma \\
&\frac{\partial \tilde{p}_i^{n+1}}{\partial \nu_i} - \mu \tilde{w}_i^{n+1} = \tilde{h}_i^n \;\text{ on }\; \Gamma \\
&\tilde{w}_i^n = 0 = \tilde{p}_i^n \;\text{ on }\; \Gamma_0 \cap \partial\Omega =: \Gamma_{0,i} \\
&\frac{\partial \tilde{w}_i^n}{\partial \nu_i} = -\tilde{p}_i^n, \quad \frac{\partial \tilde{p}_i^n}{\partial \nu_i} = 0 \;\text{ on }\; \Gamma_1 \cap \partial\Omega_i =: \Gamma_{1,i}
\end{aligned} \tag{4.3.4.1}$$

together with

$$\tilde{g}_i^{n+1} = 2\mu \tilde{p}_j^n - \tilde{g}_j^n, \quad \tilde{h}_i^{n+1} = -2\mu \tilde{w}_j^n - h_j^n \tag{4.3.4.2}$$

The variational form of (4.3.4.1) is obtained as usual:

$$\begin{aligned}
0 &= -\int_{\Omega_i} \Delta \tilde{w}_i^n v_i \, dx = -\int_{\partial\Omega_i} \frac{\partial}{\partial \nu_i} \tilde{w}_i^n v_i \, d\gamma + \int_{\Omega_i} \nabla \tilde{w}_i^n \cdot \nabla v_i \, dx \\
&= \int_{\Gamma_{1,i}} \tilde{p}_i^n v_i \, d\gamma + \int_{\Gamma} (\mu \tilde{p}_i^n - \tilde{g}_i^n) v_i \, d\gamma + \int_{\Omega_i} \nabla \tilde{w}_i^n \cdot \nabla v_i \, dx.
\end{aligned} \tag{4.3.4.3}$$

Hence with

$$a_i(w_i, v_i) := \int_{\Omega_i} \nabla w_i \cdot \nabla v_i \, dx, \qquad \langle w_i, v_i \rangle_\Gamma := \int_\Gamma w_i \cdot v_i \, d\gamma,$$

equation (4.3.4.3) has the form

$$a_i(\tilde{w}_i^n, v_i) + \langle \tilde{p}_i^n, v_i\rangle_{\Gamma_{1,i}} + \langle \mu\, \tilde{p}_i^n, v_i\rangle_\Gamma = \langle \tilde{g}_i^n, v_i\rangle, \quad \forall\, v_i \in V_i. \tag{4.3.4.4}$$

Accordingly

$$\int\limits_{\Omega_i} \kappa\, \tilde{w}_i^n\, v_i\, dx = -\int\limits_{\Omega_i} \Delta\, \tilde{p}_i^n\, v_i\, dx = \int\limits_{\Gamma} (-\mu\, \tilde{w}_i^n - \tilde{h}_i^n)\, v_i\, d\gamma + \int\limits_{\Omega_i} \nabla\, \tilde{p}_i^n\, \nabla\, v_i\, dx$$

or

$$a_i(\tilde{p}_i^n, v_i) - \langle \mu\, \tilde{w}_i^n, v_i\rangle_\Gamma - \kappa(\tilde{w}_i^n, v_i) = \langle \tilde{h}_i^n, v_i\rangle_\Gamma, \quad \forall\, v_i \in V_i. \tag{4.3.4.5}$$

With $W = V_i|_\Gamma$, the space of traces on Γ of functions in V_i, we have from (4.3.4.2):

$$\begin{aligned} \langle \tilde{g}_i^{n+1}, \phi\rangle_\Gamma &= 2\,\langle \mu \tilde{p}_j^n, \phi\rangle_\Gamma - \langle \tilde{g}_j^n, \phi\rangle_\Gamma, \quad \forall\, \phi \in W \\ \langle \tilde{h}_i^{n+1}, \phi\rangle_\Gamma &= -2\,\langle \mu \tilde{w}_j^n, \phi\rangle_\Gamma - \langle \tilde{h}_j^n, \phi\rangle, \quad \forall\, \phi \in W. \end{aligned} \tag{4.3.4.6}$$

We use the notation of Section 2.3.6. Thus, $v_i = tr_i^{-1}\, \phi$ for some $\phi \in W$. Therefore, (4.3.4.4), (4.3.4.5) can be expressed as

$$\begin{aligned} a_i(\tilde{p}_i^n, tr_i^{-1}\, \phi) - \langle \mu, \tilde{w}_i^n,\, \phi\rangle_\Gamma - \kappa(\tilde{w}_i^n, tr_i^{-1}\, \phi) &= \langle \tilde{h}_i^n, \phi\rangle_\Gamma, \quad \forall\, \phi \in W \\ a_i(\tilde{w}_i^n, tr_i^{-1}\, \phi) + \langle \tilde{p}_i^n,\, tr|_\Gamma, tr_i^{-1}\phi\rangle_{\Gamma_{1,i}} + \langle \mu\, \tilde{p}_i^n, \phi\rangle_\Gamma & \\ &= \langle \tilde{g}_i^n, \phi\rangle_\Gamma, \quad \forall\, \phi \in W. \end{aligned} \tag{4.3.4.7}$$

This gives

$$\begin{aligned} \langle \tilde{g}_i^n - \mu\, \tilde{p}_i^n, \phi\rangle_\Gamma &= \langle \tilde{p}_i^n, tr|_{\Gamma_{1,i}}\, tr_i^{-1}\, \phi\rangle_{\Gamma_{1,i}} + a_i(\tilde{w}_i^n, tr_i^{-1}\, \phi) \\ &\le C\, \|tr_i^{-1}\|\, \|\phi\|_W\, (\|\tilde{w}_i^n\|_{V_i} + \|\tilde{p}_i^n\|_{V_i}) \end{aligned} \tag{4.3.4.8}$$

and

$$\langle \tilde{h}_i^n + \mu\, \tilde{w}_i^n, \phi\rangle_\Gamma \le \tilde{C}\|tr_i^{-1}\|\, \|\phi\|_W\, (\|\tilde{p}_i^n\|_{V_i} + \|\tilde{w}_i^n\|_{V_i}) \tag{4.3.4.9}$$

Moreover,

$$\begin{aligned} \langle \mu\, \phi, \tilde{w}_i^n\rangle_\Gamma &\le \tilde{C}\, \|\mu\|\, \|\phi\|_W\, \|\tilde{w}_i^n\|_{V_i} \\ \langle \mu\, \phi, \tilde{p}_i^n\rangle_\Gamma &\le \tilde{C}\, \|\mu\|\, \|\phi\|_W\, \|\tilde{p}_i^n\|_{V_i}. \end{aligned} \tag{4.3.4.10}$$

For the sake of simplicity, in (4.3.4.8)–(4.3.4.10), we replace the constants C by a generic one. We note, however, that all constants involve just $|\beta|$, the geometry of Ω_i and given problem data. The values of those constants can be given explicitly.

Consider

$$
\begin{aligned}
a_1(\tilde w_1^{n+1}, v_1)&+a_2(\tilde w_2^n, v_2) + a_1(\tilde p_1^{n+1}, u_1) + a_2(\tilde p_2^n, u_2)\\
&= -\langle \tilde p_1^{n+1}, v_1\rangle_{\Gamma_{1,1}} + \langle \tilde g_1^{n+1} - \mu\,\tilde p_1^{n+1}, v_1\rangle_\Gamma\\
&\quad - \langle \tilde p_2^n, v_2\rangle_{\Gamma_{1,2}} + \langle \tilde g_2^n - \mu\,\tilde p_2^n, v_2\rangle_\Gamma\\
&\quad + \kappa(\tilde w_1^{n+1}, u_1) + \langle \tilde h_1^{n+1} + \mu\,\tilde w_1^{n+1}, u_1\rangle_\Gamma\\
&\quad + \kappa(\tilde w_2^n, u_2) + \langle \tilde h_2^{n+1} + \mu\,\tilde w_2^n, u_2\rangle_\Gamma\\
&= \langle \tilde p_1^{n+1}, v_1\rangle_{\Gamma_{1,1}} - \langle \tilde p_2^n, v_2\rangle_{\Gamma_{1,2}}\\
&\quad + \kappa(\tilde w_1^{n+1}, u_1) + \kappa(\tilde w_1^n, u_2)\\
&\quad + \langle \tilde g_1^{n+1} - \mu\,\tilde p_2^n, v_1\rangle_\Gamma + \langle \mu(\tilde p_2^n - \tilde p_1^{n+1}), v_1\rangle_\Gamma \qquad (4.3.4.11)\\
&\quad + \langle \tilde g_2^n - \mu\,\tilde p_2^n, v_2 - v_1\rangle + \langle \tilde g_2^n - \mu\,\tilde p_2^n, u_1\rangle_\Gamma\\
&\quad + \langle \tilde h_1^{n+1} + \mu\,\tilde w_2^n, u_1\rangle_\Gamma - \langle \mu(\tilde w_2^n - \tilde w_1^{n+1}), u_1\rangle_\Gamma\\
&\quad + \langle \tilde h_2^n + \mu\,\tilde w_2^n, u_2 - u_1\rangle + \langle \tilde h_2^n + \mu\,\tilde w_2^n, u_1\rangle_\Gamma\\
&= -\langle \tilde p_1^{n+1}, v_1\rangle_{\Gamma_{1,1}} - \langle \tilde p_2^n, v_2\rangle_{\Gamma_{1,2}}\\
&\quad + \kappa(\tilde w_1^{n+1}, u_1) + \kappa(\tilde w_2^n, u_2)\\
&\quad + \langle \mu(\tilde p_2^n - \tilde p_1^{n+1}), v_1\rangle_\Gamma + \langle \mu(\tilde w_1^{n+1} - \tilde w_2^n), u_1\rangle_\Gamma\\
&\quad + \langle \tilde g_2^n - \mu\,\tilde p_2^n, v_2 - v_1\rangle_\Gamma + \langle \tilde h_2^n - \mu\,\tilde w_2^n, u_2 - u_1\rangle_\Gamma.
\end{aligned}
$$

We set $v_1 := \tilde w_1^{n+1}$, $v_2 = \tilde w_2^n$, $u_1 := \tilde p_1^{n+1}$, $u_2 := \tilde p_2^n$ and obtain from (4.3.4.11)

$$
\begin{aligned}
a_1(\tilde w_1^{n+1}, \tilde w_2^{n+1}) + a_2(\tilde w_2^n, \tilde w_2^n)&+ a_1(\tilde p_1^{n+1}, \tilde p_1^{n+1}) + a_2(\tilde p_2^n, \tilde p_2^n)\\
&= -\left(\langle \tilde p_1^{n+1}, \tilde w_1^{n+1}\rangle_{\Gamma_{1,1}} + \langle \tilde p_2^n, \tilde w_2^n\rangle_{\Gamma_{1,2}}\right)\\
&\quad + \kappa\left((\tilde w_1^{n+1}, \tilde p_1^{n+1})_{\Omega_1} + (\tilde w_2^n, \tilde p_2^n)_{\Omega_2}\right)\\
&\quad + \langle \mu(\tilde p_2^n - \tilde p_1^{n+1}),\, \tilde w_1^{n+1}\rangle_\Gamma \qquad (4.3.4.12)\\
&\quad + \langle \mu(\tilde w_1^{n+1} - \tilde w_2^n),\, \tilde p_1^{n+1}\rangle_\Gamma\\
&\quad + \langle \tilde g_2^n - \mu\,\tilde p_2^n,\, \tilde w_2^n - \tilde w_1^{n+1}\rangle_\Gamma\\
&\quad + \langle \tilde h_2^n - \mu\,\tilde w_2^n,\, \tilde p_2^n - \tilde p_1^{n+1}\rangle_\Gamma.
\end{aligned}
$$

We now choose $v_1 = \tilde p_1^{n+1}$, $v_2 = \tilde p_2^n$, $u_1 = -\tilde w_1^{n+1}$, $u_2 = -\tilde w_2^n$. Then (4.3.4.11) gives

$$
\begin{aligned}
\|\tilde p_1^{n+1}\|^2_{\Gamma_{1,1}} &+ \|\tilde p_2^n\|^2_{\Gamma_{1,2}} + \kappa\,\|\tilde w_1^{n+1}\|^2_{\Omega_1} + \kappa\,\|\tilde w_2^n\|^2_{\Omega_2}\\
&= \langle \mu(\tilde p_2^n - \tilde p_1^{n+1}),\, \tilde p_1^{n+1}\rangle_\Gamma - \langle \mu(\tilde w_1^{n+1} - \tilde w_2^n),\, \tilde w_1^{n+1}\rangle_\Gamma\\
&\quad - \langle \tilde g_2^n - \mu\,\tilde p_2^n,\, \tilde p_1^{n+1} - \tilde p_2^n\rangle_\Gamma + \langle \tilde h_2^n - \mu\,\tilde w_2^n,\, \tilde w_1^{n+1} - \tilde w_2^n\rangle_\Gamma. \qquad (4.3.4.13)
\end{aligned}
$$

On the other hand, the mixed terms in (4.3.4.12), namely

$$
\langle \tilde p_1^{n+1}, \tilde w_1^{n+1}\rangle,\ \langle \tilde p_2^n, \tilde w_2^n\rangle, \kappa\,(\tilde w_1^{n+1}, \tilde p_1^{n+1}),\ \kappa\,(\tilde w_2^n, \tilde p_2^n),
$$

can be estimated by the Cauchy-Schwarz inequality as follows:

$$\begin{aligned}
\langle \tilde{p}_1^{n+1}, \tilde{w}_1^{n+1} \rangle &\leq \nu \, \|\tilde{w}_1^{n+1}\|^2_{\Gamma_{1,1}} + \frac{1}{\nu} \, \|\tilde{p}_1^{n+1}\|^2_{\Gamma_{1,1}}, \\
\langle \tilde{p}_2^{n}, \tilde{w}_2^{n} \rangle &\leq \nu \, \|\tilde{w}_2^{n}\|^2_{\Gamma_{1,2}} + \frac{1}{\nu} \, \|\tilde{p}_2^{n}\|^2_{\Gamma_{1,2}}, \\
\kappa \, (\tilde{w}_1^{n+1}, \tilde{p}_1^{n+1}) &\leq \frac{\kappa}{\mu} \, \|\tilde{w}_1^{n+1}\|^2_{\Omega_1} + \kappa \, \mu \, \|\tilde{p}_1^{n+1}\|^2_{\Omega_1}, \\
\kappa \, (\tilde{w}_2^{n}, \tilde{p}_2^{n}) &\leq \frac{\kappa}{\mu} \, \|\tilde{w}_2^{n}\|^2_{\Omega_2} + \kappa \, \mu \, \|\tilde{p} - 2^{n}\|^2_{\Omega_1}.
\end{aligned} \tag{4.3.4.14}$$

As the terms $\|\tilde{w}_1^{n+1}\|^2_{\Gamma_{1,1}}$, $\|\tilde{w}_2^{n}\|^2_{\Gamma_{1,2}}$, $\|\tilde{p}_1^{n+1}\|^2_{\Omega_1}$, $\|\tilde{p}_2^{n}\|^2_{\Omega_2}$ are not present in (4.3.4.13), we need to compensate them by taking ν, μ sufficiently small. The size of ν, μ will depend on Poincaré's inequality and, hence, on the geometry of Ω_1, Ω_2.

Indeed, for the sake of symmetry, we may equally well begin in (4.3.4.11) with

$$a_1 \, (\tilde{w}_1^{n}, v_1) + a_2 \, (\tilde{w}_2^{n+1}, v_2) + a_1 \, (\tilde{p}_1^{n}, u_1) + a_2 \, (\tilde{p}_2^{n+1}, u_2). \tag{4.3.4.15}$$

By putting things together and using estimates (4.3.4.8), (4.3.4.9), (4.3.4.10) on the right-hand side of (4.3.4.12), (4.3.4.15) and the corresponding estimates on the right-hand sides that would appear in (4.3.4.15), we obtain a constant $C > 0$ depending on the geometry only such that

$$\begin{aligned}
&\|\tilde{w}_1^{n+1}\|^2_{V_1} + \|\tilde{w}_1^{n}\|^2_{V_1} + \|\tilde{p}_1^{n+1}\|^2_{V_1} + \|\tilde{p}_1^{n}\|^2_{V_1} \\
&\qquad + \|\tilde{w}_2^{n+1}\|^2_{V_2} + \|\tilde{w}_2^{n}\|^2_{V_2} + \|\tilde{p}_2^{n+1}\|^2_{V_2} + \|\tilde{p}_2^{n}\|^2_{V_2} \\
&\quad \leq C[\|w_1^{n+1} - w_2^{n}\|_\Gamma + \|w_2^{n+1} - w_1^{n}\|_\Gamma + \|p_1^{n+1} - p_2^{n}\|_\Gamma + \|p_2^{n+1} - p_1^{n}\|_\Gamma] \\
&\qquad \times \, [\|w_1^{n+1}\|_{V_1} + \|p_1^{n+1}\|_{V_1} + \|w_2^{n+1}\|_{V_2} + \|p_2^{n+1}\|_{V_2} \\
&\qquad\qquad + \|w_1^{n}\|_{V_1} + \|p_1^{n}\|_{V_1} + \|w_2^{n}\|_{V_2} + \|p_2^{n}\|_{V_2}].
\end{aligned} \tag{4.3.4.16}$$

We may then introduce the total error

$$e^n := \sum_{i=1}^{2} [\|\tilde{w}_i^{n}\|_{V_i} + \|\tilde{p}_i^{n}\|_{V_i}] \tag{4.3.4.17}$$

and rewrite (4.3.4.16) using again the Cauchy-Schwarz inequality and another constant C to the effect that

$$(e^n + e^{n+1})^2 \leq C \sum_{\substack{i,j=1 \\ i \neq j}} \{\|w_i^{n+1} - w_j^{n}\|_\Gamma + \|p_i^{n+1} - p_j^{n}\|_\Gamma\}(e^n + e^{n+1}). \tag{4.3.4.18}$$

If we introduce the errors on the interface Γ by

$$\mathcal{E}_{ij}^{n,n+1} := \|w_i^{n+1} - w_j^{n}\|^2_\Gamma + \|p_i^{n+1} - p_j^{n}\|^2_\Gamma \tag{4.3.4.19}$$

then (4.3.4.18) becomes

$$e^n + e^{n+1} \leq C\,(\mathcal{E}_{12}^{n,n+1} + \mathcal{E}_{21}^{n,n+1})^{1/2}. \tag{4.3.4.20}$$

Then, with

$$e^{n,n+1} := e^n + e^{n+1} \tag{4.3.4.21}$$

we obtain the following a posteriori error estimate.

Theorem 4.3.4.1 *Let measure of error $e^{n,n+1}$ be defined by (4.3.4.21) and (4.3.4.17). Let the mismatch of the iterates along the interface Γ be defined by (4.3.4.20). Then there is a constant C, depending solely on the geometry of Ω_1, Ω_2 and the parameter $\mu > 0$ along the interface, such that*

$$e^{n,n+1} \leq C\,(\mathcal{E}_{12}^{n,n+1} + \mathcal{E}_{21}^{n,n+1})^{1/2}. \tag{4.3.4.22}$$

Chapter 5

Optimal Control of One-Dimensional Partial Differential Equations on Graphs

5.1 Introduction

In this chapter we continue the discussion of domain decomposition of optimal control problem for partial differential equations started in Chapter 4. However, here we focus on partial differential equations on graphs and, more precisely, on elliptic and hyperbolic equations. We first discuss elliptic optimal control problems on graphs. The global problem is defined and domain decomposition procedures are formulated in the cases of both distributed and boundary controls. Next we consider optimal control problems for hyperbolic systems, in which final value control is emphasized. Higher-dimensional analogs of the methods and results obtained here are discussed in Chapters 6 to 9.

5.2 Elliptic Problems

5.2.1 The Global Optimal Control Problem on a Graph

We recall the notation of Section 3.2. Let ω_i denote a subinterval of $(0, \ell_i)$. We define

$$J(f) := \frac{1}{2} \sum_{i \in C} |f_i|^2_{L^2(\omega_i)^{p_i}} + \frac{1}{2} \sum_{J \in \mathcal{J}_N^c} |f_J|^2_{\mathbb{R}^{p_J}} + \frac{\kappa}{2} \|r - r^0\|^2_{\mathcal{H}} \tag{5.2.1.1}$$

and consider the optimal control problem

$$\inf_{f \in \mathcal{U}} J(f), \quad \mathcal{U} := \Pi_{i \in C} L^2(\omega_i)^{p_i} \times \Pi_{J \in \mathcal{J}_n^c} \mathbb{R}^{p_J} \tag{5.2.1.2}$$

subject to

$$
\begin{aligned}
-[K_i(r_i' + R_i\, r_i)]' + R_i^T\, K_i(r_i' + R_i r_i) + S_i\, r_i \\
= \begin{cases} \mathbf{1}_{\omega_i} f_i, & i \in \mathcal{C} \\ 0, & i \in \mathcal{I}\backslash\mathcal{C} \end{cases}, \quad x \in (0, \ell_i) \\
r_i(v_J) = 0 \quad J \in \mathcal{J}_D, \quad i \in \mathcal{I}_J \\
\varepsilon_{iJ}\, [K_i(r_i' + R_i r_i)]\, (v_J) = f_J, \quad J \in \mathcal{J}_N^c, \quad i \in \mathcal{I}_J \\
\varepsilon_{iJ}\, [K_i(r_i' + R_i r_i)]\, (v_j) = 0, \quad J \in \mathcal{J}_N \backslash \mathcal{J}_N^c, \quad i \in \mathcal{I}_J \\
\varepsilon_{iJ}\, \Pi_{iJ}\, [K_i(r_i' + R_i\, r_i)]\, (v_J) = 0, \quad J \in \mathcal{J}_M, i \in \mathcal{I}_J \\
\sum_{i \in I_J} \varepsilon_{iJ} (C_{iJ}^+)^T\, [K_i(r_i' + R_i\, r_i)](v_J) = 0, \; J \in \mathcal{J}_M.
\end{aligned}
\tag{5.2.1.3}
$$

In (5.2.1.1)–(5.2.1.3), $\mathcal{C}$ denotes the set of control-active edges and $\mathbf{1}_{\omega_i}$ the characteristic function of ω_i, $J_N^c \subset \mathcal{J}_N$ is the index set of control-active Neumann nodes and $\kappa > 0$ is a penalty term.

The optimal control problem (5.2.1.2), (5.2.1.3) can be put into variational form as follows

$$
\inf_{f \in \mathcal{U}} J(f) \;\text{ subject to }\; a(r, \varphi) = b(f, \varphi), \quad \forall\, \varphi \in \mathcal{V}, \tag{5.2.1.4}
$$

where

$$
\begin{aligned}
b(f, \varphi) &:= \sum_{i \in \mathcal{C}} b_i(f_i, \varphi_i) + \sum_{J \in \mathcal{J}_N^c} b_{iJ}(f_J, \varphi_i) \\
&:= \sum_{i \in \mathcal{C}} \int_{\omega_i} f_i \cdot \varphi_i\, dx + \sum_{J \in \mathcal{J}_N^c} f_J \cdot C_{iJ}\, \varphi_i(v_J).
\end{aligned}
\tag{5.2.1.5}
$$

By standard arguments, problem (5.2.1.4), (5.2.1.5) admits a unit optimal control f^{opt} given by

$$
f_i^{\text{opt}} = -p_i|_{\omega_i},\; i \in \mathcal{C}, \quad f_J^{\text{opt}} = -p_i(v_J),\; J \in \mathcal{J}_M,\; i \in \mathcal{I}_J \tag{5.2.1.6}
$$

where $p = \{p_i\}_{I \in \mathcal{I}}$ is the solution of the adjoint system

$$
\begin{aligned}
-[K_i(p_i' + R_i\, p_i)]' + R_i^T\, K_i(p_i' + R_i\, p_i) + S_i\, p_i \\
= \kappa(r_i - r_i^0), \quad x \in (0, \ell_i) \\
p_i(v_J) = 0, \quad J \in \mathcal{J}_D, \quad i \in \mathcal{I}_J \\
\varepsilon_{iJ}[K_i(p_i' + R_i\, p_i)](v_J) = 0, \; J \in \mathcal{J}_N, \; i \in \mathcal{I}_J \\
\Pi_{iJ}\, [K_i(p_i' + R_i\, p_i)]\, (v_J) = 0, \; J \in \mathcal{J}_M, \; i \in \mathcal{I}_J \\
\sum_{i \in \mathcal{I}_J} \varepsilon_{iJ}\, C_{iJ}^+\, [K_i(p_i' + R_i\, p_i)](v_J) = 0, J \in \mathcal{J}_M, \; i \in \mathcal{I}_J.
\end{aligned}
\tag{5.2.1.7}
$$

(There is an obvious duplication of notation in (5.2.1.7), since the symbol p_i has been used previously for the dimension of the system on edge i. However, the intended meaning of p_i should be clear from the context in which it is used.)

The adjoint problem (5.2.1.7) can be expressed in variational form as follows

$$a(p,\varphi) = \kappa(r - r^0, \varphi), \quad \forall\, \varphi \in \mathcal{V}. \tag{5.2.1.8}$$

The optimality system for the optimal control problem (5.2.1.2), (5.2.1.3) consists of (5.2.1.3), (5.2.1.6) and (5.2.1.7), which can be conveniently expressed in variational form as

$$\begin{cases} a(r,\varphi) + b(p,\varphi) = 0 \\ a(p,\varphi) = \kappa((r - r^0), \varphi), \quad \forall\, \varphi \in \mathcal{V}. \end{cases} \tag{5.2.1.9}$$

System (5.2.1.9) is the standard format encountered in Chapter 4 with the notation of Section 3.2 of Chapter 3. Consequently, the domain decomposition technique which we apply to the optimality system (5.2.1.7) is completely analogous to the one discussed in Chapter 3 and applied to the network model of Section 3.2.

5.2.2 Domain Decomposition

We consider the following system on a single edge i at an iteration number $n+1$:

$$\begin{cases} -[K_i\,((r_i^{n+1})' + R_i\, r_i^{n+1})]' + R_i^T\, K_i((r_i^{n+1})' + R_i\, r_i^{n+1}) + S_i(r_i^{n+1}) \\ \qquad\qquad\qquad = \begin{cases} -p_i^{n+1}\mathbf{1}_{\omega_i} \text{ if } i \in \mathcal{C} \\ 0 \text{ if } i \in \mathcal{I}\backslash\mathcal{C} \end{cases} \\ -[K_i((p_i^{n+1})' + R_i\, p_i^{n+1})]' + R_i^T\, K_i((p_i^{n+1})' + R_i\, p_i^{n+1}) + S_i\, p_i^{n+1} \\ \qquad\qquad\qquad = \kappa(r_i^{n+1} - r_i^0), \quad x \in (0,\ell_i) \end{cases} \tag{5.2.2.1}$$

$$r_i^{n+1}(v_J) = p_i^{n+1}(v_J) = 0, \quad j \in \mathcal{J}_D, \quad i \in \mathcal{I}_J \tag{5.2.2.2}$$

$$\begin{cases} \varepsilon_{iJ}[K_i((r_i^{n+1})' + R_i\, r_i^{n+1})]\,(v_J) = -p_i^{n+1}(v_J), \\ \varepsilon_{iJ}[K_i((p_i^{n+1})' + R_i\, p_i^{n+1})](v_J) = 0, \ J \in \mathcal{J}_N^c,\, i \in \mathcal{I}_J \end{cases} \tag{5.2.2.3}$$

$$\begin{cases} \varepsilon_{iJ}[K_i((r_i^{n+1})' + R_i\, r_i^{n+1})]\,(v_J) = 0, \\ \varepsilon_{iJ}[K_i((p_i^{n+1})' + R_i\, p_i^{n+1})]\,(v_J) = 0, \quad J \in \mathcal{J}_N\backslash\mathcal{J}_N^c,\, i \in \mathcal{I}_J \end{cases} \tag{5.2.2.4}$$

$$\begin{cases} \varepsilon_{iJ}\,\Pi_{iJ}\,[K_i((r_i^{n+1})' + R_i\, r_i^{n+1})]\,(v_J) = 0, \\ \varepsilon_{iJ}\,\Pi_{iJ}\,[K_i((p_i^{n+1})' + R_i\, p_i^{n+1})]\,(v_J) = 0, \quad J \in \mathcal{J}_M,\, i \in \mathcal{I}_J \end{cases} \tag{5.2.2.5}$$

$$\begin{cases} \varepsilon_{iJ}(C_{iJ}^+)^T\,[K_i((r_i^{n+1})' + R_i\, r_i^{n+1})](v_J) + \alpha_J\, C_{iJ}\, r_i^{n+1}(v_J) \\ \qquad +\beta_J\, C_{iJ}\, p_i^{n+1}\,(v_J) = \lambda_{iJ}^n, \\ \varepsilon_{iJ}(C_{iJ}^+)^T\,[K_i((p_i^{n+1})' + R_i\, p_i^{n+1})](v_J) + \alpha_J\, C_{iJ}\, p_i^{n+1}(v_J) \\ \qquad -\beta_J\, C_{iJ}\, r_i^{n+1}\,(v_J) = \mu_{iJ}^n, \ J \in \mathcal{J}_M,\, i \in \mathcal{I}_J. \end{cases} \tag{5.2.2.6}$$

The inhomogeneities λ_{ij}^n, μ_{ij}^n are given by

$$\begin{aligned}\lambda_{iJ}^n := \alpha_J\Big(\frac{2}{d_J}\sum_{j\in\mathcal{I}_J} C_{jJ}\, r_j^n(v_J) - C_{iJ}\, r_i^n(v_J)\Big)\\ + \beta_J\Big(\frac{2}{d_J}\sum_{j\in\mathcal{I}_J} C_{jJ}\, p_j^n(v_J) - C_{iJ}\, p_i^n(v_J)\Big)\\ - \Big(\frac{2}{d_J}\sum_{j\in\mathcal{I}_J} \varepsilon_{jJ}(C_{jJ}^+)^T\,[K_j((r_j^n)' + R_j\, r_j^n)]\,(v_J)\\ - \varepsilon_{iJ}\,(C_{iJ}^+)^T\,[K_i((r_i^n)' + R_i\, r_i^n)](v_J)\Big)\end{aligned} \quad (5.2.2.7)$$

and

$$\begin{aligned}\mu_{iJ}^n := \alpha_J\,\Big(\frac{2}{d_J}\sum_{j\in\mathcal{I}_J} C_{jJ}\, p_j^n(v_J) - C_{iJ}\, p_i^n(v_J)\Big)\\ - \beta_J\,\Big(\frac{2}{d_J}\sum_{j\in\mathcal{I}_J} C_{jJ}\, r_j^n(v_J) - C_{iJ}\, r_i^n(v_J)\Big)\\ - \Big(\frac{2}{d_J}\sum_{j\in\mathcal{I}_J} \varepsilon_{jJ}\,(C_{jJ}^+)^T\,[K_j((p_j^n)' + R_j\, p_j^n)]\,(v_J)\\ - (C_{iJ}^+)^T\,[K_i((p_i^n)' + R_i\, p_i^n)]\,(v_J)\Big).\end{aligned} \quad (5.2.2.8)$$

According to Section 3.2 we can express (5.2.2.7) and (5.2.2.8) in a more compact form

$$\begin{aligned}Q_J(r)_i &:= \varepsilon_{iJ}\,(C_{iJ}^+)^T\,[K_i(r_i' + r_i\, r_i)]\,(v_J)\\ P_J(r)_i &:= C_{iJ}\, r_i\,(v_J), \quad J\in\mathcal{J}_M, \quad i\in\mathcal{I}_J\end{aligned} \quad (5.2.2.9)$$

and $Q_J(r) = (Q_j(r)_i)_{i\in\mathcal{I}_J}$, $P_J(r) = (P_J(r)_i)_{i\in\mathcal{I}_J}$. We also recall (3.3.2.1):

$$\mathcal{S}_J(w)_i = \frac{2}{d_J}\sum_{j\in\mathcal{I}_J} w_j - w_i, \quad w = (w_i)_{i\in\mathcal{I}_J}.$$

Thus, λ_{iJ}^n and μ_{iJ}^n defined in (5.2.2.7), (5.2.2.8) can be expressed as

$$\begin{aligned}\lambda_{iJ}^n &= \mathcal{S}_J\,(\alpha_J\, P_J(r^n) + \beta_J\, P_J(p^n) - Q_J(r^n))_i\\ \mu_{iJ}^n &= \mathcal{S}_J\,(\alpha_J\, P_J(p^n) - \alpha_J\, P_J(r^n) - Q_J(p^n))_i.\end{aligned} \quad (5.2.2.10)$$

We introduce

$$g_{iJ}^{n+1} := \lambda_{iJ}^n, \quad h_{iJ}^{n+1} := \mu_{iJ}^n. \quad (5.2.2.11)$$

Then g_{iJ}^{n+1}, h_{iJ}^{n+1} satisfy the relations

$$\begin{aligned}g_{iJ}^{n+1} &=: (g_J^{n+1})_i = \mathcal{S}_J\,(2\alpha_J\, P_J(r^n) + 2\beta_J\, P_J(p^n) - g_J^n)_i\\ h_{iJ}^{n+1} &=: (h_J^{n+1})_i = \mathcal{S}_J\,(2\alpha_J\, P_J(p^n) - 2\beta_J\, P_J(r^n) - h_J^n)_i.\end{aligned} \quad (5.2.2.12)$$

We also introduce the nodal forms

$$\begin{aligned} a_{iJ}(r_i, \varphi_i) &:= C_{iJ}\, r_i(v_J) \cdot C_{iJ}\, \varphi_i(v_J) \\ b_{iJ}(f_J, \varphi_i) &:= f_J \cdot C_{iJ}\, \varphi_i(v_J), \quad i \in \mathcal{I}_J. \end{aligned} \tag{5.2.2.13}$$

In fact, a_{iJ}, b_{iJ} do not explicitly depend on i but we keep the index in order to indicate that we deal with the edge i. If we also set

$$b_i(p_i, \varphi) = \begin{cases} \int_{\omega_i} p_i \cdot \varphi\, dx, & i \in \mathcal{C} \\ 0, & i \in \mathcal{I} \backslash \mathcal{C} \end{cases}$$

the local system (5.2.2.1)–(5.2.2.8) may be expressed in the variational form

$$\begin{aligned} a_i(r_i^n, \varphi_i) &+ \alpha_J\, a_{iJ}(r_i^n, \varphi_i) + \beta_J\, a_{iJ}(p_i^n, \varphi_i) \\ &+ b_{iJ}(p_i^n, \varphi) + b_i(p_i^n, \varphi) = b_{iJ}(g_{iJ}^n, \varphi_i), \quad \forall\, \varphi \in \mathcal{V}_i \\ a_i(p_i^n, \varphi_i) &+ \alpha_J\, a_{iJ}(p_i^n, \varphi_i) - \beta_j\, a_{iJ}(r_i^n, \varphi_i) \\ &= b_{iJ}(h_{iJ}^n, \varphi_i) + \kappa(r_i^n - r_i^0, \varphi_i), \quad \forall\, \varphi_i \in \mathcal{V}_i \\ g_{iJ}^{n+1} &= \mathcal{S}_J(2(\alpha_J\, P_J(r^n) + \beta_J\, P_J(p^n)) - g_J^n)_i \\ h_{iJ}^{n+1} &= \mathcal{S}_J(2(\alpha_J\, P_J(p^n) - \beta_J\, P_J(r^n)) - h_J^n)_i. \end{aligned} \tag{5.2.2.14}$$

We may also introduce a relaxation parameter $\varepsilon \in [0, 1)$ such that

$$\begin{aligned} g_{iJ}^{n+1} &= (1 - \varepsilon)\, \mathcal{S}_J\, (2(\alpha_j\, P_J(r^n) + \beta_J\, P_J(p^n)) - g_J^n)_i + \varepsilon\, g_{iJ}^n \\ h_{iJ}^{n+1} &= (1 - \varepsilon)\, \mathcal{S}_J\, (2(\alpha_J\, P_J(p^n) - \beta_J\, P_J(r^n)) - h_J^n)_i + \varepsilon\, h_{iJ}^n. \end{aligned} \tag{5.2.2.15}$$

5.2.3 Distributed Controls

In this section we assume that $\mathcal{J}_N^c = \emptyset$ and that $\omega_i = (0, \ell_i)$, $\forall i \in \mathcal{I}$. Then $b_i(p_i, \varphi) = (p_i, \varphi)$ for all $i \in \mathcal{I}$. Let us write (5.2.2.15) in complex form. We define $z := r + \mathbf{i}\, p$, where now $\mathbf{i}$ denotes the complex unit in order to not confuse the notation with an edge index. We define $\delta_J = \alpha_J - \mathbf{i}\, \beta_J$, and set

$$z := r + \mathbf{i}\, \frac{1}{\sqrt{\kappa}}\, p$$

and hence

$$z_i = r_i + \mathbf{i}\, \frac{1}{\sqrt{\kappa}}\, p_i.$$

We extend all real bilinear forms $a_i(\cdot,\cdot)$, $a_{iJ}(\cdot,\cdot)$, $b_{iJ}(\cdot,\cdot)$ to complex-valued functions and set

$$a_i^c(z, \psi) := a_i(z, \bar{\psi}) \tag{5.2.3.1}$$

$$c_i(z, \psi) := -\mathbf{i}\, \sqrt{\kappa}\, (z, \bar{\psi}) = k(z, \bar{\psi}). \tag{5.2.3.2}$$

Then

$$\begin{aligned} a_i^c(z_i,\psi_i) &+ c_i(z_i,\psi_i) \\ &= a_i(r_i,\bar{\psi}_i) + \frac{\mathbf{i}}{\sqrt{\kappa}}\, a_i(p_i,\bar{\psi}_i) - \mathbf{i}\sqrt{\kappa}\,(r_i,\bar{\psi}_i) + (p_i,\bar{\psi}_i) \\ &= a_i(r_i,\bar{\psi}_i) + (p_i,\bar{\psi}_i) + \frac{\mathbf{i}}{\sqrt{\kappa}}\,(a_i(p_i,\bar{\psi}_i) - \kappa(r_i,\bar{\psi}_i)). \end{aligned} \tag{5.2.3.3}$$

Thus with $a^c(\cdot,\cdot) = \sum_{I\in\mathcal{I}} a_i(\cdot,\cdot)$ and so forth, we see that

$$a^c(r,\psi) + c(r,\psi) = k(r_d,\psi), \quad \forall\, \psi \in \mathcal{V}^c, \tag{5.2.3.4}$$

where $\mathcal{V}^c$ contains now complex-valued elements, is equivalent to

$$\begin{aligned} a(r,\varphi) + b(p,\varphi) &= 0, \quad \forall\, \varphi \in \mathcal{V}, \\ a(p,\varphi) &= \kappa(r - r_d,\, \varphi), \quad \forall\, \varphi \in \mathcal{V}. \end{aligned}$$

which is the same as (5.2.1.9).

As for the local system (5.2.2.14), we proceed as follows: Define

$$\begin{aligned} F_J^n &:= g_J^n + \mathbf{i}\, h_J^n, \quad G = k\, r_d \\ a_i^c(z_i,\psi_i) &+ c_i(z_i,\psi_i) + \delta_J\, a_{iJ}^c(z_i,\psi_i) \\ &= b_{iJ}^c(F_{iJ}^n,\psi_i) + (G_i,\psi_i), \quad \forall\, \psi_i \in \mathcal{V}. \end{aligned} \tag{5.2.3.5}$$

Then

$$F_{iJ}^{n+1} = \mathcal{S}_J\,(2\delta_J\, P_J(z^n) - F_J^n)_i.$$

This is equivalent to

$$\begin{aligned} a_i\,(r_i,\varphi_i) &+ b_i\,(p_i,\varphi_i) + \alpha_J\, a_{iJ}\,(r_i,\varphi_i) \\ &+ \beta_J\, a_{iJ}\,(p_i,\varphi_i) = b_{iJ}\,(g_{iJ}^n,\varphi_i), \quad \forall\, \varphi_i \in \mathcal{V}_i \\ a_i\,(p_i,\varphi_i) &- \kappa\,(r_i,\varphi_i) + \alpha_J\, a_{iJ}\,(p_i,\varphi_i) \\ &- \beta_J\, a_{iJ}\,(r_i,\varphi_i) = b_{iJ}\,(h_{iJ}^n,\varphi_i) - \kappa\,(r_i^d,\varphi_i), \quad \forall\, \varphi \in \mathcal{V}_i. \end{aligned} \tag{5.2.3.6}$$

$$\begin{aligned} g_{iJ}^{n+1} &= \mathcal{S}_J\,(2\,(x_J\, P_J(r^n) + \beta_J\, P_J\,(p1n)) - g_J^n)_i \\ h_{iJ}^{n+1} &= \mathcal{S}_J\,(2(\alpha_J\, P_J(p^n) - \beta_j\, P_J(r^n)) - h_J^n)_i. \end{aligned}$$

Thus (5.2.3.5) reduces to (5.2.2.14) for $r_i,, p_i$ replaced with r_i^n, p_i^n.

As for consistency of the above scheme, upon convergence we have

$$F_J + \mathcal{S}_J\, F_J = 2\,\delta_J\, P_J(z). \tag{5.2.3.7}$$

If we apply $\mathcal{S}_J$ to (5.2.3.7) we obtain

$$F_J + \mathcal{S}_J\, F_J = 2\,\delta_J\, S_J(z)$$

since $\mathcal{S}_J^2 = I$.

This implies

$$\mathcal{S}_J\, P_J(z) = P_J(z)$$

and in turn

$$C_{iJ}\, z_i(v_J) = C_{iJ}\, z_j(v_J), \quad \forall\, i,\, j \in \mathcal{I}_J. \tag{5.2.3.8}$$

This is the continuity required for r_i and p_i in the global system. On the other hand

$$F_J + \mathcal{S}_J\, F_J = \frac{2}{d_J} \sum_{i\in\mathcal{I}_J} F_{iJ}$$

which implies

$$\sum_{i\in\mathcal{I}_J} F_{iJ} = \delta_J \sum_{i\in\mathcal{I}_J} P_J(z)_i. \tag{5.2.3.9}$$

In order to check consistency with the global model we have to apply test functions $\psi_i = \psi|_{e_i}$, $\psi \in \mathcal{V}$. They satisfy (5.2.3.8). As a result we obtain

$$\begin{aligned}\sum_{i\in\mathcal{I}_J} b^v_{iJ}\,(F^n_{iJ}, \psi_i) &= \sum_{i\in\mathcal{I}_J} F^n_{iJ} \cdot C_{iJ}\, \bar{\psi}_i(v_J)\\ &= (\sum_{i\in\mathcal{I}_J} F^n_{iJ}) \cdot C_{i_0 J}\, \bar{\psi}_{i_0}(v_J)\\ &= \delta_J \sum_{i\in\mathcal{I}_J} P_J(z)_i \cdot C_{iJ}\, \bar{\psi}_i(v_J).\end{aligned} \tag{5.2.3.10}$$

If we sum (5.2.3.5) over $i \in \mathcal{I}_J$ we obtain

$$\sum_{i\in\mathcal{I}_J} [a^c_i(z_i, \psi_i) + c_i(z_i, \psi_i)]$$

$$= \sum_{i\in\mathcal{I}_J} (G_i, \psi_i) + \sum_{i\in\mathcal{I}_J} b^c_{iJ}(F_{iJ}, \psi_i) - \sum_{i\in\mathcal{I}_J} b_J\, a^c_{iJ}(z_i, \psi_i).$$

But

$$a^c_{iJ}\,(z_i, \psi_i) = C_{iJ}\, z_i(v_J)\, C_{iJ}\, \bar{\psi}_i(v_J) = C_{iJ}\, z_i(v_J)\, C_{iJ}\, \bar{\psi}_{i_0}(v_J)$$

for some $i_0 \in \mathcal{I}_J$, as $\psi_i = \psi|_{e_i}$, $\psi \in \mathcal{V}$. Therefore, and because of (5.2.3.10),

$$\sum_{i\in\mathcal{I}_J} b^c_{iJ}\,(F_{iJ}, \psi_i) - \sum_{i\in\mathcal{I}_J} \delta_J\, a^c_{iJ}\,(z_i, \psi_i) = 0,$$

and, hence, (5.2.3.11) reduces to

$$\sum_{i\in\mathcal{I}_J} a^c_i\,(z_i, \psi_i) + c_i(z_i, \psi_i) = \sum_{i\in\mathcal{I}_J} (G_i, \psi_i). \tag{5.2.3.11}$$

controlled node, the form c_i in (5.2.3.2) reduces Now, each edge connects precisely two nodes. Therefore, if we add (5.2.3.12) over all $J \in \mathcal{J}$, we obtain

$$a^c(z, \psi) + c(z, \psi) = (G, \psi), \quad \forall\, \psi \in \mathcal{J}_J, \tag{5.2.3.12}$$

which is (5.2.3.4).

Convergence

According to the last section, in order to prove convergence of the iterative process (5.2.2.1)–(5.2.2.8), we may concentrate on its complex form (5.2.3.5), which is

$$\begin{aligned} a_i^c\,(z_i^n, \psi_i) + c_i\,(z_i^n, \psi_i) + \delta_J\, a_{iJ}^c\,(z_i^n, \psi_i) & \\ = b_{iJ}^c\,(F_{iJ}^n, \psi_i) + (G_i, \psi_i), \quad \forall\, \psi_i \in \mathcal{V}_i^2 & \\ F_{iJ}^{n+1} = \mathcal{S}_J\,(2\,\delta_J\, P_J\,(z^n) - F_J^n)_i. & \end{aligned} \tag{5.2.3.13}$$

System (5.2.3.13) is very much in the spirit of (4.2.3.3) and (4.2.3.4). However, as we are dealing with a graph rather than with two or even more serially connected domains, we have to give a proof of convergence. To this end we introduce the local errors

$$e_i^n = z_i^n - z_i \quad i \in \mathcal{I}, \tag{5.2.3.14}$$

where z_i^n solves (5.2.3.13) and $z_i = z|_{e_i}$ with z solving the global system (5.2.3.4). For all $J \in \mathcal{J}_M$ we have $F_J := (F_{iJ})_{i \in \mathcal{I}_J}$. We introduce $F := (F_J)_{J \in \mathcal{J}_M}$ and define a space $\mathcal{X} = \mathbb{C}^\ell$, where $\ell := \sum\limits_{J \in \mathcal{J}_M} d_J \sum\limits_{I \in \mathcal{I}_J} p_i$, with the standard norm

$$\|F\|_{\mathcal{X}} = \left(\sum_{J \in \mathcal{J}_M} \sum_{i \in \mathcal{I}_J} |F_{iJ}|^2 \right)^{1/2}. \tag{5.2.3.15}$$

In (5.2.3.13) we may now eliminate (G_i, ψ_i) and replace z_n^i by e_i^n. We then define the map

$$\begin{aligned} \mathcal{T} : \mathcal{X} \to \mathcal{X} \\ \mathcal{T}(F) = (\mathcal{S}_J\,(2\,\delta_J\, P_J(e) - F_J))_{J \in \mathcal{J}_M}. \end{aligned} \tag{5.2.3.16}$$

As we have proved consistency of (5.2.3.13) with the global problem (5.2.3.4), a fixed point F of $\mathcal{T}$ 'solves' the global system in the sense that the corresponding e in

$$F = \mathcal{S}_J(2\,\delta_J\, P_J(e) - F_J)_{J \in \mathcal{J}_M} = \mathcal{T}(F)$$

solves (5.2.3.4). We are thus led to consider the fixed point iteration

$$F^{n+1} = \mathcal{T}(F^n). \tag{5.2.3.17}$$

In fact, we could as well consider the relaxed version

$$F^{n+1} = (1 - \varepsilon)\,\mathcal{T}(F^n) + \varepsilon\, F^n. \tag{5.2.3.18}$$

However, we do not want to obscure the proof by adding yet another complication.

We first investigate under which conditions $\mathcal{T}$ is non expansive. We have

$$\begin{aligned}\|\mathcal{T}(F)\|_{\mathcal{X}}^2 &= \sum_{J\in\mathcal{J}_M}\sum_{i\in\mathcal{I}_J} |\mathcal{S}_J(2\delta_J\, P_J)(e) - F_J)_i|^2 \\ &= \sum_{J\in\mathcal{J}_M}\sum_{i\in\mathcal{I}_J} \{|\mathcal{S}_J(2\delta_J\, P_J(e))_i|^2 + |\mathcal{S}_J(F_J)_i|^2 \\ &\qquad - 2\, Re\,(\mathcal{S}_J(\overline{2\delta_J\, P_J(e)_i} \cdot F_{iJ})\} \\ &= \|F\|_{\mathcal{X}}^2 + 4\sum_{J\in\mathcal{J}_M}\sum_{i\in\mathcal{I}_J} Re\,(\mathcal{S}_J\,(\overline{\delta_j\, P_J(e)})_i \\ &\qquad \cdot [\mathcal{S}_J\,(\delta_j\, p_J(e) - F_J)_i].\end{aligned} \tag{5.2.3.19}$$

In (5.2.3.19) we have used

$$\sum_{i\in\mathcal{I}_J} |\mathcal{S}_J(F_J)_i|^2 = \langle \mathcal{S}_J\, F_J, \mathcal{S}_J\, F_J\rangle = |F_J|^2.$$

Now

$$\begin{aligned}&\sum_{i\in\mathcal{I}_J} \mathcal{S}_J\,(\overline{\delta_J\, P_J(e)})_i \cdot \mathcal{S}_J(\delta_J\, P_J(e) - F_J)_i \\ &\qquad = \sum_{i\in\mathcal{I}_J} \overline{\mathcal{S}_J(\delta_J\, P_J(e))_i} \cdot \mathcal{S}_J(\delta_J\, P_J(e) - F_J)_i \\ &\qquad = \sum_{i\in\mathcal{I}_J} (\overline{\delta_J\, P_J(e)})_i \cdot (\delta_J\, P_J(e) - F_J)_i \\ &\qquad = \bar{\delta}_J \sum_{i\in\mathcal{I}_J} G_{iJ}\, \bar{e}_i(v_J) \cdot (\delta_J\, C_{iJ}\, e_i(v_J) - F_{iJ}) \\ &\qquad = \bar{\delta}_J \sum_{i\in\mathcal{I}_J} (\delta_J\, a_{iJ}^c(e_i, e_i) - b_{iJ}^c(F_{iJ}, e_i)) \\ &\qquad = \bar{\delta}_J \sum_{i\in\mathcal{I}_J} (-a_i^c(e_i, e_i) - c_i(e_i, e_i)) \\ &\qquad = -\bar{\delta}_J \sum_{i\in\mathcal{I}_J} a_i^c(e_i, e_i) - \sum_{i\in\mathcal{I}_J} \bar{\delta}_J\, c_i(e_i, e_i).\end{aligned} \tag{5.2.3.20}$$

According to (5.2.3.2)

$$c_i(e_i, e_i) = k \int_{\omega_i} |e_i|^2\, dx. \tag{5.2.3.21}$$

Thus, with (5.2.3.21) in (5.2.3.20) we obtain from (5.2.3.19)

$$\begin{aligned}\|\mathcal{T}(F)\|_{\mathcal{X}}^2 = \|F\|_{\mathcal{X}}^2 &- 4\sum_{J\in\mathcal{J}_M} Re(\bar{\delta}_J) \sum_{i\in\mathcal{I}_J} a_i^c(e_i, e_i) \\ &- 4\sum_{J\in\mathcal{J}_M} Re(\bar{\delta}_J\, k) \sum_{i\in\mathcal{I}_J} c_i(e_i, e_i).\end{aligned} \tag{5.2.3.22}$$

This is precisely the situation met in (4.2.4.9) except for the fact that we took $\delta_J = \alpha_J - \mathbf{i}\,\beta_J$ here. For simplicity, we take $\delta_J = \delta$ at all nodes, even though the more general case might give some additional freedom in practice. There are again three cases to consider.

i) $Re(\delta) > 0,\ Re(\bar{\delta}\, k) = 0.$

We infer $a_i(e_i^n, e_i^n) \to 0,\ \ \forall\, i \in \mathcal{I}$ as $n \to \infty$. Moreover, (F_{iJ}^n), $J \in \mathcal{J}_M$, $i \in \mathcal{I}_J$ are bounded. But then, according to $(5.2.3.13)_2$ (after applying $\mathcal{S}_J$) the traces $C_{iJ}\, e_i^n(v_J)$ are uniformly bounded in i, J. Thus there exists a subsequence n_j such that $C_{iJ}\, e_i^{n_j}(v_J) \to \xi$. Together with $a_i(e_i^n, e_i^n) \to 0$ this implies that $e_i^{n_j} \to u_i =$ const. $i \in \mathcal{I}$ and, in turn, that $c_i(e_i^{n_j}, e_i^{n_j})$ converges to $c_i(u_i, u_i)$. The first equation in (5.2.3.13) (always with e_i^n's and $G_i \equiv 0$) then implies that $F_{iJ}^{n_j}$ converges to F_{iJ}. But then the second equation in (5.2.3.13) shows that $F_{iJ}^{n_j+1}$ is also a convergent sequence whose limit $\tilde{F}_{iJ}$ may conceivably differ from F_{iJ}, $\forall i \in \mathcal{I}_J$. As u_i is a constant solution of (5.2.3.13), necessarily $u_i = 0$ and, hence, $c_{iJ}\, u_i(v_J) = \xi_{iJ} = 0$ and $C_{iM}\, u_i(v_M) = \xi_{iM} = 0$, where J, M are the bounding vertices of the edge i. We may then infer $F_{iJ} = F_{iM} = 0$. The same argument applies to all edges. Thus $\tilde{F}_{iJ} = 0$ according to $(5.2.3.13)_2$.

In conclusion, on the subsequence n_j, the errors $e_i^{n_j}$ converges to zero strongly in $\mathcal{V}_i^c$. It follows that $e_i^n \to 0$ strongly in $\mathcal{V}_i^c$, $\forall i \in \mathcal{I}$.

ii) $Re(\delta) = 0,\ Re(\bar{\delta}\, k) > 0.$

In this case $c_i(e_i^n, e_i^n) \to 0,\ \forall i \in \mathcal{I}$, and F_J^n is bounded according to (5.2.3.22). But then (5.2.3.13) implies that $a_i^c(e_i^n, e_i^n)$ stays bounded. Hence, the traces $e_i^n(v_J)$ converge to zero, which in turn implies (again going back to (5.2.3.13)) ($G_i = 0$, $e_i \leftrightarrow z_i$ etc.)) that $a_i^c(e_i^n, e_i^n) \to 0$, which shows that $e_i^n \to 0$ strongly in $\mathcal{V}_i^c$.

iii) $Re(\delta) > 0,\ Re(\delta\, \bar{k}) > 0.$

In this case it is obvious that $e_n^i \to 0$ strongly in $\mathcal{V}_i^c$.

We have thus proved the following theorem.

Theorem 5.2.3.1 *Consider the optimal control problem* (5.2.1.2), (5.2.1.3) *with* $\omega_i = (0, \ell_i)$, $\forall i \in \mathcal{I}$, *and* $\mathcal{J}_N^c = \emptyset$. *Then there exists a unique family of optimal controls* $f = (f_i)_{i\in\mathcal{I}}$ *and a unique corresponding weak solution* $r = (r_i)_{i\in\mathcal{I}}$ *solving* (5.2.1.2), (5.2.3.14).

Furthermore, let $r^n = (r_i)_{i\in\mathcal{I}}^n$, $p^n = (p_i^n)_{i\in\mathcal{I}}$ *be the unique solutions of* (5.2.2.1), (5.2.2.6) *and* (5.2.2.8) *with* $\alpha_J = \alpha$, $\beta_J = \beta$, $\ \forall\ J \in \mathcal{J}_M$. *Then under the assumption* $\alpha \geq 0$, $\beta \geq 0$ *and* $\alpha + \beta > 0$, *the solutions* r_i^n, p_i^n *converge strongly in* $\mathcal{V}_i$ *to the solutions* r_i, p_i *of the optimality system* (5.2.1.3), (5.2.1.6) *and* (5.2.1.7) *such that* r_i, p_i *are the restrictions to edge* i *of* r, p *and the controls are computed according to* (5.2.1.6).

5.2.4 Boundary Controls

We recall the decomposed system (5.2.2.10)–(5.2.3.2) or equivalently (5.2.3.8), (5.2.3.9). As usual, we introduce the errors

$$\tilde{r}_i^n := r_i^n - r_i,\ \tilde{p}_i^n := p_i^n - p_i, \tag{5.2.4.1}$$

where r_i^{n+1}, p_i^{n+1} solve (5.2.2.10)–(5.2.3.2) and r_i, p_i solve (5.2.1.4), (5.2.1.7) and (5.2.1.8), with $f_i^{opt} \equiv 0$. Thus we consider the case of boundary controls, only, that is, $\mathcal{C} = \emptyset$. We will repeatedly use the representation (5.2.3.6) for the history-coupling terms. We follow the general philosophy as usual and introduce a state vector

$$X := ((\tilde{g}_{iJ})_{i\in\mathcal{I}_J},\ (\tilde{h}_{iJ})_{i\in\mathcal{I}_J})_J \in \mathbb{R}^d = \mathcal{X} \tag{5.2.4.2}$$

with $d = \sum_{J\in\mathcal{J}_M} \sum_{i\in\mathcal{I}_J} p_i$. The space $\mathcal{X}$ is endowed with the standard norm

$$\|X\|_{\mathcal{X}}^2 = \sum_{J\in V_M} \sum_{i\in\mathcal{I}_J} |X_{iJ}|^2. \tag{5.2.4.3}$$

The iteration, in terms of the errors, can now be written as follows:

$$\begin{aligned} &\mathcal{T} : \mathcal{X} \mapsto \mathcal{X} \\ \mathcal{T}(X) = \big(&\mathcal{S}_J(2(\alpha_j\, P_J(\tilde{r}) + \beta_J\, P_J(\tilde{p})) - \tilde{g}_J)_{i\in\mathcal{I}_J}, \\ &\mathcal{S}_J(2(\alpha_J\, P_J(\tilde{p}) - \beta_J\, P_J(\tilde{r})) - \tilde{h}_J)_{i\in\mathcal{I}_j}\big)_{J\in\mathcal{J}_M}. \end{aligned} \tag{5.2.4.4}$$

The fixed point iteration

$$X^{n+1} = \mathcal{T}(X^n) \tag{5.2.4.5}$$

is then equivalent to (5.2.2.10)–(5.2.3.2) for the errors $\tilde{r}_i^m$, $\tilde{p}_i^n$, $\tilde{h}_{iJ}^n$, $\tilde{g}_{iJ}^n$. We can of course also consider a relaxation as usual, but for the sake of simplicity we stay with (5.2.4.5).

The following calculation shows that $\mathcal{T}$ is non-expansive.

$$\begin{aligned} \|\mathcal{T}X\|_{\mathcal{X}}^2 = &\sum_{J\in\mathcal{J}_M} \sum_{i\in\mathcal{I}_J} |\mathcal{S}_J(2(\alpha_J\, P_J(\tilde{r}) + \beta_J\, P_J(\tilde{p})) - \tilde{g}_J)_i|^2 \\ &+ \sum_{j\in\mathcal{J}_M} \sum_{i\in\mathcal{I}_J} |\mathcal{S}_J(2(\alpha_J\, P_J(\tilde{p}) - \beta_J\, P_J(\tilde{r})) - \tilde{h}_J)_i|^2 \\ =&I + II \end{aligned} \tag{5.2.4.6}$$

where

$$\begin{aligned} I = &\sum_{J\in\mathcal{J}_M} \sum_{i\in\mathcal{I}_J} [4|\mathcal{S}_J(\alpha_J\, P_J(\tilde{r}) + \beta_J\, P_J(\tilde{p}))_i|^2 \\ &- 4(\mathcal{S}_J(\alpha_J\, P_J(\tilde{r}) + \beta_J\, P_J(\tilde{p}))_i\, (\mathcal{S}_J\tilde{g}_J)_i + |(\mathcal{S}_J\tilde{g}_J)_i|^2] \\ = &\sum_{J\in\mathcal{J}_M} \sum_{i\in\mathcal{I}_J} \{4\, \mathcal{S}_J(\alpha_J\, P_J(\tilde{r}) + \beta_J\, P_J(\tilde{p}))_i\} \\ &\cdot \mathcal{S}_J((\alpha_J\, P_J(\tilde{r}) + \beta_J\, P_J(\tilde{p}))_i - \tilde{g}_{iJ}) + |(\mathcal{S}_J\tilde{g}_J)_i|^2 \end{aligned}$$

and

$$II = \sum_{J\in\mathcal{J}_M} \sum_{i\in\mathcal{I}_J} \{4\mathcal{S}_J(\alpha_J\, P_J(\tilde{p}) - \beta_J\, P_J(\tilde{r}))_i \\ \cdot \mathcal{S}_J((\alpha_J\, P_J(\tilde{p}) - \beta_j\, P_J(\tilde{r}))_i - \tilde{h}_{iJ}) + |(\mathcal{S}_J \tilde{h}_J)_i|^2\}.$$

We can rewrite I, II as

$$I = \sum_{J\in\mathcal{J}_M} \sum_{i\in\mathcal{I}_J} \{|\tilde{g}_{iJ}|^2 + 4(\alpha_J\, C_{iJ}\tilde{r}_i(v_J) + \beta_J\, C_{iJ}\tilde{p}(v_J)) \\ \cdot (\alpha_J\, C_{iJ}\tilde{r}_i(v_J) + \beta_J\, C_{iJ}\, \tilde{p}_J(v_J) - \tilde{g}_{iJ})\}$$

$$II = \sum_{J\in\mathcal{J}_M} \sum_{i\in\mathcal{I}_J} \{|\tilde{h}_{iJ}|^2 + 4\,\alpha_J\, C_{iJ}\, \tilde{p}_i(v_J) - \beta_J\, C_{iJ}\, \tilde{r}_i(v_J)) \\ \cdot (\alpha_J\, C_{iJ}\, \tilde{p}_i(v_J) - \beta_J\, C_{iJ}\, \tilde{r}_i(v_J) - \tilde{h}_{iJ})\}.$$

According to the boundary conditions at multiple nodes (5.2.2.15) (taken without index because of (5.2.3.5))

$$I = \sum_{J\in\mathcal{J}_M} \sum_{i\in\mathcal{I}_J} \{|\tilde{g}_{iJ}|^2 - 4(\varepsilon_{iJ}(C_{iJ}^{+})^T[K_i(\tilde{r}_i' + R_i\,\tilde{r}_i)](v_J)) \\ \cdot (\alpha_j\, C_{iJ}\, \tilde{r}_i(v_J) + \beta_J\, C_{iJ}\, \tilde{p}_i(v_J))\}$$

$$II = \sum_{J\in\mathcal{J}_M} \sum_{i\in\mathcal{I}_J} \{|h_{iJ}|^2 - 4(\varepsilon_{iJ}\, C_{iJ}^{+})^T[K_i(\tilde{p}_i' + R_i\,\tilde{p}_i)](v_J)) \\ \cdot (\alpha_J\, C_{iJ}\, \tilde{p}_i(v_J) - \beta_J\, C_{iJ}\, \tilde{r}_i(v_J))\}$$

Now

$$\begin{aligned} 0 &= \int_0^{\ell_i} \{-[K_i(\tilde{r}_i' + R_i\,\tilde{r}_i)]' + R_i^T\, K_i(\tilde{r}_i' + R_i\,\tilde{r}_i) + S_i\,\tilde{r}_i\} \cdot \tilde{r}_i\, dx \\ &= -K_i(\tilde{r}_i' + R_i\,\tilde{r}_i) \cdot r_i|_0^{\ell_i} + \int_0^{\ell_i} \{K_i(\tilde{r}_i' + R_i\,\tilde{r}_i) \cdot (\tilde{r}_i' + R_i\,\tilde{r}_i) + S_i\,\tilde{r}_i \cdot \tilde{r}_i\}\, dx \\ &= -B_i(\tilde{r}_i, \tilde{r}_i)|_0^{\ell_i} + a_i(\tilde{r}_i, \tilde{r}_i), \end{aligned}$$

and

$$\begin{aligned} 0 &= \sum_{i\in\mathcal{I}} a_i(\tilde{r}_i, \tilde{r}_i) - \sum_{i\in\mathcal{I}} B_i(\tilde{r}_i, \tilde{r}_i)|_0^{\ell_i} \\ &= \sum_{i\in\mathcal{I}} a_i(\tilde{r}_i, \tilde{r}_i) - \sum_{J\in\mathcal{J}_M} \sum_{i\in\mathcal{I}_J} \varepsilon_{iJ}(C_{iJ}^{+})^T\, K_i(\tilde{r}_i' + R_i r - i)(v_J) \cdot C_{iJ}\, \tilde{r}_i(v_J) \\ &\qquad\qquad + \sum_{J\in\mathcal{J}_N^c} \tilde{p}_i(v_J)\, \tilde{r}_i(v_J), \end{aligned}$$

that is,

$$\sum_{J\in\mathcal{J}_M}\sum_{i\in\mathcal{I}_J}\varepsilon_{iJ}(C_{iJ}^{+})^T K_i(\tilde r_i'+R_i\,\tilde r_i)(v_J)\cdot C_{iJ}\,\tilde r_i(v_J)$$
$$=\sum_{i\in\mathcal{I}} a_i(\tilde r_i,\tilde r_i)+\sum_{j\in\mathcal{J}_N^c}\tilde p_i(v_J)\cdot\tilde r_i(v_J). \quad (5.2.4.7)$$

Accordingly,

$$\sum_{J\in\mathcal{J}_M}\sum_{i\in\mathcal{I}_J}\varepsilon_{iJ}(C_{iJ}^{+})^T K_i(\tilde p_i'+R_i\,\tilde p_i)(v_J)\cdot C_{iJ}\,\tilde p_i(v_J)$$
$$=\sum_{i\in\mathcal{I}} a_i(\tilde p_i,\tilde p_i)-\kappa\sum_{i\in\mathcal{I}}\int_0^{\ell_i}\tilde r_i\cdot\tilde p_i\,dx, \quad (5.2.4.8)$$

$$\sum_{J\in\mathcal{J}_M}\sum_{i\in\mathcal{I}_J}\varepsilon_{iJ}(C_{iJ}^{+})^T K_i(\tilde r_i'+R_i\,\tilde r_i)(v_J)\cdot C_{iJ}\,\tilde p_i(v_J)$$
$$=\sum_{i\in\mathcal{I}} a_i(\tilde r_i,\tilde p_i)+\sum_{J\in\mathcal{J}_n^c}|p_i(v_J)|^2, \quad (5.2.4.9)$$

$$\sum_{J\in\mathcal{J}_M}\sum_{i\in\mathcal{I}_J}\varepsilon_{iJ}(C_{iJ}^{+})^T K_i(\tilde p_i'+R_i\,\tilde p_i)(v_J)\cdot C_{iJ}\,\tilde r_i(v_J)$$
$$=\sum_{i\in\mathcal{I}} a_i(\tilde r_i,\tilde p_i)-\kappa\sum_{i\in\mathcal{I}}\int_0^{\ell_i}|\tilde r_i|^2\,dx. \quad (5.2.4.10)$$

Let us for simplicity assume that $\alpha_J=\alpha$, $\beta_J=\beta$. Then using (5.2.4.7) and (5.2.4.9) we find

$$\begin{aligned}\sum_{J\in\mathcal{J}_M}\sum_{i\in\mathcal{I}_J}\varepsilon_{iJ}(C_{iJ}^{+})^T\,[K_i(\tilde r_i'+R_i\,\tilde r_i)(v_J)]\cdot(\alpha\,C_{iJ}\,\tilde r_i(v_J)+\beta\,C_{iJ}\,\tilde p_i(v_J))\\ =\alpha\sum_{i\in\mathcal{I}} a_i(\tilde r_i,\tilde r_i)+\beta\sum_{i\in\mathcal{I}} a_i(\tilde r_i,\tilde p_i)\\ +\alpha\sum_{J\in\mathcal{J}_N^c}\tilde p_i(v_J)\cdot\tilde r_i(v_J)\\ +\beta\sum_{J\in\mathcal{J}_N^c}|\tilde p_i(v_J)|^2,\end{aligned} \quad (5.2.4.11)$$

while using (5.2.4.8) and (5.2.4.10) we obtain

$$\sum_{J\in\mathcal{J}_M}\sum_{i\in\mathcal{I}_J}\varepsilon_{iJ}(C_{iJ}^+)^T\,[K_i(\tilde{p}_i'+R_i\,\tilde{p}_i)(v_J)]\cdot(\alpha\,C_{iJ}\,\tilde{p}_i(v_J)-\beta\,C_{iJ}\,\tilde{r}_i(v_J))$$
$$=\alpha\sum_{i\in\mathcal{I}}a_i(\tilde{p}_i,\tilde{p}_i)-\kappa\,\alpha\sum_{i\in\mathcal{I}}\int_0^{\ell_i}\tilde{r}_i\cdot\tilde{p}_i\,dx$$
$$-\beta\sum_{i\in\mathcal{I}}a_i(\tilde{r}_i,\tilde{p}_i)+\beta\,\kappa\sum_{i\in\mathcal{I}}\int_0^{\ell_i}|\tilde{r}_i|^2\,dx. \qquad (5.2.4.12)$$

Therefore

$$\|\mathcal{T}X\|_{\mathcal{X}}^2=\|X\|_{\mathcal{X}}^2-4\alpha(a(\tilde{r},\tilde{r})+a(\tilde{p},\tilde{p}))$$
$$-4(\alpha\sum_{J\in\mathcal{J}_N^c}\tilde{p}_{\bar{i}}(v_J)\cdot\tilde{r}_{\bar{i}}(v_J)+\beta\sum_{J\in\mathcal{J}_N^c}|\tilde{p}_{\bar{i}}(v_J)|^2$$
$$-\kappa\,\alpha\sum_{i\in\mathcal{I}}\int_0^{\ell_i}\tilde{r}_i\cdot\tilde{p}_i\,dx+\beta\,\kappa\sum_{i\in\mathcal{I}}\int_0^{\ell_i}|\tilde{r}_i|^2\,dx) \qquad (5.2.4.13)$$

There are three cases to consider:

i) $\alpha=0,\ \beta>0$

ii) $\alpha>0,\ \beta=0$

iii) $\alpha>0,\ \beta>0$

i) In this case (5.2.4.13) reduces to

$$\|\mathcal{T}X\|_{\mathcal{X}}^2=\|X\|_{\mathcal{X}}^2-4\beta(\sum_{j\in\mathcal{J}_N^c}|p_{\bar{i}}(v_J)|^2+\kappa\sum_{i\in\mathcal{I}}\int_0^{\ell_i}|\tilde{r}_i|^2\,dx)$$

and, obviously, $\mathcal{T}$ is non-expansive. The iteration of errors satisfies

$$X^{n+1}=\mathcal{T}(X^n)$$

$$\|X^{n+1}\|_{\mathcal{X}}^2=\|X^n\|_{\mathcal{X}}^2-4\beta(\sum_{J\in\mathcal{J}_N^c}|\tilde{p}_{\bar{i}}^n(v_J)|^2+\kappa\sum_{i\in\mathcal{I}}\int_0^{\ell_i}|\tilde{r}_i^n|^2\,dx),$$

from which we infer as usual

$$(X^n)_{n\in\mathbb{N}}\quad\text{is bounded} \qquad (5.2.4.14)$$
$$\tilde{r}_i^n\to 0\quad\text{strongly in}\quad L^2(0,\ell_i) \qquad (5.2.4.15)$$
$$p_{\bar{i}}^n(v_J)\to 0,\quad \bar{i}\in\mathcal{I}_J,\quad J\in\mathcal{J}_N^c. \qquad (5.2.4.16)$$

According to (5.2.4.15), $\tilde{p}_i^n \to 0$ strongly in $L^2(0,\ell_i)$, $i \in \mathcal{I}_J$, $J \in \mathcal{J}_N^c$, since $\tilde{p}_i^n$, in the limit, has to solve the elliptic problem on $(0,\ell_i)$ with Cauchy data on v_J, $J \in \mathcal{J}_N^c$. Equation (5.2.4.8) has as its local analog

$$a_i(\tilde{p}_i^n, \tilde{p}_i^n) = \beta\, C_{iJ}\, \tilde{r}_i^n(v_J) \cdot C_{iJ}\, \tilde{p}_i^n(v_J) + \tilde{h}_{iJ}^n \cdot C_{iJ}\, \tilde{p}_i^n(v_J). \tag{5.2.4.17}$$

But (5.2.4.14) implies that $\tilde{r}_i^n(v_J)$, $\tilde{p}_i^n(v_J)$ are bounded. Thus $a(\tilde{p}_i^n, \tilde{p}_i^n)$ is bounded for all $i = 1, \ldots, n$.

Similarly,

$$\begin{aligned} a(\tilde{r}_i^n, \tilde{r}_i^n) = \tilde{g}_{iJ}^n \cdot C_{iJ}\, \tilde{r}_i^n(v_J) - \beta\, C_{iJ}\, \tilde{p}_i^n(v_J) \cdot C_{iJ}\, \tilde{r}_i^n(v_J) \\ - \tilde{p}_i^n(v_J) \cdot \tilde{r}_i^n(v_J) \end{aligned} \tag{5.2.4.18}$$

implies that $a(\tilde{r}_i^n, \tilde{r}_i^n)$ is bounded for all $i \in \mathcal{I}$. This implies that $\tilde{r}_i^n(v_J) \to 0$ at all nodes, and $\tilde{p}_i^n(v_J) \to 0$ at all nodes to which a controlled edge is incident. In turn, according to (5.2.4.17), (5.2.4.18), we have $\tilde{r}_i^n \to 0$ strongly in $H^1(0,\ell_i)$ for all indices, while $\tilde{p}_i^n \to 0$ strongly in $H^1(0,\ell_i)$ for all edges i incident to a controlled node. Assume now that G is a tree with all but possibly one simple node satisfying (controlled) Robin conditions. Then we can proceed from the controlled simple nodes, where the limiting $\tilde{r}_i$ and $\tilde{p}_i$ are all identical to zero. Then, at the penultimate nodes we transmit zero Robin boundary data across the vertices. This procedure can be repeated until we reach the clamped node. In effect, all solutions $\tilde{r}_i^n$, $\tilde{p}_i^n$ will strongly tend to zero in $H^1(0,\ell_i)$.

ii) In this case (5.2.4.13) does not allow one to infer that $\mathcal{T}$ is nonexpansive.

iii) In this case (5.2.4.13) and the Cauchy-Schwarz inequality applied to the mixed terms show that for β/α sufficiently large, we have nonexpansivity and the same arguments as in case i) apply.

Note that all this holds true for the unrelaxed iteration. Analysis of the relaxed iteration is left to the reader. That case is handled as in Chapter 3. See also Chapter 9 for the 2-dimensional case.

To summarize, we have proved the following result.

Theorem 5.2.4.1 *Assume that $\mathcal{C} = \emptyset$ and that the graph G is a tree with all but possibly one simple node satisfying controlled Robin conditions. In the cases $\alpha = 0$, $\beta > 0$, and α, $\beta > 0$ with β/α sufficiently large, the errors of the iterative scheme strongly converge to zero in the H^1-sense.*

5.3 Hyperbolic Problems

5.3.1 The Global Optimal Control Problem on a Graph

We consider the following standard final value optimal control problem

$$\min_{f\in\mathcal{U}}\{J(f):=\frac{1}{2}\sum_{i\in\mathcal{C}}\int_0^T |f_i|^2\,dt+\frac{1}{2}\sum_{J\in\mathcal{J}_N}\int_0^T |g_J|^2\,dt + \frac{\kappa}{2}\|(r(T),\dot r(T))-(r_0,r_1)\|_{\mathcal{V}\times\mathcal{H}}\} \quad (5.3.1.1)$$

subject to (3.4.1.1)–(3.4.1.7). Note that this formulation tacitly assumes that all simple Neumann nodes are controlled. For this reason we have written $\mathcal{J}_N$, the index set of all simple Neumann nodes, in place of $\mathcal{J}_N^c$.

In what follows we concentrate on the more difficult case of boundary controls and leave the case of distributed controls to the reader, that is, we take $\mathcal{C}=\emptyset$. We also make the notational change $g_J\mapsto f_J$ in (3.4.1.4) and (5.3.1.1). There exists a unique set of optimal boundary controls f_J, $J\in\mathcal{J}_N$ such that

$$f_J(t)=-p_{\hat{i}}(v_J,t),\; t\in(0,T),\; J\in\mathcal{J}_N,\; \hat{i}\in\mathcal{I}_J, \quad (5.3.1.2)$$

where p satisfies the backwards running adjoint system

$$M_i\ddot p_i=[K_i(p_i'+R_i\,p_i)]'-R_i^T\,K_i(p_i'+R_i\,p_i)-S_i\,p_i \text{ in } Q_i \quad (5.3.1.3)$$

$$p_i(v_J,t)=0,\; J\in\mathcal{J}_D,\; i\in\mathcal{I}_J,\; t\in(0,T), \quad (5.3.1.4)$$

$$C_{iJ}\,p_i(v_J,t)=C_{jJ}\,p_j(v_J,t),\; J\in\mathcal{J}_M,\; i,j\in\mathcal{I}_J,\; t\in(0,T), \quad (5.3.1.5)$$

$$\varepsilon_{iJ}\,K_i(p_i'+R_i\,p_i)(v_J,t)-\alpha_i\,\overset{\circ}{p}_i\,(v_J,t)=0,\; J\in\mathcal{J}_N,\; i\in\mathcal{I}_J,\; t\in(0,T), \quad (5.3.1.6)$$

$$\sum_{i\in\mathcal{I}_J}\varepsilon_{iJ}(C_{iJ}^+)^T\,[K_i(p_i'+R_i\,p_i)]\,(v_J,t)=0,\; J\in\mathcal{J}_M,\; t\in(0,T), \quad (5.3.1.7)$$

$$\varepsilon_{iJ}\,\Pi_{iJ}\,K_i(p_i'+R_i\,p_i)(v_J,t)=0,\; J\in\mathcal{J}_M,\; i\in\mathcal{I}_J, t\in(0,T), \quad (5.3.1.8)$$

$$p(T)=\kappa(\dot r(T)-r_1),\; \dot p(T)=-\kappa\,\mathcal{A}(r(T)-r_0), \quad (5.3.1.9)$$

where again $\mathcal{A}$ denotes the Riesz isomorphism between $\mathcal{V}$ and its dual $\mathcal{V}^*$. The superimposed 'o' denotes the linear map

$$\begin{aligned} &d_t: L^2(0,T;\mathbb{R}^{p_i})\to\left(H^1(0,T;\mathbb{R}^{p_i})\right)^*\\ &d_t\,p_i:=\overset{\circ}{p}_i\!:\; \langle\overset{\circ}{p}_i,\varphi\rangle=-\int_0^T p_i\cdot\dot\varphi\,dt, \quad \forall\phi\in H^1(0,T;\mathbb{R}^{p_i}). \end{aligned} \quad (5.3.1.10)$$

where, in abuse of notation, p_i denotes the dimension of the system on edge i and not the adjoint variable p_i; we use these notations in parallel as the intended meaning should be clear from the context in which the symbol p_i is used.

The forward system (3.4.1.1)–(3.4.1.7) can be expressed by (3.4.1.19) as

$$\begin{aligned} \ddot r + \mathcal{A} r + \mathcal{B}\mathcal{B}^* \dot r &= \mathcal{B} g \\ r(0) = r_0,\ \dot r(0) &= r_1, \end{aligned} \tag{5.3.1.11}$$

where instead of (3.4.1.14) we take $\mathcal{B} \leftrightarrow \sqrt{\alpha}\,\mathcal{B}$, and $g \leftrightarrow \frac{1}{\sqrt{\alpha}}\, g$, or in variational form as

$$\begin{aligned} &(\ddot r, \varphi)_V + a(r,\varphi) + \alpha \sum_{J\in\mathcal{J}_N} \dot r_{\hat i}(v_J)\cdot \varphi_{\hat i}(v_J) \\ &\quad = \sum_{J\in\mathcal{J}_N} f_J \cdot \varphi_{\hat i}(v_J),\ \forall\, \varphi \in \mathcal{V},\ t\in(0,T) \\ &\quad r(0) = r_0,\ \dot r(0) = r_1, \end{aligned} \tag{5.3.1.12}$$

The solution has regularity

$$(r,\dot r) \in C([0,T]; \mathcal{V}\times\mathcal{H}), \quad r_{\hat i}(v_J) \in H^1(0,T;\mathbb{R}^{p_{\hat i}}).$$

Similarly, the backward system (5.3.1.3)–(5.3.1.9) can be expressed as

$$\begin{aligned} \ddot p + \mathcal{A} p - \mathcal{B}\,(\overset{\circ}{\mathcal{B}^* p}) &= 0 \\ p(T) = \kappa(\dot r(T) - r_1),\ \ \dot p(T) &= -\kappa\,\mathcal{A}(r(T) - r_0) \end{aligned} \tag{5.3.1.13}$$

or, with the transformation

$$p = \dot q, \quad \dot p = -\mathcal{A} q, \tag{5.3.1.14}$$

as

$$\begin{aligned} \ddot q + \mathcal{A} q - \mathcal{B}\mathcal{B}^*\,\dot q &= 0 \\ q(T) = \kappa(r(T) - r_0),\ \ \dot q(T) &= \kappa(\dot r(T) - r_1). \end{aligned} \tag{5.3.1.15}$$

Equation (5.3.1.15), in its variational form, reads as follows

$$\begin{aligned} &(\ddot q, \psi)_V + a(q,\psi) - \alpha \sum_{J\in\mathcal{J}_N} \dot q_{\hat i}(v_J)\,\psi_{\hat i}(v_J) = 0,\ \forall\,\psi\in\mathcal{V},\ t\in(0,T) \\ &q(T) = \kappa(r(T) - r_0),\ \dot q(T) = \kappa(\dot r(T) - r_1). \end{aligned} \tag{5.3.1.16}$$

Existence and uniqueness of solutions r, q follow as in Section 2.5.1. See also Section 9.2 for the higher-dimensional setting.

5.3.2 The Domain Decomposition Procedure

We introduce the following boundary form on J_N, J_M.

$$b_i(r_i,\varphi_i) := \sum_{\substack{J\in\mathcal{J}_N:\\ i\in\mathcal{I}_J}} \alpha_i r_i(v_J)\cdot\varphi_i(v_J) + \sum_{\substack{J\in\mathcal{J}_M:\\ i\in\mathcal{I}_J}} \beta_J C_{iJ}\, r_i(v_J)\cdot C_{iJ}\,\varphi_i(v_J). \quad (5.3.2.1)$$

The local energy form is given by

$$a_i(r_i,\varphi_i) := \int_0^{\ell_i} \{K_i(r_i' + R_i\, r_i)\cdot(\varphi_i' + R_i\,\varphi_i) + S_i\, r_i\,\varphi_i\}\, dx. \quad (5.3.2.2)$$

For simplicity, we neglect M_i (it is hidden in the inner product otherwise) and introduce, first in the strong formulation, the typical local problem on edge e_i:

$$\ddot{r}_i = [K_i(r_i' + R_i\, r_i)]' - R_i^T\, K_i(r_i' + R_i\, r_i) - S_i\, r_i = F_i \;\text{ in }\; Q_i, \quad (5.3.2.3)$$

$$r_i(v_J,t) = 0,\; J\in\mathcal{J}_D,\; i\in\mathcal{I}_J,\; t\in(0,T), \quad (5.3.2.4)$$

$$\begin{aligned}\varepsilon_{iJ}\, K_i(r_i' + R_i\, r_i)(v_J,t) &+ \alpha_i\,\dot{r}_i(v_J,t)\\ &= f_J(t),\; J\in\mathcal{J}_N,\; i\in\mathcal{I}_J,\; t\in(0,T),\end{aligned} \quad (5.3.2.5)$$

$$\varepsilon_{iJ}\,\Pi_{iJ}\, K_i(r_i' + R_i\, r_i)(v_J,t) = 0,\; J\in\mathcal{J}_M,\; i\in\mathcal{I}_J,\; t\in(0,T), \quad (5.3.2.6)$$

$$\begin{aligned}\varepsilon_{iJ}(C_{iJ}^{+})^T\, K_i(r_i' + R_i\, r_i)(v_J,t) &+ \beta_J\, C_{iJ}\,\dot{r}_i(v_J,t)\\ &= g_{iJ}(t),\; J\in\mathcal{J}_M,\; i\in\mathcal{I}_J,\; t\in(0,T),\end{aligned} \quad (5.3.2.7)$$

$$r_i(\cdot,0) = r_{i0},\quad \dot{r}_i(0,0) = r_i. \quad (5.3.2.8)$$

Note that if edge e_i connects two multiple nodes, both (5.3.2.4) and (5.3.2.5) are vacuous, while if e_i joins a simple node to a multiple node, either (5.3.2.4) or (5.3.2.5) is vacuous.

The variational form of (5.3.2.3)–(5.3.2.7) can be written using (5.3.2.1) and (5.3.2.2) as follows

$$\begin{aligned}&(\ddot{r}_i,\varphi_i)_{\mathcal{V}_i} + a_i(r_i,\varphi_i) + b_i(\dot{r}_i,\varphi_i)\\ &= (F_i,\varphi_i)_{\mathcal{H}_i} + \sum_{\substack{J\in\mathcal{J}_N:\\ i\in\mathcal{I}_J}} f_J\cdot\varphi_i(v_J) + \sum_{\substack{J\in\mathcal{J}_M:\\ i\in\mathcal{I}_J}} g_{iJ}\cdot C_{iJ}\,\varphi_i(v_J),\;\; \forall\,\varphi_i\in\mathcal{V}_i\end{aligned} \quad (5.3.2.9)$$

We may view the quantities f_J as real controls and the parameters λ_{iJ} as virtual controls. Thus for each index i we introduce a control space

$$U_i := \prod_{\substack{J\in\mathcal{J}_N:\\ i\in\mathcal{I}_J}} \mathbb{R}^{p_i} \times \prod_{\substack{J\in\mathcal{J}_M:\\ i\in\mathcal{I}_J}} \mathbb{R}^{p_i}. \quad (5.3.2.10)$$

More concretely, $U_i = \mathbb{R}^{p_i} \times \mathbb{R}^{p_i}$ if e_i connects two multiple nodes or a Neumann node to a multiple node; else $U_i = \mathbb{R}^{p_i}$. We define the bounded linear mapping $\mathcal{B}_i : U_i \to \mathcal{V}_i^*$ by

$$(\mathcal{B}_i(f_J, g_{iJ}), \varphi_i)_{\mathcal{V}_i} = \sum_{\substack{J\in\mathcal{J}_N:\\ i\in\mathcal{I}_J}} \sqrt{\alpha_i}\, f_J\cdot\varphi_i(v_J) + \sum_{\substack{J\in\mathcal{J}_M:\\ i\in\mathcal{I}_J}} \sqrt{\beta_J}\, g_{iJ}\cdot C_{iJ}\,\varphi_i(v_J). \quad (5.3.2.11)$$

With this notation (5.3.2.9) may be expressed in the variational form

$$(\ddot{r}_i, \varphi_i)_{\mathcal{V}_i} + a_i(r_i, \varphi_i) + b_i(\dot{r}_i, \varphi_i) = (F_i, \varphi_i)_{\mathcal{H}_i} + (\mathcal{B}_i(f_J, g_{iJ}), \varphi_i)_{\mathcal{V}_i} \quad (5.3.2.12)$$

for all $\varphi_i \in \mathcal{V}_i$. The variational equation (5.3.2.12) may equally be reinterpreted in the strong formulation as

$$\ddot{r}_i + \mathcal{A}\, r_i + \mathcal{B}\,\mathcal{B}^*\,\dot{r}_i = F_i + \mathcal{B}_i(f_J, g_{iJ}) \ \text{ in } \ \mathcal{V}_i^*. \quad (5.3.2.13)$$

This is the standard format of Section 2.5.1, and, therefore, for f_J, $\lambda_{iJ} \in L^2(0,T;\mathbb{R}^{p_i})$ we obtain (3.4.1.25)–(3.4.1.27). Similarly, we obtain the analog of (3.4.2.12)

$$\begin{aligned}
&\|(r_i(T), \dot{r}_i(T))\|^2_{\mathcal{V}_i\times\mathcal{H}_i} + \sum_{\substack{J\in\mathcal{J}_N:\\ i\in\mathcal{I}_J}} \int_0^T \{\alpha_i\, |\dot{r}_i(v_J,t)|^2 \\
&\quad + \frac{1}{\alpha_i}\, |(C_{iJ}^+)^T\, K_i(r_i' + R_i\, r_i)(v_J,t)|^2\}\, dt + \sum_{\substack{J\in\mathcal{J}_M:\\ i\in\mathcal{I}_J}} \int_0^T \{\beta_j\, |C_{iJ}\,\dot{r}_i(v_J,t)|^2 \\
&\qquad + \frac{1}{\beta_J}\, |(C_{iJ}^+)^T\, K_i(r_i' + R_i\, r_i)(v_J,t)|^2\}\, dt \\
&= \|(r_0, r_1)\|^2_{\mathcal{V}_i\times\mathcal{H}_i} + \sum_{\substack{J\in\mathcal{J}_N:\\ i\in\mathcal{I}_J}} \int_0^T \frac{1}{\alpha_i}\, |f_J|^2\, dt + \sum_{\substack{J\in\mathcal{J}_M:\\ i\in\mathcal{I}_J}} \int_0^T \frac{1}{\beta_J}\, |g_{iJ}|^2\, dt. \quad (5.3.2.14)
\end{aligned}$$

We proceed to derive a similar standard formulation for problems associated with adjoint variables . Therefore we consider the system

$$\ddot{p}_i = [K_i(p_i' + R_i\, p_i)]' - R_i^T\, K_i(p_i' + R_i\, p_i) + S_i\, p_i \ \text{ in } \ Q_i, \quad (5.3.2.15)$$

$$p_i(v_J, t) = 0,\ J \in \mathcal{J}_D,\ i \in \mathcal{I}_J,\ t \in (0,T), \quad (5.3.2.16)$$

$$\begin{aligned}
\varepsilon_{iJ}\, K_i(p_i' + R_i\, p_i)(v_J,t) - \alpha_i\, \overset{\circ}{p}_i\, (v_J,t) &= \overset{\circ}{\tilde{f}}_J\, (t), \\
J \in \mathcal{J}_N,\ i \in \mathcal{I}_J,\ t &\in (0,T),
\end{aligned} \quad (5.3.2.17)$$

$$\varepsilon_{iJ}\, \Pi_{iJ}\, K_i(p_i' + R_i\, p_i)(v_J,t) = 0,\ J \in \mathcal{J}_M,\ i \in \mathcal{I}_J,\ t \in (0,T), \quad (5.3.2.18)$$

$$\varepsilon_{iJ}(C_{iJ}^{+})^T K_i(p_i' + R_i p_i)(v_J,t) - \beta_J C_{iJ} \overset{\circ}{p}_i (v_{J,t}) = \overset{\circ}{\tilde{g}}_{iJ}(t), \quad J \in \mathcal{J}_M,\ i \in \mathcal{I}_J,\ t \in (0,T), \tag{5.3.2.19}$$

$$p_i(T) = p_{0i}, \quad \dot{p}_i(T) = q_{0i}, \tag{5.3.2.20}$$

with $(p_{0i}, q_{0i}) \in \mathcal{H}_i \times \mathcal{V}_i^*$, $(\tilde{f}_J, \tilde{g}_{iJ}) \in U_i$.

If we introduce

$$p_i = \dot{q}_i,\ \dot{p}_i = -\mathcal{A}_i q_i \tag{5.3.2.21}$$

and

$$f_J^\alpha := \frac{1}{\sqrt{\alpha_i}} f_J,\ g_{iJ}^\beta := \frac{1}{\sqrt{\beta_J}} g_{iJ}, \tag{5.3.2.22}$$

then (5.3.2.15)–(5.3.2.20) can be written as

$$\begin{aligned} \ddot{q}_i + \mathcal{A}_i q_i - \mathcal{B}_i \mathcal{B}_i^* \dot{q}_i &= \mathcal{B}_i(f_J^\alpha, g_{iJ}^\beta) \\ q_i(T) = -\mathcal{A}_i^{-1} q_{0i}, &\quad \dot{q}_i(T) = p_{0i}, \end{aligned} \tag{5.3.2.23}$$

which, in turn, can be written as a forward problem by the transformation $t \to T-t$. Thus, the well-posedness of (5.3.2.15)–(5.3.2.20) is governed by transposition of Theorem 3.4.1.1 of Section 3.4.1 and (5.3.2.21), (5.3.2.22), (5.3.2.23).

As a consequence the solution of (5.3.2.15)–(5.3.2.20) has regularity

$$\begin{aligned} &(p_i, \dot{p}_i) \in C(0,T; \mathcal{H}_i \times \mathcal{V}_i^*), \\ &p_i(v_J, \cdot) \in L^2(0,T; \mathbb{R}^{p_i}), \quad J \in \mathcal{J}_N,\ i \in \mathcal{I}_J, \\ &C_{iJ}\, p_i(v_J, \cdot) \in L^2(0,T; \mathbb{R}^{p_i}), \quad J \in \mathcal{J}_M, i \in \mathcal{I}_M. \end{aligned} \tag{5.3.2.24}$$

Just as in Section 3.4.2 we want to employ absorbing conditions at the interfaces v_J, $J \in \mathcal{J}_M$, or, if possible, even transparent conditions. This choice will enhance the speed of convergence. We set:

$$\begin{aligned} &\varepsilon_{iJ}(C_{iJ}^{+})^T K_i((r_i^{n+1})' + r_i^{n+1})(v_J,t) \\ &\qquad + \beta_J C_{iJ} \dot{r}_i^{n+1}(v_J,t) + \gamma_J C_{iJ} p_i^{n+1}(v_J,t) = \lambda_{iJ}^n(t) \end{aligned} \tag{5.3.2.25}$$

and

$$\begin{aligned} &\varepsilon_{iJ}(C_{iJ}^{+})^T K_i((p_i^{n+1})' + R_i p_i^{n+1})(v_J,t) \\ &\qquad - \beta_j C_{iJ} \overset{\circ}{p}_i (v_J,t) + \gamma_J C_{iJ} \overset{\circ}{\dot{r}}_i (v_J,t) = \mu_{iJ}^n(t) \end{aligned} \tag{5.3.2.26}$$

for $t \in (0,T)$, $J \in \mathcal{J}_M$, $i \in \mathcal{I}_J$. The coupling parameters $\lambda^n_{iJ}(\cdot)$, $\mu^n_{iJ}(\cdot)$ are given as follows:

$$\begin{aligned}\lambda^n_{iJ}(t) =& \beta_j(\frac{2}{d_J}\sum_{j\in\mathcal{I}_J} C_{jJ}\,\dot r^n_j(v_J,t) - C_{iJ}\,\dot r^n_i(v_J,t))\\ &+\gamma_J\,(\frac{2}{d_J}\sum_{j\in\mathcal{I}_J} C_{iJ}\,p^n_j(v_J,t) - C_{iJ}\,p^n_i(v_J,t))\\ &-(\frac{2}{d_J}\sum_{j\in\mathcal{I}_J}\varepsilon_{jJ}(C^+_{jJ})^T K_j((r^n_j)' + R_j\,r^n_j)(v_J,t)\\ &-\varepsilon_{iJ}\,(C^+_{iJ})^T K_i((r^n_i)'+R_i\,r^n_i)(v_J,t))\end{aligned} \tag{5.3.2.27}$$

and

$$\begin{aligned}\mu^n_{iJ}(t) =& -\beta_J(\frac{2}{d_J}\sum_{j\in\mathcal{I}_J} C_{iJ}\,\overset{\circ}{p}{}^n_j(v_J,t) - C_{iJ}\,\overset{\circ}{p}{}^n_i(v_J,t))\\ &+\gamma_J(\frac{2}{d_J}\sum_{j\in\mathcal{I}_J} C_{iJ}\,\overset{\circ}{\dot r}{}^n_j(v_J,t) - C_{iJ}\,\overset{\circ}{\dot r}{}^n_i(v_J,t))\\ &-(\frac{2}{d_J}\sum_{j\in\mathcal{I}_J}\varepsilon_{iJ}(C^+_{jJ})^T K_j((p^n_j)' + R_j\,p^n_j)(v_J,t)\\ &-\varepsilon_{iJ}\,(C^+_{iJ})^T K_i((p^n_i)'+R_i\,p^n_i)(v_J,t)).\end{aligned} \tag{5.3.2.28}$$

As for consistency with the original transmission conditions (3.4.1.3), (5.3.1.5), (3.4.1.5), (5.3.1.7), we sum (5.3.2.25), (5.3.2.26), (5.3.2.27) and (5.3.2.28) over $i \in \mathcal{I}_J$, and assume convergence as $n \to \infty$.

We obtain

$$\begin{aligned}\sum_{i\in\mathcal{I}_J}&\varepsilon_{iJ}\,(C^+_{iJ})^T K_i(r'_i + R_i\,r_i)(v_J,t)\\ &+\beta_J\sum_{i\in\mathcal{I}_J} C_{iJ}\,\dot r_i(v_J,t) + \gamma_J\sum_{i\in\mathcal{I}_J} C_{iJ}\,p_i(v_J,t)\\ &=\beta_J\sum_{i\in\mathcal{I}_J} C_{iJ}\,\dot r_i(v_J,t) + \gamma_J\sum_{i\in\mathcal{I}_J} C_{iJ}\,p_i(v_J,t)\\ &\quad-\sum_{i\in\mathcal{I}_J}\varepsilon_{iJ}\,(C^+_{iJ})^T K_i(r'_i+R_i\,r_i)(v_J,t).\end{aligned}$$

This provides first of all

$$\sum_{i\in\mathcal{I}_J}\varepsilon_{iJ}\,(C^+_{iJ})^T K_i(r'_i + R_i\,r_i)(v_J,t) = 0, \tag{5.3.2.29}$$

and secondly, utilizing (5.3.2.25) and (5.3.2.27),

$$\beta_J\,(C_{iJ}\,\dot{r}_i(v_J,t) - \frac{1}{d_J}\sum_{j\in\mathcal{I}_J} C_{iJ}\,\dot{r}_j(v_J,t)) + \gamma_J(C_{iJ}\,p_i(v_J,t) - \frac{1}{d_J}\sum_{j\in\mathcal{I}_J} C_{iJ}\,p_j(v_J,t)) = 0. \quad (5.3.2.30)$$

A similar procedure, applied to (5.3.2.26), (5.3.2.28), leads to

$$\sum_{i\in\mathcal{I}_J} \varepsilon_{iJ}\,(C_{iJ}^{+})^T\,K_i(p_i' + R_i\,p_i)(v_J,t) = 0 \quad (5.3.2.31)$$

and

$$-\,\beta_J\,(C_{iJ}\,\overset{\circ}{p}_i\,(v_J,t) - \frac{1}{d_J}\sum_{j\in\mathcal{I}_J} C_{iJ}\,\overset{\circ}{p}_j\,(v_J,t)) + \gamma_J\,(C_{iJ}\,\overset{\circ}{\dot{r}}_i\,(v_J,t) - \frac{1}{d_J}\sum_{j\in\mathcal{I}_J} C_{jJ}\,\overset{\circ}{\dot{r}}_j\,(v_J,t)) = 0. \quad (5.3.2.32)$$

As the kernel of d_t is $\{0\}$ we infer from (5.3.2.30), (5.3.2.32) that

$$C_{iJ}\,\dot{r}_i(v_J,t) = \frac{1}{d_J}\sum_{j\in\mathcal{I}_J} C_{iJ}\,\dot{r}_j(v_J,t),$$
$$C_{iJ}\,p_i(v_J,t) = \frac{1}{d_J}\sum_{j\in\mathcal{I}_J} C_{iJ}\,p_j(v_J,t),$$

$\forall\, i\in\mathcal{I}_J$ and $t\in[0,T]$.

This implies

$$C_{iJ}\,r_i(v_J,t) = C_{jJ}\,r_j(v_J,t), \quad C_{iJ}\,p_i(v_J,t) = C_{jJ}\,p_j(v_J,t), \quad (5.3.2.33)$$

because of the compatibility conditions of the initial data. Equations (5.3.2.29), (5.3.2.31), (5.3.2.33) constitute the global transmission conditions conditions on the graph.

For easier reference we write down the iteratively coupled system

$$\begin{cases} \ddot{r}_i^{n+1} = [K_i((r_i^{n+1})' + R_i r_i^{n+1})]' - R_i^T K_i((r_i^{n+1})' + R_i r_i^{n+1}) - S_i r_i^{n+1} \\ \ddot{p}_i^{n+1} = [K_i((p_i^{n+1})' + R_i p_i^{n+1})]' - R_i^T K_i((p_i^{n+1})' + R_i\,p_i^{n+1}) - S_i p_i^{n+1} \end{cases} \quad \text{in } Q_i \quad (5.3.2.34)$$

$$r_i^{n+1}(v_J,t) = p_i^{n+1}(v_J,t) = 0,\ J\in\mathcal{J}_D,\ i\in\mathcal{I}_J,\ t\in(0,T) \quad (5.3.2.35)$$

$$\begin{cases} \varepsilon_{iJ}\,\Pi_{iJ}(C_{iJ}^{+})^T\,K_i((r_i^{n+1})' + R_i\,r_i^{n+1})(v_J,t) = 0, \\ \varepsilon_{iJ}\,\Pi_{iJ}(C_{iJ}^{+})^T\,K_i((p_i^{n+1})' + R_i\,p_i^{n+1})(v_J,t) = 0, \end{cases}$$
$$J \in \mathcal{J}_M,\ i \in \mathcal{I}_J,\ t \in (0,T) \quad (5.3.2.36)$$

$$\begin{cases} \varepsilon_{iJ}\,K_i\,((r_i^{n+1})' + R_i\,r_i^{n+1})(v_J,t) + \alpha_i\,\dot r_i^{n+1}(v_J,t) + p_i^{n+1}(v_J,t) = 0, \\ \varepsilon_{iJ}\,K_i\,((p_i^{n+1})' + R_i\,p_i^{n+1})(v_J,t) - \alpha_i\,\overset{\circ}{p}{}_i^{n+1}(v_J,t) = 0, \end{cases}$$
$$J \in \mathcal{J}_N,\ i \in \mathcal{I}_J,\ t \in (0,T) \quad (5.3.2.37)$$

$$\begin{cases} \varepsilon_{iJ}\,K_i\,((r_i^{n+1})' + R_i\,r_i^{n+1})(v_J,t) + \beta_J\,C_{iJ}\,\dot r_i^{n+1}(v_J,t) \\ \qquad + \gamma_J\,C_{iJ}\,p_i^{n+1}(v_J,t) = \lambda_{iJ}^n(t), \\ \varepsilon_{iJ}\,K_i\,((p_i^{n+1})' + R_i\,p_i^{n+1})(v_J,t) - \beta_J\,C_{iJ}\,\overset{\circ}{r}{}_i^{n+1} \\ \qquad + \gamma_J\,C_{iJ}\,\overset{\circ}{\dot r}{}_i^{n+1}(v_J,t) = \mu_{iJ}^n(t), \end{cases}$$
$$J \in \mathcal{J}_M,\ i \in \mathcal{I}_J,\ t \in (0,T) \quad (5.3.2.38)$$

$$\begin{aligned} r_i^{n+1}(0) &= r_{0i}, \quad \dot r_i^{n+1}(0) = r_{1i}, \\ p_i^{n+1}(T) &= \kappa\,(\dot r_i^{n+1}(T) - r_{1i}), \\ \dot p_i^{n+1}(T) &= -\kappa\,\mathcal{A}_i^{-1}(r_i^{n+1}(T) - r_{0i}), \end{aligned} \quad (5.3.2.39)$$

together with (5.3.2.27) and (5.3.2.28).

The question of well-posedness of this coupled system is not easily answered directly. However, and this is precisely one of the main reasons for this class of domain decompositions, the decoupled optimality system (5.3.2.34)–(5.3.2.39) and (5.3.2.27), (5.3.2.28) can be interpreted as an optimality system of a decomposed optimal control problem. We consider only the case where edge e_i connects a Neumann node to a multiple node. The other cases are handled similarly. We define

$$\begin{aligned} J_i^n(f,g) = \frac{1}{2} \sum_{\substack{J\in\mathcal{J}_N:\\ i\in\mathcal{I}_J}} \int_0^T |f_J|^2\,dt + \frac{\kappa}{2}\,\|(r_i(T),\dot r_i(T)) - (r_{i0},r_{i1})\|^2_{\mathcal{V}_i\times\mathcal{H}_i} \\ + \frac{1}{2} \sum_{\substack{J\in\mathcal{J}_M:\\ i\in\mathcal{I}_J}} \frac{1}{\gamma_J} \int_0^T (|g_{iJ}|^2 + |\gamma_J\,C_{iJ}\,\dot r_i(v_J) - \rho_{iJ}^n|^2)\,dt, \end{aligned} \quad (5.3.2.40)$$

where ρ_{iJ}^n is such that $\mu_{iJ}^n = \overset{\circ}{\rho}{}_{iJ}^n$, ρ_{iJ}^n, $\lambda_{iJ}^n \in L^2(0,T;\mathbb{R}^{p_i})$.

The optimal control problem on edge i is given by

$$\begin{aligned}
&\inf_{(f,g)\in U_i} \; J(f,g) \;\text{ such that}\\
&\ddot r_i = [K_i(r_i' + R_i r_i)]' - R_i^T K_i(r_i' + R_i r_i) + S_i r_i \;\text{ in }\; Q_i\\
&r_i(v_J,t) = 0,\; J \in \mathcal{J}_D,\, i \in \mathcal{I}_J,\, t \in (0,T),\\
&\varepsilon_{iJ} K_i(r_i' + R_i r_i)(v_J,t) + \alpha_i \dot r_i(v_J,t) = f_J,\; J \in \mathcal{J}_N,\, i \in \mathcal{I}_J,\, t \in (0,T)\\
&\Pi_{iJ}(C_{iJ}^+)^T K_i(r_i' + R_i r_i)(v_J,t) = 0,\; J \in \mathcal{J}_M,\, i \in \mathcal{I}_J,\, t \in (0,T)\\
&\varepsilon_{iJ} K_i(r_i' + R_i r_i)(v_J,t) + \beta_j C_{iJ} \dot r_i(v_J,t) = \lambda_{iJ}^n(t) + g_{iJ}(t),\\
&\qquad J \in \mathcal{J}_M,\, i \in \mathcal{I}_J,\, t \in (0,T),\\
&r_i(0) = r_{i0},\; \dot r_i(0) = r_{1i}.
\end{aligned} \tag{5.3.2.41}$$

Problem (5.3.2.41) is a standard quadratic optimal control problem of tracking type, for which there exist a unique solution. Exploiting the optimality conditions for (5.3.2.41) one reveals, after some standard calculations, the system (5.3.2.34)–(5.3.2.39) with indices $n+1$ placed on r_i, p_i. Therefore, iterating (5.3.2.34)–(5.3.2.39) together with (5.3.2.27), (5.3.2.28) is equivalent to iterating (5.3.2.41) together with (5.3.2.27), (5.3.2.28). Thus, the decomposition of the globally defined optimal control problem into local problems which is coupled upon iteration, is equivalent to a decomposition into locally defined optimal control problems coupled upon iteration. In addition, this decoupling procedure is completely parallel.

As for the analysis of convergence of this iteration we refer to Chapter 9, as the technical details of the proofs are rather similar.

We conclude this chapter with the display of some numerical results for a planar grid of 40 elastic strings, see the set of Figures 5.1. We begin with the initial configuration. The next figure shows the free evolution after 50 time units. This 'forward' numerical simulation has been done using the domain decomposition procedure of this chapter. The next two figures (all within Figure 5.1) show the result of the action of boundary controls acting at the end of the strings (with the exception of those excited at the beginning) at 10 and 50 time units, respectively. As a matter of fact, we display even real time capable suboptimal results which have been obtained using the so-called *instantaneous control paradigm* which, in turn, consists of a Newmark discretization with respect to time of the partial differential equations and a corresponding discretization of the cost function. Optimization then is done with respect to just one time step, at every step. This strategy has also come to be known as a *one-step moving horizon or rolling horizon* control law. This paradigm has been considered by Hundhammer and Leugering [47] and Hundhammer [46] in the context of wave equations. In these papers more numerical experiments and additional references are given. See Hinze and Kunisch [44] and Choi, Hinze and Kunisch [12] for instantaneous controls applied to parabolic problems. As a matter of fact, the method of instantaneous controls is a very active field of research. As this strategy, which also uses the domain decomposition developed in this chapter, is mainly of numerical interest at this point, we dispense with its detailed description here. It is very remarkable that instantaneous controls perform almost identically to those obtained by exploiting exact (and hereafter discretized) necessary optimality conditions. This phenomenon is visualized by Figures 5.2, 5.3 following the example of a grid.

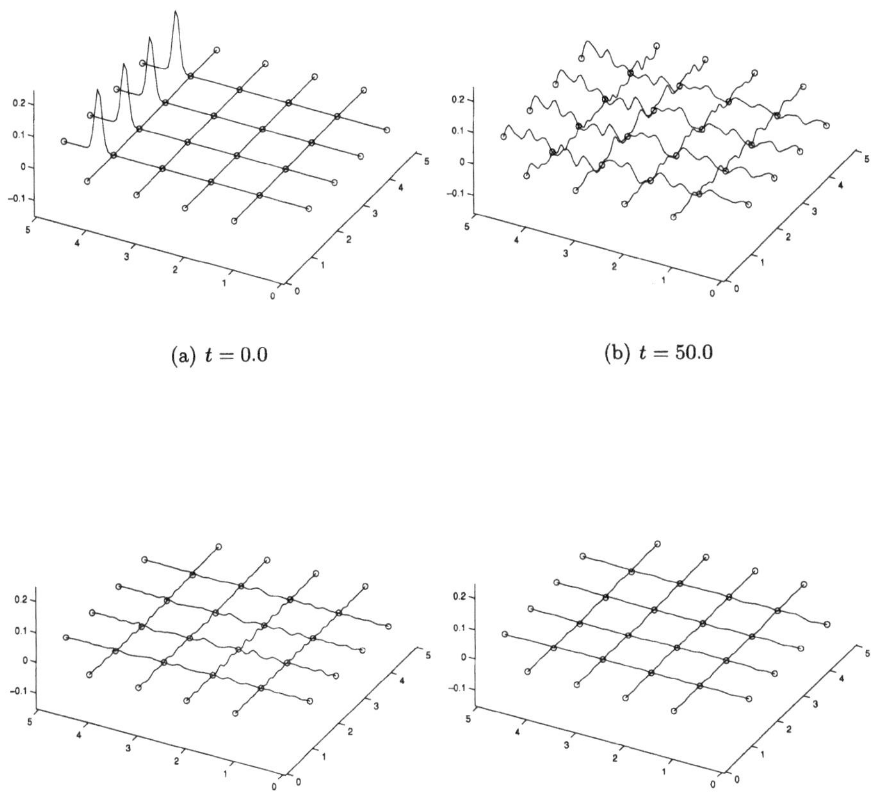

(a) $t = 0.0$ (b) $t = 50.0$

(c) $t = 10.0$ (d) $t = 50.0$

Figure 5.1: Grid of 40 elastic strings (a,b) uncontrolled at $t = 0$ and $t = 50$, (c,d) controlled at $t = 10$ and $t = 50$ time units

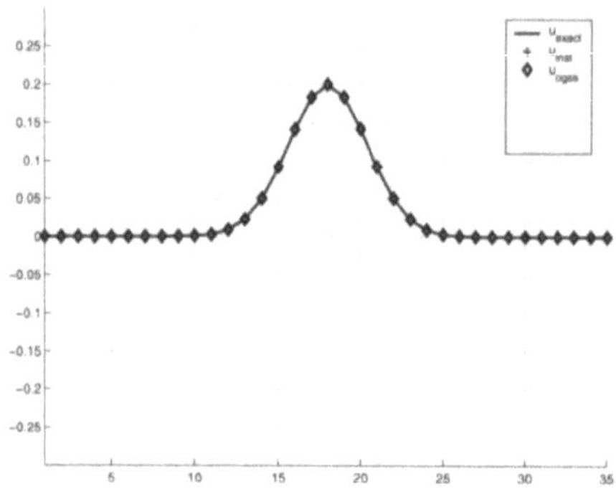

(a) $t = 0.0$

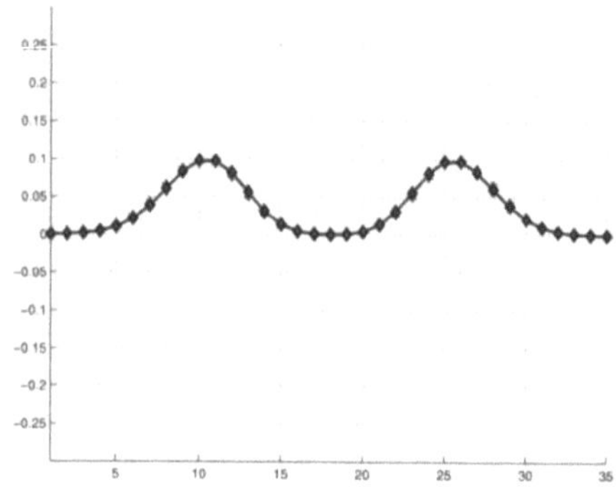

(b) $t = 0.25$

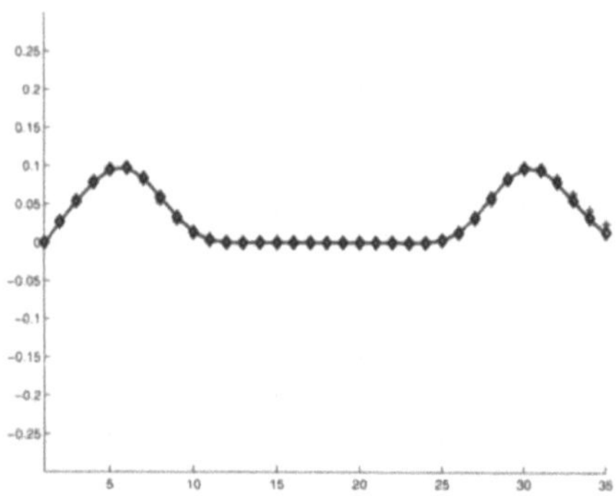

(c) $t = 0.4$

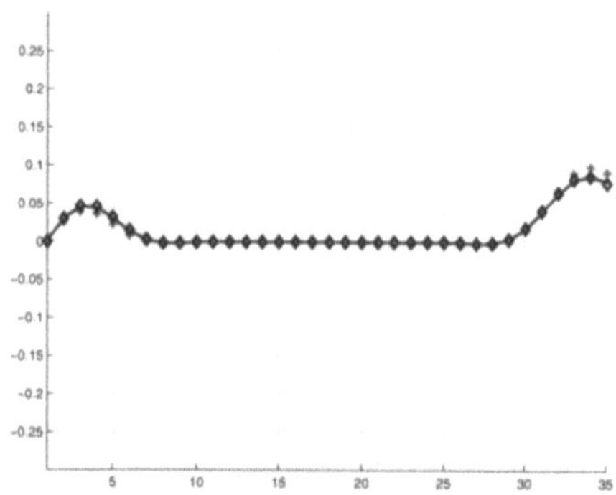

(d) $t = 0.5$

Figure 5.2: Comparison of optimal control, instantaneous control and one-step gradient method

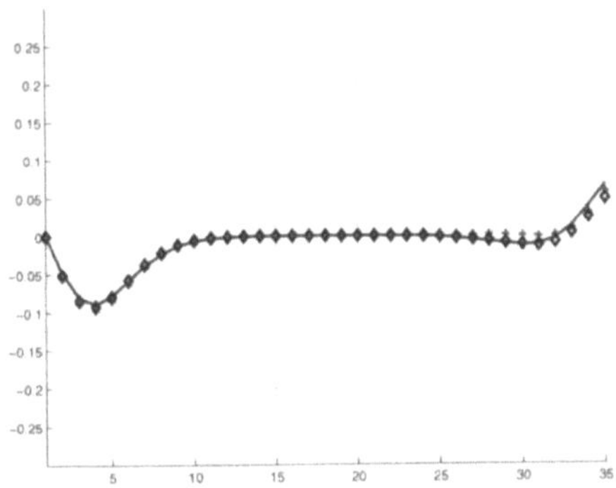

(a) $t = 0.6$

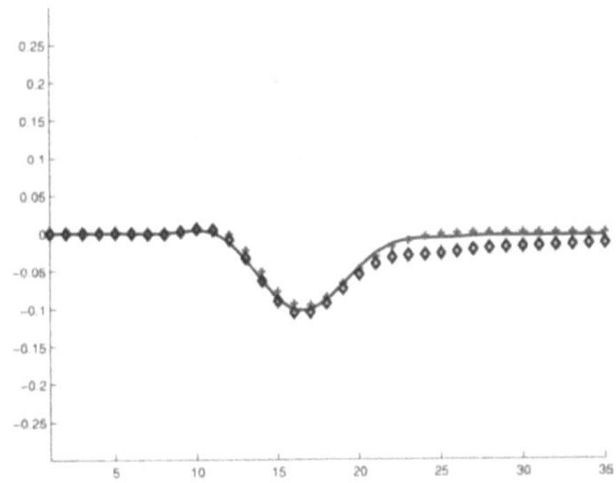

(b) $t = 1.0$

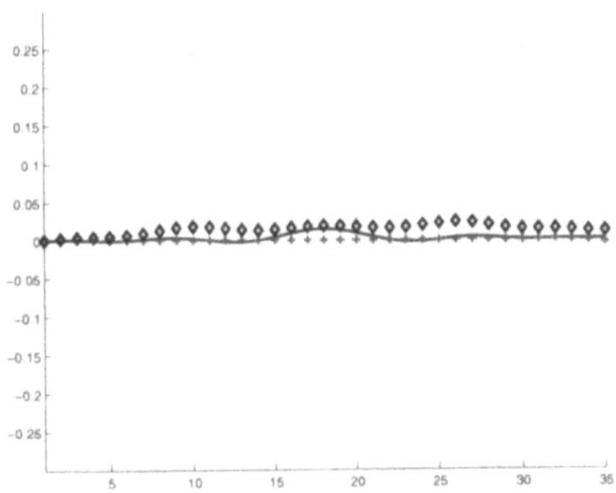

(c) $t = 2.0$

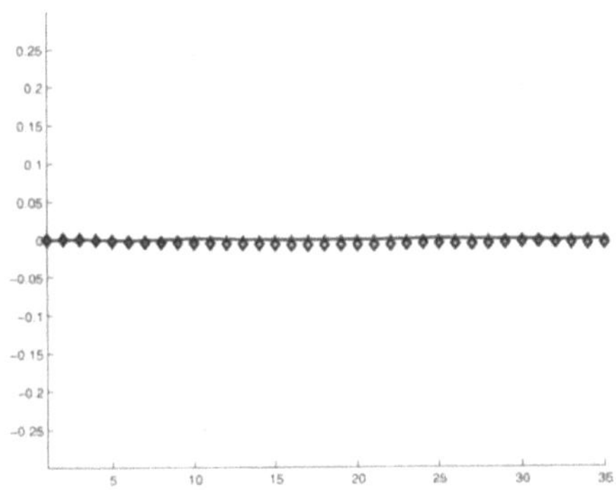

(d) $t = 4.0$

Figure 5.3: Comparison of optimal control, instantaneous control and one-step gradient method

Chapter 6

Domain Decomposition in Optimal Final Value Control of Dissipative Wave Equations

6.1 Introduction

This chapter is concerned with domain decomposition methods for the computation of the solution of the optimality system associated with final value optimal control of second-order hyperbolic equations in which the control action enters through a dissipative boundary condition. More specifically, we consider optimal final value control of solutions of hyperbolic systems of the form

$$\begin{aligned} \frac{\partial^2 w}{\partial t^2} - \nabla\cdot(A\nabla w) + cw = F \text{ in } Q := \Omega\times(0,T) \\ w = 0 \ \text{ on } \Sigma^D := \Gamma^D\times(0,T) \\ \frac{\partial w}{\partial \nu_A} + \alpha\frac{\partial w}{\partial t} = f \ \text{ on } \Sigma^N := \Gamma^N\times(0,T) \\ w(0) = w_0, \quad \frac{\partial w}{\partial t}(0) = v_0 \ \text{ in } \Omega. \end{aligned} \tag{6.1.1.1}$$

In (6.1.1.1), Ω is a bounded domain in $\mathbb{R}^d$ with boundary Γ, A is a positive definite symmetric matrix and c is a nonnegative scalar function defined on Ω, $\Gamma = \overline{\Gamma}^D \cup \overline{\Gamma}^N$ where Γ^D, Γ^N are relatively open and disjoint subsets of Γ, ν is the exterior pointing unit normal vector to Γ and $\partial w/\partial\nu_A = \nu\cdot(A\nabla w)$. Also, α is a nonnegative scalar function defined on Γ^N, and F, f are given external forces defined on $\Omega\times(0,T)$ and $\Gamma^N\times(0,T)$, respectively. Control of the system is exercised through the boundary force f.

The natural energy functional associated with equation (6.1.1.1(i)) is

$$E(t) = \int_\Omega \left[\left| \frac{\partial w}{\partial t}(x,t)\right|^2 + A\nabla w(x,t)\cdot\nabla w(x,t) + c|w(x,t)|^2 \right] dx.$$

The system (6.1.1.1) has the following important properties: (a) in the absence of external forces it is *dissipative* if $\alpha(x) > 0$ on a set of positive measure, that is

$E(t) \le E(0)$ and $E(t) \to 0$ as $t \to \infty$; (b) solutions of (6.1.1.1) have *finite energy* for all $t > 0$ provided $E(0)$ is finite and $F \in L^1(\Omega \times (0,T))$, $f \in L^2(\Gamma^N \times (0,T))$, and $f(x,t) = 0$ on the set $\{(x,t) : \alpha(x) = 0,\, 0 < t < T\}$. The latter requirement is crucial for the conclusion. Indeed, if $\alpha \equiv 0$ and $f \in L^2(\Gamma^N \times (0,T))$ there is a loss of regularity; solutions of (6.1.1.1) will in general not have finite energy for any $t > 0$ regardless of the regularity of the other data entering into the problem.

With regard to optimal final value control, the improved regularity that dissipation affords allows one to penalize in the natural energy norm the deviation of the solution at the final time from its target state. This choice of norm naturally affects the resulting optimality system which, in the case of spatial domain decomposition, leads to an iterative algorithm that is quite distinct from the classical one based on the nonoverlapping Schwarz alternating algorithm as adapted to optimal control problems and that does not seem to have been previously considered in the literature.

In addition to spatial domain decomposition, we also present iterative time domain and combined space-time domain decomposition algorithms. In each case, convergence of the algorithm is established and a posteriori error estimates in terms of the mismatch of the iterates at the interfaces of the subdomains are derived.

6.2 Optimal Dissipative Boundary Control

6.2.1 Setting the Problem

We consider the system (6.1.1.1) where $T > 0$, Ω is a bounded domain in $\mathbb{R}^d$ with piecewise smooth, Lipschitz boundary $\Gamma := \Gamma^D \cup \Gamma^N$, $\Gamma^N \neq \emptyset$ and is relatively open in Γ, $\overline{\Gamma}^D \cap \overline{\Gamma}^N = \emptyset$, ν is the unit exterior normal vector to Γ. We assume that the matrix $A = (a_{ij})$ is symmetric and uniformly positive definite in Ω with $L^\infty(\Omega)$ entries that are piecewise C^1 in Ω. That is, there are open, connected sets $\Omega_i \subset \Omega$, $i = 1, \ldots, I$, such that $\overline{\Omega} = \cup_{i=1}^I \overline{\Omega}_i$, $\overline{\Omega}_i \cap \Omega_j = \emptyset$, and such that $A|_{\Omega_i} \in C^1(\Omega_i)$ and has a C^1 extension to $\overline{\Omega}_i$. Thus (6.1.1.1i) means that

$$\frac{\partial^2 w}{\partial t^2} - \nabla \cdot (A|_{\Omega_i} \nabla w) + c|_{\Omega_i} w = F \text{ in } \Omega_i \times (0,T),\, i = 1, \ldots, I,$$

together with certain *transmission conditions* on the interfaces $\partial\Omega_i \cap \partial\Omega_j$ (see Section 6.4.1 below).

We further assume that $c \in L^\infty(\Omega)$ with $c(x) \ge 0$ a.e., and that $\alpha \in L^\infty(\Gamma^N)$ such that $\alpha_0 := \inf_{\Gamma^N} \alpha(x) > 0$. The boundary term $\alpha(\partial w/\partial t)$ makes the above system dissipative and improves the regularity of the solution. Indeed, as will be shown in the next section, if

$$(w_0, v_0) \in \mathcal{H} := H_D^1(\Omega) \times L^2(\Omega), \quad F \in L^1(0,T; L^2(\Omega)), \quad f \in L^2(\Sigma^N), \tag{6.2.1.1}$$

then (6.1.1.1) has a unique solution with regularity

$$(w, \frac{\partial w}{\partial t}) \in C([0,T];\mathcal{H}), \quad \frac{\partial w}{\partial t}|_{\Sigma^N} \in L^2(\Sigma^N), \tag{6.2.1.2}$$

neither property of which is true if $\alpha = 0$. Here

$H^1_D(\Omega)$ is the closure in $H^1(\Omega)$ of $C^\infty(\overline{\Omega})$ functions that vanish on Γ_D.

Therefore, for any $(z_0, z_1) \in \mathcal{H}$ one may formulate the optimal control problem

$$\inf_{f \in L^2(\Sigma^N)} J(f) \text{ subject to (6.1.1.1)}, \tag{6.2.1.3}$$

where

$$J(f) = \frac{1}{2}\int_{\Sigma^N} |f|^2 d\Sigma + \frac{\kappa}{2}\|(w(T), \frac{\partial w}{\partial t}(T)) - (z_0, z_1)\|^2_{\mathcal{H}}. \tag{6.2.1.4}$$

This problem has a unique optimal control given by

$$f_{\text{opt}} = -p|_{\Sigma^N} \tag{6.2.1.5}$$

where p is the solution of the backwards running adjoint system

$$\begin{gathered} \frac{\partial^2 p}{\partial t^2} - \nabla \cdot (A\nabla p) + cp = 0 \text{ in } Q \\ p = 0 \text{ on } \Sigma^D \\ \frac{\partial p}{\partial \nu_A} - \alpha d_t p = 0 \text{ on } \Sigma^N \\ p(T) = \kappa(\frac{\partial w}{\partial t}(T) - z_1), \quad \frac{\partial p}{\partial t}(T) = -\kappa \mathcal{A}(w(T) - z_0) \text{ in } \Omega \end{gathered} \tag{6.2.1.6}$$

(see Section 6.2.3 below), $\mathcal{A}$ is the Riesz isomorphism of $H^1_D(\Omega)$ onto its dual space, and d_t is a certain time differentiation operator that is specified in the next section. The *optimality system* for the optimal control problem (6.2.1.3) consists of the coupled system (6.1.1.1), (6.2.1.6), (6.2.1.5). In what follows, we shall describe and analyze various domain decompositions procedures for the numerical calculation of this system.

Remark 6.2.1.1 Instead of $L^2(\Sigma^N)$ one may choose as the control space those $L^2(\Sigma^N)$ functions that are supported in $\widetilde{\Gamma}^N \times (0,T)$, where $\widetilde{\Gamma}^N$ is a subset of Γ^N of positive $(n-1)$-dimensional measure. The analysis presented below may easily be adapted to this more general setting with only minor modifications.

Remark 6.2.1.2 Instead of controlling the system (6.1.1.1) through the boundary function f, one may treat f as a given input to the system and exercise control through the distributed force F. More generally, one may consider locally distributed controls $F\mathbf{1}_\omega$, where ω is a nonempty open subset of Ω and $\mathbf{1}_\omega$ is the characteristic function of ω. If the control space is taken to be $L^2(\omega \times (0,T))$ and if, in the cost functional, F is penalized in the natural norm in this space, then the optimal control is given by

$$F = -p\big|_{\omega\times(0,T)} \tag{6.2.1.7}$$

where p is the solution of (6.2.1.6), so that the optimality system in this case consists of (6.1.1.1), (6.2.1.6) and (6.2.1.7). The domain decomposition algorithms, and their analysis, that are presented in this chapter may be adapted to this optimality system with only minor changes. One particular modification of note is the following. In order to prove convergence of the spatial domain decomposition considered in Section 6.4 in the case of this optimality system, it is necessary to modify the dynamics $(6.1.1.1)_1$ by introducing locally distributed damping $a(\partial w/\partial t)$ into $(6.1.1.1)_1$, and the corresponding term $-a(\partial p/\partial t)$ into $(6.2.1.6)_1$, where $a \in L^\infty(\Omega)$ satisfies $a(x) \geq 0$ and $\inf_\omega a(x) > 0$; cf. Section 6.3.4, especially Example 6.3.4.1.

6.2.2 Existence and Regularity of Solutions

If (w_0, v_0), F and f satisfy (6.2.1.1), the existence of a unique solution of (6.1.1.1) with regularity (6.2.1.2) and depending continuously on the data in the relevant spaces follows, for example, from [63, Theorem 7.6.2.2], which contains the system (6.1.1.1) as a special case. However, for the sake of completeness and to set up the notation that we use in what follows, we shall outline proofs of these facts here.

The space $H := L^2(\Omega)$ denotes the standard space of Lebesgue square integrable real-valued functions defined on Ω. The standard inner product and norm on H are denoted by $(\cdot,\cdot)$ and $\|\cdot\|$, respectively. If either

(i) $c(x) > 0$ on a set in Ω of positive measure, or

(ii) $\Gamma^D \neq \emptyset$

then

$$\|\phi\|_V = \left[\int_\Omega (A\nabla\phi\cdot\nabla\phi + c|\phi|^2)\,dx\right]^{1/2}. \tag{6.2.2.1}$$

defines a norm on $H^1_D(\Omega)$ that is *equivalent* to the standard $H^1(\Omega)$ norm. In what follows we shall assume that either (i) or (ii) or both are satisfied, and for the remainder of this chapter assume that $V := H^1_D(\Omega)$ is endowed with the norm (6.2.2.1). We identify H with its dual space and denote by V^* be the dual space of V with respect to H, so that $V \subset H \subset V^*$ with dense and continuous injections. The duality pairing between elements $\phi^* \in V^*$ and $\phi \in V$ is written $\langle\phi^*,\phi\rangle_V$. When $\phi^* \in H$ we have $\langle\phi^*,\phi\rangle_V = (\phi^*,\phi)$ and one should interpret (ϕ^*,ϕ) as

$\langle\phi^*,\phi\rangle_V$ when $\phi^* \in V^*$ and $\phi \in V$. We denote by $\mathcal{A}$ the canonical isomorphism of V onto V^*. Define

$$D_{\mathcal{A}} = \{\phi \in V : \mathcal{A}\phi \in H\} \supseteq H^2(\Omega) \cap H^1_D(\Omega), \quad \|\phi\|_{D_{\mathcal{A}}} = \|\mathcal{A}\phi\|. \tag{6.2.2.2}$$

$\mathcal{A}$ is then an isomorphism of $D_{\mathcal{A}}$ onto H and, formally, functions in $D_{\mathcal{A}}$ satisfy the boundary condition $\partial\phi/\partial\nu_{\mathcal{A}} = 0$ on Γ^N.

Set $U := L^2(\Gamma^N)$ and let us define a bounded linear operator $\mathcal{B} : U \mapsto V^*$ by

$$\langle \mathcal{B}g, \phi\rangle_V = \int_{\Gamma^N} \sqrt{\alpha}\, g\phi\, d\Gamma, \quad \forall\phi \in V.$$

Its dual operator $\mathcal{B}^* \in \mathcal{L}(V, U)$ (identifying U with its dual space) is then given by

$$\mathcal{B}^*\phi = \sqrt{\alpha}\,\phi|_{\Gamma^N}.$$

With this notation, the system (6.1.1.1) may be written as the abstract equation

$$\begin{aligned} \frac{d^2w}{dt^2} + \mathcal{B}\mathcal{B}^*\frac{dw}{dt} + \mathcal{A}w &= F + \mathcal{B}(f/\sqrt{\alpha}) \text{ in } V^*, \\ w(0) = w_0,\ \frac{dw}{dt}(0) &= v_0, \end{aligned} \tag{6.2.2.3}$$

where $f \in L^2(0,T;U)$ and where we may take $F \in L^1(0,T;H)$, for example. Indeed, if we multiply (6.1.1.1(i)) by $\phi \in V$ and formally integrate the product over Ω we obtain by Green's formula

$$\langle \frac{\partial^2 w}{\partial t^2} + \mathcal{A}w, \phi\rangle_V + \int_{\Gamma^N} \alpha\frac{\partial w}{\partial t}\phi\, d\Gamma = \int_\Omega F\phi\, dx + \int_{\Gamma^N} f\phi\, d\Gamma, \quad \forall\phi \in V.$$

Equivalently, by introducing

$$\begin{aligned} \Lambda &= \begin{pmatrix} 0 & I \\ -\mathcal{A} & -\mathcal{B}\mathcal{B}^* \end{pmatrix}, \quad Bf = \begin{pmatrix} 0 \\ \mathcal{B}(f/\sqrt{\alpha}) \end{pmatrix}, \\ D(\Lambda) &= \{(\phi_1,\phi_2) \in V \times V : \mathcal{A}\phi_1 + \mathcal{B}\mathcal{B}^*\phi_2 \in H\}, \\ B^*\begin{pmatrix}\phi_1\\ \phi_2\end{pmatrix} &= \frac{1}{\sqrt{\alpha}}\mathcal{B}^*\phi_2 = \phi_2|_{\Gamma^N}, \\ W &= \begin{pmatrix} w \\ dw/dt \end{pmatrix}, \quad W_0 = \begin{pmatrix} w_0 \\ v_0 \end{pmatrix}, \quad \mathcal{F} = \begin{pmatrix} 0 \\ F \end{pmatrix}, \end{aligned}$$

we may write (6.2.2.3) as the first-order problem

$$\frac{dW}{dt} = \Lambda W + \mathcal{F} + Bf, \quad W(0) = W_0. \tag{6.2.2.4}$$

The following is well known and not difficult to prove.

Proposition 6.2.2.1 *Λ is the infinitesimal generator of a C_0 semigroup $\{S(t) : t \geq 0\}$ of contractions on $\mathcal{H} = V \times H$.*

The unique mild solution of (6.2.2.4) is therefore given formally by

$$W(t) = S(t)W_0 + \int_0^t S(t-s)[\mathcal{F}(s) + Bf(s)]\,ds, \quad 0 < t < T. \tag{6.2.2.5}$$

However, the control-to-state operator defined by

$$Lf(t) := \int_0^t S(t-s)Bf(s)\,ds$$

needs to be properly understood since Bf does not necessarily take values in $\mathcal{H}$. This is clarified in [63, Theorem 7.6.2.2] where it is proved that that the linear map $(W_0, \mathcal{F}, f) \mapsto (W, B^*W)$ is continuous from $\mathcal{H} \times L^1(0,T;\mathcal{H}) \times L^2(0,T;U)$ into $C([0,T];\mathcal{H}) \times L^2(0,T;U)$, a property that may be deduced from the a priori estimate

$$\begin{aligned}(1-\delta)\|W\|^2_{L^\infty(0,T;\mathcal{H})} &+ \left\|\frac{1}{\sqrt{\alpha}}f - \sqrt{\alpha}B^*W\right\|^2_{L^2(0,T;U)} + \|\sqrt{\alpha}B^*W\|^2_{L^2(0,T;U)} \\ &\leq \|W_0\|^2_{\mathcal{H}} + \frac{1}{\delta}\|\mathcal{F}\|^2_{L^1(0,T;\mathcal{H})} + \left\|\frac{1}{\sqrt{\alpha}}f\right\|^2_{L^2(0,T;U)}, \quad 0 < \delta < 1, \end{aligned}\tag{6.2.2.6}$$

or, equivalently,

$$\begin{aligned}(1-\delta)\left\|(w, \frac{\partial w}{\partial t})\right\|^2_{L^\infty(0,T;\mathcal{H})} &+ \int_{\Sigma^N}\left[\frac{1}{\alpha}\left|\frac{\partial w}{\partial \nu_A}\right|^2 + \alpha\left|\frac{\partial w}{\partial t}\right|^2\right] d\Sigma \\ \leq \|(w_0, v_0)\|^2_{\mathcal{H}} &+ \frac{1}{\delta}\left(\int_0^T\left(\int_\Omega |F(x,t)|^2\,dx\right)^{1/2} dt\right)^2 + \int_{\Sigma^N}\frac{1}{\alpha}|f|^2 d\Sigma. \end{aligned}\tag{6.2.2.7}$$

The estimate (6.2.2.6) can be proved by standard energy methods. Proceeding formally for the moment we obtain from (6.2.2.4)

$$\begin{aligned}\frac{1}{2}\|W(t)\|^2_{\mathcal{H}} - \frac{1}{2}\|W_0\|^2_{\mathcal{H}} &= -\int_0^t \left\|\mathcal{B}^*\frac{dw}{dt}(t)\right\|^2_U dt \\ &+ \int_0^T [(\mathcal{F}(t), W(t))_{\mathcal{H}} + (\mathcal{B}(f(t)/\sqrt{\alpha}), \frac{dw}{dt}(t))_H]\,dt. \end{aligned}\tag{6.2.2.8}$$

We have, or any $\delta > 0$,

$$\left|\int_0^t (\mathcal{F}(t), W(t))_{\mathcal{H}}\,dt\right| \leq \frac{\delta}{2}\|W\|^2_{L^\infty(0,T;\mathcal{H})} + \frac{1}{2\delta}\|\mathcal{F}\|^2_{L^1(0,T;\mathcal{H})},$$

$$\left\|\frac{1}{\sqrt{\alpha}}f(t) - \mathcal{B}^*\frac{dw}{dt}(t)\right\|^2_U = \left\|\frac{1}{\sqrt{\alpha}}f(t)\right\|^2_U + \left\|\mathcal{B}^*\frac{dw}{dt}(t)\right\|^2_U - 2(\mathcal{B}(f(t)/\sqrt{\alpha}), \frac{dw}{dt}(t))_H.$$

Use of the last two expressions in (6.2.2.8) yields (6.2.2.6). The above calculation can be completely justified for data $(W_0, \mathcal{F}, f)$ in $\mathcal{H} \times L^1(0,T;\mathcal{H}) \times L^2(0,T;U)$ by approximation with smoother data and passage to the limit; cf. [43, Section 5.2].

Remark 6.2.2.1 When $F = 0$ one has the *energy identity*

$$\|(w(T), \frac{\partial w}{\partial t}(T))\|_{\mathcal{H}}^2 + \int_{\Sigma^N} \left[\frac{1}{\alpha} \left| \frac{\partial w}{\partial \nu_A} \right|^2 + \alpha \left| \frac{\partial w}{\partial t} \right|^2 \right] d\Sigma$$
$$= \|(w_0, v_0)\|_{\mathcal{H}}^2 + \int_{\Sigma^N} \frac{1}{\alpha} |f|^2 d\Sigma \quad (6.2.2.9)$$

for every $T > 0$.

Let Λ^* denote the dual operator of Λ. It is an unbounded operator in $\mathcal{H}^* = V^* \times H$ given by

$$\Lambda^* = \begin{pmatrix} 0 & -\mathcal{A} \\ I & -\mathcal{B}\mathcal{B}^* \end{pmatrix},$$
$$D(\Lambda^*) = \{(\phi_1, \phi_2) \in V^* \times V : \phi_1 - \mathcal{B}\mathcal{B}^*\phi_2 \in H\}.$$

The operator Λ^* is the generator of the semigroup $S^*(t)$ dual to $S(t)$. Therefore, if $P_0 \in \mathcal{H}^*$ then $P(t) = S^*(T-t)P_0 \in C(\mathcal{H}^*)$ is the unique mild solution of

$$\frac{dP}{dt} = -\Lambda^* P, \quad P(T) = P_0. \quad (6.2.2.10)$$

By writing

$$P = \begin{pmatrix} r \\ p \end{pmatrix}, \quad P_0 = \begin{pmatrix} r_0 \\ p_0 \end{pmatrix},$$

this system is the same as

$$\frac{dr}{dt} = \mathcal{A}p, \ \frac{dp}{dt} = -r + \mathcal{B}\mathcal{B}^*p, \quad r(T) = r_0, \ p(T) = p_0, \quad (6.2.2.11)$$

from which one sees that p formally satisfies

$$\frac{d^2p}{dt^2} - \mathcal{B}\mathcal{B}^*\frac{dp}{dt} + \mathcal{A}p = 0, \quad p(T) = p_0, \ \frac{dp}{dt}(T) = -r_0 + \mathcal{B}\mathcal{B}^*p_0. \quad (6.2.2.12)$$

When $(r_0, p_0) \in D(\Lambda^*)$, then from our previous discussion (6.2.2.12) has a unique solution with $(p, dp/dt) \in C([0,T];\mathcal{H})$, $\mathcal{B}^*(dp/dt) \in L^2(\Sigma^N)$. When $(r_0, p_0) \in \mathcal{H}^*$, the solution of (6.2.2.12) is understood as the second component of the mild solution of the problem (6.2.2.11). A more useful characterization of the solution of (6.2.2.12) in this case is provided in the following result.

Proposition 6.2.2.2 *Let $(r_0, p_0) \in \mathcal{H}^*$. Then (r,p) is the solution of (6.2.2.11) if and only if $r = \mathcal{A}q$, $p = dq/dt$, where q be the solution of*

$$\frac{d^2q}{dt^2} - \mathcal{B}\mathcal{B}^*\frac{dq}{dt} + \mathcal{A}q = 0, \quad q(T) = \mathcal{A}^{-1}r_0, \quad \frac{dq}{dt}(T) = p_0. \tag{6.2.2.13}$$

Proof. If q satisfies (6.2.2.13), it is trivial to check that $r = \mathcal{A}q$, $p = dq/dt$ is the solution of (6.2.2.11). Conversely, if (r,p) is the solution of (6.2.2.11) and we set $q = \mathcal{A}^{-1}r$, then $dq/dt = p$ and

$$\frac{d^2q}{dt^2} = \frac{dp}{dt} = -r + \mathcal{B}\mathcal{B}^*\frac{dq}{dt} = -\mathcal{A}q + \mathcal{B}\mathcal{B}^*\frac{dq}{dt},$$
$$q(T) = \mathcal{A}^{-1}r_0, \quad \frac{dq}{dt}(T) = p_0.$$

Corollary 6.2.2.1 *Let $(r_0, p_0) \in \mathcal{H}^*$ and (r,p) be the solution of (6.2.2.11). Then $\mathcal{B}^*p \in L^2(0,T;U)$.*

Proposition 6.2.2.3 *Let $(r_0, p_0) \in \mathcal{H}^*$ and (r,p) be the solution of (6.2.2.11). Then $(r(0,p(0)), p$ and $\mathcal{B}^*p$ are the unique elements in $\mathcal{H}^*$, $C([0,T];H^*)$ and $L^2(0,T;U)$, respectively, satisfying*

$$\langle (r(0),p(0)),(\phi_0,\phi_1)\rangle_{\mathcal{H}} + \int_0^T (p(t),G(t))\,dt + \int_0^T (\mathcal{B}^*p(t), g(t)/\sqrt{\alpha}\,)_U\,dt$$
$$= \langle (r_0,p_0),(\phi(T),\frac{d\phi}{dt}(T))\rangle_{\mathcal{H}},$$
$$\forall(\phi_0,\phi_1) \in \mathcal{H},\ G \in L^1(0,T;H),\ g \in L^2(0,T;U), \tag{6.2.2.14}$$

where ϕ is the solution of

$$\begin{aligned} &\frac{d^2\phi}{dt^2} + \mathcal{B}\mathcal{B}^*\frac{d\phi}{dt} + \mathcal{A}\phi = G + \mathcal{B}(g/\sqrt{\alpha}) \\ &\phi(0) = \phi_0, \quad \frac{d\phi}{dt}(0) = \phi_1 \end{aligned} \tag{6.2.2.15}$$

and where $\langle\cdot,\cdot\rangle_{\mathcal{H}}$ is the scalar product in the $\mathcal{H}^* - \mathcal{H}$ duality.

Proof. We have

$$\begin{pmatrix} \phi(T) \\ \dfrac{d\phi}{dt}(T) \end{pmatrix} = S(T)\begin{pmatrix} \phi_0 \\ \phi_1 \end{pmatrix} + \int_0^T S(T-t)\left[\begin{pmatrix} 0 \\ G(t) \end{pmatrix} + \begin{pmatrix} 0 \\ \mathcal{B}(g(t)/\sqrt{\alpha}) \end{pmatrix}\right] dt \tag{6.2.2.16}$$

so that

$$\langle (r_0,p_0),(\phi(T),\frac{d\phi}{dt}(T))\rangle_{\mathcal{H}} = \langle (r(0),p(0)),(\phi_0,\phi_1)\rangle_{\mathcal{H}}$$
$$+ \int_0^T (p(t),G(t))\,dt + \int_0^T (\mathcal{B}^*p(t), g(t)/\sqrt{\alpha}\,)_U\,dt.$$

Thus $(r(0), p(0))$, p and $\mathcal{B}^*p$ satisfy (6.2.2.14). On the other hand, the mapping $((\phi_0, \phi_1), G, g) \mapsto (\phi, d\phi/dt)$ is bounded $\mathcal{H} \times L^1(0,T;H) \times L^2(0,T;U) \mapsto C([0,T];\mathcal{H})$, so that for any $(r_0, p_0) \in \mathcal{H}^*$ the right side of (6.2.2.14) is a continuous linear functional on $\mathcal{H} \times L^1(0,T;H) \times L^2(0,T;U)$.

Thus there are *unique*

$$(\hat{r}_0, \hat{p}_0) \in \mathcal{H}^*, \quad \hat{p} \in L^\infty(0,T;H), \quad \tilde{p} \in L^2(0,T;U),$$

such that

$$\begin{aligned}\langle (r_0, p_0), (\phi(T), \frac{d\phi}{dt}(T))\rangle_{\mathcal{H}} &= \langle (\hat{r}_0, \hat{p}_0), (\phi_0, \phi_1)\rangle_{\mathcal{H}} \\ &+ \int_0^T (\hat{p}(t), G(t))\, dt + \int_0^T (\tilde{p}(t), g(t)/\sqrt{\alpha})_U dt, \\ &\forall (\phi_0, \phi_1) \in \mathcal{H},\ G \in L^1(0,T;H),\ g \in L^2(0,T;U).\end{aligned}$$

It follows that necessarily

$$(\hat{r}_0, \hat{p}_0) = (r(0), p(0)), \quad \hat{p} = p, \quad \tilde{p} = \mathcal{B}^* p.$$

We shall also need to consider the inhomogeneous problem

$$\frac{dP}{dt} = -\Lambda^* P + Bf, \quad P(T) = P_0, \tag{6.2.2.17}$$

where $P_0 \in \mathcal{H}^*$ and $f \in L^2(0,T;U)$, whose unique mild solution is given formally by

$$P(t) = S^*(T-t)P_0 - L'f(t), \tag{6.2.2.18}$$

where

$$L'f(t) = \int_t^T S^*(s-t)Bf(s)\, ds. \tag{6.2.2.19}$$

One can prove as above that the mapping $(P_0, f) \mapsto (P, \mathcal{B}^*p)$ is continuous from $\mathcal{H}^* \times L^2(0,T;U)$ into $C([0,T];\mathcal{H}^*) \times L^2(0,T;U)$, so that (6.2.2.18) is indeed the mild solution of (6.2.2.17). To see this we write

$$P = \begin{pmatrix} r \\ p \end{pmatrix},$$

$$P_0 = \begin{pmatrix} r_0 \\ p_0 \end{pmatrix}.$$

Then (6.2.2.17) is the same as

$$\frac{dr}{dt} = \mathcal{A}p, \ \frac{dp}{dt} = -r + \mathcal{B}\mathcal{B}^*p + \mathcal{B}(f/\sqrt{\alpha}), \quad r(T) = r_0,\ p(T) = p_0. \tag{6.2.2.20}$$

From (6.2.2.20) we have, proceeding formally,

$$\frac{1}{2}(\|p_0\|^2 - \|p(t)\|^2) = -\int_t^T \langle r(s), p(s)\rangle_V ds + \int_t^T \|\mathcal{B}^* p(s)\|_U^2 ds + \int_t^T \langle \mathcal{B}(f(s)/\sqrt{\alpha}), p(s)\rangle_V dt,$$

$$\frac{1}{2}(\|\mathcal{A}^{-1} r_0\|_V^2 - \|\mathcal{A}^{-1} r(t)\|_V^2) = \int_t^T \langle r(s), p(s)\rangle_V ds.$$

Hence

$$\|p(t)\|^2 + \|r(t)\|_{V^*}^2 + 2\int_t^T \|\mathcal{B}^* p(s)\|_U^2 ds = \|p_0\|^2 + \|r_0\|_{V^*}^2 - 2\int_t^T \langle \mathcal{B}(f(s)/\sqrt{\alpha}), p(s)\rangle_V ds.$$

By writing

$$2\langle \mathcal{B}(f/\sqrt{\alpha}), p\rangle_V = 2(f/\sqrt{\alpha}, \mathcal{B}^* p)_U = \|f/\sqrt{\alpha} - \mathcal{B}^* p\|_U^2 - \|f/\sqrt{\alpha}\|_U^2 - \|\mathcal{B}^* p\|_U^2$$

one obtains the a priori identity

$$\|p(t)\|^2 + \|r(t)\|_{V^*}^2 + \int_t^T (\|\mathcal{B}^* p(s)\|_U^2 + \|f(s)/\sqrt{\alpha} - \mathcal{B}^* p(s)\|_U^2) ds = \|p_0\|^2 + \|r_0\|_{V^*}^2 + \int_t^T \|f(s)/\sqrt{\alpha}\|_U^2 ds. \quad (6.2.2.21)$$

The above computation is justified for sufficiently regular P_0 and f and then may be extended to $(P_0, f) \in \mathcal{H}^* \times L^2(0,T;U)$ by density. One can see from (6.2.2.21) that the mapping $(P_0, f) \mapsto (P, \mathcal{B}^* p)$ is continuous from $\mathcal{H}^* \times L^2(0,T;U)$ into $C([0,T];\mathcal{H}^*) \times L^2(0,T;U)$.

From (6.2.2.20) one formally obtains

$$\begin{aligned} &\frac{d^2 p}{dt^2} - \mathcal{B}\mathcal{B}^* \frac{dp}{dt} + \mathcal{A}p = \frac{d}{dt}\mathcal{B}(f/\sqrt{\alpha}), \\ &p(T) = p_0, \ \frac{dp}{dt}(T) = -r_0 + \mathcal{B}\mathcal{B}^* p_0 + \mathcal{B}(f(T)/\sqrt{\alpha}). \end{aligned} \quad (6.2.2.22)$$

If $(r_0, p_0) \in D(\Lambda^*)$ and f is sufficiently regular, say $f \in H^1(0,T;U)$ with $f(T) = 0$, then from our previous discussion (6.2.2.22) has a unique solution with $(p, dp/dt) \in C([0,T];\mathcal{H})$, $\mathcal{B}^*(dp/dt) \in L^2(\Sigma^N)$. If, however, $(r_0, p_0) \in \mathcal{H}^*$ and $f \in L^2(0,T;U)$, the solution of (6.2.2.22) is understood as the second component of the mild solution of (6.2.2.20).

The same argument used in the proof of Proposition 6.2.2.2 yields the following result.

Proposition 6.2.2.4 *Let $(r_0, p_0) \in \mathcal{H}^*$, $f \in L^2(0,T;U)$. Then (r,p) is the solution of* (6.2.2.20) *if and only if $r = \mathcal{A}q$, $p = dq/dt$, where q be the solution of*

$$\frac{d^2q}{dt^2} - \mathcal{B}\mathcal{B}^*\frac{dq}{dt} + \mathcal{A}q = \mathcal{B}(f/\sqrt{\alpha}), \quad q(T) = \mathcal{A}^{-1}r_0, \ \frac{dq}{dt}(T) = p_0. \qquad (6.2.2.23)$$

Corollary 6.2.2.2 *Let $(r_0, p_0) \in \mathcal{H}^*$, $f \in L^2(0,T;U)$ and (r,p) be the solution of* (6.2.2.20)*. Then $\mathcal{B}^*p \in L^2(0,T;U)$.*

The next result is analogous to Proposition 6.2.2.3 above.

Proposition 6.2.2.5 *Let $(r_0, p_0) \in \mathcal{H}^*$, $f \in L^2(0,T;U)$, and (r,p) be the solution of* (6.2.2.20)*. Then $(r(0,p(0))$, p and $\mathcal{B}^*p$ are the unique elements in $\mathcal{H}^*$, $C([0,T];H^*)$ and $L^2(0,T;U)$, respectively, satisfying*

$$\begin{aligned}
\langle (r(0),p(0)),(\phi_0,\phi_1)\rangle_{\mathcal{H}} &+ \int_0^T (p(t),G(t))\,dt + \int_0^T (\mathcal{B}^*p(t), g(t)/\sqrt{\alpha})_U\,dt \\
&= \langle (r_0,p_0),(\phi(T),\frac{d\phi}{dt}(T))\rangle_{\mathcal{H}} - \int_0^T (f(t)/\sqrt{\alpha}, \mathcal{B}^*\frac{d\phi}{dt})_U\,dt, \\
&\qquad \forall (\phi_0,\phi_1) \in \mathcal{H},\ G \in L^1(0,T;H),\ g \in L^2(0,T;U), \qquad (6.2.2.24)
\end{aligned}$$

where ϕ be the solution of (6.2.2.15).

Proof. We start with regular data (ϕ_0, ϕ_1), G, g such that the solution of (6.2.2.15) satisfies $(\phi, d\phi/dt) \in C([0,T]; D_{\mathcal{A}} \times V)$. From (6.2.2.16) We have

$$\begin{aligned}
&\left\langle \begin{pmatrix} r_0 \\ p_0 \end{pmatrix}, \begin{pmatrix} \phi(T) \\ \frac{d\phi}{dt}(T) \end{pmatrix} \right\rangle_{\mathcal{H}} - \left\langle S^*(T)\begin{pmatrix} r_0 \\ p_0 \end{pmatrix}, \begin{pmatrix} \phi_0 \\ \phi_1 \end{pmatrix} \right\rangle_{\mathcal{H}} \\
&= \int_0^T \left\langle S^*(T-t)\begin{pmatrix} r_0 \\ p_0 \end{pmatrix}, \left[\begin{pmatrix} 0 \\ G(t) \end{pmatrix} + \begin{pmatrix} 0 \\ \mathcal{B}(g(t)/\sqrt{\alpha}) \end{pmatrix}\right] \right\rangle_{\mathcal{H}} dt \\
&= \int_0^T \left\langle \begin{pmatrix} r(t) \\ p(t) \end{pmatrix} + L'f(t), \left[\begin{pmatrix} 0 \\ G(t) \end{pmatrix} + \begin{pmatrix} 0 \\ \mathcal{B}(g(t)/\sqrt{\alpha}) \end{pmatrix}\right] \right\rangle_{\mathcal{H}} dt \\
&= \int_0^T (p(t),G(t))\,dt + \int_0^T (\mathcal{B}^*p(t), g(t)/\sqrt{\alpha})_U\,dt \\
&\qquad + \int_0^T \left\langle L'f(t), \left[\begin{pmatrix} 0 \\ G(t) \end{pmatrix} + \begin{pmatrix} 0 \\ \mathcal{B}(g(t)/\sqrt{\alpha}) \end{pmatrix}\right] \right\rangle_{\mathcal{H}} dt. \qquad (6.2.2.25)
\end{aligned}$$

From (6.2.2.19) we obtain

$$\begin{aligned}
&\left\langle L'f(t), \left[\begin{pmatrix} 0 \\ G(t) \end{pmatrix} + \begin{pmatrix} 0 \\ \mathcal{B}(g(t)/\sqrt{\alpha}) \end{pmatrix}\right] \right\rangle_{\mathcal{H}} \\
&= \int_t^T \left\langle \begin{pmatrix} 0 \\ \mathcal{B}(f(s)/\sqrt{\alpha}) \end{pmatrix}, S(s-t)\left[\begin{pmatrix} 0 \\ G(t) \end{pmatrix} + \begin{pmatrix} 0 \\ \mathcal{B}(g(t)/\sqrt{\alpha}) \end{pmatrix}\right] \right\rangle_{\mathcal{H}} ds.
\end{aligned}$$

Therefore

$$\int_0^T \left\langle L'f(t), \left[\begin{pmatrix} 0 \\ G(t) \end{pmatrix} + \begin{pmatrix} 0 \\ \mathcal{B}(g(t)/\sqrt{\alpha}) \end{pmatrix}\right]\right\rangle_{\mathcal{H}} dt$$
$$= \int_0^T ds \int_0^s \left\langle \begin{pmatrix} 0 \\ \mathcal{B}(f(s)/\sqrt{\alpha}) \end{pmatrix}, S(s-t)\left[\begin{pmatrix} 0 \\ G(t) \end{pmatrix} + \begin{pmatrix} 0 \\ \mathcal{B}(g(t)/\sqrt{\alpha}) \end{pmatrix}\right]\right\rangle_{\mathcal{H}} dt$$
$$= \int_0^T \left\langle \begin{pmatrix} 0 \\ \mathcal{B}(f(s)/\sqrt{\alpha}) \end{pmatrix}, \begin{pmatrix} \phi(s) \\ \frac{d\phi}{dt}(s) \end{pmatrix} - S(s)\begin{pmatrix} \phi_0 \\ \phi_1 \end{pmatrix}\right\rangle_{\mathcal{H}} ds$$
$$= \int_0^T (f(s)/\sqrt{\alpha}, \mathcal{B}^* \frac{d\phi}{ds})_U d\sigma - \left\langle L'f(0), \begin{pmatrix} \phi_0 \\ \phi_1 \end{pmatrix}\right\rangle_{\mathcal{H}}. \tag{6.2.2.26}$$

Equation (6.2.2.24) follows immediately from (6.2.2.25) and (6.2.2.26).

The above computations are justified for solutions ϕ of (6.2.2.15) having the regularity assumed above and may be extended to data $(\phi_0, \phi_1) \in \mathcal{H}$, $G, g \in L^1(0,T;H) \times L^2(0,T;U)$ by density, using continuity properties of the mapping from data $((\phi_0, \phi_1), G, g)$ to $((\phi, d\phi/dt), \mathcal{B}^*(d\phi/dt))$. We have therefore proved that $(r(0), p(0))$, p, $\mathcal{B}^*p$ satisfies (6.2.2.24). On the other hand, using the same properties of this mapping we see that for any $(r_0, p_0) \in \mathcal{H}^*$ and $f \in L^2(0,T;U)$ the right side of (6.2.2.24) is a continuous linear functional on $\mathcal{H} \times L^1(0,T;H) \times L^2(0,T;U)$. It follows that (6.2.2.24) has one and only one solution in $\mathcal{H} \times L^\infty(0,T;\mathcal{H}) \times \in L^2(0,T;U)$ and that $(r(0), p(0))$, p, $\mathcal{B}^*p$ is that solution.

Proposition 6.2.2.5 allows us to interpret (6.2.2.20) as the abstract form of a certain boundary value problem. Set

$$H^1(0,T;U) = \{\phi \in L^2(0,T;U) : d\phi/dt \in L^2(0,T;U)\}$$
$$\|\phi\|^2_{H^1(0,T;U)} = \int_0^T (\|\phi(t)\|_U^2 + \|(d\phi/dt)(t)\|_U^2)dt.$$

Denote by $(H^1(0,T;U))^*$ the dual space of $H^1(0,T;U)$ with respect to the pivot space $L^2(0,T;U)$. A bounded linear operator $d_t : L^2(0,T;U) \mapsto (H^1(0,T;U))^*$ is defined by

$$\langle d_t h, \phi\rangle_{\Sigma^N} = -\int_0^T (h, \frac{d\phi}{dt})_U dt, \quad \forall \phi \in H^1(0,T;U), \tag{6.2.2.27}$$

where $\langle\cdot,\cdot\rangle_{\Sigma^N}$ denotes the inner product in the $(H^1(0,T;U))^*$, $H^1(0,T;U)$ duality pairing. We note that $d_t h$ *is not* the time derivative dh/dt of h in the sense of distributions; the latter is the restriction of the former to $H_0^1(0,T;U)$. The dual operator $d_t^* : H^1(0,T;U) \mapsto L^2(0,T;U)$ is given by

$$d_t^*\phi = -\frac{d\phi}{dt}.$$

It is easy to check that $\mathcal{R}(d_t^*) = L^2(0,T;U)$ and thus $\mathcal{N}(d_t) = \{0\}$.

Now consider the system

$$\begin{aligned}\frac{\partial^2 p}{\partial t^2} - \nabla\cdot(A\nabla p) + cp = 0 \text{ in } Q\\ p = 0 \text{ on } \Sigma^D\\ \frac{\partial p}{\partial \nu_A} - \alpha d_t p = d_t f \text{ on } \Sigma^N\\ p(T) = p_0, \quad \frac{\partial p}{\partial t}(T) = -r_0.\end{aligned} \tag{6.2.2.28}$$

If we multiply (6.2.2.28(i)) by the solution of (6.2.2.15) and formally integrate the product over $\Omega \times (0,T)$ we obtain (recalling the meaning of $d_t p$ and of $d_t f$)

$$\langle\langle(-\frac{\partial p}{\partial t}(0), p(0)), (\phi_0, \phi_1)\rangle\rangle_{\mathcal{H}} + \int_Q pG\,dx\,dt + \int_{\Sigma^N} pg\,d\Sigma = \langle\langle(r_0, p_0), (\phi(T), \frac{\partial \phi}{\partial t}(T))\rangle\rangle_{\mathcal{H}}$$
$$- \int_{\Sigma^N} f\frac{\partial \phi}{\partial t} d\Sigma, \quad \forall(\phi_0, \phi_1) \in \mathcal{H}, (G, g) \in L^2(Q) \times L^2(\Sigma^N), \tag{6.2.2.29}$$

which is the same variational equation as (6.2.2.24) if we identify $\partial p/\partial t$ with $-r$. Thus the solution (r,p) of (6.2.2.20) may be interpreted as the solution, of the boundary value problem (6.2.2.28) wherein $r = -\partial p/\partial t$. That is to say, *by definition* the solution $(p, \partial p/\partial t)$ of (6.2.2.28) is $(p, -r)$, where (r,p) is the solution of (6.2.2.20). Equivalently, according to Proposition 6.2.2.4 the solution of (6.2.2.28) may be expressed as

$$p = \frac{\partial q}{\partial t}, \quad \frac{\partial p}{\partial t} = -\mathcal{A}q, \tag{6.2.2.30}$$

where q is the solution of

$$\begin{aligned}\frac{\partial^2 q}{\partial t^2} - \nabla\cdot(A\nabla q) + cq = 0 \text{ in } Q\\ q = 0 \text{ on } \Sigma^D\\ \frac{\partial q}{\partial \nu_A} - \alpha\frac{\partial q}{\partial t} = f \text{ on } \Sigma^N\\ q(T) = \mathcal{A}^{-1}r_0, \quad \frac{\partial q}{\partial t}(T) = p_0.\end{aligned} \tag{6.2.2.31}$$

We summarize the above discussion in the following Proposition.

Proposition 6.2.2.6 *For any $(p_0, r_0) \in H \times V^*$ and $f \in L^2(\Sigma^N)$, (6.2.2.28) has a unique solution with $(p, \partial p/\partial t) \in C([0,T]; H \times V^*)$, $p|_{\Sigma^N} \in L^2(\Sigma^N)$. Further, $(-(\partial p/\partial t)(0), p(0))$, p and $p|_{\Sigma^N}$ are the unique elements in $V^* \times H$, $C([0,T];H)$ and $L^2(\Sigma^N)$, respectively, that satisfy (6.2.2.29). Moreover, (6.2.2.30) holds, where q is the solution of (6.2.2.31).*

6.2.3 The Global Optimality System

Let $(z_0, z_1) \in \mathcal{H}$ and consider the optimal control problem

$$\inf_{f \in L^2(\Sigma^N)} J(f) \text{ subject to (6.1.1.1)}, \tag{6.2.3.1}$$

where

$$J(f) = \frac{1}{2}\int_{\Sigma^N} |f|^2 d\Sigma + \frac{\kappa}{2}(\|w(T) - z_0\|_V^2 + \|\frac{\partial w}{\partial t}(T) - z_1\|^2). \tag{6.2.3.2}$$

where $\kappa > 0$. By virtue of properties of the map $((w_0, v_0), f) \mapsto (w, \partial w/\partial t)$ discussed in the previous section, the existence of a unique optimal control f_{opt} is standard. It is characterized as the solution of the variational equation

$$\int_{\Sigma^N} f_{\text{opt}} h \, d\Sigma + \kappa\big[\langle \mathcal{A}(w(T) - z_0), \psi(T)\rangle_V + (\frac{\partial w}{\partial t}(T) - z_1, \frac{\partial \psi}{\partial t}(T))\big] = 0,$$
$$\forall h \in L^2(\Sigma^N), \tag{6.2.3.3}$$

where ψ is the solution of

$$\frac{\partial^2 \psi}{\partial t^2} - \nabla \cdot (A\nabla\psi) + c\psi = 0 \text{ in } Q$$
$$\psi = 0 \text{ on } \Sigma^D, \quad \frac{\partial \psi}{\partial \nu_A} + \alpha\frac{\partial \psi}{\partial t} = h \text{ on } \Sigma^N$$
$$\psi(0) = \frac{\partial \psi}{\partial t}(0) = 0 \text{ in } \Omega.$$

Let p be the solution of the backwards running adjoint system

$$\begin{aligned} &\frac{\partial^2 p}{\partial t^2} - \nabla \cdot (A\nabla p) + cp = 0 \text{ in } Q \\ &p = 0 \text{ on } \Sigma^D \\ &\frac{\partial p}{\partial \nu_A} - \alpha d_t p = 0 \text{ on } \Sigma^N \\ &p(T) = \kappa(\frac{\partial w}{\partial t}(T) - z_1) \in L^2(\Omega), \\ &\frac{\partial p}{\partial t}(T) = -\kappa\mathcal{A}(w(T) - z_0) \in V^* \end{aligned} \tag{6.2.3.4}$$

In view of (6.2.2.29), p satisfies

$$\int_{\Sigma^N} ph \, d\Sigma = \kappa\big[\langle \mathcal{A}(w(T) - z_0), \psi(T)\rangle_V + (\frac{\partial w}{\partial t}(T) - z_1, \frac{\partial \psi}{\partial t}(T))\big] = 0,$$
$$\forall h \in L^2(\Sigma^N). \tag{6.2.3.5}$$

It follows immediately from (6.2.3.3) and (6.2.3.5) that

$$f_{\rm opt} = -p|_{\Sigma^N}. \tag{6.2.3.6}$$

The *global optimality system* consists of (6.1.1.1), (6.2.3.4) and (6.2.3.6).

Again we emphasize that, because of Proposition 6.2.2.6, the adjoint variable p is given by (6.2.2.30) where q is the solution of

$$\begin{gathered}\frac{\partial^2 q}{\partial t^2} - \nabla\cdot(A\nabla q) + cq = 0 \text{ in } Q\\ q = 0 \text{ on } \Sigma^D\\ \frac{\partial q}{\partial \nu_A} - \alpha\frac{\partial q}{\partial t} = 0 \text{ on } \Sigma^N\\ q(T) = \kappa(w(T) - z_0),\quad \frac{\partial q}{\partial t}(T) = \kappa\big(\frac{\partial w}{\partial t}(T) - z_1\big).\end{gathered} \tag{6.2.3.7}$$

6.3 Time Domain Decomposition

6.3.1 Description of the Algorithm

We decompose the time-span $(0,T)$ into $K+1$ subintervals I_k, $k = 0,1,\dots,K$, by setting

$$0 = T_0 < T_1 < T_2 < \cdots < T_K < T_{K+1} = T, \tag{6.3.1.1}$$

i.e., $I_k := (T_k, T_{k+1})$, and introduce locally defined functions $w_k := w|_{I_k}$, $p_k := p|_{I_k}$, and so forth. Next, the optimality system (6.1.1.1), (6.2.3.4) and (6.2.3.6) is decomposed into $K+1$ subsystems defined on I_k, $k = 0,\dots,K$, which are coupled through the time-wise transmission conditions

$$\begin{aligned} w_k(T_{k+1}) &= w_{k+1}(T_{k+1}),\\ \frac{\partial w_k}{\partial t}(T_{k+1}) &= \frac{\partial w_{k+1}}{\partial t}(T_{k+1}),\\ p_k(T_{k+1}) &= p_{k+1}(T_{k+1}),\\ \frac{\partial p_k}{\partial t}(T_{k+1}) &= \frac{\partial p_{k+1}}{\partial t}(T_{k+1})\end{aligned} \tag{6.3.1.2}$$

for $k = 0,\dots,K-1$. The coupling conditions (6.3.1.2) are then uncoupled by the following iteration in n:

$$\begin{aligned} \frac{\partial w_k^{n+1}}{\partial t}(T_{k+1}) - \beta p_k^{n+1}(T_{k+1}) &= \mu_{k,k+1}^n\\ \beta\frac{\partial p_k^{n+1}}{\partial t}(T_{k+1}) + \mathcal{A}w_k^{n+1}(T_{k+1}) &= \eta_{k,k+1}^n\\ \frac{\partial w_k^{n+1}}{\partial t}(T_k) + \beta p_k^{n+1}(T_k) &= \mu_{k,k-1}^n\\ -\beta\frac{\partial p_k^{n+1}}{\partial t}(T_k) + \mathcal{A}w_k^{n+1}(T_k) &= \eta_{k,k-1}^n\end{aligned} \tag{6.3.1.3}$$

where

$$\begin{aligned}\mu_{k,k+1}^n &= \frac{\partial w_{k+1}^n}{\partial t}(T_{k+1}) - \beta p_{k+1}^n(T_{k+1})\\ \eta_{k,k+1}^n &= \beta\frac{\partial p_{k+1}^n}{\partial t}(T_{k+1}) + Aw_{k+1}^n(T_{k+1})\\ \mu_{k,k-1}^n &= \frac{\partial w_{k-1}^n}{\partial t}(T_k) + \beta p_{k-1}^n(T_k)\\ \eta_{k,k-1}^n &= -\beta\frac{\partial p_{k-1}^n}{\partial t}(T_k) + Aw_{k-1}^n(T_k)\end{aligned} \tag{6.3.1.4}$$

for $k = 1,\ldots,K-1$, where $\beta > 0$. In I_0 or I_K the first or the second half of (6.3.1.3) and (6.3.1.4) are valid, respectively. It is readily seen that, upon convergence, (6.3.1.3) and (6.3.1.4) reduce to (6.3.1.2), hence (6.3.1.3) and (6.3.1.4) are consistent with (6.3.1.2). A similar splitting has been proposed by Benamou for a different control problem in [7]. However, no analysis is given in [7].

For the sake of reference we write down the complete set of decoupled systems:

$$\begin{cases}\dfrac{\partial^2 w_k^{n+1}}{\partial t^2} - \nabla\cdot(A\nabla w_k^{n+1}) + cw_k^{n+1} = F_k\\ \dfrac{\partial^2 p_k^{n+1}}{\partial t^2} - \nabla\cdot(A\nabla p_k^{n+1}) + cp_k^{n+1} = 0 \text{ in } Q_k := \Omega\times I_k\end{cases} \tag{6.3.1.5}$$

$$w_k^{n+1} = p_k^{n+1} = 0 \text{ on } \Sigma_k^D := \Gamma^D \times I_k, \tag{6.3.1.6}$$

$$\begin{cases}\dfrac{\partial w_k^{n+1}}{\partial \nu_A} + \alpha\dfrac{\partial w_k^{n+1}}{\partial t} + p_k^{n+1} = 0\\ \dfrac{\partial p_k^{n+1}}{\partial \nu_A} - \alpha d_t p_k^{n+1} = 0 \text{ on } \Sigma_k^N := \Gamma^N\times I_k\end{cases} \tag{6.3.1.7}$$

$$\begin{aligned}w_0^{n+1}(0) = w_0, \quad &\frac{\partial w_0^{n+1}}{\partial t}(0) = v_0 \text{ in } \Omega\\ p_K^{n+1}(T) &= \kappa\Big(\frac{\partial w_K^{n+1}}{\partial t}(T) - z_1\Big) \in H\\ \frac{\partial p_K^{n+1}}{\partial t}(T) &= -\kappa A(w_K^{n+1}(T) - z_0) \in V^*\end{aligned} \tag{6.3.1.8}$$

subject to (6.3.1.3), (6.3.1.4). In $(6.3.1.7)_2$, d_t is the bounded linear operator from $L^2(I_k;U)$ into $(H^1(I_k;U))^*$ defined analogously to (6.2.2.27) above.

We now show that, for $\{\mu_{k,k+1}^n, \eta_{k,k+1}^n)$ and $\{\mu_{k,k-1}^n, \eta_{k,k-1}^n)$ given in $H\times V^*$, the decoupled optimality system (6.3.1.3)–(6.3.1.8) corresponds, in fact, to an optimality system concentrated on the time-interval I_k. We omit the superscript $n+1$ for the time being and focus first on the internal intervals I_k, $k = 1,\ldots,K-1$.

Define the following cost functional on I_k:

$$J_k^n(f_k, g_{k,k-1}, h_{k,k-1}) := \frac{1}{2}\int_{\Sigma_k^N} f_k^2 d\Gamma dt + \frac{1}{2\beta}\{\|w_k(T_{k+1}) - \mathcal{A}^{-1}\eta_{k,k+1}^n\|_V^2$$
$$+ \|\frac{\partial}{\partial t} w_k(T_{k+1}) - \mu_{k,k+1}^n\|_H^2 + \|g_{k,k-1}\|_H^2 + \|h_{k,k-1}\|_{V^*}^2\}. \quad (6.3.1.9)$$

We then consider the following optimal control problem on I_k.

$$\inf_{\substack{f_k \in L^2(\Sigma_k^N) \\ (h_{k,k-1}, g_{k,k-1}) \in \mathcal{H}^*}} J_k^n(f_k, g_{k,k-1}, h_{k,k-1})$$

subject to

$$\begin{aligned}
\frac{\partial^2 w_k}{\partial t^2} - \nabla\cdot(A\nabla w_k) + c w_k &= F_k \text{ in } Q_k, \\
w_k &= 0 \text{ on } \Sigma_k^D, \\
\frac{\partial w_k}{\partial \nu_A} + \alpha \frac{\partial w_k}{\partial t} &= f_k \text{ on } \Sigma_k^N, \\
w_k(T_k) &= \mathcal{A}^{-1}(\eta_{k,k-1}^n + h_{k,k-1}) \text{ in } V, \\
\frac{\partial w_k}{\partial t}(T_k) &= -(\mu_{k,k-1}^n + g_{k,k-1}) \text{ in } H.
\end{aligned} \quad (6.3.1.10)$$

In order to derive the optimality system associated with the optimal control problem (6.3.1.10) we introduce the auxiliary sensitivity system

$$\begin{aligned}
\frac{\partial^2 \psi_k}{\partial t^2} - \nabla\cdot(A\nabla \psi_k) + c\psi_k &= 0 \text{ in } Q_k \\
\psi_k &= 0 \text{ on } \Sigma_k^D \\
\frac{\partial \psi_k}{\partial \nu_A} + \alpha\frac{\partial \psi_k}{\partial t} &= \bar{f}_k \text{ on } \Sigma_k^N \\
\psi_k(T_k) = \mathcal{A}^{-1}\bar{h}_{k,k-1}, \ \frac{\partial \psi_k}{\partial t}(T_k) &= -\bar{g}_{k,k-1} \text{ in } \Omega.
\end{aligned} \quad (6.3.1.11)$$

The directional derivative of J_k at f_k, $g_{k,k-1}$, $h_{k,k-1}$ into the direction of $\bar{f}_k$, $\bar{g}_{k,k-1}$, $\bar{h}_{k,k-1}$ is set to zero. We obtain as a necessary and sufficient condition for optimality the variational equation

$$0 = \int_{\Sigma_k^N} f_k\bar{f}_k d\Sigma + \frac{1}{\beta}\langle \mathcal{A}w_k(T_{k+1}) - \eta_{k,k+1}^n, \psi_k(T_{k+1})\rangle_V$$
$$+ \frac{1}{\beta}(\frac{\partial w_k}{\partial t}(T_{k+1}) - \mu_{k,k+1}^n, \frac{\partial \psi_k}{\partial t}(T_k))$$
$$+ \frac{1}{\beta}(g_{k,k-1}, \bar{g}_{k,k-1}) + \frac{1}{\beta}\langle \mathcal{A}^{-1}h_{k,k-1}, \bar{h}_{k,k-1}\rangle_V,$$
$$\forall \bar{f}_k \in L^2(\Sigma_k^N),\ (\bar{h}_{k,k-1}, \bar{g}_{k,k-1}) \in \mathcal{H}^*. \quad (6.3.1.12)$$

We now consider the following adjoint system

$$
\begin{aligned}
\frac{\partial^2 p_k}{\partial t^2} - \nabla \cdot (A \nabla p_k) + c p_k = 0 \ \text{ in } Q_k \\
p_k = 0 \ \text{ on } \Sigma_k^D \\
\frac{\partial p_k}{\partial \nu_A} - \alpha d_t p_k = 0 \ \text{ on } \Sigma_k^N \\
p_k(T_{k+1}) = \frac{1}{\beta}\Big(\frac{\partial w_k}{\partial t}(T_{k+1}) - \mu_{k,k+1}^n\Big) \in H \\
\frac{\partial p_k}{\partial t}(T_{k+1}) = -\frac{1}{\beta}(\mathcal{A} w_k(T_{k+1}) - \eta_{k,k+1}^n) \in V^*.
\end{aligned}
\tag{6.3.1.13}
$$

If $(6.3.1.13)_1$ is multiplied by the solution of (6.3.1.11) and the product formally integrated over Q_k one obtains

$$
\begin{aligned}
0 = \Big\langle \frac{\partial p_k}{\partial t}(T_{k+1}), \psi_k(T_{k+1})\Big\rangle_V - \Big\langle \frac{\partial p_k}{\partial t}(T_k), \mathcal{A}^{-1}\bar{h}_{k,k-1}\Big\rangle \\
- \Big(p_k(T_{k+1}), \frac{\partial \psi_k}{\partial t}(T_{k+1})\Big) - (p_k(T_k), \bar{g}_{k,k-1}) + \int_{\Sigma_k^N} p_k \bar{f}_k d\Sigma,
\end{aligned}
$$

or

$$
\begin{aligned}
\int_{\Sigma_k^N} p_k \bar{f}_k d\Sigma = \frac{1}{\beta}\langle \mathcal{A} w_k(T_{k+1}) - \eta_{k,k+1}^n, \psi_k(T_{k+1})\rangle_V \\
+ \frac{1}{\beta}\Big(\frac{\partial w_k}{\partial t}(T_{k+1}) - \mu_{k,k+1}^n, \frac{\partial \psi_k}{\partial t}(T_{k+1})\Big) \\
+ \Big\langle \mathcal{A}^{-1}\frac{\partial p_k}{\partial t}(T_k), \bar{h}_{k,k-1}\Big\rangle_V + (p_k(T_k), \bar{g}_{k,k-1}).
\end{aligned}
\tag{6.3.1.14}
$$

The above calculation is justified by virtue of the *definition*, by transposition, of the solution of (6.3.1.13); cf. the calculation in the previous section.

From (6.3.1.12) and (6.3.1.14) we obtain

$$
\begin{aligned}
0 = \int_{\Sigma_k^N} (f_k + p_k)\bar{f}_k d\Sigma - \Big\langle \mathcal{A}^{-1}\frac{\partial p_k}{\partial t}(T_k), \bar{h}_{k,k-1}\Big\rangle_V \\
- (p_k(T_k), \bar{g}_{k,k-1}) + \frac{1}{\beta}(g_{k,k-1}, \bar{g}_{k,k-1}) + \frac{1}{\beta}\langle \mathcal{A}^{-1} h_{k,k-1}, \bar{h}_{k,k-1}\rangle_V, \\
\forall\, \bar{f}_k \in L^2(\Sigma_k^N),\ (\bar{h}_{k,k-1}, \bar{g}_{k,k-1}) \in \mathcal{H}^*.
\end{aligned}
\tag{6.3.1.15}
$$

Taking variations in (6.3.1.15) gives

$$
f_k = -p_k\big|_{\Sigma_k^N}, \tag{6.3.1.16}
$$

$$h_{k,k-1} = \beta \frac{\partial p_k}{\partial t}(T_k) \text{ in } V^*, \tag{6.3.1.17}$$

$$g_{k,k-1} = \beta p_k(T_k) \text{ in } H. \tag{6.3.1.18}$$

If the optimality conditions (6.3.1.16)–(6.3.1.18) are inserted into (6.3.1.10) we obtain from (6.3.1.10) and (6.3.1.13)

$$\begin{aligned} w_k(T_k) &= \mathcal{A}^{-1}\big(\eta^n_{k,k-1} + \beta \frac{\partial p_k}{\partial t}(T)\big) \\ \frac{\partial w_k}{\partial t}(T_k) &= -\mu^n_{k,k+1} - \beta p_k(T_k) \\ p_k(T_{k+1}) &= \frac{1}{\beta}\big(\frac{\partial w_k}{\partial t}(T_{k+1}) - \mu^n_{k,k+1}\big) \\ \frac{\partial p_k}{\partial t}(T_{k+1}) &= -\frac{1}{\beta}\big(\mathcal{A} w_k(T_{k+1}) - \eta^n_{k,k+1}\big), \end{aligned}$$

which are precisely the conditions (6.3.1.3), where $\mu^n_{k,k+1}$, $\mu^n_{k,k-1}$, $\eta^n_{k,k+1}$, $\eta^n_{k,k-1}$ are given data.

We conclude that the iterative decoupling procedure for $k = 1,\dots,K-1$ according to (6.3.1.3)–(6.3.1.8) produces the optimality system corresponding to the optimal control problem (6.3.1.10). The interpretation of (6.3.1.10) is as follows. One introduces "artificial controls," or "virtual controls" $h_{k,k-1}$, $g_{k,k-1}$ (see Pironneau and Lions [76, 77]), which serve to match the initial and final conditions in terms of the iteration history, while optimizing also with respect to the "effective control" f_k.

Let us now consider the boundary intervals I_0 and I_K. On I_0 define the cost functional

$$J^n_0(f_0) := \frac{1}{2}\int\limits_{\Sigma^K_0} f_0^2 d\Sigma + \frac{1}{2\beta}\{\|w_0(T_1) - \mathcal{A}^{-1}\eta^n_{0,1}\|^2_V + \|\frac{\partial w_0}{\partial t}(T_1) - \mu^n_{0,1}\|^2_H\} \tag{6.3.1.19}$$

and consider the optimal control problem (which contains a slight conflict in notation between the solution and the initial data)

$$\begin{aligned} &\inf_{f_0 \in L^2(\Sigma^N_0)} J^n_0(f_0) \text{ subject to} \\ &\frac{\partial^2 w_0}{\partial t^2} - \nabla\cdot(A\nabla w_0) + c w_0 = F_0 \text{ in } Q_0 \\ &w_0 = 0 \text{ on } \Sigma^D_0 \\ &\frac{\partial w_0}{\partial \nu_A} + \alpha \frac{\partial w_0}{\partial t} = f_0 \text{ on } \Sigma^N_0 \\ &w_0(0) = w_0, \ \frac{\partial w_0}{\partial t} = v_0 \text{ in } \Omega. \end{aligned} \tag{6.3.1.20}$$

Problem (6.3.1.20) has the same structure as the global problem (6.2.3.1), (6.2.3.2). Hence, the optimality system is analogous to the one considered there. It is only necessary to look at the end conditions

$$p_0(T_1) = \frac{1}{\beta}(\frac{\partial w_0}{\partial t}(T_1) - \mu_{01}^n)$$
$$\frac{\partial p_0}{\partial t}(T_1) = -\frac{1}{\beta}(Aw_0(T_1) - \eta_{01}^n)$$

or

$$\begin{aligned}\frac{\partial w_0}{\partial t}(T_1) - \beta p_0(T_1) &= \mu_{0,1}^n\\ \beta\frac{\partial p_0}{\partial t}(T_1) + Aw_0(T_1) &= \eta_{0,1}^n.\end{aligned} \tag{6.3.1.21}$$

The necessary and sufficient condition for optimality yields

$$f_0 = -p_0\big|_{\Sigma_0^N}. \tag{6.3.1.22}$$

The conditions (6.3.1.21) are precisely the matching conditions for $w_0(t)$, $p_0(t)$ at $t = T_1$.

The cost functional associated with the final interval I_K is

$$\begin{aligned}J_K^n(f_K, h_{K,K-1}, g_{K,K-1}) := \frac{1}{2}\int_{\Sigma_K^N} f_K^2 d\Sigma + \frac{\kappa}{2}\{\|w_K(T_{K+1}) - z_0\|_V^2\\ + \|\frac{\partial w_K}{\partial t}(T_{K+1}) - z_1\|^2\} + \frac{1}{2\beta}\{\|g_{K,K-1}\|_H^2 + \|h_{K,K-1}\|_{V^*}^2\}\end{aligned} \tag{6.3.1.23}$$

and the optimal control problem is

$$\begin{aligned}&\inf_{\substack{f_K\in L^2(\Sigma_K^N)\\(h_{K,K-1},g_{K,K-1})\in\mathcal{H}^*}} J_K^n(f_K, h_{K,K-1}, g_{K,K-1}) \text{ subject to}\\ &\frac{\partial^2 w_K}{\partial t^2} - \nabla\cdot(A\nabla w_K) + cw_K = F_K \text{ in } Q_K\\ &w_K = 0 \text{ on } \Sigma_K^D,\\ &\frac{\partial w_K}{\partial \nu_A} + \alpha\frac{\partial}{\partial t}w_K = f_K \text{ on } \Sigma_K^N,\\ &w_K(T_K) = A^{-1}(\eta_{K,K-1}^n + h_{K,K-1}) \text{ in } V,\\ &\frac{\partial w_K}{\partial t}(T_K) = -(\mu_{K,K-1}^n + g_{K,K-1}) \text{ in } H.\end{aligned} \tag{6.3.1.24}$$

The optimality system for (6.3.1.24) is examined in the same way as above.

In conclusion, the decoupled system (6.3.1.3)–(6.3.1.8) is a decoupled system of optimality conditions associated with the optimal control problems (6.3.1.20), (6.3.1.10) and (6.3.1.24) on I_0, $\{I_k\}_{k=1}^{K-1}$, I_K, respectively. By observing that the conditions

$$(\mu_{k,k+1}^n, \eta_{k,k+1}^n) \in H \times V^*,\ k = 0, \dots, K-1$$

$$(\mu_{k,k-1}^n, \eta_{k,k-1}^n) \in H \times V^*,\ k = 1, \dots, K$$

propagate with n, it follows that the iterative procedure (6.3.1.3)–(6.3.1.8) is well defined for $n = 0, 1, \dots$, provided $(\mu_{k,k+1}^0, \eta_{k,k+1}^0) \in H \times V^*$ for $k = 0, \dots, K-1$ and $(\mu_{k,k-1}^0, \eta_{k,k-1}^0) \in H \times V^*$ for $k = 1, \dots, K$.

This iterative decoupling procedure (6.3.1.3)–(6.3.1.8) is analogous to and reminiscent of a Jacobi iteration. A Gauß-Seidel variant of the iteration would consist of solving from $k = 0$ to $k = K$ and using the most recent transmission values at the corresponding left end T_k of the interval $I_k = (T_k, T_{k+1})$. Moreover, the usage of "virtual controls" is reminiscent of a shooting method.

6.3.2 Convergence of the Iterates

We consider the iterative procedure with the basic step described in (6.3.1.3), (6.3.1.4). In fact, we consider a relaxation of the iteration step through the introduction of a relaxation parameter $\epsilon \in [0, 1)$ and replacement of (6.3.1.3), (6.3.1.4) with the under-relaxed iteration

$$\begin{aligned}
\frac{\partial w_k^{n+1}}{\partial t}(T_{k+1}) - \beta p_k^{n+1}(T_{k+1}) &= (1-\epsilon)\mu_{k,k+1}^n + \epsilon\Big(\frac{\partial w_k^n}{\partial t}(T_{k+1}) - \beta p_k^n(T_{k+1})\Big)\\
\beta\frac{\partial p_k^{n+1}}{\partial t}(T_{k+1}) + \mathcal{A}w_k^{n+1}(T_{k+1}) &= (1-\epsilon)\eta_{k,k+1}^n + \epsilon\Big(\beta\frac{\partial p_k^n}{\partial t}(T_{k+1}) + \mathcal{A}w_k^n(T_{k+1})\Big)\\
\frac{\partial w_k^{n+1}}{\partial t}(T_k) + \beta p_k^{n+1}(T_k) &= (1-\epsilon)\mu_{k,k-1}^n + \epsilon\Big(\frac{\partial w_k^n}{\partial t}(T_k) + \beta p_k^n(T_k)\Big)\\
-\beta\frac{\partial p_k^{n+1}}{\partial t}(T_k) + \mathcal{A}w_k^{n+1}(T_k) &= (1-\epsilon)\eta_{k,k-1}^n + \epsilon\Big(-\beta\frac{\partial p_k^n}{\partial t}(T_k) + \mathcal{A}w_k^n(T_k)\Big).
\end{aligned} \tag{6.3.2.1}$$

In order to prove convergence, we introduce the errors

$$\tilde{p}_k^{n+1} = p_k^{n+1} - p_k, \quad \tilde{w}_k^{n+1} = w_k^{n+1} - w_k.$$

These functions satisfy

$$\begin{cases}
\dfrac{\partial^2 \tilde{w}_k^{n+1}}{\partial t^2} - \nabla\cdot(A\nabla\tilde{w}_k^{n+1}) + c\tilde{w}_k^{n+1} = 0\\[2mm]
\dfrac{\partial^2 \tilde{p}_k^{n+1}}{\partial t^2} - \nabla\cdot(A\nabla\tilde{p}_k^{n+1}) + c\tilde{p}_k^{n+1} = 0 \qquad \text{in } Q_k
\end{cases} \tag{6.3.2.2}$$

$$\tilde{w}_k^{n+1} = \tilde{p}_k^{n+1} = 0 \ \text{ on } \Sigma_k^D, \tag{6.3.2.3}$$

$$\begin{cases} \dfrac{\partial \tilde{w}_k^{n+1}}{\partial \nu_A} + \alpha \dfrac{\partial \tilde{w}_k^{n+1}}{\partial t} + \tilde{p}_k^{n+1} = 0 \\ \dfrac{\partial \tilde{p}_k^{n+1}}{\partial \nu_A} - \alpha d_t \tilde{p}_i^{n+1} = 0 \qquad \text{on } \Sigma_k^N \end{cases} \tag{6.3.2.4}$$

$$\begin{cases} \tilde{w}_0^{n+1}(0) = \dfrac{\partial \tilde{w}_0^{n+1}}{\partial t}(0) = 0 \\ \tilde{p}_K^{n+1}(T) = \kappa \dfrac{\partial \tilde{w}_K^{n+1}}{\partial t}(T) \\ \dfrac{\partial \tilde{p}_K^{n+1}}{\partial t}(T) = -\kappa \mathcal{A} \tilde{w}_K^{n+1}(T) \quad \text{in } \Omega \end{cases} \tag{6.3.2.5}$$

subject to

$$\begin{aligned} \frac{\partial \tilde{w}_k^{n+1}}{\partial t}(T_{k+1}) - \beta \tilde{p}_k^{n+1}(T_{k+1}) &= \tilde{\mu}_{k,k+1}^n \\ \beta \frac{\partial \tilde{p}_k^{n+1}}{\partial t}(T_{k+1}) + \mathcal{A}\tilde{w}_k^{n+1}(T_{k+1}) &= \tilde{\eta}_{k,k+1}^n \\ \frac{\partial \tilde{w}_k^{n+1}}{\partial t}(T_k) + \beta \tilde{p}_k^{n+1}(T_k) &= \tilde{\mu}_{k,k-1}^n \\ -\beta \frac{\partial \tilde{p}_k^{n+1}}{\partial t}(T_k) + \mathcal{A}\tilde{w}_k^{n+1}(T_k) &= \tilde{\eta}_{k,k-1}^n \end{aligned} \tag{6.3.2.6}$$

where, for $n \geq 1$,

$$\begin{aligned} \tilde{\mu}_{k,k+1}^n &= (1-\epsilon)\big(\frac{\partial \tilde{w}_{k+1}^n}{\partial t}(T_{k+1}) - \beta\, \tilde{p}_{k+1}^n(T_{k+1})\big) + \epsilon\big(\frac{\partial \tilde{w}_k^n}{\partial t}(T_{k+1}) - \beta \tilde{p}_k^n(T_{k+1})\big) \\ \tilde{\eta}_{k,k+1}^n &= (1-\epsilon)\big(\beta \frac{\partial \tilde{p}_{k+1}^n}{\partial t}(T_{k+1}) + \mathcal{A}\tilde{w}_{k+1}^n(T_{k+1})\big) + \epsilon\big(\beta \frac{\partial \tilde{p}_k^n}{\partial t}(T_{k+1}) + \mathcal{A}\tilde{w}_k^n(T_{k+1})\big) \\ \tilde{\mu}_{k,k-1}^n &= (1-\epsilon)\big(\frac{\partial \tilde{w}_{k-1}^n}{\partial t}(T_k) + \beta \tilde{p}_{k-1}^n(T_k)\big) + \epsilon\big(\frac{\partial \tilde{w}_k^n}{\partial t}(T_k) + \beta \tilde{p}_k^n(T_k)\big) \\ \tilde{\eta}_{k,k-1}^n &= (1-\epsilon)\big(-\beta \frac{\partial \tilde{p}_{k-1}^n}{\partial t}(T_k) + \mathcal{A}\tilde{w}_{k-1}^n(T_k)\big) + \epsilon\big(-\beta \frac{\partial \tilde{p}_k^n}{\partial t}(T_k) + \mathcal{A}\tilde{w}_k^n(T_k)\big) \end{aligned} \tag{6.3.2.7}$$

and

$$\begin{aligned} \tilde{\mu}_{k,k+1}^0 &= \mu_{k,k+1}^0 - \big(\frac{\partial w_{k+1}}{\partial t}(T_{k+1}) - \beta p_{k+1}(T_{k+1})\big) \\ \tilde{\eta}_{k,k+1}^0 &= \eta_{k,k+1}^0 - \big(\beta \frac{\partial p_{k+1}}{\partial t}(T_{k+1}) + \mathcal{A} w_{k+1}(T_{k+1})\big) \\ \tilde{\mu}_{k,k-1}^0 &= \mu_{k,k-1}^0 - \big(\frac{\partial w_k}{\partial t}(T_k) + \beta p_k(T_k)\big) \\ \tilde{\eta}_{k,k-1}^0 &= \eta_{k,k-1}^0 - \big(-\beta \frac{\partial p_k}{\partial t}(T_k) + \mathcal{A} w_k(T_k)\big) \end{aligned} \tag{6.3.2.8}$$

We shall prove the following result.

Theorem 6.3.2.1 *Let $\beta > 0$. Then*
(i) *for any $\epsilon \in [0,1)$*

$$\tilde{p}_k^n|_{\Sigma_k^N} \to 0 \text{ strongly in } L^2(\Sigma_k^N),\ k = 0,\dots,K$$
$$(\tilde{w}_0^n, \frac{\partial \tilde{w}_0^n}{\partial t}) \to 0 \text{ in } C(I_0; V \times H)$$
$$\left.\frac{\partial \tilde{w}_0^n}{\partial t}\right|_{\Sigma_0^N} \to 0, \text{ strongly in } L^2(\Sigma_0^N)$$
$$(\tilde{p}_K^n, \frac{\partial \tilde{p}_K^n}{\partial t}) \to 0 \text{ in } C(I_K; H \times V^*).$$

(ii) *For any $\epsilon \in (0,1)$ and for $k = 0,\dots,K$*

$$(\tilde{w}_k^n, \frac{\partial \tilde{w}_k^n}{\partial t}) \to 0 \text{ in } C(I_k; V \times H)$$
$$(\tilde{p}_k^n, \frac{\partial \tilde{p}_k^n}{\partial t}) \to 0 \text{ in } C(I_k; H \times V^*)$$
$$\left.\frac{\partial \tilde{w}_k^n}{\partial t}\right|_{\Sigma_k^N} \to 0, \text{ strongly in } L^2(\Sigma_k^N).$$

Remark 6.3.2.1 Part (i) shows that, in particular, for any $\epsilon \in [0,1)$ the effective local optimal controls $\{p_k^n|_{\Sigma_k^N}\}_{k=0}^K$ converge strongly to the global optimal control $p|_{\Sigma^N}$, and the deviation $\|(w_K^n(T), (\partial w_K^n/\partial t)(T)) - (z_0, z_1)\|_{\mathcal{H}}$ of the local optimal trajectory $(w_K^n, \partial w_K^n/\partial t)$ at time T from the target state (z_0, z_1) converges to the deviation $\|(w(T), (\partial w/\partial t)(T)) - (z_0, z_1)\|_{\mathcal{H}}$ of the global optimal trajectory at time T from the target state. An a posteriori estimate of the error in the approximation is derived in the next section.

Remark 6.3.2.2 When $\epsilon = 0$ it is possible to show that

$$\begin{cases} (\tilde{w}_k^n, \dfrac{\partial \tilde{w}_k^n}{\partial t}) \to 0 \text{ weakly* in } L^\infty(0,T; V \times H) \\ \left.\dfrac{\partial \tilde{w}_k^n}{\partial t}\right|_{\Sigma_k^N} \to 0 \text{ weakly in } L^2(\Sigma_k^N),\ \ k = 1,\dots,K \end{cases}$$
$$(\tilde{p}_k^n, \frac{\partial \tilde{p}_k^n}{\partial t}) \to 0 \text{ weakly* in } L^\infty(0,T; H \times V^*),\ k = 0,\dots,K-1.$$

provided the following *backwards uniqueness property* is valid: if w satisfies (6.1.1.1) with $F = 0$ $f = 0$, and if $w(T) = (\partial w/\partial t)(T) = 0$, then necessarily $w_0 = v_0 = 0$. In dimension one it is known that this uniqueness property is true provided α is constant and $\alpha \neq 1$, but its validity otherwise seems to be an open question.

Proof of Theorem 6.3.2.1. We set

$$\mathcal{X} = (H \times V^*)^{2K}$$

with the standard product norm. We define a bounded linear operator $\mathcal{T} : \mathcal{X} \mapsto \mathcal{X}$ as follows. Let

$$X = \{(\tilde{\mu}_{k,k+1}, \tilde{\eta}_{k,k+1})_{k=0}^{K-1}, (\tilde{\mu}_{k,k-1}, \tilde{\eta}_{k,k-1})_{k=1}^{K}\} \in \mathcal{X} \tag{6.3.2.9}$$

and let $(\tilde{w}_k, \tilde{p}_k)_{k=0}^K$ be the solutions of (6.3.2.2)–(6.3.2.6), (6.3.1.3) with the superscripts n and $n+1$ removed. Set

$$\begin{aligned}\mathcal{T}X = \{&\Big(\frac{\partial \tilde{w}_{k+1}}{\partial t}(T_{k+1}) - \beta \tilde{p}_{k+1}(T_{k+1}), \beta \frac{\partial \tilde{p}_{k+1}}{\partial t}(T_{k+1}) + \mathcal{A}\tilde{w}_{k+1}(T_{k+1})\Big)_{k=0}^{K-1}, \\ &\Big(\frac{\partial \tilde{w}_{k-1}}{\partial t}(T_k) + \beta \tilde{p}_{k-1}(T_k), -\beta \frac{\partial \tilde{p}_{k-1}}{\partial t}(T_k) + \mathcal{A}\tilde{w}_{k-1}(T_k)\Big)_{k=1}^{K}\}.\end{aligned} \tag{6.3.2.10}$$

Note that X is a fixed point of $\mathcal{T}$ if and only if the transmission conditions (6.3.1.2) are satisfied, that is, if and only if $(\tilde{w}_k, \tilde{p}_k)_{k=0}^K$ is the solution of the global optimality system (6.1.1.1), (6.2.3.4) and (6.2.3.6) with vanishing data

$$F_k = w_0 = v_0 = z_0 = z_1 = 0 \tag{6.3.2.11}$$

Since, in this case, the optimal control is clearly $f_{\text{opt}} = 0$, it follows that the only fixed point of $\mathcal{T}$ is $X = 0$.

The significance of the mapping $\mathcal{T}$ is that, if we set

$$X^n = \{(\tilde{\mu}_{k,k+1}^{n-1}, \tilde{\eta}_{k,k+1}^{n-1})_{k=0}^{K-1}, (\tilde{\mu}_{k,k-1}^{n-1}, \tilde{\eta}_{k,k-1}^{n-1})_{k=1}^{K}\},$$

and let $(\tilde{w}_k^n, \tilde{p}_k^n)$ be the solutions of (6.3.2.2)–(6.3.2.6) with n replaced by $n-1$, then

$$\begin{aligned}X^n = \{&\Big(\frac{\partial \tilde{w}_k^n}{\partial t}(T_{k+1}) - \beta \tilde{p}_k^n(T_{k+1}), \beta \frac{\partial \tilde{p}_k^n}{\partial t}(T_{k+1}) + \mathcal{A}\tilde{w}_k^n(T_{k+1})\Big)_{k=0}^{K-1}, \\ &\Big(\frac{\partial \tilde{w}_k^n}{\partial t}(T_k) + \beta \tilde{p}_k^n(T_k), -\beta \frac{\partial \tilde{p}_k^n}{\partial t}(T_k) + \mathcal{A}\tilde{w}_k^n(T_k)\Big)_{k=1}^{K}\},\end{aligned}$$

$$\begin{aligned}\mathcal{T}X^n = \{&\Big(\frac{\partial \tilde{w}_{k+1}^n}{\partial t}(T_{k+1}) - \beta \tilde{p}_{k+1}^n(T_{k+1}), \beta \frac{\partial \tilde{p}_{k+1}^n}{\partial t}(T_{k+1}) + \mathcal{A}\tilde{w}_{k+1}^n(T_{k+1})\Big)_{k=0}^{K-1}, \\ &\Big(\frac{\partial \tilde{w}_{k-1}^n}{\partial t}(T_k) + \beta \tilde{p}_{k-1}^n(T_k), -\beta \frac{\partial \tilde{p}_{k-1}^n}{\partial t}(T_k) + \mathcal{A}\tilde{w}_{k-1}^n(T_k)\Big)_{k=1}^{K}\},\end{aligned}$$

hence the iteration step (6.3.1.3), (6.3.1.4) may be expressed as the fixed point iteration

$$X^{n+1} = \mathcal{T}X^n$$

while the relaxed iteration (6.3.2.1), (6.3.1.4) may be written as the relaxed fixed point iteration

$$X^{n+1} = (1-\epsilon)\mathcal{T}X^n + \epsilon X^n. \tag{6.3.2.12}$$

Let $X \in \mathcal{X}$ be given by (6.3.2.9) and $\tilde{w}_k$, $\tilde{p}_k$ be the corresponding solution as above. Set

$$\mathcal{E}_k(t) := \|\tilde{w}_k(t)\|_V^2 + \left\|\frac{\partial \tilde{w}_k}{\partial t}(t)\right\|^2 + \beta^2\left(\|\tilde{p}_k(t)\|^2 + \left\|\frac{\partial \tilde{p}_k}{\partial t}(t)\right\|_{V^*}^2\right), \tag{6.3.2.13}$$

$$\begin{aligned}\mathcal{E} :&= \sum_{k=1}^{K-1}[\mathcal{E}_k(T_{k+1}) + \mathcal{E}_k(T_k)] + \mathcal{E}_0(T_1) + \mathcal{E}_K(T_K)\\ &= \sum_{k=0}^{K-1}[\mathcal{E}_k(T_{k+1}) + \mathcal{E}_{k+1}(T_{k+1})].\end{aligned} \tag{6.3.2.14}$$

The following result shows that $\mathcal{T}$ is nonexpansive.

Lemma 6.3.2.1 *For any $X \in \mathcal{X}$ given by (6.3.2.9) we may write,*

$$\|X\|_{\mathcal{X}}^2 = \mathcal{E} + \mathcal{F}, \quad \|\mathcal{T}X\|_{\mathcal{X}}^2 = \mathcal{E} - \mathcal{F}, \tag{6.3.2.15}$$

where

$$\mathcal{F} = 2\beta\sum_{k=0}^{K}\int_{\Sigma_k^N}\tilde{p}_k^2\,d\Sigma + 2\beta\kappa\left(\|\tilde{w}_K(T)\|_V^2 + \left\|\frac{\partial\tilde{w}_K}{\partial t}(T)\right\|^2\right). \tag{6.3.2.16}$$

Proof of Lemma 6.3.2.1. We have

$$\begin{aligned}\|X\|_{\mathcal{X}}^2 = &\sum_{k=0}^{K-1}\left\{\left\|\frac{\partial\tilde{w}_k}{\partial t}(T_{k+1}) - \beta\tilde{p}_k(T_{k+1})\right\|^2 + \left\|\beta\frac{\partial\tilde{p}_k}{\partial t}(T_{k+1}) + A\tilde{w}_k(T_{k+1})\right\|_{V^*}^2\right\}\\ &+\sum_{k=1}^{K}\left\{\left\|\frac{\partial\tilde{w}_k}{\partial t}(T_k) + \beta\tilde{p}_k(T_k)\right\|^2 + \left\|-\beta\frac{\partial\tilde{p}_k}{\partial t}(T_k) + A\tilde{w}_k(T_k)\right\|_{V^*}^2\right\}.\end{aligned} \tag{6.3.2.17}$$

A straightforward calculation shows that (6.3.2.17) may be written

$$\begin{aligned}\|X\|_{\mathcal{X}}^2 = &\sum_{k=1}^{K-1}[\mathcal{E}_k(T_{k+1}) + \mathcal{E}_k(T_k)] + \mathcal{E}_0(T_1) + \mathcal{E}_K(T_K)\\ &+ 2\beta\sum_{k=0}^{K-1}\left[\langle\frac{\partial\tilde{p}_k}{\partial t}, \tilde{w}_k\rangle_V\Big|_{t=T_k}^{T_{k+1}} - (\tilde{p}_k, \frac{\partial\tilde{w}_k}{\partial t})\Big|_{t=T_k}^{T_{k+1}}\right]\\ &+ 2\beta(\tilde{p}_K(T_K), \frac{\partial\tilde{w}_K}{\partial t}(T_K)) - 2\beta\langle\frac{\partial\tilde{p}_K}{\partial t}(T_K), \tilde{w}_K(T_K)\rangle_V.\end{aligned} \tag{6.3.2.18}$$

We note that the product of $(6.3.2.2)_2$ with w_k of $(6.3.2.2)_1$ (with $F_k = 0$) gives

$$\langle\frac{\partial\tilde{p}_k}{\partial t}, \tilde{w}_k\rangle_V\Big|_{t=T_k}^{T_{k+1}} - (\tilde{p}_k, \frac{\partial\tilde{w}_k}{\partial t})\Big|_{t=T_k}^{T_{k+1}} = \int_{\Sigma_k^N}\tilde{p}_k^2 d\Sigma. \tag{6.3.2.19}$$

In addition, the last two terms of (6.3.2.18) may be written

$$2\beta\langle\frac{\partial\tilde{p}_K}{\partial t},\tilde{w}_K\rangle_V\Big|_{t=T_K}^T - 2\beta(\tilde{p}_K,\frac{\partial\tilde{w}_K}{\partial t})\Big|_{t=T_K}^T + 2\beta\kappa\left(\|\tilde{w}_K(T)\|_V^2 + \left\|\frac{\partial\tilde{w}_K}{\partial t}(T)\right\|^2\right)$$

$$= 2\beta\int_{\Sigma_K^N}\tilde{p}_K^2\,d\Sigma + 2\beta\kappa\left(\|\tilde{w}_K(T)\|_V^2 + \left\|\frac{\partial\tilde{w}_K}{\partial t}(T)\right\|^2\right). \qquad (6.3.2.20)$$

It follows that

$$\|X\|_{\mathcal{X}}^2 = \mathcal{E} + \mathcal{F}, \qquad (6.3.2.21)$$

A similar calculation yields

$$\begin{aligned}\|\mathcal{T}X\|_{\mathcal{X}}^2 = &\sum_{k=0}^{K-1}\left\{\left\|\frac{\partial\tilde{w}_{k+1}}{\partial t}(T_{k+1}) - \beta\tilde{p}_{k+1}(T_{k+1})\right\|^2 + \left\|\beta\frac{\partial\tilde{p}_{k+1}}{\partial t}(T_{k+1}) + \mathcal{A}\tilde{w}_{k+1}(T_{k+1})\right\|_{V^*}^2\right\}\\ &+\sum_{k=1}^{K}\left\{\left\|\frac{\partial\tilde{w}_{k-1}}{\partial t}(T_k)\beta\tilde{p}_{k-1}(T_k)\right\|^2 + \left\|-\beta\frac{\partial\tilde{p}_{k-1}}{\partial t}(T_k) + \mathcal{A}\tilde{w}_{k-1}(T_k)\right\|_{V^*}^2\right\} \qquad (6.3.2.22)\\ =&\,\mathcal{E} - \mathcal{F}.\end{aligned}$$

This completes the proof of Lemma 6.3.2.1.

To continue the proof of Theorem 6.3.2.1, set

$$\mathcal{E}_k^n(t) := \|\tilde{w}_k^n(t)\|_V^2 + \left\|\frac{\partial\tilde{w}_k^n}{\partial t}(t)\right\|^2 + \beta^2\left(\|\tilde{p}_k^n(t)\|^2 + \left\|\frac{\partial\tilde{p}_k^n}{\partial t}(t)\right\|_{V^*}^2\right)$$

$$\mathcal{E}^n = \sum_{k=0}^{K-1}[\mathcal{E}_k^n(T_{k+1}) + \mathcal{E}_{k+1}^n(T_{k+1})] \qquad (6.3.2.23)$$

$$\mathcal{F}^n = 2\beta\sum_{k=0}^{K}\int_{\Sigma_k^N}(\tilde{p}_k^n)^2\,d\Sigma + 2\beta\kappa\left(\|\tilde{w}_K^n(T)\|_V^2 + \left\|\frac{\partial\tilde{w}_K^n}{\partial t}(T)\right\|^2\right).$$

From (6.3.2.21) and (6.3.2.22) we have

$$\|X^n\|_{\mathcal{X}}^2 = \mathcal{E}^n + \mathcal{F}^n, \quad \|\mathcal{T}X^n\|_{\mathcal{X}}^2 = \mathcal{E}^n - \mathcal{F}^n. \qquad (6.3.2.24)$$

Lemma 6.3.2.2 *For any* $\epsilon \in [0,1)$ *we have*

$$\mathcal{E}^{n+1} + \mathcal{F}^{n+1} \le \mathcal{E}^n - (1-2\epsilon)\mathcal{F}^n, \quad n = 1,2,\ldots. \qquad (6.3.2.25)$$

Proof of Lemma 6.3.2.2. The relaxed iteration (6.3.2.1), (6.3.1.4) may be expressed as the relaxed fixed point iteration (6.3.2.12), as noted above. Hence

$$\|X^{n+1}\|_{\mathcal{X}}^2 = (1-\epsilon)^2\|\mathcal{T}X^n\|_{\mathcal{X}}^2 + \epsilon^2\|X^n\|_{\mathcal{X}}^2 + 2\epsilon(1-\epsilon)(\mathcal{T}X^n, X^n)_{\mathcal{X}}.$$

By using (6.3.2.24) this equality may be written

$$\mathcal{E}^{n+1} + \mathcal{F}^{n+1} = \left((1-\epsilon)^2 + \epsilon^2\right)\mathcal{E}^n - (1-2\epsilon)\mathcal{F}^n. + 2\epsilon(1-\epsilon)(\mathcal{T}X, X^n)_{\mathcal{X}} \quad (6.3.2.26)$$

Furthermore,

$$|(\mathcal{T}X, X^n)_{\mathcal{X}}| \le \|\mathcal{T}X^n\|_{\mathcal{X}}\|X^n\|_{\mathcal{X}} = \sqrt{(\mathcal{E}^n)^2 - (\mathcal{F}^n)^2} \le \mathcal{E}^n. \quad (6.3.2.27)$$

Inequality (6.3.2.25) follows immediately from (6.3.2.26) and (6.3.2.27).

Iterating (6.3.2.25) down to index $n = 1$ we obtain

$$\mathcal{E}^{n+1} \le \mathcal{E}^1 - \sum_{p=1}^{n+1} c_p(\epsilon)\mathcal{F}^p, \quad (6.3.2.28)$$

where

$$c_p(\epsilon) = 2(1-\epsilon),\ p = 2,\dots,n; \quad c_1(\epsilon) = 1-2\epsilon,\ c_{n+1}(\epsilon) = 1.$$

The crucial inequality (6.3.2.28) implies

$$(\tilde{w}_K^n(T), \frac{\partial \tilde{w}_K^n}{\partial t}(T)) \longrightarrow 0 \ \text{ strongly in } V \times H, \quad (6.3.2.29)$$

$$\tilde{p}_k^n|_{\Sigma_k^N} \longrightarrow 0 \ \text{ strongly in } \ L^2(\Sigma_k^N), \ \ k = 0,\dots,K, \quad (6.3.2.30)$$

$$\mathcal{E}^n \ \text{ bounded.} \quad (6.3.2.31)$$

It follows from (6.3.2.29) and $(6.3.1.8)_3$ that

$$(\tilde{p}_K^n(T), \frac{\partial \tilde{p}_K^n}{\partial t}(T)) \longrightarrow 0 \ \text{ strongly in } H \times V^*. \quad (6.3.2.32)$$

Since, for the solution of (6.1.1.1) (with $F = 0$) the mapping taking $((w_0, v_0), f)$ to $((w, \frac{\partial w}{\partial t}), \frac{\partial w}{\partial t}|_{\Sigma^N})$ is continuous from $(V \times H) \times L^2(\Sigma^N)$ into $L^\infty(0,T; V \times H) \times L^2(\Sigma^N)$, it follows from (6.3.2.30) and $\tilde{w}_0^n(0) = \frac{\partial \tilde{w}_0^n}{\partial t}(0) = 0$, that

$$(\tilde{w}_0^n, \frac{\partial \tilde{w}_0^n}{\partial t}) \longrightarrow 0 \ \text{ strongly in } C(I_0; V \times H) \quad (6.3.2.33)$$

$$\left.\frac{\partial \tilde{w}_0^n}{\partial t}\right|_{\Sigma_0^N} \to 0 \text{ strongly in } L^2(\Sigma_0^N) \quad (6.3.2.34)$$

Similarly, for the solution of (6.2.1.6) the mapping from the final data to $((p, \frac{\partial p}{\partial t}), p|_{\Sigma^n})$ is continuous $H \times V^* \mapsto L^\infty(0,T; H \times V^*) \times L^2(\Sigma^n)$ so in particular we obtain from (6.3.2.32)

$$(\tilde{p}_K^n, \frac{\partial \tilde{p}_K^n}{\partial t}) \longrightarrow 0 \text{ strongly in } C(I_K; H \times V^*). \tag{6.3.2.35}$$

This proves part (i) of Theorem 6.3.2.1

Now suppose that $0 < \epsilon < 1$. To obtain the stronger convergence results of part (ii) of Theorem 6.3.2.1 we use the result of Schaefer [94] stated in Proposition 2.3.7.1 above. Since 0 is a fixed point of $\mathcal{T}$, it follows from Proposition 2.3.7.1 that

$$\|\mathcal{T}_\epsilon^n X^1 - \mathcal{T}_\epsilon^{n-1} X^1\|_{\mathcal{X}} = (1-\epsilon)\|\mathcal{T} X^n - X^n\|_{\mathcal{X}} \to 0.$$

From (6.3.2.24) we obtain

$$\|\mathcal{T} X^n - X^n\|_{\mathcal{X}}^2 = 2\big(\mathcal{E}^n - (\mathcal{T} X^n, X^n)_{\mathcal{X}}\big). \tag{6.3.2.36}$$

We have

$$\begin{aligned}
(\mathcal{T} X, X^n)_{\mathcal{X}} = \sum_{k=0}^{K-1} \Big\{ &\big(\frac{\partial \tilde{w}_{k+1}^n}{\partial t}(T_{k+1}) - \beta \tilde{p}_{k+1}^n(T_{k+1}), \frac{\partial \tilde{w}_k^n}{\partial t}(T_{k+1}) - \beta \tilde{p}_k^n(T_{k+1})\big) \\
&+ \big(\beta \frac{\partial \tilde{p}_{k+1}^n}{\partial t}(T_{k+1}) + \mathcal{A}\tilde{w}_{k+1}^n(T_{k+1}), \beta \frac{\partial \tilde{p}_k^n}{\partial t}(T_{k+1}) + \mathcal{A}\tilde{w}_k^n(T_{k+1})\big)_{V^*} \Big\} \\
&+ \sum_{k=1}^{K} \Big\{ \big(\frac{\partial \tilde{w}_{k-1}^n}{\partial t}(T_k) + \beta \tilde{p}_{k-1}^n(T_k), \frac{\partial \tilde{w}_k^n}{\partial t}(T_k) + \beta \tilde{p}_k^n(T_k)\big) \\
&+ \big(-\beta \frac{\partial \tilde{p}_{k-1}^n}{\partial t}(T_k) + \mathcal{A}\tilde{w}_{k-1}^n(T_k), -\beta \frac{\partial \tilde{p}_k^n}{\partial t}(T_k) + \mathcal{A}\tilde{w}_k^n(T_k)\big)_{V^*} \Big\}.
\end{aligned}$$

After a little calculation one finds that

$$\begin{aligned}
(\mathcal{T} X, X^n)_{\mathcal{X}} = 2\sum_{k=0}^{K-1} \Big\{ &\big(\frac{\partial \tilde{w}_{k+1}^n}{\partial t}(T_{k+1}), \frac{\partial \tilde{w}_k^n}{\partial t}(T_{k+1})\big) + \big(\tilde{w}_{k+1}(T_{k+1}), \tilde{w}_k(T_{k+1})\big)_V \\
&\beta^2 \big(\tilde{p}_{k+1}^n(T_{k+1}), \tilde{p}_k(T_{k+1})\big) + \beta^2 \big(\frac{\partial \tilde{p}_{k+1}^n}{\partial t}(T_{k+1}), \frac{\partial \tilde{p}_k^n}{\partial t}(T_{k+1})\big)_{V^*} \Big\}.
\end{aligned} \tag{6.3.2.37}$$

It follows from (6.3.2.37) and the definition (6.3.2.23) of $\mathcal{E}^n$ that

$$\begin{aligned}
\mathcal{E}^n - (\mathcal{T} X^n, X^n)_{\mathcal{X}} = \sum_{k=0}^{K-1} \Big\{ &\|\tilde{w}_k^n(T_{k+1}) - \tilde{w}_{k+1}^n(T_{k+1})\|_V^2 \\
&+ \Big\|\frac{\partial \tilde{w}_k^n}{\partial t}(T_{k+1}) - \frac{\partial \tilde{w}_{k+1}^n}{\partial t}(T_{k+1})\Big\|^2 + \beta^2 \|\tilde{p}_k^n(T_{k+1}) - \tilde{p}_{k+1}^n(T_{k+1})\|^2 \\
&+ \beta^2 \Big\|\frac{\partial \tilde{p}_k^n}{\partial t}(T_{k+1}) - \frac{\partial \tilde{p}_{k+1}^n}{\partial t}(T_{k+1})\Big\|_{V^*}^2 \Big\}.
\end{aligned} \tag{6.3.2.38}$$

Therefore, for $k = 0, \ldots, K-1$ we have

$$\begin{aligned} &\tilde{w}_k^n(T_{k+1}) - \tilde{w}_{k+1}^n(T_{k+1}) \to 0 \text{ in } V \\ &\tilde{p}_k^n(T_{k+1}) - \tilde{p}_{k+1}^n(T_{k+1}) \to 0 \text{ in } H \\ &\frac{\partial \tilde{w}_k^n}{\partial t}(T_{k+1}) - \frac{\partial \tilde{w}_{k+1}^n}{\partial t}(T_{k+1}) \to 0 \text{ in } H \\ &\frac{\partial \tilde{p}_k^n}{\partial t}(T_{k+1}) - \frac{\partial \tilde{p}_{k+1}^n}{\partial t}(T_{k+1}) \to 0 \text{ in } V^*. \end{aligned} \tag{6.3.2.39}$$

By what has already been proved we have

$$\Big(\tilde{p}_K^n, \frac{\partial \tilde{p}_K^n}{\partial t}\Big) \to 0 \text{ in } C(I_K; H \times V^*)$$

so that, in particular,

$$\Big(\tilde{p}_K^n(T_K), \frac{\partial \tilde{p}_K^n}{\partial t}(T_K)\Big) \to 0 \text{ in } H \times V^*.$$

It then follows that from (6.3.2.39) with $k = K - 1$ that

$$\Big(\tilde{p}_{K-1}^n(T_K), \frac{\partial \tilde{p}_{K-1}^n}{\partial t}(T_K)\Big) \to 0 \text{ in } H \times V^*,$$

from which we may deduce that

$$\Big(\tilde{p}_{K-1}^n, \frac{\partial \tilde{p}_{K-1}^n}{\partial t}\Big) \to 0 \text{ in } C(I_{K-1}; H \times V^*).$$

By repeating the argument we conclude that

$$\Big(\tilde{p}_k^n, \frac{\partial \tilde{p}_k^n}{\partial t}\Big) \to 0 \text{ in } C(I_k; H \times V^*),\ k = K, K-1, \ldots, 0.$$

Similarly, since we have already proved that

$$\Big(\tilde{w}_0^n, \frac{\partial \tilde{w}_0^n}{\partial t}\Big) \to 0 \text{ in } C(I_0; V \times H)$$

we have

$$\Big(\tilde{w}_0^n(T_1), \frac{\partial \tilde{w}_0^n}{\partial t}(T_1)\Big) \to 0 \text{ in } V \times H.$$

Use of (6.3.2.39) with $k = 0$ then gives

$$\Big(\tilde{w}_1^n(T_1), \frac{\partial \tilde{w}_1^n}{\partial t}(T_1)\Big) \to 0 \text{ in } V \times H.$$

Since also, from part (i) of Theorem 6.3.2.1,

$$\tilde{p}_1^n\big|_{\Sigma_1^N} \to 0 \text{ strongly in } L^2(\Sigma_1^N)$$

we may conclude

$$(\tilde{w}_1^n, \frac{\partial \tilde{w}_1^n}{\partial t}) \to 0 \text{ in } C(I_1; V \times H)$$

and

$$\frac{\partial \tilde{w}_1^n}{\partial t}\Big|_{\Sigma_1^N} \to 0 \text{ strongly in } L^2(\Sigma_1^N).$$

In particular,

$$(\tilde{w}_1^n(T_2), \frac{\partial \tilde{w}_1^n}{\partial t}(T_2)) \to 0 \text{ in } V \times H$$

and so forth. This completes the proof of Theorem 6.3.2.1.

6.3.3 A Posteriori Error Estimates

The purpose of this section is to derive estimates on the norms of the errors $\tilde{w}_k^n = w_k^n - w_k$, $\tilde{p}_k^n = p_k^n - p_k$ in terms of the mismatch of the nth iterates w_k^n, p_k^n, or of successive iterates, across the break points $t = T_{k+1}$, $k = 0, \ldots, K-1$. In fact, for this purpose it is convenient to work with the auxiliary variables q_k, q_k^n instead of the adjoint variables p_k, p_k^n. Here $q_k = q|_{I_k}$, where q is related to p through (6.2.2.30) and where q is the solution of (6.2.3.7). Similarly, the local adjoint variable p_k^n may be expressed as

$$p_k^n = \frac{\partial q_k^n}{\partial t}, \quad \frac{\partial p_k^n}{\partial t} = -\mathcal{A} q_k^n$$

where q_k^{n+1}, $k = 0, \ldots, K$, is the solution of

$$\begin{aligned}
\frac{\partial^2 q_k^{n+1}}{\partial t^2} - \nabla \cdot (A \nabla q_k^{n+1}) + c q_k^{n+1} &= 0 \text{ in } Q_k \\
q_k^{n+1} &= 0 \text{ on } \Sigma_k^D \\
\frac{\partial q_k^{n+1}}{\partial \nu_A} - \alpha \frac{\partial q_k^{n+1}}{\partial t} &= 0 \text{ on } \Sigma_k^N \\
q_K^{n+1}(T) = \kappa(w_K^{n+1}(T) - z_0), \quad \frac{\partial q_K^{n+1}}{\partial t}(T) &= \kappa\Big(\frac{\partial w_K^{n+1}}{\partial t}(T) - z_1\Big).
\end{aligned}$$

In addition, at the break points $t = T_k$ the functions w_k^n, q_k^n must satisfy the constraints (6.3.1.3) which, in terms of these functions, take the simple forms

$$\begin{aligned}
\frac{\partial w_k^{n+1}}{\partial t}(T_{k+1}) - \beta \frac{\partial q_k^{n+1}}{\partial t}(T_{k+1}) &= \mu_{k,k+1}^n \\
-\beta q_k^{n+1}(T_{k+1}) + w_k^{n+1}(T_{k+1}) &= \eta_{k,k+1}^n \\
\frac{\partial w_k^{n+1}}{\partial t}(T_k) + \beta \frac{\partial q_k^{n+1}}{\partial t}(T_k) &= \mu_{k,k-1}^n \\
\beta q_k^{n+1}(T_k) + w_k^{n+1}(T_k) &= \eta_{k,k-1}^n
\end{aligned} \tag{6.3.3.1}$$

where

$$\begin{aligned}
\mu^n_{k,k+1} &= \frac{\partial w^n_{k+1}}{\partial t}(T_{k+1}) - \beta\frac{\partial q^n_{k+1}}{\partial t}(T_{k+1}) \\
\eta^n_{k,k+1} &= -\beta q^n_{k+1} T_{k+1}) + w^n_{k+1}(T_{k+1}) \\
\mu^n_{k,k-1} &= \frac{\partial w^n_{k-1}}{\partial t}(T_k) + \beta\frac{\partial q^n_{k-1}}{\partial t}(T_k) \\
\eta^n_{k,k-1} &= \beta q^n_{k-1}(T_k) + w^n_{k-1}(T_k).
\end{aligned} \tag{6.3.3.2}$$

Thus, w^n_k, q^n_k satisfy the decoupled systems

$$\begin{cases} \dfrac{\partial^2 w^{n+1}_k}{\partial t^2} - \nabla\cdot(A\nabla w^{n+1}_k) + c w^{n+1}_k = F_k \\ \dfrac{\partial^2 q^{n+1}_k}{\partial t^2} - \nabla\cdot(A\nabla q^{n+1}_k) + c q^{n+1}_k = 0 \ \text{ in } Q_k \end{cases} \tag{6.3.3.3}$$

$$w^{n+1}_k = q^{n+1}_k = 0 \ \text{ on } \Sigma^D_k, \tag{6.3.3.4}$$

$$\begin{cases} \dfrac{\partial w^{n+1}_k}{\partial \nu_A} + \alpha\dfrac{\partial w^{n+1}_k}{\partial t} + \dfrac{\partial q^{n+1}_k}{\partial t} = 0 \\ \dfrac{\partial q^{n+1}_k}{\partial \nu_A} - \alpha\dfrac{\partial q^{n+1}_k}{\partial t} = 0 \ \text{ on } \Sigma^N_k \end{cases} \tag{6.3.3.5}$$

$$\begin{cases} w^{n+1}_0(0) = \varphi, \quad \dfrac{\partial w^{n+1}_0}{\partial t}(0) = \psi \\ q^{n+1}_K(T) = \kappa(w^{n+1}_K(T) - z_0) \\ \dfrac{\partial q^{n+1}_K}{\partial t}(T) = \kappa\Big(\dfrac{\partial w^{n+1}_K}{\partial t}(T) - z_1\Big) \ \text{ in } \Omega \end{cases} \tag{6.3.3.6}$$

subject to (6.3.3.1), (6.3.3.2).

If we relax the iteration (6.3.3.1) as in the previous section and introduce the local errors

$$\tilde{w}^{n+1}_k = w^{n+1}_k - w_k, \quad \tilde{q}^{n+1}_k = q^{n+1}_k - q_k$$

then $\tilde{w}^{n+1}_k$, $\tilde{q}^{n+1}_k$ satisfy

$$\begin{cases} \dfrac{\partial^2 \tilde{w}^{n+1}_k}{\partial t^2} - \nabla\cdot(A\nabla \tilde{w}^{n+1}_k) + c\tilde{w}^{n+1}_k = 0 \\ \dfrac{\partial^2 \tilde{q}^{n+1}_k}{\partial t^2} - \nabla\cdot(A\nabla \tilde{q}^{n+1}_k) + c\tilde{q}^{n+1}_k = 0 \ \text{ in } Q_k \end{cases} \tag{6.3.3.7}$$

$$\tilde{w}^{n+1}_k = \tilde{q}^{n+1}_k = 0 \ \text{ on } \Sigma^D_k, \tag{6.3.3.8}$$

$$\begin{cases} \dfrac{\partial \tilde{w}^{n+1}_k}{\partial \nu_A} + \alpha\dfrac{\partial \tilde{w}^{n+1}_k}{\partial t} + \dfrac{\partial \tilde{q}^{n+1}_k}{\partial t} = 0 \\ \dfrac{\partial \tilde{q}^{n+1}_k}{\partial \nu_A} - \alpha\dfrac{\partial \tilde{q}^{n+1}_k}{\partial t} = 0 \ \text{ on } \Sigma^N_k \end{cases} \tag{6.3.3.9}$$

$$\begin{cases} \tilde{w}_0^{n+1}(0) = \dfrac{\partial \tilde{w}_0^{n+1}}{\partial t}(0) = 0 \\ \tilde{q}_K^{n+1}(T) = \kappa \tilde{w}_K^{n+1}(T) \\ \dfrac{\partial \tilde{q}_K^{n+1}}{\partial t}(T) = \kappa \dfrac{\partial \tilde{w}_K^{n+1}}{\partial t}(T) \text{ in } \Omega \end{cases} \tag{6.3.3.10}$$

subject to

$$\begin{aligned} &\frac{\partial \tilde{w}_k^{n+1}}{\partial t}(T_{k+1}) - \beta \frac{\partial \tilde{q}_k^{n+1}}{\partial t}(T_{k+1}) = \tilde{\mu}_{k,k+1}^n \\ &-\beta \tilde{q}_k^{n+1}(T_{k+1}) + \tilde{w}_k^{n+1}(T_{k+1}) = \tilde{\eta}_{k,k+1}^n \\ &\frac{\partial \tilde{w}_k^{n+1}}{\partial t}(T_k) + \beta \frac{\partial \tilde{q}_k^{n+1}}{\partial t}(T_k) = \tilde{\mu}_{k,k-1}^n \\ &\beta \tilde{q}_k^{n+1}(T_k) + \tilde{w}_k^{n+1}(T_k) = \tilde{\eta}_{k,k-1}^n \end{aligned} \tag{6.3.3.11}$$

where, for $n \geq 1$,

$$\begin{aligned} \tilde{\mu}_{k,k+1}^n &= (1-\epsilon)\Big(\frac{\partial \tilde{w}_{k+1}^n}{\partial t}(T_{k+1}) - \beta \frac{\partial \tilde{q}_{k+1}^n}{\partial t}(T_{k+1})\Big) + \epsilon\Big(\frac{\partial \tilde{w}_k^n}{\partial t}(T_{k+1}) - \beta \frac{\partial \tilde{q}_k^n}{\partial t}(T_{k+1})\Big) \\ \tilde{\eta}_{k,k+1}^n &= (1-\epsilon)\big(-\beta \tilde{q}_{k+1}^n(T_{k+1}) + \tilde{w}_{k+1}^n(T_{k+1})\big) + \epsilon\big(-\beta \tilde{q}_k^n(T_{k+1}) + \tilde{w}_k^n(T_{k+1})\big) \\ \tilde{\mu}_{k,k-1}^n &= (1-\epsilon)\Big(\frac{\partial \tilde{w}_{k-1}^n}{\partial t}(T_k) + \beta \frac{\partial \tilde{q}_{k-1}^n}{\partial t}(T_k)\Big) + \epsilon\Big(\frac{\partial \tilde{w}_k^n}{\partial t}(T_k) + \beta \frac{\partial \tilde{q}_k^n}{\partial t}(T_k)\Big) \\ \tilde{\eta}_{k,k-1}^n &= (1-\epsilon)\big(\beta \tilde{q}_{k-1}^n(T_k) + \tilde{w}_{k-1}^n(T_k)\big) + \epsilon\big(\beta \tilde{q}_k^n(T_k) + \tilde{w}_k^n(T_k)\big) \end{aligned} \tag{6.3.3.12}$$

and

$$\begin{aligned} \tilde{\mu}_{k,k+1}^0 &= \mu_{k,k+1}^0 - \Big(\frac{\partial w}{\partial t}(T_{k+1}) - \beta \frac{\partial q}{\partial t}(T_{k+1})\Big) \\ \tilde{\eta}_{k,k+1}^0 &= \eta_{k,k+1}^0 - \big(-\beta q(T_{k+1}) + w(T_{k+1})\big) \\ \tilde{\mu}_{k,k-1}^0 &= \mu_{k,k-1}^0 - \Big(\frac{\partial w}{\partial t}(T_k) + \beta \frac{\partial q}{\partial t}(T_k)\Big) \\ \tilde{\eta}_{k,k-1}^0 &= \eta_{k,k-1}^0 - \big(\beta q(T_k) + w(T_k)\big) \end{aligned} \tag{6.3.3.13}$$

From Theorem 6.3.2.1 one immediately deduces the following convergence properties of the auxiliary variables q_k^n:

(i) for any $\epsilon \in [0,1)$

$$\frac{\partial \tilde{q}_k^n}{\partial t}\Big|_{\Sigma_k^N} \to 0 \text{ strongly in } L^2(\Sigma_k^n),\ k = 0, \dots, K$$

$$\big(\tilde{q}_K^n, \frac{\partial \tilde{q}_K^n}{\partial t}\big) \to 0 \text{ in } C(I_K; \mathcal{H}).$$

(ii) For any $\epsilon \in (0,1)$ and for $= 0, \dots, K$

$$\big(\tilde{q}_k^n, \frac{\partial \tilde{q}_k^n}{\partial t}\big) \to 0 \text{ in } C(I_k; \mathcal{H}).$$

We define *the global error at the nth iteration*, e^n, by

$$e^n = \max_{0\le k\le K}\left\|\left(\tilde{w}_k^n, \frac{\partial \tilde{w}_k^n}{\partial t}\right)\right\|^2_{L^\infty(I_k;\mathcal{H})} + \max_{0\le k\le K}\left\|\left(\tilde{q}_k^n, \frac{\partial \tilde{q}_k^n}{\partial t}\right)\right\|^2_{L^\infty(I_k;\mathcal{H})}$$
$$+\sum_{k=0}^{K}\int_{\Sigma_k^N}\left[\alpha\left(\left|\frac{\partial \tilde{w}_k^n}{\partial t}\right|^2+\left|\frac{\partial \tilde{q}_k^n}{\partial t}\right|^2\right)+\frac{1}{\alpha}\left(\left|\frac{\partial \tilde{w}_k^n}{\partial \nu_A}\right|^2+\left|\frac{\partial \tilde{q}_k^n}{\partial \nu_A}\right|^2\right)\right]d\Sigma.$$

We also define, for $t\in I_k$, *the pointwise local error at the nth iteration* as

$$e_k^n(t) = \left\|\left(\tilde{w}_k^n(t), \frac{\partial \tilde{w}_k^n}{\partial t}(t)\right)\right\|^2_{\mathcal{H}} + \int_{T_k}^{t}\int_{\Gamma^N}\left(\alpha\left|\frac{\partial \tilde{w}_k^n}{\partial t}\right|^2+\frac{1}{\alpha}\left|\frac{\partial \tilde{w}_k^n}{\partial \nu_A}\right|^2\right)d\Gamma dt$$
$$+\left\|\left(\tilde{q}_k^n(t), \frac{\partial \tilde{q}_k^n}{\partial t}(t)\right)\right\|^2_{\mathcal{H}} + \int_{t}^{T_{k+1}}\int_{\Gamma^N}\left(\alpha\left|\frac{\partial \tilde{q}_k^n}{\partial t}\right|^2+\frac{1}{\alpha}\left|\frac{\partial \tilde{q}_k^n}{\partial \nu_A}\right|^2\right)d\Gamma dt$$

and we set

$$e^{n,n+1} = \max_{0\le k\le K}\|e_k^n + e_k^{n+1}\|_{L^\infty(I_k)},$$

which we refer to as *the cumulative global error at the nth and $(n+1)$st iterations.* In what follows we shall obtain estimates of e^n and $e^{n,n+1}$ in terms of the quantities

$$\mathcal{E}^n_{k,k+1}(T_{k+1}) = \left\|\left(w_k^n(T_{k+1})-w_{k+1}^n(T_{k+1}), \frac{\partial w_k^n}{\partial t}(T_{k+1})-\frac{\partial w_{k+1}^n}{\partial t}(T_{k+1})\right)\right\|^2_{\mathcal{H}}$$
$$+\left\|\left(q_k^n(T_{k+1})-q_{k+1}^n(T_{k+1}), \frac{\partial q_k^n}{\partial t}(T_{k+1})-\frac{\partial q_{k+1}^n}{\partial t}(T_{k+1})\right)\right\|^2_{\mathcal{H}},$$

$$\mathcal{E}^{n,n+1}_{k,k+1}(T_{k+1}) = \left\|\left(w_k^n(T_{k+1})-w_{k+1}^{n+1}(T_{k+1}), \frac{\partial w_k^n}{\partial t}(T_{k+1})-\frac{\partial w_{k+1}^{n+1}}{\partial t}(T_{k+1})\right)\right\|^2_{\mathcal{H}}$$
$$+\left\|\left(q_k^n(T_{k+1})-q_{k+1}^{n+1}(T_{k+1}), \frac{\partial q_k^n}{\partial t}(T_{k+1})-\frac{\partial q_{k+1}^{n+1}}{\partial t}(T_{k+1})\right)\right\|^2_{\mathcal{H}}.$$

The quantities $\mathcal{E}^n_{k,k+1}(T_{k+1})$ and $\mathcal{E}^{n,n+1}_{k,k+1}(T_{k+1})$ are measures, respectively, of the mismatch of the nth iterate and of successive iterates across the break point $t = T_{k+1}$. They may be expressed directly in terms of the adjoint variable p_k^n instead of q_n^k if desired by observing that $\|(q_k^n, \partial q_k^n/\partial t)\|_{\mathcal{H}} = \|(\partial p_k^n/\partial t, p_k^n)\|_{\mathcal{H}^*}$.

The main results of this section are the following error estimates.

Theorem 6.3.3.1 *Let $\epsilon\in[0,1)$. Then*

$$e^n \le C\sqrt{\mathcal{E}^n+\mathcal{F}^n}\Big\{\sum_{k=0}^{K-1}\mathcal{E}^n_{k,k+1}(T_{k+1})\Big\}^{1/2} \tag{6.3.3.14}$$

where

$$C = \sqrt{2}\max(1,1/\beta)\big(2+\max(\|\alpha^{-1}\|_{L^\infty(\Gamma^N)},\kappa)\big).$$

Theorem 6.3.3.2 *Let $\epsilon \in [0,1/2)$. Then*

$$e^{n,n+1} \le C_\epsilon \sqrt{\mathcal{E}^{n+1}+\mathcal{E}^n}\Big\{\sum_{k=0}^{K-1}\big(\mathcal{E}^{n,n+1}_{k,k+1}(T_{k+1})+\mathcal{E}^{n,n+1}_{k+1,k}(T_{k+1})\big)\Big\}^{1/2} \qquad (6.3.3.15)$$

where

$$C_\epsilon = \sqrt{2}\max(1,1/\beta)\big(2+(1-2\epsilon)^{-1}\max(\|\alpha^{-1}\|_{L^\infty(\Gamma^N)},\kappa)\big) \qquad (6.3.3.16)$$

It shall be shown below that $\mathcal{E}^n + \mathcal{F}^n \le \tilde{C}$ with a constant $\tilde{C}$ that is *explicitly computable* in terms of β, K and the input data to the global and local optimal control problems. We therefore have the following corollaries.

Corollary 6.3.3.1 *Let $\epsilon \in [0,1)$. There is an explicitly computable constant C, depending only on β, K and the input data to the global and local optimal control problems, such that*

$$e^n \le C\Big\{\sum_{k=0}^{K-1}\mathcal{E}^n_{k,k+1}(T_{k+1})\Big\}^{1/2}. \qquad (6.3.3.17)$$

Corollary 6.3.3.2 *Let $\epsilon \in [0,1/2)$. There is an explicitly computable constant C_ϵ, depending only on ϵ, β, K and the input data to the global and local optimal control problems, such that*

$$e^{n,n+1} \le C_\epsilon\Big\{\sum_{k=0}^{K-1}\big(\mathcal{E}^{n,n+1}_{k,k+1}(T_{k+1})+\mathcal{E}^{n,n+1}_{k+1,k}(T_{k+1})\big)\Big\}^{1/2}. \qquad (6.3.3.18)$$

Theorem 6.3.3.1 is a consequence of the following two lemmas.

Lemma 6.3.3.1 *Let $\epsilon \in [0,1)$. Then*

$$\mathcal{F}^n \le \sqrt{2}\max(1,\beta)(\mathcal{E}^n+\mathcal{F}^n)^{1/2}\Big\{\sum_{k=0}^{K-1}\mathcal{E}^n_{k,k+1}(T_{k+1})\Big\}^{1/2}. \qquad (6.3.3.19)$$

Lemma 6.3.3.2 *Let $\epsilon \in [0,1)$. Then*

$$e^n \le 2\sqrt{2}\max(1,1/\beta)\sqrt{\mathcal{E}^n}\Big\{\sum_{k=0}^{K-1}\mathcal{E}^n_{k,k+1}(T_{k+1})\Big\}^{1/2} + \frac{1}{\beta}\max(\|\alpha^{-1}\|_{L^\infty(\Gamma^N)},\kappa)\mathcal{F}^n. \qquad (6.3.3.20)$$

Similarly, Theorem 6.3.3.2 follows immediately from the following two results.

Lemma 6.3.3.3 *Let $\epsilon \in [0,1)$. Then*

$$\mathcal{F}^{n+1} + (1-2\epsilon)\mathcal{F}^n \\ \leq \sqrt{2}\max(1,\beta)\sqrt{\mathcal{E}^{n+1}+\mathcal{E}^n}\Big\{\sum_{k=0}^{K-1}\big(\mathcal{E}^{n,n+1}_{k,k+1}(T_{k+1}) + \mathcal{E}^{n,n+1}_{k+1,k}(T_{k+1})\big)\Big\}^{1/2}. \tag{6.3.3.21}$$

Lemma 6.3.3.4 *We have*

$$e^{n,n+1} \\ \leq \sqrt{2}\max(1,1/\beta)\sqrt{\mathcal{E}^{n+1}+\mathcal{E}^n}\Big\{\sum_{k=0}^{K-1}\big(\mathcal{E}^{n,n+1}_{k,k+1}(T_{k+1}) + \mathcal{E}^{n,nL^\infty(\Gamma)+1}_{k+1,k}(T_{k+1})\big)\Big\}^{1/2} \\ + \frac{1}{2\beta}\max(\|\alpha^{-1}\|_{L^\infty(\Gamma^N)},\kappa)\big(\mathcal{F}^{n+1}+\mathcal{F}^n\big). \tag{6.3.3.22}$$

Proof of Lemma 6.3.3.1. We start with the fundamental recursion (6.3.2.25), which we write as

$$2(1-\epsilon)\mathcal{F}^n \leq (\mathcal{E}^n + \mathcal{F}^n) - (\mathcal{E}^{n+1} + \mathcal{F}^{n+1}). \tag{6.3.3.23}$$

From (6.3.2.24) the right side of (6.3.3.23) may be written

$$\begin{aligned} \|X^n\|^2_{\mathcal{X}} - \|X^{n+1}\|^2_{\mathcal{X}} &\leq \|X^n - X^{n+1}\|_{\mathcal{X}}\big(\|X^n\|_{\mathcal{X}} + \|X^{n+1}\|_{\mathcal{X}}\big) \\ &\leq 2(1-\epsilon)\|X^n - \mathcal{T}X^n\|_{\mathcal{X}}\|X^n\|_{\mathcal{X}} \end{aligned} \tag{6.3.3.24}$$

since $\{\|X^n\|_{\mathcal{X}}\}$ is a nonincreasing sequence. Since

$$\tilde{w}^n_{k+1}(T_{k+1}) - \tilde{w}^n_k(T_{k+1}) = w^n_{k+1}(T_{k+1}) - w^n_k(T_{k+1})$$

and, similarly, for the other components, it follows from (6.3.3.23), (6.3.3.24), (6.3.2.36) and (6.3.2.38) that

$$\mathcal{F}^n \leq \sqrt{2}\max(1,\beta)\|X^n\|_{\mathcal{X}}\left\{\sum_{k=0}^{K-1}\mathcal{E}^n_{k,k+1}(T_{k+1})\right\}^{1/2},$$

which is the conclusion of Lemma 6.3.3.1.

Proof of Lemma 6.3.3.2. We first observe that the solution of (6.1.1.1) with $F=0$ satisfies the following a priori identity (see Remark 6.2.2.1):

$$\|(w(t),\frac{\partial w}{\partial t}(t))\|^2_{\mathcal{H}} + \int_0^t\int_{\Gamma^N}\left(\alpha\left|\frac{\partial w}{\partial t}\right|^2 + \frac{1}{\alpha}\left|\frac{\partial w}{\partial \nu_A}\right|^2\right)d\Gamma dt \\ = \|(w_0,v_0)\|^2_{\mathcal{H}} + \int_0^t\int_{\Gamma^N}\frac{1}{\alpha}|f|^2 d\Gamma dt, \quad 0\leq t\leq T. \tag{6.3.3.25}$$

We apply (6.3.3.25) to $\tilde{w}_k^n$ on the interval I_k to obtain

$$\begin{aligned}\|(\tilde{w}_k^n(t), \frac{\partial \tilde{w}_k^n}{\partial t}(t))\|_{\mathcal{H}}^2 &+ \int_{T_k}^t \int_{\Gamma^N} \left(\alpha \left| \frac{\partial \tilde{w}_k^n}{\partial t} \right|^2 + \frac{1}{\alpha} \left| \frac{\partial \tilde{w}_k^n}{\partial \nu_A} \right|^2 \right) d\Gamma dt \\ &= \|(\tilde{w}_k^n(T_k), \frac{\partial \tilde{w}_k^n}{\partial t}(T_k))\|_{\mathcal{H}}^2 + \int_{T_k}^t \int_{\Gamma^N} \frac{1}{\alpha} \left| \frac{\partial \tilde{q}_k^n}{\partial t} \right|^2 d\Gamma dt, \quad (6.3.3.26)\end{aligned}$$

valid for $T_k \le t \le T_{k+1}$ and for $k = 0, \dots, K$. The first term on the right is written

$$\begin{aligned}\|(\tilde{w}_k^n(T_k), \frac{\partial \tilde{w}_k^n}{\partial t}(T_k))\|_{\mathcal{H}}^2 &- \|(\tilde{w}_{k-1}^n(T_k), \frac{\partial \tilde{w}_{k-1}^n}{\partial t}(T_k))\|_{\mathcal{H}}^2 \\ &+ \|(\tilde{w}_{k-1}^n(T_k), \frac{\partial \tilde{w}_{k-1}^n}{\partial t}(T_k))\|_{\mathcal{H}}^2. \quad (6.3.3.27)\end{aligned}$$

Application of (6.3.3.25) to $\tilde{w}_{k-1}^n$ yields

$$\begin{aligned}\|(\tilde{w}_{k-1}^n(T_k), &\frac{\partial \tilde{w}_{k-1}^n}{\partial t}(T_k))\|_{\mathcal{H}}^2 \\ &= -\int_{T_{k-1}}^{T_k} \int_{\Gamma^N} \left(\alpha \left| \frac{\partial \tilde{w}_{k-1}^n}{\partial t} \right|^2 + \frac{1}{\alpha} \left| \frac{\partial \tilde{w}_{k-1}^n}{\partial \nu_A} \right|^2 \right) d\Gamma dt \\ &\quad + \|(\tilde{w}_{k-1}^n(T_{k-1}), \frac{\partial \tilde{w}_{k-1}^n}{\partial t}(T_{k-1}))\|_{\mathcal{H}}^2 \\ &\quad + \int_{T_{k-1}}^{T_k} \int_{\Gamma^N} \frac{1}{\alpha} \left| \frac{\partial \tilde{q}_{k-1}^n}{\partial t} \right|^2 d\Gamma dt. \quad (6.3.3.28)\end{aligned}$$

From (6.3.3.26)–(6.3.3.28) we obtain

$$\begin{aligned}\|(\tilde{w}_k^n(t), \frac{\partial \tilde{w}_k^n}{\partial t}(t))\|_{\mathcal{H}}^2 &+ \int_{T_{k-1}}^{T_k} \int_{\Gamma^N} \left(\alpha \left| \frac{\partial \tilde{w}_{k-1}^n}{\partial t} \right|^2 + \frac{1}{\alpha} \left| \frac{\partial \tilde{w}_{k-1}^n}{\partial \nu_A} \right|^2 \right) d\Gamma dt \\ &+ \int_{T_k}^t \int_{\Gamma^N} \left(\alpha \left| \frac{\partial \tilde{w}_k^n}{\partial t} \right|^2 + \frac{1}{\alpha} \left| \frac{\partial \tilde{w}_k^n}{\partial \nu_A} \right|^2 \right) d\Gamma dt \\ &= \|(\tilde{w}_k^n(T_k), \frac{\partial \tilde{w}_k^n}{\partial t}(T_k))\|_{\mathcal{H}}^2 - \|(\tilde{w}_{k-1}^n(T_k), \frac{\partial \tilde{w}_{k-1}^n}{\partial t}(T_k))\|_{\mathcal{H}}^2 \\ &\quad + \|(\tilde{w}_{k-1}^n(T_{k-1}), \frac{\partial \tilde{w}_{k-1}^n}{\partial t}(T_{k-1}))\|_{\mathcal{H}}^2 \\ &+ \int_{T_{k-1}}^{T_k} \int_{\Gamma^N} \frac{1}{\alpha} \left| \frac{\partial \tilde{q}_{k-1}^n}{\partial t} \right|^2 d\Gamma dt + \int_{T_k}^t \int_{\Gamma^N} \frac{1}{\alpha} \left| \frac{\partial \tilde{q}_k^n}{\partial t} \right|^2 d\Gamma dt. \quad (6.3.3.29)\end{aligned}$$

We repeat the argument, using (6.3.3.27), (6.3.3.28) with k replaced by $k-1$ to replace the term $\|(\tilde{w}_{k-1}^n(T_{k-1}), \frac{\partial \tilde{w}_{k-1}^n}{\partial t}(T_{k-1}))\|_{\mathcal{H}}^2$ in (6.3.3.29), and so forth down to $k = 0$.

Since

$$\tilde{w}_0^n(0) = (\partial \tilde{w}_0^n/\partial t)(0) = 0,$$

we end up with

$$\begin{aligned}
&\|(\tilde{w}_k^n(t), \frac{\partial \tilde{w}_k^n}{\partial t}(t))\|_{\mathcal{H}}^2 + \sum_{m=0}^{k-1} \int_{T_m}^{T_{m+1}} \int_{\Gamma^N} \left(\alpha \left|\frac{\partial \tilde{w}_m^n}{\partial t}\right|^2 + \frac{1}{\alpha}\left|\frac{\partial \tilde{w}_m^n}{\partial \nu_A}\right|^2 \right) d\Gamma dt \\
&\qquad + \int_{T_k}^{t} \int_{\Gamma^N} \left(\alpha \left|\frac{\partial \tilde{w}_k^n}{\partial t}\right|^2 + \frac{1}{\alpha}\left|\frac{\partial \tilde{w}_k^n}{\partial \nu_A}\right|^2 \right) d\Gamma dt \\
&= \sum_{m=0}^{k-1} \{\|(\tilde{w}_{m+1}^n(T_{m+1}), \frac{\partial \tilde{w}_{m+1}^n}{\partial t}(T_{m+1}))\|_{\mathcal{H}}^2 - \|(\tilde{w}_m^n(T_{m+1}), \frac{\partial \tilde{w}_m^n}{\partial t}(T_{m+1}))\|_{\mathcal{H}}^2\} \\
&\qquad + \sum_{m=0}^{k-1} \int_{T_m}^{T_{m+1}} \int_{\Gamma^N} \frac{1}{\alpha}\left|\frac{\partial \tilde{q}_m^n}{\partial t}\right|^2 d\Gamma dt + \int_{T_k}^{t} \int_{\Gamma^N} \frac{1}{\alpha}\left|\frac{\partial \tilde{q}_k^n}{\partial t}\right|^2 d\Gamma dt. \qquad (6.3.3.30)
\end{aligned}$$

This identity is valid for $t \in I_k$ and for $k = 0, \ldots, K$. It follows that

$$\begin{aligned}
&\max_{0 \le k \le K} \|(\tilde{w}_k^n, \frac{\partial \tilde{w}_k^n}{\partial t})\|_{L^\infty(I_k;\mathcal{H})}^2 + \sum_{k=0}^{K} \int_{\Sigma_k^N} \left(\alpha \left|\frac{\partial \tilde{w}_k^n}{\partial t}\right|^2 + \frac{1}{\alpha}\left|\frac{\partial \tilde{w}_k^n}{\partial \nu_A}\right|^2 \right) d\Sigma \\
&\le 2 \sum_{k=0}^{K-1} \|(w_{k+1}^n(T_{k+1}), \frac{\partial w_{k+1}^n}{\partial t}(T_{k+1})) - (w_k^n(T_{k+1}), \frac{\partial w_k^n}{\partial t}(T_{k+1}))\|_{\mathcal{H}} \\
&\times \{\|(\tilde{w}_{k+1}^n(T_{k+1}), \frac{\partial \tilde{w}_{k+1}^n}{\partial t}(T_{k+1}))\|_{\mathcal{H}} + \|(\tilde{w}_k^n(T_{k+1}), \frac{\partial \tilde{w}_k^n}{\partial t}(T_{k+1}))\|_{\mathcal{H}}\} \\
&\qquad + 2 \sum_{k=0}^{K} \int_{\Sigma_k^N} \frac{1}{\alpha}\left|\frac{\partial \tilde{q}_k^n}{\partial t}\right|^2 d\Sigma. \qquad (6.3.3.31)
\end{aligned}$$

By using the Cauchy-Schwarz inequality, the first sum on the right side of (6.3.3.31) may be estimated by

$$\begin{aligned}
&\sqrt{2}\{\sum_{k=0}^{K-1} \|(w_k^n(T_{k+1}) - w_{k+1}^n(T_{k+1}), \frac{\partial w_k^n}{\partial t}(T_{k+1}) - \frac{\partial w_{k+1}^n}{\partial t}(T_{k+1}))\|_{\mathcal{H}}^2\}^{1/2} \\
&\times \{\sum_{k=0}^{K-1} \|(\tilde{w}_k^n(T_{k+1}), \frac{\partial \tilde{w}_k^n}{\partial t}(T_{k+1}))\|_{\mathcal{H}}^2 + \|(\tilde{w}_{k+1}^n(T_{k+1}), \frac{\partial \tilde{w}_{k+1}^n}{\partial t}(T_{k+1}))\|_{\mathcal{H}}^2\}^{1/2}
\end{aligned}$$

and therefore

$$\max_{0\le k\le K}\|(\tilde w_k^n,\frac{\partial \tilde w_k^n}{\partial t})\|^2_{L^\infty(I_k;\mathcal H)}+\sum_{k=0}^K\int_{\Sigma_k^N}\left(\alpha\left|\frac{\partial \tilde w_k^n}{\partial t}\right|^2+\frac{1}{\alpha}\left|\frac{\partial \tilde w_k^n}{\partial \nu_A}\right|^2\right)d\Sigma$$

$$\le 2\sqrt{2}\{\sum_{k=0}^{K-1}\|(w_k^n(T_{k+1})-w_{k+1}^n(T_{k+1}),\frac{\partial w_k^n}{\partial t}(T_{k+1})-\frac{\partial w_{k+1}^n}{\partial t}(T_{k+1}))\|^2_{\mathcal H}\}^{1/2}$$

$$\times\{\sum_{k=0}^{K-1}\|(\tilde w_k^n(T_{k+1}),\frac{\partial \tilde w_k^n}{\partial t}(T_{k+1}))\|^2_{\mathcal H}+\|(\tilde w_{k+1}^n(T_{k+1}),\frac{\partial \tilde w_{k+1}^n}{\partial t}(T_{k+1}))\|^2_{\mathcal H}\}^{1/2}$$

$$+2\sum_{k=0}^K\int_{\Sigma_k^N}\frac{1}{\alpha}\left|\frac{\partial \tilde q_k^n}{\partial t}\right|^2 d\Sigma. \quad (6.3.3.32)$$

We estimate $\tilde q_k^n$ in a similar manner, starting with

$$\|(\tilde q_k^n(t),\frac{\partial \tilde q_k^n}{\partial t}(t))\|^2_{\mathcal H}+\int_t^{T_{k+1}}\int_{\Gamma^N}\left(\alpha\left|\frac{\partial \tilde q_k^n}{\partial t}\right|^2+\frac{1}{\alpha}\left|\frac{\partial \tilde q_k^n}{\partial \nu_A}\right|^2\right)d\Gamma dt$$

$$=\|(\tilde q_k^n(T_{k+1}),\frac{\partial \tilde q_k^n}{\partial t}(T_{k+1}))\|^2_{\mathcal H},\quad t\in I_k.$$

For the term on the right we have, as above,

$$\|(\tilde q_k^n(T_{k+1}),\frac{\partial \tilde q_k^n}{\partial t}(T_{k+1}))\|^2_{\mathcal H}$$

$$=\|(\tilde q_k^n(T_{k+1}),\frac{\partial \tilde q_k^n}{\partial t}(T_{k+1}))\|^2_{\mathcal H}-\|(\tilde q_{k+1}^n(T_{k+1}),\frac{\partial \tilde q_{k+1}^n}{\partial t}(T_{k+1}))\|^2_{\mathcal H}$$

$$+\|(\tilde q_{k+1}^n(T_{k+2}),\frac{\partial \tilde q_{k+1}^n}{\partial t}(T_{k+2}))\|^2_{\mathcal H}-\int_{T_{k+1}}^{T_{k+2}}\int_{\Gamma^N}\left(\alpha\left|\frac{\partial \tilde q_{k+1}^n}{\partial t}\right|^2+\frac{1}{\alpha}\left|\frac{\partial \tilde q_{k+1}^n}{\partial \nu_A}\right|^2\right)d\Gamma dt.$$

Iterating the argument leads to

$$\|(\tilde q_k^n(t),\frac{\partial \tilde q_k^n}{\partial t}(t))\|^2_{\mathcal H}+\int_t^{T_{k+1}}\int_{\Gamma^N}\left(\alpha\left|\frac{\partial \tilde q_k^n}{\partial t}\right|^2+\frac{1}{\alpha}\left|\frac{\partial \tilde q_k^n}{\partial \nu_A}\right|^2\right)d\Gamma dt$$

$$+\sum_{m=k+1}^K\int_{T_m}^{T_{m+1}}\int_{\Gamma^N}\left(\alpha\left|\frac{\partial \tilde q_m^n}{\partial t}\right|^2+\frac{1}{\alpha}\left|\frac{\partial \tilde q_m^n}{\partial \nu_A}\right|^2\right)d\Gamma dt$$

$$=\sum_{m=k}^{K-1}(\|(\tilde q_m^n(T_{m+1}),\frac{\partial \tilde q_m^n}{\partial t}(T_{m+1}))\|^2_{\mathcal H}-\|(\tilde q_{m+1}^n(T_{m+1}),\frac{\partial \tilde q_{m+1}^n}{\partial t}(T_{m+1}))\|^2_{\mathcal H})$$

$$+\kappa^2\|(\tilde w_K^n(T),\frac{\partial \tilde w_K^n}{\partial t}(T))\|^2_{\mathcal H},\quad t\in I_k,\ k=0,\dots,K. \quad (6.3.3.33)$$

Therefore

$$\max_{0\le k\le K} \|(\tilde{q}_k^n, \frac{\partial \tilde{q}_k^n}{\partial t})\|^2_{L^\infty(I_k;\mathcal{H})} + \sum_{k=0}^{K} \int_{\Sigma_k^N} \left(\alpha \left|\frac{\partial \tilde{q}_k^n}{\partial t}\right|^2 + \frac{1}{\alpha}\left|\frac{\partial \tilde{q}_k^n}{\partial \nu_A}\right|^2 \right) d\Sigma$$
$$\le 2\sqrt{2}\{\sum_{k=0}^{K-1} \|(q_k^n(T_{k+1}) - q_{k+1}^n(T_{k+1}), \frac{\partial q_k^n}{\partial t}(T_{k+1}) - \frac{\partial q_{k+1}^n}{\partial t}(T_{k+1}))\|^2_{\mathcal{H}}\}^{1/2}$$
$$\times \{\sum_{k=0}^{K-1} \|(\tilde{q}_k^n(T_{k+1}), \frac{\partial \tilde{q}_k^n}{\partial t}(T_{k+1}))\|^2_{\mathcal{H}} + \|(\tilde{q}_{k+1}^n(T_{k+1}), \frac{\partial \tilde{q}_{k+1}^n}{\partial t}(T_{k+1}))\|^2_{\mathcal{H}}\}^{1/2}$$
$$+ 2\kappa^2 \|(\tilde{w}_K^n(T), \frac{\partial \tilde{w}_K^n}{\partial t}(T))\|^2_{\mathcal{H}}. \quad (6.3.3.34)$$

From the Cauchy-Schwarz inequality and the definitions of $\mathcal{E}^n$ and $\mathcal{E}^n_{k,k+1}(T_{k+1})$ we obtain

$$\{\sum_{k=0}^{K-1} \|(w_k^n(T_{k+1}) - w_{k+1}^n(T_{k+1}), \frac{\partial w_k^n}{\partial t}(T_{k+1}) - \frac{\partial w_{k+1}^n}{\partial t}(T_{k+1}))\|^2_{\mathcal{H}}\}^{1/2}$$
$$\times \{\sum_{k=0}^{K-1} \|(\tilde{w}_k^n(T_{k+1}), \frac{\partial \tilde{w}_k^n}{\partial t}(T_{k+1}))\|^2_{\mathcal{H}} + \|(\tilde{w}_{k+1}^n(T_{k+1}), \frac{\partial \tilde{w}_{k+1}^n}{\partial t}(T_{k+1}))\|^2_{\mathcal{H}}\}^{1/2}$$
$$+ \{\sum_{k=0}^{K-1} \|(q_k^n(T_{k+1}) - q_{k+1}^n(T_{k+1}), \frac{\partial q_k^n}{\partial t}(T_{k+1}) - \frac{\partial q_{k+1}^n}{\partial t}(T_{k+1}))\|^2_{\mathcal{H}}\}^{1/2}$$
$$\times \{\sum_{k=0}^{K-1} \|(\tilde{q}_k^n(T_{k+1}), \frac{\partial \tilde{q}_k^n}{\partial t}(T_{k+1}))\|^2_{\mathcal{H}} + \|(\tilde{q}_{k+1}^n(T_{k+1}), \frac{\partial \tilde{q}_{k+1}^n}{\partial t}(T_{k+1}))\|^2_{\mathcal{H}}\}^{1/2}$$
$$\le \max(1, 1/\beta)\sqrt{\mathcal{E}^n}\{\sum_{k=0}^{K-1} \mathcal{E}^n_{k,k+1}(T_{k+1})\}^{1/2}. \quad (6.3.3.35)$$

The definition (6.3.2.23) of $\mathcal{F}^n$ gives

$$2\sum_{k=0}^{K} \int_{\Sigma_k^N} \frac{1}{\alpha}\left|\frac{\partial \tilde{q}_k^n}{\partial t}\right|^2 d\Sigma + 2\kappa^2 \|(\tilde{w}_K^n(T), \frac{\partial \tilde{w}_K^n}{\partial t}(T))\|^2_{\mathcal{H}} \le \frac{1}{\beta}\max(\|\alpha^{-1}\|_{L^\infty(\Gamma^N)}, \kappa)\mathcal{F}^n. \quad (6.3.3.36)$$

If follows from (6.3.3.32), (6.3.3.34)–(6.3.3.36) that

$$e^n \le 2\sqrt{2}\max(1, 1/\beta)\sqrt{\mathcal{E}^n}\{\sum_{k=0}^{K-1} \mathcal{E}^n_{k,k+1}(T_{k+1})\}^{1/2} + \frac{1}{\beta}\max(\|\alpha^{-1}\|_{L^\infty(\Gamma^N)}, \kappa)\mathcal{F}^n.$$

Proof of Lemma 6.3.3.3. The starting point is the recursion formula (6.3.2.25), which is written

$$\mathcal{F}^{n+1} + (1-2\epsilon)\mathcal{F}^n \le \mathcal{E}^n - \mathcal{E}^{n+1}. \quad (6.3.3.37)$$

We proceed to estimate

$$\mathcal{E}^n - \mathcal{E}^{n+1} = \sum_{k=0}^{K-1} \big\{ \big[\mathcal{E}_k^n(T_{k+1}) - \mathcal{E}_{k+1}^{n+1}(T_{k+1}) \big] + \big[\mathcal{E}_{k+1}^n(T_{k+1}) - \mathcal{E}_k^{n+1}(T_{k+1}) \big] \big\}. \tag{6.3.3.38}$$

We have

$$\begin{aligned}
&\mathcal{E}_k^n(T_{k+1}) - \mathcal{E}_{k+1}^{n+1}(T_{k+1}) \\
&\quad \le \| w_k^n(T_{k+1}) - w_{k+1}^{n+1}(T_{k+1}) \|_V \big(\| \tilde{w}_k^n(T_{k+1}) \|_V + \| \tilde{w}_{k+1}^{n+1}(T_{k+1}) \|_V \big) \\
&\quad + \Big\| \frac{\partial w_k^n}{\partial t}(T_{k+1}) - \frac{\partial w_{k+1}^{n+1}}{\partial t}(T_{k+1}) \Big\| \Big(\Big\| \frac{\partial \tilde{w}_k^n}{\partial t}(T_{k+1}) \Big\| + \Big\| \frac{\partial \tilde{w}_{k+1}^{n+1}}{\partial t}(T_{k+1}) \Big\| \Big) \\
&\quad + \beta^2 \| q_k^n(T_{k+1}) - q_{k+1}^{n+1}(T_{k+1}) \|_V \big(\| \tilde{q}_k^n(T_{k+1}) \|_V + \| \tilde{q}_{k+1}^{n+1}(T_{k+1}) \|_V \big) \\
&\quad + \beta^2 \Big\| \frac{\partial q_k^n}{\partial t}(T_{k+1}) - \frac{\partial q_{k+1}^{n+1}}{\partial t}(T_{k+1}) \Big\| \Big(\Big\| \frac{\partial \tilde{q}_k^n}{\partial t}(T_{k+1}) \Big\| + \Big\| \frac{\partial \tilde{q}_{k+1}^{n+1}}{\partial t}(T_{k+1}) \Big\| \Big).
\end{aligned}$$

By using the Cauchy-Schwarz inequality we obtain

$$\begin{aligned}
&\sum_{k=0}^{K-1} \big[\mathcal{E}_k^n(T_{k+1}) - \mathcal{E}_{k+1}^{n+1}(T_{k+1}) \big] \\
&\quad \le \sqrt{2} \Big\{ \sum_{k=0}^{K-1} \Big[\| w_k^n(T_{k+1}) - w_{k+1}^{n+1}(T_{k+1}) \|_V^2 + \Big\| \frac{\partial w_k^n}{\partial t}(T_{k+1}) - \frac{\partial w_{k+1}^{n+1}}{\partial t}(T_{k+1}) \Big\|^2 \\
&\quad + \beta^2 \| q_k^n(T_{k+1}) - q_{k+1}^{n+1}(T_{k+1}) \|_V^2 + \beta^2 \Big\| \frac{\partial q_k^n}{\partial t}(T_{k+1}) - \frac{\partial q_{k+1}^{n+1}}{\partial t}(T_{k+1}) \Big\|^2 \Big] \Big\}^{1/2} \\
&\quad \times \Big\{ \sum_{k=0}^{K-1} \Big[\big(\| \tilde{w}_k^n(T_{k+1}) \|_V^2 + \| \tilde{w}_{k+1}^{n+1}(T_{k+1}) \|_V^2 \big) + \Big(\Big\| \frac{\partial \tilde{w}_k^n}{\partial t}(T_{k+1}) \Big\|^2 + \Big\| \frac{\partial \tilde{w}_{k+1}^{n+1}}{\partial t}(T_{k+1}) \Big\|^2 \Big) \\
&\quad + \beta^2 \big(\| \tilde{q}_k^n(T_{k+1}) \|_V^2 + \| \tilde{q}_{k+1}^{n+1}(T_{k+1}) \|_V^2 \big) + \beta^2 \Big(\Big\| \frac{\partial \tilde{q}_k^n}{\partial t}(T_{k+1}) \Big\|^2 + \Big\| \frac{\partial \tilde{q}_{k+1}^{n+1}}{\partial t}(T_{k+1}) \Big\|^2 \Big) \Big] \Big\}^{1/2} \\
&\quad \le \sqrt{2} \max(1, \beta) \Big\{ \sum_{k=0}^{K-1} \mathcal{E}_{k,k+1}^{n,n+1}(T_{k+1}) \Big\}^{1/2} \Big\{ \sum_{k=0}^{K-1} B_{k,k+1}^{n,n+1}(T_{k+1}) \Big\}^{1/2},
\end{aligned} \tag{6.3.3.39}$$

where

$$\begin{aligned}
B_{k,k+1}^{n,n+1}(T_{k+1}) &= \big(\| \tilde{w}_k^n(T_{k+1}) \|_V^2 + \| \tilde{w}_{k+1}^{n+1}(T_{k+1}) \|_V^2 \big) \\
&+ \Big(\Big\| \frac{\partial \tilde{w}_k^n}{\partial t}(T_{k+1}) \Big\|^2 + \Big\| \frac{\partial \tilde{w}_{k+1}^{n+1}}{\partial t}(T_{k+1}) \Big\|^2 \Big) + \beta^2 \big(\| \tilde{q}_k^n(T_{k+1}) \|_V^2 + \| \tilde{q}_{k+1}^{n+1}(T_{k+1}) \|_V^2 \big) \\
&+ \beta^2 \Big(\Big\| \frac{\partial \tilde{q}_k^n}{\partial t}(T_{k+1}) \Big\|^2 + \Big\| \frac{\partial \tilde{q}_{k+1}^{n+1}}{\partial t}(T_{k+1}) \Big\|^2 \Big).
\end{aligned}$$

Similarly

$$\sum_{k=0}^{K-1}\left(\mathcal{E}_{k+1}^{n}(T_{k+1})-\mathcal{E}_{k}^{n+1}(T_{k+1})\right)$$
$$\leq \sqrt{2}\max(1,\beta)\Big\{\sum_{k=0}^{K-1}\mathcal{E}_{k+1,k}^{n,n+1}(T_{k+1})\Big\}^{1/2}\Big\{\sum_{k=0}^{K-1}B_{k+1,k}^{n,n+1}(T_{k+1})\Big\}^{1/2}. \quad (6.3.3.40)$$

It follows from (6.3.3.38)–(6.3.3.40) that

$$\mathcal{E}^{n}-\mathcal{E}^{n+1}\leq \sqrt{2}\max(1,\beta)\Big\{\sum_{k=0}^{K-1}\left(B_{k,k+1}^{n,n+1}(T_{k+1})+B_{k+1,k}^{n,n+1}(T_{k+1})\right)\Big\}^{1/2}$$
$$\cdot\Big\{\sum_{k=0}^{K-1}\left(\mathcal{E}_{k,k+1}^{n,n+1}(T_{k+1})+\mathcal{E}_{k+1,k}^{n,n+1}(T_{k+1})\right)\Big\}^{1/2} \quad (6.3.3.41)$$

The proof is completed by noting that

$$\sum_{k=0}^{K-1}\left(B_{k,k+1}^{n,n+1}(T_{k+1})+B_{k+1,k}^{n,n+1}(T_{k+1})\right)=\mathcal{E}^{n+1}+\mathcal{E}^{n}. \quad (6.3.3.42)$$

Proof of Lemma 6.3.3.4. From (6.3.3.30) and (6.3.3.33) we have

$$e_k^{n+1}(t)+e_k^{n}(t)\leq \Sigma_w^k+\Sigma_q^k$$
$$+\sum_{m=0}^{k-1}\int_{T_m}^{T_{m+1}}\int_{\Gamma^N}\frac{1}{\alpha}\Big(\|\frac{\partial\tilde q_m^n}{\partial t}|^2+\|\frac{\partial\tilde q_m^{n+1}}{\partial t}|^2\Big)d\Gamma dt$$
$$+\int_{T_k}^{t}\int_{\Gamma^N}\frac{1}{\alpha}\Big(\|\frac{\partial\tilde q_k^n}{\partial t}|^2+\|\frac{\partial\tilde q_k^{n+1}}{\partial t}|^2\Big)d\Gamma dt$$
$$+\kappa^2\Big(\big\|\big(\tilde w_K^n(T),\frac{\partial\tilde w_K^n}{\partial t}(T)\big)\big\|_{\mathcal H}^2+\big\|\big(\tilde w_K^{n+1}(T),\frac{\partial\tilde w_K^{n+1}}{\partial t}(T)\big)\big\|_{\mathcal H}^2\Big), \quad (6.3.3.43)$$

where

$$\Sigma_w^k=\sum_{m=0}^{k-1}\Big\{\big\|\big(\tilde w_{m+1}^{n+1}(T_{m+1}),\frac{\partial\tilde w_{m+1}^{n+1}}{\partial t}(T_{m+1})\big)\big\|_{\mathcal H}^2-\big\|\big(\tilde w_m^{n}(T_{m+1}),\frac{\partial\tilde w_m^{n}}{\partial t}(T_{m+1})\big)\big\|_{\mathcal H}^2$$
$$+\big\|\big(\tilde w_{m+1}^{n}(T_{m+1}),\frac{\partial\tilde w_{m+1}^{n}}{\partial t}(T_{m+1})\big)\big\|_{\mathcal H}^2-\big\|\big(\tilde w_m^{n+1}(T_{m+1}),\frac{\partial\tilde w_m^{n+1}}{\partial t}(T_{m+1})\big)\big\|_{\mathcal H}^2\Big\}$$

$$\Sigma_q^k=\sum_{m=k}^{K-1}\Big\{\big\|\big(\tilde q_m^{n+1}(T_{m+1}),\frac{\partial\tilde q_m^{n+1}}{\partial t}(T_{m+1})\big)\big\|_{\mathcal H}^2-\big\|\big(\tilde q_{m+1}^{n}(T_{m+1}),\frac{\partial\tilde q_{m+1}^{n}}{\partial t}(T_{m+1})\big)\big\|_{\mathcal H}^2$$
$$+\big\|\big(\tilde q_m^{n}(T_{m+1}),\frac{\partial\tilde q_m^{n}}{\partial t}(T_{m+1})\big)\big\|_{\mathcal H}^2-\big\|\big(\tilde q_{m+1}^{n+1}(T_{m+1}),\frac{\partial\tilde q_{m+1}^{n+1}}{\partial t}(T_{m+1})\big)\big\|_{\mathcal H}^2\Big\}.$$

By the Cauchy-Schwarz and other elementary inequalities we have, for all $k = 0,\dots,K$,

$$\Sigma_w^k \le \sqrt{2}\Big\{\sum_{k=0}^{K-1}\Big[\Big\|\Big(w_{k+1}^{n+1}(T_{k+1}),\frac{\partial w_{k+1}^{n+1}}{\partial t}(T_{k+1})\Big)-\Big(w_k^n(T_{k+1}),\frac{\partial w_k^n}{\partial t}(T_{k+1})\Big)\Big\|_{\mathcal{H}}^2$$

$$+\Big\|\Big(w_{k+1}^n(T_{k+1}),\frac{\partial w_{k+1}^n}{\partial t}(T_{k+1})\Big)-\Big(w_k^{n+1}(T_{k+1}),\frac{\partial w_k^{n+1}}{\partial t}(T_{k+1})\Big)\Big\|_{\mathcal{H}}^2\Big]\Big\}^{1/2}$$

$$\times\Big\{\sum_{k=0}^{K-1}\Big[\Big\|\Big(\tilde{w}_{k+1}^{n+1}(T_{k+1}),\frac{\partial \tilde{w}_{k+1}^{n+1}}{\partial t}(T_{k+1})\Big)\Big\|_{\mathcal{H}}^2+\Big\|\Big(\tilde{w}_k^n(T_{k+1}),\frac{\partial \tilde{w}_k^n}{\partial t}(T_{k+1})\Big)\Big\|_{\mathcal{H}}^2$$

$$+\Big\|\Big(\tilde{w}_{k+1}^n(T_{k+1}),\frac{\partial \tilde{w}_{k+1}^n}{\partial t}(T_{k+1})\Big)\Big\|_{\mathcal{H}}^2+\Big\|\Big(\tilde{w}_k^{n+1}(T_{k+1}),\frac{\partial \tilde{w}_k^{n+1}}{\partial t}(T_{k+1})\Big)\Big\|_{\mathcal{H}}^2\Big]\Big\}^{1/2}.$$

A similar estimate holds for Σ_q^k. It follows that

$$\max_{0\le k\le K}\|e_k^{n+1}+e_k^n\|_{L^\infty(I_k)}$$

$$\le \sqrt{2}\max(1,1/\beta)\sqrt{\mathcal{E}^n+\mathcal{E}^{n+1}}\Big\{\sum_{k=0}^{K-1}\big[\mathcal{E}_{k,k+1}^{n,n+1}(T_{k+1})+\mathcal{E}_{k+1,k}^{n,n+1}(T_{k+1})\big]\Big\}^{1/2}$$

$$+\frac{1}{2\beta}\max\big(\|\alpha^{-1}\|_{L^\infty(\Gamma^N)},\kappa\big)\big(\mathcal{F}^n+\mathcal{F}^{n+1}\big).$$

Proof of Corollary 6.3.3.1. Since $\mathcal{E}^n+\mathcal{F}^n=\|X^n\|_{\mathcal{X}}^2$ is nonincreasing, it is sufficient to show that $\mathcal{E}^1+\mathcal{F}^1\le\tilde{C}$ for some constant $\tilde{C}$ depending only on β, K and the input data to the global and local optimal control problems; that is, on (w_0,v_0), (z_0,z_1), F, κ, $\{(\mu_{k,k+1}^0,\eta_{k,k+1}^0)\}_{k=0}^{K-1}$, $\{(\mu_{k,k-1}^0,\eta_{k,k+1}^0)\}_{k=1}^{K}$, and the coefficients in the system (6.1.1.1).

From (6.3.2.8) we have

$$\mathcal{E}^1+\mathcal{F}^1=\|X^1\|_{\mathcal{X}}^2=\sum_{k=0}^{K-1}\{\|(\tilde{\mu}_{k,k+1}^0,\tilde{\eta}_{k,k+1}^0)\|_{H\times V^*}^2+\|(\tilde{\mu}_{k+1,k}^0,\tilde{\eta}_{k+1,k}^0)\|_{H\times V^*}^2\}$$

$$\le 2\sum_{k=0}^{K-1}\{\|(\mu_{k,k+1}^0,\eta_{k,k+1}^0)\|_{H\times V^*}^2+\|(\mu_{k+1,k}^0,\eta_{k+1,k}^0)\|_{H\times V^*}^2\}$$

$$+4\sum_{k=0}^{K-1}\Big\{\|w(T_{k+1})\|_V^2+\Big\|\frac{\partial w}{\partial t}(T_{k+1})\Big\|^2+\beta^2\|p(T_{k+1})\|^2+\beta^2\Big\|\frac{\partial p}{\partial t}(T_{k+1})\Big\|_{V^*}^2\Big\}.$$

Use of (6.2.2.7) with $\delta=1/2$ gives

$$\Big\|\Big(w,\frac{\partial w}{\partial t}\Big)\Big\|_{L^\infty(0,T;\mathcal{H})}^2\le 2\|(w_0,v_0)\|_{\mathcal{H}}^2+2\int_{\Sigma^N}\frac{1}{\alpha}|p|^2d\Sigma+4\|F\|_{L^1(0,T;H)}^2.$$

Similarly,

$$\|(p, \frac{\partial p}{\partial t})\|^2_{L^\infty(0,T;H\times V^*)} = \|(q, \frac{\partial q}{\partial t})\|^2_{L^\infty(0,T;\mathcal{H})} \le \kappa^2 \|(w(T) - z_0, \frac{\partial w}{\partial t}(T) - z_1)\|^2_{\mathcal{H}}.$$

Thus

$$\begin{aligned}\mathcal{E}^1 + \mathcal{F}^1 \le 2 \sum_{k=0}^{K-1} \{\|(\mu^0_{k,k+1}, \eta^0_{k,k+1})\|^2_{H\times V^*} + \|(\mu^0_{k+1,k}, \eta^0_{k+1,k})\|^2_{H\times V^*}\} \\ + 4K \max(1, \beta^2)\{2\|(w_0, v_0)\|^2_{\mathcal{H}} + 2\int_{\Sigma^N} \frac{1}{\alpha}|p|^2 d\Sigma + 4\|F\|^2_{L^1(0,T;H)} \\ + \kappa^2 \|(w(T) - z_0, \frac{\partial w}{\partial t}(T) - z_1)\|^2_{\mathcal{H}}\}. \quad (6.3.3.44)\end{aligned}$$

Since $J(p|_{\Sigma^N}) \le J(0)$ we have

$$\int_{\Sigma^N} |p|^2 + \kappa \|(w(T) - z_0, \frac{\partial w}{\partial t}(T) - z_1)\|^2_{\mathcal{H}} \le \kappa \|(\hat{w}(T) - z_0, \frac{\partial \hat{w}}{\partial t}(T) - z_1)\|^2_{\mathcal{H}}$$

where $\hat{w}$ is the solution of (6.1.1.1) corresponding to $f = 0$. From (6.2.2.7) we have

$$\|(\hat{w}(T), \frac{\partial \hat{w}}{\partial t}(T))\|^2_{\mathcal{H}} \le 2\|(w_0, v_0)\|^2_{\mathcal{H}} + 4\|F\|^2_{L^1(0,T;H)}.$$

Therefore

$$\begin{aligned}\int_{\Sigma^N} |p|^2 + \kappa \|(w(T) - z_0, \frac{\partial w}{\partial t}(T) - z_1)\|^2_{\mathcal{H}} \\ \le 2\kappa\big[2\|(w_0, v_0)\|^2_{\mathcal{H}} + 4\|F\|^2_{L^1(0,T;H)} + \|(z_0, z_1)\|^2_{\mathcal{H}}\big] \quad (6.3.3.45)\end{aligned}$$

It follows from (6.3.3.44) and (6.3.3.45) that

$$\begin{aligned}\mathcal{E}^1 + \mathcal{F}^1 \le \tilde{C} := 2 \sum_{k=0}^{K-1} \{\|(\mu^0_{k,k+1}, \eta^0_{k,k+1})\|^2_{H\times V^*} + \|(\mu^0_{k+1,k}, \eta^0_{k+1,k})\|^2_{H\times V^*}\} \\ + 8\{K \max(1, \beta^2) + \kappa \max(2\|\alpha^{-1}\|_{L^\infty(\Gamma^N)}, \kappa)\}\{\|(w_0, v_0)\|^2_{\mathcal{H}} + 2\|F\|^2_{L^1(0,T;H)}\} \\ + 2\kappa \max(2\|\alpha^{-1}\|_{L^\infty(\Gamma^N)}, \kappa)\|(z_0, z_1)\|^2_{\mathcal{H}}. \quad (6.3.3.46)\end{aligned}$$

This completes the proof of Corollary 6.3.3.1.

Corollary 6.3.3.2 follows immediately from (6.3.3.46) since $\mathcal{E}^n + \mathcal{E}^{n+1} \le 2(\mathcal{E}^n + \mathcal{F}^n)$.

6.3.4 Extension to General Dissipative Control Systems

The time domain decomposition procedure described above for the system (6.1.1.1) may be extended without difficulty to general second-order control systems with dissipation. Let V and H be Hilbert spaces such that $V \hookrightarrow H$. We denote by V^* the dual space of V with respect to the pivot space H and by $\mathcal{A}$ the Riesz isomorphism of V onto V^*. Let U be another Hilbert space and $\mathcal{B}$ be a bounded linear operator from U into V^*, and let $\mathcal{K}$ be a bounded linear operator on H. We consider systems of the form

$$\begin{gathered}\frac{d^2w}{dt^2} + \mathcal{A}w + \mathcal{B}\mathcal{B}^*\frac{dw}{dt} = \mathcal{B}(\mathcal{K}f) + F \text{ in } V^* \\ w(0) = w_0, \quad \frac{dw}{dt}(0) = v_0.\end{gathered} \tag{6.3.4.1}$$

In this system the initial data $(w_0, v_0) \in V \times H$, $F \in L^1(0,T;H)$, and the control function $f \in L^2(0,T;U)$. The system (6.1.1.1) clearly falls into this framework wherein the spaces V and H, and the operator $\mathcal{B}$, are those defined in Section 6.2.2, and the operator $\mathcal{K}f := (1/\sqrt{\alpha})f$.

Set $\mathcal{H} := V \times H$. From [63, Chapter 7], or through the arguments of Section 6.2.2, system (6.3.4.1) has a unique solution with regularity

$$\Big(w, \frac{dw}{dt}\Big) \in C([0,T];\mathcal{H}), \quad \mathcal{B}^*w \in H^1(0,T;U)$$

and the following estimate holds (cf. (6.2.2.6)):

$$\begin{aligned}(1-\delta)\Big\|\Big(w, \frac{dw}{dt}\Big)\Big\|^2_{L^\infty(0,T;\mathcal{H})} &+ \int_0^T \Big[\Big\|\mathcal{K}f - B^*\frac{dw}{dt}\Big\|^2_U + \Big\|B^*\frac{dw}{dt}\Big\|^2_U\Big]\,dt \\ &\le \|(w_0, v_0)\|^2_{\mathcal{H}} + \frac{1}{\delta}\|F\|^2_{L^1(0,T;H)} + \int_0^T \|\mathcal{K}f\|^2_U\,dt, \; 0 < \delta < 1.\end{aligned} \tag{6.3.4.2}$$

In addition, if $F = 0$ one has the identity (cf. (6.2.2.9))

$$\begin{aligned}\Big\|\Big(w(T), \frac{dw}{dt}(T)\Big)\Big\|^2_{\mathcal{H}} &+ \int_0^T \Big[\Big\|\mathcal{K}f - \mathcal{B}^*\frac{dw}{dt}\Big\|^2_U + \Big\|\mathcal{B}^*\frac{dw}{dt}\Big\|^2_U\Big]\,dt \\ &= \|(w_0, v_0)\|^2_{\mathcal{H}} + \int_0^T \|\mathcal{K}f\|^2_U dt.\end{aligned} \tag{6.3.4.3}$$

One may therefore consider the optimal control problem

$$\inf_{f \in L^2(0,T;U)} \Big[\int_0^T \|f(t)\|^2_U dt + \kappa\Big\|\Big(w(T), \frac{dw}{dt}(T)\Big) - (z_0, z_1)\Big\|^2_{\mathcal{H}}\Big] \tag{6.3.4.4}$$

subject to (6.3.4.1), where (z_0, z_1) is the desired target state. There is a unique optimal control $f = f_{\text{opt}}$ characterized as the solution of the variational equation

$$\begin{aligned}\int_0^T (f, g)_U\,dt &+ \kappa\big\langle \mathcal{A}(w(T) - z_0), \psi(T)\big\rangle_V \\ &+ \Big(\frac{dw}{dt}(T) - z_1, \frac{d\psi}{dt}(T)\Big) = 0, \quad \forall g \in L^2(0,T;U),\end{aligned} \tag{6.3.4.5}$$

where ψ is the solution of

$$\frac{d^2\psi}{dt^2} + \mathcal{A}\psi + \mathcal{B}\mathcal{B}^*\frac{d\psi}{dt} = \mathcal{B}(\mathcal{K}g),$$
$$\psi(0) = \frac{d\psi}{dt}(0) = 0.$$

Introduce the adjoint variable p as the solution of

$$\begin{aligned} &\frac{d^2p}{dt^2} + \mathcal{A}p + \mathcal{B}d_t(\mathcal{B}^*p) = 0, \\ p(T) = \kappa\Big(\frac{dw}{dt}(T) - z_1\Big), \quad &\frac{dp}{dt}(T) = -\kappa\mathcal{A}(w(T) - z_0). \end{aligned} \tag{6.3.4.6}$$

The solution of (6.3.4.5) must be properly interpreted, for example by the transposition method, along the lines of Section 6.2.2. In this interpretation the operator $d_t : L^2(0,T;U) \mapsto \big(H^1(0,T;U)\big)^*$ is defined by

$$\langle d_t h, \phi\rangle = -\int_0^T \Big(h(t), \frac{d\phi}{dt}(t)\Big)_U\, dt, \quad \forall \phi \in H^1(0,T;U).$$

System (6.3.4.6) has a unique solution with regularity

$$\Big(\frac{dp}{dt}, p\Big) \in C([0,T];\mathcal{H}^*), \quad \mathcal{B}^*p \in L^2(0,T;U).$$

A formal integration by parts in the equality

$$0 = \int_0^T \Big\langle \frac{d^2p}{dt^2} + \mathcal{A}p + \mathcal{B}d_t(\mathcal{B}^*p), \psi\Big\rangle_V\, dt$$

that may be justified by the arguments of Section 6.2.2 leads to

$$\begin{aligned} 0 = \int_0^T (g, \mathcal{K}^*\mathcal{B}^*p)_U\, dt - \kappa\big\langle \mathcal{A}(w(T) - z_0), \psi(T)\big\rangle_V \\ - \Big(\frac{dw}{dt}(T) - z_1, \frac{d\psi}{dt}(T)\Big), \quad \forall g \in L^2(0,T;U). \end{aligned} \tag{6.3.4.7}$$

It follows from (6.3.4.5) and (6.3.4.7) that

$$f_{\text{opt}} = -\mathcal{K}^*\mathcal{B}^*p. \tag{6.3.4.8}$$

The time domain decomposition of the optimality system consisting of (6.3.4.1), (6.3.4.6) and (6.3.4.8), proceeds along the same lines as in Section 6.3.1 and has the form

$$\begin{aligned} \frac{d^2w_k^{n+1}}{dt^2} + \mathcal{A}w_k^{n+1} + \mathcal{B}\mathcal{B}^*\frac{dw_k^{n+1}}{dt} + (\mathcal{B}\mathcal{K})(\mathcal{K}^*\mathcal{B}^*)p_k^{n+1} = F_k, \\ \frac{d^2p_k^{n+1}}{dt^2} + \mathcal{A}p_k^{n+1} + \mathcal{B}d_t(\mathcal{B}^*p_k^{n+1}) = 0 \quad \text{in } I_k \end{aligned} \tag{6.3.4.9}$$

$$\begin{aligned}
w_0^{n+1}(0) &= w_0, \quad \frac{dw_0^{n+1}}{dt}(0) = v_0, \\
p_K^{n+1}(T) &= \kappa\Big(\frac{dw_K^{n+1}}{dt}(T) - z_1\Big), \\
\frac{dp_K^{n+1}}{dt}(T) &= -\kappa\mathcal{A}(w_K^{n+1}(T) - z_0).
\end{aligned} \tag{6.3.4.10}$$

The initial conditions for $w_k^{n+1}(T_k)$, $k = 1, \ldots, K$, and the final conditions for $p_k^{n+1}(T_{k+1})$, $k = K-1, \ldots, 0$, are determined through the iterations (6.3.1.3), in which $\mu_{k,k\pm1}$, $\eta_{k,k\pm1}$ are given by (6.3.1.4), just as before. Once again, each local problem is itself an optimality system. The specifics can easily be inferred from the discussion in Section 6.3.1.

Convergence results for the above algorithm and a posteriori error estimates, similar in description to those established in Sections 6.3.2 and 6.3.3, may be proved by using the same arguments employed there. For instance, we once again have the fundamental recursion estimate

$$\mathcal{E}^{n+1} + \mathcal{F}^{n+1} \le \mathcal{E}^n - (1-2\epsilon)\mathcal{F}^n,$$

where $\epsilon \in [0,1)$ is a relaxation parameter,

$$\mathcal{E}^n = \sum_{k=0}^{K-1} [\mathcal{E}_k^n(T_{k+1}) + \mathcal{E}_{k+1}^n(T_{k+1})],$$

$$\mathcal{E}_k^n(t) = \big\|\big(\tilde{w}_k^n(t), \frac{d\tilde{w}_k^n}{dt}(t)\big)\big\|_{\mathcal{H}}^2 + \big\|\big(\frac{d\tilde{p}_k^n}{dt}(t), \tilde{p}_k^n(t)\big)\big\|_{\mathcal{H}^*}^2,$$

$$\mathcal{F}^n = 2\beta\sum_{k=0}^{K}\int_0^T \|\mathcal{K}^*\mathcal{B}^*\tilde{p}_k^n\|_U^2 dt + 2\beta\kappa\big\|\big(\tilde{w}_K^n(T), \frac{d\tilde{w}_K^n}{dt}(T)\big)\big\|_{\mathcal{H}}^2,$$

in which

$$\tilde{w}_k^n = w_k^n - w\big|_{I_k}, \quad \tilde{p}_k^n = p_k^n - p\big|_{I_k}$$

are the local errors. In particular, for the algorithm with or without relaxation one obtains

$$\mathcal{K}^*\mathcal{B}^*p_k^n \to \mathcal{K}^*\mathcal{B}^*p\big|_{I_k} \quad \text{strongly in } L^2(I_k, U),\ k = 0, \ldots, K$$

$$\big(w_K^n(T), \frac{dw_K^n}{dt}(T)\big) \to \big(w(T), \frac{dw}{dt}(T)\big) \quad \text{strongly in } \mathcal{H}.$$

That is, in all cases the effective local optimal controls, in aggregate, converge strongly to the global optimal control, and the final value of the local optimal trajectory $\big(w_K^n, \frac{dw_K^n}{dt}\big)$ converges strongly to the final value of the global optimal trajectory.

Example 6.3.4.1 *(Locally distributed control)* Let V and H be as in Section 6.2.2, that is $V = H_D^1(\Omega)$ with norm given by (6.2.2.1), $H = L^2(\Omega)$. Let ω be a nonempty open subset of Ω, $U = L^2(\omega)$, and $\alpha(\cdot) \in L^\infty(\Omega)$ satisfy $\alpha(x) \geq \alpha_0 > 0$ on ω. Define $\mathcal{B} \in \mathcal{L}(U, H)$ by

$$(\mathcal{B}g, \phi) = \int_\omega \sqrt{\alpha}\, g\, \phi\, dx, \quad \forall \phi \in H$$

and set

$$\mathcal{K}g = \frac{1}{\sqrt{\alpha}} g = \mathcal{K}^* g$$

Then

$$\mathcal{B}^* \phi = \sqrt{\alpha}\, \phi\big|_\omega$$

and the system (6.3.4.1) takes the form

$$\begin{gathered}
\frac{\partial^2 w}{\partial t^2} - \nabla \cdot (A \nabla w) + \mathbf{1}_\omega \alpha \frac{dw}{dt} + cw = f + F \;\text{ in } Q \\
w = 0 \;\text{ on } \Sigma^D, \quad \frac{\partial w}{\partial \nu_A} = 0 \;\text{ on } \Sigma^N \\
w(0) = w_0, \quad \frac{\partial w}{\partial t}(0) = v_0 \;\text{ in } \Omega.
\end{gathered}$$

where $\mathbf{1}_\omega$ is the characteristic function of ω. The global optimal control is

$$f_{\text{opt}} = -\mathbf{1}_\omega p$$

where p is the solution of

$$\begin{gathered}
\frac{\partial^2 p}{\partial t^2} - \nabla \cdot (A \nabla p) - \mathbf{1}_\omega \alpha \frac{dp}{dt} + cp = 0 \;\text{ in } Q \\
p = 0 \;\text{ on } \Sigma^D, \quad \frac{\partial p}{\partial \nu_A} = 0 \;\text{ on } \Sigma^N \\
p(T) = \kappa\Big(\frac{\partial w}{dt}(T) - z_1\Big), \quad \frac{dp}{dt}(T) = -\kappa \mathcal{A}(w(T) - z_0).
\end{gathered}$$

Example 6.3.4.2 *(Boundary control of thin plates)* Let $\Omega \subset \mathbb{R}^2$ be a bounded domain with piecewise smooth, Lipschitz boundary $\Gamma := \Gamma^D \cup \Gamma^N$ with $\Gamma^N \neq \emptyset$ and relatively open in Γ. Set $H := L^2(\Omega)$ with the standard norm, and set

$$V = H_D^2(\Omega) := \Big\{\phi \in H^2(\Omega) : \phi = \frac{\partial \phi}{\partial \nu} = 0 \text{ on } \Gamma^D\Big\},$$

$$\begin{aligned}
(w, \phi)_V = \int_\Omega \big[& w_{xx}\phi_{xx} + w_{yy}\phi_{yy} + \mu(w_{xx}\phi_{yy} + w_{yy}\phi_{xx}) \\
& + 2(1-\mu) w_{xy}\phi_{xy}\big]\, dxdy
\end{aligned}$$

$$\|\phi\|_V = \sqrt{(\phi, \phi)_V},$$

where $\mu \in (0,1)$ is a constant. The functional $\phi \mapsto \|\phi\|_V$ defines a norm on V if it is assumed that $\Gamma^D \neq \emptyset$, as we do, and one has the Green's Theorem

$$\int_\Omega (\Delta^2 w)\phi\, dxdy = (w,\phi)_V + \int_{\Gamma^N} \{[\frac{\partial \Delta w}{\partial \nu} + (1-\mu)B_2 w]\phi$$
$$- \left[\Delta w + (1-\mu)B_1 w\right]\frac{\partial \phi}{\partial \nu}\}\, dxdy, \quad \forall w,\phi \in V, \quad (6.3.4.11)$$

where Δ and Δ^2 are the two-dimensional harmonic and biharmonic operators, respectively,

$$B_1 w = -\frac{\partial^2 w}{\partial \tau^2} - \kappa\frac{\partial w}{\partial \nu}, \quad B_2 w = \frac{\partial}{\partial \tau}(\frac{\partial^2 w}{\partial \tau \partial \nu} - 2\kappa \frac{\partial w}{\partial \tau}),$$

$\frac{\partial}{\partial \tau}$ and $\frac{\partial}{\partial \nu}$ respectively denoting differentiation with respect to the positively oriented unit tangential vector, and the outward pointing unit normal vector to Γ, and where κ denotes the curvature of Γ. If $w \in V$ has less regularity than $H^4(\Omega)$, the integral on the left side of (6.3.4.11) may be interpreted in the duality between $\Delta^2 w \in V^*$ and $\phi \in V$. The reader is referred to [49] for information about various mathematical models of thin plates.

There are various possible choices for the control space U and control operator $\mathcal{B}$. Here we concentrate on boundary control and list three possibilities.

(i) Set $U = L^2(\Gamma^N)$, let $\alpha \in L^\infty(\Gamma^N)$ be positive such that $(1/\alpha) \in L^\infty(\Gamma^N)$, set $\mathcal{K}g := (1/\sqrt{\alpha})g$, and define $\mathcal{B} : U \mapsto V^*$ by

$$\langle \mathcal{B}g, \phi\rangle_V = \int_{\Gamma^N} \sqrt{\alpha}\, g\phi\, d\Gamma, \quad \forall g \in U,\ \phi \in V.$$

(ii) Set $U = L^2(\Gamma^N)$, let $\beta \in L^\infty(\Gamma^N)$ be positive such that $(1/\beta) \in L^\infty(\Gamma^N)$, set $\mathcal{K}h := (1/\sqrt{\beta})h$,and define $\mathcal{B} : U \mapsto V^*$ by

$$\langle \mathcal{B}h, \phi\rangle_V = \int_{\Gamma^N} \sqrt{\beta}\, h\frac{\partial \phi}{\partial \nu}\, d\Gamma, \quad \forall h \in U,\ \phi \in V.$$

(iii) Set $U = L^2(\Gamma^N) \times L^2(\Gamma^N)$, let α, β be as in (i) and (ii), set $\mathcal{K}(g,h) := \big((1/\sqrt{\alpha})g, (1/\sqrt{\beta})h\big)$, and define $\mathcal{B} : U \mapsto V^*$ by

$$\langle \mathcal{B}(g,h), \phi\rangle_V = \int_{\Gamma^N} \left[\sqrt{\alpha}\, g\phi + \sqrt{\beta}\, h\frac{\partial \phi}{\partial \nu}\right], d\Gamma, \quad \forall (g,h) \in U,\ \phi \in V.$$

With the above three choices, the control system (6.3.4.1) corresponds partial differential equation

$$\frac{\partial^2 w}{\partial t^2} + \Delta^2 w = F \quad \text{in } Q, \qquad (6.3.4.12)$$

the initial conditions

$$w(0) = w_0, \quad \frac{\partial w}{\partial t}(0) = v_0 \quad \text{in } \Omega, \qquad (6.3.4.13)$$

and respective boundary conditions: in case (i)

$$w = \frac{\partial w}{\partial \nu} = 0 \quad \text{on } \Sigma^D$$

$$\begin{cases} \Delta w + (1-\mu) B_1 w = 0 \\ \dfrac{\partial \Delta w}{\partial \nu} + (1-\mu) B_2 w - \alpha \dfrac{\partial w}{\partial t} = -f \quad \text{on } \Sigma^N; \end{cases} \tag{6.3.4.14}$$

in case (ii)

$$w = \frac{\partial w}{\partial \nu} = 0 \quad \text{on } \Sigma^D$$

$$\begin{cases} \Delta w + (1-\mu) B_1 w + \beta \dfrac{\partial}{\partial \nu} \left(\dfrac{\partial w}{\partial t} \right) = f \\ \dfrac{\partial \Delta w}{\partial \nu} + (1-\mu) B_2 w = 0 \qquad \text{on } \Sigma^N; \end{cases} \tag{6.3.4.15}$$

in case (iii)

$$w = \frac{\partial w}{\partial \nu} = 0 \quad \text{on } \Sigma^D$$

$$\begin{cases} \Delta w + (1-\mu) B_1 w + \beta \dfrac{\partial}{\partial \nu} \left(\dfrac{\partial w}{\partial t} \right) = f_1 \\ \dfrac{\partial \Delta w}{\partial \nu} + (1-\mu) B_2 w - \alpha \dfrac{\partial w}{\partial t} = -f_2 \qquad \text{on } \Sigma^N; \end{cases} \tag{6.3.4.16}$$

Equation (6.3.4.12) is the *Kirchhoff model* of a thin, homogeneous, isotropic elastic plate in which the mass density per unit area, and the modulus of flexural rigidity, have been set equal to unity. In the boundary conditions (6.3.4.14), the control f represents a force acting perpendicularly to the edge of the plate; in (6.3.4.15), f represents a bending moment about the tangent vector τ. In the case of the control problem (6.3.4.12)–(6.3.4.14), the optimal control is

$$f_{\text{opt}} = -\mathcal{K}^* \mathcal{B}^* p = -p\big|_{\Sigma^N},$$

where p is the solution of

$$\begin{aligned} &\frac{\partial^2 p}{\partial t^2} + \Delta^2 p = 0 \quad \text{in } Q, \\ &p = \frac{\partial p}{\partial \nu} = 0 \quad \text{on } \Sigma^D; \end{aligned} \tag{6.3.4.17}$$

$$\begin{aligned} p(T) &= \kappa\Big(\frac{\partial w}{\partial t}(T) - z_1\Big), \\ \frac{\partial p}{\partial t}(T) &= -\kappa \mathcal{A}\big(w(T) - z_0\big) \quad \text{in } \Omega; \end{aligned} \tag{6.3.4.18}$$

$$\begin{aligned} &\Delta p + (1-\mu)B_1 p = 0, \\ &\frac{\partial \Delta p}{\partial \nu} + (1-\mu)B_2 p + \alpha d_t p = 0 \text{on } \Sigma^N. \end{aligned} \tag{6.3.4.19}$$

For the control problem (6.3.4.12), (6.3.4.13), (6.3.4.15), the optimal control is

$$f_{\text{opt}} = -\left.\frac{\partial p}{\partial \nu}\right|_{\Sigma^N},$$

where p is the solution of (6.3.4.17), (6.3.4.18), and

$$\begin{aligned} &\Delta p + (1-\mu)B_1 p - \beta d_t \left(\frac{\partial p}{\partial \nu}\right) = 0, \\ &\frac{\partial \Delta p}{\partial \nu} + (1-\mu)B_2 p = 0 \text{on } \Sigma^N. \end{aligned} \tag{6.3.4.20}$$

6.4 Decomposition of the Spatial Domain

6.4.1 Description of the Algorithm

Let $\{\Omega_i\}_{i=1}^I$ be bounded domains in $\mathbb{R}^d$ with piecewise smooth, Lipschitz boundaries such that

$$\Omega_i \cap \Omega_j = \emptyset,\ i \neq j, \quad \Omega_i \subset \Omega,\ i = 1,\dots,I, \quad \overline{\Omega} = \bigcup_{i=1}^I \overline{\Omega}_i.$$

We set

$$\begin{aligned} \Gamma_i^D = \partial\Omega_i \cap \Gamma^D, \quad \Gamma_i^N = \partial\Omega_i \cap \Gamma^N, \\ \Gamma_{ij} = \partial\Omega_i \cap \partial\Omega_j = \Gamma_{ji},\ i \neq j, \quad \Gamma_i^{\text{int}} = \bigcup_{j\in\mathcal{J}_i} \Gamma_{ij}, \quad \mathcal{J}_i = \{j : \Gamma_{ij} \neq \emptyset\}. \end{aligned}$$

Then Γ_i^{int} is the portion of $\partial\Omega_i$ that is in the interior of Ω, $\partial\Omega_i = \Gamma_i^{\text{int}} \cup \Gamma_i^D \cup \Gamma_i^N$, and $\mathcal{J}_i$ is set of indices j of regions Ω_j adjacent to Ω_i. It is assumed that each Γ_i^D, Γ_i^N and Γ_{ij} is either empty or has a nonempty interior. We further set

$$\begin{aligned} \Sigma_{ij} = \Gamma_{ij} \times (0,T), \quad \Sigma_i^{\text{int}} = \Gamma_i^{\text{int}} \times (0,T), \\ Q_i = \Omega_i \times (0,T) \quad \Sigma_i^D = \Gamma_i^D \times (0,T), \quad \Sigma_i^N = \Gamma_i^N \times (0,T). \end{aligned}$$

Let $A_i = A|_{\Omega_i}$. $c_i = c|_{\Omega_i}$, $\alpha_i = \alpha|_{\Gamma_i^N}$. We assume that

$$\begin{aligned} &A_i \in C^1(\Omega_i) \text{ and has a } C^1 \text{ extension to } \overline{\Omega}_i \\ &c_i(x) > 0 \text{ on a set of positive measure in } \Omega_i \text{ if } \Gamma_i^D = \emptyset. \end{aligned} \tag{6.4.1.1}$$

Let ν_i denote the unit exterior normal vector to $\partial\Omega_i$ and $\partial/\partial_{\nu_{A_i}} = \nu_i \cdot (A_i \nabla)$. Set

$$H_D^1(\Omega_i) = \{\phi \in H^1(\Omega_i) : \phi|_{\Gamma_i^D} = 0\}$$

and $V_i = H_D^1(\Omega_i)$ endowed with the norm

$$\|\phi\|_{V_i} = \left[\int_{\Omega_i} (A_i \nabla\phi \cdot \nabla\phi + c_i|\phi|^2)\, dx\right]^{1/2}, \quad \phi \in H_D^1(\Omega_i).$$

Remark 6.4.1.1 The second assumption in (6.4.1.1) is imposed to assure that $\|\cdot\|_{V_i}$ is a norm on $H_D^1(\Omega_i)$. It would be desirable to remove this restriction on c but it is unfortunately crucial to the proofs of the results that follow.

Let $\phi \in V$ and $\phi_i = \phi|_{\Omega_i}$. Since

$$\|\phi\|_V^2 = \sum_{i=1}^{I} \|\phi\|_{V_i}^2,$$

if $\mathcal{A}_i$ is the canonical isomorphism of V_i onto V_i^* we have

$$\mathcal{A}\phi = (\mathcal{A}_1\phi_1, \ldots, \mathcal{A}_I\phi_I)$$

in the sense that

$$\langle \mathcal{A}\phi, \psi\rangle_V = \sum_{i=1}^{I} \langle \mathcal{A}_i\phi_i, \psi_i\rangle_{V_i}, \quad \forall \phi, \psi \in V,$$

where the angle brackets are taken in the indicated dualities.

We use the standard norm in $L^2(\Omega_i)$, and set

$$H_i = L^2(\Omega_i), \quad \mathcal{H}_i = V_i \times H_i, \quad U_i = L^2(\Gamma_i^N), \quad U_{ij} = L^2(\Gamma_{ij}) = U_{ji}.$$

We identify H_i with its dual space and set $\mathcal{H}_i^* = V_i^* \times H_i$, which is the dual space of $\mathcal{H}_i$ with respect to $H_i \times H_i$. Let (w, p) be the solution of the global optimality system (6.1.1.1), (6.2.3.4) and (6.2.3.6). This system is formally equivalent to the coupled local systems

$$\begin{cases} \dfrac{\partial^2 w_i}{\partial t^2} - \nabla \cdot (A_i \nabla w_i) + c_i w_i = F_i \\ \dfrac{\partial^2 p_i}{\partial t^2} - \nabla \cdot (A_i \nabla p_i) + c_i p_i = 0 \text{ in } Q_i \end{cases} \tag{6.4.1.2}$$

$$w_i = p_i = 0 \text{ on } \Sigma_i^D \tag{6.4.1.3}$$

$$\begin{cases} \dfrac{\partial w_i}{\partial \nu_{A_i}} + \alpha_i \dfrac{\partial w_i}{\partial t} + p_i = 0 \\ \dfrac{\partial p_i}{\partial \nu_{A_i}} - \alpha_i d_t p_i = 0 \text{ on } \Sigma_i^N \end{cases} \tag{6.4.1.4}$$

$$\begin{cases} w_i(0) = w_{0i}, & \dfrac{\partial w_i}{\partial t}(0) = v_{0i} \\ p_i(T) = \kappa(\dfrac{\partial w_i}{\partial t}(T) - z_{1i}) \\ \dfrac{\partial p_i}{\partial t}(T) = -\kappa \mathcal{A}_i(w_i(T) - z_{0i}) \text{ in } \Omega_i \end{cases} \tag{6.4.1.5}$$

where $F_i = F|_{Q_i}$, together with the following *transmission conditions* on the interfaces Σ_{ij}:

$$\begin{aligned} \frac{\partial w_i}{\partial \nu_{A_i}} + \beta \frac{\partial w_i}{\partial t} + \gamma p_i &= -\frac{\partial w_j}{\partial \nu_{A_j}} + \beta \frac{\partial w_j}{\partial t} + \gamma p_j \\ \frac{\partial p_i}{\partial \nu_{A_i}} - \beta d_t p_i + \gamma d_t \frac{\partial w_i}{\partial t} &= -\frac{\partial p_j}{\partial \nu_{A_j}} - \beta d_t p_j + \gamma d_t \frac{\partial w_j}{\partial t} \end{aligned} \tag{6.4.1.6}$$

where β and γ are nonzero constants. In (6.4.1.4), d_t is the bounded linear mapping $L^2(0,T;U_i) \mapsto \left(H^1(0,T;U_i)\right)^*$ defined as in (6.2.2.27). Similarly, in (6.4.1.6), $d_t : L^2(0,T;U_{ij}) \mapsto \left(H^1(0,T;U_{ij})\right)^*$. Note that by interchanging i and j in (6.4.1.6) and adding the result to (6.4.1.6) we obtain

$$\frac{\partial w_i}{\partial \nu_{A_i}} = -\frac{\partial w_j}{\partial \nu_{A_j}}, \quad \frac{\partial p_i}{\partial \nu_{A_i}} = -\frac{\partial p_j}{\partial \nu_{A_j}} \text{ on } \Sigma_{ij}$$

and then

$$\beta \frac{\partial w_i}{\partial t} + \gamma p_i = \beta \frac{\partial w_j}{\partial t} + \gamma p_j, \quad \gamma d_t \frac{\partial w_j}{\partial t} - \beta d_t p_i = \gamma d_t \frac{\partial w_j}{\partial t} - \beta d_t p_j$$

on Σ_{ij}. Upon applying d_t to the first equation one finds that

$$d_t \frac{\partial w_i}{\partial t} = d_t \frac{\partial w_j}{\partial t}, \quad d_t p_i = d_t p_j \text{ on } \Sigma_{ij}.$$

Thus

$$p_i = p_j, \quad \frac{\partial w_i}{\partial t} = \frac{\partial w_j}{\partial t} \text{ on } \Sigma_{ij}$$

in the $L^2(\Sigma_{ij})$ sense since $\mathcal{N}(d_t) = \{0\}$, and then $w_i = w_j$ on Σ_{ij} since $w_{0i} = w_{0j}$ there.

Set

$$\begin{aligned} \lambda_{ij} &= (1-\epsilon)\left(-\frac{\partial w_j}{\partial \nu_{A_j}} + \beta \frac{\partial w_j}{\partial t} + \gamma p_j\right) + \epsilon \left(\frac{\partial w_i}{\partial \nu_{A_i}} + \beta \frac{\partial w_i}{\partial t} + \gamma p_i\right)\Bigg|_{\Sigma_{ij}} \\ \mu_{ij} &= (1-\epsilon)\left(-\frac{\partial p_j}{\partial \nu_{A_j}} - \beta d_t p_j + \gamma d_t \frac{\partial w_j}{\partial t}\right) \\ &\quad + \epsilon \left(\frac{\partial p_i}{\partial \nu_{A_i}} - \beta d_t p_i + \gamma d_t \frac{\partial w_i}{\partial t}\right)\Bigg|_{\Sigma_{ij}}, \end{aligned} \tag{6.4.1.7}$$

where $\epsilon \in [0,1)$ is a relaxation parameter in the iterative scheme to follow. Then the transmission conditions (6.4.1.6) may be written

$$\begin{aligned} \frac{\partial w_i}{\partial \nu_{A_i}} + \beta \frac{\partial w_i}{\partial t} + \gamma p_i &= \lambda_{ij} \text{ on } \Sigma_{ij} \\ \frac{\partial p_i}{\partial \nu_{A_i}} - \beta d_t p_i + \gamma d_t \frac{\partial w_i}{\partial t} &= \mu_{ij} \text{ on } \Sigma_{ij} \end{aligned} \tag{6.4.1.8}$$

Lemma 6.4.1.1 *Assume that $\beta > 0$, $\gamma > 0$, that $\lambda_{ij}, \rho_{ij} \in L^2(\Sigma_{ij})$, and that $\mu_{ij} = d_t \rho_{ij}$, $\forall j : \Gamma_{ij} \neq \emptyset$. Then (6.4.1.2)–(6.4.1.5), (6.4.1.8) is well posed. The solution has regularity*

$$\begin{aligned} &(w_i, \frac{\partial w_i}{\partial t}) \in C([0,T]; \mathcal{H}_i), \quad \frac{\partial w_i}{\partial t}|_{\Sigma_i^N \cup \Sigma_i^{int}} \in L^2(\Sigma_i^N \cup \Sigma_i^{int}), \\ &(\frac{\partial p_i}{\partial t}, p_i) \in C([0,T]; \mathcal{H}_i^*), \quad p_i|_{\Sigma_i^N \cup \Sigma_i^{int}} \in L^2(\Sigma_i^N \cup \Sigma_i^{int}). \end{aligned}$$

Proof. The lemma follows from the fact that (6.4.1.2)–(6.4.1.5), (6.4.1.6) is the optimality system for the LQR problem

$$\inf_{f_i, g_{ij}} \{ \int_{\Sigma_i^N} |f_i|^2 d\Sigma + \frac{1}{\gamma} \sum_{j \in \mathcal{J}_i} \int_{\Sigma_{ij}} (|g_{ij}|^2 + |\gamma \frac{\partial w_i}{\partial t} - \rho_{ij}|^2) d\Sigma$$
$$+ \kappa \|(w_i(T), \frac{\partial w_i}{\partial t}(T)) - (z_{0i}, z_{1i})\|^2_{\mathcal{H}_i} \}$$

subject to

$$\begin{aligned} \frac{\partial^2 w_i}{\partial t^2} - \nabla \cdot (A_i \nabla w_i) + c_i w_i &= F_i \text{ in } Q_i \\ w_i &= 0 \text{ on } \Sigma_i^D \\ \frac{\partial w_i}{\partial \nu_{A_i}} + \alpha_i \frac{\partial w_i}{\partial t} &= f_i \text{ on } \Sigma_i^N \\ \frac{\partial w_i}{\partial \nu_{A_i}} + \beta \frac{\partial w_i}{\partial t} &= \lambda_{ij} + g_{ij} \text{ on } \Sigma_{ij}, \forall j : \Gamma_{ij} \neq \emptyset \\ w_i(0) = w_{0i}, \quad \frac{\partial w_i}{\partial t}(0) &= v_{0i} \text{ in } \Omega_i, \end{aligned}$$

where $f_i \in L^2(\Sigma_i^N)$, $g_{ij} \in L^2(\Sigma_{ij})$. This problem has the same structure as the global optimal control problem. So for $f_i \in L^2(\Sigma_i^N)$ and $g_{ij} \in L^2(\Sigma_{ij})$, the system has a unique solution w_i with the indicated regularity and the optimal controls are given by

$$f_i = -p_i|_{\Sigma_i^N}, \quad g_{ij} = -\gamma p_i|_{\Sigma_{ij}},$$

where p_i is the solution of

$$\begin{aligned}
\frac{\partial^2 p_i}{\partial t^2} - \nabla\cdot(A_i\nabla p_i) + c_i p_i = 0 \text{ in } Q_i\\
p_i = 0 \text{ on } \Sigma_i^D\\
\frac{\partial p_i}{\partial \nu_{A_i}} - \alpha_i d_t p_i = 0 \text{ on } \Sigma_i^N\\
\frac{\partial p_i}{\partial \nu_{A_i}} - \beta d_t p = -\gamma d_t \frac{\partial w_i}{\partial t} + d_t \rho_{ij} \text{ on } \Sigma_{ij}\\
p_i(T) = \kappa(\frac{\partial w_i}{\partial t}(T) - z_{1i}) \in H_i\\
\frac{\partial p_i}{\partial t}(T) = -\kappa A_i(w_i(T) - z_{0i}) \in V_i^*
\end{aligned}$$

Since $-\gamma \partial w/\partial t\big|_{\Sigma_{ij}} + \rho_{ij} \in L^2(\Sigma_{ij})$, the last problem has a unique solution p_i with the indicated regularity.

As in Proposition 6.2.2.6 we have

$$\begin{aligned}
p_i &= \frac{\partial q_i}{\partial t},\\
\frac{\partial p_i}{\partial t} &= -A_i q_i
\end{aligned}$$

where q_i is the solution of

$$\begin{aligned}
\frac{\partial^2 q_i}{\partial t^2} - \nabla\cdot(A_i\nabla q_i) + c_i q_i = 0 \text{ in } Q_i\\
q_i = 0 \text{ on } \Sigma_i^D\\
\frac{\partial q_i}{\partial \nu_{A_i}} - \alpha_i \frac{\partial q_i}{\partial t} = 0 \text{ on } \Sigma_i^N\\
\frac{\partial q_i}{\partial \nu_{A_i}} - \beta\frac{\partial q_i}{\partial t} + \gamma\frac{\partial w_i}{\partial t} = \rho_{ij} \text{ on } \Sigma_{ij}\\
q_i(T) = \kappa(w_i(T) - z_{0i}) \in V_i\\
\frac{\partial q_i}{\partial t}(T) = \kappa(\frac{\partial w_i}{\partial t}(T) - z_{1i}) \in H_i.
\end{aligned} \tag{6.4.1.9}$$

Further, p_i satisfies the variational equation

$$\begin{aligned}
&\langle\langle(-\frac{\partial p_i}{\partial t}(0), p_i(0)), \phi_{i0}, \phi_{i1}\rangle\rangle + \int_{Q_i} p_i G_i\, dxdt + \left(\int_{\Sigma_i^N} + \int_{\Sigma_i^{\text{int}}}\right) p_i h_i\, d\Sigma\\
&= \langle(r_{0i}, p_{0i}), (\phi_i(T), \frac{\partial \phi_i}{\partial t}(T))\rangle_{\mathcal{H}_i} - \sum_{j\in\mathcal{J}_i}\int_{\Sigma_{ij}} g_{ij}\frac{\partial \phi_i}{\partial t}\, d\Sigma,\\
&\forall\,(\phi_{i0}, \phi_{i1}) \in \mathcal{H}_i,\ G_i \in L^2(Q_i),\ h_i \in L^2(\Sigma_i^N \cup \Sigma_i^{\text{int}}),
\end{aligned} \tag{6.4.1.10}$$

where

$$p_{0i} = \kappa(\frac{\partial w_i}{\partial t}(T) - z_{1i}), \quad r_{0i} = \kappa \mathcal{A}_i(w_i(T) - z_{0i}), \quad g_{ij} = \rho_{ij} - \gamma \frac{\partial w_i}{\partial t}|_{\Sigma_{ij}}$$

and where ϕ_i is the solution of

$$\begin{aligned} \frac{\partial^2 \phi_i}{\partial t^2} - \nabla \cdot (A_i \nabla \phi_i) + c_i \phi_i &= G_i \text{ in } Q_i \\ \phi_i &= 0 \text{ on } \Sigma_i^D \\ \frac{\partial \phi_i}{\partial \nu_{A_i}} + \alpha_i \frac{\partial \phi_i}{\partial t} &= h_i \text{ on } \Sigma_i^N \\ \frac{\partial \phi_i}{\partial \nu_{A_i}} + \beta \frac{\partial \phi_i}{\partial t} &= h_i \text{ on } \Sigma_i^{\text{int}} \\ \phi_i(0) = \phi_{i0}, \quad \frac{\partial \phi_i}{\partial t}(0) &= \phi_{i1} \text{ in } \Omega_i. \end{aligned} \tag{6.4.1.11}$$

The transmission conditions (6.4.1.6) suggest the following domain decomposition iteration of the optimality system:

$$\begin{cases} \dfrac{\partial^2 w_i^{n+1}}{\partial t^2} - \nabla \cdot (A_i \nabla w_i^{n+1}) + c_i w_i^{n+1} = F_i \\ \dfrac{\partial^2 p_i^{n+1}}{\partial t^2} - \nabla \cdot (A_i \nabla p_i^{n+1}) + c_i p_i^{n+1} = 0 \text{ in } Q_i \end{cases} \tag{6.4.1.12}$$

$$w_i^{n+1} = p_i^{n+1} = 0 \text{ on } \Sigma_i^D \tag{6.4.1.13}$$

$$\begin{cases} \dfrac{\partial w_i^{n+1}}{\partial \nu_{A_i}} + \alpha_i \dfrac{\partial w_i^{n+1}}{\partial t} + p_i^{n+1} = 0 \\ \dfrac{\partial p_i^{n+1}}{\partial \nu_{A_i}} - \alpha_i d_t p_i^{n+1} = 0 \text{ on } \Sigma_i^N \end{cases} \tag{6.4.1.14}$$

$$\begin{cases} \dfrac{\partial w_i^{n+1}}{\partial \nu_{A_i}} + \beta \dfrac{\partial w_i^{n+1}}{\partial t} + \gamma p_i^{n+1} = \lambda_{ij}^n \\ \dfrac{\partial p_i^{n+1}}{\partial \nu_{A_i}} - \beta d_t p_i^{n+1} + \gamma d_t \dfrac{\partial w_i^{n+1}}{\partial t} = \mu_{ij}^n \text{ on } \Sigma_{ij} \end{cases} \tag{6.4.1.15}$$

$$\begin{cases} w_i^{n+1}(0) = w_{0i}, \quad \dfrac{\partial w_i^{n+1}}{\partial t}(0) = v_{0i} \\ p_i^{n+1}(T) = \kappa(\dfrac{\partial w_i^{n+1}}{\partial t}(T) - z_{1i}) \\ \dfrac{\partial p_i^{n+1}}{\partial t}(T) = -\kappa \mathcal{A}_i(w_i^{n+1}(T) - z_{0i}) \text{ in } \Omega_i \end{cases} \tag{6.4.1.16}$$

where λ_{ij}^0, $\rho_{ij}^0 \in L^2(\Sigma_{ij})$ are arbitrary, $\mu_{ij}^0 = d_t \rho_{ij}^0$ and, for $n \geq 1$,

$$\lambda_{ij}^n = (1-\epsilon)\left(-\frac{\partial w_j^n}{\partial \nu_{A_j}} + \beta \frac{\partial w_j^n}{\partial t} + \gamma p_j^n\right) + \epsilon\left(\frac{\partial w_i^n}{\partial \nu_{A_i}} + \beta \frac{\partial w_i^n}{\partial t} + \gamma p_i^n\right)\Bigg|_{\Sigma_{ij}}$$

$$\mu_{ij}^n = (1-\epsilon)\left(-\frac{\partial p_j^n}{\partial \nu_{A_j}} - \beta d_t p_j^n + \gamma d_t \frac{\partial w_j^n}{\partial t}\right) \tag{6.4.1.17}$$

$$+\epsilon\left(\frac{\partial p_i^n}{\partial \nu_{A_i}} - \beta d_t p_i^n + \gamma d_t \frac{\partial w_i^n}{\partial t}\right)\Bigg|_{\Sigma_{ij}}.$$

Lemma 6.4.1.2 *Assume that $\beta > 0$, $\gamma > 0$, that $\lambda_{ij}^0 \in L^2(\Sigma_{ij})$ and $\mu_{ij}^0 = d_t \rho_{ij}^0$ where $\rho_{ij}^0 \in L^2(\Sigma_{ij})$. Then for $n = 0, 1, \ldots$, $\lambda_{ij}^n \in L^2(\Sigma_{ij})$ and $\mu_{ij}^n = d_t \rho_{ij}^n$ where $\rho_{ij}^n \in L^2(\Sigma_{ij})$. As a consequence, for each $i = 1, \ldots, I$ the iteration* (6.4.1.12)–(6.4.1.17) *is well posed and*

$$(w_i^{n+1}, \frac{\partial w_i^{n+1}}{\partial t}) \in C([0,T]; \mathcal{H}_i), \quad \frac{\partial w_i^{n+1}}{\partial t}|_{\Sigma_i^N \cup \Sigma_i^{int}} \in L^2(\Sigma_i^N \cup \Sigma_i^{int}),$$

$$(\frac{\partial p_i^{n+1}}{\partial t}, p_i^{n+1}) \in C([0,T]; \mathcal{H}_i^*), \quad p_i^{n+1}|_{\Sigma_i^N \cup \Sigma_i^{int}} \in L^2(\Sigma_i^N \cup \Sigma_i^{int}).$$

Proof. From Lemma 6.4.1.1, for $n = 0$ and each $i = 1, \ldots, I$ the system has a unique solution with the indicated regularity. Suppose the Lemma is true for the index n. It follows immediately that $\lambda_{ij}^{n+1} \in L^2(\Sigma_{ij})$. We show that $\mu_{ij}^{n+1} = d_t \rho_{ij}^{n+1}$ for some $\rho_{ij}^{n+1} \in L^2(\Sigma_{ij})$. The conclusion of the Lemma will then follow by induction.

We may write

$$p_i^{n+1} = \frac{\partial q_i^{n+1}}{\partial t}, \quad \frac{\partial p_i^{n+1}}{\partial t} = -A_i q_i^{n+1}, \quad i = 1, \ldots, I,$$

where q_i^{n+1} is the solution of

$$\begin{gathered}
\frac{\partial^2 q_i^{n+1}}{\partial t^2} - \nabla \cdot (A_i \nabla q_i^{n+1}) + c_i q_i^{n+1} = 0 \text{ in } Q_i \\
q_i^{n+1} = 0 \text{ on } \Sigma_i^D \\
\frac{\partial q_i^{n+1}}{\partial \nu_{A_i}} - \alpha_i \frac{\partial q_i^{n+1}}{\partial t} = 0 \text{ on } \Sigma_i^N \\
\frac{\partial q_i^{n+1}}{\partial \nu_{A_i}} - \beta \frac{\partial q_i^{n+1}}{\partial t} + \gamma \frac{\partial w_i^{n+1}}{\partial t} = \rho_{ij}^n \text{ on } \Sigma_{ij} \\
q_i^{n+1}(T) = \kappa(w_i^{n+1}(T) - z_{0i}), \quad \frac{\partial q_i^{n+1}}{\partial t}(T) = k(\frac{\partial w_i^{n+1}}{\partial t}(T) - z_{1i}) \text{ in } \Omega.
\end{gathered} \tag{6.4.1.18}$$

Set

$$\rho_{ij}^{n+1} = (1-\epsilon)\left(-\frac{\partial q_j^{n+1}}{\partial \nu_{A_j}} - \beta \frac{\partial q_j^{n+1}}{\partial t} + \gamma \frac{\partial w_j^{n+1}}{\partial t}\right)\Bigg|_{\Sigma_{ij}} + \epsilon \rho_{ij}^n.$$

One has

$$\frac{\partial q_j^{n+1}}{\partial \nu_{A_j}} = \beta \frac{\partial q_j^{n+1}}{\partial t} - \gamma \frac{\partial w_j^{n+1}}{\partial t} + \rho_{ji}^n \text{ on } \Sigma_{ji} = \Sigma_{ij}.$$

Therefore if $\langle \cdot, \cdot \rangle_{\Sigma_{ij}}$ denotes the inner product in the $(H^1(0,T;U_{ij}))^*$, $H^1(0,T;U_{ij})$ duality, we have

$$\begin{aligned}
\langle d_t \frac{\partial q_j^{n+1}}{\partial \nu_{A_j}}, \phi \rangle_{\Sigma_{ij}} &:= -\int_{\Sigma_{ij}} \frac{\partial q_j^{n+1}}{\partial \nu_{A_j}} \frac{\partial \phi}{\partial t}\, d\Sigma = -\int_{\Sigma_{ij}} \left(\beta p_j^{n+1} - \gamma \frac{\partial w_j^{n+1}}{\partial t} + \rho_{ji}^n\right) \frac{\partial \phi}{\partial t}\, d\Sigma \\
&= \langle \beta d_t p_j^{n+1} - \gamma d_t \frac{\partial w_j^{n+1}}{\partial t} + \mu_{ji}^n, \phi \rangle_{\Sigma_{ij}} \\
&= \langle \frac{\partial p_j^{n+1}}{\partial \nu_{A_j}}, \phi \rangle_{\Sigma_{ij}}, \quad \forall \phi \in H^1(0,T;U_{ij}).
\end{aligned}$$

It follows that

$$d_t \rho_{ij}^{n+1} = (1-\epsilon)\left(-\frac{\partial p_j^{n+1}}{\partial \nu_{A_j}} - \beta d_t p_j^{n+1} + \gamma d_t \frac{\partial w_j^{n+1}}{\partial t}\right)\Bigg|_{\Sigma_{ij}} + \epsilon d_t \rho_{ij}^n = \mu_{ij}^{n+1}.$$

Remark 6.4.1.2 In the nonoverlapping Schwarz alternating method of P.-L. Lions as adapted to optimal control problems by Benamou and others, the interface conditions (6.4.1.15) are replaced by

$$\frac{\partial w_i^{n+1}}{\partial \nu_{A_i}} + \gamma p_i^{n+1} = \tilde{\lambda}_{ij}^n, \quad \frac{\partial p_i^{n+1}}{\partial \nu_{A_i}} - \gamma w_i^{n+1} = \tilde{\mu}_{ij}^n \text{ on } \Sigma_{ij} \tag{6.4.1.19}$$

where $\gamma > 0$ and

$$\tilde{\lambda}_{ij}^n = -\frac{\partial w_j^n}{\partial \nu_{A_j}} + \gamma p_j^n, \quad \tilde{\mu}_{ij}^n = -\frac{\partial p_j^n}{\partial \nu_{A_j}} - \gamma w_j^n \tag{6.4.1.20}$$

(or, more generally, by a relaxation of (6.4.1.19), (6.4.1.20)). However, if it is assumed that $\tilde{\lambda}_{ij}^n$, $\tilde{\mu}_{ij}^n \in L^2(\Sigma_{ij})$ and we use (6.4.1.19) in place of (6.4.1.15), the local problems are then not necessarily well posed. Indeed, the system consisting of (6.4.1.12)–(6.4.1.14), (6.4.1.16) and (6.4.1.19) is formally the optimality system associated with the optimal control problem

$$\begin{aligned}
\inf_{f_i, g_{ij}} \Bigg\{ \int_{\Sigma_i^N} |f_i|^2 d\Sigma &+ \frac{1}{\gamma} \sum_{j \in \mathcal{J}_i} \int_{\Sigma_{ij}} (|g_{ij}|^2 + |\gamma w_i + \tilde{\mu}_{ij}^n|^2)\, d\Sigma \\
&+ \kappa \|(w_i(T), \frac{\partial w_i}{\partial t}(T)) - (z_{0i}, z_{1i})\|_{\mathcal{H}_i}^2 \Bigg\}
\end{aligned} \tag{6.4.1.21}$$

subject to

$$\begin{aligned}
\frac{\partial^2 w_i}{\partial t^2} - \nabla \cdot (A_i \nabla w_i) + c_i w_i &= F_i \text{ in } Q_i \\
w_i &= 0 \text{ on } \Sigma_i^D \\
\frac{\partial w_i}{\partial \nu_{A_i}} + \alpha_i \frac{\partial w_i}{\partial t} &= f_i \text{ on } \Sigma_i^N \\
\frac{\partial w_i}{\partial \nu_{A_i}} &= \tilde{\lambda}_{ij} + g_{ij} \text{ on } \Sigma_{ij}, \forall j : \Gamma_{ij} \neq \emptyset \\
w_i(0) = w_{0i}, \quad \frac{\partial w_i}{\partial t}(0) &= v_{0i} \text{ in } \Omega_i,
\end{aligned}$$

where the controls $f_i \in L^2(\Sigma_i^N)$, $g_{ij} \in L^2(\Sigma_{ij})$. Because of the absence of dissipation in the boundary condition on Σ_{ij}, the state space for the above system will be less regular than $\mathcal{H}_i$ so that the cost functional will not be defined for all admissible controls. If the space of admissible controls is restricted to those for which the cost functional is well defined, there will not be an optimal control, in general. On the other hand, if the cost functional is changed to penalize the deviation of the final state from its target in a appropriate weaker norm than the $\mathcal{H}_i$ norm, there will be a unique optimal control. However, in this case the system (6.4.1.12)–(6.4.1.14), (6.4.1.16), (6.4.1.19), will no longer be the optimality system for the optimal control problem.

6.4.2 Convergence of the Iterates

To establish convergence of the solutions $\{(w_i^{n+1}, p_i^{n+1})\}_{i=1}^I$ of the local optimality systems to the solution $\{w_i, p_i\}$ of the global optimality system, we consider the local errors

$$\tilde{w}_i^{n+1} := w_i^{n+1} - w_i, \quad \tilde{p}_i^{n+1} := p_i^{n+1} - p_i.$$

These satisfy

$$\begin{aligned}
\frac{\partial^2 \tilde{w}_i^{n+1}}{\partial t^2} - \nabla \cdot (A_i \nabla \tilde{w}_i^{n+1}) + c_i \tilde{w}_i^{n+1} &= 0 \text{ in } Q_i \\
\tilde{w}_i^{n+1} &= 0 \text{ on } \Sigma_i^D \\
\frac{\partial \tilde{w}_i^{n+1}}{\partial \nu_{A_i}} + \alpha_i \frac{\partial \tilde{w}_i^{n+1}}{\partial t} + \tilde{p}_i^{n+1} &= 0 \text{ on } \Sigma_i^N \\
\frac{\partial \tilde{w}_i^{n+1}}{\partial \nu_{A_i}} + \beta \frac{\partial \tilde{w}_i^{n+1}}{\partial t} + \gamma \tilde{p}_i^{n+1} &= \tilde{\lambda}_{ij}^n \text{ on } \Sigma_{ij} \\
\tilde{w}_i^{n+1}(0) = \frac{\partial \tilde{w}_i^{n+1}}{\partial t}(0) &= 0 \text{ in } \Omega_i,
\end{aligned} \tag{6.4.2.1}$$

$$
\begin{aligned}
\frac{\partial^2 \tilde{p}_i^{n+1}}{\partial t^2} - \nabla \cdot (A_i \nabla \tilde{p}_i^{n+1}) + c_i \tilde{p}_i^{n+1} &= 0 \text{ in } Q_i \\
\tilde{p}_i^{n+1} &= 0 \text{ on } \Sigma_i^D \\
\frac{\partial \tilde{p}_i^{n+1}}{\partial \nu_{A_i}} - \alpha_i d_t \tilde{p}_i^{n+1} &= 0 \text{ on } \Sigma_i^N \\
\frac{\partial \tilde{p}_i^{n+1}}{\partial \nu_{A_i}} - \beta d_t \tilde{p}_i^{n+1} + \gamma d_t \frac{\partial \tilde{w}_i^{n+1}}{\partial t} &= \tilde{\mu}_{ij}^n \text{ on } \Sigma_{ij} \\
\tilde{p}_i^{n+1}(T) = \kappa \frac{\partial \tilde{w}_i^{n+1}}{\partial t}(T), \quad \frac{\partial \tilde{p}_i^{n+1}}{\partial t}(T) &= -\kappa A_i \tilde{w}_i^{n+1}(T) \text{ in } \Omega_i,
\end{aligned}
\tag{6.4.2.2}
$$

where

$$
\tilde{\lambda}_{ij}^n = \lambda_{ij}^n - \lambda_{ij}, \quad \tilde{\mu}_{ij}^n = \mu_{ij}^n - \mu_{ij} \tag{6.4.2.3}
$$

and where λ_{ij}^n, μ_{ij}^n, λ_{ij}, μ_{ij} are given by (6.4.1.17) and (6.4.1.7), respectively.

As above, we may express $\tilde{p}_i^{n+1}$ as

$$
\tilde{p}_i^{n+1} = \frac{\partial \tilde{q}_i^{n+1}}{\partial t}, \quad \frac{\partial \tilde{p}_i^{n+1}}{\partial t} = -A_i q_i^{n+1}, \tag{6.4.2.4}
$$

where $\tilde{q}_i^{n+1}$ is the solution of

$$
\begin{aligned}
\frac{\partial^2 \tilde{q}_i^{n+1}}{\partial t^2} - \nabla \cdot (A_i \nabla \tilde{q}_i^{n+1}) + c_i \tilde{q}_i^{n+1} &= 0 \text{ in } Q_i \\
\tilde{q}_i^{n+1} &= 0 \text{ on } \Sigma_i^D \\
\frac{\partial \tilde{q}_i^{n+1}}{\partial \nu_{A_i}} - \alpha_i \frac{\partial \tilde{q}_i^{n+1}}{\partial t} &= 0 \text{ on } \Sigma_i^N \\
\frac{\partial \tilde{q}_i^{n+1}}{\partial \nu_{A_i}} - \beta \frac{\partial \tilde{q}_i^{n+1}}{\partial t} + \gamma \frac{\partial \tilde{w}_i^{n+1}}{\partial t} &= \tilde{\rho}_{ij}^n \text{ on } \Sigma_{ij} \\
\tilde{q}_i^{n+1}(T) = \kappa \tilde{w}_i^{n+1}(T), \quad \frac{\partial \tilde{q}_i^{n+1}}{\partial t}(T) &= \kappa \frac{\partial \tilde{w}_i^{n+1}}{\partial t}(T) \text{ in } \Omega,
\end{aligned}
\tag{6.4.2.5}
$$

where

$$
\begin{aligned}
\tilde{\rho}_{ij}^n = (1-\epsilon)\Big(-\frac{\partial \tilde{q}_j^n}{\partial \nu_{A_j}} - \beta \frac{\partial \tilde{q}_j^n}{\partial t} + \gamma \frac{\partial \tilde{w}_j^n}{\partial t} \Big) \\
+ \epsilon \Big(\frac{\partial \tilde{q}_i^n}{\partial \nu_{A_i}} - \beta \frac{\partial \tilde{q}_i^n}{\partial t} + \gamma \frac{\partial \tilde{w}_i^n}{\partial t} \Big).
\end{aligned}
\tag{6.4.2.6}
$$

We have the following convergence result.

Theorem 6.4.2.1 *In addition to the assumptions of Lemma* 6.4.1.2, *suppose that*

$$
\alpha_0 := \inf_{\Gamma^N} \alpha(x) > \frac{1}{2}\left(-\frac{\gamma}{\beta} + \sqrt{\frac{\gamma^2}{\beta^2} + 1} \right), \quad \frac{\gamma}{\beta} > \frac{\kappa^2 - 1}{2\kappa}. \tag{6.4.2.7}
$$

Then for any $\epsilon \in [0,1)$ *and for* $i = 1, \ldots, I$

$$\frac{\partial \tilde{q}_i^n}{\partial t}\Big|_{\Sigma_i^N} \to 0, \quad \frac{\partial \tilde{w}_i^n}{\partial t}\Big|_{\Sigma_i^N} \to 0 \text{ strongly in } L^2(\Sigma_i^N)$$

$$\begin{cases} \dfrac{\partial \tilde{q}_i^n}{\partial t}\Big|_{\Sigma_i^{int}} \to 0, \quad \dfrac{\partial \tilde{w}_i^n}{\partial t}\Big|_{\Sigma_i^{int}} \to 0 \\ \dfrac{\partial \tilde{q}_i^n}{\partial \nu_{A_i}}\Big|_{\Sigma_i^{int}} \to 0, \quad \dfrac{\partial \tilde{w}_i^n}{\partial \nu_{A_i}}\Big|_{\Sigma_i^{int}} \to 0 \quad \text{weakly in } L^2(\Sigma_i^{int}). \end{cases}$$

In addition,

(i) *if* $\epsilon \in (0,1)$ *then*

$$(\tilde{w}_i^n, \frac{\partial \tilde{w}_i^n}{\partial t}) \to 0, \quad (\tilde{q}_i^n, \frac{\partial \tilde{q}_i^n}{\partial t}) \to 0 \text{ strongly in } L^\infty(0,T;\mathcal{H}_i)$$

$$\begin{cases} \dfrac{\partial w_i^n}{\partial t} - \dfrac{\partial w_j^n}{\partial t}\Big|_{\Sigma_{ij}} \to 0, \quad \dfrac{\partial w_i^n}{\partial \nu_{A_i}} + \dfrac{\partial w_j^n}{\partial \nu_{A_j}}\Big|_{\Sigma_{ij}} \to 0 \\ \dfrac{\partial q_i^n}{\partial t} - \dfrac{\partial q_j^n}{\partial t}\Big|_{\Sigma_{ij}} \to 0, \quad \dfrac{\partial q_i^n}{\partial \nu_{A_i}} + \dfrac{\partial q_j^n}{\partial \nu_{A_j}}\Big|_{\Sigma_{ij}} \to 0 \quad \text{strongly in } L^2(\Sigma_{ij}). \end{cases}$$

(ii) *If* $\epsilon = 0$ *then*

$$(\tilde{w}_i^n, \frac{\partial \tilde{w}_i^n}{\partial t}) \to 0, \quad (\tilde{q}_i^n, \frac{\partial \tilde{q}_i^n}{\partial t}) \to 0 \text{ weakly* in } L^\infty(0,T;\mathcal{H}_i)$$

$$(\tilde{w}_i^n(T), \frac{\partial \tilde{w}_i^n}{\partial t}(T)) \to 0, \quad (\tilde{q}_i^n(0), \frac{\partial \tilde{q}_i^n}{\partial t}(0)) \to 0 \text{ strongly in } \mathcal{H}_i.$$

Remark 6.4.2.1 Note that for a given $\kappa > 1$ and $\alpha_0 > 0$, (6.4.2.7) can always be satisfied by choosing γ/β sufficiently large.

Remark 6.4.2.2 One immediately deduces from Theorem 6.4.2.1 and Corollary 6.2.2.2 the following convergence properties of the adjoint variables $\tilde{p}_i^n = \dfrac{\partial \tilde{q}_i^n}{\partial t}$, $\dfrac{\partial \tilde{p}_i^n}{\partial t} = -\mathcal{A}_i \tilde{q}_i^n$: for any $\epsilon \in [0,1)$

$$\tilde{p}_i^n|_{\Sigma_i^N} \to 0 \text{ strongly in } L^2(\Sigma_i^N)$$
$$\tilde{p}_i^n|_{\Sigma_{ij}} \to 0 \text{ weakly in } L^2(\Sigma_{ij}).$$

In addition, if $\epsilon \in (0,1)$ then

$$(\tilde{p}_i^n, \frac{\partial \tilde{p}_i^n}{\partial t}) \to 0 \text{ strongly in } L^\infty(0,T;H_i \times V_i^*)$$
$$p_i^n - p_j^n|_{\Sigma_{ij}} \to 0 \text{ strongly in } L^2(\Sigma_{ij})$$

while, if $\epsilon = 0$, then

$$(\tilde{p}_i^n, \frac{\partial \tilde{p}_i^n}{\partial t}) \to 0 \text{ weakly}^* \text{ in } L^\infty(0,T; H_i \times V_i^*)$$
$$(\tilde{p}_i^n(T), \frac{\partial \tilde{p}_i^n}{\partial t}(T)) \to 0 \text{ strongly in } H_i \times V_i^*.$$

In particular, for all $\epsilon \in [0,1)$ and for $i = 1,\dots,I$ the effective local optimal control $p_i^n|_{\Sigma_i^N}$ converges strongly in $L^2(\Sigma_i^N)$ to the restriction $p|_{\Sigma_i^N}$ of the global optimal control $p|_{\Sigma^N}$ to Σ_i^N. Further, the deviation of the local trajectory $(w_i^n(T), \partial w_i^n(T)/\partial t)$ at time T from its target state (z_{0i}, z_{1i}) converges strongly in $\mathcal{H}_i$ to the deviation of the global trajectory $(w(T), \partial w(T)/\partial t)$ at time T from its target (z_0, z_1), restricted to Ω_i.

Proof of Theorem 6.4.2.1. We introduce the space

$$\mathcal{X} = \prod_{i=1}^I L^2(\Sigma_i^{\text{int}}) \times L^2(\Sigma_i^{\text{int}})$$

with norm defined by

$$\|X\|_{\mathcal{X}}^2 = \sum_{i=1}^I \frac{1}{2\gamma} \int_{\Sigma_i^{\text{int}}} (|\lambda_i|^2 + |\rho_i|^2)\, d\Sigma, \quad X = \{(\lambda_i, \rho_i) : i = 1,\dots,I\}.$$

For $X = \{(\lambda_i, \rho_i)\}_{i=1}^I \in \mathcal{X}$ let $\{(w_i, q_i)\}_{i=1}^I$ be the solutions of

$$\begin{cases} \dfrac{\partial^2 w_i}{\partial t^2} - \nabla \cdot (A_i \nabla w_i) + c_i w_i = 0 \\ \dfrac{\partial^2 q_i}{\partial t^2} - \nabla \cdot (A_i \nabla q_i) + c_i q_i = 0 \text{ in } Q_i \end{cases}$$

$$w_i = q_i = 0 \text{ on } \Sigma_i^D$$

$$\begin{cases} \dfrac{\partial w_i}{\partial \nu_{A_i}} + \alpha_i \dfrac{\partial w_i}{\partial t} + \dfrac{\partial q_i}{\partial t} = 0 \\ \dfrac{\partial q_i}{\partial \nu_{A_i}} - \alpha_i \dfrac{\partial q_i}{\partial t} = 0 \text{ on } \Sigma_i^N \end{cases}$$

$$\begin{cases} \dfrac{\partial w_i}{\partial \nu_{A_i}} + \beta \dfrac{\partial w_i}{\partial t} + \gamma \dfrac{\partial q_i}{\partial t} = \lambda_i \\ \dfrac{\partial q_i}{\partial \nu_{A_i}} - \beta \dfrac{\partial q_i}{\partial t} + \gamma \dfrac{\partial w_i}{\partial t} = \rho_i \text{ on } \Sigma_i^{\text{int}} \end{cases}$$

$$\begin{cases} q_i(T) = \kappa w_i(T), \ \dfrac{\partial q_i}{\partial t}(T) = \kappa \dfrac{\partial w_i}{\partial t}(T) \\ w_i(0) = \dfrac{\partial w_i}{\partial t}(0) = 0 \text{ in } \Omega_i \end{cases}$$

This system has a unique solution with regularity

$$(w_i, \frac{\partial w_i}{\partial t}), (q_i, \frac{\partial q_i}{\partial t}) \in C([0,T];\mathcal{H}_i), \quad \frac{\partial w_i}{\partial t}\Big|_{\Sigma_i^N \cup \Sigma_i^{\text{int}}}, \frac{\partial q_i}{\partial t}\Big|_{\Sigma_i^N \cup \Sigma_i^{\text{int}}} \in L^2(\Sigma_i^N \cup \Sigma_i^{\text{int}}).$$

In fact, this system is nothing but the system (6.4.2.1), (6.4.2.5) with the superscripts suppressed.

We define a linear mapping $\mathcal{T} : \mathcal{X} \mapsto \mathcal{X}$ as follows. Set

$$(\mathcal{T}X)_{ij} = \left((-\frac{\partial w_j}{\partial \nu_{A_j}} + \beta\frac{\partial w_j}{\partial t} + \gamma\frac{\partial q_j}{\partial t})\Big|_{\Sigma_{ij}}, (-\frac{\partial q_j}{\partial \nu_{A_j}} - \beta\frac{\partial q_j}{\partial t} + \gamma\frac{\partial w_j}{\partial t})\Big|_{\Sigma_{ij}} \right)$$

$$(\mathcal{T}X)_i = \{(\mathcal{T}X)_{ij} : j \in \mathcal{J}_i\}, \quad \mathcal{T}X = \{(\mathcal{T}X)_i : i = 1, \dots, I\}. \tag{6.4.2.8}$$

Note that X is a fixed point of $\mathcal{T}$ if and only if (w_i, p_i), where $p_i = \partial q_i/\partial t$, is the solution of the global optimality system (6.4.1.2)–(6.4.1.6) corresponding to $w_{0i} = v_{0i} = z_{0i} = z_{1i} = 0$, $F_i = 0$. Since the solution of the global optimal control problem corresponding to such vanishing data is clearly $f_{\text{opt}} = 0$, it follows that $X = 0$ is the unique fixed point of $\mathcal{T}$.

The main significance of the mapping $\mathcal{T}$ is that if, for $n = 1, 2, \dots$, we set

$$X^n = \{(\tilde{\lambda}_{ij}^{n-1}, \tilde{\rho}_{ij}^{n-1}) : i = 1, \dots, I;\ j \in \mathcal{J}_i\} \tag{6.4.2.9}$$

and let $(\tilde{w}_i^n, \tilde{q}_i^n)$ be the solution of (6.4.2.1), (6.4.2.5) with n replace by $n-1$, then

$$X^n = \Big\{ \Big((\frac{\partial \tilde{w}_i^n}{\partial \nu_{A_i}} + \beta\frac{\partial \tilde{w}_i^n}{\partial t} + \gamma\frac{\partial \tilde{q}_i^n}{\partial t})\Big|_{\Sigma_{ij}}, (\frac{\partial \tilde{q}_i^n}{\partial \nu_{A_i}} - \beta\frac{\partial \tilde{q}_i^n}{\partial t} + \gamma\frac{\partial \tilde{w}_i^n}{\partial t})\Big|_{\Sigma_{ij}} \Big) : \\ i = 1, \dots, I;\ j \in \mathcal{J}_i \Big\} \tag{6.4.2.10}$$

$$\mathcal{T}X^n = \Big\{ \Big((-\frac{\partial \tilde{w}_j^n}{\partial \nu_{A_j}} + \beta\frac{\partial \tilde{w}_j^n}{\partial t} + \gamma\frac{\partial \tilde{q}_j^n}{\partial t})\Big|_{\Sigma_{ij}}, (-\frac{\partial \tilde{q}_j^n}{\partial \nu_{A_j}} - \beta\frac{\partial \tilde{q}_j^n}{\partial t} + \gamma\frac{\partial \tilde{w}_j^n}{\partial t})\Big|_{\Sigma_{ij}} \Big) : \\ i = 1, \dots, I;\ j \in \mathcal{J}_i \Big\} \tag{6.4.2.11}$$

and the interface conditions on Σ_{ij} in (6.4.2.1), (6.4.2.5) may be expressed as the fixed point iteration

$$X^{n+1} = (1-\epsilon)\mathcal{T}X^n + \epsilon X^n. \tag{6.4.2.12}$$

The proof of Theorem 6.4.2.1 is based on the following property of $\mathcal{T}$.

Lemma 6.4.2.1 *For any $X = \{(\lambda_i, \rho_i) : i = i, \dots, I\} \in \mathcal{X}$, we may write*

$$\|X\|_{\mathcal{X}}^2 = \mathcal{E} + \mathcal{F}, \quad \|\mathcal{T}X\|_{\mathcal{X}}^2 = \mathcal{E} - \mathcal{F} \tag{6.4.2.13}$$

where

$$\mathcal{E} = \sum_{i=1}^{I} \int_{\Sigma_i^{int}} \left\{ \frac{1}{2\gamma} \left(\left| \frac{\partial w_i}{\partial \nu_{A_i}} \right|^2 + \left| \frac{\partial q_i}{\partial \nu_{A_i}} \right|^2 \right) + \frac{\beta^2 + \gamma^2}{2\gamma} \left(\left| \frac{\partial w_i}{\partial t} \right|^2 + \left| \frac{\partial q_i}{\partial t} \right|^2 \right) \right\} d\Sigma, \quad (6.4.2.14)$$

$$\mathcal{F} = \sum_{i=1}^{I} \left\{ [\kappa + \frac{\beta}{2\gamma}(1-\kappa^2)] \|(w_i(T), \frac{\partial w_i}{\partial t}(T))\|_{\mathcal{H}_i}^2 + \frac{\beta}{2\gamma} \|(q_i(0), \frac{\partial q_i}{\partial t}(0))\|_{\mathcal{H}_i}^2 \right\} + \sum_{i=1}^{I} \int_{\Sigma_i^N} \left\{ \left(1 + \frac{\alpha_i \beta}{\gamma}\right) \left| \frac{\partial q_i}{\partial t} \right|^2 + \frac{\beta}{\gamma} \frac{\partial q_i}{\partial t} \frac{\partial w_i}{\partial t} + \frac{\alpha_i \beta}{\gamma} \left| \frac{\partial w_i}{\partial t} \right|^2 \right\} d\Sigma. \quad (6.4.2.15)$$

Corollary 6.4.2.1 *If β, γ satisfy* (6.4.2.7), *then $X \mapsto \mathcal{T}X : \mathcal{X} \mapsto \mathcal{X}$ is nonexpansive.*

Proof of Lemma 6.4.2.1. The proof follows by direct calculation. One has

$$\begin{aligned}
\|X\|_{\mathcal{X}}^2 &= \sum_{i=1}^{I} \frac{1}{2\gamma} \int_{\Sigma_i^{int}} \left(\left| \frac{\partial w_i}{\partial \nu_{A_i}} + \beta \frac{\partial w_i}{\partial t} + \gamma \frac{\partial q_i}{\partial t} \right|^2 + \left| \frac{\partial q_i}{\partial \nu_{A_i}} - \beta \frac{\partial q_i}{\partial t} + \gamma \frac{\partial w_i}{\partial t} \right|^2 \right) d\Sigma \\
&= \sum_{i=1}^{I} \int_{\Sigma_i^{int}} \left\{ \frac{1}{2\gamma} \left(\left| \frac{\partial w_i}{\partial \nu_{A_i}} \right|^2 + \left| \frac{\partial q_i}{\partial \nu_{A_i}} \right|^2 \right) + \frac{\beta^2 + \gamma^2}{2\gamma} \left(\left| \frac{\partial w_i}{\partial t} \right|^2 + \left| \frac{\partial q_i}{\partial t} \right|^2 \right) \right. \\
&\quad \left. + \frac{\beta}{\gamma} \left(\frac{\partial w_i}{\partial \nu_{A_i}} \frac{\partial w_i}{\partial t} - \frac{\partial q_i}{\partial \nu_{A_i}} \frac{\partial q_i}{\partial t} \right) + \left(\frac{\partial w_i}{\partial \nu_{A_i}} \frac{\partial q_i}{\partial t} + \frac{\partial q_i}{\partial \nu_{A_i}} \frac{\partial w_i}{\partial t} \right) \right\} d\Sigma \\
&= \mathcal{E} + \sum_{i=1}^{I} \int_{\Sigma_i^{int}} \left\{ \frac{\beta}{\gamma} \left(\frac{\partial w_i}{\partial \nu_{A_i}} \frac{\partial w_i}{\partial t} - \frac{\partial q_i}{\partial \nu_{A_i}} \frac{\partial q_i}{\partial t} \right) + \left(\frac{\partial w_i}{\partial \nu_{A_i}} \frac{\partial q_i}{\partial t} + \frac{\partial q_i}{\partial \nu_{A_i}} \frac{\partial w_i}{\partial t} \right) \right\} d\Sigma.
\end{aligned} \quad (6.4.2.16)$$

Similarly, since $\sum_{i=1}^{I} \sum_{j \in \mathcal{J}_i} = \sum_{j=1}^{I} \sum_{i \in \mathcal{J}_j}$, $\|\mathcal{T}X\|_{\mathcal{X}}^2$ may be written

$$\|\mathcal{T}X\|_{\mathcal{X}}^2 = \mathcal{E} - \sum_{i=1}^{I} \int_{\Sigma_i^{int}} \left\{ \frac{\beta}{\gamma} \left(\frac{\partial w_i}{\partial \nu_{A_i}} \frac{\partial w_i}{\partial t} - \frac{\partial q_i}{\partial \nu_{A_i}} \frac{\partial q_i}{\partial t} \right) + \left(\frac{\partial w_i}{\partial \nu_{A_i}} \frac{\partial q_i}{\partial t} + \frac{\partial q_i}{\partial \nu_{A_i}} \frac{\partial w_i}{\partial t} \right) \right\} d\Sigma. \quad (6.4.2.17)$$

We will show that

$$\sum_{i=1}^{I} \int_{\Sigma_i^{int}} \left\{ \frac{\beta}{\gamma} \left(\frac{\partial w_i}{\partial \nu_{A_i}} \frac{\partial w_i}{\partial t} - \frac{\partial q_i}{\partial \nu_{A_i}} \frac{\partial q_i}{\partial t} \right) + \left(\frac{\partial w_i}{\partial \nu_{A_i}} \frac{\partial q_i}{\partial t} + \frac{\partial q_i}{\partial \nu_{A_i}} \frac{\partial w_i}{\partial t} \right) \right\} d\Sigma = \mathcal{F}. \quad (6.4.2.18)$$

It then follows from (6.4.2.16) and (6.4.2.17) that

$$\begin{aligned} \|X\|_{\mathcal{X}}^2 &= \mathcal{E} + \mathcal{F}, \\ \|\mathcal{T}X\|_{\mathcal{X}}^2 &= \mathcal{E} - \mathcal{F}. \end{aligned} \tag{6.4.2.19}$$

From

$$0 = \int_{Q_i} (\frac{\partial^2 w_i}{\partial t^2} - \nabla \cdot (A_i \nabla w_i) + c_i w_i) \frac{\partial w_i}{\partial t}\, dxdt$$

we easily obtain

$$\frac{1}{2}\|(w_i(T), \frac{\partial w_i}{\partial t}(T))\|_{\mathcal{H}_i}^2 = \left(\int_{\Sigma_i^N} + \int_{\Sigma_i^{\text{int}}}\right) \frac{\partial w_i}{\partial \nu_{A_i}} \frac{\partial w_i}{\partial t}\, d\Sigma,$$

hence

$$\begin{aligned} &\int_{\Sigma_i^{\text{int}}} \frac{\partial w_i}{\partial \nu_{A_i}} \frac{\partial w_i}{\partial t}\, d\Sigma \\ &\qquad = \frac{1}{2}\|(w_i(T), \frac{\partial w_i}{\partial t}(T))\|_{\mathcal{H}_i}^2 + \int_{\Sigma_i^N} \left(\alpha_i \left|\frac{\partial w_i}{\partial t}\right|^2 + \frac{\partial q_i}{\partial t} \frac{\partial w_i}{\partial t}\right) d\Sigma. \end{aligned}$$

Similarly

$$\begin{aligned} \int_{\Sigma_i^{\text{int}}} \frac{\partial q_i}{\partial \nu_{A_i}} \frac{\partial q_i}{\partial t}\, d\Sigma &= \frac{\kappa^2}{2}\|(w_i(T), \frac{\partial w_i}{\partial t}(T))\|_{\mathcal{H}_i}^2 \\ &\quad - \frac{1}{2}\|(q_i(0), \frac{\partial q_i}{\partial t}(0))\|_{\mathcal{H}_i}^2 - \int_{\Sigma_i^N} \alpha_i \left|\frac{\partial q_i}{\partial t}\right|^2 d\Sigma. \end{aligned}$$

Therefore

$$\begin{aligned} &\int_{\Sigma_i^{\text{int}}} \left(\frac{\partial w_i}{\partial \nu_{A_i}} \frac{\partial w_i}{\partial t} - \frac{\partial q_i}{\partial \nu_{A_i}} \frac{\partial q_i}{\partial t}\right) d\Sigma \\ &\qquad = \frac{1}{2}(1-\kappa^2)\|(w_i(T), \frac{\partial w_i}{\partial t}(T))\|_{\mathcal{H}_i}^2 \\ &\qquad\quad + \frac{1}{2}\|(q_i(0), \frac{\partial q_i}{\partial t}(0))\|_{\mathcal{H}_i}^2 \\ &\qquad\quad + \int_{\Sigma_i^N} \left\{\alpha_i \left(\left|\frac{\partial w_i}{\partial t}\right|^2 + \left|\frac{\partial q_i}{\partial t}\right|^2\right) + \frac{\partial q_i}{\partial t} \frac{\partial w_i}{\partial t}\right\} d\Sigma. \end{aligned} \tag{6.4.2.20}$$

To evaluate the last term on the right side of (6.4.2.18), we write

$$
\begin{aligned}
0 = \int_{Q_i} \Big[&\big(\frac{\partial^2 w_i}{\partial t^2} - \nabla\cdot(A_i \nabla w_i) + c_i w_i\big)\frac{\partial q_i}{\partial t} \\
&+ \big(\frac{\partial^2 q_i}{\partial t^2} - \nabla\cdot(A_i \nabla q_i) + c_i q_i\big)\frac{\partial w_i}{\partial t}\Big]\, dxdt \\
&= \big(\frac{\partial w_i}{\partial t}(T), \frac{\partial q_i}{\partial t}(T)\big) + (w_i(T), q_i(T))_{V_i} \\
&- \Big(\int_{\Sigma_i^{\text{int}}} + \int_{\Sigma_i^N}\Big)\Big(\frac{\partial w_i}{\partial t}\frac{\partial q_i}{\partial \nu_{A_i}} + \frac{\partial q_i}{\partial t}\frac{\partial w_i}{\partial \nu_{A_i}}\Big)\, d\Sigma \\
&= \kappa \|\big(w_i(T), \frac{\partial w_i}{\partial t}(T)\big)\|^2_{\mathcal{H}_i} + \int_{\Sigma_i^N}\Big|\frac{\partial q_i}{\partial t}\Big|^2 d\Sigma \\
&\qquad - \int_{\Sigma_i^{\text{int}}}\Big(\frac{\partial w_i}{\partial t}\frac{\partial q_i}{\partial \nu_{A_i}} + \frac{\partial q_i}{\partial t}\frac{\partial w_i}{\partial \nu_{A_i}}\Big)\, d\Sigma.
\end{aligned}
$$

Therefore

$$
\begin{aligned}
\int_{\Sigma_i^{\text{int}}} &\Big(\frac{\partial w_i}{\partial t}\frac{\partial q_i}{\partial \nu_{A_i}} + \frac{\partial q_i}{\partial t}\frac{\partial w_i}{\partial \nu_{A_i}}\Big)\, d\Sigma \\
&= \kappa \|\big(w_i(T), \frac{\partial w_i}{\partial t}(T)\big)\|^2_{\mathcal{H}_i} + \int_{\Sigma_i^N}\Big|\frac{\partial q_i}{\partial t}\Big|^2 d\Sigma. \qquad (6.4.2.21)
\end{aligned}
$$

The equality (6.4.2.18) now follows from (6.4.2.20) and (6.4.2.21).

Now let X^n be given by (6.4.2.9) and set

$$
\begin{aligned}
\mathcal{E}^n = \frac{1}{2}\sum_{i=1}^{I}\int_{\Sigma_i^{\text{int}}} \Big\{ &\frac{\beta^2+\gamma^2}{\gamma}\Big(\Big|\frac{\partial \tilde{q}_i^n}{\partial t}\Big|^2 + \Big|\frac{\partial \tilde{w}_i^n}{\partial t}\Big|^2\Big) \\
&+ \frac{1}{\gamma}\Big(\Big|\frac{\partial \tilde{w}_i^n}{\partial \nu_{A_i}}\Big|^2 + \Big|\frac{\partial \tilde{q}_i^n}{\partial \nu_{A_i}}\Big|^2\Big)\Big\}\, d\Sigma, \qquad (6.4.2.22)
\end{aligned}
$$

$$
\begin{aligned}
\mathcal{F}^n = &\sum_{i=1}^{I}\Big\{[\kappa + \frac{\beta}{2\gamma}(1-\kappa^2)]\|\big(\tilde{w}_i^n(T), \frac{\partial \tilde{w}_i^n}{\partial t}(T)\big)\|^2_{\mathcal{H}_i} + \frac{\beta}{2\gamma}\|\big(\tilde{q}_i^n(0), \frac{\partial \tilde{q}_i^n}{\partial t}(0)\big)\|^2_{\mathcal{H}_i}\Big\} \\
&+ \sum_{i=1}^{I}\int_{\Sigma_i^N}\Big\{\Big(1+\frac{\alpha_i\beta}{\gamma}\Big)\Big|\frac{\partial \tilde{q}_i^n}{\partial t}\Big|^2 + \frac{\beta}{\gamma}\frac{\partial \tilde{q}_i^n}{\partial t}\frac{\partial \tilde{w}_i^n}{\partial t} + \frac{\alpha_i\beta}{\gamma}\Big|\frac{\partial \tilde{w}_i^n}{\partial t}\Big|^2\Big\}\, d\Sigma. \qquad (6.4.2.23)
\end{aligned}
$$

From (6.4.2.19) we may write

$$
\|X^n\|^2_{\mathcal{X}} = \mathcal{E}^n + \mathcal{F}^n, \qquad \|\mathcal{T}X^n\|^2_{\mathcal{X}} = \mathcal{E}^n - \mathcal{F}^n. \qquad (6.4.2.24)
$$

If we take the norm of each side of (6.4.2.12) and use (6.4.2.19) we obtain

$$\mathcal{E}^{n+1} + \mathcal{F}^{n+1} = \left((1-\epsilon)^2 + \epsilon^2\right)\mathcal{E}^n + \left(\epsilon^2 - (1-\epsilon)^2\right)\mathcal{F}^n + 2\epsilon(1-\epsilon)(\mathcal{T}X^n, X^n)_{\mathcal{X}}.$$

From (6.4.2.19) and the Cauchy-Schwarz inequality we have

$$|(\mathcal{T}X^n, X^n)_{\mathcal{X}}| \le \sqrt{(\mathcal{E}^n)^2 - (\mathcal{F}^n)^2} \le \mathcal{E}^n,$$

hence we deduce the following *fundamental recursion formula:*

$$\mathcal{E}^{n+1} + \mathcal{F}^{n+1} \le \mathcal{E}^n - (1-2\epsilon)\mathcal{F}^n, \tag{6.4.2.25}$$

By iterating (6.4.2.25) we find that

$$\mathcal{E}^{n+1} \le \mathcal{E}^1 - \sum_{p=1}^{n+1} c_p(\epsilon)\mathcal{F}^p$$

where

$$c_1(\epsilon) = 1 - 2\epsilon, \quad c_{n+1}(\epsilon) = 1, \quad c_p(\epsilon) = 2(1-\epsilon),\ p = 2, \dots, n. \tag{6.4.2.26}$$

Since $\alpha_i \ge \alpha_0$, under the stated condition on α_0 the quadratic form

$$\left(1 + \frac{\alpha_i\beta}{\gamma}\right)\left|\frac{\partial \tilde{q}_i^n}{\partial t}\right|^2 + \frac{\beta}{\gamma}\frac{\partial \tilde{q}_i^n}{\partial t}\frac{\partial \tilde{w}_i^n}{\partial t} + \frac{\alpha_i\beta}{\gamma}\left|\frac{\partial \tilde{w}_i^n}{\partial t}\right|^2$$

is positive definite. If $\kappa > 1$, $\epsilon \in [0,1)$ and if we further restrict β/γ according to (6.4.2.7), then

$$\sum_{p=1}^{\infty} \mathcal{F}^p \text{ converges and } \{\mathcal{E}^n\}_{n=1}^{\infty} \text{ is a bounded sequence.}$$

The convergence of $\sum \mathcal{F}^p$ then implies that

$$\begin{aligned} &(\tilde{w}_i^n(T), \frac{\partial \tilde{w}_i^n}{\partial t}(T)) \to 0, \quad (\tilde{q}_i^n(0), \frac{\partial \tilde{q}_i^n}{\partial t}(0)) \to 0 \text{ strongly in } \mathcal{H}_i \\ &\left.\frac{\partial \tilde{q}_i^n}{\partial t}\right|_{\Sigma_i^N} \to 0, \quad \left.\frac{\partial \tilde{w}_i^n}{\partial t}\right|_{\Sigma_i^N} \to 0 \text{ strongly in } L^2(\Sigma_i^N). \end{aligned} \tag{6.4.2.27}$$

From (6.4.1.14) and (6.4.2.5) we then have

$$\begin{aligned} &\left.\frac{\partial \tilde{w}_i^n}{\partial \nu_{A_i}}\right|_{\Sigma_i^N} \to 0, \quad \left.\frac{\partial \tilde{q}_i^n}{\partial \nu_{A_i}}\right|_{\Sigma_i^N} \to 0 \text{ strongly in } L^2(\Sigma_i^N) \\ &(\tilde{q}_i^n(T), \frac{\partial \tilde{q}_i^n}{\partial t}(T)) \to 0 \text{ strongly in } \mathcal{H}_i. \end{aligned} \tag{6.4.2.28}$$

The boundedness of $\mathcal{E}^n$ implies that λ_{ij}^n, ρ_{ij}^n are bounded in $L^2(\Sigma_{ij})$. Therefore

$$(\tilde{w}_i^{n+1}, \frac{\partial \tilde{w}_i^{n+1}}{\partial t}), \quad (\tilde{q}_i^{n+1}, \frac{\partial \tilde{q}_i^{n+1}}{\partial t}) \text{ are bounded in } L^\infty(0,T;\mathcal{H}_i)$$

which implies, in particular, that $\{\tilde{w}_i^{n+1}\}$, $\{\tilde{q}_i^{n+1}\}$ are uniformly bounded and equicontinuous on $[0,T]$ in the $L^2(\Omega_i)$ topology. It follows that on a subsequence $n = n_k$ of the positive integers,

$$\begin{aligned}
&\tilde{w}_i^{n+1} \to \tilde{w}_i, \quad \tilde{q}_i^{n+1} \to \tilde{q}_i \text{ weakly* in } L^\infty(0,T;V_i) \\
&\qquad\qquad\qquad\qquad \text{and strongly in } C([0,T];H_i) \\
&\frac{\partial \tilde{w}_i^{n+1}}{\partial t} \to \frac{\partial \tilde{w}_i}{\partial t}, \quad \frac{\partial \tilde{q}_i^{n+1}}{\partial t} \to \frac{\partial \tilde{q}_i}{\partial t} \text{ weakly* in } L^\infty(0,T;H_i) \\
&\left.\frac{\partial \tilde{w}_i^{n+1}}{\partial t}\right|_{\Sigma_i^{\text{int}}} \to f_i, \quad \left.\frac{\partial \tilde{w}_i^{n+1}}{\partial \nu_{A_i}}\right|_{\Sigma_i^{\text{int}}} \to g_i \text{ weakly in } L^2(\Sigma_i^{\text{int}}) \\
&\left.\frac{\partial \tilde{q}_i^{n+1}}{\partial t}\right|_{\Sigma_i^{\text{int}}} \to F_i, \quad \left.\frac{\partial \tilde{q}_i^{n+1}}{\partial \nu_{A_i}}\right|_{\Sigma_i^{\text{int}}} \to G_i \text{ weakly in } L^2(\Sigma_i^{\text{int}})
\end{aligned} \tag{6.4.2.29}$$

for some f_i, g_i, F_i, G_i in $L^2(\Sigma_i^{\text{int}})$, where both $\tilde{w}_i$ and $\tilde{q}_i$ satisfy

$$\begin{gathered}
\frac{\partial^2 u_i}{\partial t^2} - \nabla \cdot (A_i \nabla u_i) + c_i u_i = 0 \text{ in } Q_i \\
u_i|_{\Sigma_i^D} = 0 \\
\left.\frac{\partial u_i}{\partial \nu_{A_i}}\right|_{\Sigma_i^N} = \left.\frac{\partial u_i}{\partial t}\right|_{\Sigma_i^N} = 0 \\
u_i(0) = \frac{\partial u_i}{\partial t}(0) = u_i(T) = \frac{\partial u_i}{\partial t}(T) = 0 \text{ in } \Omega_i,
\end{gathered} \tag{6.4.2.30}$$

and

$$\left.\frac{\partial \tilde{w}_i}{\partial t}\right|_{\Sigma_i^{\text{int}}} = f_i, \quad \left.\frac{\partial \tilde{w}_i}{\partial \nu_{A_i}}\right|_{\Sigma_i^{\text{int}}} = g_i, \quad \left.\frac{\partial \tilde{q}_i}{\partial t}\right|_{\Sigma_i^{\text{int}}} = F_i, \quad \left.\frac{\partial \tilde{q}_i}{\partial \nu_{A_i}}\right|_{\Sigma_i^{\text{int}}} = G_i \tag{6.4.2.31}$$

Since $\tilde{w}_i$ and $\tilde{q}_i$ vanish at $t = 0$ we also have $\tilde{w}_i|_{\Sigma_i^N} = \tilde{q}_i|_{\Sigma_i^N} = 0$. Thus $\tilde{w}_i$ and $\tilde{q}_i$ have vanishing Cauchy on Σ_i^N. Note also that $\tilde{w}_i$ and $\tilde{q}_i$ are in $H^1(\Omega_i \times (0,T))$ since, in particular, $\tilde{w}_i, \tilde{q}_i \in H^1(0,T;L^2(\Omega_i)) \cap L^2(0,T;H_D^1(\Omega_i)) \subseteq H^1(\Omega_i \times (0,T))$.

Case 1: $\epsilon = 0$. In this case we can use a unique continuation argument to conclude that $\tilde{w}_i = \tilde{q}_i = 0$ in Q_i for $i = 1, \ldots, I$. We start with regions Ω_i adjacent to the global boundary Γ such that $\Gamma_i^N \neq \emptyset$. Since $\tilde{w}_i$, $\tilde{q}_i \in H^1(\Omega_i \times (0,T))$ and have vanishing Cauchy on Σ_i^N and at $t = 0$ and $t = T$, they may be continued by zero to $\Omega \times \{t < 0\}$ and to $\Omega \times \{t > T\}$ as $H^1(\Omega_i \times (-\infty, \infty))$ solutions of the hyperbolic equation (6.4.2.30a) that have vanishing Cauchy data on $\Gamma_i^N \times (-\infty, +\infty)$. It then follows by unique continuation (see, for example, Hörmander [45]) that $\tilde{w}_i = \tilde{q}_i = 0$

in $\Omega \times (-\infty, +\infty)$. In particular, $f_i = g_i = F_i = G_i = 0$ and therefore the convergence in (6.4.2.29) is through the entire sequence of positive integers.

Now suppose that Ω_j is a region adjacent to a boundary region Ω_i such that $\Gamma_i^N \neq \emptyset$. Then $\Sigma_{ij} \neq \emptyset$ and from (6.4.1.15), (6.4.1.17), (6.4.2.4)–(6.4.2.6) we have on Σ_{ij} (since $\epsilon = 0$)

$$\frac{\partial \tilde{w}_i^{n+2}}{\partial \nu_{A_i}} + \beta \frac{\partial \tilde{w}_i^{n+2}}{\partial t} + \gamma \frac{\partial \tilde{q}_i^{n+2}}{\partial t} = -\frac{\partial \tilde{w}_j^{n+1}}{\partial \nu_{A_j}} + \beta \frac{\partial \tilde{w}_j^{n+1}}{\partial t} + \gamma \frac{\partial \tilde{q}_j^{n+1}}{\partial t} \tag{6.4.2.32}$$

$$\frac{\partial \tilde{q}_i^{n+2}}{\partial \nu_{A_i}} - \beta \frac{\partial \tilde{q}_i^{n+2}}{\partial t} + \gamma \frac{\partial \tilde{w}_i^{n+2}}{\partial t} = -\frac{\partial \tilde{q}_j^{n+1}}{\partial \nu_{A_j}} - \beta \frac{\partial \tilde{q}_j^{n+1}}{\partial t} + \gamma \frac{\partial \tilde{w}_j^{n+1}}{\partial t} \tag{6.4.2.33}$$

$$\frac{\partial \tilde{w}_j^{n+1}}{\partial \nu_{A_j}} + \beta \frac{\partial \tilde{w}_j^{n+1}}{\partial t} + \gamma \frac{\partial \tilde{q}_j^{n+1}}{\partial t} = -\frac{\partial \tilde{w}_i^{n}}{\partial \nu_{A_i}} + \beta \frac{\partial \tilde{w}_i^{n}}{\partial t} + \gamma \frac{\partial \tilde{q}_i^{n}}{\partial t} \tag{6.4.2.34}$$

$$\frac{\partial \tilde{q}_j^{n+1}}{\partial \nu_{A_j}} - \beta \frac{\partial \tilde{q}_j^{n+1}}{\partial t} + \gamma \frac{\partial \tilde{w}_j^{n+1}}{\partial t} = -\frac{\partial \tilde{q}_i^{n}}{\partial \nu_{A_i}} - \beta \frac{\partial \tilde{q}_i^{n}}{\partial t} + \gamma \frac{\partial \tilde{w}_i^{n}}{\partial t}. \tag{6.4.2.35}$$

Since, for the index i, convergence is through the entire sequence of positive integers, if we pass to the weak $L^2(\Sigma_{ij})$ limit in (6.4.2.32)–(6.4.2.35) through the subsequence $n = n_k$ we obtain

$$\begin{aligned} -g_j + \beta f_j + \gamma F_j = 0, &\quad -G_j - \beta F_j + \gamma f_j = 0 \\ g_j + \beta f_j + \gamma F_j = 0, &\quad G_j - \beta F_j + \gamma f_j = 0 \end{aligned} \tag{6.4.2.36}$$

on Σ_{ij}. Therefore $f_j = g_j = F_j = G_j = 0$, that is

$$\frac{\partial \tilde{w}_j}{\partial t} = \frac{\partial \tilde{w}_j}{\partial \nu_{A_j}} = \frac{\partial \tilde{q}_j}{\partial t} = \frac{\partial \tilde{q}_j}{\partial \nu_{A_j}} = 0 \text{ on } \Sigma_{ij}.$$

The same unique continuation argument as above gives $\tilde{w}_j = \tilde{q}_j = 0$ in Q_j. One may now proceed step-by-step into the remaining regions Ω_j and conclude that $\tilde{w}_j = \tilde{q}_j = 0$ in Q_j and $f_j = g_j = F_j = G_j = 0$ for $j = 1, \dots, I$. This completes the proof for the case $\epsilon = 0$.

Case 2: $0 < \epsilon < 1$. It remains true that $\mathcal{F}^n \to 0$ and therefore (6.4.2.27) and (6.4.2.28) continue to hold. However, the unique continuation argument used above fails in this case. But we can obtain even stronger convergence results than in the case $\epsilon = 0$ by utilizing Proposition 2.3.7.1 as in Section 6.3.1, together with Opial's Lemma [89] as stated in Proposition 2.3.7.3 above. In fact, we first use Opial's Lemma to deduce

$$\begin{cases} \left.\dfrac{\partial \tilde{w}_i^n}{\partial t}\right|_{\Sigma_i^{\text{int}}} \to 0, & \left.\dfrac{\partial \tilde{q}_i^n}{\partial t}\right|_{\Sigma_i^{\text{int}}} \to 0 \\ \left.\dfrac{\partial \tilde{w}_i^n}{\partial \nu_{A_i}}\right|_{\Sigma_i^{\text{int}}} \to 0, & \left.\dfrac{\partial \tilde{q}_i^n}{\partial \nu_{A_i}}\right|_{\Sigma_i^{\text{int}}} \to 0 \quad \text{weakly in } L^2(\Sigma_i^{\text{int}}). \end{cases} \tag{6.4.2.37}$$

Indeed, since the only fixed point of $\mathcal{T}$ is $X = 0$, we may conclude that $X^n \to 0$ and $\mathcal{T}X^n \to 0$ weakly in $L^2(\Sigma_i^{\text{int}})$. Referring to the definitions (6.4.2.10) and (6.4.2.11) of X^n and $\mathcal{T}X^n$, respectively, one immediately infers (6.4.2.37).

To obtain the remaining conclusions of Theorem 6.4.2.1, we apply Proposition 2.3.7.1 to obtain $\|X^{n+1} - X^n\|_{\mathcal{X}} \to 0$, or, equivalently, that $\|\mathcal{T}X^n - X^n\|_{\mathcal{X}} \to 0$ since $X^{n+1} - X^n = (1-\epsilon)(\mathcal{T}X^n - X^n)$. From (6.4.2.24) we have

$$\|\mathcal{T}X^n - X^n\|_{\mathcal{X}}^2 = 2\mathcal{E}^n - 2\text{Re}(\mathcal{T}X^n, X^n) \to 0. \tag{6.4.2.38}$$

We calculate

$$\begin{aligned}(\mathcal{T}X^n, X^n)_{\mathcal{X}} = \sum_{i=1}^{I}\sum_{j\in\mathcal{J}_i}\frac{1}{2\gamma}\int_{\Sigma_{ij}}\Big\{&\Big(-\frac{\partial\tilde{w}_j^n}{\partial\nu_{A_j}} + \beta\frac{\partial\tilde{w}_j^n}{\partial t} + \gamma\frac{\partial\tilde{q}_j^n}{\partial t}\Big)\\ &\cdot\Big(\frac{\partial\tilde{w}_i^n}{\partial\nu_{A_i}} + \beta\frac{\partial\tilde{w}_i^n}{\partial t} + \gamma\frac{\partial\tilde{q}_i^n}{\partial t}\Big)\\ &+\Big(-\frac{\partial\tilde{q}_j^n}{\partial\nu_{A_j}} - \beta\frac{\partial\tilde{q}_j^n}{\partial t} + \gamma\frac{\partial\tilde{w}_j^n}{\partial t}\Big)\Big(\frac{\partial\tilde{q}_i^n}{\partial\nu_{A_i}} - \beta\frac{\partial\tilde{q}_i^n}{\partial t} + \gamma\frac{\partial\tilde{w}_i^n}{\partial t}\Big)\Big\}d\Sigma\end{aligned}$$

The last expression may be written

$$\begin{aligned}(\mathcal{T}X^n, X^n)_{\mathcal{X}} = \sum_{i=1}^{I}\sum_{j\in\mathcal{J}_i}\frac{1}{2\gamma}\int_{\Sigma_{ij}}\Big\{&-\Big(\frac{\partial\tilde{w}_i^n}{\partial\nu_{A_i}}\frac{\partial\tilde{w}_j^n}{\partial\nu_{A_j}} + \frac{\partial\tilde{q}_i^n}{\partial\nu_{A_i}}\frac{\partial\tilde{q}_j^n}{\partial\nu_{A_j}}\Big)\\ &+(\gamma^2+\beta^2)\Big(\frac{\partial\tilde{w}_i^n}{\partial t}\frac{\partial\tilde{w}_j^n}{\partial t} + \frac{\partial\tilde{q}_i^n}{\partial t}\frac{\partial\tilde{q}_j^n}{\partial t}\Big)\\ +\beta\Big(\frac{\partial\tilde{w}_i^n}{\partial\nu_{A_i}}\frac{\partial\tilde{w}_j^n}{\partial t} - \frac{\partial\tilde{w}_j^n}{\partial\nu_{A_j}}\frac{\partial\tilde{w}_i^n}{\partial t}\Big) &+ \beta\Big(\frac{\partial\tilde{q}_j^n}{\partial\nu_{A_j}}\frac{\partial\tilde{q}_i^n}{\partial t} - \frac{\partial\tilde{q}_i^n}{\partial\nu_{A_i}}\frac{\partial\tilde{q}_j^n}{\partial t}\Big)\\ +\gamma\Big(\frac{\partial\tilde{w}_i^n}{\partial\nu_{A_i}}\frac{\partial\tilde{q}_j^n}{\partial t} - \frac{\partial\tilde{w}_j^n}{\partial\nu_{A_j}}\frac{\partial\tilde{q}_i^n}{\partial t}\Big) &+ \gamma\Big(\frac{\partial\tilde{q}_i^n}{\partial\nu_{A_i}}\frac{\partial\tilde{w}_j^n}{\partial t} - \frac{\partial\tilde{q}_j^n}{\partial\nu_{A_j}}\frac{\partial\tilde{w}_i^n}{\partial t}\Big)\Big\}d\Sigma.\end{aligned}$$

All integrals of the form

$$\sum_{i=1}^{I}\sum_{j\in\mathcal{J}_i}\int_{\Sigma_{ij}}(f_ig_j - f_jg_i)\,d\Sigma$$

vanish. Therefore

$$\begin{aligned}(\mathcal{T}X^n, X^n)_{\mathcal{X}} = \sum_{i=1}^{I}\sum_{j\in\mathcal{J}_i}\frac{1}{2\gamma}\int_{\Sigma_{ij}}\Big\{&-\Big(\frac{\partial\tilde{w}_i^n}{\partial\nu_{A_i}}\frac{\partial\tilde{w}_j^n}{\partial\nu_{A_j}} + \frac{\partial\tilde{q}_i^n}{\partial\nu_{A_i}}\frac{\partial\tilde{q}_j^n}{\partial\nu_{A_j}}\Big)\\ &+(\gamma^2+\beta^2)\Big(\frac{\partial\tilde{w}_i^n}{\partial t}\frac{\partial\tilde{w}_j^n}{\partial t} + \frac{\partial\tilde{q}_i^n}{\partial t}\frac{\partial\tilde{q}_j^n}{\partial t}\Big)\Big\}d\Sigma.\end{aligned} \tag{6.4.2.39}$$

We may write

$$\sum_{i=1}^{I}\int_{\Sigma_i^{\rm int}}|f_i|^2 d\Sigma=\sum_{i=1}^{I-1}\sum_{\substack{j\in\mathcal{J}_i\\ j>i}}\int_{\Sigma_{ij}}(|f_i|^2+|f_j|^2)\,d\Sigma$$
$$\sum_{i=1}^{I}\sum_{j\in\mathcal{J}_i}\int_{\Sigma_{ij}}f_if_j=2\sum_{i=1}^{I-1}\sum_{\substack{j\in\mathcal{J}_i\\ j>i}}\int_{\Sigma_{ij}}f_if_j\,d\Sigma. \tag{6.4.2.40}$$

From (6.4.2.38)–(6.4.2.40), and observing that

$$\frac{\partial\tilde{w}_i^n}{\partial t}-\frac{\partial\tilde{w}_j^n}{\partial t}=\frac{\partial w_i^n}{\partial t}-\frac{\partial w_j^n}{\partial t},\quad \frac{\partial\tilde{w}_i^n}{\partial\nu_{A_i}}+\frac{\partial\tilde{w}_j^n}{\partial\nu_{A_j}}=\frac{\partial w_i^n}{\partial\nu_{A_i}}+\frac{\partial w_j^n}{\partial\nu_{A_j}}$$

and, similarly, for q, we obtain

$$\|\mathcal{T}X^n-X^n\|_{\mathcal{X}}^2=\sum_{i=1}^{I-1}\sum_{\substack{j\in\mathcal{J}_i\\ j>i}}\int_{\Sigma_{ij}}\frac{\beta^2+\gamma^2}{\gamma}\left(\left|\frac{\partial w_i^n}{\partial t}-\frac{\partial w_j^n}{\partial t}\right|^2+\left|\frac{\partial q_i^n}{\partial t}-\frac{\partial q_j^n}{\partial t}\right|^2\right)$$
$$+\frac{1}{\gamma}\left(\left|\frac{\partial w_i^n}{\partial\nu_{A_i}}+\frac{\partial w_j^n}{\partial\nu_{A_j}}\right|^2+\left|\frac{\partial q_i^n}{\partial\nu_{A_i}}+\frac{\partial q_j^n}{\partial\nu_{A_j}}\right|^2\right)d\Sigma. \tag{6.4.2.41}$$

It now follows from (6.4.2.42) that

$$\left\|\frac{\partial w_i^n}{\partial t}-\frac{\partial w_j^n}{\partial t}\right\|_{L^2(\Sigma_{ij})}\to 0,\quad \left\|\frac{\partial q_i^n}{\partial t}-\frac{\partial q_j^n}{\partial t}\right\|_{L^2(\Sigma_{ij})}\to 0$$
$$\left\|\frac{\partial w_i^n}{\partial\nu_{A_i}}+\frac{\partial w_j^n}{\partial\nu_{A_j}}\right\|_{L^2(\Sigma_{ij})}\to 0,\quad \left\|\frac{\partial q_i^n}{\partial\nu_{A_i}}+\frac{\partial q_j^n}{\partial\nu_{A_j}}\right\|_{L^2(\Sigma_{ij})}\to 0. \tag{6.4.2.42}$$

To complete the proof of Theorem 6.4.2.1 for the case $\epsilon\in(0,1)$ it remains to establish the strong convergence in $L^\infty(0,T;\mathcal{H}_i)$ of $(\tilde{w}_i^n,\partial\tilde{w}_i^n/\partial t)$ and of $(\tilde{q}_i^n,\partial\tilde{q}_i^n/\partial t)$. For this we appeal to the following result.

Proposition 6.4.2.1 *For $i=1,\dots,I$ let ϕ_i^n satisfy*

$$\frac{\partial^2\phi_i^n}{\partial t^2}-\nabla\cdot(A\nabla\phi_i^n)+c\phi_i^n=F_i^n \text{ in } Q_i$$
$$\phi_i^n=0 \text{ on } \Sigma_i^D$$
$$\frac{\partial\phi_i^n}{\partial\nu_{A_i}}+\alpha_i\frac{\partial\phi_i^n}{\partial t}=f_i^n \text{ on } \Sigma_i^N$$
$$\phi_i^n(0)=\phi_{i0}^n,\quad \frac{\partial\phi_i^n}{\partial t}(0)=\psi_{i0}^n \text{ in } \Omega_i$$

where $F_i^n \in L^1(0,T;H_i)$, $f_i^n \in L^2(\Sigma_i^N)$ *and* $(\phi_{i0}^n, \psi_{i0}^n) \in \mathcal{H}_i$. *Assume also that*

$$\frac{\partial \phi_i^n}{\partial t}\bigg|_{\Sigma_i^{int}}, \quad \frac{\partial \phi_i^n}{\partial \nu_{A_i}}\bigg|_{\Sigma_i^{int}} \quad \textit{are bounded in } L^2(\Sigma_i^{int}).$$

If

$$\|(\phi_{i0}^n, \psi_{i0}^n)\|_{\mathcal{H}_i} \to 0, \quad \|F_i^n\|_{L^1(0,T;H_i)} \to 0, \quad \|f_i^n\|_{L^2(\Sigma_i^N)} \to 0$$
$$\left\|\frac{\partial \phi_i^n}{\partial t} - \frac{\partial \phi_j^n}{\partial t}\right\|_{L^2(\Sigma_{ij})} \to 0, \quad \left\|\frac{\partial \phi_i^n}{\partial \nu_{A_i}} + \frac{\partial \phi_j^n}{\partial \nu_{A_j}}\right\|_{L^2(\Sigma_{ij})} \to 0, \quad \forall j \in \mathcal{J}_i,$$

then

$$(\phi_i^n, \frac{\partial \phi_i^n}{\partial t}) \to 0 \textit{ strongly in } L^\infty(0,T;\mathcal{H}_i).$$

The proof of this proposition is given below.

We apply Proposition 6.4.2.1 first with $\phi_i^n = \tilde{w}_i^n$, in which case $F_i^n = 0$, $f_i^n = -p_i^n|_{\Sigma_i^N}$, $\phi_{i0}^n = \psi_{i0}^n = 0$. By (6.4.2.27), (6.4.2.42) and the boundedness of the sequence $\mathcal{E}^n$ we may conclude that

$$(\tilde{w}_i^n, \frac{\partial \tilde{w}_i^n}{\partial t}) \to 0 \text{ strongly in } L^\infty(0,T;\mathcal{H}_i).$$

Next, by the change of variable $t \mapsto T-t$, we may apply Proposition 6.4.2.1 with $\phi_i^n = q_i^n$, in which case $(\phi_{i0}^n, \psi_{i0}^n) = (\tilde{q}_i^n(T), \partial \tilde{q}_i^n(T)/\partial t)$, $F_i^n = 0$ and $f_i^n = 0$. By appealing to (6.4.2.28), (6.4.2.42) and the boundedness of $\mathcal{E}^n$ we deduce that

$$(\tilde{q}_i^n, \frac{\partial \tilde{q}_i^n}{\partial t}) \to 0 \text{ strongly in } L^\infty(0,T;\mathcal{H}_i).$$

The proof of Theorem 6.4.2.1 is now complete, except for proving Proposition 6.4.2.1.

Proof of Proposition 6.4.2.1. First assume that $F_i^n = 0$. We multiply the partial differential equation by $\partial \phi_i^n/\partial t$, integrate over $\Omega_i \times (0,t)$ and use Green's Theorem to obtain

$$\frac{1}{2}\left\|(\phi_i^n(t), \frac{\partial \phi_i^n(t)}{\partial t})\right\|_{\mathcal{H}_i}^2 + \int_0^t \int_{\Gamma_i^N} \alpha_i \left|\frac{\partial \phi_i^n}{\partial t}\right|^2 d\Gamma dt$$
$$= \frac{1}{2}\|(\phi_{i0}^n, \psi_{i0}^n)\|_{\mathcal{H}_i}^2 + \int_0^t \int_{\Gamma_i^N} f_i^n \frac{\partial \phi_i^n}{\partial t}\, d\Gamma dt - \int_0^t \int_{\Gamma_i^{\text{int}}} \frac{\partial \phi_i^n}{\partial t} \frac{\partial \phi_i^n}{\partial \nu_{A_i}}\, d\Gamma dt.$$

Since

$$f_i^n \frac{\partial \phi_i^n}{\partial t} = \frac{1}{2\alpha_i}|f_i^n|^2 - \frac{1}{2\alpha_i}\left|\frac{\partial \phi_i^n}{\partial \nu_{A_i}}\right|^2 + \frac{\alpha_i}{2}\left|\frac{\partial \phi_i^n}{\partial t}\right|^2$$

we obtain

$$\begin{aligned}&\left\|(\phi_i^n(t), \frac{\partial \phi_i^n(t)}{\partial t})\right\|_{\mathcal{H}_i}^2 + \int_0^t \int_{\Gamma_i^N} \left(\alpha_i \left|\frac{\partial \phi_i^n}{\partial t}\right|^2 + \frac{1}{\alpha_i}\left|\frac{\partial \phi_i^n}{\partial \nu_{A_i}}\right|^2\right) d\Gamma dt \\ &= \|(\phi_{i0}^n, \psi_{i0}^n)\|_{\mathcal{H}_i}^2 + \int_0^t \int_{\Gamma_i^N} \frac{1}{\alpha_i}|f_i^n|^2 d\Gamma dt - 2\int_0^t \int_{\Gamma_i^{\mathrm{int}}} \frac{\partial \phi_i^n}{\partial t}\frac{\partial \phi_i^n}{\partial \nu_{A_i}} d\Gamma dt. \quad (6.4.2.43)\end{aligned}$$

Next, we write

$$\begin{aligned}\sum_{i=1}^{I} \int_0^t \int_{\Gamma_i^{\mathrm{int}}} \frac{\partial \phi_i^n}{\partial t}\frac{\partial \phi_i^n}{\partial \nu_{A_i}} d\Gamma dt = \sum_{i=1}^{I-1} \sum_{\substack{j\in\mathcal{J}_i \\ j>i}} \int_0^t \int_{\Gamma_{ij}} &\left(\frac{\partial \phi_i^n}{\partial t} - \frac{\partial \phi_j^n}{\partial t}\right)\frac{\partial \phi_i^n}{\partial \nu_{A_i}} \\ &+ \frac{\partial \phi_j^n}{\partial t}\left(\frac{\partial \phi_j^n}{\partial \nu_{A_j}} + \frac{\partial \phi_i^n}{\partial \nu_{A_i}}\right) d\Gamma dt. \quad (6.4.2.44)\end{aligned}$$

It follows immediately from (6.4.2.43), (6.4.2.44) and the hypotheses of the proposition that

$$\sup_{0\le t\le T} \sum_{i=1}^{I} \left\|(\phi_i^n(t), \frac{\partial \phi_i^n(t)}{\partial t})\right\|_{\mathcal{H}_i}^2 \to 0.$$

This proves the proposition if $F_i^n = 0$.

Now assume that $\phi_{i0}^n = \psi_{i0}^n = 0$ and $f_i^n = 0$. One easily finds in the same manner as above that for any $\delta \in (0,1)$,

$$\begin{aligned}&\frac{1}{2}\left\|(\phi_i^n(t), \frac{\partial \phi_i^n(t)}{\partial t})\right\|_{\mathcal{H}_i}^2 + \int_0^t \int_{\Gamma_i^N} \alpha_i \left|\frac{\partial \phi_i^n}{\partial t}\right|^2 d\Gamma dt \\ &\qquad \le \frac{\delta}{2}\left\|\frac{\partial \phi_i^n}{\partial t}\right\|_{L^\infty(0,T;H_i)}^2 + \frac{1}{2\delta}\|F_i^n\|_{L^1(0,T;H_i)}^2 - \int_0^t \int_{\Gamma_i^{\mathrm{int}}} \frac{\partial \phi_i^n}{\partial t}\frac{\partial \phi_i^n}{\partial \nu_{A_i}} d\Gamma dt.\end{aligned}$$

Therefore

$$\begin{aligned}&(1-\delta)\left\|(\phi_i^n, \frac{\partial \phi_i^n}{\partial t})\right\|_{L^\infty(0,T;\mathcal{H}_i)}^2 \\ &\qquad \le \frac{1}{\delta}\|F_i^n\|_{L^1(0,T;H_i)}^2 + 2\sup_{0\le t\le T}\left|\int_0^t \int_{\Gamma_i^{\mathrm{int}}} \frac{\partial \phi_i^n}{\partial t}\frac{\partial \phi_i^n}{\partial \nu_{A_i}} d\Gamma dt\right|.\end{aligned}$$

By using (6.4.2.44) we then obtain

$$(1-\delta)\sum_{i=1}^{I}\left\|(\phi_i^n,\frac{\partial\phi_i^n}{\partial t})\right\|^2_{L^\infty(0,T;\mathcal{H}_i)} \le \frac{1}{\delta}\sum_{i=1}^{I}\|F_i^n\|^2_{L^1(0,T;H_i)}$$
$$+2\sum_{i=1}^{I-1}\sum_{\substack{j\in\mathcal{J}_i\\ j>i}}\left\{\left\|\frac{\partial\phi_i^n}{\partial t}-\frac{\partial\phi_j^n}{\partial t}\right\|_{L^2(\Sigma_{ij})}\left\|\frac{\partial\phi_i^n}{\partial\nu_{A_i}}\right\|_{L^2(\Sigma_{ij})}\right.$$
$$\left.+\left\|\frac{\partial\phi_j^n}{\partial t}\right\|_{L^2(\Sigma_{ij})}\left\|\frac{\partial\phi_j^n}{\partial\nu_{A_j}}-\frac{\partial\phi_i^n}{\partial\nu_{A_i}}\right\|_{L^2(\Sigma_{ij})}\right\},$$

hence

$$\left\|(\phi_i^n,\frac{\partial\phi_i^n}{\partial t})\right\|^2_{L^\infty(0,T;\mathcal{H}_i)}\to 0,\quad i=1,\ldots,I.$$

6.4.3 A Posteriori Error Estimates

In this section we shall derive *a posteriori* estimates of the difference between the solutions of the local optimality systems and the solution of the global optimality system in terms of the mismatch of the time derivatives and normal derivatives of the nth iterates, or of successive iterates, across the interfaces Σ_{ij}. We define *the pointwise global error $e^n(t)$ at the nth iteration at time t* by

$$e^n(t)=\sum_{i=1}^{I}\{\|(\tilde{w}_i^n(t),\frac{\partial\tilde{w}_i^n}{\partial t}(t))\|^2_{\mathcal{H}_i}+\|(\tilde{q}_i^n(t),\frac{\partial\tilde{q}_i^n}{\partial t}(t))\|^2_{\mathcal{H}_i}$$
$$+\int_0^t\int_{\Gamma_i^N}(\frac{1}{\alpha_i}\left|\frac{\partial\tilde{w}_i^n}{\partial\nu_{A_i}}\right|^2+\alpha_i\left|\frac{\partial\tilde{w}_i^n}{\partial t}\right|^2)\,d\Gamma dt+\int_t^T\int_{\Gamma_i^N}(\frac{1}{\alpha_i}\left|\frac{\partial\tilde{q}_i^n}{\partial\nu_{A_i}}\right|^2+\alpha_i\left|\frac{\partial\tilde{q}_i^n}{\partial t}\right|^2)d\Gamma dt\}.$$

We further define

$$\mathcal{E}_{ij}^n(\Sigma_{ij}):=\int_{\Sigma_{ij}}\left(\left|\frac{\partial w_i^n}{\partial t}-\frac{\partial w_j^n}{\partial t}\right|^2+\left|\frac{\partial q_i^n}{\partial t}-\frac{\partial q_j^n}{\partial t}\right|^2\right.$$
$$\left.+\left|\frac{\partial w_i^n}{\partial\nu_{A_i}}+\frac{\partial w_j^n}{\partial\nu_{A_j}}\right|^2+\left|\frac{\partial q_i^n}{\partial\nu_{A_i}}+\frac{\partial q_j^n}{\partial\nu_{A_j}}\right|^2\right)d\Sigma$$

and

$$\mathcal{E}_{ij}^{n,n+1}(\Sigma_{ij}):=\int_{\Sigma_{ij}}\left(\left|\frac{\partial w_i^n}{\partial t}-\frac{\partial w_j^{n+1}}{\partial t}\right|^2+\left|\frac{\partial q_i^n}{\partial t}-\frac{\partial q_j^{n+1}}{\partial t}\right|^2\right.$$
$$\left.+\left|\frac{\partial w_i^n}{\partial\nu_{A_i}}+\frac{\partial w_j^{n+1}}{\partial\nu_{A_j}}\right|^2+\left|\frac{\partial q_i^n}{\partial\nu_{A_i}}+\frac{\partial q_j^{n+1}}{\partial\nu_{A_j}}\right|^2\right)d\Sigma.$$

The quantity $\mathcal{E}^n_{ij}(\Sigma_{ij})$ is a measure of the mismatch of the time and normal derivatives of the nth iterates across the interface Σ_{ij} while $\mathcal{E}^{n,n+1}_{ij}(\Sigma_{ij})$ measures the mismatch of successive iterates.

For the remainder of this section, we assume that α, β, γ and κ satisfy (6.4.2.7) of Theorem 6.4.2.1.

Theorem 6.4.3.1 *Let $\epsilon \in [0,1)$. Then*

$$\sup_{0\le t\le T} e^n(t) \le C\sqrt{\mathcal{E}^n+\mathcal{F}^n}\Big\{\sum_{i=1}^{I-1}\sum_{\substack{j\in\mathcal{J}_i\\ j>i}} \mathcal{E}^n_{ij}(\Sigma_{ij})\Big\}^{1/2} \tag{6.4.3.1}$$

where C depends only on α_0, β, γ and κ.

Theorem 6.4.3.2 *Let $\epsilon \in [0,1/2)$. Then*

$$\sup_{0\le t\le T} [e^{n+1}(t)+e^n(t)]$$

$$\le C_\epsilon\sqrt{\mathcal{E}^{n+1}+\mathcal{E}^n}\Big\{\sum_{i=1}^{I-1}\sum_{\substack{j\in\mathcal{J}_i\\ j>i}} \big(\mathcal{E}^{n,n+1}_{ij}(\Sigma_{ij})+\mathcal{E}^{n+1,n}_{ij}(\Sigma_{ij})\big)\Big\}^{1/2} \tag{6.4.3.2}$$

where C_ϵ depends only on α_0, β, γ, κ and on ϵ.

Remark 6.4.3.1 The constants C and C_ϵ will be explicitly calculated in the proofs of Theorems 6.4.3.1 and 6.4.3.2; see (6.4.3.6), (6.4.3.9) and (6.4.3.5) below. Note also that $\mathcal{E}^n+\mathcal{F}^n = \|X^n\|^2_{\mathcal{X}}$ is nonincreasing. Therefore

$$\mathcal{E}^n+\mathcal{F}^n \le \|X^1\|^2_{\mathcal{X}} = \sum_{i=1}^{I}\sum_{j\in\mathcal{J}_i}\frac{1}{2\gamma}\int_{\Sigma_{ij}}(|\tilde\lambda^0_{ij}|^2+|\tilde\rho^0_{ij}|^2)\,d\Sigma.$$

The right side can be estimated in terms of the starting inputs λ^0_{ij}, ρ^0_{ij} to the local problems, and the time and normal derivatives of the solution of the global optimality system on the interfaces Σ_{ij}. We do not, however, have estimates of these traces in terms of the input data (w_0,v_0), (z_0,z_1), F and κ to the global optimal control problem.

Theorem 6.4.3.1 is an immediate consequence of the following two lemmas.

Lemma 6.4.3.1 *Let $\epsilon \in [0,1)$. Then*

$$\mathcal{F}^n \le \Big\{\max\Big(\frac{1}{\gamma},\frac{\beta^2+\gamma^2}{\gamma}\Big)\Big\}^{1/2}\sqrt{\mathcal{E}^n+\mathcal{F}^n}\Big\{\sum_{i=1}^{I-1}\sum_{\substack{j\in\mathcal{J}_i\\ j>i}} \mathcal{E}^n_{ij}(\Sigma_{ij})\Big\}^{1/2}. \tag{6.4.3.3}$$

Lemma 6.4.3.2 *Let* $\epsilon \in [0,1)$. *Then*

$$\sup_{0\leq t\leq T} e^n(t) \leq C_1\mathcal{F}^n + C_2\sqrt{\mathcal{E}^n}\{\sum_{i=1}^{I-1}\sum_{\substack{j\in\mathcal{J}_i\\ j>i}} \mathcal{E}^n_{ij}(\Sigma_{ij})\}^{1/2} \tag{6.4.3.4}$$

where

$$C_1 = \max\left(\frac{\kappa^2}{\kappa+\frac{\beta}{2\gamma}(1-\kappa^2)}, \frac{1}{\alpha_0+\alpha_0^2\frac{\beta}{\gamma}-\frac{\beta}{4\gamma}}\right),$$
$$C_2 = 2\sqrt{2}\{\max(\frac{\gamma}{\beta^2+\gamma^2},\gamma)\}^{1/2}. \tag{6.4.3.5}$$

Note that $C_1 > 0$ by virtue of assumption (6.4.2.7). From (6.4.3.3) and (6.4.3.4) one obtains (6.4.3.1) with

$$C = C_1\{\max(\frac{1}{\gamma},\frac{\beta^2+\gamma^2}{\gamma})\}^{1/2} + C_2. \tag{6.4.3.6}$$

Similarly, Theorem 6.4.3.2 follows from the following two results.

Lemma 6.4.3.3 *Let* $\epsilon \in [0,1)$. *then*

$$\mathcal{F}^{n+1} + (1-2\epsilon)\mathcal{F}^n \leq \{\max(\frac{1}{\gamma},\frac{\beta^2+\gamma^2}{\gamma})\}^{1/2}\sqrt{\mathcal{E}^{n+1}+\mathcal{E}^n}$$
$$\times\{\sum_{i=1}^{I-1}\sum_{\substack{j\in\mathcal{J}_i\\ j>i}}(\mathcal{E}^{n,n+1}_{ij}(\Sigma_{ij})+\mathcal{E}^{n+1,n}_{ij}(\Sigma_{ij}))\}^{1/2}. \tag{6.4.3.7}$$

Lemma 6.4.3.4 *Let* $\epsilon \in [0,1)$. *Then*

$$\sup_{0\leq t\leq T}[e^{n+1}(t)+e^n(t)] \leq C_1(\mathcal{F}^{n+1}+\mathcal{F}^n)$$
$$+ C_2\sqrt{\mathcal{E}^{n+1}+\mathcal{E}^n}\{\sum_{i=1}^{I-1}\sum_{\substack{j\in\mathcal{J}_i\\ j>i}}(\mathcal{E}^{n,n+1}_{ij}(\Sigma_{ij})+\mathcal{E}^{n+1,n}_{ij}(\Sigma_{ij}))\}^{1/2} \tag{6.4.3.8}$$

where C_1, C_2 *are given by* (6.4.3.5).

It follows that the constant C_ϵ in Theorem 6.4.3.2 is given by

$$C_\epsilon = \frac{C_1}{1-2\epsilon}\{\max(\frac{1}{\gamma},\frac{\beta^2+\gamma^2}{\gamma})\}^{1/2} + C_2. \tag{6.4.3.9}$$

Proof of Lemma 6.4.3.1. We start with the fundamental recursion (6.4.2.25), which we write as

$$2(1-\epsilon)\mathcal{F}^n_\beta \leq (\mathcal{E}^n_\beta+\mathcal{F}^n_\beta) - (\mathcal{E}^{n+1}_\beta+\mathcal{F}^{n+1}_\beta). \tag{6.4.3.10}$$

From (6.4.2.24) the right side of (6.4.3.10) may be written

$$\begin{aligned}\|X^n\|_{\mathcal{X}}^2 - \|X^{n+1}\|_{\mathcal{X}}^2 &\leq \|X^n - X^{n+1}\|_{\mathcal{X}}(\|X^n\|_{\mathcal{X}} + \|X^{n+1}\|_{\mathcal{X}}) \\ &\leq 2(1-\epsilon)\|X^n - \mathcal{T}X^n\|_{\mathcal{X}}\|X^n\|_{\mathcal{X}}\end{aligned} \tag{6.4.3.11}$$

since $\{\|X^n\|_{\mathcal{X}}\}$ is a nonincreasing sequence. In addition, it follows from (6.4.2.41) that

$$\|X^n - \mathcal{T}X^n\|_{\mathcal{X}}^2 \leq \max(\frac{1}{\gamma}, \frac{\beta^2+\gamma^2}{\gamma}) \sum_{i=1}^{I-1} \sum_{\substack{j\in\mathcal{J}_i \\ j>i}} \mathcal{E}_{ij}^n(\Sigma_{ij}). \tag{6.4.3.12}$$

The conclusion of Lemma 6.4.3.1 now follows from (6.4.3.10)–(6.4.3.12).

Proof of Lemma 6.4.3.2. We begin with the energy identity

$$\begin{aligned}\|(\tilde{w}_i^n(t), \frac{\partial \tilde{w}_i^n}{\partial t}(t))\|_{\mathcal{H}_i}^2 &+ \int_0^t \int_{\Gamma_i^N} \left(\frac{1}{\alpha_i}\left|\frac{\partial \tilde{w}_i^n}{\partial \nu_{A_i}}\right|^2 + \alpha_i \left|\frac{\partial \tilde{w}_i^n}{\partial t}\right|^2\right) d\Gamma dt \\ &= \int_0^t \int_{\Gamma_i^N} \frac{1}{\alpha_i}\left|\frac{\partial \tilde{q}_i^n}{\partial t}\right|^2 d\Gamma dt + 2\int_0^t \int_{\Gamma_i^{\text{int}}} \frac{\partial \tilde{w}_i^n}{\partial t}\frac{\partial \tilde{w}_i^n}{\partial \nu_{A_i}} d\Gamma dt,\end{aligned} \tag{6.4.3.13}$$

which follows from

$$0 = \int_0^t \int_{\Omega_i} \frac{\partial \tilde{w}_i^n}{\partial t}(\frac{\partial^2 \tilde{w}_i^n}{\partial t^2} - \nabla\cdot(A_i\nabla\tilde{w}_i^n) + c_i\tilde{w}_i^n)\, dxdt$$

by integration by parts and a little algebra. By using (6.4.2.44) and the Cauchy-Schwarz inequality we then obtain, for $0 \leq t \leq T$,

$$\begin{aligned}\sum_{i=1}^{I}\Big\{\|(\tilde{w}_i^n(t), \frac{\partial \tilde{w}_i^n}{\partial t}(t))\|_{\mathcal{H}_i}^2 &+ \int_0^t \int_{\Gamma_i^N} \left(\frac{1}{\alpha_i}\left|\frac{\partial \tilde{w}_i^n}{\partial \nu_{A_i}}\right|^2 + \alpha_i \left|\frac{\partial \tilde{w}_i^n}{\partial t}\right|^2\right) d\Gamma dt\Big\} \\ &\leq \sum_{i=1}^{I} \int_{\Sigma_i^N} \frac{1}{\alpha_i}\left|\frac{\partial \tilde{q}_i^n}{\partial t}\right|^2 d\Sigma \\ &+ 2\sum_{i=1}^{I-1}\sum_{\substack{j\in\mathcal{J}_i \\ j>i}}\Big\{\Big(\int_{\Sigma_{ij}}\left|\frac{\partial w_i^n}{\partial t} - \frac{\partial w_j^n}{\partial t}\right|^2 d\Sigma\Big)^{1/2}\Big(\int_{\Sigma_{ij}}\left|\frac{\partial \tilde{w}_i^n}{\partial \nu_{A_i}}\right|^2 d\Sigma\Big)^{1/2} \\ &+ \Big(\int_{\Sigma_{ij}}\left|\frac{\partial w_i^n}{\partial \nu_{A_i}} + \frac{\partial w_j^n}{\partial \nu_{A_j}}\right|^2 d\Sigma\Big)^{1/2}\Big(\int_{\Sigma_{ij}}\left|\frac{\partial \tilde{w}_j^n}{\partial t}\right|^2 d\Sigma\Big)^{1/2}\Big\}.\end{aligned} \tag{6.4.3.14}$$

Similarly, from

$$\begin{aligned}\|(\tilde{q}_i^n(t), \frac{\partial \tilde{q}_i^n}{\partial t}(t))\|_{\mathcal{H}_i}^2 &+ \int_t^T \int_{\Gamma_i^N} \left(\frac{1}{\alpha_i}\left|\frac{\partial \tilde{q}_i^n}{\partial \nu_{A_i}}\right|^2 + \alpha_i \left|\frac{\partial \tilde{q}_i^n}{\partial t}\right|^2\right) d\Gamma dt \\ &= \|(\tilde{q}_i^n(T), \frac{\partial \tilde{q}_i^n}{\partial t}(T))\|_{\mathcal{H}_i}^2 + 2\int_t^T \int_{\Gamma_i^{\text{int}}} \frac{\partial \tilde{q}_i^n}{\partial t}\frac{\partial \tilde{q}_i^n}{\partial \nu_{A_i}} d\Gamma dt\end{aligned} \tag{6.4.3.15}$$

we obtain, for $0 \leq t \leq T$,

$$\sum_{i=1}^{I}\Big\{\|(\tilde q_i^n(t), \frac{\partial \tilde q_i^n}{\partial t}(t))\|^2_{\mathcal{H}_i} + \int_t^T\int_{\Gamma_i^N}\Big(\frac{1}{\alpha_i}\Big|\frac{\partial \tilde q_i^n}{\partial \nu_{A_i}}\Big|^2 + \alpha_i\Big|\frac{\partial \tilde q_i^n}{\partial t}\Big|^2\Big)\,d\Gamma dt\Big\}$$

$$\leq \sum_{i=1}^{I}\|(\tilde q_i^n(T), \frac{\partial \tilde q_i^n}{\partial t}(T))\|^2_{\mathcal{H}_i}$$

$$+2\sum_{i=1}^{I-1}\sum_{\substack{j\in\mathcal{J}_i\\ j>i}}\Big\{\Big(\int_{\Sigma_{ij}}\Big|\frac{\partial q_i^n}{\partial t}-\frac{\partial q_j^n}{\partial t}\Big|^2 d\Sigma\Big)^{1/2}\Big(\int_{\Sigma_{ij}}\Big|\frac{\partial \tilde q_i^n}{\partial \nu_{A_i}}\Big|^2 d\Sigma\Big)^{1/2}$$

$$+\Big(\int_{\Sigma_{ij}}\Big|\frac{\partial q_i^n}{\partial \nu_{A_i}}+\frac{\partial q_j^n}{\partial \nu_{A_j}}\Big|^2 d\Sigma\Big)^{1/2}\Big(\int_{\Sigma_{ij}}\Big|\frac{\partial \tilde q_j^n}{\partial t}\Big|^2 d\Sigma\Big)^{1/2}\Big\}. \quad (6.4.3.16)$$

It follows from (6.4.3.14) and (6.4.3.16) that

$$\sup_{0\leq t\leq T} e^n(t) \leq \sum_{i=1}^{I}\Big\{\|(\tilde q_i^n(T), \frac{\partial \tilde q_i^n}{\partial t}(T))\|^2_{\mathcal{H}_i} + \int_{\Sigma_i^N}\frac{1}{\alpha_i}\Big|\frac{\partial \tilde q_i^n}{\partial t}\Big|^2 d\Sigma\Big\}$$

$$+2\Big\{\sum_{i=1}^{I-1}\sum_{\substack{j\in\mathcal{J}_i\\ j>i}}\mathcal{E}_{ij}^n(\Sigma_{ij})\Big\}^{1/2}\Big\{\sum_{i=1}^{I-1}\sum_{\substack{j\in\mathcal{J}_i\\ j>i}}\int_{\Sigma_{ij}}\Big(\Big|\frac{\partial \tilde w_i^n}{\partial \nu_{A_i}}\Big|^2+\Big|\frac{\partial \tilde w_j^n}{\partial t}\Big|^2$$

$$+\Big|\frac{\partial \tilde q_i^n}{\partial \nu_{A_i}}\Big|^2+\Big|\frac{\partial \tilde q_j^n}{\partial t}\Big|^2\Big)\,d\Sigma\Big\}^{1/2}, \quad (6.4.3.17)$$

where we have used the Cauchy-Schwarz inequality once again. We have

$$\sum_{i=1}^{I-1}\sum_{\substack{j\in\mathcal{J}_i\\ j>i}}\int_{\Sigma_{ij}}\Big(\Big|\frac{\partial \tilde w_i^n}{\partial \nu_{A_i}}\Big|^2+\Big|\frac{\partial \tilde w_j^n}{\partial t}\Big|^2+\Big|\frac{\partial \tilde q_i^n}{\partial \nu_{A_i}}\Big|^2+\Big|\frac{\partial \tilde q_j^n}{\partial t}\Big|^2\Big)\,d\Sigma$$

$$\leq\sum_{i=1}^{I}\int_{\Sigma_i^{\rm int}}\Big(\Big|\frac{\partial \tilde w_i^n}{\partial \nu_{A_i}}\Big|^2+\Big|\frac{\partial \tilde q_i^n}{\partial \nu_{A_i}}\Big|^2+\Big|\frac{\partial \tilde w_i^n}{\partial t}\Big|^2+\Big|\frac{\partial \tilde q_i^n}{\partial t}\Big|^2\Big)\,d\Sigma$$

$$\leq 2\max(\frac{\gamma}{\beta^2+\gamma^2},\gamma)\mathcal{E}^n. \quad (6.4.3.18)$$

From (6.4.3.17) and (6.4.3.18) we therefore obtain

$$\sup_{0\leq t\leq T} e^n(t) \leq \sum_{i=1}^{I}\Big\{\|(\tilde q_i^n(T), \frac{\partial \tilde q_i^n}{\partial t}(T))\|^2_{\mathcal{H}_i} + \int_{\Sigma_i^N}\frac{1}{\alpha_i}\Big|\frac{\partial \tilde q_i^n}{\partial t}\Big|^2 d\Sigma\Big\}$$

$$+2\sqrt{2}\Big\{\max(\frac{\gamma}{\beta^2+\gamma^2},\gamma)\mathcal{E}^n\Big\}^{1/2}\Big\{\sum_{i=1}^{I-1}\sum_{\substack{j\in\mathcal{J}_i\\ j>i}}\mathcal{E}_{ij}^n(\Sigma_{ij})\Big\}^{1/2}. \quad (6.4.3.19)$$

Thus it remains to show that

$$\sum_{i=1}^{I}\Big\{\big\|(\tilde q_i^n(T),\frac{\partial \tilde q_i^n}{\partial t}(T))\big\|^2_{\mathcal{H}_i}+\int_{\Sigma_i^N}\frac{1}{\alpha_i}\Big|\frac{\partial \tilde q_i^n}{\partial t}\Big|^2 d\Sigma\Big\}\le C_1\mathcal{F}^n. \tag{6.4.3.20}$$

We have

$$\sum_{i=1}^{I}\Big\{\big\|(\tilde q_i^n(T),\frac{\partial \tilde q_i^n}{\partial t}(T))\big\|^2_{\mathcal{H}_i}+\int_{\Sigma_i^N}\frac{1}{\alpha_i}\Big|\frac{\partial \tilde q_i^n}{\partial t}\Big|^2 d\Sigma\Big\}$$
$$=\sum_{i=1}^{I}\Big\{\kappa^2\big\|(\tilde w_i^n(T),\frac{\partial \tilde w_i^n}{\partial t}(T))\big\|^2_{\mathcal{H}_i}+\int_{\Sigma_i^N}\frac{1}{\alpha_i}\Big|\frac{\partial \tilde q_i^n}{\partial t}\Big|^2 d\Sigma\Big\}. \tag{6.4.3.21}$$

By using the definition (6.4.2.20) of $\mathcal{F}^n$ and a little algebra, one finds that

$$\mathcal{F}^n\ge\frac{1}{\kappa^2}\big(\kappa+\frac{\beta}{2\gamma}(1-\kappa^2)\big)\sum_{i=1}^{I}\kappa^2\big\|(\tilde w_i^n(T),\frac{\partial \tilde w_i^n}{\partial t}(T))\big\|^2_{\mathcal{H}_i}$$
$$+\sum_{i=1}^{I}\big(\alpha_0+\alpha_0^2\frac{\beta}{\gamma}-\frac{\beta}{4\gamma}\big)\int_{\Sigma_i^N}\frac{1}{\alpha_i}\Big|\frac{\partial \tilde q_i^n}{\partial t}\Big|^2 d\Sigma. \tag{6.4.3.22}$$

The estimate (6.4.3.20) follows immediately from (6.4.3.21) and (6.4.3.22).

Proof of Lemma 6.4.3.3. We again start from the recursion formula (6.4.2.25), which may be written

$$\mathcal{F}^{n+1}+(1-2\epsilon)\mathcal{F}^n\le\mathcal{E}^n-\mathcal{E}^{n+1},$$

and proceed to estimate the right-hand side. From the definition (6.4.2.22) of $\mathcal{E}^n$ we have

$$\mathcal{E}^n-\mathcal{E}^{n+1}=\frac12\sum_{i=1}^{I}\int_{\Sigma_i^{\text{int}}}\Big\{\frac{\beta^2+\gamma^2}{\gamma}\Big(\Big|\frac{\partial \tilde w_i^n}{\partial \nu_{A_i}}\Big|^2-\Big|\frac{\partial \tilde w_i^{n+1}}{\partial \nu_{A_i}}\Big|^2+\Big|\frac{\partial \tilde q_i^n}{\partial \nu_{A_i}}\Big|^2-\Big|\frac{\partial \tilde q_i^{n+1}}{\partial \nu_{A_i}}\Big|^2\Big)$$
$$+\frac{1}{\gamma}\Big(\Big|\frac{\partial \tilde w_i^n}{\partial t}\Big|^2-\Big|\frac{\partial \tilde w_i^{n+1}}{\partial t}\Big|^2+\Big|\frac{\partial \tilde q_i^n}{\partial t}\Big|^2-\Big|\frac{\partial \tilde q_i^{n+1}}{\partial t}\Big|^2\Big)\Big\}d\Sigma. \tag{6.4.3.23}$$

We rewrite the right-hand side using (6.4.2.44). So, for example, we have

$$\sum_{i=1}^{I}\int_{\Sigma_i^{\text{int}}}\Big\{\Big|\frac{\partial \tilde w_i^n}{\partial \nu_{A_i}}\Big|^2-\Big|\frac{\partial \tilde w_i^{n+1}}{\partial \nu_{A_i}}\Big|^2\Big\}d\Sigma$$
$$\le\sum_{i=1}^{I-1}\sum_{\substack{j\in\mathcal{J}_i\\ j>i}}\int_{\Sigma_{ij}}\Big(\Big|\frac{\partial w_i^n}{\partial \nu_{A_i}}+\frac{\partial w_j^{n+1}}{\partial \nu_{A_j}}\Big|\,\Big|\frac{\partial \tilde w_i^n}{\partial \nu_{A_i}}-\frac{\partial \tilde w_j^{n+1}}{\partial \nu_{A_j}}\Big|$$
$$+\Big|\frac{\partial w_j^n}{\partial \nu_{A_j}}+\frac{\partial w_i^{n+1}}{\partial \nu_{A_i}}\Big|\,\Big|\frac{\partial \tilde w_j^n}{\partial \nu_{A_j}}-\frac{\partial \tilde w_i^{n+1}}{\partial \nu_{A_i}}\Big|\Big)\,d\Sigma$$

and, similarly, for the remaining three differences in (6.4.3.23). An application of the Cauchy-Schwarz inequality, together with the trivial estimate $(a \pm b)^2 \leq 2(a^2 + b^2)$, yields

$$\mathcal{E}^n - \mathcal{E}^{n+1} \leq \frac{1}{2}\{\max(\frac{\beta^2+\gamma^2}{\gamma}, \frac{1}{\gamma})\}^{1/2} I^{n,n+1} \times \{\sum_{i=1}^{I-1} \sum_{\substack{j\in\mathcal{J}_i \\ j>i}} (\mathcal{E}_{ij}^{n,n+1}(\Sigma_{ij}) + \mathcal{E}_{ij}^{n+1,n}(\Sigma_{ij}))\}^{1/2} \tag{6.4.3.24}$$

where

$$I^{n,n+1} = \Bigg\{\sum_{i=1}^{I-1} \sum_{\substack{j\in\mathcal{J}_i \\ j>i}} 2\int_{\Sigma_{ij}} \Bigg\{\frac{\beta^2+\gamma^2}{\gamma}\Bigg(\Big|\frac{\partial \tilde{w}_i^n}{\partial \nu_{A_i}}\Big|^2 + \Big|\frac{\partial \tilde{w}_j^n}{\partial \nu_{A_j}}\Big|^2 + \Big|\frac{\partial \tilde{w}_i^{n+1}}{\partial \nu_{A_i}}\Big|^2 + \Big|\frac{\partial \tilde{w}_j^{n+1}}{\partial \nu_{A_j}}\Big|^2 + \Big|\frac{\partial \tilde{q}_i^n}{\partial \nu_{A_i}}\Big|^2 + \Big|\frac{\partial \tilde{q}_j^n}{\partial \nu_{A_j}}\Big|^2 + \Big|\frac{\partial \tilde{q}_i^{n+1}}{\partial \nu_{A_i}}\Big|^2 + \Big|\frac{\partial \tilde{q}_j^{n+1}}{\partial \nu_{A_j}}\Big|^2\Bigg)$$
$$+ \frac{1}{\gamma}\Bigg(\Big|\frac{\partial \tilde{w}_i^n}{\partial t}\Big|^2 + \Big|\frac{\partial \tilde{w}_j^n}{\partial t}\Big|^2 + \Big|\frac{\partial \tilde{w}_i^{n+1}}{\partial t}\Big|^2 + \Big|\frac{\partial \tilde{w}_j^{n+1}}{\partial t}\Big|^2 + \Big|\frac{\partial \tilde{q}_i^n}{\partial t}\Big|^2 + \Big|\frac{\partial \tilde{q}_j^n}{\partial t}\Big|^2 + \Big|\frac{\partial \tilde{q}_i^{n+1}}{\partial t}\Big|^2 + \Big|\frac{\partial \tilde{q}_j^{n+1}}{\partial t}\Big|^2\Bigg)\Bigg\} d\Sigma\Bigg\}^{1/2} = 2\sqrt{\mathcal{E}^n + \mathcal{E}^{n+1}}$$

Use of the last expression in (6.4.3.24) completes the proof of Lemma 6.4.3.3.

Proof of Lemma 6.4.3.4. From (6.4.3.13), (6.4.3.15) and the formula (6.4.2.44) we have

$$\begin{aligned} e^{n+1}(t) + e^n(t) \leq \sum_{i=1}^{I} \Bigg\{ &\|(\tilde{q}_i^n(T), \frac{\partial \tilde{q}_i^n}{\partial t}(T))\|^2_{\mathcal{H}_i} + \|(\tilde{q}_i^{n+1}(T), \frac{\partial \tilde{q}_i^{n+1}}{\partial t}(T))\|^2_{\mathcal{H}_i} \\ &+ \int_{\Sigma_i^N} \Big\{\alpha_i \Big|\frac{\partial \tilde{q}_i^n}{\partial \nu_{A_i}}\Big|^2 + \alpha_i \Big|\frac{\partial \tilde{q}_i^{n+1}}{\partial \nu_{A_i}}\Big|^2\Big\} d\Sigma\Bigg\} \\ &+ I_1(w^n, \tilde{w}^n)(t) + I_1(w^{n+1}, \tilde{w}^{n+1})(t) \\ &+ I_2(q^n, \tilde{q}^n)(t) + I_2(q^{n+1}, \tilde{q}^{n+1})(t), \end{aligned} \tag{6.4.3.25}$$

where

$$I_1(w^n, \tilde{w}^n)(t) = 2\sum_{i=1}^{I-1} \sum_{\substack{j\in\mathcal{J}_i \\ j>i}} \int_0^t \int_{\Gamma_{ij}} \Bigg\{\Big(\frac{\partial w_i^n}{\partial t} - \frac{\partial w_j^n}{\partial t}\Big) \frac{\partial \tilde{w}_i^n}{\partial \nu_{A_i}} + \frac{\partial \tilde{w}_j^n}{\partial t}\Big(\frac{\partial w_i^n}{\partial \nu_{A_i}} + \frac{\partial w_j^n}{\partial \nu_{A_j}}\Big)\Bigg\} d\Gamma dt,$$

$$I_2(q^n,\tilde{q}^n)(t) = 2\sum_{i=1}^{I-1}\sum_{\substack{j\in\mathcal{J}_i\\ j>i}}\int_t^T\int_{\Gamma_{ij}}\Big\{\Big(\frac{\partial q_i^n}{\partial t}-\frac{\partial q_j^n}{\partial t}\Big)\frac{\partial \tilde{q}_i^n}{\partial \nu_{A_i}} + \frac{\partial \tilde{q}_j^n}{\partial t}\Big(\frac{\partial q_i^n}{\partial \nu_{A_i}}+\frac{\partial q_j^n}{\partial \nu_{A_j}}\Big)\Big\}d\Gamma dt.$$

It may be verified that

$$\begin{aligned} I_1(w^n,\tilde{w}^n)(t) + I_1(w^{n+1},\tilde{w}^{n+1})(t) &= 2\sum_{i=1}^{I-1}\sum_{\substack{j\in\mathcal{J}_i\\ j>i}}\int_0^t\int_{\Gamma_{ij}}\Big\{\Big(\frac{\partial w_i^n}{\partial t}-\frac{\partial w_j^{n+1}}{\partial t}\Big)\frac{\partial \tilde{w}_i^n}{\partial \nu_{A_i}} \\ &+ \Big(\frac{\partial w_i^{n+1}}{\partial t}-\frac{\partial w_j^n}{\partial t}\Big)\frac{\partial \tilde{w}_i^{n+1}}{\partial \nu_{A_i}} + \frac{\partial \tilde{w}_j^n}{\partial t}\Big(\frac{\partial w_i^{n+1}}{\partial \nu_{A_i}}+\frac{\partial w_j^n}{\partial \nu_{A_j}}\Big) \\ &+ \frac{\partial \tilde{w}_j^{n+1}}{\partial t}\Big(\frac{\partial w_i^n}{\partial \nu_{A_i}}+\frac{\partial w_j^{n+1}}{\partial \nu_{A_j}}\Big)\Big\}d\Gamma dt. \end{aligned}$$

A similar expression holds for $I_2(q^n,\tilde{q}^n)(t)+I_2(q^{n+1},\tilde{q}^{n+1})(t)$. It follows that

$$\begin{aligned} &|I_1(w^n,\tilde{w}^n)(t) + I_1(w^{n+1},\tilde{w}^{n+1})(t) + I_2(q^n,\tilde{q}^n)(t) + I_2(q^{n+1},\tilde{q}^{n+1})(t)| \\ &\le 2\Big\{\sum_{i=1}^{I-1}\sum_{\substack{j\in\mathcal{J}_i\\ j>i}}[\mathcal{E}_{ij}^{n,n+1}(\Sigma_{ij})+\mathcal{E}_{ij}^{n+1,n}(\Sigma_{ij})]\Big\}^{1/2} \\ &\times\Big\{\sum_{i=1}^{I-1}\sum_{\substack{j\in\mathcal{J}_i\\ j>i}}\int_{\Sigma_{ij}}\Big\{\Big|\frac{\partial \tilde{w}_i^n}{\partial \nu_{A_i}}\Big|^2+\Big|\frac{\partial \tilde{w}_i^{n+1}}{\partial \nu_{A_i}}\Big|^2+\Big|\frac{\partial \tilde{w}_j^n}{\partial t}\Big|^2+\Big|\frac{\partial \tilde{w}_j^{n+1}}{\partial t}\Big|^2 \\ &+\Big|\frac{\partial \tilde{q}_i^n}{\partial \nu_{A_i}}\Big|^2+\Big|\frac{\partial \tilde{q}_i^{n+1}}{\partial \nu_{A_i}}\Big|^2+\Big|\frac{\partial \tilde{q}_j^n}{\partial t}\Big|^2+\Big|\frac{\partial \tilde{q}_j^{n+1}}{\partial t}\Big|^2\Big\}d\Sigma\Big\}^{1/2} \\ &\le 2\sqrt{2}\Big\{\max\big(\frac{\gamma}{\beta^2+\gamma^2},\gamma\big)\Big\}^{1/2}\sqrt{\mathcal{E}^n+\mathcal{E}^{n+1}} \\ &\times\Big\{\sum_{i=1}^{I-1}\sum_{\substack{j\in\mathcal{J}_i\\ j>i}}[\mathcal{E}_{ij}^{n,n+1}(\Sigma_{ij})+\mathcal{E}_{ij}^{n+1,n}(\Sigma_{ij})]\Big\}^{1/2} \\ &= C_2\sqrt{\mathcal{E}^n+\mathcal{E}^{n+1}}\Big\{\sum_{i=1}^{I-1}\sum_{\substack{j\in\mathcal{J}_i\\ j>i}}[\mathcal{E}_{ij}^{n,n+1}(\Sigma_{ij})+\mathcal{E}_{ij}^{n+1,n}(\Sigma_{ij})]\Big\}^{1/2}. \qquad (6.4.3.26) \end{aligned}$$

Lemma 6.4.3.4 now follows from (6.4.3.25), (6.4.3.26) and (6.4.3.20).

6.5 Space and Time Domain Decomposition

We partition Ω into subregions Ω_i, $i = i, \ldots, I$, as in Section 6.4.1, and the time interval $(0, T)$ into subintervals $I_k = (T_k, T_{k+1})$, $k = 0, \ldots, K$, as in Section 6.3.1. In this section we consider three algorithms for the approximation of the solution of the global optimality system (6.1.1.1), (6.2.3.4), (6.2.3.6) through iterative, parallel computations concentrated on subdomains $\Omega_i \times I_k$ of the space-time domain $\Omega \times (0, T)$. Two candidates for such algorithms come immediately to mind. In the first, a time domain decomposition of the global optimality system into local optimality systems concentrated on the cylinder $\Omega \times I_k$, $k = 0, \ldots, K$ as described in Section 6.3.1, is followed by a spatial domain decomposition of each of the $K + 1$ local optimality system, as in Section 6.4.1. We shall refer to such an approach as *sequential space-time domain decomposition.* In the second approach, referred to as *sequential time-space domain decomposition*, the spatial domain decomposition of the global optimality system into local optimality systems concentrated on the cylinders $\Omega_i \times (0, T)$, $i = 1, \ldots, I$, is followed by a time domain decomposition of each of the I local optimality systems.

Each of these two approaches involves two iteration parameters: a parameter m associated with time domain decomposition, and an iteration parameter n occurring in the spatial domain decomposition. In the sequential space-time domain decomposition, the solution of the global optimality system is to be obtained in the repeated limit $\lim_{m\to\infty} \lim_{n\to\infty}$, while it is to be obtained in the repeated limit $\lim_{n\to\infty} \lim_{m\to\infty}$ in the case of sequential time-space domain decomposition.

It is natural to ask whether one can find a convergent, iterative scheme for the calculation of the solution of the global optimality system that involves only one iteration parameter and parallel computations on the subregions $\Omega_i \times I_k$ for each value of that parameter. An obvious approach to this question is to directly combine the time domain and space domain decomposition algorithms presented in Sections 6.3.1 and 6.4.1 above. The combined algorithm essentially looks for the limit along the diagonal $m = n$. However, convergence of this scheme has not been established and remains an open, and important, question.

In what follows we use the notation

$$Q_{ik} = \Omega_i \times I_k, \;\; \Sigma_{ik}^D = \Sigma_i^D \times I_k, \;\; \Sigma_{ik}^N = \Sigma_i^N \times I_k,$$
$$\Sigma_{ijk} = \Sigma_{ij} \times I_k, \quad \Sigma_{ik}^{\text{int}} = \Gamma_i^{\text{int}} \times I_k = \bigcup_{j\in\mathcal{J}_i} \Sigma_{ijk},$$

6.5.1 Sequential Space-Time Domain Decomposition

We begin with the time domain decomposition of the global optimality system described in Section 6.3.1. We write m in place of the iteration parameter $n+1$ of that section, and θ in place of the parameter β in the iteration step. Thus for

$k = 0, \dots, K$ we consider functions w_k^m, p_k^m that satisfy

$$\begin{aligned} \frac{\partial^2 w_k^m}{\partial t^2} - \nabla \cdot (A \nabla w_k^m) + c w_k^m &= F_k \ \text{ in } Q_k, \\ w_k^m \ \text{ on } \Sigma_k^D, \\ \frac{\partial w_k^m}{\partial \nu_A} + \alpha \frac{\partial w_k^m}{\partial t} + p_k^m &= 0 \ \text{ on } \Sigma_k^N; \end{aligned} \tag{6.5.1.1}$$

$$\begin{aligned} \frac{\partial^2 p_k^m}{\partial t^2} - \nabla \cdot (A \nabla p_k^m) + c p_k^m &= 0 \ \text{ in } Q_k, \\ p_k^m &= 0 \ \text{ on } \Sigma_k^D, \\ \frac{\partial p_k^m}{\partial \nu_A} - \alpha d_t p_k^m &= 0 \ \text{ on } \Sigma_k^N. \end{aligned} \tag{6.5.1.2}$$

Initial values for $(w_k^m, \partial w_k^m/\partial t)$ at $t = T_k$ and final values for $(p_k^m, \partial p_k^m/\partial t)$ at $t = T_{k+1}$ are determined from the iteration step (6.3.1.3). On the interval I_0 these conditions are

$$\begin{aligned} w_0^m(0) = w_0, \quad & \frac{\partial w_0^m}{\partial t}(0) = v_0, \\ p_0^m(T_1) &= \frac{1}{\theta}\Big(\frac{\partial w_0^m}{\partial t}(T_1) - \mu_{0,1}^{m-1}\Big), \\ \frac{\partial p_0^m}{\partial t}(T_1) &= -\frac{1}{\theta}\big(\mathcal{A} w_0^m(T_1) - \eta_{0,1}^{m-1}\big). \end{aligned} \tag{6.5.1.3}$$

On the intervals I_k, $k = 1, \dots, K-1$ the conditions are

$$\begin{aligned} w_k^m(T_k) &= \mathcal{A}^{-1}\Big(\theta \frac{\partial p_k^m}{\partial t}(T_k) + \eta_{k,k-1}^{m-1}\Big) \\ \frac{\partial w_k^m}{\partial t}(T_k) &= -\theta p_k^m(T_k) + \mu_{k,k-1}^{m-1} \\ p_k^m(T_{k+1}) &= \frac{1}{\theta}\Big(\frac{\partial w_k^m}{\partial t}(T_{k+1}) - \mu_{k,k+1}^{m-1}\Big), \\ \frac{\partial p_k^m}{\partial t}(T_{k+1}) &= -\frac{1}{\theta}\big(\mathcal{A} w_k^m(T_{k+1}) - \eta_{k,k+1}^{m-1}\big). \end{aligned} \tag{6.5.1.4}$$

On the final interval I_K the initial values for $(w_K^m, \partial w_K^m/\partial t)$ at $t = T_K$ and the final values for $(p_K^m, \partial p_K^m/\partial t)$ at $t = T_{K+1} = T$ are

$$\begin{aligned} w_K^m(T_K) &= \mathcal{A}^{-1}\Big(\theta \frac{\partial p_K^m}{\partial t}(T_K) + \eta_{K,K-1}^{m-1}\Big) \\ \frac{\partial w_K^m}{\partial t}(T_K) &= -\theta p_K^m(T_K) + \mu_{K,K-1}^{m-1} \\ p_K^m(T_{K+1}) &= \kappa\Big(\frac{\partial w_K^m}{\partial t}(T_{K+1}) - z_1\Big), \\ \frac{\partial p_K^m}{\partial t}(T_{K+1}) &= -\kappa\big(\mathcal{A} w_K^m(T_{K+1}) - z_0\big). \end{aligned} \tag{6.5.1.5}$$

In (6.5.1.4)–(6.5.1.6) the quantities $\mu_{k,k\pm1}^{m-1} \in H$ and $\eta_{k,k\pm1}^{m-1} \in V^*$ are given data to the problem.

The system (6.5.1.1), (6.5.1.2) with initial/final values given by (6.5.1.3), (6.5.1.4) or (6.5.1.5) as appropriate, is an optimality system concentrated on the cylinder $Q_k = \Omega \times I_k$, as was shown in Section 6.3.1. Let us note that on the cylinder I_0 this local optimality system has a structure identical to that of the global optimality system (6.1.1.1), (6.2.3.4), (6.2.3.6), the only superficial differences being that $(0,T)$ is replaced by $(0,T_1)$, κ by $1/\theta$ and (z_0, z_1) by $(\mu_{0,1}^{m-1}, \mathcal{A}^{-1}\eta_{0,1}^{m-1})$. On the intervals I_k, $k = 1,\ldots,K$, the local optimality system differs in structure from the global optimality system only in that the initial conditions for $(w_k^m, \partial w_k^m/\partial t)$, rather than being a given element of $V \times H$, involve also the adjoint state $(p_k^m, \partial p_k^m/\partial t)$.

We now consider the following spatial domain decomposition of (6.5.1.1)–(6.5.1.5), which is based on the same interface conditions used in the domain decomposition algorithm in Section 6.4.1: for $n = 0,1,\ldots,$

$$\begin{cases} \dfrac{\partial^2 w_{ik}^{m,n+1}}{\partial t^2} - \nabla\cdot(A_i \nabla w_{ik}^{m,n+1}) + c_i w_{ik}^{m,n+1} = F_{ik} \\ \dfrac{\partial^2 p_{ik}^{m,n+1}}{\partial t^2} - \nabla\cdot(A_i \nabla p_{ik}^{m,n+1}) + c_i p_{ik}^{m,n+1} = 0 \text{ in } Q_{ik}; \end{cases} \tag{6.5.1.6}$$

$$w_{ik}^{m,n+1} = p_{ik}^{m,n+1} = 0 \text{ on } \Sigma_{ik}^D; \tag{6.5.1.7}$$

$$\begin{cases} \dfrac{\partial w_{ik}^{m,n+1}}{\partial \nu_{A_i}} + \alpha_i \dfrac{\partial w_{ik}^{m,n+1}}{\partial t} + p_{ik}^{m,n+1} = 0 \\ \dfrac{\partial p_{ik}^{m,n+1}}{\partial \nu_{A_i}} - \alpha_i d_t p_{ik}^{m,n+1} = 0 \text{ on } \Sigma_{ik}^N; \end{cases} \tag{6.5.1.8}$$

$$\begin{cases} \dfrac{\partial w_{ik}^{m,n+1}}{\partial \nu_{A_i}} + \beta \dfrac{\partial w_{ik}^{m,n+1}}{\partial t} + \gamma p_{ik}^{m,n+1} = \lambda_{ijk}^{m,n} \\ \dfrac{\partial p_{ik}^{m,n+1}}{\partial \nu_{A_i}} - \beta d_t p_{ik}^{m,n+1} + \gamma d_t \dfrac{\partial w_{ik}^{m,n+1}}{\partial t} = \zeta_{ijk}^{m,n} \text{ on } \Sigma_{ijk}; \end{cases} \tag{6.5.1.9}$$

$$\begin{cases} w_{i0}^{m,n+1}(0) = w_{0i}, \quad \dfrac{\partial w_{i0}^{m,n+1}}{\partial t}(0) = v_{0i} \\ p_{iK}^{m,n+1}(T) = \kappa\Big(\dfrac{\partial w_{iK}^{m,n+1}}{\partial t}(T) - z_{1i}\Big) \\ \dfrac{\partial p_{iK}^{m,n+1}}{\partial t}(T) = -\kappa \mathcal{A}_i (w_{iK}^{m,n+1}(T) - z_{0i}) \text{ in } \Omega_i; \end{cases} \tag{6.5.1.10}$$

$$\begin{cases} w_{ik}^{m,n+1}(T_k) = \mathcal{A}_i^{-1}(\theta \dfrac{\partial p_{ik}^{m,n+1}}{\partial t}(T_k) + \eta_{i,k,k-1}^{m-1}) \\ \dfrac{\partial w_{ik}^{m,n+1}}{\partial t}(T_k) = -\theta p_{ik}^{m,n+1}(T_k) + \mu_{i,k,k-1}^{m-1} \\ p_{ik}^{m,n+1}(T_{k+1}) = \dfrac{1}{\theta}(\dfrac{\partial w_{ik}^{m,n+1}}{\partial t}(T_{k+1}) - \mu_{i,k,k+1}^{m-1}) \\ \dfrac{\partial p_{ik}^{m,n+1}}{\partial t}(T_{k+1}) = -\dfrac{1}{\theta}\mathcal{A}_i(w_{ik}^{m,n+1}(T_{k+1}) - \mathcal{A}_i^{-1}\eta_{i,k,k+1}^{m-1}) \quad \text{in } \Omega_i. \end{cases} \tag{6.5.1.11}$$

where β and γ are positive constants, $F_{ik} = F|_{Q_{ik}}$, $\mu_{i,k,k\pm1}^{m-1} = \mu_{k,k+1}^{m-1}|_{\Omega_i}$, $\eta_{i,k,k\pm1}^{m-1} = \mathcal{A}_i(\mathcal{A}^{-1}\eta_{k,k+1}^{m-1})_{\Omega_i}$. The issue we wish to investigate is that of convergence, for each fixed index m, of the system (6.5.1.6)–(6.5.1.11) as $n \to \infty$ to the solution of the optimality system (6.5.1.1)–(6.5.1.5). In (6.5.1.11), the first two conditions hold for $k = 1, \ldots, K$ while the last two hold for $k = 0, \ldots, K-1$. In (6.5.1.9), $\lambda_{ijk}^0, \rho_{ijk}^0 \in L^2(\Sigma_{ijk})$ are arbitrary, $\zeta_{ijk}^0 = d_t\rho_{ijk}^0$, while for $n = 1, \ldots$ we set

$$\begin{aligned} \lambda_{ijk}^{m,n} &= (1-\epsilon)(-\frac{\partial w_{jk}^{m,n}}{\partial \nu_{A_j}} + \beta\frac{\partial w_{jk}^{m,n}}{\partial t} + \gamma p_{jk}^{m,n}) \\ &\qquad + \epsilon(\frac{\partial w_{ik}^{m,n+1}}{\partial \nu_{A_i}} + \beta\frac{\partial w_{ik}^{m,n}}{\partial t} + \gamma p_{ik}^{m,n}) \\ \zeta_{ijk}^{m,n} &= (1-\epsilon)(-\frac{\partial p_{jk}^{m,n}}{\partial \nu_{A_j}} - \beta d_t p_{jk}^{m,n} + \gamma d_t \frac{\partial w_{jk}^{m,n}}{\partial t}) \\ &\qquad + \epsilon(\frac{\partial p_{ik}^{m,n}}{\partial \nu_{A_i}} - \beta d_t p_{ik}^{m,n+1} + \gamma d_t \frac{\partial w_{ik}^{m,n}}{\partial t}) \end{aligned} \tag{6.5.1.12}$$

where $\epsilon \in [0,1)$ is a relaxation parameter. The operator d_t appearing in (6.5.1.9) is the bounded linear operator $L^2(I_k; U_{ij}) \mapsto \big(H^1(I_k; U_{ij})\big)^*$ defined by

$$\langle d_t h, \phi\rangle = -\int_{I_k} \big(h(t), \frac{d\phi}{dt}(t)\big)_{U_{ij}}\, dt, \quad \forall \phi \in H^1(I_k; U_{ij}).$$

The operator d_t in (6.5.1.8) is similarly defined, but with U_i in place of U_{ij}.

If $(\mu_{i,k,k\pm1}^{m-1}, \eta_{i,k,k\pm1}^{m-1} \in H_i \times V_i^*$, if $\lambda_{ijk}^{m,n}, \rho_{ijk}^{m,n} \in L^2(\Sigma_{ijk})$ and $\zeta_{ijk}^{m,n} = d_t\rho_{ijk}^{m,n}$, then the system (6.5.1.1)–(6.5.1.6) is an optimality system concentrated on the cylinder Q_{ik}. Indeed, for $i = 1, \ldots, I$ and $k = 1, \ldots, K-1$ it is the optimality system for the problem

$$\inf J_{ik}^{m,n}\big(f_{ik}, g_{ijk}, \phi_{i,k,k-1}, \psi_{i,k,k-1}\big),$$

taken over

$$f_{ik} \in L^2(\Sigma_{ik}^N), \quad g_{ijk} \in L^2(\Sigma_{ijk}), \quad \phi_{i,k,k-1} \in H_i, \quad \psi_{i,k,k-1} \in V_i^*,$$

subject to

$$\begin{aligned}
\frac{\partial^2 w_{ik}}{\partial t^2} - \nabla\cdot(A_i\nabla w_{ik}) + c_i w_{ik} &= F_{ik} \text{ in } Q_{ik}\\
w_{ik} &= 0 \quad \text{on } \Sigma_{ik}^D\\
\frac{\partial w_{ik}}{\partial \nu_{A_i}} + \alpha_i\frac{\partial w_{ik}}{\partial t} &= f_{ik} \quad \text{on } \Sigma_{ik}^N\\
\frac{\partial w_{ik}}{\partial \nu_{A_i}} + \beta\frac{\partial w_{ik}}{\partial t} &= \lambda_{ijk}^{m,n} + g_{ijk} \quad \text{on } \Sigma_{ijk},\ \forall j\in\mathcal{J}_i\\
&\begin{cases} w_{ik}(T_k) = \mathcal{A}_i^{-1}\big(\eta_{i,k,k-1}^{m-1} + \psi_{i,k,k-1}\big)\\ \dfrac{\partial w_{ik}}{\partial t}(T_k) = \mu_{i,k,k-1}^{m-1} + \phi_{i,k,k-1} \qquad \text{in } \Omega_i,\end{cases}
\end{aligned}\tag{6.5.1.13}$$

where

$$\begin{aligned}
J_{ik}^{m,n}\big(f_{ik}, g_{ijk}, \phi_{i,k,k-1}, \psi_{i,k,k-1}\big) &= \frac{1}{2}\int_{\Sigma_{ik}^N} |f_{ik}|^2 d\Sigma\\
&+ \frac{1}{2\gamma}\sum_{j\in\mathcal{J}_i}\int_{\Sigma_{ijk}} \Big\{|g_{ijk}|^2 + \Big|\gamma\frac{\partial w_{ik}}{\partial t} - \rho_{ijk}^{m,n}\Big|^2\Big\}\, d\Sigma\\
+ \frac{1}{2\theta}\Big\{\big\|w_{ik}(T_{k+1}) - \mathcal{A}_i^{-1}\eta_{i,k,k+1}^{m-1}\big\|_{V_i}^2 &+ \Big\|\frac{\partial w_{ik}}{\partial t}(T_{k+1}) - \mu_{i,k,k+1}^{m-1}\Big\|_{H_i}^2\\
&+ \|\phi_{i,k.k-1}\|_{H_i}^2 + \|\psi_{i,k.k-1}\|_{V_i^*}^2.
\end{aligned}\tag{6.5.1.14}$$

For the cylinder $Q_{i0} = \Omega_i \times I_0$, the appropriate cost functional is given by (6.5.1.14) with $k = 0$ but with the last two terms omitted, subject to (6.5.1.13) with $k = 0$ and with the initial conditions at $t = T_0 = 0$ replaced by

$$w_{i0}(0) = w_{0i}, \qquad \frac{\partial w_{i0}}{\partial t}(0) = v_{0i} \quad \text{in } \Omega_i.$$

For the cylinder Q_{iK} the appropriate cost functional is given by (6.5.1.14) with $k = K$ but with the terms

$$\frac{1}{2\theta}\Big\{\big\|w_{ik}(T_{k+1}) - \mathcal{A}_i^{-1}\eta_{i,k,k+1}^{m-1}\big\|_{V_i}^2 + \Big\|\frac{\partial w_{ik}}{\partial t}(T_{k+1}) - \zeta_{i,k,k+1}^{n}\Big\|^2\Big\}$$

replaced by

$$\frac{\kappa}{2}\Big\{\big\|w_{iK}(T) - z_{0i}\big\|_{V_i}^2 + \Big\|\frac{\partial w_{iK}}{\partial t}(T) - z_{1i}\Big\|^2\Big\}.$$

Verification of these statements follows the same lines as the calculations carried out in Sections 6.3.1 and 6.4.1, and are omitted. From the discussion of Section

6.2.2 it follows that, for $i = 1,\dots,I$ and $k = 0,\dots,K$, the solution of (6.5.1.6)–(6.5.1.11) has regularity

$$(w_{ik}^{m,n+1}, \frac{\partial w_{ik}^{m,n+1}}{\partial t}) \in C(I_k; V_i \times H_i), \qquad \frac{\partial w_{ik}^{m,n+1}}{\partial t}\Big|_{\Sigma_{ik}^N \cup \Sigma_{ik}^{\rm int}} \in L^2(\Sigma_{ik}^N \cup \Sigma_{ik}^{\rm int}),$$

$$(p_{ik}^{m,n+1}, \frac{\partial p_{ik}^{n+1}}{\partial t}) \in C(I_k; H_i \times V_i^*), \quad p_{ik}^{m,n+1}|_{\Sigma_{ik}^N \cup \Sigma_{ik}^{\rm int}} \in L^2(\Sigma_{ik}^N \cup \Sigma_{ik}^{\rm int}).$$

As was done in Section 6.4.1, we replace the adjoint variable $p_{ik}^{m,n+1}$ by the more regular variable $q_{ik}^{m,n+1}$ through the substitution

$$p_{ik}^{m,n+1} = \frac{\partial q_{ik}^{m,n+1}}{\partial t}, \qquad \frac{\partial p_{ik}^{m,n+1}}{\partial t} = -A_i q_{ik}^{m,n+1}.$$

In terms of the variables $w_{ik}^{m,n+1}$, $q_{ik}^{m,n+1}$ the system (6.5.1.6)–(6.5.1.11) takes the form

$$\begin{cases} \dfrac{\partial^2 w_{ik}^{m,n+1}}{\partial t^2} - \nabla\cdot(A_i \nabla w_{ik}^{m,n+1}) + c_i w_{ik}^{m,n+1} = F_{ik} \\ \dfrac{\partial^2 q_{ik}^{m,n+1}}{\partial t^2} - \nabla\cdot(A_i \nabla q_{ik}^{m,n+1}) + c_i q_{ik}^{m,n+1} = 0 \text{ in } Q_{ik}; \end{cases} \tag{6.5.1.15}$$

$$w_{ik}^{m,n+1} = q_{ik}^{m,n+1} = 0 \text{ on } \Sigma_{ik}^D; \tag{6.5.1.16}$$

$$\begin{cases} \dfrac{\partial w_{ik}^{m,n+1}}{\partial \nu_{A_i}} + \alpha_i \dfrac{\partial w_{ik}^{m,n+1}}{\partial t} + \dfrac{\partial q_{ik}^{m,n+1}}{\partial t} = 0 \\ \dfrac{\partial q_{ik}^{m,n+1}}{\partial \nu_{A_i}} - \alpha_i \dfrac{\partial q_{ik}^{m,n+1}}{\partial t} = 0 \text{ on } \Sigma_{ik}^N; \end{cases} \tag{6.5.1.17}$$

$$\begin{cases} \dfrac{\partial w_{ik}^{m,n+1}}{\partial \nu_{A_i}} + \beta \dfrac{\partial w_{ik}^{m,n+1}}{\partial t} + \gamma \dfrac{\partial q_{ik}^{m,n+1}}{\partial t} = \lambda_{ijk}^{m,n} \\ \dfrac{\partial q_{ik}^{m,n+1}}{\partial \nu_{A_i}} - \beta \dfrac{\partial q_{ik}^{m,n+1}}{\partial t} + \gamma \dfrac{\partial w_{ik}^{m,n+1}}{\partial t} = \rho_{ijk}^{m,n} \text{ on } \Sigma_{ijk}; \end{cases} \tag{6.5.1.18}$$

$$\begin{cases} w_{i0}^{m,n+1}(0) = w_{0i}, \quad \dfrac{\partial w_{i0}^{m,n+1}}{\partial t}(0) = v_{0i} \\ q_{iK}^{m,n+1}(T) = \kappa(w_{iK}^{m,n+1}(T) - z_{0i}) \\ \dfrac{\partial q_{iK}^{m,n+1}}{\partial t}(T) = \kappa(\dfrac{\partial w_{ik}^{m,n+1}}{\partial t}(T) - z_{1i}) \text{ in } \Omega_i; \end{cases} \tag{6.5.1.19}$$

$$\begin{cases} w_{ik}^{m,n+1}(T_k) = -\theta q_{ik}^{m,n+1}(T_k) + \mathcal{A}_i^{-1}\eta_{i,k,k-1}^{m-1} \\ \dfrac{\partial w_{ik}^{m,n+1}}{\partial t}(T_k) = -\theta \dfrac{\partial q_{ik}^{m,n+1}}{\partial t}(T_k) + \mu_{i,k,k-1}^{m-1} \\ q_{ik}^{m,n+1}(T_{k+1}) = \dfrac{1}{\theta}\big(w_{ik}^{m,n+1}(T_{k+1}) - \mathcal{A}_i^{-1}\eta_{i,k,k+1}^{m-1}\big) \\ \dfrac{\partial q_{ik}^{m,n+1}}{\partial t}(T_{k+1}) = \dfrac{1}{\theta}\Big(\dfrac{\partial w_{ik}^{m,n+1}}{\partial t}(T_{k+1}) - \mu_{i,k,k+1}^{m-1}\Big) \quad \text{in } \Omega_i. \end{cases} \tag{6.5.1.20}$$

Similarly, it is convenient to write the optimality system (6.5.1.1)–(6.5.1.5) in terms of w_{ik}^m, q_{ik}^m where

$$p_k^m = \frac{\partial q_k^m}{\partial t}, \quad \frac{\partial p_k^m}{\partial t} = -\mathcal{A}q_k^m, \qquad w_{ik}^m = w_k^m|_{Q_{ik}}, \quad q_{ik}^m = q_k^m|_{Q_{ik}}.$$

This system then takes the form

$$\begin{cases} \dfrac{\partial^2 w_{ik}^m}{\partial t^2} - \nabla\cdot(A\nabla w_{ik}^m) + c w_{ik}^m = F_{ik} \\ \dfrac{\partial^2 q_{ik}^m}{\partial t^2} - \nabla\cdot(A\nabla q_{ik}^m) + c q_{ik}^m = 0 \qquad \text{in } Q_{ik}; \end{cases} \tag{6.5.1.21}$$

$$w_{ik}^m = q_{ik}^m = 0 \;\text{ on } \Sigma_{ik}^D; \tag{6.5.1.22}$$

$$\begin{cases} \dfrac{\partial w_{ik}^m}{\partial \nu_A} + \alpha \dfrac{\partial w_{ik}^m}{\partial t} + q_{ik}^m = 0 \\ \dfrac{\partial q_{ik}^m}{\partial \nu_A} - \alpha \dfrac{\partial q_{ik}^m}{\partial t} = 0 \qquad \text{on } \Sigma_{ik}^N; \end{cases} \tag{6.5.1.23}$$

$$\begin{cases} w_{i0}^m(0) = w_{0i}, \quad \dfrac{\partial w_{i0}^m}{\partial t}(0) = v_{0i} \\ q_{iK}^m(T) = \kappa\big(w_{iK}^m(T) - z_{0i}\big) \\ \dfrac{\partial q_{iK}^m}{\partial t}(T) = \kappa\Big(\dfrac{\partial w_{iK}^m}{\partial t}(T) - z_{1i}\Big) \quad \text{in } \Omega_i; \end{cases} \tag{6.5.1.24}$$

$$\begin{cases} w_{ik}^m(T_k) = -\theta q_{ik}^m(T_k) + \mathcal{A}_i^{-1}\eta_{i,k,k-1}^{m-1} \\ \dfrac{\partial w_{ik}^m}{\partial t}(T_k) = -\theta \dfrac{\partial q_{ik}^m}{\partial t}(T_k) + \mu_{i,k,k-1}^{m-1} \\ q_{ik}^m(T_{k+1}) = \dfrac{1}{\theta}\big(w_{ik}^m(T_{k+1}) - \mathcal{A}_i^{-1}\eta_{i,k,k+1}^{m-1}\big) \\ \dfrac{\partial q_{ik}^m}{\partial t}(T_{k+1}) = \dfrac{1}{\theta}\Big(\dfrac{\partial w_{ik}^m}{\partial t}(T_{k+1}) - \mu_{i,k,k+1}^{m-1}\Big) \;\text{ in } \Omega_i, \end{cases} \tag{6.5.1.25}$$

together with the *transmission conditions*

$$\begin{aligned} &w_{ik}^m\big|_{\Sigma_{ijk}} = w_{jk}^m\big|_{\Sigma_{ijk}}, \quad q_{ik}^m\big|_{\Sigma_{ijk}} = q_{jk}^m\big|_{\Sigma_{ijk}} \\ &\frac{\partial w_{ik}^m}{\partial \nu_{A_i}}\bigg|_{\Sigma_{ijk}} = -\frac{\partial w_{jk}^m}{\partial \nu_{A_j}}\bigg|_{\Sigma_{ijk}}, \quad \frac{\partial q_{ik}^m}{\partial \nu_{A_i}}\bigg|_{\Sigma_{ijk}} = -\frac{\partial q_{jk}^m}{\partial \nu_{A_j}}\bigg|_{\Sigma_{ijk}}. \end{aligned} \tag{6.5.1.26}$$

In (6.5.1.24), the first two conditions are imposed for $k = 1, \dots, K$ and the latter two for $k = 0, \dots, K-1$.

Convergence of the solutions of (6.5.1.15)–(6.5.1.20) to the solution of (6.5.1.21)–(6.5.1.25) is proved along the lines of Theorem 6.4.2.1, so we will not provide all details. One introduces the local errors

$$\tilde{w}_{ik}^{m,n+1} = w_{ik}^{m,n+1} - w_{ik}^{m}, \qquad \tilde{q}_{ik}^{m,n+1} = q_{ik}^{m,n+1} = q_{ik}^{m},$$

which satisfy (6.5.1.15)–(6.5.1.20) with homogeneous data

$$F_{ik} = 0, \quad w_{0i} = v_{0i} = z_{0i} = z_{1i} = 0, \quad \mu_{i,k,k\pm1}^{m-1} = \eta_{i,k,k\pm1}^{m-1} = 0 \tag{6.5.1.27}$$

and with $\lambda_{ijk}^{m,n}$, $\rho_{ijk}^{m,n}$ replaced by

$$\begin{aligned}
\tilde{\lambda}_{ijk}^{m,n} &= (1-\epsilon)\Big(-\frac{\partial \tilde{w}_{jk}^{m,n}}{\partial \nu_{A_j}} + \beta\frac{\partial \tilde{w}_{jk}^{m,n}}{\partial t} + \gamma\frac{\partial \tilde{q}_{jk}^{m,n}}{\partial t}\Big)\\
&\qquad + \epsilon\Big(\frac{\partial \tilde{w}_{ik}^{m,n}}{\partial \nu_{A_i}} + \beta\frac{\partial \tilde{w}_{ik}^{m,n}}{\partial t} + \gamma\frac{\partial \tilde{q}_{ik}^{m,n}}{\partial t}\Big)\\
\tilde{\rho}_{ijk}^{m,n} &= (1-\epsilon)\Big(-\frac{\partial \tilde{q}_{jk}^{m,n}}{\partial \nu_{A_j}} - \beta\frac{\partial \tilde{q}_{jk}^{m,n}}{\partial t} + \gamma\frac{\partial \tilde{w}_{jk}^{m,n}}{\partial t}\Big)\\
&\qquad + \epsilon\Big(\frac{\partial \tilde{q}_{ik}^{m,n}}{\partial \nu_{A_i}} - \beta\frac{\partial \tilde{q}_{ik}^{m,n}}{\partial t} + \gamma\frac{\partial \tilde{w}_{ik}^{m,n}}{\partial t}\Big).
\end{aligned} \tag{6.5.1.28}$$

The local errors therefore satisfy

$$\begin{cases}
\dfrac{\partial^2 \tilde{w}_{ik}^{m,n+1}}{\partial t^2} - \nabla\cdot(A_i \nabla \tilde{w}_{ik}^{m,n+1}) + c_i \tilde{w}_{ik}^{m,n+1} = 0\\
\dfrac{\partial^2 \tilde{q}_{ik}^{m,n+1}}{\partial t^2} - \nabla\cdot(A_i \nabla \tilde{q}_{ik}^{m,n+1}) + c_i \tilde{q}_{ik}^{m,n+1} = 0 \text{ in } Q_{ik};
\end{cases} \tag{6.5.1.29}$$

$$\tilde{w}_{ik}^{m,n+1} = \tilde{q}_{ik}^{m,n+1} = 0 \text{ on } \Sigma_{ik}^{D}; \tag{6.5.1.30}$$

$$\begin{cases}
\dfrac{\partial \tilde{w}_{ik}^{m,n+1}}{\partial \nu_{A_i}} + \alpha_i \dfrac{\partial \tilde{w}_{ik}^{m,n+1}}{\partial t} + \dfrac{\partial \tilde{q}_{ik}^{m,n+1}}{\partial t} = 0\\
\dfrac{\partial \tilde{q}_{ik}^{m,n+1}}{\partial \nu_{A_i}} - \alpha_i \dfrac{\partial \tilde{q}_{ik}^{m,n+1}}{\partial t} = 0 \qquad \text{on } \Sigma_{ik}^{N};
\end{cases} \tag{6.5.1.31}$$

$$\begin{cases}
\dfrac{\partial \tilde{w}_{ik}^{m,n+1}}{\partial \nu_{A_i}} + \beta\dfrac{\partial \tilde{w}_{ik}^{m,n+1}}{\partial t} + \gamma\dfrac{\partial \tilde{q}_{ik}^{m,n+1}}{\partial t} = \tilde{\lambda}_{ijk}^{m,n}\\
\dfrac{\partial \tilde{q}_{ik}^{m,n+1}}{\partial \nu_{A_i}} - \beta\dfrac{\partial \tilde{q}_{ik}^{m,n+1}}{\partial t} + \gamma\dfrac{\partial w_{ik}^{m,n+1}}{\partial t} = \tilde{\rho}_{ijk}^{m,n} \text{ on } \Sigma_{ijk};
\end{cases} \tag{6.5.1.32}$$

$$\begin{cases}
\tilde{w}_{i0}^{m,n+1}(0) = \dfrac{\partial \tilde{w}_{i0}^{m,n+1}}{\partial t}(0) = 0\\
\tilde{q}_{iK}^{m,n+1}(T) = \kappa \tilde{w}_{iK}^{m,n+1}(T), \quad \dfrac{\partial \tilde{q}_{iK}^{m,n+1}}{\partial t}(T) = \kappa\dfrac{\partial \tilde{w}_{ik}^{m,n+1}}{\partial t}(T) \quad \text{in } \Omega_i;
\end{cases} \tag{6.5.1.33}$$

$$\begin{cases} \tilde{w}_{ik}^{m,n+1}(T_k) = -\theta \tilde{q}_{ik}^{m,n+1}(T_k) \\ \dfrac{\partial \tilde{w}_{ik}^{m,n+1}}{\partial t}(T_k) = -\theta \dfrac{\partial \tilde{q}_{ik}^{m,n+1}}{\partial t}(T_k) \\ \tilde{q}_{ik}^{m,n+1}(T_{k+1}) = \dfrac{1}{\theta} \tilde{w}_{ik}^{m,n+1}(T_{k+1}) \\ \dfrac{\partial \tilde{q}_{ik}^{m,n+1}}{\partial t}(T_{k+1}) = \dfrac{1}{\theta} \dfrac{\partial \tilde{w}_{ik}^{m,n+1}}{\partial t}(T_{k+1}) \quad \text{in } \Omega_i. \end{cases} \tag{6.5.1.34}$$

The following result describes the convergence of the local errors.

Theorem 6.5.1.1 *Suppose that β, γ satisfy the hypotheses of Theorem* 6.4.2.1, *that* $\theta > 0$ *and*

$$\frac{\gamma}{\beta} > \frac{|1-\theta^2|}{2\theta}. \tag{6.5.1.35}$$

Then as $n \to \infty$ we have, for $i = 1, \ldots, I$; $k = 0, \ldots, K$, and $m = 1, 2, \ldots$:
(i) *for any $\epsilon \in [0, 1)$*

$$(\tilde{w}_{ik}^{m,n}(T_{k+1}), \frac{\partial \tilde{w}_{ik}^{m,n}}{\partial t}(T_{k+1})) \to 0 \ \textit{strongly in } \mathcal{H}_i,$$

$$(\tilde{q}_{ik}^{m,n}(T_k), \frac{\partial \tilde{q}_{ik}^{m,n}}{\partial t}(T_k)) \to 0 \ \textit{strongly in } \mathcal{H}_i,$$

$$(\tilde{w}_{ik}^{m,n}, \frac{\partial \tilde{w}_{ik}^{m,n}}{\partial t}) \to 0, \ (\tilde{q}_i^{m,n}, \frac{\partial \tilde{q}_i^{m,n}}{\partial t}) \to 0 \ \textit{weakly* in } L^\infty(I_k; \mathcal{H}_i),$$

$$\left.\frac{\partial \tilde{w}_{ik}^{m,n}}{\partial t}\right|_{\Sigma_{ik}^N} \to 0, \quad \left.\frac{\partial \tilde{q}_{ik}^{m,n}}{\partial t}\right|_{\Sigma_{ik}^N} \to 0 \ \textit{strongly in } L^2(\Sigma_{ik}^N),$$

$$\left.\frac{\partial \tilde{w}_{ik}^{m,n}}{\partial t}\right|_{\Sigma_{ik}^{int}} \to 0, \quad \left.\frac{\partial \tilde{q}_{ik}^{m,n}}{\partial t}\right|_{\Sigma_{ik}^{int}} \to 0 \ \textit{weakly in } L^2(\Sigma_{ik}^{int}),$$

$$\left.\frac{\partial \tilde{w}_{ik}^{m,n}}{\partial \nu_{A_i}}\right|_{\Sigma_{ik}^{int}} \to 0, \quad \left.\frac{\partial \tilde{q}_{ik}^{m,n}}{\partial \nu_{A_i}}\right|_{\Sigma_{ik}^{int}} \to 0 \ \textit{weakly in } L^2(\Sigma_{ik}^{int}).$$

(ii) *In addition, if $\epsilon > 0$,*

$$(\tilde{w}_{ik}^{m,n}, \frac{\partial \tilde{w}_{ik}^{m,n}}{\partial t}) \to 0, \ (\tilde{q}_{ik}^{m,n}, \frac{\partial \tilde{q}_{ik}^{m,n}}{\partial t}) \to 0 \ \textit{strongly in } L^\infty(I_k; \mathcal{H}_i)$$

$$\begin{cases} \left.\dfrac{\partial w_{ik}^{m,n}}{\partial t} - \dfrac{\partial w_{jk}^{m,n}}{\partial t}\right|_{\Sigma_{ijk}} \to 0, \quad \left.\dfrac{\partial w_{ik}^{m,n}}{\partial \nu_{A_i}} + \dfrac{\partial w_{jk}^{m,n}}{\partial \nu_{A_j}}\right|_{\Sigma_{ijk}} \to 0 \\ \left.\dfrac{\partial q_{ik}^{m,n}}{\partial t} - \dfrac{\partial q_{jk}^{m,n}}{\partial t}\right|_{\Sigma_{ijk}} \to 0, \quad \left.\dfrac{\partial q_{ik}^{m,n}}{\partial \nu_{A_i}} + \dfrac{\partial q_{jk}^{m,n}}{\partial \nu_{A_j}}\right|_{\Sigma_{ijk}} \to 0 \end{cases}$$

strongly in $L^2(\Sigma_{ijk})$, $\forall j \in \mathcal{J}_i$.

Remark 6.5.1.1 For given α, κ and θ, all of the hypotheses of Theorem 6.5.1.1 may be satisfied by choosing γ/β sufficiently large.

For the proof one writes the relaxed iteration step for the local errors as the relaxed fixed point iteration of a nonexpansive mapping $\mathcal{T}$. Convergence of the algorithm is then deduced from properties of $\mathcal{T}$. To this end, we set

$$\mathcal{X} = \prod_{i=1}^{I} \prod_{k=0}^{K} L^2(\Sigma_{ik}^{\text{int}}) \times L^2(\Sigma_{ik}^{\text{int}})$$

with norm defined by

$$\|X\|_{\mathcal{X}}^2 = \sum_{i=1}^{I} \sum_{k=0}^{K} \frac{1}{2\gamma} \int_{\Sigma_{ik}^{\text{int}}} (|\lambda_{ik}|^2 + |\rho_{ik}|^2)\, d\Sigma.$$

Let $X = \{(\lambda_{ik}, \rho_{ik}) : i = 1, \ldots, I;\ k = 0, \ldots, K\} \in \mathcal{X}$ and let (w_{ik}, q_{ik}) be the solution of (6.5.1.29)–(6.5.1.34) with $\tilde{\lambda}_{ijk}^{m,n}$, $\tilde{\rho}_{ijk}^{m,n}$ replaced by $\lambda_{ijk} := \lambda_{ik}|_{\Sigma_{ijk}}$ and $\rho_{ijk} := \rho_{ik}|_{\Sigma_{ijk}}$, respectively. Define $\mathcal{T} : \mathcal{X} \mapsto \mathcal{X}$ according to

$$(\mathcal{T}X)_{ijk} = \left(\left(-\frac{\partial w_{jk}}{\partial \nu_{A_j}} + \beta \frac{\partial w_{jk}}{\partial t} + \gamma \frac{\partial q_{jk}}{\partial t}\right)\Big|_{\Sigma_{ijk}}, \left(-\frac{\partial q_{jk}}{\partial \nu_{A_j}} - \beta \frac{\partial q_{jk}}{\partial t} + \gamma \frac{\partial w_{jk}}{\partial t}\right)\Big|_{\Sigma_{ijk}} \right)$$

$$\begin{aligned}(\mathcal{T}X)_{ik} &= \{(\mathcal{T}X)_{ijk} : j \in \mathcal{J}_i\},\\ \mathcal{T}X &= \{(\mathcal{T}X)_{i,k} : i = 1, \ldots, I;\ k = 0, \ldots, K\}.\end{aligned}$$

Then X is a fixed point of $\mathcal{T}$ if and only if $\{w_{ik},\, q_{ik}\}_{i=1}^{I}$ satisfy (6.5.1.21)–(6.5.1.26) with homogeneous data (6.5.1.27). This means that $(w_k,\, p_k) := \{w_{ik},\, p_{ik}\}_{i=1}^{I}$ is the solution of the optimality system (6.5.1.1)–(6.5.1.5) corresponding to homogeneous data (6.5.1.27). Therefore $X = 0$ is the only fixed point of $\mathcal{T}$.

In analogy with Lemma 6.4.2.1, we have the following result.

Lemma 6.5.1.1 *For any* $X = \{(\lambda_{ik}, \rho_{ik}) : i = i, \ldots, I\} \in \mathcal{X}$, *we may write*

$$\|X\|_{\mathcal{X}}^2 = \mathcal{E} + \mathcal{F}, \quad \|\mathcal{T}X\|_{\mathcal{X}}^2 = \mathcal{E} - \mathcal{F} \tag{6.5.1.36}$$

where

$$\mathcal{E} = \sum_{i=1}^{I} \sum_{k=0}^{K} \int_{\Sigma_{ik}^{\text{int}}} \left\{ \frac{1}{2\gamma} \left(\left|\frac{\partial w_{ik}}{\partial \nu_{A_i}}\right|^2 + \left|\frac{\partial q_{ik}}{\partial \nu_{A_i}}\right|^2 \right) + \frac{\beta^2 + \gamma^2}{2\gamma} \left(\left|\frac{\partial w_{ik}}{\partial t}\right|^2 + \left|\frac{\partial q_{ik}}{\partial t}\right|^2 \right) \right\} d\Sigma, \tag{6.5.1.37}$$

$$\mathcal{F} = \sum_{i=1}^{I} \Big\{ [\kappa + \frac{\beta}{2\gamma}(1-\kappa^2)] \|(w_{iK}(T), \frac{\partial w_{iK}}{\partial t}(T))\|^2_{\mathcal{H}_i} + \frac{\beta}{2\gamma} \|(q_{i0}(0), \frac{\partial q_{i0}}{\partial t}(0))\|^2_{\mathcal{H}_i}$$
$$+ \sum_{k=0}^{K-1} \frac{1}{\theta^2}[\theta + \frac{\beta}{2\gamma}(\theta^2 - 1)] \|(w_{ik}(T_{k+1}), \frac{\partial w_{ik}}{\partial t}(T_{k+1}))\|^2_{\mathcal{H}_i}$$
$$+ \sum_{k=1}^{K} [\theta + \frac{\beta}{2\gamma}(1-\theta^2)] \|(q_{ik}(T_k), \frac{\partial q_{ik}}{\partial t}(T_k))\|^2_{\mathcal{H}_i} \Big\}$$
$$+ \sum_{i=1}^{I}\sum_{k=0}^{K} \int_{\Sigma_{ik}^N} \left\{ \left(1 + \frac{\alpha_i \beta}{\gamma}\right) \left|\frac{\partial q_{ik}}{\partial t}\right|^2 + \frac{\beta}{\gamma}\frac{\partial q_{ik}}{\partial t}\frac{\partial w_{ik}}{\partial t} + \frac{\alpha_i\beta}{\gamma}\left|\frac{\partial w_{ik}}{\partial t}\right|^2 \right\} d\Sigma. \tag{6.5.1.38}$$

It is thus seen that $\mathcal{T}$ is a nonexpansive mapping if the parameters β, γ satisfy the conditions of Theorem 6.4.2.1 and if (6.5.1.35) holds.

Proof of Lemma 6.5.1.1. The proof follows the lines of the proof of Lemma 6.4.2.1. Indeed, direct calculation gives

$$\|X\|^2_{\mathcal{X}} =$$
$$\mathcal{E} + \sum_{i=1}^{I}\sum_{k=0}^{K} \int_{\Sigma_{ik}^{\text{int}}} \left\{ \frac{\beta}{\gamma}\left(\frac{\partial w_{ik}}{\partial \nu_{A_i}}\frac{\partial w_{ik}}{\partial t} - \frac{\partial q_{ik}}{\partial \nu_{A_i}}\frac{\partial q_{ik}}{\partial t} \right) + \left(\frac{\partial w_{ik}}{\partial \nu_{A_i}}\frac{\partial q_{ik}}{\partial t} + \frac{\partial q_{ik}}{\partial \nu_{A_i}}\frac{\partial w_{ik}}{\partial t} \right) \right\} d\Sigma, \tag{6.5.1.39}$$

$$\|\mathcal{T}X\|^2_{\mathcal{X}} =$$
$$\mathcal{E} - \sum_{i=1}^{I}\sum_{k=0}^{K} \int_{\Sigma_{ik}^{\text{int}}} \left\{ \frac{\beta}{\gamma}\left(\frac{\partial w_{ik}}{\partial \nu_{A_i}}\frac{\partial w_{ik}}{\partial t} - \frac{\partial q_{ik}}{\partial \nu_{A_i}}\frac{\partial q_{ik}}{\partial t} \right) + \left(\frac{\partial w_{ik}}{\partial \nu_{A_i}}\frac{\partial q_{ik}}{\partial t} + \frac{\partial q_{ik}}{\partial \nu_{A_i}}\frac{\partial w_{ik}}{\partial t} \right) \right\} d\Sigma. \tag{6.5.1.40}$$

By the same arguments used in the proof of Lemma 6.4.2.1 one obtains

$$\sum_{i=1}^{I}\sum_{k=0}^{K} \frac{\beta}{\gamma} \int_{\Sigma_{ik}^{\text{int}}} \left(\frac{\partial w_{ik}}{\partial \nu_{A_i}}\frac{\partial w_{ik}}{\partial t} - \frac{\partial q_{ik}}{\partial \nu_{A_i}}\frac{\partial q_{ik}}{\partial t} \right) d\Sigma$$
$$= \sum_{i=1}^{I} \frac{\beta}{\gamma} \Big\{ \frac{1}{2}(1-\kappa^2) \big\| \big(w_{iK}(T), \frac{\partial w_{iK}}{\partial t}(T)\big) \big\|^2_{\mathcal{H}_i} + \frac{1}{2} \big\| \big(q_{i0}(0), \frac{\partial q_{i0}}{\partial t}(0)\big) \big\|^2_{\mathcal{H}_i}$$
$$+ \sum_{k=0}^{K-1} \frac{1}{2\theta^2}(\theta^2 - 1) \big\| \big(w_{ik}(T_{k+1}), \frac{\partial w_{ik}}{\partial t}(T_{k+1})\big) \big\|^2_{\mathcal{H}_i}$$
$$+ \sum_{k=1}^{K} \frac{1}{2}(1-\theta^2) \big\| \big(q_{ik}(T_k), \frac{\partial q_{ik}}{\partial t}(T_k)\big) \big\|^2_{\mathcal{H}_i} \Big\}$$
$$+ \sum_{i=1}^{I}\sum_{k=0}^{K} \frac{\beta}{\gamma} \int_{\Sigma_i^N} \left(\alpha_i \left|\frac{\partial q_i}{\partial t}\right|^2 + \frac{\partial q_i}{\partial t}\frac{\partial w_i}{\partial t} + \alpha_i \left|\frac{\partial w_i}{\partial t}\right|^2 \right) d\Sigma, \tag{6.5.1.41}$$

$$\sum_{i=1}^{I}\sum_{k=0}^{K}\int_{\Sigma_{ik}^{\rm int}}\left(\frac{\partial w_{ik}}{\partial \nu_{A_i}}\frac{\partial q_{ik}}{\partial t}+\frac{\partial q_{ik}}{\partial \nu_{A_i}}\frac{\partial w_{ik}}{\partial t}\right)d\Sigma$$

$$=\sum_{i=1}^{I}\Big\{\kappa\Big\|\Big(w_{iK}(T),\frac{\partial w_{iK}}{\partial t}(T)\Big)\Big\|_{\mathcal{H}_i}^2+\sum_{k=0}^{K-1}\frac{1}{\theta}\Big\|\Big(w_{ik}(T_{k+1}),\frac{\partial w_{ik}}{\partial t}(T_{k+1})\Big)\Big\|_{\mathcal{H}_i}^2$$

$$+\sum_{k=1}^{K}\theta\Big\|\Big(q_{ik}(T_k),\frac{\partial q_{ik}}{\partial t}(T_k)\Big)\Big\|_{\mathcal{H}_i}^2\Big\}+\sum_{i=1}^{I}\sum_{k=0}^{K}\int_{\Sigma_i^N}\left|\frac{\partial q_i}{\partial t}\right|^2 d\Sigma. \quad (6.5.1.42)$$

Lemma 6.5.1.1 follows immediately from (6.5.1.39)–(6.5.1.42).

Proof of Theorem 6.5.1.1. Set

$$(\tilde\lambda_{ik}^{m,n-1},\rho_{ik}^{m,n-1}):=\{(\tilde\lambda_{ijk}^{m,n-1},\rho_{ijk}^{m,n-1}):\tilde\lambda_{ijk}^{m,n-1},\rho_{ijk}^{m,n-1})\in L^2(\Sigma_{ijk},\,j\in\mathcal{J}_i\},$$
$$X^{m,n}=\{(\tilde\lambda_{ik}^{m,n-1},\rho_{ik}^{m,n-1}):i=1,\dots,I;\ k=0,\dots,K\}\in\mathcal{X},$$

and let $w_{ik}^{m,n}$, $q_{ik}^{m,n}$ be the solution of (6.5.1.29)–(6.5.1.34) with n replaced by $n-1$. Then the interface conditions (6.5.1.32) may be expressed as the relaxed fixed point iteration in n

$$X^{m,n+1}=(1-\epsilon)\mathcal{T}X^{m,n}+\epsilon X^{m,n}. \quad (6.5.1.43)$$

Just as in Section 6.4.2, from (6.5.1.43) and Lemma 6.5.1.1 we obtain the recursion formula

$$\mathcal{E}^{m,n+1}+\mathcal{F}^{m,n+1}\le\mathcal{E}^{m,n}+(1-2\epsilon)\mathcal{F}^{m,n},$$

where $\mathcal{E}^{m,n}$, $\mathcal{F}^{m,n}$ are defined by (6.5.1.37), (6.5.1.38) with w_{ik}, q_{ik} replaced by the local errors $\tilde w_{ik}^{m,n}$, $\tilde q_{ik}^{m,n}$. Then, by iterating in the index n, we obtain

$$\mathcal{E}^{m,n+1}\le\mathcal{E}^{m,1}-\sum_{p=1}^{n+1}c_p(\epsilon)\mathcal{F}^{m,p},\quad m=1,2,\dots, \quad (6.5.1.44)$$

where the coefficients $c_p(\epsilon)$ are defined in (6.4.2.26). The remainder of the proof of Theorem 6.5.1.1 is based on (6.5.1.44) and is essentially identical to that of Theorem 6.4.2.1 and is therefore omitted.

6.5.2 Sequential Time-Space Domain Decomposition

In this approach we begin with the spatial domain decomposition algorithm of the global optimality system, as described in Section 6.4.1, which we write as

$$\begin{cases}\dfrac{\partial^2 w_i^n}{\partial t^2}-\nabla\cdot(A_i\nabla w_i^n)+c_iw_i^n=F_i\\[2mm] \dfrac{\partial^2 p_i^n}{\partial t^2}-\nabla\cdot(A_i\nabla p_i^n)+c_ip_i^n=0\text{ in }Q_i\end{cases} \quad (6.5.2.1)$$

$$w_i^n = p_i^n = 0 \text{ on } \Sigma_i^D \tag{6.5.2.2}$$

$$\begin{cases} \dfrac{\partial w_i^n}{\partial \nu_{A_i}} + \alpha_i \dfrac{\partial w_i^n}{\partial t} + p_i^n = 0 \\ \dfrac{\partial p_i^n}{\partial \nu_{A_i}} - \alpha_i d_t p_i^n = 0 \text{ on } \Sigma_i^N \end{cases} \tag{6.5.2.3}$$

$$\begin{cases} \dfrac{\partial w_i^n}{\partial \nu_{A_i}} + \beta \dfrac{\partial w_i^n}{\partial t} + \gamma p_i^n = \lambda_i^{n-1} \\ \dfrac{\partial p_i^n}{\partial \nu_{A_i}} - \beta d_t p_i^n + \gamma d_t \dfrac{\partial w_i^n}{\partial t} = \mu_i^{n-1} \text{ on } \Sigma_i^{\text{int}} \end{cases} \tag{6.5.2.4}$$

$$\begin{cases} w_i^n(0) = w_{0i}, \quad \dfrac{\partial w_i^n}{\partial t}(0) = v_{0i} \\ p_i^n(T) = \kappa(\dfrac{\partial w_i^n}{\partial t}(T) - z_{1i}) \\ \dfrac{\partial p_i^n}{\partial t}(T) = -\kappa \mathcal{A}_i(w_i^n(T) - z_{0i}) \text{ in } \Omega_i. \end{cases} \tag{6.5.2.5}$$

The localized systems (6.5.2.1)–(6.5.2.5) are, for each value of the iteration index n, optimality systems concentrated on the cylinders Q_i, $i = 1, \dots, I$. We then apply our time domain decomposition algorithm of Section 6.3.1 to each of these local optimality systems. We are thus led to consider the following collection of problems concentrated on the cylinders Q_{ik}:

$$\begin{cases} \dfrac{\partial^2 w_{ik}^{m+1,n}}{\partial t^2} - \nabla \cdot (A_i \nabla w_{ik}^{m+1,n}) + c_i w_{ik}^{m+1,n} = F_{ik} \\ \dfrac{\partial^2 p_{ik}^{m+1,n}}{\partial t^2} - \nabla \cdot (A_i \nabla p_{ik}^{m+1,n}) + c_i p_{ik}^{m+1,n} = 0 \text{ in } Q_{ik}; \end{cases} \tag{6.5.2.6}$$

$$w_{ik}^{m+1,n} = p_{ik}^{m+1,n} = 0 \text{ on } \Sigma_{ik}^D; \tag{6.5.2.7}$$

$$\begin{cases} \dfrac{\partial w_{ik}^{m+1,n}}{\partial \nu_{A_i}} + \alpha_i \dfrac{\partial w_{ik}^{m+1,n}}{\partial t} + p_{ik}^{m+1,n} = 0 \\ \dfrac{\partial p_{ik}^{m+1,n}}{\partial \nu_{A_i}} - \alpha_i d_t p_{ik}^{m+1,n} = 0 \text{ on } \Sigma_{ik}^N; \end{cases} \tag{6.5.2.8}$$

$$\begin{cases} \dfrac{\partial w_{ik}^{m+1,n}}{\partial \nu_{A_i}} + \beta \dfrac{\partial w_{ik}^{m+1,n}}{\partial t} + \gamma p_{ik}^{m+1,n} = \lambda_{ik}^{n-1} \\ \dfrac{\partial p_{ik}^{m+1,n}}{\partial \nu_{A_i}} - \beta d_t p_{ik}^{m+1,n} + \gamma d_t \dfrac{\partial w_{ik}^{m+1,n}}{\partial t} = \zeta_{ik}^{n-1} \text{ on } \Sigma_{ik}^{\text{int}}; \end{cases} \tag{6.5.2.9}$$

$$\begin{cases} w_{i0}^{m+1,n}(0) = w_{0i}, \quad \dfrac{\partial w_{i0}^{m+1,n}}{\partial t}(0) = v_{0i} \\ p_{iK}^{m+1,n}(T) = \kappa(\dfrac{\partial w_{iK}^{m+1,n}}{\partial t}(T) - z_{1i}) \\ \dfrac{\partial p_{iK}^{m+1,n}}{\partial t}(T) = -\kappa \mathcal{A}_i(w_{iK}^{m+1,n}(T) - z_{0i}) \text{ in } \Omega_i; \end{cases} \tag{6.5.2.10}$$

$$\begin{cases} w_{ik}^{m+1,n}(T_k) = \mathcal{A}_i^{-1}\big(\theta \dfrac{\partial p_{ik}^{m+1,n}}{\partial t}(T_k) + \eta_{i,k,k-1}^{m,n}\big) \\ \dfrac{\partial w_{ik}^{m+1,n}}{\partial t}(T_k) = -\theta p_{ik}^{m+1,n}(T_k) + \mu_{i,k,k-1}^{m,n} \\ p_{ik}^{m+1,n}(T_{k+1}) = \dfrac{1}{\theta}\big(\dfrac{\partial w_{ik}^{m+1,n}}{\partial t}(T_{k+1}) - \mu_{i,k,k+1}^{m,n}\big) \\ \dfrac{\partial p_{ik}^{m+1,n}}{\partial t}(T_{k+1}) = -\dfrac{1}{\theta}\mathcal{A}_i\big(w_{ik}^{m+1,n}(T_{k+1}) - \mathcal{A}_i^{-1}\eta_{i,k,k+1}^{m,n}\big) \quad \text{in } \Omega_i. \end{cases} \tag{6.5.2.11}$$

It is seen that, except for a notational change in indexing, the system (6.5.2.6)–(6.5.2.11) is identical to the system (6.5.1.6)–(6.5.1.11). However, in the present context the index $n \geq 1$ is fixed, λ_{ik}^{n-1}, μ_{ik}^{n-1} are taken as given data, and iteration is in the index m, where $\mu_{i,k,k\pm1}^{m,n}$, $\eta_{i,k,k\pm1}^{m,n}$ are given by

$$\begin{aligned} \mu_{i,k,k+1}^{m,n} &= (1-\epsilon)\big(\frac{\partial w_{i,k+1}^{m,n}}{\partial t}(T_{k+1}) - \theta\, p_{i,k+1}^{m,n}(T_{k+1})\big) \\ &\qquad + \epsilon\big(\frac{\partial w_{ik}^{m,n}}{\partial t}(T_{k+1}) - \theta p_{ik}^{m,n}(T_{k+1})\big), \\ \eta_{i,k,k+1}^{m,n} &= (1-\epsilon)\big(\theta \frac{\partial p_{k+1}^{m,n}}{\partial t}(T_{k+1}) + \mathcal{A}_i w_{k+1}^{m,n}(T_{k+1})\big) \\ &\qquad + \epsilon\big(\theta \frac{\partial p_{k}^{m,n}}{\partial t}(T_{k+1}) + \mathcal{A}_i w_{k}^{m,n}(T_{k+1})\big), \\ \mu_{i,k,k-1}^{m,n} &= (1-\epsilon)\big(\frac{\partial w_{i,k-1}^{m,n}}{\partial t}(T_{k}) + \theta p_{i,k-1}^{m,n}(T_{k})\big) \\ &\qquad + \epsilon\big(\frac{\partial w_{ik}^{m,n}}{\partial t}(T_{k}) + \theta p_{ik}^{m,n}(T_{k})\big), \\ \eta_{i,k,k-1}^{m,n} &= (1-\epsilon)\big(-\theta \frac{\partial p_{i,k-1}^{m,n}}{\partial t}(T_{k}) + \mathcal{A}_i w_{i,k-1}^{m,n}(T_{k})\big) \\ &\qquad + \epsilon\big(-\theta \frac{\partial p_{ik}^{m,n}}{\partial t}(T_{k}) + \mathcal{A}_i w_{ik}^{m,n}(T_{k})\big), \end{aligned} \tag{6.5.2.12}$$

where $\epsilon \in [0,1)$ is a relaxation parameter. The above are valid for $m \geq 1$; the quantities $\mu_{i,k,k\pm1}^{0,n}$, $\eta_{i,k,k\pm1}^{0,n}$ are arbitrary. The question to be investigated is the convergence of $w_{ik}^{m+1,n}$, $p_{ik}^{m+1,n}$ as $m \to \infty$.

As noted above, the system (6.5.2.6)–(6.5.2.11) is an optimality system concentrated on the cylinder Q_{ik}. It is well posed for fixed n and for $m = 0, 1, \ldots$ provided $\mu_{i,k,k\pm1}^{0,n} \in H_i$, $\eta_{i,k,k\pm1}^{0,n} \in V_i^*$, λ_{ik}^{n-1}, $\rho_{ik}^{n-1} \in L^2(\Sigma_{ik}^{\text{int}})$, and $\zeta_{ik}^{n-1} = d_t\rho_{ik}^{n-1}$.

Let w_{ik}^n, p_{ik}^n be the restriction to Q_{ik} of the solution w_i^n, p_i^n of (6.5.2.1)–(6.5.2.5). Then w_{ik}^n, q_{ik}^n satisfy (6.5.2.6)–(6.5.2.10) together with the *transmission conditions*

$$\begin{aligned} w_{ik}^n(T_{k+1}) &= w_{i,k+1}^n(T_{k+1}), \quad p_{ik}^n(T_{k+1}) = p_{i,k+1}^n(T_{k+1}), \\ \frac{\partial w_{ik}^n}{\partial t}(T_{k+1}) &= \frac{\partial w_{i,k+1}^n}{\partial t}(T_{k+1}), \quad \frac{\partial p_{ik}^n}{\partial t}(T_{k+1}) = \frac{\partial p_{i,k+1}^n}{\partial t}(T_{k+1}). \end{aligned} \tag{6.5.2.13}$$

Therefore the *local errors*

$$\tilde{w}_{ik}^{m,n} := w_{ik}^{m,n} - w_{ik}^n, \quad \tilde{p}_{ik}^{m,n} := p_{ik}^{m,n} - p_{ik}^n$$

satisfy (6.5.2.6)–(6.5.2.11) with homogeneous data

$$F_{ik} = 0, \quad w_{0i} = v_{0i} = z_{0i} = z_{1i} = 0, \quad \lambda_{ik}^{n-1} = \zeta_{ik}^{n-1} = 0 \tag{6.5.2.14}$$

and with $\mu_{i,k,k\pm1}^{m,n}$, $\eta_{i,k,k\pm1}^{m,n}$ replaced by

$$\begin{aligned} \tilde{\mu}_{i,k,k+1}^{m,n} &= (1-\epsilon)\Big(\frac{\partial \tilde{w}_{i,k+1}^{m,n}}{\partial t}(T_{k+1}) - \theta\, \tilde{p}_{i,k+1}^{m,n}(T_{k+1})\Big) \\ &\qquad + \epsilon\Big(\frac{\partial \tilde{w}_{ik}^{m,n}}{\partial t}(T_{k+1}) - \theta \tilde{p}_{ik}^{m,n}(T_{k+1})\Big), \\ \tilde{\eta}_{i,k,k+1}^{m,n} &= (1-\epsilon)\Big(\theta \frac{\partial \tilde{p}_{k+1}^{m,n}}{\partial t}(T_{k+1}) + \mathcal{A}_i \tilde{w}_{k+1}^{m,n}(T_{k+1})\Big) \\ &\qquad + \epsilon\Big(\theta \frac{\partial \tilde{p}_{k}^{m,n}}{\partial t}(T_{k+1}) + \mathcal{A}_i \tilde{w}_{k}^{m,n}(T_{k+1})\Big), \\ \tilde{\mu}_{i,k,k-1}^{m,n} &= (1-\epsilon)\Big(\frac{\partial \tilde{w}_{i,k-1}^{m,n}}{\partial t}(T_k) + \theta \tilde{p}_{i,k-1}^{m,n}(T_k)\Big) \\ &\qquad + \epsilon\Big(\frac{\partial \tilde{w}_{ik}^{m,n}}{\partial t}(T_k) + \theta \tilde{p}_{ik}^{m,n}(T_k)\Big), \\ \tilde{\eta}_{i,k,k-1}^{m,n} &= (1-\epsilon)\Big(-\theta \frac{\partial \tilde{p}_{i,k-1}^{m,n}}{\partial t}(T_k) + \mathcal{A}_i \tilde{w}_{i,k-1}^{m,n}(T_k)\Big) \\ &\qquad + \epsilon\Big(-\theta \frac{\partial \tilde{p}_{ik}^{m,n}}{\partial t}(T_k) + \mathcal{A}_i \tilde{w}_{ik}^{m,n}(T_k)\Big), \end{aligned} \tag{6.5.2.15}$$

It is worth noting that the system satisfied by the local errors $\tilde{w}_{ik}^{m+1,n}$, $\tilde{p}_{ik}^{m+1,n}$ has the same structure as the system (6.3.2.2)–(6.3.2.7) for the local errors in the pure time domain decomposition algorithm, save for the additional term $\gamma d_t\big(\partial \tilde{w}_{ik}^{m+1,n}/\partial t\big)$ appearing in the boundary condition on Σ_{ik}^{int}. Therefore, it is only a matter of checking that this additional term does not deleteriously affect the convergence to zero of the local errors. The following result shows that this is the case.

Theorem 6.5.2.1 *Let β, γ, θ be positive constants. Then as $m \to \infty$ we have, for $i = 1, \ldots, I$ and $n = 1, 2, \ldots$:*

(i) *for any* $\epsilon \in [0,1)$

$$\tilde{p}_{ik}^{m,n}\big|_{\Sigma_{ik}^N \cup \Sigma_{ik}^{int}} \to 0 \ \textit{strongly in } L^2(\Sigma_{ik}^N \cup \Sigma_{ik}^{int}),\ k = 0,\dots,K$$

$$\frac{\partial \tilde{w}_{ik}^{m,n}}{\partial t}\bigg|_{\Sigma_{ik}^{int}} \to 0 \ \textit{strongly in } L^2(\Sigma_{ik}^{int}),\ k = 0,\dots,K$$

$$(\tilde{p}_{iK}^{m,n}, \frac{\partial \tilde{p}_{iK}^{m,n}}{\partial t}) \to 0 \ \textit{in } C(I_K; H_i \times V_i^*)$$

$$\big(\tilde{w}_{ik}^{m,n}(T), \frac{\partial \tilde{w}_{ik}^{m,n}}{\partial t}(T)\big) \to 0 \ \textit{strongly in } V_i \times H_i$$

$$(\tilde{w}_{i0}^{m,n}, \frac{\partial \tilde{w}_{i0}^{m,n}}{\partial t}) \to 0 \ \textit{in } C(I_0; V_i \times H_i).$$

(ii) *In addition, if* $\epsilon > 0$ *then for* $k = 0,\dots,K$

$$(\tilde{w}_{ik}^{m,n}, \frac{\partial \tilde{w}_{ik}^{m,n}}{\partial t}) \to 0 \ \textit{in } C(I_k; V_i \times H_i)$$

$$(\tilde{p}_{ik}^{m,n}, \frac{\partial \tilde{p}_{ik}^{m,n}}{\partial t}) \to 0 \ \textit{in } C(I_k; H_i \times V_i^*)$$

$$\frac{\partial \tilde{w}_{ik}^{m,n}}{\partial t}\bigg|_{\Sigma_{ik}^N} \to 0 \ \textit{strongly in } L^2(\Sigma_{ik}^N).$$

Theorem 6.5.2.1 is deduced from properties of the mapping $\mathcal{T}_\epsilon$ associated with the relaxed iteration step (6.5.2.11), (6.5.2.12). Thus, proceeding along the lines of Section 6.3.2, we set

$$\mathcal{X} = \prod_{i=1}^{I} (H_i \times V_i^*)^{2K}$$

with the standard product norm. Let

$$X = \{(\mu_{i,k,k+1}, \eta_{i,k,k+1})_{k=0}^{K-1}, (\mu_{i,k,k-1}, \eta_{i,k,k-1})_{k=1}^{K}\} \in \mathcal{X} \tag{6.5.2.16}$$

and let w_{ik}, p_{ik} be the solution of (6.5.2.6)–(6.5.2.11) with homogeneous data (6.5.2.14) and all superscripts removed. Define $\mathcal{T} : \mathcal{X} \mapsto \mathcal{X}$ by setting

$$(\mathcal{T}X)_i = \{ (\frac{\partial w_{i,k+1}}{\partial t}(T_{k+1}) - \theta p_{i,k+1}(T_{k+1}), \theta \frac{\partial p_{i,k+1}}{\partial t}(T_{k+1}) + \mathcal{A}_i w_{i,k+1}(T_{k+1}))_{k=0}^{K=1},$$
$$(\frac{\partial w_{i,k-1}}{\partial t}(T_k) + \theta p_{i,k-1}(T_k), -\theta \frac{\partial p_{i,k-1}}{\partial t}(T_k) + \mathcal{A}_i w_{i,k-1}(T_k))_{k=1}^{K} \} \in \mathcal{X},$$

$$\mathcal{T}X = ((\mathcal{T}X)_1, \dots, (\mathcal{T}X)_I).$$

If we then set

$$X^{m,n} = \{(\tilde{\mu}_{i,k,k+1}^{m-1,n}, \tilde{\eta}_{i,k,k+1}^{m-1,n})_{k=0}^{K-1}, (\tilde{\mu}_{i,k,k-1}^{m-1,n}, \tilde{\eta}_{i,k,k-1}^{m-1,n})_{k=1}^{K}\} \tag{6.5.2.17}$$

and let $\tilde{w}_{ik}^{m,n}$, $\tilde{p}_{ik}^{m,n}$ be the solution of (6.5.2.6)–(6.5.2.11) corresponding to $X^{m,n}$ and homogeneous data (6.5.2.14), the relaxed iteration step for the local errors is given by the relaxed fixed point iteration

$$X^{m+1,n} = (1-\epsilon)\mathcal{T}X^{m,n} + \epsilon X^{m,n} := \mathcal{T}_\epsilon X^{m,n}.$$

For $X \in \mathcal{X}$ given by (6.5.2.16) and w_{ik}, p_{ik} the solution of (6.5.2.6)–(6.5.2.11), (6.5.2.14) corresponding to X, let us set

$$\mathcal{E}_{ik}(t) := \big\|\big(w_{ik}(t), \frac{\partial w_{ik}}{\partial t}(t)\big)\big\|^2_{V_i\times H_i} + \theta^2\big\|\big(p_{ik}(t), \frac{\partial p_{ik}}{\partial t}(t)\big)\big\|^2_{H_i\times V_i^*}, \tag{6.5.2.18}$$

$$\mathcal{E} = \sum_{i=1}^{I}\sum_{k=0}^{K-1}[\mathcal{E}_{ik}(T_{k+1}) + \mathcal{E}_{i,k+1}(T_{k+1})], \tag{6.5.2.19}$$

$$\begin{aligned}\mathcal{F} = 2\theta\sum_{i=1}^{I}\sum_{k=0}^{K}\Big\{\int_{\Sigma_{ik}^N}|p_{ik}|^2 d\Sigma + \int_{\Sigma_{ik}^{\text{int}}}\gamma\big(|p_{ik}|^2 + \big|\frac{\partial w_{ik}}{\partial t}\big|^2\big)\,d\Sigma\Big\}\\ + 2\theta\kappa\sum_{i=1}^{I}\big\|\big(w_{iK}(T), \frac{\partial w_{iK}}{\partial t}(T)\big)\big\|^2.\end{aligned} \tag{6.5.2.20}$$

Lemma 6.5.2.1 *For any $X \in \mathcal{X}$ We may write*

$$\|X\|^2_{\mathcal{X}} = \mathcal{E} + \mathcal{F}, \quad \|\mathcal{T}X\|^2_{\mathcal{X}} = \mathcal{E} - \mathcal{F},$$

where $\mathcal{E}$, $\mathcal{F}$ are defined by (6.5.2.18)–(6.5.2.20).

Thus $\mathcal{T}$ is nonexpansive provided $\gamma > 0$ and $\theta > 0$.

Proof of Lemma 6.5.2.1. The computation follows the lines of the proof of Lemma 6.3.2.1. One has

$$\begin{aligned}\|X\|^2_{\mathcal{X}} =& \sum_{i=1}^{I}\Big\{\sum_{k=0}^{K-1}\Big[\big\|-\theta p_{ik}(T_{k+1}) + \frac{\partial w_{ik}}{\partial t}(T_{k+1})\big\|^2_{H_i}\\ &+ \big\|\theta\frac{\partial p_{ik}}{\partial t}(T_{k+1}) + A_i w_{ik}(T_{k+1})\big\|^2_{V_i^*}\Big]\\ &+ \sum_{k=1}^{K}\Big[\big\|\theta p_{ik}(T_k) + \frac{\partial w_{ik}}{\partial t}(T_k)\big\|^2_{H_i} + \big\|-\theta\frac{\partial p_{ik}}{\partial t}(T_k) + A_i w_{ik}(T_k)\big\|^2_{V_i^*}\Big]\Big\}\\ =& \sum_{i=1}^{I}\Big\{\sum_{k=1}^{K}[\mathcal{E}_{ik}(T_{k+1}) + \mathcal{E}_{i,k+1}(T_{k+1})]\\ &+ 2\theta\sum_{k=0}^{K}\Big[\big\langle\frac{\partial p_{ik}}{\partial t}(t), w_{ik}(t)\big\rangle_{V_i} - \big(p_{ik}(t), \frac{\partial w_{ik}}{\partial t}(t)\big)_{H_i}\Big]_{t=T_k}^{t=T_{k+1}}\\ &+ 2\theta\kappa\big\|\big(w_{iK}(T), \frac{\partial w_{iK}}{\partial t}(T)\big)\big\|^2_{V_i\times H_i}\Big\}.\end{aligned} \tag{6.5.2.21}$$

From the definition of the solution p_{ik} of the adjoint system we have (cf. (6.2.2.29); this is also obtainable from a formal calculation as in Lemma 6.3.2.1)

$$\Big[\big\langle \frac{\partial p_{ik}}{\partial t}(t), w_{ik}(t)\big\rangle_{V_i} - \big(p_{ik}(t), \frac{\partial w_{ik}}{\partial t}(t)\big)_{H_i}\Big]_{t=T_k}^{t=T_{k+1}} \\ = \int_{\Sigma_{ik}^N} |p_{ik}|^2 d\Sigma + \int_{\Sigma_{ik}^{\mathrm{int}}} \gamma\big(|p_{ik}|^2 + |\frac{\partial w_{ik}}{\partial t}|^2\big) d\Sigma. \quad (6.5.2.22)$$

It follows from (6.5.2.21) and (6.5.2.22) that $\|X\|_{\mathcal{X}}^2 = \mathcal{E} + \mathcal{F}$. By a similar calculation we obtain

$$\begin{aligned} \|\mathcal{T}X\|_{\mathcal{X}}^2 =& \sum_{i=1}^{I}\Big\{\sum_{k=0}^{K-1}\Big[\big\|\theta p_{ik}(T_{k+1}) + \frac{\partial w_{ik}}{\partial t}(T_{k+1})\big\|_{H_i}^2 \\ &+ \big\|-\theta\frac{\partial p_{ik}}{\partial t}(T_{k+1}) + \mathcal{A}_i w_{ik}(T_{k+1})\big\|_{V_i^*}^2\Big] \\ &+ \sum_{k=1}^{K}\Big[\big\|-\theta p_{ik}(T_k) + \frac{\partial w_{ik}}{\partial t}(T_k)\big\|_{H_i}^2 + \big\|\theta\frac{\partial p_{ik}}{\partial t}(T_k) + \mathcal{A}_i w_{ik}(T_k)\big\|_{V_i^*}^2\Big]\Big\} \\ =& \sum_{i=1}^{I}\Big\{\sum_{k=1}^{K}[\mathcal{E}_{ik}(T_{k+1}) + \mathcal{E}_{i,k+1}(T_{k+1})] \\ &- 2\theta\sum_{k=0}^{K}\Big[\big\langle\frac{\partial p_{ik}}{\partial t}(t), w_{ik}(t)\big\rangle_{V_i} - \big(p_{ik}(t), \frac{\partial w_{ik}}{\partial t}(t)\big)_{H_i}\Big]_{t=T_k}^{t=T_{k+1}} \\ &- 2\theta\kappa\big\|\big(w_{iK}(T), \frac{\partial w_{iK}}{\partial t}(T)\big)\big\|_{V_i\times H_i}^2\Big\}. \end{aligned} \quad (6.5.2.23)$$

It follows from (6.5.2.22) and (6.5.2.23) that $\|\mathcal{T}X\|_{\mathcal{X}}^2 = \mathcal{E} - \mathcal{F}$. As a consequence of Lemma 6.5.2.1 we obtain the fundamental recursion formula

$$\mathcal{E}^{m+1,n} + \mathcal{F}^{m+1,n} \le \mathcal{E}^{m,n} + (1-2\epsilon)\mathcal{F}^{m,n} \quad (6.5.2.24)$$

where $\mathcal{E}^{m,n}$, $\mathcal{F}^{m,n}$ are defined by (6.5.2.18)–(6.5.2.20) with w_{ik}, p_{ik} replaced by the local errors $\tilde{w}_{ik}^{m,n}$, $\tilde{p}_{ik}^{m,n}$. By iterating (6.5.2.24) in the index m we obtain

$$\mathcal{E}^{m+1,n} \le \mathcal{E}^{1,n} - \sum_{p=0}^{m+1} c_p(\epsilon)\mathcal{F}^{p,n}, \quad n = 1, 2, \ldots, \quad (6.5.2.25)$$

where the coefficients $c_p(\epsilon)$ are given by (6.4.2.26). The proof of Theorem 6.5.2.1 is based on the (6.5.2.25) and is essentially identical to that of Theorem 6.3.2.1 and is therefore omitted.

Remark 6.5.2.1 One may derive a posteriori estimates for the local errors both in the sequential space-time and the sequential time-space domain decomposition algorithms, of a form similar to those obtained in Sections 6.4.3 and 6.3.3, respectively, by means of the arguments put forth therein.

Chapter 7

Domain Decomposition in Optimal Final Value Boundary Control of Maxwell's System

7.1 Introduction

This chapter is concerned with domain decomposition in optimal final value control of the heterogeneous Maxwell system

$$\begin{aligned} &\varepsilon E' - \operatorname{rot} H + \sigma E = F, \\ &\mu H' + \operatorname{rot} E = G \quad \text{in } Q := \Omega \times (0,T) \\ &H_\tau - \alpha(\nu \wedge E) = J \text{ on } \Sigma := \Gamma \times (0,T) \\ &E(0) = E_0, \quad H(0) = H_0 \text{ in } \Omega \end{aligned} \tag{7.1.1.1}$$

where $' = \partial/\partial t$, $\wedge$ denotes the vector product operation, ν denotes the exterior pointing unit normal vector to Γ and H_τ is the tangential component of H, that is,

$$H_\tau = H - (H \cdot \nu)\nu = \nu \wedge (H \wedge \nu).$$

E and H represent, respectively, the temporally dependent electric and magnetic fields in the bounded region Ω. The electric permittivity ε and magnetic permeability μ are positive definite 3×3 Hermitian matrices, and the resistivity σ is a 3×3 nonnegative Hermitian matrix, all defined on a bounded region $\Omega \subset \mathbb{R}^3$, α is a nonnegative scalar function defined on Γ, and F, G, J are given vector-valued inputs defined on $\Omega \times (0,T)$ and $\Gamma \times (0,T)$ respectively, of which J is taken as the control variable. In (7.1.1.1) the term σE represents internal damping while $\alpha(\nu \wedge E)$ represents boundary damping.

When $J = 0$ and $\alpha = 1$, the boundary condition $(7.1.1.1)_3$ is known as the Silver-Müller boundary condition. It is the first approximation to the so-called transparent boundary condition, which corresponds to the transmission of electromagnetic waves through the boundary without reflections. In general, when $\alpha > 0$ on a set on positive measure in Γ, the boundary condition $(7.1.1.1)_3$ is *dissipative.* That is, with the electromagnetic energy defined by

$$\mathcal{E}(t) = \int_\Omega (\varepsilon E \cdot E + \mu H \cdot H)\, dx,$$

in the absence of external inputs F, G, J, the functional $\mathcal{E}(t)$ is nonincreasing. Moreover, the boundary condition $(7.1.1.1)_3$ is also *regularizing*. That is to say, if $\mathcal{E}(0) < \infty$, if F, G and J have regularity assumed above, and if the support of J is contained in the support of α, the solution of (7.1.1.1) satisfies $\mathcal{E}(t) < \infty$ for all $t \geq 0$. On the other hand, if $\alpha \equiv 0$ and $J \not\equiv 0$ then, in general, solutions of (7.1.1.1) will have less regularity. Thus the term $\alpha \nu \wedge (H \wedge \nu)$ in (7.1.1.1 plays the same role in the system (7.1.1.1) as does the term $\alpha \partial w / \partial t$ in (6.1.1.1, namely, it improves the regularity of the solution relative to that which holds when $\alpha = 0$ (when $J \neq 0$). We may therefore consider optimal final value control problems in which, in the cost functional, penalization of the deviation of the final state from its target state is taken in the natural energy norm for the problem. This problem has a unique solution J_{opt} that is given in terms of the adjoint variable in the optimality system. In this chapter we shall present iterative domain decomposition algorithms for this system that, as shall be seen, parallel those for the optimality system associated with optimal control of the system (6.1.1.1). Once again, we shall establish convergence of the algorithms and derive a posteriori error estimates in terms of the mismatch of the iterates at the interfaces of the subdomains.

7.2 Optimal Dissipative Boundary Control

7.2.1 Setting the Problem

Let Ω be a bounded, open, connected set in $\mathbb{R}^3$ with piecewise smooth, Lipschitz boundary Γ, and let $T > 0$. We consider the Maxwell system (7.1.1.1) in which $\alpha \in L^\infty(\Gamma)$ such that $\alpha_0 := \inf_\Gamma \alpha(x) > 0$. Further, $\varepsilon = (\varepsilon^{jk}(x))$, $\mu = (\mu^{jk}(x))$ and $\sigma = (\sigma^{jk}(x))$ are 3×3 Hermitian matrices with $L^\infty(\Omega)$ entries such that ε and μ are uniformly positive definite and $\sigma \geq 0$ in Ω. The functions F, $G \in L^1(0,T;\mathcal{L}^2(\Omega))$ are given while J is a control input and is taken from the class

$$\mathcal{U} = \mathcal{L}^2_\tau(\Sigma) := \{J | \, J \in L^2(0,T;\mathcal{L}^2(\Gamma)), \, \nu \cdot J(t) = 0 \quad \text{for a.a. } x \in \Gamma \text{ and a.a. } t \in (0,T)\}$$

Function spaces of $\mathcal{C}$-valued functions are denoted by capital Roman letters, while function spaces of $\mathcal{C}^3$-valued functions are denoted by capital script letters. We use $\alpha \cdot \beta$ to denote the natural scalar product in $\mathcal{C}^3$, i.e., $\alpha \cdot \beta = \sum_{j=1}^3 \alpha_j \overline{\beta}_j$, and write $\langle \cdot, \cdot \rangle$ for the natural scalar product in various function spaces such as $L^2(\Omega)$ and $\mathcal{L}^2(\Omega)$. A subscript may sometimes be added to avoid confusion. The spaces $L^2(\Omega)$ and $\mathcal{L}^2(\Omega)$ denote the usual spaces of Lebesgue square integrable $\mathcal{C}$-valued functions and $\mathcal{C}^3$-valued functions, respectively, and, similarly, for the Sobolev space $H^s(\Omega)$, $\mathcal{H}^s(\Omega)$.

We set $\mathcal{H} = \mathcal{L}^2(\Omega) \times \mathcal{L}^2(\Omega)$ with weight matrix $M = \text{diag}(\varepsilon, \mu)$. Thus

$$\|(\phi, \psi)\|^2_{\mathcal{H}} = \langle \varepsilon \phi, \phi \rangle + \langle \mu \psi, \psi \rangle$$

It will be proved below that for $J \in \mathcal{U}$ and $(E_0, H_0) \in \mathcal{H}$, (7.1.1.1) has a unique solution $(E, H) \in C([0,T]; \mathcal{H})$ such that $\nu \wedge E|_\Sigma \in \mathcal{L}^2_\tau(\Sigma)$. We shall consider the optimal control problem

$$\inf_{J \in \mathcal{U}} \int_\Sigma |J|^2 d\Sigma + \kappa \|(E(T), H(T)) - (E_T, H_T)\|^2_{\mathcal{H}}, \quad \kappa > 0, \tag{7.2.1.1}$$

subject to (7.1.1.1), where $(E_T, H_T) \in \mathcal{H}$ is given. Problem (7.1.1.1), (7.2.1.1) admits a unique optimal control J_{opt} given by

$$J_{\mathrm{opt}} = \nu \wedge P|_\Sigma, \tag{7.2.1.2}$$

where (P, Q) is the solution of the backwards running adjoint system

$$\begin{aligned} &\begin{cases} \varepsilon P' - \operatorname{rot} Q - \sigma P = 0 \\ \mu Q' + \operatorname{rot} P = 0 & \text{in } \mathcal{Q} \end{cases} \\ &\quad Q_\tau + \alpha(\nu \wedge P) = 0 \text{ on } \Sigma \\ &\begin{cases} P(T) = \kappa(E(T) - E_T) \\ Q(T) = \kappa(H(T) - H_T) & \text{in } \Omega \end{cases} \end{aligned} \tag{7.2.1.3}$$

(see Section 7.2.3). The *optimality system* for the optimal control problem (7.2.1.1) consists of (7.1.1.1), (7.2.1.3), (7.2.1.2). The main purpose of this chapter is to present and analyze certain domain decomposition algorithms for the numerical calculation of this system.

Remark 7.2.1.1 Instead of $\mathcal{L}^2_\tau(\Sigma)$ one may choose as the control space those $\mathcal{L}^2_\tau(\Sigma)$ functions that are supported in $\widetilde{\Gamma} \times (0,T)$, where $\widetilde{\Gamma}$ is a subset of Γ of positive 2-dimensional measure. The analysis presented below may easily be adapted to this more general setting with only minor modifications.

Remark 7.2.1.2 Instead of controlling the system (7.1.1.1) through a tangentially flowing electric current J, one may treat J as a given input to the system and exercise control through the distributed electromagnetic force F or, more generally, through a locally distributed force $F\mathbf{1}_\omega$, where ω is a nonempty open subset of Ω and $\mathbf{1}_\omega$ is the characteristic function of ω. If the control space is taken to be $\mathcal{L}^2(\omega \times (0,T))$ and if, in the cost functional, F is penalized in the natural norm in this space, then the optimal control is given by

$$F = P\big|_{\omega \times (0,T)} \tag{7.2.1.4}$$

where (P, Q) is the solution of (7.2.1.3), so that the optimality system in this case consists of (7.1.1.1), (7.2.1.3) and (7.2.1.4). The domain decomposition algorithms, and their analysis, that are presented in this chapter may be adapted to this optimality system with only minor changes one of which, in the case decomposition of the spatial domain, is that it is necessary to require that $\sigma|_\omega$ be uniformly positive definite (cf. Section 6.3.4). esp. Example 6.3.4.1.

7.2.2 Existence and Uniqueness of Solution

We set

$$V = \{\phi \in \mathcal{L}^2(\Omega) : \operatorname{rot}\phi \in \mathcal{L}^2(\Omega),\ \nu\wedge\phi \in \mathcal{L}^2_\tau(\Gamma)\},$$
$$\|\phi\|_V^2 = \int_\Omega (|\phi|^2 + |\operatorname{rot}\phi|^2)dx + \int_\Gamma |\nu\wedge\phi|^2 d\Gamma.$$

It is proved in [5] that $\mathcal{H}^1(\Omega)$ is dense in V. Therefore the classical Green's formula

$$\begin{aligned}\langle \operatorname{rot}\phi, \psi\rangle &= \langle \phi, \operatorname{rot}\psi\rangle + \int_\Gamma (\nu\wedge\phi)\cdot\psi_\tau\, d\Gamma \\ &= \langle \phi, \operatorname{rot}\psi\rangle - \int_\Gamma \phi_\tau\cdot(\nu\wedge\psi)d\Gamma\end{aligned} \tag{7.2.2.1}$$

is valid for all functions in $V\times V$.

Let us consider the problem

$$\begin{cases}\varepsilon\phi' - \operatorname{rot}\psi + \sigma\phi = \varepsilon f \\ \mu\psi' + \operatorname{rot}\phi = \mu g \qquad \text{in } Q\end{cases}$$
$$\psi_\tau - \alpha(\nu\wedge\phi) = 0 \ \text{ on } \Sigma \tag{7.2.2.2}$$
$$\phi(0) = \phi_0, \quad \psi(0) = \psi_0 \text{ in } \Omega$$

Set

$$A = M^{-1}\begin{pmatrix} -\sigma & \operatorname{rot} \\ -\operatorname{rot} & 0\end{pmatrix},$$
$$\operatorname{Dom}(A) = \{(\phi,\psi)\in V\times V : \psi_\tau - \alpha(\nu\wedge\phi) = 0 \text{ on } \Gamma\}.$$

The system (7.2.2.2) may be formally written

$$\begin{aligned} U' = AU + F, \ \text{where } U = \begin{pmatrix}\phi\\ \psi\end{pmatrix}, \ F = \begin{pmatrix} f \\ g\end{pmatrix} \\ U(0) = U_0 := \begin{pmatrix}\phi_0 \\ \psi_0\end{pmatrix}\end{aligned} \tag{7.2.2.3}$$

Lemma 7.2.2.1 *A is an m-dissipative operator in $\mathcal{H}$.*

Proof. A is densely defined and from the Green's formula (7.2.2.1) we obtain

$$\langle AU, U\rangle_{\mathcal{H}} = -\langle\sigma\phi,\phi\rangle - \int_\Gamma \alpha|\nu\wedge\phi|^2 d\Gamma + 2\sqrt{-1}\operatorname{Im}\langle\psi, \operatorname{rot}\phi\rangle, \quad U\in D(A),$$

so $\operatorname{Re}\langle AU, U\rangle_{\mathcal{H}} \le 0$.

Let $(f,g) \in \mathcal{H}$ and consider the variational equation

$$\langle(\varepsilon+\sigma)\phi,\chi\rangle + \langle\mu^{-1}\operatorname{rot}\phi,\operatorname{rot}\chi\rangle + \int_\Gamma \alpha(\nu\wedge\phi)\cdot(\nu\wedge\chi)\,d\Gamma$$
$$= \langle g,\operatorname{rot}\chi\rangle + \langle\varepsilon f,\chi\rangle, \quad \forall\chi\in V. \quad (7.2.2.4)$$

The left side of (7.2.2.4) defines a strictly coercive bilinear form on $V\times V$ and the right side defines a continuous linear functional on V. Thus (7.2.2.4) has a unique solution $\phi\in V$. Set $\psi = g-\mu^{-1}\operatorname{rot}\phi \in \mathcal{L}^2(\Omega)$. Then (7.2.2.4) reads

$$\langle\psi,\operatorname{rot}\chi\rangle = \langle(\varepsilon+\sigma)\phi,\chi\rangle - \langle\varepsilon f,\chi\rangle + \int_\Gamma \alpha(\nu\wedge\phi)\cdot(\nu\wedge\chi)\,d\Gamma, \quad \forall\chi\in V.$$

It follows that $\operatorname{rot}\psi\in\mathcal{L}^2(\Omega)$ and that

$$(\varepsilon+\sigma)\phi - \operatorname{rot}\psi = \varepsilon f \text{ in } \Omega, \quad \psi_\tau = \alpha(\nu\wedge\phi) \text{ on } \Gamma.$$

Therefore $\begin{pmatrix}\phi\\ \psi\end{pmatrix}\in D(A)$ and $(I-A)\begin{pmatrix}\phi\\ \psi\end{pmatrix} = \begin{pmatrix}f\\ g\end{pmatrix}$. This completes the proof.

The following result follows from Lemma 7.2.2.1 and standard semigroup theory.

Corollary 7.2.2.1 (1) *If $(\phi_0,\psi_0)\in\mathcal{H}$ and $(f,g)\in L^1(0,T;\mathcal{H})$, then (7.2.2.2) has a unique mild solution $(\phi,\psi)\in C([0,T];\mathcal{H})$ and*

$$\|(\phi(t),\psi(t))\|_{L^\infty(0,T;\mathcal{H})} \le C\big(\|(\phi_0,\psi_0)\|_{\mathcal{H}} + \|(f,g)\|_{L^1(0,T;\mathcal{H})}\big).$$

(2) *If $(\phi_0,\psi_0)\in D(A)$ and $(f,g)\in C^1([0,T];\mathcal{H})$, then $(\phi,\psi)\in C([0,T];D(A))$.*

Lemma 7.2.2.2 *Let $(\phi_0,\psi_0)\in\mathcal{H}$ and $(f,g)\in L^1(0,T;\mathcal{H})$. Then the solution of (7.2.2.2) satisfies $\nu\wedge\phi|_\Sigma\in\mathcal{L}^2_\tau(\Sigma)$. Moreover, the mapping $((\phi_0,\psi_0),(f,g))\mapsto \nu\wedge\phi|_\Sigma$ is continuous from $\mathcal{H}\times L^1(0,T;\mathcal{H})$ into $\mathcal{L}^2_\tau(\Sigma)$.*

Proof. First suppose that $(\phi_0,\psi_0)\in D(A)$ and $(f,g)\in C^1([0,T];\mathcal{H})$. From

$$\int_0^t ((f,g),(\phi,\psi))_{\mathcal{H}}dt = \int_0^t [\langle\varepsilon\phi',\phi\rangle + \langle\mu\psi',\psi\rangle - \langle\operatorname{rot}\psi,\phi\rangle + \langle\operatorname{rot}\phi,\psi\rangle + \langle\sigma\phi,\phi\rangle]\,dt$$

we have

$$\frac12\|(\phi(t),\psi(t))\|^2_{\mathcal{H}} - \frac12\|(\phi_0,\psi_0)\|^2_{\mathcal{H}} + \int_0^t\langle\sigma\phi,\phi\rangle\,dt + \int_0^t\int_\Gamma \alpha|\nu\wedge\phi|^2 d\Gamma dt$$
$$-2\sqrt{-1}\operatorname{Im}\int_0^t\langle\psi,\operatorname{rot}\phi\rangle\,dt = \int_0^t ((f,g),(\phi,\psi))_{\mathcal{H}}dt.$$

Take the real part of the last equation and use

$$\left|\int_0^t \langle (f,g),(\phi,\psi)\rangle_{\mathcal{H}} dt\right| \le \epsilon\|(\phi,\psi)\|^2_{L^\infty(0,T;\mathcal{H})} + C_\epsilon\|(f,g)\|^2_{L^1(0,T;\mathcal{H})}, \quad \forall\epsilon > 0,$$

to obtain

$$\|(\phi,\psi)\|^2_{L^\infty(0,T;\mathcal{H})} + \int_0^T \langle\sigma\phi,\phi\rangle\, dt + \int_0^T\int_\Gamma \alpha|\nu\wedge\phi|^2 d\Gamma dt$$
$$\le C\left\{\|(\phi_0,\psi_0)\|^2_{\mathcal{H}} + \|(f,g)\|^2_{L^1(0,T;\mathcal{H})}\right\}.$$

The result now follows by density.

We now turn to the nonhomogeneous boundary value problem (7.1.1.1). Solutions of (7.1.1.1) are defined by the transposition method. Indeed, a formal calculation using Green's formula (7.2.2.1) shows that if (E,H) satisfies (7.1.1.1) then (E,H) satisfies the variational equation

$$\langle (E(T),H(T)),(\phi_0,\psi_0)\rangle_{\mathcal{H}} - \int_0^T \langle (E(t),H(t)),(f(t),g(t))\rangle_{\mathcal{H}} dx dt$$
$$= \langle (E_0,H_0),(\phi(0),\psi(0))\rangle_{\mathcal{H}} + \int_0^T \langle (F(t),G(t)),(\phi(t),\psi(t))\rangle dt - \int_\Sigma J\cdot(\nu\wedge\phi)\, d\Sigma,$$
$$\forall(\phi_0,\psi_0)\in\mathcal{H},\ (f,g)\in L^1(0,T;\mathcal{H}), \qquad (7.2.2.5)$$

where (ϕ,ψ) is the solution of

$$\begin{cases} \varepsilon\phi' - \operatorname{rot}\psi - \sigma\phi = \varepsilon f \\ \mu\psi' + \operatorname{rot}\psi = \mu g \qquad \text{in } Q \end{cases}$$
$$\psi_\tau + \alpha(\nu\wedge\phi) = 0 \text{ on } \Sigma \qquad (7.2.2.6)$$
$$\phi(T) = \phi_0, \quad \psi(T) = \psi_0 \text{ in } \Omega.$$

A solution of (7.1.1.1) is, *by definition*, a solution of (7.2.2.5). We have the following result.

Theorem 7.2.2.1 *Suppose that $(E_0,H_0)\in\mathcal{H}$, $(F,G)\in L^1(0,T;\mathcal{H})$ and $J\in\mathcal{L}^2_\tau(\Sigma)$. Then (7.1.1.1) has a unique solution $(E,H)\in C([0,T];\mathcal{H})$. Moreover $\nu\wedge E|_\Sigma\in\mathcal{L}^2_\tau(\Sigma)$ and the linear map from the data to $((E,H),\nu\wedge E|_\Sigma)$ is continuous in the indicated topologies.*

Proof. Since the mapping $(\phi_0,\psi_0),(f,g))\mapsto((\phi,\psi),\nu\wedge\phi|_\Sigma)$ is continuous from $\mathcal{H}\times L^1(0,T;\mathcal{H})$ into $C([0,T];\mathcal{H})\times\mathcal{L}^2_\tau(\Sigma)$, the right side of (7.2.2.5) defines a continuous linear functional on $\mathcal{H}\times L^1(0,T;\mathcal{H})$. Therefore, there exists a *unique* pair (E_T, H_T), $(E, H) \in L^\infty(0,T;\mathcal{H})$ satisfying

$$\langle (E_T, H_T), (\phi_0, \psi_0)\rangle_{\mathcal{H}} - \int_0^T \langle (E(t), H(t)), (f(t), g(t))\rangle_{\mathcal{H}} dx dt$$

$$= \langle (E_0, H_0), (\phi(0), \psi(0))\rangle_{\mathcal{H}} + \int_0^T \langle (F(t), G(t)), (\phi(t), \psi(t))\rangle dt - \int_\Sigma J \cdot (\nu \wedge \phi)\, d\Sigma,$$
$$\forall (\phi_0, \psi_0) \in \mathcal{H},\ (f, g) \in L^1(0,T;\mathcal{H}),$$

where (ϕ, ψ) is the solution of (7.2.2.6). *By definition*, (E_T, H_T) is the value of (E, H) at $t = T$. This definition is justified because it is true if $(E, H) \in C([0,T];\mathcal{H})$ (see, e.g., [50]). Thus it remains to show that $(E, H) \in C([0,T];\mathcal{H})$, $\nu \wedge E|_\Sigma \in \mathcal{L}^2_\tau(\Sigma)$, and that the mapping from the data to $((E, H), \nu \wedge E|_\Sigma)$ is continuous in the indicated topologies. If $J = 0$ this follows from Corollary 7.2.2.1 and Lemma 7.2.2.2. Indeed, in this case it is easy to verify that the solution given according to semigroup theory satisfies (7.2.2.5). Therefore, by superposition, we may assume that $F = G = 0$ and $E_0 = H_0 = 0$.

Let $\mathcal{T}_\tau$ denote the tangential trace operator on Γ, that is,

$$\mathcal{T}_\tau \phi = \nu \wedge (\phi \wedge \nu)|_\Gamma,$$

which is well defined for $\phi \in V$. We suppose that

$$J(t) = \mathcal{T}_\tau \hat{J}(t), \quad 0 \le t \le T, \tag{7.2.2.7}$$

where

$$\hat{J} \in C^1([0,T]; V) \cap C^2([0,T]; \mathcal{L}^2(\Omega)), \quad \hat{J}(0) = 0. \tag{7.2.2.8}$$

In (7.1.1.1) (with $F = 0$ and $E_0 = H_0 = 0$) we set

$$\hat{E} = E, \quad \hat{H} = H - \hat{J}, \tag{7.2.2.9}$$

which formally leads to

$$\begin{cases} \varepsilon \hat{E}' - \operatorname{rot} \hat{H} + \sigma \hat{E} = \operatorname{rot} \hat{J} \\ \mu \hat{H}' + \operatorname{rot} \hat{E} = -\hat{J}' \qquad \text{in } Q \end{cases}$$
$$\hat{H}_\tau - \alpha(\nu \wedge \hat{E}) = 0 \text{ on } \Sigma \tag{7.2.2.10}$$
$$\hat{E}(0) = \hat{H}(0) = 0 \text{ in } \Omega$$

Since $(\operatorname{rot} \hat{J}, -\hat{J}') \in C^1([0,T];\mathcal{H})$, system (7.2.2.10) has a unique solution $(\hat{E}, \hat{H}) \in C([0,T]; D(A)) \cap C^1([0,T];\mathcal{H})$. Thus (E, H), defined by (7.2.2.9), satisfies $(E, H) \in C([0,T]; V \times V) \cap C^1([0,T];\mathcal{H})$ and obviously satisfies (7.1.1.1) (with $F = G = 0$ and $E_0 = H_0 = 0$) in the pointwise sense, and therefore also satisfies (7.2.2.5), provided that J is given by (7.2.2.7).

We observe that functions J of the form (7.2.2.7) are dense in $\mathcal{L}^2_\tau(\Sigma)$. Indeed, $C_0^\infty(0,T;\mathcal{H}^{1/2}_\tau(\Gamma))$ is dense in $\mathcal{L}^2_\tau(\Sigma)$ and each $J \in C_0^\infty(0,T;\mathcal{H}^{1/2}_\tau(\Gamma))$ may be represented, according to trace theory, by some $\hat{J} \in C_0^\infty(0,T;\mathcal{H}^1(\Omega))$ in the sense of (7.2.2.7). Such $\hat{J}$ obviously satisfy (7.2.2.8).

We have the following a priori energy identity for solutions of (7.1.1.1) with $F = G = 0$ and J given by (7.2.2.7):

$$\|(E(t),H(t))\|^2_{\mathcal{H}} + 2\int_0^t \langle\sigma E,E\rangle\,dt + \int_0^t\int_\Gamma (\alpha|\nu\wedge E|^2 + \frac{1}{\alpha}|H_\tau|^2)\,d\Gamma dt$$
$$= \|(E_0,H_0)\|^2_{\mathcal{H}} + \int_0^t\int_\Gamma \frac{1}{\alpha}|J|^2 d\Gamma dt. \quad (7.2.2.11)$$

Assume for the moment that (7.2.2.11) holds, and suppose $J \in \mathcal{L}^2_\tau(\Sigma)$. Let $\{J_n\}$ be a sequence of functions of the form (7.2.2.7) for some corresponding $\hat{J}_n$ satisfying (7.2.2.8) such that $J_n \to J$ strongly in $\mathcal{L}^2_\tau(\Sigma)$, and let $\{(E_n,H_n)\}$ be the corresponding solutions of (7.1.1.1). By applying (7.2.2.11) to the differences $(E_n - E_m, H_n - H_m)$ and $J_n - J_m$ we find that

$$(E_n,H_n) \text{ converges to } (E,H) \text{ in } C([0,T];\mathcal{H})$$
$$\nu\wedge E_n \text{ converges to } \nu\wedge E \text{ strongly in } \mathcal{L}^2_\tau(\Sigma)$$

for some $(E,H) \in C([0,T];\mathcal{H})$ such that $\nu\wedge E \in \mathcal{L}^2_\tau(\Sigma)$. Since, for each n, (E_n,H_n) satisfies (7.2.2.5) with $J = J_n$, upon passing to the limit we see that the limit pair (E,H) satisfies (7.2.2.5). Moreover, we note that (7.2.2.11) *holds in the limit*, from which follows that the mapping $((E_0,H_0),J) \mapsto ((E,H),\nu\wedge\phi|_\Sigma)$ is continuous from $\mathcal{H}\times\mathcal{L}^2_\tau(\Sigma)$ into $C([0,T];\mathcal{H})\times\mathcal{L}^2_\tau(\Sigma)$. This completes the proof of Theorem 7.2.2.1, except for establishing (7.2.2.11). This is done by the standard energy method.

From

$$0 = \operatorname{Re}\int_0^t [\langle\varepsilon E' - \operatorname{rot} H + \sigma E, E\rangle + \langle\mu H' + \operatorname{rot} E, H\rangle]\,dt$$

we obtain by Green's formula and integration by parts

$$\frac{1}{2}\|(E(t),H(t))\|^2_{\mathcal{H}} + \int_0^t \langle\sigma E,E\rangle\,dt + \int_0^t\int_\Gamma \alpha|\nu\wedge E|^2 d\Gamma dt$$
$$= \frac{1}{2}\|(E_0,H_0)\|^2_{\mathcal{H}} - \operatorname{Re}\int_0^t\int_\Gamma J\cdot(\nu\wedge E)\,d\Gamma dt. \quad (7.2.2.12)$$

From the boundary condition we have

$$-\operatorname{Re} J\cdot(\nu\wedge E) = \frac{1}{2}(\alpha|\nu\wedge E|^2 + \frac{1}{\alpha}|J|^2 - \frac{1}{\alpha}|H_\tau|^2). \quad (7.2.2.13)$$

The identity (7.2.2.11) follows upon substitution of (7.2.2.13) into (7.2.2.12).

Later on we shall need the following result, related to Theorem 7.2.2.1 above.

Theorem 7.2.2.2 *Suppose that* $(E_0, H_0) \in \mathcal{H}$, $(F, G) \in L^1(0, T; \mathcal{H})$ *and* $J \in \mathcal{L}^2_\tau(\Sigma)$. *Then the solution* (E, H) *of* (7.1.1.1) *is the (unique) solution of the variational equation*

$$\langle (E(T), H(T)), (\phi_0, \psi_0)\rangle_{\mathcal{H}} - \int_0^T \langle (E(t), H(t)), (f(t), g(t))\rangle_{\mathcal{H}} dx dt + \int_\Sigma (\nu \wedge E) \cdot K \, d\Sigma$$
$$= \langle (E_0, H_0), (\phi(0), \psi(0))\rangle_{\mathcal{H}} + \int_0^T \langle (F(t), G(t)), (\phi(t), \psi(t))\rangle dt - \int_\Sigma J \cdot (\nu \wedge \phi) d\Sigma,$$
$$\forall (\phi_0, \psi_0) \in \mathcal{H}, (f, g) \in L^1(0, T; \mathcal{H}), K \in \mathcal{L}^2_\tau(\Sigma) \tag{7.2.2.14}$$

where (ϕ, ψ) *is the solution of*

$$\begin{cases} \varepsilon\phi' - \operatorname{rot}\psi - \sigma\phi = \varepsilon f \\ \mu\psi' + \operatorname{rot}\psi = \mu g \qquad \text{in } Q \end{cases}$$
$$\psi_\tau + \alpha(\nu \wedge \phi) = K \quad \text{on } \Sigma \tag{7.2.2.15}$$
$$\phi(T) = \phi_0, \quad \psi(T) = \psi_0 \text{ in } \Omega.$$

Proof. Since, by Theorem 7.2.2.1, the linear mapping $((\phi_0, \psi_0), (f, g), K) \mapsto ((\phi, \psi), (\nu \wedge \phi)|_\Sigma)$ is continuous from $\mathcal{H} \times L^1(0, T; \mathcal{H}) \times \mathcal{L}^2_\tau(\Sigma)$ into $C([0, T]; \mathcal{H}) \times \mathcal{L}^2_\tau(\Sigma)$, the right side of (7.2.2.14) defines a continuous linear functional on $\mathcal{H} \times L^1(0, T\mathcal{H}) \times \mathcal{L}^2_\tau(\Sigma)$. Thus (7.2.2.14) has a unique solution $(E, H) \in L^\infty(0, T; \mathcal{H})$, $(E(T), H(T)) \in \mathcal{H}$, $(\nu \wedge E)|_\Sigma \in \mathcal{L}^2_\tau(\Sigma)$. On the other hand, if (E, H) is the solution of (7.1.1.1) and (ϕ, ψ) the solution of (7.2.2.15), then the use of Green's formula and integration by parts in

$$\int_0^T \langle (F(t), G(t)), (\phi(t), \psi(t))\rangle \, dt$$
$$= \int_0^T \langle (\varepsilon E' - \operatorname{rot} H + \sigma E, \mu H' + \operatorname{rot} E), (\phi, \psi)\rangle \, dt$$

shows that (E, H) satisfies (7.2.2.14). This calculation is justified if the data in (7.2.2.15) is sufficiently regular, e.g., $(\phi_0, \psi_0) \in D(A)$, $(f, g) \in C^1([0, T]; \mathcal{H})$, and $K = \mathcal{T}_\tau \tilde{K}$ as in (7.2.2.7) and (7.2.2.8) above. The formula (7.2.2.14) then follows for general data by a standard density argument.

7.2.3 The Global Optimality System

We consider the optimal control problem (7.2.1.1) subject to (7.1.1.1).

Since the cost functional is convex and in view of the properties of the mapping $((E_0, H_0), (F, G), J) \mapsto (E, H)$ discussed in the last section, there is a unique optimal control J_{opt} that is characterized by the variational equation

$$\int_\Sigma J_{\text{opt}} \cdot \hat{J}\, d\Sigma + \kappa \langle (E(T) - E_T, H(T) - H_T), (\hat{E}(T), \hat{H}(T)) \rangle_{\mathcal{H}} = 0, \qquad \forall \hat{J} \in \mathcal{L}^2_\tau(\Sigma), \quad (7.2.3.1)$$

where $(\hat{E}, \hat{H})$ is the solution of

$$\begin{cases} \varepsilon \hat{E}' - \operatorname{rot} \hat{H} + \sigma \hat{E} = 0 \\ \mu \hat{H}' + \operatorname{rot} \hat{E} = 0 \qquad \text{in } Q \end{cases}$$
$$\hat{H}_\tau - \alpha(\nu \wedge \hat{E}) = \hat{J} \text{ on } \Sigma$$
$$\hat{E}(0) = \hat{H}(0) = 0 \text{ in } \Omega.$$

The pair $(\hat{E}, \hat{H})$ is the unique solution of the variational equation

$$\langle (\hat{E}(T), \hat{H}(T)), (\phi_0, \psi_0) \rangle_{\mathcal{H}} - \int_0^T \langle (\hat{E}(t), \hat{H}(t)), (f(t), g(t)) \rangle_{\mathcal{H}} dx dt$$
$$= - \int_\Sigma \hat{J} \cdot (\nu \wedge \phi)\, d\Sigma, \quad \forall (\phi_0, \psi_0) \in \mathcal{H},\ (f, g) \in L^1(0, T; \mathcal{H}), \quad (7.2.3.2)$$

where (ϕ, ψ) is the solution of (7.2.2.6)

Let (P, Q) be the solution of(7.2.1.3). If in (7.2.3.2) we replace (ϕ, ψ) by (P, Q) we obtain

$$\kappa \langle (\hat{E}(T), \hat{H}(T)), (E(T) - E_T, H(T) - H_T) \rangle_{\mathcal{H}} = - \int_\Sigma \hat{J} \cdot (\nu \wedge P)\, d\Sigma, \qquad \forall \hat{J} \in \mathcal{L}^2_\tau(\Sigma). \quad (7.2.3.3)$$

It follows from (7.2.3.1) and (7.2.3.3) that

$$J_{\text{opt}} = \nu \wedge P|_\Sigma. \quad (7.2.3.4)$$

Thus the global optimality system consists of the Maxwell system (7.1.1.1) and the backwards running adjoint system (7.2.1.3), with $J = J_{\text{opt}}$ given by (7.2.3.4).

7.3 Time Domain Decomposition

7.3.1 Description of the Algorithm

We introduce a partition of the time interval $[0, T]$ by setting

$$0 = T_0 < T_1 < \cdots < T_K < T_{K+1} = T \quad (7.3.1.1)$$

and thereby decompose $[0,T]$ into $K+1$ subintervals $I_k := [T_k, T_{k+1}]$, $k = 0, \dots, K$. We further introduce locally defined functions $E_k = E|_{I_k}$, $H_k = H|_{I_k}$, and so forth. We proceed to decompose the optimality system (7.1.1.1), (7.2.1.3), (7.2.1.2) into the following local systems defined on I_k, $k = 0, \dots, K$:

$$\begin{cases} \varepsilon E_k' - \operatorname{rot} H_k + \sigma E_k = F_k \\ \mu H_k' + \operatorname{rot} E_k = G_k \qquad \text{in } \mathcal{Q}_k := \Omega \times I_k \end{cases} \tag{7.3.1.2}$$
$$H_{k\tau} - \alpha(\nu \wedge E_k) = \nu \wedge P_k \text{ on } \Sigma_k := \Gamma \times I_k,$$

$$\begin{cases} \varepsilon P_k' - \operatorname{rot} Q_k - \sigma P_k = 0 \\ \mu Q_k' + \operatorname{rot} P_k = 0 \qquad \text{in } \mathcal{Q}_k \end{cases} \tag{7.3.1.3}$$
$$Q_{k\tau} + \alpha(\nu \wedge P_k) = 0 \text{ on } \Sigma_k.$$

The "initial" values $(E_k(T_k), H_k(T_k))$ and $(P_k(T_{k+1}), Q_k(T_{k+1}))$ for (7.3.1.2) and (7.3.1.3) are given through the coupling conditions

$$\begin{aligned} E_k(T_k) &= E_{k-1}(T_k), \quad H_k(T_k) = H_{k-1}(T_k), \quad k = 1, \dots, K, \\ P_k(T_{k+1}) &= P_{k+1}(T_{k+1}), \quad Q_k(T_{k+1}) = Q_{k+1}(T_{k+1}), \quad k = K-1, \dots, 0, \end{aligned} \tag{7.3.1.4}$$

together with

$$\begin{aligned} E_0(0) &= E_0, \quad H_0(0) = H_0, \\ P_K(T) &= \kappa(E_K(T) - E_T), \quad Q_K(T) = \kappa(H_K(T) - H_T). \end{aligned} \tag{7.3.1.5}$$

We now uncouple the local problems by uncoupling the transmission conditions (7.3.1.4) through an iteration as follows:

$$\begin{aligned} &\begin{cases} \beta E_k^{n+1}(T_{k+1}) - P_k^{n+1}(T_{k+1}) = \mu_{k,k+1}^n \\ \beta H_k^{n+1}(T_{k+1}) - Q_k^{n+1}(T_{k+1}) = \eta_{k,k+1}^n, \quad k = 0, \dots, K-1 \end{cases} \\ &\begin{cases} \beta E_k^{n+1}(T_k) + P_k^{n+1}(T_k) = \mu_{k,k-1}^n \\ \beta H_k^{n+1}(T_k) + Q_k^{n+1}(T_k) = \eta_{k,k-1}^n, \quad k = 1, \dots, K, \end{cases} \end{aligned} \tag{7.3.1.6}$$

where $\beta > 0$ and

$$\begin{aligned} \mu_{k,k+1}^n &= \beta E_{k+1}^n(T_{k+1}) - P_{k+1}^n(T_{k+1}) \\ \eta_{k,k+1}^n &= \beta H_{k+1}^n(T_{k+1}) - Q_{k+1}^n(T_{k+1}) \\ \mu_{k,k-1}^n &= \beta E_{k-1}^n(T_k) + P_{k-1}^n(T_k) \\ \eta_{k,k-1}^n &= \beta H_{k-1}^n(T_k) + Q_{k-1}^n(T_k). \end{aligned} \tag{7.3.1.7}$$

It is readily seen that in the limit of (7.3.1.6), (7.3.1.7) one recovers (7.3.1.4), so that (7.3.1.6), (7.3.1.7) are consistent with (7.3.1.4).

For the convenience of the reader we write down the complete set of uncoupled systems:

$$\begin{cases} \varepsilon(E_k^{n+1})' - \operatorname{rot} H_k^{n+1} + \sigma E_k^{n+1} = F_k \\ \mu(H_k^{n+1})' + \operatorname{rot} E_k^{n+1} = G_k & \text{in } \mathcal{Q}_k \end{cases}$$
$$\begin{cases} \varepsilon(P_k^{n+1})' - \operatorname{rot} Q_k^{n+1} - \sigma P_k^{n+1} = 0 \\ \mu(Q_k^{n+1})' + \operatorname{rot} P_k^{n+1} = 0 & \text{in } \mathcal{Q}_k \end{cases} \tag{7.3.1.8}$$

$$\begin{cases} H_{k\tau}^{n+1} - \alpha(\nu \wedge E_k^{n+1}) = \nu \wedge P_k^{n+1} \\ Q_{k\tau}^{n+1} + \alpha(\nu \wedge P_k^{n+1}) = 0 & \text{on } \Sigma_k \end{cases} \tag{7.3.1.9}$$

$$E_0^{n+1}(0) = E_0, \quad H_0^{n+1}(0) = H_0$$
$$\begin{cases} P_K^{n+1}(T) = \kappa(E_K^{n+1}(T) - E_T) \\ Q_K^{n+1}(T) = \kappa(H_K^{n+1}(T) - H_T) & \text{in } \Omega \end{cases} \tag{7.3.1.10}$$

subject to (7.3.1.6), (7.3.1.7).

We wish to show that for $(\mu_{k,k+1}^n, \eta_{k,k+1}^n)$ and $(\mu_{k,k-1}^n, \eta_{k,k-1}^n)$ given in $\mathcal{H}$, the local problems are well posed for $k = 0, \ldots, K$. To do so we shall show that the local problem with index k is in fact an optimality system concentrated on the interval I_k, $k = 0, \ldots, K$.

Proposition 7.3.1.1 *Assume that* $(\mu_{k,k+1}^n, \eta_{k,k+1}^n) \in \mathcal{H}$, $(\mu_{k,k-1}^n, \eta_{k,k-1}^n) \in \mathcal{H}$. *Set*

$$\begin{aligned} \mathcal{J}_k(J_k, h_{k,k-1}, g_{k,k-1}) &= \frac{1}{2}\int_{\Sigma_k} |J_k|^2 d\Sigma + \frac{1}{2\beta}\{\|\beta E_k(T_{k+1}) - \mu_{k,k+1}^n\|_\varepsilon^2 \\ &+ \|\beta H_k(T_{k+1}) - \eta_{k,k+1}^n\|_\mu^2 + \|(h_{k,k-1}, g_{k,k-1})\|_{\mathcal{H}}^2\}, \quad k = 1, \ldots, K-1, \end{aligned} \tag{7.3.1.11}$$

$$\begin{aligned} \mathcal{J}_K(J_K, h_{K,K-1}, g_{K,K-1}) &= \frac{1}{2}\int_{\Sigma_K} |J_K|^2 d\Sigma \\ &+ \frac{\kappa}{2}\|(E_K(T), H_K(T)) - (E_T, H_T)\|_{\mathcal{H}}^2 + \frac{1}{2\beta}\|(h_{K,K-1}, g_{K,K-1})\|_{\mathcal{H}}^2, \end{aligned} \tag{7.3.1.12}$$

$$\mathcal{J}_0(J_0) = \frac{1}{2}\int_{\Sigma_0} |J_0|^2 d\Sigma + \frac{1}{2\beta}\{\|\beta E_0(T_1) - \mu_{0,1}^n\|_\varepsilon^2 + \|\beta H_0(T_1) - \eta_{0,1}^n\|_\mu^2\}. \tag{7.3.1.13}$$

For $k = 0, \ldots, K$, *the system* (7.3.1.8)–(7.3.1.10), (7.3.1.6) *is the optimality system for the optimal control problem* $\inf_{\mathcal{U}_k} \mathcal{J}_k$, *where* $\mathcal{U}_k = \mathcal{L}_\tau^2(\Sigma_k) \times \mathcal{H}$ *for* $k = 1, \ldots, K$, $\mathcal{U}_0 = \mathcal{L}_\tau^2(\Sigma_0)$, *subject to*

$$\begin{cases} \varepsilon E_k' - \operatorname{rot} H_k + \sigma E_k = F_k \\ \mu H_k' + \operatorname{rot} E_k = G_k & \textit{in } \mathcal{Q}_k \end{cases} \tag{7.3.1.14}$$
$$H_{k\tau} - \alpha(\nu \wedge E_k) = J_k \textit{ on } \Sigma_k$$

and

$$\begin{cases} E_k(T_k) = \dfrac{1}{\beta}(h_{k,k-1} + \mu^n_{k,k-1}) \\ H_k(T_k) = \dfrac{1}{\beta}(g_{k,k-1} + \eta^n_{k,k-1}) \ \textit{in } \Omega, \quad k = 1, \ldots, K, \end{cases} \tag{7.3.1.15}$$

$$E_0(0) = E_0, \quad H_0(0) = H_0 \ \textit{in } \Omega. \tag{7.3.1.16}$$

Remark 7.3.1.1 In the local optimal control problem $\inf_{\mathcal{U}_k} \mathcal{J}_k(J_k, h_{k,k-1}, g_{k,k-1})$, the control J_k is, in the terminology of J.-L. Lions and O. Pironneau, the *effective control* and the controls $h_{k,k-1}$, $g_{k,k-1}$ are *virtual*, or *artificial controls*, as discussed in Chapter 2.

Proof. We give the proof only for $k = 1, \ldots, K-1$ since the proofs in the two remaining cases are similar. The necessary and sufficient condition that J_k, $h_{k,k-1}$, $g_{k,k-1}$ be optimal is that the directional derivative of $\mathcal{J}_k$ at J_k, $h_{k,k-1}$, $g_{k,k-1}$ in the direction of $\hat{J}_k$, $\hat{h}_{k,k-1}$, $\hat{g}_{k,k-1}$ be zero. This leads to the variational equation

$$\begin{aligned} 0 = &\int_{\Sigma_k} J_k \cdot \hat{J}_k \, d\Sigma + \langle \beta E_k(T_{k+1}) - \mu^n_{k,k+1}, \hat{E}_k(T_{k+1})\rangle_\varepsilon \\ &+ \langle \beta H_k(T_{k+1}) - \eta^n_{k,k+1}, \hat{H}_k(T_{k+1})\rangle_\mu + \frac{1}{\beta}\langle (h_{k,k-1}, g_{k,k-1}), (\hat{h}_{k,k-1}, \hat{g}_{k,k-1})\rangle_{\mathcal{H}}, \\ &\qquad\qquad \forall (\hat{J}_k, \hat{h}_{k,k-1}, \hat{g}_{k,k-1}) \in \mathcal{U}_k, \end{aligned} \tag{7.3.1.17}$$

where

$$\begin{gathered} \begin{cases} \varepsilon \hat{E}'_k - \operatorname{rot} \hat{H}_k + \sigma \hat{E}_k = 0 \\ \mu \hat{H}'_k + \operatorname{rot} \hat{E}_k = 0 \qquad \text{in } \mathcal{Q}_k \end{cases} \\ \hat{H}_{k\tau} - \alpha(\nu \wedge \hat{E}_k) = \hat{J}_k \text{ on } \Sigma_k \\ \hat{E}_k(T_k) = \frac{1}{\beta}\hat{h}_{k,k-1}, \quad \hat{H}_k(T_k) = \frac{1}{\beta}\hat{g}_{k,k-1} \text{ in } \Omega. \end{gathered} \tag{7.3.1.18}$$

Introduce (P_k, Q_k) as the solution of

$$\begin{gathered} \begin{cases} \varepsilon P'_k - \operatorname{rot} Q_k - \sigma P_k = 0 \\ \mu Q'_k + \operatorname{rot} P_k = 0 \qquad \text{in } \mathcal{Q}_k \end{cases} \\ Q_{k\tau} + \alpha(\nu \wedge P_k) = 0 \text{ on } \Sigma_k \\ \begin{cases} P_k(T_{k+1}) = \beta E_k(T_{k+1}) - \mu^n_{k,k+1} \\ Q_k(T_{k+1}) = \beta H_k(T_{k+1}) - \eta^n_{k,k+1} \quad \text{in } \Omega. \end{cases} \end{gathered} \tag{7.3.1.19}$$

We have

$$0 = \int_{T_k}^{T_{k+1}} \{\langle \varepsilon P_k' - \operatorname{rot} Q_k - \sigma P_k, \hat{E}_k\rangle + \langle \mu Q_k' + \operatorname{rot} P_k, \hat{H}_k\rangle\}\, dt \tag{7.3.1.20}$$

$$= \langle (P_k(t), Q_k(t)), (\hat{E}_k(t), \hat{H}_k(t))\rangle_{\mathcal{H}} \Big|_{t=T_k}^{T_{k+1}}$$

$$+ \int_{\Sigma_k} [Q_{k\tau} \cdot (\nu \wedge \hat{E}_k) + (\nu \wedge P_k) \cdot \hat{H}_{k\tau}] d\Sigma$$

where we have used (7.2.2.1). By utilizing (7.3.1.18) and (7.3.1.19) it is seen that (7.3.1.20) may be written

$$0 = \langle (\beta E_k(T_{k+1}) - \mu^n_{k,k+1}, \beta H_k(T_{k+1}) - \eta^n_{k,k+1}), (\hat{E}_k(T_{k+1}), \hat{H}_k(T_{k+1}))\rangle_{\mathcal{H}}$$

$$- \langle (P_k(T_k), Q_k(T_k)), (\frac{1}{\beta}\hat{h}_{k,k-1}, \frac{1}{\beta}\hat{g}_{k,k-1})\rangle_{\mathcal{H}} + \int_{\Sigma_k} (\nu \wedge P_k) \cdot \hat{J}_k \, d\Sigma. \tag{7.3.1.21}$$

It now follows from (7.3.1.21) and (7.3.1.17) that

$$J_k = \nu \wedge P_k|_{\Sigma_k}, \quad h_{k,k-1} = -P_k(T_k), \quad g_{k,k-1} = -Q_k(T_k),$$

from which the conclusion of Proposition 7.3.1.1 follows immediately.

Corollary 7.3.1.1 *Let* $(\mu^0_{k,k+1}, \eta^0_{k,k+1})|_{k=0}^{K-1}$ *and* $(\mu^0_{k,k-1}, \eta^0_{k,k-1})|_{k=1}^{K}$ *be given arbitrarily in* $\mathcal{H}$. *Then the iterative procedure described by* (7.3.1.6)–(7.3.1.10) *is well defined for* $n = 0, 1, \ldots$.

7.3.2 Convergence of the Iterates

We consider the iterative procedure with the basic step given by (7.3.1.6)–(7.3.1.10). In fact, as we have done previously, we consider a relaxation of the iteration step (7.3.1.6), (7.3.1.7). Thus we introduce a relaxation parameter $\epsilon \in [0, 1)$ and consider the iterative step (7.3.1.6), (7.3.1.7) with under relaxation:

$$\begin{aligned}
\beta E_k^{n+1}(T_{k+1}) - P_k^{n+1}(T_{k+1}) &= (1-\epsilon)\mu^n_{k,k+1} + \epsilon(\beta E_k^n(T_{k+1}) - P_k^n(T_{k+1})) \\
\beta H_k^{n+1}(T_{k+1}) - Q_k^{n+1}(T_{k+1}) &= (1-\epsilon)\eta^n_{k,k+1} + \epsilon(\beta H_k^n(T_{k+1}) - Q_k^n(T_{k+1})) \\
\beta E_k^{n+1}(T_k) + P_k^{n+1}(T_k) &= (1-\epsilon)\mu^n_{k,k-1} + \epsilon(\beta E_k^n(T_k) + P_k^n(T_k)) \\
\beta H_k^{n+1}(T_k) + Q_k^{n+1}(T_k) &= (1-\epsilon)\eta^n_{k,k-1} + \epsilon(\beta H_k^n(T_k) + Q_k^n(T_k)).
\end{aligned} \tag{7.3.2.1}$$

We introduce the errors

$$\widetilde{E}_k^n = E_k^n - E_k, \quad \widetilde{H}_k^n = H_k^n - H_k, \quad \widetilde{P}_k^n = P_k^n - P_k, \quad \widetilde{Q}_k^n = Q_k^n - Q_k.$$

These satisfy

$$\begin{cases} \varepsilon(\widetilde{E}_k^{n+1})' - \operatorname{rot} \widetilde{H}_k^{n+1} + \sigma \widetilde{E}_k^{n+1} = 0 \\ \mu(\widetilde{H}_k^{n+1})' + \operatorname{rot} \widetilde{E}_k^{n+1} = 0 \qquad \text{in } Q_k \end{cases}$$
$$\begin{cases} \varepsilon(\widetilde{P}_k^{n+1})' - \operatorname{rot} \widetilde{Q}_k^{n+1} - \sigma \widetilde{P}_k^{n+1} = 0 \\ \mu(\widetilde{Q}_k^{n+1})' + \operatorname{rot} \widetilde{P}_k^{n+1} = 0 \qquad \text{in } Q_k \end{cases} \tag{7.3.2.2}$$

$$\begin{cases} \widetilde{H}_{k\tau}^{n+1} - \alpha(\nu \wedge \widetilde{E}_k^{n+1}) = \nu \wedge \widetilde{P}_k^{n+1} \\ Q_{k\tau}^{n+1} + \alpha(\nu \wedge P_k^{n+1}) = 0 \qquad \text{on } \Sigma_k \end{cases} \tag{7.3.2.3}$$

$$\begin{gathered} \widetilde{E}_0^{n+1}(0) = \widetilde{H}_0^{n+1}(0) = 0 \\ \widetilde{P}_K^{n+1}(T) = \kappa \widetilde{E}_K^{n+1}(T), \quad \widetilde{Q}_K^{n+1}(T) = \kappa \widetilde{H}_K^{n+1}(T) \text{ in } \Omega \end{gathered} \tag{7.3.2.4}$$

subject to

$$\begin{cases} \beta \widetilde{E}_k^{n+1}(T_{k+1}) - \widetilde{P}_k^{n+1}(T_{k+1}) = \tilde{\mu}_{k,k+1}^n \\ \beta \widetilde{H}_k^{n+1}(T_{k+1}) - \widetilde{Q}_k^{n+1}(T_{k+1}) = \tilde{\eta}_{k,k+1}^n, \quad k = 0, \dots, K-1 \end{cases}$$
$$\begin{cases} \beta \widetilde{E}_k^{n+1}(T_k) + \widetilde{P}_k^{n+1}(T_k) = \tilde{\mu}_{k,k-1}^n \\ \beta \widetilde{H}_k^{n+1}(T_k) + \widetilde{Q}_k^{n+1}(T_k) = \tilde{\eta}_{k,k-1}^n, \quad k = 1, \dots, K \end{cases} \tag{7.3.2.5}$$

where, for $n \geq 1$,

$$\begin{aligned} \tilde{\mu}_{k,k+1}^n &= (1-\epsilon)(\beta \widetilde{E}_{k+1}^n(T_{k+1}) - \widetilde{P}_{k+1}^n(T_{k+1})) + \epsilon(\beta \widetilde{E}_k^n(T_{k+1}) - \widetilde{P}_k^n(T_{k+1})) \\ \tilde{\eta}_{k,k+1}^n &= (1-\epsilon)(\beta \widetilde{H}_{k+1}^n(T_{k+1}) - \widetilde{Q}_{k+1}^n(T_{k+1})) + \epsilon(\beta \widetilde{H}_k^n(T_{k+1}) - \widetilde{Q}_k^n(T_{k+1})) \\ \tilde{\mu}_{k,k-1}^n &= (1-\epsilon)(\beta \widetilde{E}_{k-1}^n(T_k) + \widetilde{P}_{k-1}^n(T_k)) + \epsilon(\beta \widetilde{E}_k^n(T_k) + \widetilde{P}_k^n(T_k)) \\ \tilde{\eta}_{k,k-1}^n &= (1-\epsilon)(\beta \widetilde{H}_{k-1}^n(T_k) + \widetilde{Q}_{k-1}^n(T_k)) + \epsilon(\beta \widetilde{H}_k^n(T_k) + \widetilde{Q}_k^n(T_k)), \end{aligned} \tag{7.3.2.6}$$

$$\begin{aligned} \tilde{\mu}_{k,k+1}^0 &= \mu_{k,k+1}^0 - (\beta E_{k+1}(T_{k+1}) - P_{k+1}(T_{k+1})) \\ \tilde{\eta}_{k,k+1}^0 &= \eta_{k,k+1}^0 - (\beta H_{k+1}(T_{k+1}) - Q_{k+1}(T_{k+1})) \\ \tilde{\mu}_{k,k-1}^0 &= \mu_{k,k-1}^0 - (\beta E_k(T_k) + P_k(T_k)) \\ \tilde{\eta}_{k,k-1}^0 &= \eta_{k,k-1}^0 - (\beta H_k(T_k) + Q_k(T_k)). \end{aligned} \tag{7.3.2.7}$$

We shall prove the following convergence result.

Theorem 7.3.2.1 *Let $\beta > 0$. Then*

(i) *for any $\epsilon \in [0, 1)$*

$$\begin{gathered} \nu \wedge \widetilde{P}_k^n|_{\Sigma_k} \to 0 \ \textit{strongly in } \mathcal{L}_\tau^2(\Sigma_k),\ k = 0, \dots, K \\ (\widetilde{P}_K^n, \widetilde{Q}_K^n) \to 0 \ \textit{in } C(I_K; \mathcal{H}) \\ (\widetilde{E}_0^n, \widetilde{H}_0^n) \to 0 \ \textit{in } C(I_0; \mathcal{H}) \\ \nu \wedge \widetilde{E}_0^n|_{\Sigma_0} \to 0 \ \textit{strongly in } \mathcal{L}_\tau^2(\Sigma_0). \end{gathered}$$

(ii) *For any* $\epsilon \in (0,1)$ *and for* $k = 0, \dots, K$,

$$(\widetilde{E}_k^n, \widetilde{H}_k^n) \to 0 \ \ \textit{and} \ \ (\widetilde{P}_k^n, \widetilde{Q}_k^n) \to 0 \ \ \textit{in} \ C(I_k; \mathcal{H})$$
$$\nu \wedge \widetilde{E}_k^n|_{\Sigma_k} \to 0 \ \ \textit{strongly in} \ \mathcal{L}_\tau^2(\Sigma_k).$$

Remark 7.3.2.1 Part $(\mathrm{i})_{1,2}$ states that for any $\epsilon \in [0,1)$ the effective local optimal controls $\{\nu \wedge P_k^n|_{\Sigma_k}\}_{k=0}^K$ converge strongly to the global optimal control $\nu \wedge P|_\Sigma$, and the deviation $\|(E_K^n(T), H_K^n(T)) - (E_T, H_T)\|_\mathcal{H}$ of the local optimal trajectory (E_K^n, H_K^n) at time T from the target state (E_T, H_T) converges to the deviation $\|(E(T), H(T)) - (E_T, H_T)\|_\mathcal{H}$ of the global optimal trajectory at time T from the target state. An a posteriori estimate of the error in the approximation is derived in the next section.

Remark 7.3.2.2 When $\epsilon = 0$ it is possible to show that

$$\begin{cases} (\widetilde{E}_k^n, \widetilde{H}_k^n) \to 0 \ \ \text{weakly* in } L^\infty(0,T;\mathcal{H}) \\ \nu \wedge \widetilde{E}_k^n|_{\Sigma_k} \to 0 \ \ \text{weakly in } \mathcal{L}_\tau^2(\Sigma_k), \qquad k = 1, \dots, K \end{cases} \tag{7.3.2.8}$$
$$(\widetilde{P}_k^n, \widetilde{Q}_k^n) \to 0 \ \ \text{weakly* in } L^\infty(0,T;\mathcal{H}),\ k = 0, \dots, K-1.$$

provided the following *backwards uniqueness property* is valid: if (E, H) satisfies the Maxwell system $(7.1.1.1)_{1,2}$ with $F = G = 0$ and the boundary condition $(7.1.1.1)_3$ with $J = 0$, and if $E(T) = H(T) = 0$, then $E(0) = H(0) = 0$. Whether or not this uniqueness property holds seems to be an open question.

Proof of Theorem 7.3.2.1. Set $\mathcal{X} = \mathcal{H}^{2K}$ with the standard product norm. Let

$$X = \{(\mu_{k,k+1}, \eta_{k,k+1})_{k=0}^{K-1}, (\mu_{k,k-1}, \eta_{k,k-1})_{k=1}^K\} \in \mathcal{X}$$

and let (E_k, H_k), (P_k, Q_k) be the solution of (7.3.1.8)–(7.3.1.10), (7.3.1.6) with the superscripts $n+1$ and n removed and with zero data

$$F_k = G_k = 0, \quad E_0 = H_0 = E_T = H_T = 0. \tag{7.3.2.9}$$

Define the linear mapping $\mathcal{T} : \mathcal{X} \mapsto \mathcal{X}$ by

$$\begin{aligned} \mathcal{T}X = \{&(\beta E_{k+1}(T_{k+1}) - P_{k+1}(T_{k+1}), \beta H_{k+1}(T_{k+1}) - Q_{k+1}(T_{k+1}))_{k=0}^{K-1}, \\ &(\beta E_{k-1}(T_k) + P_{k-1}(T_k), \beta H_{k-1}(T_k) + Q_{k-1}(T_k))_{k=1}^K\}. \end{aligned}$$

Note that X is a fixed point of $\mathcal{T}$ if and only if the transmission conditions (7.3.1.4) are satisfied, that is, if and only if $\{(E_k, H_k), (P_k, Q_k)\}_{k=0}^K$ is the solution of the global optimality system with vanishing data (7.3.2.9). Since, in this case, the optimal control is obviously $J_{\text{opt}} = 0$, it follows that the only fixed point of $\mathcal{T}$ is $X = 0$.

As before, the significance of the map $\mathcal{T}$ is that, if we set

$$X^n = \{(\tilde{\mu}^{n-1}_{k,k+1}, \tilde{\eta}^{n-1}_{k,k+1})_{k=0}^{K-1}, (\tilde{\mu}^{n-1}_{k,k-1}, \tilde{\eta}^{n-1}_{k,k-1})_{k=1}^{K}\} \tag{7.3.2.10}$$

the relaxed iteration step (7.3.2.5), (7.3.2.6) may be expressed as the relaxed fixed point iteration

$$X^{n+1} = (1-\epsilon)\mathcal{T}X^n + \epsilon X^n := \mathcal{T}_\epsilon X^n. \tag{7.3.2.11}$$

Indeed, let $\{(\widetilde{E}^n_k, \widetilde{H}^n_k), (\widetilde{P}^n_k, \widetilde{Q}^n_k)\}_{k=0}^{K}$ be the solution of (7.3.2.2)–(7.3.2.5) with n replaced by $n-1$. Then

$$\begin{aligned} X^n = \{&(\beta\widetilde{E}^n_k(T_{k+1}) - \widetilde{P}^n_k(T_{k+1}), \beta\widetilde{H}^n_k(T_{k+1}) - \widetilde{Q}^n_k(T_{k+1}))_{k=0}^{K-1}, \\ &(\beta\widetilde{E}^n_k(T_k) + \widetilde{P}^n_k(T_k), \beta\widetilde{H}^n_k(T_k) + \widetilde{Q}^n_k(T_k))_{k=1}^{K}\} \end{aligned}$$

and

$$\begin{aligned} \mathcal{T}X^n = \{&(\beta\widetilde{E}^n_{k+1}(T_{k+1}) - \widetilde{P}^n_{k+1}(T_{k+1}), \beta\widetilde{H}^n_{k+1}(T_{k+1}) - \widetilde{Q}^n_{k+1}(T_{k+1}))_{k=0}^{K-1}, \\ &(\beta\widetilde{E}^n_{k-1}(T_k) + \widetilde{P}^n_{k-1}(T_k), \beta\widetilde{H}^n_{k-1}(T_k) + \widetilde{Q}^n_{k-1}(T_k))_{k=1}^{K}\}, \end{aligned}$$

from which (7.3.2.11) follows.

The following result shows that $\mathcal{T}$ is nonexpansive.

Lemma 7.3.2.1 *For any $X \in \mathcal{X}$,*

$$\|\mathcal{T}X\|^2_{\mathcal{X}} = \|X\|^2_{\mathcal{X}} - 2\mathcal{F} \tag{7.3.2.12}$$

where

$$\mathcal{F} = 2\beta\sum_{k=0}^{K}\int_{\Sigma_k} |\nu \wedge P_k|^2\, d\Sigma + 2\beta\kappa\|(E_K(T_{K+1}), H_K(T_{K+1}))\|^2_{\mathcal{H}}. \tag{7.3.2.13}$$

Proof. We have

$$\begin{aligned} \|X\|^2_{\mathcal{X}} = &\sum_{k=0}^{K-1} \|(\beta E_k(T_{k+1}) - P_k(T_{k+1}), \beta H_k(T_{k+1}) - Q_k(T_{k+1}))\|^2_{\mathcal{H}} \\ &+ \sum_{k=1}^{K} \|(\beta E_k(T_k) + P_k(T_k), \beta H_k(T_k) + Q_k(T_k))\|^2_{\mathcal{H}}. \end{aligned} \tag{7.3.2.14}$$

Let us define

$$\mathcal{E}_k(t) = \beta^2(\|E_k(t)\|^2_\varepsilon + \|H_k(t)\|^2_\mu) + \|P_k(t)\|^2_\varepsilon + \|Q_k(t)\|^2_\mu$$

$$\begin{aligned} \mathcal{E} &= \sum_{k=1}^{K-1}\big[\mathcal{E}_k(T_{k+1}) + \mathcal{E}_k(T_k)\big] + \mathcal{E}_0(T_1) + \mathcal{E}_K(T_K) \\ &= \sum_{k=0}^{K-1}\big[\mathcal{E}_k(T_{k+1}) + \mathcal{E}_{k+1}(T_{k+1})\big]. \end{aligned} \tag{7.3.2.15}$$

One finds after a little calculation that (7.3.2.14) may be written

$$\|X\|_{\mathcal{X}}^2 = \mathcal{E} - 2\beta \sum_{k=0}^{K-1} \mathrm{Re}\big[\langle E_k(t), P_k(t)\rangle_\varepsilon + \langle H_k(t), Q_k(t)\rangle_\mu\big]_{t=T_k}^{T_{k+1}} + 2\beta\, \mathrm{Re}\big[\langle E_K(T_K), P_K(T_K)\rangle_\varepsilon + \langle H_K(T_K), Q_K(T_K)\rangle_\mu\big]. \quad (7.3.2.16)$$

Form

$$0 = \int_{T_k}^{T_{k+1}} \big[\langle \varepsilon E_k' - \mathrm{rot}\, H_k + \sigma E_k, P_k\rangle + \langle \mu H_k' + \mathrm{rot}\, E_k, Q_k\rangle\big]\, dt = \langle (E_k(t), H_k(t)), (P_k(t), Q_k(t))\rangle_{\mathcal{H}}\big|_{t=T_k}^{T_{k+1}} + \int_{\Sigma_k} |\nu \wedge P_k|^2\, d\Sigma, \quad (7.3.2.17)$$

where we have used (7.2.2.1). In particular we have

$$\begin{aligned} &\langle (E_K(T_K), H_K(T_K)), (P_K(T_K), Q_K(T_K))\rangle_{\mathcal{H}} \\ &= \langle (E_K(T_{K+1}), H_K(T_{K+1})), (P_K(T_{K+1}), Q_K(T_{K+1}))\rangle_{\mathcal{H}} + \int_{\Sigma_K} |\nu \wedge P_K|^2\, d\Sigma \\ &= \kappa \|(E_K(T_{K+1}), H_K(T_{K+1}))\|_{\mathcal{H}}^2 + \int_{\Sigma_K} |\nu \wedge P_K|^2\, d\Sigma. \end{aligned} \quad (7.3.2.18)$$

It follows from (7.3.2.16)–(7.3.2.18) that

$$\|X\|_{\mathcal{X}}^2 = \mathcal{E} + \mathcal{F}. \quad (7.3.2.19)$$

A similar calculation yields

$$\begin{aligned} \|\mathcal{T}X\|_{\mathcal{X}}^2 &= \sum_{k=0}^{K-1} \|(\beta E_{k+1}(T_{k+1}) - P_{k+1}(T_{k+1}), \beta H_{k+1}(T_{k+1}) - Q_{k+1}(T_{k+1}))\|_{\mathcal{H}}^2 \\ &\quad + \sum_{k=1}^{K} \|(\beta E_{k-1}(T_k) + P_{k-1}(T_k), \beta H_{k-1}(T_k) + Q_{k-1}(T_k))\|_{\mathcal{H}}^2 \\ &= \mathcal{E} + 2\beta \sum_{k=0}^{K-1} \mathrm{Re}\big[\langle E_k(t), P_k(t)\rangle_\varepsilon + \langle H_k(t), Q_k(t)\rangle_\mu\big]_{t=T_k}^{T_{k+1}} \\ &\quad - 2\beta\, \mathrm{Re}\big[\langle E_K(T_K), P_K(T_K)\rangle_\varepsilon + \langle H_K(T_K), Q_K(T_K)\rangle_\mu\big] \\ &= \mathcal{E} - \mathcal{F}. \end{aligned} \quad (7.3.2.20)$$

Thus (7.3.2.12) follows from (7.3.2.19) and (7.3.2.20).

To continue with the proof of Theorem 7.3.2.1, we set

$$\mathcal{E}_k^n(t) = \beta^2(\|\widetilde{E}_k^n(t)\|_\varepsilon^2 + \|\widetilde{H}_k^n(t)\|_\mu^2) + \|\widetilde{P}_k^n(t)\|_\varepsilon^2 + \|\widetilde{Q}_k^n(t)\|_\mu^2$$

$$\mathcal{E}^n = \sum_{k=0}^{K-1} \left[\mathcal{E}_k^n(T_{k+1}) + \mathcal{E}_{k+1}^n(T_{k+1})\right]$$

$$\mathcal{F}^n = 2\beta \sum_{k=0}^{K} \int_{\Sigma_k} |\nu \wedge \widetilde{P}_k^n|^2 \, d\Sigma + 2\beta\kappa \|(\widetilde{E}_K^n(T_{K+1}), \widetilde{H}_K^n(T_{K+1}))\|_{\mathcal{H}}^2.$$

From (7.3.2.19) and (7.3.2.20) we then have

$$\|X^n\|_{\mathcal{X}}^2 = \mathcal{E}^n + \mathcal{F}^n, \quad \|\mathcal{T}X^n\|_{\mathcal{X}}^2 = \mathcal{E}^n - \mathcal{F}^n. \tag{7.3.2.21}$$

Lemma 7.3.2.2 *For any* $\epsilon \in [0,1)$ *we have*

$$\mathcal{E}^{n+1} + \mathcal{F}^{n+1} \leq \mathcal{E}^n - (1-2\epsilon)\mathcal{F}^n. \tag{7.3.2.22}$$

Proof. As the relaxed iteration step (7.3.2.5), (7.3.2.6) may be expressed as the relaxed fixed point iteration (7.3.2.11), from Lemma 7.3.2.1 and (7.3.2.21) we have

$$\begin{aligned} \|X^{n+1}\|_{\mathcal{X}}^2 &= \mathcal{E}^{n+1} + \mathcal{F}^{n+1} \\ &= ((1-\epsilon)^2 + \epsilon^2)\|X^n\|_{\mathcal{X}}^2 - 2(1-\epsilon)^2\mathcal{F}^n + 2\epsilon(1-\epsilon)\mathrm{Re}(\mathcal{T}X^n, X^n)_{\mathcal{X}} \\ &= ((1-\epsilon)^2 + \epsilon^2)\mathcal{E}^n - (1-2\epsilon)\mathcal{F}^n + 2\epsilon(1-\epsilon)\mathrm{Re}(\mathcal{T}X^n, X^n)_{\mathcal{X}}, \end{aligned} \tag{7.3.2.23}$$

From (7.3.2.21) and the Cauchy-Schwarz inequality we have

$$|(\mathcal{T}X^n, X^n)_{\mathcal{X}}| \leq \sqrt{(\mathcal{E}^n)^2 - (\mathcal{F}^n)^2} \leq \mathcal{E}^n. \tag{7.3.2.24}$$

Lemma 7.3.2.2 follows immediately from (7.3.2.23) and (7.3.2.24).

By iterating (7.3.2.22) down to $n = 1$ we obtain

$$\mathcal{E}^{n+1} \leq \mathcal{E}^1 - \sum_{p=1}^{n+1} c_p(\epsilon)\mathcal{F}^p, \tag{7.3.2.25}$$

where

$$c_p(\epsilon) = 2(1-\epsilon),\ p = 2, \ldots, n; \quad c_1(\epsilon) = 1 - 2\epsilon,\ c_{n+1}(\epsilon) = 1.$$

The crucial inequality (7.3.2.25) implies, in particular, that $\mathcal{F}^n \to 0$ as $n \to \infty$, hence

$$\nu \wedge \widetilde{P}_k^n|_{\Sigma_k} \to 0 \ \text{ strongly in } \mathcal{L}_\tau^2(\Sigma_k),\ k = 0, \ldots, K \tag{7.3.2.26}$$

$$(\widetilde{E}_K^n(T_{K+1}), \widetilde{H}_K^n(T_{K+1})) \to 0 \ \text{ strongly in } \mathcal{H}. \tag{7.3.2.27}$$

From (7.3.2.27) and $(7.3.2.4)_2$ we then obtain

$$(\widetilde{P}_K^n(T_{K+1}), \widetilde{Q}_K^n(T_{K+1})) \to 0 \ \text{ strongly in } \mathcal{H}. \tag{7.3.2.28}$$

Since, for the solution of (7.1.1.1) (with $F = G = 0$) the mapping taking $((E_0, H_0), J)$ to $((E, H), \nu \wedge E|_\Sigma)$ is continuous from $\mathcal{H} \times \mathcal{L}^2_\tau(\Sigma)$ into $L^\infty(0, T; \mathcal{H}) \times \mathcal{L}^2_\tau(\Sigma)$, it follows from (7.3.2.26) and $\widetilde{E}^n_0(0) = \widetilde{H}^n_0(0) = 0$, that

$$(\widetilde{E}^n_0, \widetilde{H}^n_0) \to 0 \text{ strongly in } C(I_0; \mathcal{H}) \tag{7.3.2.29}$$

$$\nu \wedge \widetilde{E}^n_0|_{\Sigma_0} \to 0 \text{ strongly in } \mathcal{L}^2_\tau(\Sigma_0) \tag{7.3.2.30}$$

Similarly, for the solution of (7.2.1.3) the mapping from the final data $(P(T), Q(T))$ to $((P, Q), \nu \wedge P|_\Sigma)$ is continuous $\mathcal{H} \mapsto L^\infty(0, T; \mathcal{H}) \times \mathcal{L}^2_\tau(\Sigma)$ so, in particular, we obtain from (7.3.2.28)

$$(\widetilde{P}^n_K, \widetilde{Q}^n_K) \to 0 \text{ strongly in } C(I_K; \mathcal{H}). \tag{7.3.2.31}$$

This proves part (i) of Theorem 7.3.2.1.

To prove part (ii) we invoke Proposition 2.3.7.1 to deduce that $\|X^{n+1} - X^n\|_{\mathcal{X}} \to 0$ or, equivalently,

$$\|\mathcal{T}X^n - X^n\|_{\mathcal{X}} \to 0 \tag{7.3.2.32}$$

since $X^{n+1} - X^n = (1 - \epsilon)(\mathcal{T}X^n - X^n)$. We have

$$\begin{aligned}\mathcal{T}X^n - X^n = \big\{&\big(\beta(\widetilde{E}^n_{k+1}(T_{k+1}) - \widetilde{E}^n_k(T_{k+1})) - (\widetilde{P}^n_{k+1}(T_{k+1}) - \widetilde{P}^n_k(T_{k+1})),\\ &\beta(\widetilde{H}^n_{k+1}(T_{k+1}) - \widetilde{H}^n_k(T_{k+1})) - (\widetilde{Q}^n_{k+1}(T_{k+1}) - \widetilde{Q}^n_k(T_{k+1}))\big)_{k=0}^{K-1},\\ &\big(\beta(\widetilde{E}^n_{k-1}(T_k) - \widetilde{E}^n_k(T_k)) + (\widetilde{P}^n_{k-1}(T_k) - \widetilde{P}^n_k(T_k)),\\ &\beta(\widetilde{H}^n_{k-1}(T_k) - \widetilde{H}^n_k(T_k)) + (\widetilde{Q}^n_{k-1}(T_k) - \widetilde{Q}^n_k(T_k))\big)_{k=1}^{K}\big\}.\end{aligned}$$

A straightforward calculation shows that

$$\begin{aligned}\|\mathcal{T}X^n - X^n\|^2_{\mathcal{X}} = 2\sum_{k=0}^{K-1}\big\{&\beta^2\|(\widetilde{E}^n_{k+1}(T_{k+1}) - \widetilde{E}^n_k(T_{k+1}), \widetilde{H}^n_{k+1}(T_{k+1}) - \widetilde{H}^n_k(T_{k+1}))\|^2_{\mathcal{H}}\\ &+ \|(\widetilde{P}^n_{k+1}(T_{k+1}) - \widetilde{P}^n_k(T_{k+1}), \widetilde{Q}^n_{k+1}(T_{k+1}) - \widetilde{Q}^n_k(T_{k+1}))\|^2_{\mathcal{H}}\big\}.\end{aligned} \tag{7.3.2.33}$$

Therefore, for $k = 0, \ldots, K-1$,

$$\|(\widetilde{E}^n_{k+1}(T_{k+1}) - \widetilde{E}^n_k(T_{k+1}), \widetilde{H}^n_{k+1}(T_{k+1}) - \widetilde{H}^n_k(T_{k+1}))\|_{\mathcal{H}} \to 0 \tag{7.3.2.34}$$

$$\|(\widetilde{P}^n_{k+1}(T_{k+1}) - \widetilde{P}^n_k(T_{k+1}), \widetilde{Q}^n_{k+1}(T_{k+1}) - \widetilde{Q}^n_k(T_{k+1}))\|_{\mathcal{H}} \to 0. \tag{7.3.2.35}$$

By what has already been proved we have

$$(\widetilde{P}^n_K, \widetilde{Q}^n_K) \to 0 \text{ in } C(I_K; \mathcal{H})$$

so that, in particular,

$$\|(\widetilde{P}^n_K(T_K), \widetilde{Q}^n_K(T_K))\|_{\mathcal{H}} \to 0.$$

It then follows from (7.3.2.35) with $k = K - 1$ that

$$\|(\widetilde{P}^n_{K-1}(T_K), \widetilde{Q}^n_{K-1}(T_K))\|_{\mathcal{H}} \to 0,$$

from which follows that

$$(\widetilde{P}^n_{K-1}, \widetilde{Q}^n_{K-1}) \to 0 \text{ in } C(I_{K-1}; \mathcal{H})$$

and, in particular, that

$$\|(\widetilde{P}^n_{K-1}(T_{K-1}), \widetilde{Q}^n_{K-1}(T_{K-1}))\|_{\mathcal{H}} \to 0.$$

We may now repeat the argument to conclude that

$$(\widetilde{P}^n_k, \widetilde{Q}^n_k) \to 0 \text{ in } C(I_k; \mathcal{H}),\ k = K, K-1, \dots, 0.$$

In a similar way, since we have already proved that

$$(\widetilde{E}^n_0, \widetilde{H}^n_0) \to 0 \text{ in } C(I_0; \mathcal{H})$$

we have

$$\|(\widetilde{E}^n_0(T_1), \widetilde{H}^n_0(T_1))\|_{\mathcal{H}} \to 0.$$

Use of (7.3.2.34) with $k = 0$ gives

$$\|(\widetilde{E}^n_1(T_1), \widetilde{H}^n_1(T_1))\|_{\mathcal{H}} \to 0.$$

Also, by what was proved earlier,

$$\nu \wedge \widetilde{P}^n_1\big|_{\Sigma_1} \to 0 \text{ strongly in } \mathcal{L}^2_\tau(\Sigma_1),$$

hence

$$(\widetilde{E}^n_1, \widetilde{H}^n_1) \to 0 \text{ in } C(I_1; \mathcal{H})$$
$$\nu \wedge \widetilde{E}^n_1|_{\Sigma_1} \to 0 \text{ strongly in } \mathcal{L}^2_\tau(\Sigma_1),$$

and so forth. This completes the proof of Theorem 7.3.2.1.

7.3.3 A Posteriori Error Estimates

The purpose of this section is to derive *a posteriori* estimates of the difference between the solutions of the local optimality systems and the solution of the global optimality system in terms of the mismatch of the nth iterates, or of successive iterates, across the interfaces $t = T_k$, $k = 1, \dots, K$. Although the arguments closely parallel those of Section 6.3.3, for the sake of completeness they are included except when the arguments are identical up to minor details.

We introduce *the global error at the nth iteration*, e^n, by

$$e^n := \max_{0\le k\le K} \|(\widetilde{E}_k^n, \widetilde{H}_k^n)\|^2_{L^\infty(I_k;\mathcal{H})} + \max_{0\le k\le K} \|(\widetilde{P}_k^n, \widetilde{Q}_k^n)\|^2_{L^\infty(I_k;\mathcal{H})}$$
$$+\sum_{k=0}^{K}\int_{\Sigma_k}\{\alpha(|\nu\wedge\widetilde{E}_k^n|^2+|\nu\wedge\widetilde{P}_k^n|^2)+\frac{1}{\alpha}(|\widetilde{H}_{k\tau}^n|^2+|\widetilde{Q}_{k\tau}^n|^2)\}\,d\Sigma$$

and further define *the local pointwise error at the nth iteration*

$$e_k^n(t) = \|(\widetilde{E}_k^n(t), \widetilde{H}_k^n(t))\|^2_{\mathcal{H}} + \int_{T_k}^t\int_\Gamma (\alpha|\nu\wedge\widetilde{E}_k^n|^2+\frac{1}{\alpha}|\widetilde{H}_{k\tau}^n|^2)d\Gamma dt$$
$$+\|(\widetilde{P}_k^n(t), \widetilde{Q}_k^n(t))\|^2_{\mathcal{H}} + \int_t^{T_{k+1}}\int_\Gamma\left(\alpha|\nu\wedge\widetilde{P}_k^n|^2+\frac{1}{\alpha}|\widetilde{Q}_{k\tau}^n|^2\right)d\Gamma dt$$

for $t\in I_k$. The *cumulative global error at the nth and* $(n+1)$*st iterations* is then defined to be

$$e^{n,n+1} = \max_{0\le k\le K}\|e_k^n+e_k^{n+1}\|_{L^\infty(I_k)}.$$

In what follows we shall obtain a posteriori estimates of e^n and $e^{n,n+1}$ in terms of the quantities

$$\mathcal{E}_{k,k+1}^n(T_{k+1}) = \|(E_k^n(T_{k+1})-E_{k+1}^n(T_{k+1}), H_k^n(T_{k+1})-H_{k+1}^n(T_{k+1}))\|^2_{\mathcal{H}}$$
$$+\|(P_k^n(T_{k+1})-P_{k+1}^n(T_{k+1}), Q_k^n(T_{k+1})-Q_{k+1}^n(T_{k+1}))\|^2_{\mathcal{H}}$$

and

$$\mathcal{E}_{k,k+1}^{n,n+1}(T_{k+1}) = \|E_k^n(T_{k+1})-E_{k+1}^{n+1}(T_{k+1}), H_k^n(T_{k+1})-H_{k+1}^{n+1}(T_{k+1}))\|^2_{\mathcal{H}}$$
$$+\|(P_k^n(T_{k+1})-P_{k+1}^{n+1}(T_{k+1}), Q_k^n(T_{k+1})-Q_{k+1}^{n+1}(T_{k+1}))\|^2_{\mathcal{H}},$$

which are, respectively, measures of the mismatch of the nth iterates, and of successive iterates, across the break point $t=T_{k+1}$. The main results of this section are the following error estimates.

Theorem 7.3.3.1 *Let* $\epsilon\in[0,1)$*. Then*

$$e^n \le C\sqrt{\mathcal{E}^n+\mathcal{F}^n}\Big\{\sum_{k=0}^{K-1}\mathcal{E}_{k,k+1}^n(T_{k+1})\Big\}^{1/2} \qquad (7.3.3.1)$$

where

$$C=\sqrt{2}\max(1,1/\beta)\big(2+\max(\|\alpha^{-1}\|_{L^\infty(\Gamma)},\kappa)\big).$$

Theorem 7.3.3.2 *Let* $\epsilon \in [0, 1/2)$. *Then*

$$e^{n,n+1} \leq C_\epsilon \sqrt{\mathcal{E}^{n+1} + \mathcal{E}^n} \Big\{ \sum_{k=0}^{K-1} \big(\mathcal{E}_{k,k+1}^{n,n+1}(T_{k+1}) + \mathcal{E}_{k+1,k}^{n,n+1}(T_{k+1}) \big) \Big\}^{1/2} \qquad (7.3.3.2)$$

where

$$C_\epsilon = \sqrt{2} \max(1, 1/\beta) \big(2 + (1 - 2\epsilon)^{-1} \max(\|\alpha^{-1}\|_{L^\infty(\Gamma)}, \kappa) \big) \qquad (7.3.3.3)$$

It shall be shown below that $\mathcal{E}^n + \mathcal{F}^n \leq \tilde{C}$ with a constant $\tilde{C}$ that is *explicitly computable* in terms of β, K and the input data to the global and local optimal control problems. We therefore have the following corollaries.

Corollary 7.3.3.1 *Let* $\epsilon \in [0, 1)$. *There is an explicitly computable constant* C, *depending only on* β, K *and the input data to the global and local optimal control problems, such that*

$$e^n \leq C \Big\{ \sum_{k=0}^{K-1} \mathcal{E}_{k,k+1}^n(T_{k+1}) \Big\}^{1/2}. \qquad (7.3.3.4)$$

Corollary 7.3.3.2 *Let* $\epsilon \in [0, 1/2)$. *There is an explicitly computable constant* C_ϵ, *depending only on* ϵ, β, K *and the input data to the global and local optimal control problems, such that*

$$e^{n,n+1} \leq C_\epsilon \Big\{ \sum_{k=0}^{K-1} \big(\mathcal{E}_{k,k+1}^{n,n+1}(T_{k+1}) + \mathcal{E}_{k+1,k}^{n,n+1}(T_{k+1}) \big) \Big\}^{1/2}. \qquad (7.3.3.5)$$

Theorem 7.3.3.1 is a consequence of the following two lemmas.

Lemma 7.3.3.1 *Let* $\epsilon \in [0, 1)$. *Then*

$$\mathcal{F}^n \leq \sqrt{2} \max(1, \beta) (\mathcal{E}^n + \mathcal{F}^n)^{1/2} \Big\{ \sum_{k=0}^{K-1} \mathcal{E}_{k,k+1}^n(T_{k+1}) \Big\}^{1/2}. \qquad (7.3.3.6)$$

Lemma 7.3.3.2 *Let* $\epsilon \in [0, 1)$. *Then*

$$e^n \leq 2\sqrt{2} \max(1, 1/\beta) \sqrt{\mathcal{E}^n} \Big\{ \sum_{k=0}^{K-1} \mathcal{E}_{k,k+1}^n(T_{k+1}) \Big\}^{1/2} + \frac{1}{\beta} \max(\|\alpha^{-1}\|_{L^\infty(\Gamma)}, \kappa) \mathcal{F}^n. \qquad (7.3.3.7)$$

Similarly, Theorem 7.3.3.2 follows immediately from the following two results.

Lemma 7.3.3.3 *Let* $\epsilon \in [0,1)$. *Then*

$$\mathcal{F}^{n+1} + (1-2\epsilon)\mathcal{F}^n \le \sqrt{2}\max(1,\beta)\sqrt{\mathcal{E}^{n+1}+\mathcal{E}^n}\Big\{\sum_{k=0}^{K-1}\big(\mathcal{E}^{n,n+1}_{k,k+1}(T_{k+1}) + \mathcal{E}^{n,n+1}_{k+1,k}(T_{k+1})\big)\Big\}^{1/2}. \quad (7.3.3.8)$$

Lemma 7.3.3.4 *We have*

$$e^{n,n+1} \le \frac{1}{\beta}\max(\|\alpha^{-1}\|_{L^\infty(\Gamma)},\kappa)\big(\mathcal{F}^{n+1}+\mathcal{F}^n\big) + \sqrt{2}\max(1,1/\beta)\sqrt{\mathcal{E}^{n+1}+\mathcal{E}^n}\Big\{\sum_{k=0}^{K-1}\big(\mathcal{E}^{n,n+1}_{k,k+1}(T_{k+1}) + \mathcal{E}^{n,n+1}_{k+1,k}(T_{k+1})\big)\Big\}^{1/2}. \quad (7.3.3.9)$$

Proof of Lemma 7.3.3.1. The proof follows exactly the proof of Lemma 6.3.3.1 and is omitted.

Proof of Lemma 7.3.3.2. The starting point is the identity (7.2.2.7), which we apply to the solution $(\widetilde{E}^n_k, \widetilde{H}^n_k)$ on the interval I_k to obtain

$$\|(\widetilde{E}^n_k(t), \widetilde{H}^n_k(t))\|^2_{\mathcal{H}} + \int_{T_k}^t\int_\Gamma \big(\alpha|\nu\wedge\widetilde{E}^n_k|^2 + \frac{1}{\alpha}|\widetilde{H}^n_{k\tau}|^2\big)\,d\Gamma dt \le \|(\widetilde{E}^n_k(T_k), \widetilde{H}^n_k(T_k))\|^2_{\mathcal{H}} + \int_{T_k}^t\int_\Gamma \frac{1}{\alpha}|\nu\wedge\widetilde{P}^n_k|^2 d\Gamma dt, \quad (7.3.3.10)$$

valid for $T_k \le t \le T_{k+1}$ and $k = 0,\dots,K$. We write the first term on the right side of (7.3.3.10) as

$$\|(\widetilde{E}^n_k(T_k), \widetilde{H}^n_k(T_k))\|^2_{\mathcal{H}} - \|(\widetilde{E}^n_{k-1}(T_k), \widetilde{H}^n_{k-1}(T_k))\|^2_{\mathcal{H}} + \|(\widetilde{E}^n_{k-1}(T_k), \widetilde{H}^n_{k-1}(T_k))\|^2_{\mathcal{H}} \quad (7.3.3.11)$$

and then use (7.2.2.7) again to estimate the last term as follows:

$$\|(\widetilde{E}^n_{k-1}(T_k), \widetilde{H}^n_{k-1}(T_k))\|^2_{\mathcal{H}} \le -\int_{T_{k-1}}^{T_k}\int_\Gamma \big(\alpha|\nu\wedge\widetilde{E}^n_{k-1}|^2 + \frac{1}{\alpha}|\widetilde{H}^n_{k-1,\tau}|^2\big)\,d\Gamma dt + \|(\widetilde{E}^n_{k-1}(T_{k-1}), \widetilde{H}^n_{k-1}(T_{k-1}))\|^2_{\mathcal{H}} + \int_{T_{k-1}}^{T_k}\int_\Gamma \frac{1}{\alpha}|\nu\wedge\widetilde{P}^n_{k-1}|^2 d\Gamma dt, \quad (7.3.3.12)$$

From (7.3.3.10)–(7.3.3.12) we then obtain

$$
\begin{aligned}
&\|(\widetilde{E}_k^n(t), \widetilde{H}_k^n(t))\|_{\mathcal{H}}^2 + \int_{T_{k-1}}^{T_k} \int_\Gamma \big(\alpha|\nu \wedge \widetilde{E}_{k-1}^n|^2 + \frac{1}{\alpha}|\widetilde{H}_{k-1,\tau}^n|^2\big)\, d\Gamma dt \\
&\qquad + \int_{T_k}^{t} \int_\Gamma \big(\alpha|\nu \wedge \widetilde{E}_k^n|^2 + \frac{1}{\alpha}|\widetilde{H}_{k\tau}^n|^2\big)\, d\Gamma dt \\
&\quad \le \big(\|(\widetilde{E}_k^n(T_k), \widetilde{H}_k^n(T_k))\|_{\mathcal{H}}^2 - \|(\widetilde{E}_{k-1}^n(T_k), \widetilde{H}_{k-1}^n(T_k))\|_{\mathcal{H}}^2\big) \\
&\qquad + \|(\widetilde{E}_{k-1}^n(T_{k-1}), \widetilde{H}_{k-1}^n(T_{k-1}))\|_{\mathcal{H}}^2 \\
&\qquad + \int_{T_{k-1}}^{T_k} \int_\Gamma \frac{1}{\alpha}|\nu \wedge \widetilde{P}_{k-1}^n|^2 d\Gamma dt \\
&\qquad + \int_{T_k}^{t} \int_\Gamma \frac{1}{\alpha}|\nu \wedge \widetilde{P}_k^n|^2 d\Gamma dt. \qquad (7.3.3.13)
\end{aligned}
$$

We now repeat the argument, using (7.3.3.11)–(7.3.3.13) with k replaced by $k-1$, and so forth down to $k = 0$. Since $\widetilde{E}_0^n(0) = \widetilde{H}_0^n(0) = 0$ we end up with

$$
\begin{aligned}
&\|(\widetilde{E}_k^n(t), \widetilde{H}_k^n(t))\|_{\mathcal{H}}^2 + \sum_{m=0}^{k-1} \int_{T_m}^{T_{m+1}} \int_\Gamma \big(\alpha|\nu \wedge \widetilde{E}_m^n|^2 + \frac{1}{\alpha}|\widetilde{H}_{m\tau}^n|^2\big)\, d\Gamma dt \\
&\qquad + \int_{T_k}^{t} \int_\Gamma \big(\alpha|\nu \wedge \widetilde{E}_k^n|^2 + \frac{1}{\alpha}|\widetilde{H}_{k\tau}^n|^2\big)\, d\Gamma dt \\
&\le \sum_{m=0}^{k-1} \big(\|(\widetilde{E}_{m+1}^n(T_{m+1}), \widetilde{H}_{m+1}^n(T_{m+1}))\|_{\mathcal{H}}^2 - \|(\widetilde{E}_m^n(T_{m+1}), \widetilde{H}_m^n(T_{m+1}))\|_{\mathcal{H}}^2\big) \\
&\qquad + \sum_{m=0}^{k-1} \int_{T_m}^{T_{m+1}} \int_\Gamma \frac{1}{\alpha}|\nu \wedge \widetilde{P}_m^n|^2 d\Gamma dt \\
&\qquad + \int_{T_k}^{t} \int_\Gamma \frac{1}{\alpha}|\nu \wedge \widetilde{P}_k^n|^2 d\Gamma dt. \qquad (7.3.3.14)
\end{aligned}
$$

This estimate is valid for $t \in I_k$ and $k = 0, \dots, K$.

By the Cauchy-Schwarz inequality we have

$$
\begin{aligned}
&\sum_{m=0}^{k-1} \big(\|(\widetilde{E}_{m+1}^n(T_{m+1}), \widetilde{H}_{m+1}^n(T_{m+1}))\|_{\mathcal{H}}^2 - \|(\widetilde{E}_m^n(T_{m+1}), \widetilde{H}_m^n(T_{m+1}))\|_{\mathcal{H}}^2\big) \\
&\le \sqrt{2}\Big\{\sum_{m=0}^{k-1} \|(E_{m+1}^n(T_{m+1}), H_{m+1}^n(T_{m+1})) - (E_m^n(T_{m+1}), H_m^n(T_{m+1}))\|_{\mathcal{H}}^2\Big\}^{1/2} \\
&\times \Big\{\sum_{m=0}^{k-1} \big(\|(\widetilde{E}_{m+1}^n(T_{m+1}), \widetilde{H}_{m+1}^n(T_{m+1}))\|_{\mathcal{H}}^2 + \|(\widetilde{E}_m^n(T_{m+1}), \widetilde{H}_m^n(T_{m+1}))\|_{\mathcal{H}}^2\big)\Big\}^{1/2}.
\end{aligned}
$$

(7.3.3.15)

It follows that

$$\max_{0\le k\le K}\|(\widetilde{E}_k^n,\widetilde{H}_k^n)\|^2_{L^\infty(I_k;\mathcal{H})}$$
$$\le\sqrt{2}\Big\{\sum_{k=0}^{K-1}\|(E_{k+1}^n(T_{k+1}),H_{k+1}^n(T_{k+1}))-(E_k^n(T_{k+1}),H_k^n(T_{k+1}))\|^2_{\mathcal{H}}\Big\}^{1/2}$$
$$\times\Big\{\sum_{k=0}^{K-1}\big(\|(\widetilde{E}_{k+1}^n(T_{k+1}),\widetilde{H}_{k+1}^n(T_{k+1}))\|^2_{\mathcal{H}}+\|(\widetilde{E}_k^n(T_{k+1}),\widetilde{H}_k^n(T_{k+1}))\|^2_{\mathcal{H}}\big)\Big\}^{1/2}$$
$$+\sum_{k=0}^{K}\int_{\Sigma_k}\frac{1}{\alpha}|\nu\wedge\widetilde{P}_k^n|^2d\Sigma \quad (7.3.3.16)$$

and that

$$\sum_{k=0}^{K}\int_{\Sigma_k}\big(\alpha|\nu\wedge\widetilde{E}_k^n|^2+\frac{1}{\alpha}|\widetilde{H}_{k\tau}^n|^2\big)\,d\Sigma$$
$$\le\sqrt{2}\Big\{\sum_{k=0}^{K-1}\|(E_{k+1}^n(T_{k+1}),H_{k+1}^n(T_{k+1}))-(E_k^n(T_{k+1}),H_k^n(T_{k+1}))\|^2_{\mathcal{H}}\Big\}^{1/2}$$
$$\times\Big\{\sum_{k=0}^{K-1}\big(\|(\widetilde{E}_{k+1}^n(T_{k+1}),\widetilde{H}_{k+1}^n(T_{k+1}))\|^2_{\mathcal{H}}+\|(\widetilde{E}_k^n(T_{k+1}),\widetilde{H}_k^n(T_{k+1}))\|^2_{\mathcal{H}}\big)\Big\}^{1/2}$$
$$+\sum_{k=0}^{K}\int_{\Sigma_k}\frac{1}{\alpha}|\nu\wedge\widetilde{P}_k^n|^2d\Sigma. \quad (7.3.3.17)$$

We estimate the solutions $(\widetilde{P}_k^n,\widetilde{Q}_k^n)$ in a similar manner, starting with the estimate

$$\|(\widetilde{P}_k^n(t),\widetilde{Q}_k^n(t))\|^2_{\mathcal{H}}+\int_t^{T_{k+1}}\int_\Gamma\big(\alpha|\nu\wedge\widetilde{P}_k^n|^2+\frac{1}{\alpha}|\widetilde{Q}_{k\tau}^n|^2\big)\,d\Gamma dt$$
$$\le\|(\widetilde{P}_k^n(T_{k+1}),\widetilde{Q}_k^n(T_{k+1}))\|^2_{\mathcal{H}},\quad t\in I_k. \quad (7.3.3.18)$$

The term on the right side is written

$$\|(\widetilde{P}_k^n(T_{k+1}),\widetilde{Q}_k^n(T_{k+1}))\|^2_{\mathcal{H}}-\|(\widetilde{P}_{k+1}^n(T_{k+1}),\widetilde{Q}_{k+1}^n(T_{k+1}))\|^2_{\mathcal{H}}$$
$$+\|(\widetilde{P}_{k+1}^n(T_{k+1}),\widetilde{Q}_{k+1}^n(T_{k+1}))\|^2_{\mathcal{H}}$$
$$\le\|(\widetilde{P}_k^n(T_{k+1}),\widetilde{Q}_k^n(T_{k+1}))\|^2_{\mathcal{H}}-\|(\widetilde{P}_{k+1}^n(T_{k+1}),\widetilde{Q}_{k+1}^n(T_{k+1}))\|^2_{\mathcal{H}}$$
$$+\|(\widetilde{P}_{k+1}^n(T_{k+2}),\widetilde{Q}_{k+1}^n(T_{k+2}))\|^2_{\mathcal{H}}-\int_{T_{k+1}}^{T_{k+2}}\int_\Gamma\big(\alpha|\nu\wedge\widetilde{P}_{k+1}^n|^2+\frac{1}{\alpha}|\widetilde{Q}_{k+1,\tau}^n|^2\big)\,d\Gamma dt.$$
$$(7.3.3.19)$$

From (7.3.3.18) and (7.3.3.19) we thus obtain

$$\begin{aligned}
&\|(\widetilde{P}_k^n(t), \widetilde{Q}_k^n(t))\|_{\mathcal{H}}^2 \\
&\quad + \int_t^{T_{k+1}} \int_\Gamma \Big(\alpha|\nu \wedge \widetilde{P}_k^n|^2 + \frac{1}{\alpha}|\widetilde{Q}_{k\tau}^n|^2\Big)\, d\Gamma dt \\
&\quad + \int_{T_{k+1}}^{T_{k+2}} \int_\Gamma \Big(\alpha|\nu \wedge \widetilde{P}_{k+1}^n|^2 + \frac{1}{\alpha}|\widetilde{Q}_{k+1,\tau}^n|^2\Big)\, d\Gamma dt \\
&\le \|(\widetilde{P}_k^n(T_{k+1}), \widetilde{Q}_k^n(T_{k+1}))\|_{\mathcal{H}}^2 \\
&\quad - \|(\widetilde{P}_{k+1}^n(T_{k+1}), \widetilde{Q}_{k+1}^n(T_{k+1}))\|_{\mathcal{H}}^2 \\
&\quad + \|(\widetilde{P}_{k+1}^n(T_{k+2}), \widetilde{Q}_{k+1}^n(T_{k+2}))\|_{\mathcal{H}}^2.
\end{aligned}$$

Iterating the argument leads finally to

$$\begin{aligned}
&\|(\widetilde{P}_k^n(t), \widetilde{Q}_k^n(t))\|_{\mathcal{H}}^2 \\
&\quad + \int_t^{T_{k+1}} \int_\Gamma \Big(\alpha|\nu \wedge \widetilde{P}_k^n|^2 + \frac{1}{\alpha}|\widetilde{Q}_{k\tau}^n|^2\Big)\, d\Gamma dt \\
&\quad + \sum_{m=k+1}^{K} \int_{T_m}^{T_{m+1}} \int_\Gamma \Big(\alpha|\nu \wedge \widetilde{P}_m^n|^2 + \frac{1}{\alpha}|\widetilde{Q}_{m\tau}^n|^2\Big)\, d\Gamma dt \\
&\le \sum_{m=k}^{K-1} \Big(\|(\widetilde{P}_m^n(T_{m+1}), \widetilde{Q}_m^n(T_{m+1}))\|_{\mathcal{H}}^2 \\
&\quad - \|(\widetilde{P}_{m+1}^n(T_{m+1}), \widetilde{Q}_{m+1}^n(T_{m+1}))\|_{\mathcal{H}}^2\Big) \\
&\quad + \kappa^2 \|(\widetilde{E}_K^n(T), \widetilde{H}_K^n(T))\|_{\mathcal{H}}^2
\end{aligned} \tag{7.3.3.20}$$

since $(\widetilde{P}_K^n(T), \widetilde{Q}_K^n(T)) = \kappa(\widetilde{E}_K^n(T), \widetilde{H}_K^n(T))$. It follows that

$$\begin{aligned}
&\max_{0\le k\le K} \|(\widetilde{P}_k^n, \widetilde{Q}_k^n)\|_{L^\infty(I_k;\mathcal{H})}^2 \\
&\le \sqrt{2}\Big\{ \sum_{k=0}^{K-1} \|(P_k^n(T_{k+1}), Q_k^n(T_{k+1})) \\
&\qquad\qquad - (P_{k+1}^n(T_{k+1}), Q_{k+1}^n(T_{k+1}))\|_{\mathcal{H}}^2 \Big\}^{1/2} \\
&\quad \times \Big\{ \sum_{k=0}^{K-1} \Big(\|(\widetilde{P}_k^n(T_{k+1}), \widetilde{Q}_k^n(T_{k+1}))\|_{\mathcal{H}}^2 \\
&\qquad\qquad + \|(\widetilde{P}_{k+1}^n(T_{k+1}), \widetilde{Q}_{k+1}^n(T_{k+1}))\|_{\mathcal{H}}^2\Big) \Big\}^{1/2} \\
&\qquad\qquad + \kappa^2 \|(\widetilde{E}_K^n(T), \widetilde{H}_K^n(T))\|_{\mathcal{H}}^2
\end{aligned} \tag{7.3.3.21}$$

and that

$$\sum_{k=0}^{K}\int_{\Sigma_k}\big(\alpha|\nu\wedge\widetilde{P}_k^n|^2+\frac{1}{\alpha}|\widetilde{Q}_{k\tau}^n|^2\big)\,d\Sigma$$
$$\le\sqrt{2}\Big\{\sum_{k=0}^{K-1}\|(P_k^n(T_{k+1}),Q_k^n(T_{k+1}))-(P_{k+1}^n(T_{k+1}),Q_{k+1}^n(T_{k+1}))\|_{\mathcal H}^2\Big\}^{1/2}$$
$$\times\Big\{\sum_{k=0}^{K-1}\big(\|(\widetilde{P}_k^n(T_{k+1}),\widetilde{Q}_k^n(T_{k+1}))\|_{\mathcal H}^2+\|(\widetilde{P}_{k+1}^n(T_{k+1}),\widetilde{Q}_{k+1}^n(T_{k+1}))\|_{\mathcal H}^2\big)\Big\}^{1/2}$$
$$+\kappa^2\|(\widetilde{E}_K^n(T),\widetilde{H}_K^n(T))\|_{\mathcal H}^2. \quad (7.3.3.22)$$

We therefore obtain from (7.3.3.16), (7.3.3.17), (7.3.3.21) and (7.3.3.22)

$$e^n\le 2\sqrt{2}\Big\{\sum_{k=0}^{K-1}\|(E_{k+1}^n(T_{k+1}),H_{k+1}^n(T_{k+1}))-(E_k^n(T_{k+1}),H_k^n(T_{k+1}))\|_{\mathcal H}^2\Big\}^{1/2}$$
$$\times\Big\{\sum_{k=0}^{K-1}\big(\|(\widetilde{E}_{k+1}^n(T_{k+1}),\widetilde{H}_{k+1}^n(T_{k+1}))\|_{\mathcal H}^2+\|(\widetilde{E}_k^n(T_{k+1}),\widetilde{H}_k^n(T_{k+1}))\|_{\mathcal H}^2\big)\Big\}^{1/2}$$
$$+2\sqrt{2}\Big\{\sum_{k=0}^{K-1}\|(P_k^n(T_{k+1}),Q_k^n(T_{k+1}))-(P_{k+1}^n(T_{k+1}),Q_{k+1}^n(T_{k+1}))\|_{\mathcal H}^2\Big\}^{1/2}$$
$$\times\Big\{\sum_{k=0}^{K-1}\big(\|(\widetilde{P}_k^n(T_{k+1}),\widetilde{Q}_k^n(T_{k+1}))\|_{\mathcal H}^2+\|(\widetilde{P}_{k+1}^n(T_{k+1}),\widetilde{Q}_{k+1}^n(T_{k+1}))\|_{\mathcal H}^2\big)\Big\}^{1/2}$$
$$+2\sum_{k=0}^{K}\int_{\Sigma_k}\frac{1}{\alpha}|\nu\wedge\widetilde{P}_k^n|^2d\Sigma+2\kappa^2\|(\widetilde{E}_K^n(T),\widetilde{H}_K^n(T))\|_{\mathcal H}^2. \quad (7.3.3.23)$$

By applying the Cauchy-Schwarz inequality once again and recalling the definitions of $\mathcal{E}^n$ and $\mathcal{E}_{k,k+1}^n(T_{k+1})$ we obtain

$$e^n\le 2\sqrt{2}\max(1,1/\beta)\sqrt{\mathcal{E}^n}\Big\{\sum_{k=0}^{K-1}\mathcal{E}_{k,k+1}^n(T_{k+1})\Big\}^{1/2}$$
$$+2\sum_{k=0}^{K}\int_{\Sigma_k}\frac{1}{\alpha}|\nu\wedge\widetilde{P}_k^n|^2d\Sigma+2\kappa^2\|(\widetilde{E}_K^n(T),\widetilde{H}_K^n(T))\|_{\mathcal H}^2. \quad (7.3.3.24)$$

From the definition of $\mathcal{F}^n$ we obtain

$$2\sum_{k=0}^{K}\int_{\Sigma_k}\frac{1}{\alpha}|\nu\wedge\widetilde{P}_k^n|^2d\Sigma+2\kappa^2\|(\widetilde{E}_K^n(T),\widetilde{H}_K^n(T))\|_{\mathcal H}^2\le\frac{1}{\beta}\max(\|\alpha^{-1}\|_{L^\infty(\Gamma)},\kappa)\,\mathcal{F}^n. \quad (7.3.3.25)$$

Lemma 7.3.3.2 follows from (7.3.3.24) and (7.3.3.25).

Proof of Lemma 7.3.3.3. We start with our fundamental recursion which we now write as

$$\mathcal{F}^{n+1} + (1-2\epsilon)\mathcal{F}^n \le \mathcal{E}^n - \mathcal{E}^{n+1}. \tag{7.3.3.26}$$

We proceed to estimate

$$\mathcal{E}^n - \mathcal{E}^{n+1} = \sum_{k=0}^{K-1} \big\{ \big[\mathcal{E}_k^n(T_{k+1}) - \mathcal{E}_{k+1}^{n+1}(T_{k+1})\big] + \big[\mathcal{E}_{k+1}^n(T_{k+1}) - \mathcal{E}_k^{n+1}(T_{k+1})\big] \big\}. \tag{7.3.3.27}$$

We have

$$\begin{aligned}
&\mathcal{E}_k^n(T_{k+1}) - \mathcal{E}_{k+1}^{n+1}(T_{k+1}) \\
&\quad = \beta^2 \big(\|(\widetilde{E}_k^n(T_{k+1}), \widetilde{H}_k^n(T_{k+1}))\|_{\mathcal{H}} - \|(\widetilde{E}_{k+1}^{n+1}(T_{k+1}), \widetilde{H}_{k+1}^{n+1}(T_{k+1}))\|_{\mathcal{H}} \big) \\
&\qquad \cdot \big(\|(\widetilde{E}_k^n(T_{k+1}), \widetilde{H}_k^n(T_{k+1}))\|_{\mathcal{H}} + \|(\widetilde{E}_{k+1}^{n+1}(T_{k+1}), \widetilde{H}_{k+1}^{n+1}(T_{k+1}))\|_{\mathcal{H}} \big) \\
&\qquad + \big(\|(\widetilde{P}_k^n(T_{k+1}), \widetilde{Q}_k^n(T_{k+1}))\|_{\mathcal{H}} - \|(\widetilde{P}_{k+1}^{n+1}(T_{k+1}), \widetilde{Q}_{k+1}^{n+1}(T_{k+1}))\|_{\mathcal{H}} \big) \\
&\qquad\qquad \cdot \big(\|(\widetilde{P}_k^n(T_{k+1}), \widetilde{Q}_k^n(T_{k+1}))\|_{\mathcal{H}} + \|(\widetilde{P}_{k+1}^{n+1}(T_{k+1}), \widetilde{Q}_{k+1}^{n+1}(T_{k+1}))\|_{\mathcal{H}} \big).
\end{aligned}$$

Therefore by the Cauchy-Schwarz inequality we obtain

$$\begin{aligned}
&\sum_{k=0}^{K-1} \big(\mathcal{E}_k^n(T_{k+1}) - \mathcal{E}_{k+1}^{n+1}(T_{k+1}) \big) \\
&\quad \le \sqrt{2} \Big\{ \sum_{k=0}^{K-1} \big[\beta^2 \|(E_k^n(T_{k+1}) - E_{k+1}^{n+1}(T_{k+1}), H_k^n(T_{k+1}) - H_{k+1}^{n+1}(T_{k+1}))\|_{\mathcal{H}}^2 \\
&\qquad + \|(P_k^n(T_{k+1}) - P_{k+1}^{n+1}(T_{k+1}), Q_k^n(T_{k+1}) - Q_{k+1}^{n+1}(T_{k+1}))\|_{\mathcal{H}}^2 \big] \Big\}^{1/2} \\
&\quad \cdot \Big\{ \sum_{k=0}^{K-1} \big[\beta^2 \big(\|(\widetilde{E}_k^n(T_{k+1}), \widetilde{H}_k^n(T_{k+1}))\|_{\mathcal{H}}^2 + \|(\widetilde{E}_{k+1}^{n+1}(T_{k+1}), \widetilde{H}_{k+1}^{n+1}(T_{k+1}))\|_{\mathcal{H}}^2 \big) \\
&\qquad + \|(\widetilde{P}_k^n(T_{k+1}), \widetilde{Q}_k^n(T_{k+1}))\|_{\mathcal{H}}^2 + \|(\widetilde{P}_{k+1}^{n+1}(T_{k+1}), \widetilde{Q}_{k+1}^{n+1}(T_{k+1}))\|_{\mathcal{H}}^2 \big] \Big\}^{1/2} \\
&\quad \le \sqrt{2} \max(1,\beta) \Big\{ \sum_{k=0}^{K-1} \mathcal{E}_{k,k+1}^{n,n+1}(T_{k+1}) \Big\}^{1/2} \Big\{ \sum_{k=0}^{K-1} B_{k,k+1}^{n,n+1}(T_{k+1}) \Big\}^{1/2},
\end{aligned} \tag{7.3.3.28}$$

where

$$\begin{aligned}
&B_{k,k+1}^{n,n+1}(T_{k+1}) \\
&\quad = \beta^2 \big(\|(\widetilde{E}_k^n(T_{k+1}), \widetilde{H}_k^n(T_{k+1}))\|_{\mathcal{H}}^2 + \|(\widetilde{E}_{k+1}^{n+1}(T_{k+1}), \widetilde{H}_{k+1}^{n+1}(T_{k+1}))\|_{\mathcal{H}}^2 \big) \\
&\qquad + \|(\widetilde{P}_k^n(T_{k+1}), \widetilde{Q}_k^n(T_{k+1}))\|_{\mathcal{H}}^2 + \|(\widetilde{P}_{k+1}^{n+1}(T_{k+1}), \widetilde{Q}_{k+1}^{n+1}(T_{k+1}))\|_{\mathcal{H}}^2.
\end{aligned}$$

Similarly

$$\sum_{k=0}^{K-1}\big(\mathcal{E}_{k+1}^{n}(T_{k+1})-\mathcal{E}_{k}^{n+1}(T_{k+1})\big)$$
$$\leq \sqrt{2}\max(1,\beta)\Big\{\sum_{k=0}^{K-1}\mathcal{E}_{k+1,k}^{n,n+1}(T_{k+1})\Big\}^{1/2}\Big\{\sum_{k=0}^{K-1}B_{k+1,k}^{n,n+1}(T_{k+1})\Big\}^{1/2}. \quad (7.3.3.29)$$

It follows from (7.3.3.27)–(7.3.3.29) that

$$\mathcal{E}^{n}-\mathcal{E}^{n+1}\leq \sqrt{2}\max(1,\beta)\Big\{\sum_{k=0}^{K-1}\big(B_{k,k+1}^{n,n+1}(T_{k+1})+B_{k+1,k}^{n,n+1}(T_{k+1})\big)\Big\}^{1/2}$$
$$\cdot\Big\{\sum_{k=0}^{K-1}\big(\mathcal{E}_{k,k+1}^{n,n+1}(T_{k+1})+\mathcal{E}_{k+1,k}^{n,n+1}(T_{k+1})\big)\Big\}^{1/2} \quad (7.3.3.30)$$

The proof is completed by noting that

$$\sum_{k=0}^{K-1}\big(B_{k,k+1}^{n,n+1}(T_{k+1})+B_{k+1,k}^{n,n+1}(T_{k+1})\big)=\mathcal{E}^{n+1}+\mathcal{E}^{n}. \quad (7.3.3.31)$$

Proof of Lemma 7.3.3.4. From (7.3.3.14), (7.3.3.20) and (7.3.3.25) we have

$$\begin{aligned}
&e_k^{n+1}(t)+e_k^{n}(t)\\
&\leq \sum_{m=0}^{k-1}\big(\|(\widetilde{E}_{m+1}^{n+1}(T_{m+1}),\widetilde{H}_{m+1}^{n+1}(T_{m+1}))\|_{\mathcal{H}}^{2}\\
&\qquad -\|(\widetilde{E}_{m}^{n}(T_{m+1}),\widetilde{H}_{m}^{n}(T_{m+1}))\|_{\mathcal{H}}^{2}\\
&\qquad +\|(\widetilde{E}_{m+1}^{n}(T_{m+1}),\widetilde{H}_{m+1}^{n}(T_{m+1}))\|_{\mathcal{H}}^{2}\\
&\qquad -\|(\widetilde{E}_{m}^{n+1}(T_{m+1}),\widetilde{H}_{m}^{n+1}(T_{m+1}))\|_{\mathcal{H}}^{2}\big)\\
&\quad +\sum_{m=k}^{K-1}\big(\|(\widetilde{P}_{m}^{n+1}(T_{m+1}),\widetilde{Q}_{m}^{n+1}(T_{m+1}))\|_{\mathcal{H}}^{2}\\
&\qquad -\|(\widetilde{P}_{m+1}^{n}(T_{m+1}),\widetilde{Q}_{m+1}^{n}(T_{m+1}))\|_{\mathcal{H}}^{2}\\
&\qquad +\|(\widetilde{P}_{m}^{n}(T_{m+1}),\widetilde{Q}_{m}^{n}(T_{m+1}))\|_{\mathcal{H}}^{2}\\
&\qquad -\|(\widetilde{P}_{m+1}^{n+1}(T_{m+1}),\widetilde{Q}_{m+1}^{n+1}(T_{m+1}))\|_{\mathcal{H}}^{2}\big)\\
&\quad +\frac{1}{\beta}\max(\|\alpha^{-1}\|_{L^{\infty}(\Gamma)},\kappa)(\mathcal{F}^{n+1}+\mathcal{F}^{n}).
\end{aligned} \quad (7.3.3.32)$$

The sums on the right are estimated as in (7.3.3.15), resulting in

$$\begin{aligned}
&\max_{0\le k\le K} \|e_k^{n+1}+e_k^n\|_{L^\infty(I_k)} \\
&\le \sqrt{2}\Big\{\sum_{k=0}^{K-1}\big(\|(E_{k+1}^{n+1}(T_{k+1}),H_{k+1}^{n+1}(T_{k+1}))-(E_k^n(T_{k+1}),H_k^n(T_{k+1}))\|_{\mathcal H}^2 \\
&\quad+\|(E_{k+1}^n(T_{k+1}),H_{k+1}^n(T_{k+1}))-(E_k^{n+1}(T_{k+1}),H_k^{n+1}(T_{k+1}))\|_{\mathcal H}^2\big)\Big\}^{1/2} \\
&\quad\times\Big\{\sum_{k=0}^{K-1}\big(\|(\widetilde E_{k+1}^{n+1}(T_{k+1}),\widetilde H_{k+1}^{n+1}(T_{k+1}))\|_{\mathcal H}^2+\|(\widetilde E_k^n(T_{k+1}),\widetilde H_k^n(T_{k+1}))\|_{\mathcal H}^2 \\
&\quad+\|(\widetilde E_{k+1}^n(T_{k+1}),\widetilde H_{k+1}^n(T_{k+1}))\|_{\mathcal H}^2+\|(\widetilde E_k^{n+1}(T_{k+1}),\widetilde H_k^{n+1}(T_{k+1}))\|_{\mathcal H}^2\big)\Big\}^{1/2} \\
&\quad+\Big\{\sum_{k=0}^{K-1}\big(\|(P_k^{n+1}(T_{k+1}),Q_k^{n+1}(T_{k+1}))-(P_{k+1}^n(T_{k+1}),Q_{k+1}^n(T_{k+1}))\|_{\mathcal H}^2 \\
&\quad+\|(P_k^n(T_{k+1}),Q_k^n(T_{k+1}))-(P_{k+1}^{n+1}(T_{k+1}),Q_{k+1}^{n+1}(T_{k+1}))\|_{\mathcal H}^2\big)\Big\}^{1/2} \\
&\quad\times\Big\{\sum_{k=0}^{K-1}\big(\|\widetilde P_k^{n+1}(T_{k+1}),\widetilde Q_k^{n+1}(T_{k+1}))\|_{\mathcal H}^2+\|(\widetilde P_{k+1}^n(T_{k+1}),\widetilde Q_{k+1}^n(T_{k+1}))\|_{\mathcal H}^2 \\
&\quad+\|\widetilde P_k^n(T_{k+1}),\widetilde Q_k^n(T_{k+1}))\|_{\mathcal H}^2+\|(\widetilde P_{k+1}^{n+1}(T_{k+1}),\widetilde Q_{k+1}^{n+1}(T_{k+1}))\|_{\mathcal H}^2\Big\}^{1/2} \\
&\qquad\qquad+\frac{1}{\beta}\max(\|\alpha^{-1}\|_{L^\infty(\Gamma)},\kappa)(\mathcal F^{n+1}+\mathcal F^n).
\end{aligned}$$

After one more application of the Cauchy-Schwarz inequality we obtain the conclusion of Lemma 7.3.3.4.

Proof of Corollary 7.3.3.1. Since $\mathcal E^n+\mathcal F^n=\|X^n\|_{\mathcal X}^2$ is nonincreasing, it suffices to estimate $\mathcal E^1+\mathcal F^1$. From (7.3.2.21) and (7.3.2.10) we have

$$\mathcal E_\beta^1+\mathcal F_\beta^1=\sum_{k=0}^{K-1}\big(\|(\tilde\mu_{k,k+1}^0,\tilde\eta_{k,k+1}^0)\|_{\mathcal H}^2+\|(\tilde\mu_{k+1,k}^0,\tilde\eta_{k+1,k}^0)\|_{\mathcal H}^2\big). \tag{7.3.3.33}$$

By (7.3.2.7) the right side of (7.3.3.33) is bounded above by

$$\begin{aligned}
&2\sum_{k=0}^{K-1}\big(\|(\mu_{k,k+1}^0,\eta_{k,k+1}^0)\|_{\mathcal H}^2+\|(\mu_{k+1,k}^0,\eta_{k+1,k}^0)\|_{\mathcal H}^2\big) \\
&\quad+4\sum_{k=0}^{K-1}\big(\beta^2\|(E(T_{k+1}),H(T_{k+1}))\|_{\mathcal H}^2+\|(P(T_{k+1}),Q(T_{k+1}))\|_{\mathcal H}^2\big).
\end{aligned}\tag{7.3.3.34}$$

The following a priori estimate for the solution of (7.1.1.1) can be proved by standard multiplier methods:

$$(1-\delta)\|(E,H)\|^2_{L^\infty(0,T;\mathcal{H})} \le \|(E_0,H_0)\|^2_{\mathcal{H}} + \int_\Sigma \frac{1}{\alpha}|J|^2 d\Sigma + \frac{1}{\delta}\|(\varepsilon^{-1}F,\mu^{-1}G)\|^2_{L^1(0,T;\mathcal{H})}, \quad 0<\delta<1. \tag{7.3.3.35}$$

Thus for every $k=0,\dots,K-1$ we have

$$\|(E(T_{k+1}),H(T_{k+1}))\|^2_{\mathcal{H}} \le 2\|(E_0,H_0)\|^2_{\mathcal{H}} + 2\int_\Sigma \frac{1}{\alpha}|\nu\wedge P|^2 d\Sigma + 4\|(\varepsilon^{-1}F,\mu^{-1}G)\|^2_{L^1(0,T;\mathcal{H})}.$$

In addition,

$$\|(P(T_{k+1}),Q(T_{k+1}))\|^2_{\mathcal{H}} \le \|(P(T),Q(T))\|^2_{\mathcal{H}} = \kappa^2\|(E(T)-E_T,H(T)-H_T)\|^2_{\mathcal{H}}.$$

Thus the right side of (7.3.3.34) is bounded above by

$$\begin{aligned} &2\sum_{k=0}^{K-1}\left(\|(\mu^0_{k,k+1},\eta^0_{k,k+1})\|^2_{\mathcal{H}} + \|(\mu^0_{k+1,k},\eta^0_{k+1,k})\|^2_{\mathcal{H}}\right) \\ &\quad + 4K\big\{2\beta^2\|(E_0,H_0)\|^2_{\mathcal{H}} + 4\beta^2\|(\varepsilon^{-1}F,\mu^{-1}G)\|^2_{L^1(0,T;\mathcal{H})} \\ &+ \max(2\beta^2\|\alpha^{-1}\|_{L^\infty(\Gamma)},\kappa)\Big(\int_\Sigma |\nu\wedge P|^2 d\Sigma + \kappa\|(E(T)-E_T,H(T)-H_T)\|^2_{\mathcal{H}}\Big)\big\}. \end{aligned} \tag{7.3.3.36}$$

Since $\mathcal{J}(\nu\wedge P|_\Sigma) \le \mathcal{J}(0)$, we have

$$\int_\Sigma |\nu\wedge P|^2 d\Sigma + \kappa\|(E(T)-E_T,H(T)-H_T)\|^2_{\mathcal{H}} \le \kappa\|(\hat{E}(T)-E_T,\hat{H}(T)-H_T)\|^2_{\mathcal{H}},$$

where $(\hat{E},\hat{H})$ is the solution of (7.1.1.1) corresponding to $J=0$. Therefore

$$\|(\hat{E}(T),\hat{H}(T))\|^2_{\mathcal{H}} \le 2\|(E_0,H_0)\|^2_{\mathcal{H}} + 4\|(\varepsilon^{-1}F,\mu^{-1}G)\|^2_{L^1(0,T;\mathcal{H})},$$

hence

$$\begin{aligned} &\int_\Sigma |\nu\wedge P|^2 d\Sigma + \kappa\|(E(T)-E_T,H(T)-H_T)\|^2_{\mathcal{H}} \\ &\le 2\kappa\big(2\|(E_0,H_0)\|^2_{\mathcal{H}} + 4\|(\varepsilon^{-1}F,\mu^{-1}G)\|^2_{L^1(0,T;\mathcal{H})} + \|(E_T,H_T)\|^2_{\mathcal{H}}\big). \end{aligned} \tag{7.3.3.37}$$

It follows from (7.3.3.33)–(7.3.3.37) that

$$\mathcal{E}^1 + \mathcal{F}^1 \le \tilde{C} \tag{7.3.3.38}$$

where

$$\tilde{C} = 2\sum_{k=0}^{K-1}\left(\|(\mu^0_{k,k+1},\eta^0_{k,k+1})\|^2_{\mathcal{H}} + \|(\mu^0_{k+1,k},\eta^0_{k+1,k})\|^2_{\mathcal{H}}\right)$$
$$+ 4\big(K\beta^2 + \kappa\max(\beta^2\|\alpha^{-1}\|_{L^\infty(\Gamma)},\kappa)\big)\big(\|(E_0,H_0)\|^2_{\mathcal{H}}$$
$$+ 2\|(\varepsilon^{-1}F,\mu^{-1}G)\|^2_{L^1(0,T;\mathcal{H})}\big) + \kappa\max(\beta^2\|\alpha^{-1}\|_{L^\infty(\Gamma)},\kappa)\|(E_T,H_T)\|^2_{\mathcal{H}}\big). \tag{7.3.3.39}$$

The proof of Corollary 7.3.3.1 is complete.

Corollary 7.3.3.2 is an immediate consequence of (7.3.3.38), since $\mathcal{E}^n + \mathcal{E}^{n+1} \le 2(\mathcal{E}^n + \mathcal{F}^n)$.

7.4 Decomposition of the Spatial Domain

7.4.1 Description of the Algorithm

In this section we describe an iterative spatial domain decomposition algorithm for the optimality system (7.1.1.1), (7.2.1.3), (7.2.1.2). Let $\{\Omega_i\}_{i=1}^I$ be bounded domains in $\mathbb{R}^3$ with piecewise smooth, Lipschitz boundaries such that

$$\Omega_i \cap \Omega_j = \emptyset,\ i \ne j, \quad \Omega_i \subset \Omega,\ i = 1,\dots,I, \quad \overline{\Omega} = \bigcup_{i=1}^I \overline{\Omega}_i.$$

We set

$$\Gamma_{ij} = \partial\Omega_i \cap \partial\Omega_j = \Gamma_{ji},\ i \ne j,\ \Gamma_i = \partial\Omega_i \cap \Gamma,\ \Gamma_i^{\text{int}} = \bigcup_{j\in\mathcal{J}_i}\Gamma_{ij},\ \mathcal{J}_i = \{j : \Gamma_{ij} \ne \emptyset\}.$$

Then $\partial\Omega_i = \Gamma_i^{\text{int}} \cup \Gamma_i$. It is assumed that each Γ_i and Γ_{ij} is either empty or has a nonempty interior. We further set

$$\mathcal{Q}_i = \Omega_i \times (0,T), \quad \Sigma_{ij} = \Gamma_{ij} \times (0,T), \quad \Sigma_i = \Gamma_i \times (0,T), \quad \Sigma_i^{\text{int}} = \Gamma_i^{\text{int}} \times (0,T).$$

Let (E,H), (P,Q) be the solution of the optimality system (7.1.1.1), (7.2.1.3), (7.2.1.2) with $(E_0,H_0) \in \mathcal{H}$. The global optimality system may be formally expressed as the coupled system

$$\begin{aligned}
&\begin{cases} \varepsilon_i E_i' - \operatorname{rot} H_i + \sigma_i E_i = F_i \\ \mu_i H_i' + \operatorname{rot} E_i = G_i \qquad \text{in } \mathcal{Q}_i \end{cases}\\
&H_{i\tau} - \alpha_i(\nu_i \wedge E_i) = \nu \wedge P_i \text{ on } \Sigma_i\\
&E_i(0) = E_{0i}, \quad H_i(0) = H_{0i} \text{ in } \Omega_i,
\end{aligned} \tag{7.4.1.1}$$

$$\begin{cases} \varepsilon_i P_i' - \operatorname{rot} Q_i - \sigma_i P_i = 0 \\ \mu_i Q_i' + \operatorname{rot} P_i = 0 \qquad \text{in } \mathcal{Q}_i \end{cases}$$
$$Q_{i\tau} + \alpha_i(\nu_i \wedge P_i) = 0 \text{ on } \Sigma_i \tag{7.4.1.2}$$
$$(P_i(T), Q_i(T)) = \kappa((E_i(T), H_i(T)) - (E_{iT}, H_{iT})) \text{ in } \Omega_i,$$

together with the *transmission conditions*

$$\begin{cases} \nu_i \wedge E_i = -\nu_j \wedge E_j, \quad \nu_i \wedge H_i = -\nu_j \wedge H_j, \\ \nu_i \wedge P_i = -\nu_j \wedge P_j, \quad \nu_i \wedge Q_i = -\nu_j \wedge Q_j \quad \text{on } \Sigma_{ij}. \end{cases} \tag{7.4.1.3}$$

The subscript i on E, H, P, Q, F, G indicates restriction to $\mathcal{Q}_i$; α_i is the restriction of α to Γ_i. For the remaining coefficients and data the subscript indicates restriction to Ω_i. The vector ν_i is the unit exterior pointing normal vector to $\partial\Omega_i$. The interface conditions (7.4.1.3) are realized in a weak sense by the solution of the global optimality system and will hold in the sense of traces if

$$\operatorname{rot} E(t),\ \operatorname{rot} H(t),\ \operatorname{rot} P(t),\ \operatorname{rot} Q(t) \in \mathcal{L}^2(\Omega),\ 0 < t < T.$$

Remark 7.4.1.1 Although the tangential components of E, H, P, Q are required to be continuous across an interface, in general there may be a discontinuity in their normal components.

It is easy to see that (7.4.1.3) will hold if and only if $E_{i\tau} = E_{j\tau}$ on Σ_{ij} and similarly for the remaining three interface conditions. We further note that (7.4.1.3) is equivalent to

$$\begin{aligned} H_{i\tau} - \beta(\nu_i \wedge E_i) - \gamma(\nu_i \wedge P_i) &= H_{j\tau} + \beta(\nu_j \wedge E_j) + \gamma(\nu_j \wedge P_j) \\ Q_{i\tau} + \beta(\nu_i \wedge P_i) - \gamma(\nu_i \wedge E_i) &= Q_{j\tau} - \beta(\nu_j \wedge P_j) + \gamma(\nu_j \wedge E_j), \end{aligned} \tag{7.4.1.4}$$

where β, γ are nonzero constants. Indeed, by interchanging i and j in (7.4.1.4) and subtracting the results from (7.4.1.4) we find that $H_{i\tau} = H_{j\tau}$, $Q_{i\tau} = Q_{j\tau}$, and that

$$\begin{aligned} \beta(\nu_i \wedge E_i) + \gamma(\nu_i \wedge P_i) &= -\beta(\nu_j \wedge E_j) - \gamma(\nu_j \wedge P_j) \\ \beta(\nu_i \wedge P_i) - \gamma(\nu_i \wedge E_i) &= -\beta(\nu_j \wedge P_j) + \gamma(\nu_j \wedge E_j), \end{aligned}$$

hence $\nu_i \wedge E_i = -\nu_j \wedge E_j$ and $\nu_i \wedge P_i = -\nu_j \wedge P_j$.

Let us now consider the local iterations

$$\begin{cases} \varepsilon_i (E_i^{n+1})' - \operatorname{rot} H_i^{n+1} + \sigma_i E_i^{n+1} = F_i \\ \mu_i (H_i^{n+1})' + \operatorname{rot} E_i^{n+1} = G_i \qquad \text{in } \mathcal{Q}_i \end{cases}$$
$$H_{i\tau}^{n+1} - \alpha_i(\nu_i \wedge E_i^{n+1}) = \nu_i \wedge P_i^{n+1} \text{ on } \Sigma_i \tag{7.4.1.5}$$
$$E_i^{n+1}(0) = E_{0i}, \quad H_i^{n+1}(0) = H_{0i} \text{ in } \Omega_i,$$

$$\begin{cases} \varepsilon_i (P_i^{n+1})' - \operatorname{rot} Q_i^{n+1} - \sigma_i P_i = 0 \\ \mu_i (Q_i^{n+1})' + \operatorname{rot} P_i^{n+1} = 0 \qquad \text{in } \mathcal{Q}_i \end{cases}$$
$$Q_{i\tau}^{n+1} + \alpha_i(\nu_i \wedge P_i^{n+1}) = 0 \text{ on } \Sigma_i \tag{7.4.1.6}$$
$$(P_i^{n+1}(T), Q_i^{n+1}(T)) = \kappa((E_i^{n+1}(T), H_i^{n+1}(T)) - (E_{iT}, H_{iT})) \text{ in } \Omega_i$$

$$\begin{cases} H_{i\tau}^{n+1} - \beta(\nu_i \wedge E_i^{n+1}) - \gamma(\nu_i \wedge P_i^{n+1}) = \lambda_{ij}^n \\ Q_{i\tau}^{n+1} + \beta(\nu_i \wedge P_i^{n+1}) - \gamma(\nu_i \wedge E_i^{n+1}) = \rho_{ij}^n \quad \text{on } \Sigma_{ij} \end{cases} \tag{7.4.1.7}$$

where

$$\lambda_{ij}^n = (1-\epsilon)[H_{j\tau}^n + \beta(\nu_j \wedge E_j^n) + \gamma(\nu_j \wedge P_j^n)] \\ +\epsilon[H_{i\tau}^n - \beta(\nu_i \wedge E_i^n) - \gamma(\nu_i \wedge P_i^n)]\big|_{\Sigma_{ij}} \tag{7.4.1.8}$$
$$\rho_{ij}^n = (1-\epsilon)[Q_{j\tau}^n - \beta(\nu_j \wedge P_j^n) + \gamma(\nu_j \wedge E_j^n)] \\ +\epsilon[Q_{i\tau}^n + \beta(\nu_i \wedge P_i^n) - \gamma(\nu_i \wedge E_i^n)]\big|_{\Sigma_{ij}}. \tag{7.4.1.9}$$

In (7.4.1.8), (7.4.1.9), $\epsilon \in [0,1)$ is a relaxation parameter.

Let $\mathcal{H}_i = \mathcal{L}^2(\Omega_i) \times \mathcal{L}^2(\Omega_i)$ with weight matrix $M_i = \operatorname{diag}(\varepsilon_i, \mu_i)$. We denote by $\langle \cdot, \cdot \rangle_i$ the natural scalar product in $\mathcal{L}^2(\Omega_i)$.

Theorem 7.4.1.1 *Assume that $\beta > 0$, $\gamma > 0$. If λ_{ij}^n, $\rho_{ij}^n \in \mathcal{L}_\tau^2(\Sigma_{ij})$, $\forall j \in \mathcal{J}_i$, problem (7.4.1.5)–(7.4.1.7) has a unique solution such that $(E_i^{n+1}, H_i^{n+1}) \in C([0,T];\mathcal{H}_i)$, $(P_i^{n+1}, Q_i^{n+1}) \in C([0,T];\mathcal{H}_i)$, and all of the traces appearing in the boundary conditions in (7.4.1.5)–(7.4.1.7) are in $\mathcal{L}_\tau^2$.*

Proof. This theorem is a consequence of the fact that (7.4.1.5)–(7.4.1.7) is the optimality system for the local optimal control problem

$$\int_{\Sigma_i} |J_i|^2 d\Sigma + \frac{1}{\gamma} \sum_{j\in\mathcal{J}_i} \int_{\Sigma_{ij}} \{|J_{ij}|^2 + |\gamma(\nu_i \wedge E_i) + \rho_{ij}^n|^2\} d\Sigma \\ + \kappa \|(E_i(T) - E_{iT}, H_i(T) - H_{iT})\|_{\mathcal{H}_i}^2 \to \inf_{J_i, J_{ij}}$$

subject to

$$\begin{cases} \varepsilon_i E_i' - \operatorname{rot} H_i + \sigma_i E_i = F_i \\ \mu_i H_i' + \operatorname{rot} E_i = G_i \qquad \text{in } \mathcal{Q}_i \end{cases}$$
$$H_{i\tau} - \alpha_i(\nu_i \wedge E_i) = J_i \text{ on } \Sigma_i \tag{7.4.1.10}$$
$$H_{i\tau} - \beta(\nu_i \wedge E_i) = J_{ij} + \lambda_{ij}^n \text{ on } \Sigma_{ij}$$
$$E_i(0) = E_{0i}, \quad H_i(0) = H_{0i} \text{ in } \Omega_i,$$

where $J_i \in \mathcal{L}_\tau^2(\Sigma_i)$, $J_{ij} \in \mathcal{L}_\tau^2(\Sigma_{ij})$. Since $\beta > 0$, problem (7.4.1.10) has the same structure as (7.1.1.1) and, therefore, from Theorem 7.2.2.1 its solution satisfies

$(E_i, H_i) \in C([0,T]; \mathcal{H}_i)$, $\nu_i \wedge E_i$, $H_{i\tau} \in \mathcal{L}^2_\tau(\Sigma_i) \cup \mathcal{L}^2_\tau(\Sigma_i^{\text{int}})$. The optimal functions J_i, J_{ij}, $j \in \mathcal{J}_i$, are characterized by the variational equation

$$\int_{\Sigma_i} J_i \cdot \hat{J}_i d\Sigma + \frac{1}{\gamma} \sum_{j\in\mathcal{J}_i} \int_{\Sigma_{ij}} \{J_{ij} \cdot \hat{J}_{ij} + (\gamma(\nu_i \cdot E_i) + \rho^n_{ij}) \cdot (\nu_i \cdot \hat{E}_i)\} d\Sigma$$
$$+ \kappa \langle (E_i(T) - E_{iT}, H_i(T) - H_{iT}), (\hat{E}_i(T), \hat{H}_i(T)) \rangle_{\mathcal{H}_i} = 0,$$
$$\forall \hat{J}_i \in \mathcal{L}^2_\tau(\Sigma_i),\ \hat{J}_{ij} \in \mathcal{L}^2_\tau(\Sigma_{ij}), \quad (7.4.1.11)$$

where $(\hat{E}_i, \hat{H}_i)$ is the solution of

$$\begin{cases} \varepsilon_i \hat{E}'_i - \operatorname{rot} \hat{H}_i + \sigma_i \hat{E}_i = 0 \\ \mu_i \hat{H}'_i + \operatorname{rot} \hat{E}_i = 0 \qquad \text{in } Q_i \end{cases}$$
$$\hat{H}_{i\tau} - \alpha_i(\nu_i \wedge \hat{E}_i) = \hat{J}_i \text{ on } \Sigma_i \quad (7.4.1.12)$$
$$\hat{H}_{i\tau} - \beta(\nu_i \wedge \hat{E}_i) = \hat{J}_{ij} \text{ on } \Sigma_{ij}$$
$$\hat{E}_i(0) = \hat{H}_i(0) = 0 \text{ in } \Omega_i.$$

Let (P_i, Q_i) be the solution of

$$\begin{cases} \varepsilon_i P'_i - \operatorname{rot} Q_i - \sigma_i P_i = 0 \\ \mu_i Q'_i + \operatorname{rot} P_i = 0 \qquad \text{in } Q_i \end{cases}$$
$$Q_{i\tau} + \alpha_i(\nu_i \wedge P_i) = 0 \text{ on } \Sigma_i \quad (7.4.1.13)$$
$$Q_{i\tau} + \beta(\nu_i \wedge P_i) = \gamma(\nu_i \wedge E_i) + \rho^n_{ij} \text{ on } \Sigma_{ij}$$
$$(P_i(T), Q_i(T)) = \kappa(E_i(T) - E_{iT}, H_i(T) - H_{iT}) \text{ in } \Omega_i.$$

Since $\gamma(\nu_i \wedge E_i) + \rho^n_{ij} \in \mathcal{L}^2_\tau(\Sigma_{ij})$, $\forall j \in \mathcal{J}_i$, (7.4.1.13) has a unique solution $(P_i, Q_i) \in C([0,T]; \mathcal{H}_i)$ and $\nu_i \wedge P_i$, $Q_{i\tau} \in \mathcal{L}^2_\tau(\Sigma_i) \cup \mathcal{L}^2_\tau(\Sigma_i^{\text{int}})$. If we apply Theorem 7.2.2.2 to (7.4.1.12) and (7.4.1.13) we obtain

$$\kappa \langle (E_i(T) - E_{iT}, H_i(T) - H_{iT}), (\hat{E}_i(T), \hat{H}_i(T)) \rangle_{\mathcal{H}_i}$$
$$+ \sum_{j\in\mathcal{J}_i} \int_{\Sigma_{ij}} (\gamma(\nu_i \cdot E_i) + \rho^n_{ij}) \cdot (\nu_i \cdot \hat{E}_i) d\Sigma$$
$$= - \int_{\Sigma_i} (\nu_i \cdot P_i) \cdot \hat{J}_i d\Sigma - \sum_{j\in\mathcal{J}_i} \int_{\Sigma_{ij}} (\nu_i \cdot P_i) \cdot \hat{J}_{ij} d\Sigma. \quad (7.4.1.14)$$

It follows from (7.4.1.11) and (7.4.1.14) that the optimal solution is given by

$$J_i = \nu_i \wedge P_i|_{\Sigma_i}, \quad J_{ij} = \gamma(\nu_i \wedge P_i)|_{\Sigma_{ij}}.$$

This completes the proof.

Remark 7.4.1.2 In the local optimal control problem on the region Q_i the control $J_i = \nu_i \wedge P_i|_{\Sigma_i}$ is the *effective control* and the controls J_{ij} are *virtual controls*; see [76, 77, 73, 74].

As a consequence of Theorem 7.4.1.1, it follows that the iterations (7.4.1.5)–(7.4.1.7) are well defined for $n = 0, 1, \ldots$, if $\beta > 0$, $\gamma > 0$, and $\lambda_{ij}^0, \rho_{ij}^0 \in \mathcal{L}_\tau^2(\Sigma_{ij})$, $\forall j \in \mathcal{J}_i$.

7.4.2 Convergence of the Iterates

In this section we prove that the solutions $\{(E_i^{n+1}, H_i^{n+1})\}_{i=1}^I$, $\{(P_i^{n+1}, Q_i^{n+1})\}_{i=1}^I$ of the local optimality systems (7.4.1.5)–(7.4.1.7) converge to the solution $\{(E_i, H_i)\}_{i=1}^I$, $\{(P_i, Q_i)\}_{i=1}^I$ of the global optimality system (7.4.1.1)–(7.4.1.3). We introduce the local errors

$$(\widetilde{E}_i^n, \widetilde{H}_i^n) := (E_i^n - E_i, H_i^n - H_i), \quad (\widetilde{P}_i^n, \widetilde{Q}_i^n) := (P_i^n - P_i, Q_i^n - Q_i), \quad n \geq 1.$$

Then $(\widetilde{E}_i^n, \widetilde{H}_i^n)$, $(\widetilde{P}_i^n, \widetilde{Q}_i^n)$ satisfy

$$\begin{cases} \varepsilon_i(\widetilde{E}_i^{n+1})' - \operatorname{rot} \widetilde{H}_i^{n+1} + \sigma_i \widetilde{E}_i^{n+1} = 0 \\ \mu_i(\widetilde{H}_i^{n+1})' + \operatorname{rot} \widetilde{E}_i^{n+1} = 0 \qquad \text{in } Q_i \end{cases}$$
$$\widetilde{H}_{i\tau}^{n+1} - \alpha_i(\nu_i \wedge \widetilde{E}_i^{n+1}) = \nu_i \wedge \widetilde{P}_i^{n+1} \text{ on } \Sigma_i \tag{7.4.2.1}$$
$$\widetilde{E}_i^{n+1}(0) = \widetilde{H}_i^{n+1}(0) = 0 \text{ in } \Omega_i,$$

$$\begin{cases} \varepsilon_i(\widetilde{P}_i^{n+1})' - \operatorname{rot} \widetilde{Q}_i^{n+1} - \sigma_i \widetilde{P}_i = 0 \\ \mu_i(\widetilde{Q}_i^{n+1})' + \operatorname{rot} \widetilde{P}_i^{n+1} = 0 \qquad \text{in } Q_i \end{cases}$$
$$\widetilde{Q}_{i\tau}^{n+1} + \alpha_i(\nu_i \wedge \widetilde{P}_i^{n+1}) = 0 \text{ on } \Sigma_i \tag{7.4.2.2}$$
$$(\widetilde{P}_i^{n+1}(T), \widetilde{Q}_i^{n+1}(T)) = \kappa(\widetilde{E}_i^{n+1}(T), \widetilde{H}_i^{n+1}(T)) \text{ in } \Omega_i,$$

$$\begin{cases} \widetilde{H}_{i\tau}^{n+1} - \beta(\nu_i \wedge \widetilde{E}_i^{n+1}) - \gamma(\nu_i \wedge \widetilde{P}_i^{n+1}) = \tilde{\lambda}_{ij}^n, \\ \widetilde{Q}_{i\tau}^{n+1} + \beta(\nu_i \wedge \widetilde{P}_i^{n+1}) - \gamma(\nu_i \wedge \widetilde{E}_i^{n+1}) = \tilde{\rho}_{ij}^n \qquad \text{on } \Sigma_{ij} \end{cases} \tag{7.4.2.3}$$

where, for $n \geq 1$,

$$\tilde{\lambda}_{ij}^n = (1-\epsilon)[\widetilde{H}_{j\tau}^n + \beta(\nu_j \wedge \widetilde{E}_j^n) + \gamma(\nu_j \wedge \widetilde{P}_j^n)] \\ + \epsilon[\widetilde{H}_{i\tau}^n - \beta(\nu_i \wedge \widetilde{E}_i^n) - \gamma(\nu_i \wedge \widetilde{P}_i^n)]|_{\Sigma_{ij}} \tag{7.4.2.4}$$

$$\tilde{\rho}_{ij}^n = (1-\epsilon)[\widetilde{Q}_{j\tau}^n - \beta(\nu_j \wedge \widetilde{P}_j^n) + \gamma(\nu_j \wedge \widetilde{E}_j^n)] \\ + \epsilon[\widetilde{Q}_{i\tau}^n + \beta(\nu_i \wedge \widetilde{P}_i^n) - \gamma(\nu_i \wedge \widetilde{E}_i^n)]|_{\Sigma_{ij}}, \tag{7.4.2.5}$$

$$\tilde{\lambda}_{ij}^0 = \lambda_{ij}^0 - [H_{j\tau} + \beta(\nu_j \wedge E_j) + \gamma(\nu_j \wedge P_j)]|_{\Sigma_{ij}} \tag{7.4.2.6}$$
$$\tilde{\rho}_{ij}^0 = \rho_{ij}^0 - [Q_{j\tau} - \beta(\nu_j \wedge P_j) + \gamma(\nu_j \wedge E_j)]|_{\Sigma_{ij}}. \tag{7.4.2.7}$$

The main convergence result of this section is the following.

Theorem 7.4.2.1 *Assume that* $\tilde{\lambda}^0_{ij}, \tilde{\rho}^0_{ij} \in \mathcal{L}^2_\tau(\Sigma_{ij})$, $i = 1, \dots, I$, $j \in \mathcal{J}_i$, *that* $\beta > 0$, $\gamma > 0$ *and that*

$$\alpha_0 := \inf_\Gamma \alpha(x) > \frac{1}{2}\left(-\frac{\gamma}{\beta} + \sqrt{\frac{\gamma^2}{\beta^2} + 1}\right), \quad \frac{\gamma}{\beta} > \frac{\kappa^2 - 1}{2\kappa}. \tag{7.4.2.8}$$

Then for any $\epsilon \in [0, 1)$ *and for* $i = 1, \dots, I$,

$$(\widetilde{E}^n_i(T), \widetilde{H}^n_i(T)) \to 0,\ (\widetilde{P}^n_i(0), \widetilde{Q}^n_i(0)) \to 0 \text{ strongly in } \mathcal{H}_i$$

$$\begin{cases} \nu_i \wedge \widetilde{E}^n_i|_{\Sigma_i} \to 0,\ \widetilde{H}^n_{i\tau}|_{\Sigma_i} \to 0 \\ \nu_i \wedge \widetilde{P}^n_i|_{\Sigma_i} \to 0,\ \widetilde{Q}^n_{i\tau}|_{\Sigma_i} \to 0 \quad \text{strongly in } \mathcal{L}^2_\tau(\Sigma_i) \end{cases}$$

$$\begin{cases} \nu_i \wedge \widetilde{E}^n_i|_{\Sigma_i^{int}} \to 0,\ \widetilde{H}^n_{i\tau}|_{\Sigma_i^{int}} \to 0 \\ \nu_i \wedge \widetilde{P}^n_i|_{\Sigma_i^{int}} \to 0,\ \widetilde{Q}^n_{i\tau}|_{\Sigma_i^{int}} \to 0 \quad \text{weakly in } \mathcal{L}^2_\tau(\Sigma_i^{int}). \end{cases}$$

In addition,

(i) *if* $\epsilon \in (0, 1)$,

$$(\widetilde{E}^n_i, \widetilde{H}^n_i) \to 0,\ (\widetilde{P}^n_i, \widetilde{Q}^n_i) \to 0 \text{ strongly in } L^\infty(0, T; \mathcal{H}_i)$$

$$\begin{cases} H^n_{i\tau} - H^n_{j\tau}|_{\Sigma_{ij}} \to 0,\ \nu_i \wedge E^n_i + \nu_j \wedge E^n_j|_{\Sigma_{ij}} \to 0 \\ Q^n_{i\tau} - Q^n_{j\tau}|_{\Sigma_{ij}} \to 0,\ \nu_i \wedge P^n_i + \nu_j \wedge P^n_j|_{\Sigma_{ij}} \to 0 \quad \text{strongly in } \mathcal{L}^2_\tau(\Sigma_{ij}). \end{cases}$$

(ii) *If* $\epsilon = 0$, *and if it is assumed that for each* $i = 1, \dots, I$, *either (i)* ε_i, μ_i *are* $C^2(\overline{\Omega}_i)$ *positive scalar functions; or (ii)* ε_i, μ_i *are* $C^\infty(\overline{\Omega}_i)$ *positive definite matrices such that* $\mu_i = f_i \varepsilon_i$ *for some* $C^\infty(\overline{\Omega}_i)$ *positive scalar function* f_i, *then*

$$(\widetilde{E}^n_i, \widetilde{H}^n_i) \to 0,\ (\widetilde{P}^n_i, \widetilde{Q}^n_i) \to 0 \text{ weakly* in } L^\infty(0, T; \mathcal{H}_i).$$

Remark 7.4.2.1 We see from Theorem 7.4.2.1 that, for any $\epsilon \in [0, 1)$, the local optimal control $\nu_i \wedge P^n_i|_{\Sigma_i}$ converges strongly in $\mathcal{L}^2_\tau(\Sigma_i)$ to the restriction to Σ_i of the global optimal control $\nu \wedge P$. Moreover, the deviation $(E^n_i(T) - E_{iT}, H^n_i(T) - H_{iT})$ of the local optimal trajectory at time T from its target state (E_{iT}, H_{iT}) converges strongly in $\mathcal{H}_i$ to the restriction to Ω_i of the deviation of the global optimal trajectory at time T from its target state.

Proof of Theorem 7.4.2.1. We introduce the space

$$\begin{aligned} \mathcal{X} &= \prod_{i=1}^I \mathcal{L}^2_\tau(\Sigma_i^{\rm int}) \times \mathcal{L}^2_\tau(\Sigma_i^{\rm int}) \\ \|X\|^2_{\mathcal{X}} &= \sum_{i=1}^I \frac{1}{2\gamma} \int_{\Sigma_i^{\rm int}} (|\lambda_i|^2 + |\rho_i|^2)\, d\Sigma, \quad X = \{(\lambda_i, \rho_i)\}_{i=1}^I. \end{aligned} \tag{7.4.2.9}$$

For $X = \{(\lambda_i, \rho_i)\}_{i=1}^I \in \mathcal{X}$ let $\{(E_i, H_i), (P_i, Q_i)\}_{i=1}^I$ be the solutions of the family of problems

$$\begin{cases} \varepsilon E_i' - \operatorname{rot} H_i + \sigma E_i = 0, & \mu H_i' + \operatorname{rot} E_i = 0 \\ \varepsilon P_i' - \operatorname{rot} Q_i - \sigma P_i = 0, & \mu Q_i' + \operatorname{rot} P_i = 0 \quad \text{in } \mathcal{Q}_i \end{cases}$$

$$\begin{cases} H_{i\tau} - \alpha_i(\nu_i \wedge E_i) - \nu_i \wedge P_i = 0 \\ Q_{i\tau} + \alpha_i(\nu_i \wedge P_i) = 0 \qquad \text{on } \Sigma_i \end{cases}$$

$$\begin{cases} H_{i\tau} - \beta(\nu_i \wedge E_i) - \gamma(\nu_i \wedge P_i) = \lambda_i \\ Q_{i\tau} + \beta(\nu_i \wedge P_i) - \gamma(\nu_i \wedge E_i) = \rho_i \quad \text{on } \Sigma_i^{\text{int}} \end{cases}$$

$$E_i(0) = H_i(0) = 0, \quad P_i(T) = \kappa E_i(T), \quad Q_i(T) = \kappa H_i(T) \text{ in } \Omega_i.$$

This system, which is nothing but the system (7.4.2.1)–(7.4.2.3) above with the superscripts removed, has a unique solution with (E_i, H_i), $(P_i, Q_i) \in C([0,T]; \mathcal{H})$, and all of the boundary terms are in $\mathcal{L}_\tau^2$. Define the linear mapping $\mathcal{T} : \mathcal{X} \mapsto \mathcal{X}$ as follows. Set

$$\begin{aligned} (\mathcal{T}X)_{ij} &= \big(H_{j\tau} + \beta(\nu_j \wedge E_j) + \gamma(\nu_j \wedge P_j), Q_{j\tau} - \beta(\nu_j \wedge P_j) + \gamma(\nu_j \wedge E_j)\big)\big|_{\Sigma_{ij}} \\ (\mathcal{T}X)_i &= \{(\mathcal{T}X)_{ij} : j \in \mathcal{J}_i\}, \quad \mathcal{T}X = \{(\mathcal{T}X)_i : i = 1, \dots, I\}. \end{aligned} \tag{7.4.2.10}$$

We see that X is a fixed point of $\mathcal{T}$ if and only if (E_i, H_i), (P_i, Q_i), $i = 1, \dots, I$, is the solution of the global optimality system corresponding to vanishing data $E_0 = H_0 = E_T = H_T = 0$, $F = G = 0$. Since for vanishing data the optimal control is $J = 0$, it follows that the only fixed point of $\mathcal{T}$ is $X = 0$.

The mapping $\mathcal{T}$ is useful for the following reason. If, for $n = 1, 2, \dots$, we set

$$X^n = \{(\tilde{\lambda}_{ij}^{n-1}, \tilde{\rho}_{ij}^{n-1}) : i = 1, \dots, I;\ j \in \mathcal{J}_i\} \tag{7.4.2.11}$$

and let $(\widetilde{E}^n, \widetilde{H}^n)$, $(\widetilde{P}^n, \widetilde{Q}^n)$ be the solution of (7.4.2.1)–(7.4.2.3) with n replaced by $n-1$, then

$$\begin{aligned} X^n = \{\big(\widetilde{H}_{i\tau}^n - \beta(\nu_i \wedge \widetilde{E}_i^n) - \gamma(\nu_i \wedge \widetilde{P}_i^n), \widetilde{Q}_{i\tau}^n + \beta(\nu_i \wedge \widetilde{P}_i^n) - \gamma(\nu_i \wedge \widetilde{E}_i^n)\big)\big|_{\Sigma_{ij}} : \\ i = 1, \dots, I;\ j \in \mathcal{J}_i\}, \end{aligned} \tag{7.4.2.12}$$

$$\begin{aligned} \mathcal{T}X^n = \{\big(\widetilde{H}_{j\tau}^n + \beta(\nu_j \wedge \widetilde{E}_j^n) + \gamma(\nu_j \wedge \widetilde{P}_j^n), \widetilde{Q}_{j\tau}^n - \beta(\nu_j \wedge \widetilde{P}_j^n) + \gamma(\nu_j \wedge \widetilde{E}_j^n)\big)\big|_{\Sigma_{ij}} : \\ i = 1, \dots, I;\ j \in \mathcal{J}_i\}, \end{aligned} \tag{7.4.2.13}$$

and the interface conditions on Σ_{ij} in (7.4.2.3) may be expressed as the fixed point iteration

$$X^{n+1} = (1 - \epsilon)\mathcal{T}X^n + \epsilon X^n. \tag{7.4.2.14}$$

The following result shows that $\mathcal{T}$ is nonexpansive if β, γ satisfy (7.4.2.8).

Lemma 7.4.2.1 *For any $X \in \mathcal{X}$ one may write*

$$\|X\|_{\mathcal{X}}^2 = \mathcal{E} + \mathcal{F}, \quad \|\mathcal{T}X\|_{\mathcal{X}}^2 = \mathcal{E} - \mathcal{F} \tag{7.4.2.15}$$

where

$$\mathcal{E} = \frac{1}{2}\sum_{i=1}^{I}\int_{\Sigma_i^{int}}\{\frac{\beta^2+\gamma^2}{\gamma}(|\nu_i \wedge E_i|^2 + |\nu_i \wedge P_i|^2) + \frac{1}{\gamma}(|H_{i\tau}|^2 + |Q_{i\tau}|^2)\}d\Sigma. \tag{7.4.2.16}$$

$$\begin{aligned}\mathcal{F} = \sum_{i=1}^{I}\Big\{&\Big(\kappa + \frac{\beta}{2\gamma}(1-\kappa^2)\Big)\|(E_i(T), H_i(T))\|_{\mathcal{H}_i}^2 + \frac{\beta}{2\gamma}\|(P_i(0), Q_i(0))\|_{\mathcal{H}_i}^2 \\ &+ \frac{\beta}{\gamma}\int_0^T (\langle\sigma_i E_i, E_i\rangle_i + \langle\sigma_i P_i, P_i\rangle_i)\,dt + \int_{\Sigma_i}\{\alpha_i\frac{\beta}{\gamma}|\nu_i \wedge E_i|^2 \\ &+ (1+\alpha_i\frac{\beta}{\gamma})|\nu_i \wedge P_i|^2 + \frac{\beta}{\gamma}\mathrm{Re}\,(\nu_i \wedge E_i)\cdot(\nu_i \wedge P_i)\}d\Sigma\Big\}.\end{aligned} \tag{7.4.2.17}$$

One has $\mathcal{F} \geq 0$ if β, γ satisfy (7.4.2.8).

Proof of Lemma 7.4.2.1. We have

$$\begin{aligned}\|X\|_{\mathcal{X}}^2 &= \sum_{i=1}^{I}\frac{1}{2\gamma}\int_{\Sigma_i^{\rm int}}\{|H_{i\tau} - \beta(\nu_i \wedge E_i) - \gamma(\nu_i \wedge P_i)|^2 \\ &\qquad + |Q_{i\tau} + \beta(\nu_i \wedge P_i) - \gamma(\nu_i \wedge E_i)|^2\}d\Sigma \\ &= \sum_{i=1}^{I}\int_{\Sigma_i^{\rm int}}\Big\{\frac{1}{2\gamma}(|H_{i\tau}|^2 + |Q_{i\tau}|^2) + \frac{\beta^2+\gamma^2}{2\gamma}(|\nu \wedge E_i|^2 + |\nu \wedge P_i|^2) \\ &\qquad + \frac{\beta}{\gamma}\mathrm{Re}\,\big(Q_{i\tau}\cdot(\nu_i \wedge P_i) - H_{i\tau}\cdot(\nu_i \wedge E_i)\big) \\ &\qquad - \mathrm{Re}\,\big(H_{i\tau}\cdot(\nu_i \wedge P_i) + Q_{i\tau}\cdot(\nu_i \wedge E_i)\big)\Big\}d\Sigma.\end{aligned} \tag{7.4.2.18}$$

From

$$0 = \int_0^T \{\langle\varepsilon E_i' - \mathrm{rot}\, H_i + \sigma_i E_i, E_i\rangle_i + \langle\mu_i H_i' + \mathrm{rot}\, E_i, H_i\rangle_i\}dt$$

we obtain

$$0 = \frac{1}{2}\|(E_i(T), H_i(T))\|_{\mathcal{H}_i}^2 + \int_0^T \langle\sigma_i E_i, E_i\rangle_i dt + \int_{\Sigma_i \cup \Sigma_i^{\rm int}} \mathrm{Re}\, H_{i\tau}\cdot(\nu_i \wedge E_i)d\Sigma$$

and, therefore,

$$\begin{aligned}\int_{\Sigma_i^{\text{int}}} &\operatorname{Re} H_{i\tau} \cdot (\nu_i \wedge E_i) d\Sigma \\ &= -\frac{1}{2}\|(E_i(T), H_i(T))\|_{\mathcal{H}_i}^2 - \int_0^T \langle \sigma_i E_i, E_i \rangle_i dt \\ &\quad - \int_{\Sigma_i} \{\alpha_i |\nu_i \wedge E_i|^2 + \operatorname{Re}(\nu_i \wedge P_i) \cdot (\nu_i \wedge E_i)\} d\Sigma. \quad (7.4.2.19)\end{aligned}$$

Similarly,

$$0 = \int_0^T \{\langle \varepsilon P_i' - \operatorname{rot} Q_i - \sigma_i P_i, P_i \rangle_i + \langle \mu_i Q_i' + \operatorname{rot} P_i, Q_i \rangle_i\} dt$$

leads to

$$\begin{aligned}\int_{\Sigma_i^{\text{int}}} &\operatorname{Re} Q_{i\tau} \cdot (\nu_i \wedge P_i) d\Sigma \\ &= -\frac{\kappa^2}{2}\|(E_i(T), H_i(T))\|_{\mathcal{H}_i}^2 + \frac{1}{2}\|(P_i(0), Q_i(0))\|_{\mathcal{H}_i}^2 \\ &\quad + \int_0^T \langle \sigma_i P_i, P_i \rangle_i \, dt + \int_{\Sigma_i} \alpha_i |\nu_i \wedge P_i|^2 d\Sigma. \quad (7.4.2.20)\end{aligned}$$

Next we calculate

$$\begin{aligned}0 &= \operatorname{Re} \int_0^T \{\langle \varepsilon E_i' - \operatorname{rot} H_i + \sigma_i E_i, P_i \rangle_i + \langle \mu_i H_i' + \operatorname{rot} E_i, Q_i \rangle_i\} dt \\ &= \kappa \|(E_i(T), H_i(T))\|_{\mathcal{H}_i}^2 \\ &\quad + \int_{\Sigma_i \cup \Sigma_i^{\text{int}}} \operatorname{Re}\{H_{i\tau} \cdot (\nu_i \wedge P_i) + (\nu_i \wedge E_i) \cdot Q_{i\tau}\} d\Sigma \\ &= \kappa \|(E_i(T), H_i(T))\|_{\mathcal{H}_i}^2 \\ &\quad + \int_{\Sigma_i} |\nu_i \wedge P_i|^2 d\Sigma \\ &\quad + \int_{\Sigma_i^{\text{int}}} \operatorname{Re}\left(H_{i\tau} \cdot (\nu_i \wedge P_i) + (\nu_i \wedge E_i) \cdot Q_{i\tau}\right) d\Sigma \quad (7.4.2.21)\end{aligned}$$

It follows from (7.4.2.18)–(7.4.2.21) that

$$\|X\|_{\mathcal{X}}^2 = \mathcal{E} + \mathcal{F}. \quad (7.4.2.22)$$

Similarly,

$$\begin{aligned}\|\mathcal{T}X\|_{\mathcal{X}}^2 &= \sum_{i=1}^{I} \frac{1}{2\gamma} \int_{\Sigma_i^{\text{int}}} \{|H_{i\tau} + \beta(\nu_i \wedge E_i) + \gamma(\nu_i \wedge P_i)|^2 \\ &\quad + |Q_{i\tau} - \beta(\nu_i \wedge P_i) + \gamma(\nu_i \wedge E_i)|^2\} d\Sigma \\ &= \sum_{i=1}^{I} \int_{\Sigma_i^{\text{int}}} \left\{ \frac{1}{2\gamma}(|H_{i\tau}|^2 + |Q_{i\tau}|^2) + \frac{\beta^2+\gamma^2}{2\gamma}(|\nu \wedge E_i|^2 + |\nu \wedge P_i|^2) \right. \\ &\quad - \frac{\beta}{\gamma}\mathrm{Re}\left(Q_{i\tau} \cdot (\nu_i \wedge P_i) - H_{i\tau} \cdot (\nu_i \wedge E_i)\right) \\ &\quad \left. + \mathrm{Re}\left(H_{i\tau} \cdot (\nu_i \wedge P_i) + Q_{i\tau} \cdot (\nu_i \wedge E_i)\right) \right\} d\Sigma \\ &= \mathcal{E} - \mathcal{F}. \end{aligned} \tag{7.4.2.23}$$

Inequality (7.4.2.15) now follows from (7.4.2.22) and (7.4.2.23).

Under the stated condition on the parameters β, γ, we have

$$\left(\kappa + \frac{\beta}{2\gamma}(1-\kappa^2)\right) > 0.$$

and, moreover, the quadratic form

$$\alpha_i \frac{\beta}{\gamma}|\nu_i \wedge \widetilde{E}_i^{n+1}|^2 + (1 + \alpha_i \frac{\beta}{\gamma})|\nu_i \wedge \widetilde{P}_i^{n+1}|^2 + \frac{\beta}{\gamma}\mathrm{Re}\,(\nu_i \wedge \widetilde{E}_i^{n+1}) \cdot (\nu_i \wedge \widetilde{P}_i^{n+1})$$

is positive definite. Therefore $\mathcal{F} \geq 0$ if β, γ satisfy (7.4.2.8). This completes the proof of Lemma 7.4.2.1.

Let X^n be given by (7.4.2.11), so that $\mathcal{T}X^n$ is given by (7.4.2.13) and the interface conditions are expressible as the fixed point iteration (7.4.2.14). From (7.4.2.22) and (7.4.2.23) we have

$$\|X^n\|_{\mathcal{X}}^2 = \mathcal{E}^n + \mathcal{F}^n, \quad \|\mathcal{T}X^n\|_{\mathcal{X}}^2 = \mathcal{E}^n - \mathcal{F}^n \tag{7.4.2.24}$$

where

$$\begin{aligned}\mathcal{E}^n = \frac{1}{2}\sum_{i=1}^{I} \int_{\Sigma_i^{\text{int}}} \{ &\frac{\beta^2+\gamma^2}{\gamma}(|\nu_i \wedge \widetilde{E}_i^n|^2 + |\nu_i \wedge \widetilde{P}_i^n|^2) \\ &+ \frac{1}{\gamma}(|\widetilde{H}_{i\tau}^n|^2 + |\widetilde{Q}_{i\tau}^n|^2)\} d\Sigma, \end{aligned} \tag{7.4.2.25}$$

$$\mathcal{F}^n = \sum_{i=1}^{I} \Big\{ \big(\kappa + \frac{\beta}{2\gamma}(1-\kappa^2)\big) \|(\widetilde{E}_i^n(T), \widetilde{H}_i^n(T))\|_{\mathcal{H}_i}^2 + \frac{\beta}{2\gamma} \|(\widetilde{P}_i^n(0), \widetilde{Q}_i^n(0))\|_{\mathcal{H}_i}^2$$

$$+ \frac{\beta}{\gamma} \int_0^T (\langle \sigma_i \widetilde{E}_i^n, \widetilde{E}_i^n \rangle_i + \langle \sigma_i \widetilde{P}_i^n, \widetilde{P}_i^n \rangle_i)\, dt + \int_{\Sigma_i} \{ \alpha_i \frac{\beta}{\gamma} |\nu_i \wedge \widetilde{E}_i^n|^2 + (1+\alpha_i \frac{\beta}{\gamma}) |\nu_i \wedge \widetilde{P}_i^n|^2$$

$$+ \frac{\beta}{\gamma} \mathrm{Re}\,(\nu_i \wedge \widetilde{E}_i^n) \cdot (\nu_i \wedge \widetilde{P}_i^n) \} d\Sigma \Big\}. \quad (7.4.2.26)$$

From Lemma 7.4.2.1, (7.4.2.24) and (7.4.2.14) we have

$$\mathcal{E}^{n+1} + \mathcal{F}^{n+1} = ((1-\epsilon)^2 + \epsilon^2)\mathcal{E}^n + 2\epsilon(1-\epsilon)\mathrm{Re}\,(\mathcal{T}X^n, X^n)_{\mathcal{X}} - (1-2\epsilon)\mathcal{F}^n \quad (7.4.2.27)$$

From (7.4.2.24) and the Cauchy-Schwarz inequality we have

$$|(\mathcal{T}X^n, X^n)_{\mathcal{X}}| \le \sqrt{(\mathcal{E}^n)^2 - (\mathcal{F}^n)^2} \le \mathcal{E}^n$$

and, consequently,

$$((1-\epsilon)^2 + \epsilon^2)\mathcal{E}^n + 2\epsilon(1-\epsilon)\mathrm{Re}\,(\mathcal{T}X^n, X^n)_{\mathcal{X}} \le \mathcal{E}^n. \quad (7.4.2.28)$$

The following *fundamental recursion relation* now follows from (7.4.2.27) and (7.4.2.28):

$$\mathcal{E}^{n+1} + \mathcal{F}^{n+1} \le \mathcal{E}^n - (1-2\epsilon)\mathcal{F}^n, \quad n = 1, 2, \ldots, \quad (7.4.2.29)$$

If (7.4.2.29) is iterated down to $n = 1$ we obtain

$$\mathcal{E}^{n+1} \le \mathcal{E}^1 - 2\sum_{p=1}^{n+1} c_p(\epsilon)\mathcal{F}^p, \quad (7.4.2.30)$$

where the $c_p(\epsilon)$ are defined in (6.4.2.26). Since $\mathcal{F}^p \ge 0$ for all p it follows from (7.4.2.30) that

$$\sum_{p=1}^{\infty} \mathcal{F}^p \text{ converges and } \{\mathcal{E}^n\}_{n=1}^{\infty} \text{ is a bounded sequence.}$$

The convergence of $\sum \mathcal{F}^p$ implies that

$$\begin{aligned} &(\widetilde{E}_i^n(T), \widetilde{H}_i^n(T)) \to 0 \text{ and } (\widetilde{P}_i^n(0), \widetilde{Q}_i^n(0)) \to 0 \text{ strongly in } \mathcal{H}_i \\ &\nu_i \wedge \widetilde{E}_i^n|_{\Sigma_i} \to 0 \text{ and } \nu_i \wedge \widetilde{P}_i^n|_{\Sigma_i} \to 0 \text{ strongly in } \mathcal{L}_\tau^2(\Sigma_i), \\ &\sigma_i \widetilde{E}_i^{n+1} \to 0,\ \sigma_i \widetilde{P}_i^{n+1} \to 0 \text{ weakly in } L^2(\mathcal{Q}_i),\ i = 1, \ldots, I. \end{aligned} \quad (7.4.2.31)$$

From (7.4.2.1) and (7.4.2.2),

$$\begin{aligned} \widetilde{H}_{i\tau}^{n+1}|_{\Sigma_i} \to 0,\ \widetilde{Q}_{i\tau}^{n+1}|_{\Sigma_i} \to 0 \text{ strongly in } \mathcal{L}_\tau^2(\Sigma_i) \\ (\widetilde{P}_i^n(T), \widetilde{Q}_i^n(T)) \to 0 \text{ strongly in } \mathcal{H}_i. \end{aligned} \tag{7.4.2.32}$$

The boundedness of $\mathcal{E}^n$ implies that $\tilde{\lambda}_{ij}^n$, $\tilde{\rho}_{ij}^n$ are bounded in $\mathcal{L}_\tau^2(\Sigma_{ij})$. Therefore

$$(\widetilde{E}_i^{n+1}, \widetilde{H}_i^{n+1}),\ (\widetilde{P}_i^{n+1}, \widetilde{Q}_i^{n+1}) \text{ are bounded in } L^\infty(0,T;\mathcal{H}_i),\ i = 1,\dots,I$$

and all of the boundary traces appearing in the interface iteration (7.4.1.7) are bounded in $\mathcal{L}_\tau^2(\Sigma_{ij})$. It follows that, on a subsequence $n = n_k$ of the positive integers,

$$\begin{aligned} &(\widetilde{E}_i^{n+1}, \widetilde{H}_i^{n+1}) \to (\widetilde{E}_i, \widetilde{H}_i) \text{ weakly* in } L^\infty(0,T;\mathcal{H}_i) \\ &(\widetilde{P}_i^{n+1}, \widetilde{Q}_i^{n+1}) \to (\widetilde{P}_i, \widetilde{Q}_i) \text{ weakly* in } L^\infty(0,T;\mathcal{H}_i) \\ &\nu_i \wedge \widetilde{E}_i^{n+1} \to A_i,\ \widetilde{H}_{i\tau}^{n+1} \to B_i \text{ weakly in } \mathcal{L}_\tau^2(\Sigma_i^{\text{int}}) \\ &\nu_i \wedge \widetilde{P}_i^{n+1} \to C_i,\ \widetilde{Q}_{i\tau}^{n+1} \to D_i \text{ weakly in } \mathcal{L}_\tau^2(\Sigma_i^{\text{int}}) \end{aligned} \tag{7.4.2.33}$$

for some A_i, B_i, C_i, $D_i \in \mathcal{L}_\tau^2(\Sigma_i^{\text{int}})$, $(\widetilde{E}_i, \widetilde{H}_i)$, $(\widetilde{P}_i, \widetilde{Q}_i) \in L^\infty(0,T;\mathcal{H}_i)$.

Let $(\phi, \psi) \in \mathcal{C}^\infty(\overline{\Omega_i} \times [0,T])$. We have

$$\begin{aligned} 0 &= \int_0^T [\langle \varepsilon_i(\widetilde{E}_i^{n+1})' - \operatorname{rot} \widetilde{H}_i^{n+1} + \sigma_i \widetilde{E}_i^{n+1}, \phi\rangle_i + \langle \mu_i(\widetilde{H}_i^{n+1})' + \operatorname{rot} \widetilde{E}_i^{n+1}, \psi\rangle_i]\, dt \\ &= \langle (\widetilde{E}_i^{n+1}(T), \widetilde{H}_i^{n+1}(T)), (\phi(T), \psi(T))\rangle_{\mathcal{H}_i} - \int_0^T [\langle \widetilde{E}_i^{n+1}, \varepsilon_i\phi' - \operatorname{rot}\psi - \sigma_i\phi\rangle_i \\ &\quad + \langle \widetilde{H}_i^{n+1}, \mu_i\psi' + \operatorname{rot}\phi\rangle_i]\, dt + \int_{\Sigma_i} \{\widetilde{H}_{i\tau}^{n+1} \cdot (\nu_i \wedge \phi) + (\nu_i \wedge \widetilde{E}_i^{n+1}) \cdot \psi_\tau\} d\Sigma \\ &\qquad + \int_{\Sigma_i^{\text{int}}} \{\widetilde{H}_{i\tau}^{n+1} \cdot (\nu_i \wedge \phi) + (\nu_i \wedge \widetilde{E}_i^{n+1}) \cdot \psi_\tau\} d\Sigma \end{aligned}$$

Upon passing to the limit through the subsequence $n = n_k$ we obtain

$$\begin{aligned} &\int_0^T [\langle \widetilde{E}_i, \varepsilon_i\phi' - \operatorname{rot}\psi\rangle_i + \langle \widetilde{H}_i, \mu_i\psi' + \operatorname{rot}\phi\rangle_i]\, dt \\ &\qquad = \int_{\Sigma_i^{\text{int}}} \{B_i \cdot (\nu_i \wedge \phi) + A_i \cdot \psi_\tau\} d\Sigma, \quad \forall (\phi,\psi) \in \mathcal{C}^\infty(\overline{\Omega_i} \times [0,T]). \end{aligned}$$

Therefore

$$\begin{aligned} &\begin{cases} \varepsilon_i \widetilde{E}_i' - \operatorname{rot} \widetilde{H}_i = 0 \\ \mu_i \widetilde{H}_i' + \operatorname{rot} \widetilde{E}_i = 0 \quad \text{in } Q_i \end{cases} \\ &\nu_i \wedge \widetilde{E}_i = \widetilde{H}_{i\tau} = 0 \text{ on } \Sigma_i \\ &\nu_i \wedge \widetilde{E}_i = A_i,\ \widetilde{H}_{i\tau} = B_i \text{ on } \Sigma_i^{\text{int}} \\ &\widetilde{E}_i(0) = \widetilde{H}_i(0) = \widetilde{E}_i(T) = \widetilde{H}_i(T) = 0 \text{ in } \Omega_i. \end{aligned} \tag{7.4.2.34}$$

Similarly,

$$\begin{cases} \varepsilon_i \widetilde{P}_i' - \operatorname{rot} \widetilde{Q}_i = 0 \\ \mu_i \widetilde{Q}_i' + \operatorname{rot} \widetilde{P}_i = 0 \quad \text{in } \mathcal{Q}_i \end{cases}$$
$$\nu_i \wedge \widetilde{P}_i = \widetilde{Q}_{i\tau} = 0 \ \text{ on } \Sigma_i \tag{7.4.2.35}$$
$$\nu_i \wedge \widetilde{P}_i = C_i, \ \widetilde{Q}_{i\tau} = D_i \ \text{ on } \Sigma_i^{\text{int}}$$
$$\widetilde{P}_i(0) = \widetilde{Q}_i(0) = \widetilde{P}_i(T) = \widetilde{Q}_i(T) = 0 \ \text{ in } \Omega_i.$$

Case 1: $\epsilon = 0$. In this case we use a unique continuation argument to conclude that $\widetilde{E}_i = \widetilde{H}_i = 0$ and $\widetilde{P}_i = \widetilde{Q}_i = 0$ in $\mathcal{Q}_i$ for $i = 1, \dots, I$. Together with (7.4.2.31) and (7.4.2.32), this will complete the proof of part (ii) of Theorem 7.4.2.1. However, the unique continuation argument requires additional hypotheses on the coefficients ε_i, μ_i as indicated in part (ii) of Theorem 7.4.2.1. Suppose first of all that Ω_i is a region adjacent to Γ, that is $\Gamma_i \neq \emptyset$. From (7.4.2.34) we have

$$\varepsilon_i \widetilde{E}_i(t) = \int_0^t \operatorname{rot} \widetilde{H}_i(s)\, ds, \quad \mu_i \widetilde{H}_i(t) = -\int_0^t \operatorname{rot} \widetilde{E}_i(s)\, ds,$$

hence

$$\operatorname{div}(\varepsilon_i \widetilde{E}_i) = \operatorname{div}(\mu_i \widetilde{H}_i) = 0 \ \text{ in } \mathcal{Q}_i, \tag{7.4.2.36}$$
$$\nu_i \cdot (\varepsilon_i \widetilde{E}_i) = \nu_i \cdot (\mu_i \widetilde{H}_i) = 0 \ \text{ on } \Sigma_i,$$

since $\nu_i \cdot \operatorname{rot} \phi$ is a tangential differential operation on $\nu_i \wedge \phi$ on Σ_i. Since

$$0 = \widetilde{H}_{i\tau} = \widetilde{H}_i - (\widetilde{H}_i \cdot \nu_i)\nu_i \ \text{ on } \Sigma_i$$

we have

$$0 = \nu_i \cdot (\mu_i \widetilde{H}_i) = (\widetilde{H}_i \cdot \nu_i)\nu_i \cdot (\mu_i \nu_i) \ \text{ on } \Sigma_i,$$

which implies that $\nu_i \cdot \widetilde{H}_i = 0$ on Σ_i since μ_i is positive definite. Therefore $\widetilde{H}_i|_{\Sigma_i} = 0$ and, similarly, $\widetilde{E}_i|_{\Sigma_i} = 0$. Hence $(\widetilde{E}_i, \widetilde{H}_i)$ is a solution of the dynamic Maxwell system and of (7.4.2.36) in $\mathcal{Q}_i$ that vanishes on Σ_i. Moreover, since $(\widetilde{E}_i, \widetilde{H}_i)$ vanishes at $t = 0$ and $t = T$, this pair may be continued by zero to $\Omega \times \{t < 0\}$ and to $\Omega \times \{t > T\}$ as $\mathcal{D}'(-\infty, \infty); \mathcal{H}_i)$ solutions of the Maxwell system satisfying (7.4.2.36) and vanishing on $\Gamma_i \times (-\infty, \infty)$. It then follows from results of Eller (see [21, Corollary 50] when ε_i, μ_i are $C^2(\overline{\Omega}_i)$ positive scalar functions, or [20, Theorem 1.1] when ε_i, μ_i are $C^\infty(\overline{\Omega}_i)$ positive definite matrices and $\mu_i = f_i \varepsilon_i$ for some positive $C^\infty(\overline{\Omega}_i)$ scalar function) that $\widetilde{E}_i = \widetilde{H}_i = 0$ in $\mathcal{Q}_i$. Therefore, $\widetilde{E}_i = \widetilde{H}_i = 0$ in $\mathcal{Q}_i$ for every index i such that $\Gamma_i \neq \emptyset$. The argument is the same for $(\widetilde{P}_i, \widetilde{Q}_i)$. It follows that for such i we have $A_i = B_i = C_i = D_i = 0$ and therefore the convergence in (7.4.2.33) is through the entire sequence of positive integers.

Now suppose Ω_j is a region adjacent to a boundary region, i.e., to a region Ω_i such that $\Gamma_i \neq \emptyset$. Then $\Sigma_{ij} = \partial\Omega_i \cap \partial\Omega_j$ has a nonempty interior. From the interface conditions on Σ_{ij} we have

$$\widetilde{H}_{i\tau}^{n+2} - \beta(\nu_i \wedge \widetilde{E}_i^{n+2}) - \gamma(\nu_i \wedge \widetilde{P}_i^{n+2}) \\ = \widetilde{H}_{j\tau}^{n+1} + \beta(\nu_j \wedge \widetilde{E}_j^{n+1}) + \gamma(\nu_i \wedge \widetilde{P}_j^{n+1}), \quad (7.4.2.37)$$

$$\widetilde{H}_{j\tau}^{n+1} - \beta(\nu_j \wedge \widetilde{E}_j^{n+1}) - \gamma(\nu_i \wedge \widetilde{P}_j^{n+1}) \\ = \widetilde{H}_{i\tau}^{n} + \beta(\nu_i \wedge \widetilde{E}_i^{n}) + \gamma(\nu_i \wedge \widetilde{P}_i^{n}) \text{ on } \Sigma_{ij}. \quad (7.4.2.38)$$

Since, for the index i, convergence is through the entire sequence of positive integers, if we pass to the weak $\mathcal{L}_\tau^2(\Sigma_{ij})$ limit in (7.4.2.37) and (7.4.2.38) through the subsequence $n = n_k$ we obtain

$$B_j + \beta A_j + \gamma C_j = 0, \quad B_j - \beta A_j - \gamma C_j = 0 \text{ on } \Sigma_{ij}.$$

Similarly,

$$D_j - \beta C_j + \gamma A_j = 0, \quad D_j + \beta C_j - \gamma A_j = 0 \text{ on } \Sigma_{ij},$$

hence $A_j = B_j = C_j = D_j = 0$, that is to say

$$\nu_j \wedge \widetilde{E}_j = \widetilde{H}_{j\tau} = \nu_j \wedge \widetilde{P}_j = \widetilde{Q}_{j\tau} = 0 \text{ on } \Sigma_{ij}, \quad \forall j \in \mathcal{J}_i. \quad (7.4.2.39)$$

The same unique continuation argument as above gives $\widetilde{E}_j = \widetilde{H}_j = \widetilde{P}_j = \widetilde{Q}_j = 0$ in $\mathcal{Q}_j$. One may now proceed step-by-step into the remaining interior regions Ω_j and conclude that $\widetilde{E}_j = \widetilde{H}_j = \widetilde{P}_j = \widetilde{Q}_j = 0$ in $\mathcal{Q}_j$ for $j = 1, \dots, I$.

Case 2: $0 < \epsilon < 1$. In this case we can obtain stronger convergence results than in the case $\epsilon = 0$ by invoking Proposition 2.3.7.1 to deduce that $\|X^{n+1} - X^n\|_{\mathcal{X}} \to 0$, or, equivalently, that $\|\mathcal{T}X^n - X^n\|_{\mathcal{X}} \to 0$. From (7.4.2.24) we have

$$\|\mathcal{T}X^n - X^n\|_{\mathcal{X}}^2 = 2\mathcal{E}^n - 2\mathrm{Re}(\mathcal{T}X^n, X^n) \to 0. \quad (7.4.2.40)$$

We calculate

$$(\mathcal{T}X^n, X^n)_{\mathcal{X}} = \sum_{i=1}^{I} \sum_{j \in \mathcal{J}_i} \frac{1}{2\gamma} \int_{\Sigma_{ij}} \{(\widetilde{H}_{j\tau}^n + \beta(\nu_j \wedge \widetilde{E}_j^n) + \gamma(\nu_j \wedge \widetilde{P}_j)) \\ \cdot (\widetilde{H}_{i\tau}^n - \beta(\nu_i \wedge \widetilde{E}_i^n) - \gamma(\nu_i \wedge \widetilde{P}_i)) \\ + (\widetilde{Q}_{j\tau}^n - \beta(\nu_j \wedge \widetilde{P}_j^n) + \gamma(\nu_j \wedge \widetilde{E}_j)) \cdot (\widetilde{Q}_{i\tau}^n + \beta(\nu_i \wedge \widetilde{P}_i^n) - \gamma(\nu_i \wedge \widetilde{E}_i))\} d\Sigma.$$

The integrand equals

$$\widetilde{H}_{j\tau}^n \cdot \widetilde{H}_{i\tau}^n + \widetilde{Q}_{j\tau}^n \cdot \widetilde{Q}_{i\tau}^n - (\beta^2 + \gamma^2)[(\nu_j \wedge \widetilde{E}_j^n) \cdot (\nu_i \wedge \widetilde{E}_i^n) + (\nu_j \wedge \widetilde{P}_j^n) \cdot (\nu_i \wedge \widetilde{P}_i^n)] \\ - \beta[\widetilde{H}_{j\tau}^n \cdot (\nu_i \wedge \widetilde{E}_i^n) - (\nu_j \wedge \widetilde{E}_j) \cdot \widetilde{H}_{i\tau}] - \gamma[\widetilde{H}_{j\tau}^n \cdot (\nu_i \wedge \widetilde{P}_i^n) - (\nu_j \wedge \widetilde{P}_j) \cdot \widetilde{H}_{i\tau}] \\ + \beta[\widetilde{Q}_{j\tau}^n \cdot (\nu_i \wedge \widetilde{P}_i^n) - (\nu_j \wedge \widetilde{P}_j) \cdot \widetilde{Q}_{i\tau}] - \gamma[\widetilde{Q}_{j\tau}^n \cdot (\nu_i \wedge \widetilde{E}_i^n) - (\nu_j \wedge \widetilde{E}_j) \cdot \widetilde{Q}_{i\tau}].$$

Integrals of the form

$$\sum_{i=1}^{I}\sum_{j\in\mathcal{J}_i}\int_{\Sigma_{ij}}(F_j\cdot G_i - G_j\cdot F_i)d\Sigma$$

may be written

$$\sum_{i=1}^{I-1}\sum_{\substack{j\in\mathcal{J}_i\\ j>i}}\int_{\Sigma_{ij}}\left[(F_j\cdot G_i - G_j\cdot F_i)+(F_i\cdot G_j - G_i\cdot F_j)\right]d\Sigma$$

$$= 2\sum_{i=1}^{I-1}\sum_{\substack{j\in\mathcal{J}_i\\ j>i}}\int_{\Sigma_{ij}}\sqrt{-1}\,\mathrm{Im}\,(F_j\cdot G_i - G_j\cdot F_i)\,d\Sigma$$

and, similarly,

$$\sum_{i=1}^{I}\sum_{j\in\mathcal{J}_i}\int_{\Sigma_{ij}}F_j\cdot F_i\,d\Sigma = 2\sum_{i=1}^{I-1}\sum_{\substack{j\in\mathcal{J}_i\\ j>i}}\int_{\Sigma_{ij}}\mathrm{Re}\,(F_j\cdot F_i)\,d\Sigma.$$

Therefore

$$\mathrm{Re}\,(\mathcal{T}X^n, X^n)_{\mathcal{X}} = \sum_{i=1}^{I-1}\sum_{\substack{j\in\mathcal{J}_i\\ j>i}}\int_{\Sigma_{ij}}\mathrm{Re}\left\{\frac{1}{\gamma}(\widetilde{H}^n_{j\tau}\cdot\widetilde{H}^n_{i\tau}+\widetilde{Q}^n_{j\tau}\cdot\widetilde{Q}^n_{i\tau})\right.$$
$$\left.-\frac{\beta^2+\gamma^2}{\gamma}\left[(\nu_j\wedge\widetilde{E}^n_j)\cdot(\nu_i\wedge\widetilde{E}^n_i)+(\nu_j\wedge\widetilde{P}^n_j)\cdot(\nu_i\wedge\widetilde{P}^n_i)\right]\right\}d\Sigma. \quad (7.4.2.41)$$

By writing

$$\sum_{i=1}^{I}\int_{\Sigma_i^{\mathrm{int}}}|F_i|^2 d\Sigma = \sum_{i=1}^{I-1}\sum_{\substack{j\in\mathcal{J}_i\\ j>i}}\int_{\Sigma_{ij}}(|F_i|^2+|F_j|^2)\,d\Sigma \quad (7.4.2.42)$$

we obtain from the definition of $\mathcal{E}^n$ and (7.4.2.41)

$$2\mathcal{E}^n - 2\mathrm{Re}\,(\mathcal{T}X^n, X^n)_{\mathcal{X}} = \sum_{i=1}^{I-1}\sum_{\substack{j\in\mathcal{J}_i\\ j>i}}\int_{\Sigma_{ij}}\{\frac{1}{\gamma}(|\widetilde{H}^n_{i\tau}-\widetilde{H}^n_{j\tau}|^2+|\widetilde{Q}^n_{i\tau}-\widetilde{Q}^n_{j\tau}|^2)$$
$$+\frac{\beta^2+\gamma^2}{\gamma}\left(|\nu_i\wedge\widetilde{E}^n_i+\nu_j\wedge\widetilde{E}^n_j|^2+|\nu_i\wedge\widetilde{P}^n_i+\nu_j\wedge\widetilde{P}^n_j|^2\right)\}\,d\Sigma \quad (7.4.2.43)$$

Since $\widetilde{H}^n_{i\tau} - \widetilde{H}^n_{j\tau} = H^n_{i\tau} - H^n_{j\tau}$ and, similarly, for the other terms in (7.4.2.43), it follows from (7.4.2.40) and (7.4.2.43) that

$$\|\mathcal{T}X^n - X^n\|^2_{\mathcal{X}} = \sum_{i=1}^{I-1} \sum_{\substack{j\in\mathcal{J}_i \\ j>i}} \int_{\Sigma_{ij}} \{\frac{1}{\gamma}(|H^n_{i\tau} - H^n_{j\tau}|^2 + |Q^n_{i\tau} - Q^n_{j\tau}|^2)$$

$$+ \frac{\beta^2+\gamma^2}{\gamma}(|\nu_i \wedge E^n_i + \nu_j \wedge E^n_j|^2 + |\nu_i \wedge P^n_i + \nu_j \wedge P^n_j|^2)\} \, d\Sigma. \quad (7.4.2.44)$$

Thus, for $i = 1, \ldots, I$ and $j \in \mathcal{J}_i$,

$$\begin{aligned} &\nu_i \wedge E^n_i + \nu_j \wedge E^n_j|_{\Sigma_{ij}} \to 0, \quad \nu_i \wedge P^n_i + \nu_j \wedge P^n_j|_{\Sigma_{ij}} \to 0 \\ &H^n_{i\tau} - H^n_{j\tau}|_{\Sigma_{ij}} \to 0, \quad Q^n_{i\tau} - Q^n_{j\tau}|_{\Sigma_{ij}} \text{ strongly in } \mathcal{L}^2_\tau(\Sigma_{ij}). \end{aligned} \quad (7.4.2.45)$$

We now show that (7.4.2.45), together with

$$(\widetilde{P}^n_i(T), \widetilde{Q}^n_i(T)) \to 0 \text{ strongly in } \mathcal{H}_i, \; \nu_i \wedge \widetilde{P}_i|_{\Sigma_i} \to 0 \text{ strongly in } \mathcal{L}^2_\tau(\Sigma_i) \quad (7.4.2.46)$$

for $i = 1, \ldots, I$ (both of which hold, according to $(7.4.2.31)_2$ and $(7.4.2.32)_2$, for $\epsilon \in [0,1)$) imply that

$$(\widetilde{E}^n_i, \widetilde{H}^n_i) \to 0, \quad (\widetilde{P}^n_i, \widetilde{Q}^n_i) \to 0 \text{ strongly in } L^\infty(0,T;\mathcal{H}_i), \; i = 1, \ldots, I. \quad (7.4.2.47)$$

Part (i) of Theorem 7.4.2.1 will then follow from (7.4.2.47), (7.4.2.45) and $(7.4.2.31)_2$.

To prove (7.4.2.47) we use the following result.

Proposition 7.4.2.1 *For $i = 1, \ldots, I$, let (Φ^n_i, Ψ^n_i) satisfy*

$$\begin{cases} \varepsilon_i(\Phi^n_i)' - \operatorname{rot}\Psi^n_i + \sigma_i\Phi^n_i = F^n_i \\ \mu_i(\Psi^n_i)' + \operatorname{rot}\Phi^n_i = G^n_i \qquad \text{in } Q_i \end{cases}$$

$$\Psi^n_{i\tau} - \alpha_i(\nu_i \wedge \Phi^n_i) = J^n_i \text{ on } \Sigma_i$$

$$\Phi^n_i(0) = \Phi^n_{i0}, \quad \Psi^n_i(0) = \Psi^n_{i0} \text{ in } \Omega_i$$

where $(F^n_i, G^n_i) \in L^1(0,T;\mathcal{H}_i)$, $J^n_i \in \mathcal{L}^2_\tau(\Sigma_i)$ and $(\Phi^n_{i0}, \Psi^n_{i0}) \in \mathcal{H}_i$. Assume also that

$$\Psi^n_{i\tau}|_{\Sigma^{int}_i}, \quad \nu_i \wedge \Phi^n_i|_{\Sigma^{int}_i} \text{ are bounded in } \mathcal{L}^2_\tau(\Sigma^{int}_i).$$

If

$$\|(\Phi^n_{i0}, \Psi^n_{i0})\|_{\mathcal{H}_i} \to 0, \quad \|(F^n_i, G^n_i)\|_{L^1(0,T;\mathcal{H}_i)} \to 0, \quad \|J^n_i\|_{\mathcal{L}^2_\tau(\Sigma_i)} \to 0,$$
$$\|\Psi^n_{i\tau} - \Psi^n_{j\tau}\|_{\mathcal{L}^2_\tau(\Sigma_{ij})} \to 0, \quad \|\nu_i \wedge \Phi^n_i + \nu_j \wedge \Phi^n_j\|_{\mathcal{L}^2_\tau(\Sigma_{ij})} \to 0, \quad \forall j \in \mathcal{J}_i,$$

then $(\Phi^n_i, \Psi^n_i) \to 0$ strongly in $L^\infty(0,T;\mathcal{H}_i)$, $i = 1, \ldots, I$.

The proof of this proposition is given immediately below.

We apply Proposition 7.4.2.1 with $(\Phi_i^n, \Psi_i^n) = (\widetilde{E}_i^n, \widetilde{H}_i^n)$, in which case $J_i^n = \nu_i \wedge \widetilde{P}_i^n$, $(\Phi_{i0}^n, \Psi_{i0}^n) = (0,0)$ and $F_i^n = G_i^n = 0$. In view of (7.4.2.45) and (7.4.2.46) we may conclude that

$$(\widetilde{E}_i^n, \widetilde{H}_i^n) \to 0 \text{ strongly in } L^\infty(0,T;\mathcal{H}_i),\ i = 1,\dots,I.$$

Similarly, by the change of variable $t \mapsto T - t$, we may apply the Proposition 7.4.2.1 with $(\Phi_i^n, \Psi_i^n) = (\widetilde{P}_i^n, \widetilde{Q}_i^n)$, in which case $(\Phi_{i0}^n, \Psi_{i0}^n) := (\widetilde{P}_i^n(T), \widetilde{Q}_i^n(T))$, $F_i^n = G_i^n = 0$, $J_i^n = 0$, and conclude that

$$(\widetilde{P}_i^n, \widetilde{Q}_i^n) \to 0 \text{ strongly in } L^\infty(0,T;\mathcal{H}_i),\ i = 1,\dots,I.$$

Proof of Proposition 7.4.2.1. The hypotheses imply that $(\Phi_i^n, \Psi_i^n) \in C([0,T];\mathcal{H}_i)$ and that $\Psi_i^n|_{\Sigma_i}$, $\nu_i \wedge \Phi_i^n|_{\Sigma_i} \in \mathcal{L}_\tau^2(\Sigma_i)$. Suppose first of all that $F_i^n = G_i^n = 0$. Then (Φ_i^n, Ψ_i^n) satisfies the following identity, whose proof follows the same lines as the proof of (7.2.2.11):

$$\begin{aligned}
&\|(\Phi_i^n(t), \Psi_i^n(t))\|_{\mathcal{H}_i}^2 + 2\int_0^t \int_{\Omega_i} \sigma_i \Phi_i^n \cdot \Phi_i^n \, dx dt \\
&+ \int_0^t \int_{\Gamma_i} (\alpha_i |\nu \wedge \Phi_i^n|^2 + \frac{1}{\alpha_i} |\Psi_{i\tau}^n|^2) \, d\Gamma dt + 2\,\mathrm{Re} \int_0^t \int_{\Gamma_i^{\mathrm{int}}} \Psi_{i\tau}^n \cdot (\nu_i \wedge \Phi_i^n) \, d\gamma dt \\
&\qquad = \|(\Phi_{i0}^n, \Psi_{i0}^n)\|_{\mathcal{H}}^2 + \int_0^t \int_{\Gamma_i} \frac{1}{\alpha_i} |J_i^n|^2 d\Gamma dt. \qquad (7.4.2.48)
\end{aligned}$$

We have

$$\begin{aligned}
\sum_{i=1}^{I} \int_0^t \int_{\Gamma_i^{\mathrm{int}}} \Psi_{i\tau}^n \cdot (\nu_i \wedge \Phi_i^n) \, d\gamma dt = \sum_{i=1}^{I-1} \sum_{\substack{j \in \mathcal{J}_i \\ j > i}} \int_0^t \int_{\Gamma_{ij}} \big[(\Psi_{i\tau}^n - \Psi_{j\tau}^n) \cdot (\nu_i \wedge \Phi_i^n) \\
+ \Psi_{j\tau}^n \cdot (\nu_j \wedge \Phi_j^n + \nu_i \wedge \Phi_i^n) \big] \, d\Gamma dt. \qquad (7.4.2.49)
\end{aligned}$$

It follows from (7.4.2.48), (7.4.2.49) and the hypotheses of the proposition that

$$\|(\Phi_i^n, \Psi_i^n)\|_{L^\infty(0,T;\mathcal{H}_i)} \to 0, \quad i = 1,\dots,I.$$

Now assume that $\Phi_{i0}^n = \Psi_{i0}^n = 0$ and $J_i^n = 0$. If we multiply the two Maxwell equations by Φ_i^n and Ψ_i^n, respectively, and integrate by parts we obtain

$$\begin{aligned}
&\frac{1}{2} \|(\Phi_i^n(t), \Psi_i^n(t))\|_{\mathcal{H}_i}^2 + \int_0^t \int_{\Omega_i} \sigma_i \Phi_i^n \cdot \Phi_i^n \, dx dt + \int_0^t \int_{\Gamma_i} \alpha_i |\nu_i \wedge \Phi_i^n|^2 d\Gamma dt \\
&\quad = \mathrm{Re} \int_0^t \int_{\Omega_i} (F_i^n \cdot \Phi_i^n + G_i^n \cdot \Psi_i^n) \, dx dt - \mathrm{Re} \int_0^t \int_{\Gamma_i^{\mathrm{int}}} \Psi_{i\tau}^n \cdot (\nu_i \wedge \Phi_i^n) \, d\gamma dt.
\end{aligned}$$

It follows that for any $\delta \in (0,1)$,

$$(1-\delta)\|(\Phi_i^n,\Psi_i^n)\|^2_{L^\infty(0,T;\mathcal{H}_i)} \le \frac{1}{\delta}\|(\varepsilon_i^{-1}F_i^n,\mu_i^{-1}G_i^n)\|^2_{L^1(0,T;\mathcal{H}_i)} + 2\int_{\Sigma_i^{\mathrm{int}}} |\Psi_{i\tau}^n\cdot(\nu_i\wedge\Phi_i^n)|\,d\Sigma.$$

With the aid of (7.4.2.49) we therefore obtain

$$\begin{aligned}(1-\delta)\sum_{i=1}^I \|(\Phi_i^n,\Psi_i^n)\|^2_{L^\infty(0,T;\mathcal{H}_i)} &\le \frac{1}{\delta}\sum_{i=1}^I \|(\varepsilon_i^{-1}F_i^n,\mu_i^{-1}G_i^n)\|^2_{L^1(0,T;\mathcal{H}_i)}\\ &+2\sum_{i=1}^{I-1}\sum_{\substack{j\in\mathcal{J}_i\\ j>i}}\{\|\Psi_{i\tau}^n-\Psi_{j\tau}^n\|_{\mathcal{L}^2_\tau(\Sigma_{ij})}\|\nu_i\wedge\Phi_i^n\|_{\mathcal{L}^2_\tau(\Sigma_{ij})}\\ &+\|\Psi_{j\tau}^n\|_{\mathcal{L}^2_\tau(\Sigma_{ij})}\|\nu_j\wedge\Phi_j^n+\nu_i\wedge\Phi_i^n\|_{\mathcal{L}^2_\tau(\Sigma_{ij})}\}.\end{aligned}$$

7.4.3 A Posteriori Error Estimates

In this section we shall derive *a posteriori* estimates of the difference between the solutions of the local optimality systems and the solution of the global optimality system in terms of the mismatch of the tangential components of the nth iterates, or of successive iterates, across the interfaces Σ_{ij}. The analysis of this section parallels that of Section 6.4.3. Nonetheless, we provide it for the sake of completeness.

The *error* $e^n(t)$ *at the* n*th iteration at time* t is defined by

$$\begin{aligned}e^n(t) &= \sum_{i=1}^I\{\|(\widetilde{E}_i^n(t),\widetilde{H}_i^n(t))\|^2_{\mathcal{H}_i}+\|(\widetilde{P}_i^n(t),\widetilde{Q}_i^n(t))\|^2_{\mathcal{H}_i}\\ &+\int_0^t\int_{\Gamma_i}(\alpha_i|\nu_i\wedge\widetilde{E}_i^n|^2+\frac{1}{\alpha_i}|\widetilde{H}_{i\tau}^n|^2)\,d\Gamma dt+\int_t^T\int_{\Gamma_i}(\alpha_i|\nu_i\wedge\widetilde{P}_i^n|^2+\frac{1}{\alpha_i}|\widetilde{Q}_{i\tau}^n|^2)d\Gamma dt\}.\end{aligned}$$

We further define

$$\begin{aligned}\mathcal{E}_{ij}^n(\Sigma_{ij}) := \int_{\Sigma_{ij}}&(|H_{i\tau}^n-H_{j\tau}^n|^2+|Q_{i\tau}^n-Q_{j\tau}^n|^2\\ &+|\nu_i\wedge E_i^n+\nu_j\wedge E_j^n|^2+|\nu_i\wedge P_i^n+\nu_j\wedge P_j^n|^2)\,d\Sigma\end{aligned}$$

and

$$\begin{aligned}\mathcal{E}_{ij}^{n,n+1}(\Sigma_{ij}) := \int_{\Sigma_{ij}}&(|H_{i\tau}^n-H_{j\tau}^{n+1}|^2+|Q_{i\tau}^n-Q_{j\tau}^{n+1}|^2\\ &+|\nu_i\wedge E_i^n+\nu_j\wedge E_j^{n+1}|^2+|\nu_i\wedge P_i^n+\nu_j\wedge P_j^{n+1}|^2)\,d\Sigma.\end{aligned}$$

The quantity $\mathcal{E}_{ij}^n(\Sigma_{ij})$ is a measure of the mismatch of the tangential components of the nth iterates across the interface Σ_{ij} while $\mathcal{E}_{ij}^{n,n+1}(\Sigma_{ij})$ measures the mismatch of successive iterates.

For the remainder of this section, we assume that α, β, γ and κ satisfy (7.4.2.8) of Theorem 7.4.2.1.

Theorem 7.4.3.1 *Let $\epsilon \in [0,1)$. Then*

$$\|e^n\|_{L^\infty(0,T)} \leq C\sqrt{\mathcal{E}^n + \mathcal{F}^n}\Big\{\sum_{i=1}^{I-1}\sum_{\substack{j\in\mathcal{J}_i\\ j>i}} \mathcal{E}_{ij}^n(\Sigma_{ij})\Big\}^{1/2} \tag{7.4.3.1}$$

where C depends only on α_0, β, γ and κ.

Theorem 7.4.3.2 *Let $\epsilon \in [0,1/2)$. Then*

$$\|e^{n+1} + e^n\|_{L^\infty(0,T)} \leq C_\epsilon\sqrt{\mathcal{E}^{n+1} + \mathcal{E}^n}\Big\{\sum_{i=1}^{I-1}\sum_{\substack{j\in\mathcal{J}_i\\ j>i}} (\mathcal{E}_{ij}^{n,n+1}(\Sigma_{ij}) + \mathcal{E}_{ij}^{n+1,n}(\Sigma_{ij}))\Big\}^{1/2} \tag{7.4.3.2}$$

where C_ϵ depends only on α_0, β, γ, κ and on ϵ.

Remark 7.4.3.1 The constants C and C_ϵ will be explicitly calculated in the proofs of Theorems 7.4.3.1 and 7.4.3.2; see (7.4.3.6), (7.4.3.9) and (7.4.3.5) below. Note also that $\mathcal{E}^n + \mathcal{F}^n = \|X^n\|_{\mathcal{X}}^2$ is nonincreasing. Therefore

$$\mathcal{E}^n + \mathcal{F}^n \leq \|X^1\|_{\mathcal{X}}^2 = \sum_{i=1}^{I}\sum_{j\in\mathcal{J}_i}\frac{1}{2\gamma}\int_{\Sigma_{ij}}(|\tilde{\lambda}_{ij}^0|^2 + |\tilde{\rho}_{ij}^0|^2)d\Sigma.$$

By using (7.4.2.6) and (7.4.2.7), the right side can be estimated in terms of the starting inputs λ_{ij}^0, ρ_{ij}^0 to the local problems, and the tangential traces of the solution of the global optimality system on the interfaces Σ_{ij}. We do not, however, have estimates of these traces in terms of the input data (E_0, H_0), (E_T, H_T), (F, G) and κ to the global optimal control problem.

Theorem 7.4.3.1 is an immediate consequence of the following two lemmas.

Lemma 7.4.3.1 *Let $\epsilon \in [0,1)$. Then*

$$\mathcal{F}^n \leq \Big\{\max\big(\frac{1}{\gamma}, \frac{\beta^2+\gamma^2}{\gamma}\big)\Big\}^{1/2}\sqrt{\mathcal{E}^n + \mathcal{F}^n}\Big\{\sum_{i=1}^{I-1}\sum_{\substack{j\in\mathcal{J}_i\\ j>i}} \mathcal{E}_{ij}^n(\Sigma_{ij})\Big\}^{1/2}. \tag{7.4.3.3}$$

Lemma 7.4.3.2 *Let* $\epsilon \in [0,1)$. *Then*

$$\sup_{0\le t\le T} e^n(t) \le C_1\mathcal{F}^n + C_2\sqrt{\mathcal{E}^n}\Big\{\sum_{i=1}^{I-1}\sum_{\substack{j\in\mathcal{J}_i\\ j>i}} \mathcal{E}^n_{ij}(\Sigma_{ij})\Big\}^{1/2} \tag{7.4.3.4}$$

where

$$C_1 = \max\left(\frac{\kappa^2}{\kappa + \frac{\beta}{2\gamma}(1-\kappa^2)}, \frac{1}{\alpha_0 + \alpha_0^2\frac{\beta}{\gamma} - \frac{\beta}{4\gamma}}\right),$$
$$C_2 = 2\sqrt{2}\Big\{\max\Big(\frac{\gamma}{\beta^2+\gamma^2}, \gamma\Big)\Big\}^{1/2}. \tag{7.4.3.5}$$

Note that $C_1 > 0$ by virtue of assumption (7.4.2.8). From (7.4.3.3) and (7.4.3.4) one obtains (7.4.3.1) with

$$C = C_1\Big\{\max\Big(\frac{1}{\gamma}, \frac{\beta^2+\gamma^2}{\gamma}\Big)\Big\}^{1/2} + C_2. \tag{7.4.3.6}$$

Similarly, Theorem 7.4.3.2 follows from the following two results.

Lemma 7.4.3.3 *Let* $\epsilon \in [0,1)$. *then*

$$\mathcal{F}^{n+1} + (1-2\epsilon)\mathcal{F}^n \le \Big\{\max\Big(\frac{1}{\gamma}, \frac{\beta^2+\gamma^2}{\gamma}\Big)\Big\}^{1/2}\sqrt{\mathcal{E}^{n+1}+\mathcal{E}^n}$$
$$\times \Big\{\sum_{i=1}^{I-1}\sum_{\substack{j\in\mathcal{J}_i\\ j>i}} \big(\mathcal{E}^{n,n+1}_{ij}(\Sigma_{ij}) + \mathcal{E}^{n+1,n}_{ij}(\Sigma_{ij})\big\}^{1/2}. \tag{7.4.3.7}$$

Lemma 7.4.3.4 *Let* $\epsilon \in [0,1)$. *Then*

$$\sup_{0\le t\le T}[e^{n+1}(t) + e^n(t)] \le C_1(\mathcal{F}^{n+1} + \mathcal{F}^n)$$
$$+ C_2\sqrt{\mathcal{E}^{n+1}+\mathcal{E}^n}\Big\{\sum_{i=1}^{I-1}\sum_{\substack{j\in\mathcal{J}_i\\ j>i}} \big(\mathcal{E}^{n,n+1}_{ij}(\Sigma_{ij}) + \mathcal{E}^{n+1,n}_{ij}(\Sigma_{ij})\big\}^{1/2} \tag{7.4.3.8}$$

where C_1, C_2 *are given by* (7.4.3.5).

It follows that the constant C_ϵ in Theorem 7.4.3.2 is given by

$$C_\epsilon = \frac{C_1}{1-2\epsilon}\Big\{\max\Big(\frac{1}{\gamma}, \frac{\beta^2+\gamma^2}{\gamma}\Big)\Big\}^{1/2} + C_2. \tag{7.4.3.9}$$

Proof of Lemma 7.4.3.1. The proof follows exactly the proof of Lemma 6.4.3.1 and is omitted.

Proof of Lemma 7.4.3.2. We begin with the energy estimate

$$\|(\widetilde{E}_i^n(t),\widetilde{H}_i^n(t))\|_{\mathcal{H}_i}^2+\int_0^t\int_{\Gamma_i}\big(\alpha_i\big(|\nu_i\wedge\widetilde{E}_i^n|^2+\frac{1}{\alpha_i}|\widetilde{H}_{i\tau}^n|^2\big)\,d\Gamma dt$$
$$\le\int_0^t\int_{\Gamma_i}\frac{1}{\alpha_i}|\nu_i\wedge\widetilde{P}_i^n|^2d\Gamma dt-2\mathrm{Re}\int_0^t\int_{\Gamma_i^{\mathrm{int}}}\widetilde{H}_{i\tau}^n\cdot(\nu_i\wedge\widetilde{E}_i^n)\,d\Gamma dt \quad (7.4.3.10)$$

whose simple proof follows the lines of the proof of (7.2.2.11). By using (7.4.2.49) and the Cauchy-Schwarz inequality we then obtain, for $0\le t\le T$,

$$\sum_{i=1}^{I}\Big\{\|(\widetilde{E}_i^n(t),\widetilde{H}_i^n(t))\|_{\mathcal{H}_i}^2+\int_0^t\int_{\Gamma_i}\big(\alpha_i\big(|\nu_i\wedge\widetilde{E}_i^n|^2+\frac{1}{\alpha_i}|\widetilde{H}_{i\tau}^n|^2\big)\,d\Gamma dt\Big\}$$
$$\le\sum_{i=1}^{I}\int_{\Sigma_i}\frac{1}{\alpha_i}|\nu_i\wedge\widetilde{P}_i^n|^2d\Sigma$$
$$+2\sum_{i=1}^{I-1}\sum_{\substack{j\in\mathcal{J}_i\\ j>i}}\Big\{\Big(\int_{\Sigma_{ij}}|H_{i\tau}^n-H_{j\tau}^n|^2d\Sigma\Big)^{1/2}\Big(\int_{\Sigma_{ij}}|\nu_i\wedge\widetilde{E}_i^n|^2d\Sigma\Big)^{1/2}$$
$$+\Big(\int_{\Sigma_{ij}}|\nu_i\wedge E_i^n+\nu_j\wedge E_j^n|^2d\Sigma\Big)^{1/2}\Big(\int_{\Sigma_{ij}}|\widetilde{H}_{j\tau}^n|^2d\Sigma\Big)^{1/2}\Big\}. \quad (7.4.3.11)$$

Similarly, from

$$\|(\widetilde{P}_i^n(t),\widetilde{Q}_i^n(t))\|_{\mathcal{H}_i}^2+\int_t^T\int_{\Gamma_i}\big(\alpha_i\big(|\nu_i\wedge\widetilde{P}_i^n|^2+\frac{1}{\alpha_i}|\widetilde{Q}_{i\tau}^n|^2\big)\,d\Gamma dt$$
$$\le\|(\widetilde{P}_i^n(T),\widetilde{Q}_i^n(T))\|_{\mathcal{H}_i}^2-2\mathrm{Re}\int_t^T\int_{\Gamma_i^{\mathrm{int}}}\widetilde{Q}_{i\tau}^n\cdot(\nu_i\wedge\widetilde{P}_i^n)\,d\Gamma dt \quad (7.4.3.12)$$

we obtain, for $0\le t\le T$,

$$\sum_{i=1}^{I}\Big\{\|(\widetilde{P}_i^n(t),\widetilde{Q}_i^n(t))\|_{\mathcal{H}_i}^2+\int_t^T\int_{\Gamma_i}\big(\alpha_i\big(|\nu_i\wedge\widetilde{P}_i^n|^2+\frac{1}{\alpha_i}|\widetilde{Q}_{i\tau}^n|^2\big)\,d\Gamma dt\Big\}$$
$$\le\sum_{i=1}^{I}\|(\widetilde{P}_i^n(T),\widetilde{Q}_i^n(T))\|_{\mathcal{H}_i}^2$$
$$+2\sum_{i=1}^{I-1}\sum_{\substack{j\in\mathcal{J}_i\\ j>i}}\Big\{\Big(\int_{\Sigma_{ij}}|Q_{i\tau}^n-Q_{j\tau}^n|^2d\Sigma\Big)^{1/2}\Big(\int_{\Sigma_{ij}}|\nu_i\wedge\widetilde{P}_i^n|^2d\Sigma\Big)^{1/2}$$
$$+\Big(\int_{\Sigma_{ij}}|\nu_i\wedge P_i^n+\nu_j\wedge P_j^n|^2d\Sigma\Big)^{1/2}\Big(\int_{\Sigma_{ij}}|\widetilde{Q}_{j\tau}^n|^2d\Sigma\Big)^{1/2}\Big\}. \quad (7.4.3.13)$$

It follows from (7.4.3.11) and (7.4.3.13) that

$$\sup_{0\le t\le T} e^n(t) \le \sum_{i=1}^{I}\Big\{\|(\widetilde{P}_i^n(T),\widetilde{Q}_i^n(T))\|^2_{\mathcal{H}_i} + \int_{\Sigma_i}\frac{1}{\alpha_i}|\nu_i\wedge \widetilde{P}_i^n|^2 d\Sigma\Big\}$$
$$+2\{\sum_{i=1}^{I-1}\sum_{\substack{j\in\mathcal{J}_i\\ j>i}}\mathcal{E}_{ij}^n(\Sigma_{ij})\}^{1/2}\{\sum_{i=1}^{I-1}\sum_{\substack{j\in\mathcal{J}_i\\ j>i}}\int_{\Sigma_{ij}}(|\nu_i\wedge\widetilde{E}_i^n|^2+|\widetilde{H}_{j\tau}^n|^2$$
$$+|\nu_i\wedge\widetilde{P}_i^n|^2+|\widetilde{Q}_{j\tau}^n|^2)\,d\Sigma\}^{1/2}, \quad (7.4.3.14)$$

where we have used the Cauchy-Schwarz inequality once again. We have

$$\sum_{i=1}^{I-1}\sum_{\substack{j\in\mathcal{J}_i\\ j>i}}\int_{\Sigma_{ij}}(|\nu_i\wedge\widetilde{E}_i^n|^2+|\widetilde{H}_{j\tau}^n|^2+|\nu_i\wedge\widetilde{P}_i^n|^2+|\widetilde{Q}_{j\tau}^n|^2)\,d\Sigma$$
$$\le\sum_{i=1}^{I}\int_{\Sigma_i^{\rm int}}(|\nu_i\wedge\widetilde{E}_i^n|^2+|\nu_i\wedge\widetilde{P}_i^n|^2+|\widetilde{H}_{i\tau}^n|^2+|\widetilde{Q}_{i\tau}^n|^2)d\Sigma$$
$$\le 2\max(\frac{\gamma}{\beta^2+\gamma^2},\gamma)\mathcal{E}^n. \quad (7.4.3.15)$$

From (7.4.3.14) and (7.4.3.15) we therefore obtain

$$\sup_{0\le t\le T} e^n(t) \le \sum_{i=1}^{I}\Big\{\|(\widetilde{P}_i^n(T),\widetilde{Q}_i^n(T))\|^2_{\mathcal{H}_i} + \int_{\Sigma_i}\frac{1}{\alpha_i}|\nu_i\wedge \widetilde{P}_i^n|^2 d\Sigma\Big\}$$
$$+2\sqrt{2}\{\max(\frac{\gamma}{\beta^2+\gamma^2},\gamma)\mathcal{E}^n\}^{1/2}\{\sum_{i=1}^{I-1}\sum_{\substack{j\in\mathcal{J}_i\\ j>i}}\mathcal{E}_{ij}^n(\Sigma_{ij})\}^{1/2}. \quad (7.4.3.16)$$

Thus it remains to show that

$$\sum_{i=1}^{I}\Big\{\|(\widetilde{P}_i^n(T),\widetilde{Q}_i^n(T))\|^2_{\mathcal{H}_i} + \int_{\Sigma_i}\frac{1}{\alpha_i}|\nu_i\wedge \widetilde{P}_i^n|^2 d\Sigma\Big\}\le C_1\mathcal{F}^n. \quad (7.4.3.17)$$

We have

$$\sum_{i=1}^{I}\Big\{\|(\widetilde{P}_i^n(T),\widetilde{Q}_i^n(T))\|^2_{\mathcal{H}_i} + \int_{\Sigma_i}\frac{1}{\alpha_i}|\nu_i\wedge \widetilde{P}_i^n|^2 d\Sigma\Big\}$$
$$=\sum_{i=1}^{I}\Big\{\kappa^2\|(\widetilde{E}_i^n(T),\widetilde{H}_i^n(T))\|^2_{\mathcal{H}_i} + \int_{\Sigma_i}\frac{1}{\alpha_i}|\nu_i\wedge \widetilde{P}_i^n|^2 d\Sigma\Big\}. \quad (7.4.3.18)$$

By using the definition (7.4.2.26) of $\mathcal{F}^n$ and a little algebra, one finds that

$$\mathcal{F}^n \geq \frac{1}{\kappa^2}\big(\kappa + \frac{\beta}{2\gamma}(1-\kappa^2)\big)\sum_{i=1}^{I}\kappa^2\|(\widetilde{E}_i^n(T),\widetilde{H}_i^n(T))\|_{\mathcal{H}_i}^2 + \sum_{i=1}^{I}\big(\alpha_0 + \alpha_0^2\frac{\beta}{\gamma} - \frac{\beta}{4\gamma}\big)\int_{\Sigma_i}\frac{1}{\alpha_i}|\nu_i \wedge \widetilde{P}_i^n|^2 d\Sigma. \quad (7.4.3.19)$$

The estimate (7.4.3.17) follows immediately from (7.4.3.18) and (7.4.3.19).

Proof of Lemma 7.4.3.3. We again start from the recursion formula (7.4.2.28), which may be written

$$\mathcal{F}^{n+1} + (1-2\epsilon)\mathcal{F}^n \leq \mathcal{E}^n - \mathcal{E}^{n+1},$$

and proceed to estimate the right-hand side. From the definition (7.4.2.25) of $\mathcal{E}^n$ we have

$$\mathcal{E}^n - \mathcal{E}^{n+1} = \frac{1}{2}\sum_{i=1}^{I}\int_{\Sigma_i^{\text{int}}}\Big\{\frac{\beta^2+\gamma^2}{\gamma}\big(|\nu_i\wedge\widetilde{E}_i^n|^2 - |\nu_i\wedge\widetilde{E}_i^{n+1}|^2 + |\nu_i\wedge\widetilde{P}_i^n|^2 - |\nu_i\wedge\widetilde{P}_i^{n+1}|^2\big) + \frac{1}{\gamma}\big(|\widetilde{H}_{i\tau}^n|^2 - |\widetilde{H}_{i\tau}^{n+1}|^2 + |\widetilde{Q}_{i\tau}^n|^2 - |\widetilde{Q}_{i\tau}^{n+1}|^2\big)\Big\}d\Sigma. \quad (7.4.3.20)$$

We rewrite the right-hand side using (7.4.2.42). So, for example, we have

$$\sum_{i=1}^{I}\int_{\Sigma_i^{\text{int}}}\big\{|\nu_i\wedge\widetilde{E}_i^n|^2 - |\nu_i\wedge\widetilde{E}_i^{n+1}|^2\big\}d\Sigma \leq \sum_{i=1}^{I-1}\sum_{\substack{j\in\mathcal{J}_i\\ j>i}}\int_{\Sigma_{ij}}\big(|\nu_i\wedge E_i^n + \nu_j\wedge E_j^{n+1}||\nu_i\wedge\widetilde{E}_i^n - \nu_j\wedge\widetilde{E}_j^{n+1}| + |\nu_j\wedge E_j^n + \nu_i\wedge E_i^{n+1}||\nu_j\wedge\widetilde{E}_j^n - \nu_i\wedge\widetilde{E}_i^{n+1}|\big)\,d\Sigma$$

and, similarly, for the remaining three differences in (7.4.3.20). An application of the Cauchy-Schwarz inequality, together with the trivial estimate $(a\pm b)^2 \leq 2(a^2+b^2)$, yields

$$\mathcal{E}^n - \mathcal{E}^{n+1} \leq \frac{1}{2}\Big\{\max\big(\frac{\beta^2+\gamma^2}{\gamma},\frac{1}{\gamma}\big)\Big\}^{1/2} I^{n,n+1} \times \Big\{\sum_{i=1}^{I-1}\sum_{\substack{j\in\mathcal{J}_i\\ j>i}}\big(\mathcal{E}_{ij}^{n,n+1}(\Sigma_{ij}) + \mathcal{E}_{ij}^{n+1,n}(\Sigma_{ij})\big)\Big\}^{1/2} \quad (7.4.3.21)$$

where

$$I^{n,n+1} = \{\sum_{i=1}^{I-1}\sum_{\substack{j\in\mathcal{J}_i\\ j>i}} 2\int_{\Sigma_{ij}}\{\frac{\beta^2+\gamma^2}{\gamma}(|\nu_i\wedge\widetilde{E}_i^n|^2+|\nu_j\wedge\widetilde{E}_j^n|^2+|\nu_i\wedge\widetilde{E}_i^{n+1}|^2+|\nu_j\wedge\widetilde{E}_j^{n+1}|^2$$
$$+|\nu_i\wedge\widetilde{P}_i^n|^2+|\nu_j\wedge\widetilde{P}_j^n|^2+|\nu_i\wedge\widetilde{P}_i^{n+1}|^2+|\nu_j\wedge\widetilde{P}_j^{n+1}|^2)$$
$$+\frac{1}{\gamma}(|\widetilde{H}_{i\tau}^n|^2+|\widetilde{H}_{j\tau}^n|^2+|\widetilde{H}_{i\tau}^{n+1}|^2+|\widetilde{H}_{j\tau}^{n+1}|^2$$
$$+|\widetilde{Q}_{i\tau}^n|^2+|\widetilde{Q}_{j\tau}^n|^2+|\widetilde{Q}_{i\tau}^{n+1}|^2+|\widetilde{Q}_{j\tau}^{n+1}|^2)\}\,d\Sigma\}^{1/2} = 2\sqrt{\mathcal{E}^n+\mathcal{E}^{n+1}}$$

Use of the last expression in (7.4.3.21) completes the proof of Lemma 7.4.3.3.

Proof of Lemma 7.4.3.4. From (7.4.3.10), (7.4.3.12) and the formula (7.4.2.49) we have

$$e^{n+1}(t)+e^n(t) \le \sum_{i=1}^{I}\Big\{\|(\widetilde{P}_i^n(T),\widetilde{Q}_i^n(T))\|^2_{\mathcal{H}_i}+\|(\widetilde{P}_i^{n+1}(T),\widetilde{Q}_i^{n+1}(T))\|^2_{\mathcal{H}_i}$$
$$+\int_{\Sigma_i}\{\frac{1}{\alpha_i}|\nu_i\wedge\widetilde{P}_i^n|^2+\frac{1}{\alpha_i}|\nu_i\wedge\widetilde{P}_i^{n+1}|^2\}d\Sigma\Big\}$$
$$-I_1(\widetilde{E}^n,\widetilde{H}^n)(t)-I_1(\widetilde{E}^{n+1},\widetilde{H}^{n+1})(t)$$
$$-I_2(\widetilde{P}^n,\widetilde{Q}^n)(t)-I_2(\widetilde{P}^{n+1},\widetilde{Q}^{n+1})(t), \quad (7.4.3.22)$$

where

$$I_1(\widetilde{E}^n,\widetilde{H}^n)(t) = 2\sum_{i=1}^{I-1}\sum_{\substack{j\in\mathcal{J}_i\\ j>i}}\mathrm{Re}\int_0^t\int_{\Gamma_{ij}}\{(H_{i\tau}^n-H_{j\tau}^n)\cdot(\nu_i\wedge\widetilde{E}_i^n)$$
$$+\widetilde{H}_{j\tau}^n(\nu_i\wedge E_i^n+\nu_j\wedge E_j^n)\}d\Gamma dt,$$

$$I_2(\widetilde{P}^n,\widetilde{Q}^n)(t) = 2\sum_{i=1}^{I-1}\sum_{\substack{j\in\mathcal{J}_i\\ j>i}}\mathrm{Re}\int_t^T\int_{\Gamma_{ij}}\{(Q_{i\tau}^n-Q_{j\tau}^n)\cdot(\nu_i\wedge\widetilde{P}_i^n)$$
$$+\widetilde{Q}_{j\tau}^n(\nu_i\wedge P_i^n+\nu_j\wedge P_j^n)\}d\Gamma dt.$$

It may be verified that

$$I_1(\widetilde{E}^n,\widetilde{H}^n)(t)+I_1(\widetilde{E}^{n+1},\widetilde{H}^{n+1})(t) = 2\sum_{i=1}^{I-1}\sum_{\substack{j\in\mathcal{J}_i\\ j>i}}\mathrm{Re}\int_0^t\int_{\Gamma_{ij}}\{(H_{i\tau}^n-H_{j\tau}^{n+1})\cdot(\nu_i\wedge\widetilde{E}_i^n)$$
$$+(H_{i\tau}^{n+1}-H_{j\tau}^n)\cdot(\nu_i\wedge\widetilde{E}_i^{n+1})+\widetilde{H}_{j\tau}^n\cdot(\nu_i\wedge E_i^{n+1}+\nu_j\wedge E_j^n)$$
$$+\widetilde{H}_{j\tau}^{n+1}\cdot(\nu_i\wedge E_i^n+\nu_j\wedge E_j^{n+1})\}d\Gamma dt.$$

A similar expression holds for $I_2(\widetilde{P}^n, \widetilde{Q}^n)(t) + I_2(\widetilde{P}^{n+1}, \widetilde{Q}^{n+1})(t)$. It follows that

$$
\begin{aligned}
&|I_1(\widetilde{E}^n, \widetilde{H}^n)(t) + I_1(\widetilde{E}^{n+1}, \widetilde{H}^{n+1})(t) + I_2(\widetilde{P}^n, \widetilde{Q}^n)(t) + I_2(\widetilde{P}^{n+1}, \widetilde{Q}^{n+1})(t)| \\
&\qquad \leq 2\Big\{\sum_{i=1}^{I-1} \sum_{\substack{j\in\mathcal{J}_i \\ j>i}} [\mathcal{E}_{ij}^{n,n+1}(\Sigma_{ij}) + \mathcal{E}_{ij}^{n+1,n}(\Sigma_{ij})]\Big\}^{1/2} \\
&\qquad \times \Big\{\sum_{i=1}^{I-1} \sum_{\substack{j\in\mathcal{J}_i \\ j>i}} \int_{\Sigma_{ij}} \{|\nu_i \wedge \widetilde{E}_i^n|^2 + |\nu_i \wedge \widetilde{E}_i^{n+1}|^2 + |\widetilde{H}_{j\tau}^n|^2 + |\widetilde{H}_{j\tau}^{n+1}|^2 \\
&\qquad + |\nu_i \wedge \widetilde{P}_i^n|^2 + |\nu_i \wedge \widetilde{P}_i^{n+1}|^2 + |\widetilde{Q}_{j\tau}^n|^2 + |\widetilde{Q}_{j\tau}^{n+1}|^2\} d\Sigma\Big\}^{1/2} \\
&\qquad \leq 2\sqrt{2}\Big\{\max\Big(\frac{\gamma}{\beta^2+\gamma^2}, \gamma\Big)\Big\}^{1/2} \sqrt{\mathcal{E}^n + \mathcal{E}^{n+1}} \\
&\qquad \times \Big\{\sum_{i=1}^{I-1} \sum_{\substack{j\in\mathcal{J}_i \\ j>i}} [\mathcal{E}_{ij}^{n,n+1}(\Sigma_{ij}) + \mathcal{E}_{ij}^{n+1,n}(\Sigma_{ij})]\Big\}^{1/2} \\
&\qquad = C_2\sqrt{\mathcal{E}^n + \mathcal{E}^{n+1}} \Big\{\sum_{i=1}^{I-1} \sum_{\substack{j\in\mathcal{J}_i \\ j>i}} [\mathcal{E}_{ij}^{n,n+1}(\Sigma_{ij}) + \mathcal{E}_{ij}^{n+1,n}(\Sigma_{ij})]\Big\}^{1/2}. \qquad (7.4.3.23)
\end{aligned}
$$

Lemma 7.4.3.4 now follows from (7.4.3.22), (7.4.3.23) and (7.4.3.17).

7.5 Time and Space Domain Decomposition

7.5.1 Sequential Space-Time Domain Decomposition

As in Section 6.5.1, one may write a sequential space-time domain decomposition scheme for the global optimality system (6.1.1.1), (6.2.3.4), (6.2.3.6) by applying the spatial domain decomposition algorithm of Section 7.4.1 to each of the time-local problems arising in the time domain decomposition scheme of Section 7.3. The resulting local problems on the cylinders $\mathcal{Q}_{ik}$ are then given by

$$
\begin{cases}
\varepsilon_i(E_{ik}^{m,n+1})' - \operatorname{rot} H_{ik}^{m,n+1} + \sigma_i E_{ik}^{m,n+1} = F_{ik} \\
\mu_i(H_{ik}^{m,n+1})' + \operatorname{rot} E_{ik}^{m,n+1} = G_{ik} \qquad \text{in } \mathcal{Q}_{ik}
\end{cases} \qquad (7.5.1.1)
$$

$$
H_{ik\tau}^{m,n+1} - \alpha_i(\nu_i \wedge E_{ik}^{m,n+1}) = \nu_i \wedge P_{ik}^{m,n+1} \text{ on } \Sigma_{ik}
$$

$$
\begin{cases}
\varepsilon_i(P_{ik}^{m,n+1})' - \operatorname{rot} Q_{ik}^{m,n+1} - \sigma_i P_{ik}^{m,n+1} = 0 \\
\mu_i(Q_{ik}^{m,n+1})' + \operatorname{rot} P_{ik}^{m,n+1} = 0 \qquad \text{in } \mathcal{Q}_{ik}
\end{cases} \qquad (7.5.1.2)
$$

$$
Q_{ik\tau}^{m,n+1} + \alpha_i(\nu_i \wedge P_{ik}^{m,n+1}) = 0 \text{ on } \Sigma_{ik}
$$

$$\begin{cases} H_{ik\tau}^{m,n+1} - \beta(\nu_i \wedge E_{ik}^{m,n+1}) - \gamma(\nu_i \wedge P_{ik}^{m,n+1}) = \lambda_{ijk}^{m,n} \\ Q_{ik\tau}^{m,n+1} + \beta(\nu_i \wedge P_{ik}^{m,n+1}) - \gamma(\nu_i \wedge E_{ik}^{m,n+1}) = \rho_{ijk}^{m,n} \quad \text{on } \Sigma_{ijk} \end{cases} \tag{7.5.1.3}$$

$$\begin{cases} E_{i0}^{m,n+1}(0) = E_{i0}, \quad H_{i0}^{m,n+1}(0) = H_{i0} \\ E_{ik}^{m,n+1}(T_k) = \dfrac{1}{\theta}\big(\mu_{i,k,k-1}^{m-1} - P_{ik}^{m,n+1}(T_k)\big) \\ H_{ik}^{m,n+1}(T_k) = \dfrac{1}{\theta}\big(\eta_{i,k,k-1}^{m-1} - Q_{ik}^{m,n+1}(T_k)\big) \quad \text{in } \Omega_i \end{cases} \tag{7.5.1.4}$$

$$\begin{cases} P_{ik}^{m,n+1}(T_{k+1}) = \theta E_{ik}^{m,n+1}(T_{k+1}) - \mu_{i,k,k+1}^{m-1} \\ Q_{ik}^{m,n+1}(T_{k+1}) = \theta H_{ik}^{m,n+1}(T_{k+1}) - \eta_{i,k,k+1}^{m-1} \\ P_{iK}^{m,n+1}(T) = \kappa(E_{iK}^{m,n+1}(T) - E_{iT}) \\ Q_{iK}^{m,n+1}(T) = \kappa(H_{iK}^{m,n+1}(T) - H_{iT}) \quad \text{in } \Omega_i \end{cases} \tag{7.5.1.5}$$

where β, γ, θ are positive constants and

$$\mathcal{Q}_{ik} = \mathcal{Q}_i \times I_k, \quad \Sigma_{ik} = \Gamma_i \times I_k, \quad \Sigma_{ijk} = \Gamma_{ij} \times I_k.$$

Further, $\mu_{i,k,k\pm1}^{m-1}$, $\eta_{i,k,k\pm1}^{m-1}$ and $\lambda_{ijk}^{m,0}$, $\rho_{ijk}^{m,0}$ are taken as given data while, for $n \geq 1$,

$$\begin{aligned} \lambda_{ijk}^{m,n} =& (1-\epsilon)[H_{jk\tau}^{m,n} + \beta(\nu_j \wedge E_{jk}^{m,n}) + \gamma(\nu_j \wedge P_{jk}^{m,n})] \\ &+ \epsilon[H_{ik\tau}^{m,n} - \beta(\nu_i \wedge E_{ik}^{m,n}) - \gamma(\nu_i \wedge P_{ik}^{m,n})]\big|_{\Sigma_{ij}} \\ \rho_{ij}^{m,n} =& (1-\epsilon)[Q_{jk\tau}^{m,n} - \beta(\nu_j \wedge P_{jk}^{m,n}) + \gamma(\nu_j \wedge E_{jk}^{m,n})] \\ &+ \epsilon[Q_{ik\tau}^{m,n} + \beta(\nu_i \wedge P_{ik}^{m,n}) - \gamma(\nu_i \wedge E_{ik}^{n})]\big|_{\Sigma_{ijk}}, \end{aligned} \tag{7.5.1.6}$$

where $\epsilon \in [0,1)$ is a relaxation parameter. In (7.5.1.4) the index k runs from 1 to K, while k runs from 0 to $K-1$ in (7.5.1.5).

The system (7.5.1.1)–(7.5.1.5) is well posed under our standing assumptions on the global data and provided $(\mu_{i,k,k\pm1}^{m-1}, \eta_{i,k,k\pm1}^{m-1}) \in \mathcal{H}_i$, $\lambda_{ijk}^{m,n}$, $\rho_{ijk}^{m,n} \in L^2(\Sigma_{ijk})$ for all $j \in \mathcal{J}_i$. Indeed, for the regions $\mathcal{Q}_{ik}$ with $k = 1, \dots, K-1$ it is the optimality system for the problem

$$\inf \mathcal{C}_{ik}^{m,n}(J_{ik}, J_{ijk}, h_{i,k,k-1}, g_{i,k,k-1})$$

taken over

$$(h_{i,k,k-1}, g_{i,k,k-1}) \in \mathcal{H}_i, \quad J_{ik} \in L^2(\Sigma_{ik}), \quad J_{ijk} \in L^2(\Sigma_{ijk}), \ \forall j \in \mathcal{J}_i,$$

subject to

$$\begin{aligned} &\begin{cases} \varepsilon_i(E_{ik})' - \operatorname{rot} H_{ik} + \sigma_i E_{ik} = F_{ik} \\ \mu_i(H_{ik})' + \operatorname{rot} E_{ik} = G_{ik} \quad \text{in } \mathcal{Q}_{ik} \end{cases} \\ &\quad H_{ik\tau} - \alpha_i(\nu_i \wedge E_{ik}) = J_{ik} \text{ on } \Sigma_{ik} \\ &H_{ik\tau} - \beta(\nu_i \wedge E_{ik}) = J_{ijk} + \lambda_{ijk}^{m,n} \text{ on } \Sigma_{ijk}, \ j \in \mathcal{J}_i \end{aligned} \tag{7.5.1.7}$$

$$\begin{cases} E_{ik}(T_k) = \frac{1}{\theta}\big(\mu^{m-1}_{i,k,k-1} + h_{i,k,k-1}\big) \\ H_{ik}(T_k) = \frac{1}{\theta}\big(\eta^{m-1}_{i,k,k-1} + g_{i,k,k-1}\big) \quad \text{in } \Omega_i \end{cases} \tag{7.5.1.8}$$

where

$$\begin{aligned} \mathcal{C}^{m,n}_{ik} = \frac{1}{2}\int_{\Sigma_{ik}} |J_{ik}|^2 d\Sigma + \frac{1}{2\gamma}\sum_{j\in\mathcal{J}_i}\int_{\Sigma_{ijk}} \{|J_{ijk}|^2 + |\gamma(\nu_i \wedge E_{ik}) + \rho^{m,n}_{ijk}|^2\}\, d\Sigma \\ + \frac{1}{2\theta}\{\|(\theta E_{ik}(T_{k+1}) - \mu^{m-1}_{i,k,k+1}, \theta H_{ik}(T_{k+1}) - \eta^{m-1}_{i,k,k+1})\|^2_{\mathcal{H}_i} \\ + \|(h_{i,k,k-1}, g_{i,k,k-1})\|^2_{\mathcal{H}_i}\}. \end{aligned}$$

For the cylinder Q_{i0} the appropriate cost functional is $\mathcal{C}^{m,n}_{i0}$ but with the last term omitted, subject to (7.5.1.7) with $k = 0$ and with (7.5.1.8) replaced by

$$E_{i0}(0) = E_{0i}, \quad H_{i0}(0) = H_{0i}.$$

For the cylinder Q_{iK} the appropriate cost functional is $\mathcal{C}^{m,n}_{iK}$ with the second to last term replaced by

$$\frac{\kappa}{2}\|(E_{iK}(T), H_{iK}(T)) - (E_{iT}, H_{iT})\|^2_{\mathcal{H}_i},$$

subject to (7.5.1.7), (7.5.1.8) with $k = K$.

Since the conditions $\lambda^{m,n}_{ijk}, \rho^{m,n}_{ijk} \in L^2(\Sigma_{ijk})$ propagate with the index n, it follows that the iteration (7.5.1.1)–(7.5.1.6) is well defined under the standing assumptions on the global data and provided $(\mu^{m-1}_{i,k,k\pm1}, \eta^{m-1}_{i,k,k\pm1}) \in \mathcal{H}_i$, $\lambda^{m,0}_{ijk}, \rho^{m,0}_{ijk} \in L^2(\Sigma_{ijk})$ for all $j \in \mathcal{J}_i$. The goal of this section is to verify that as $n \to \infty$ we have

$$\begin{aligned} (E^{m,n}_{ik}, H^{m,n}_{ik}) &\to (E^m_{ik}, H^m_{ik}) := (E^m_k, H^m_k)\big|_{\Omega_i} \\ (P^{m,n}_{ik}, Q^{m,n}_{ik}) &\to (P^m_{ik}, Q^m_{ik}) := (P^m_k, Q^m_k)\big|_{\Omega_i} \end{aligned}$$

where (E^m_k, H^m_k), (P^m_k, Q^m_k) is the solution on I_k of the time domain decomposition scheme (7.3.1.8)–(7.3.1.10), (7.3.1.6) at iteration level m.

In fact, this conclusion is almost obvious since, on each interval I_k, the iteration (7.5.1.1)–(7.5.1.6) has the exact structure of the spatial domain decomposition scheme (7.4.1.5)–(7.4.1.9) except for the presence of the terms $P^{m,n+1}_{ik}(T_k)$ and $Q^{m,n+1}_{ik}(T_k)$ in the initial conditions for $E^{m,n+1}_{ik}$, $H^{m,n+1}_{ik}$ on the interval I_k. However, these additional terms to not compromise convergence. The arguments used to confirm this statement are the same as those used in the proof of Theorem 7.4.2.1 and lead to the following result (cf. Theorem 6.5.1.1). Introduce the local errors

$$\begin{aligned} \widetilde{E}^{m,n}_{ik} &= E^{m,n}_{ik} - E^m_{ik}, & \widetilde{H}^{m,n}_{ik} &= H^{m,n}_{ik} - H^m_{ik} \\ \widetilde{P}^{m,n}_{ik} &= P^{m,n}_{ik} - P^m_{ik}, & \widetilde{Q}^{m,n}_{ik} &= Q^{m,n}_{ik} - Q^m_{ik}. \end{aligned}$$

Theorem 7.5.1.1 *Assume that β, γ satisfy (7.4.2.8) and that*

$$\frac{\gamma}{\beta} > \frac{|1-\theta^2|}{2\theta}.$$

Then as $n \to \infty$ we have, for $i = 1, \ldots, I$; $k = 0, \ldots, K$; and $m = 1, 2, \ldots$:

(i) *for any $\epsilon \in [0, 1)$,*

$$\begin{cases} (\widetilde{E}_{ik}^{m,n}(T_{k+1}), \widetilde{H}_{ik}^{m,n}(T_{k+1})) \to 0 \\ (\widetilde{P}_{ik}^{m,n}(T_k), \widetilde{Q}_{ik}^{m,n}(T_k)) \to 0 \qquad \textit{strongly in } \mathcal{H}_i \end{cases}$$

$$\begin{cases} \nu_i \wedge \widetilde{E}_{ik}^{m,n}\big|_{\Sigma_{ik}} \to 0, \quad \widetilde{H}_{ik\tau}^{m,n}\big|_{\Sigma_{ik}} \to 0 \\ \nu_i \wedge \widetilde{P}_{ik}^{m,n}\big|_{\Sigma_{ik}} \to 0, \quad \widetilde{Q}_{ik\tau}^{m,n}\big|_{\Sigma_{ik}} \to 0 \quad \textit{strongly in } L^2(\Sigma_{ik}) \end{cases}$$

$$\begin{cases} \nu_i \wedge \widetilde{E}_{ik}^{m,n}\big|_{\Sigma_{ik}^{int}} \to 0, \quad \widetilde{H}_{ik\tau}^{m,n}\big|_{\Sigma_{ik}^{int}} \to 0 \\ \nu_i \wedge \widetilde{P}_{ik}^{m,n}\big|_{\Sigma_{ik}^{int}} \to 0, \quad \widetilde{Q}_{ik\tau}^{m,n}\big|_{\Sigma_{ik}^{int}} \to 0 \quad \textit{weakly in } L^2(\Sigma_{ik}^{int}). \end{cases}$$

(ii) *In addition, if $\epsilon > 0$, then*

$$(\widetilde{E}_{ik}^{m,n}, \widetilde{H}_{ik}^{m,n}) \to 0, \quad (\widetilde{P}_{ik}^{m,n}, \widetilde{Q}_{ik}^{m,n}) \to 0 \quad \textit{in } L^\infty(I_k; \mathcal{H}_i)$$

$$\begin{cases} \nu_i \wedge E_{ik}^{m,n} + \nu_j \wedge E_{jk}^{m,n}\big|_{\Sigma_{ijk}} \to 0, \quad H_{ik\tau}^{m,n} - H_{jk\tau}^{m,n}\big|_{\Sigma_{ijk}} \to 0 \\ \nu_i \wedge P_{ik}^{m,n} + \nu_j \wedge P_{jk}^{m,n}\big|_{\Sigma_{ijk}} \to 0, \quad Q_{ik\tau}^{m,n} - Q_{jk\tau}^{m,n}\big|_{\Sigma_{ijk}} \to 0 \end{cases}$$
$$\textit{strongly in } L^2(\Sigma_{ijk}),\ \forall j \in \mathcal{J}_i.$$

Indeed, by proceeding as in the proof of Theorem 7.4.2.1 we arrive at the fundamental recursion

$$\mathcal{E}^{m,n+1} + \mathcal{F}^{m,n+1} \le \mathcal{E}^{m,n} + (1-2\epsilon)\mathcal{F}^{m,n},$$

valid for every $\epsilon \in [0, 1)$, where

$$\mathcal{E}^{m,n} = \frac{1}{2}\sum_{i=1}^{I}\sum_{k=0}^{K}\int_{\Sigma_{ik}^{\text{int}}}\Big\{\frac{\beta^2+\gamma^2}{\gamma}(|\nu_i \wedge \widetilde{E}_{ik}^{m,n}|^2 + |\nu_i \wedge \widetilde{P}_{ik}^{m,n}|^2)$$
$$+ \frac{1}{\gamma}(|\widetilde{H}_{ik\tau}^{m,n}|^2 + |\widetilde{Q}_{ik\tau}^{m,n}|^2)\Big\}d\Sigma$$

and

$$\mathcal{F}^{m,n} = \sum_{i=1}^{I} \Big\{ \sum_{k=1}^{K} \Big[\big(\theta + \frac{\beta}{2\gamma}(1-\theta^2)\big) \|(\widetilde{E}_{ik}^{m,n}(T_k), \widetilde{H}_{ik}^{m,n}(T_k))\|_{\mathcal{H}_i}^2$$

$$+ \frac{1}{\theta^2}\big(\theta + \frac{\beta}{2\gamma}(\theta^2 - 1)\big) \|(\widetilde{P}_{ik}^{m,n}(T_k), \widetilde{Q}_{ik}^{m,n}(T_k))\|_{\mathcal{H}_i}^2 \Big]$$

$$+ \big(\kappa + \frac{\beta}{2\gamma}(1-\kappa^2)\big) \|(\widetilde{E}_{iK}^{m,n}(T), \widetilde{H}_{iK}^{m,n}(T))\|_{\mathcal{H}_i}^2 + \frac{\beta}{2\gamma} \|(\widetilde{P}_{i0}^{m,n}(0), \widetilde{Q}_{i0}^{m,n}(0))\|_{\mathcal{H}_i}^2$$

$$+ \sum_{k=0}^{K} \frac{\beta}{\gamma} \int_{T_k}^{T_{k+1}} \{ \langle \sigma_i \widetilde{E}_{ik}^{m,n}, \widetilde{E}_{ik}^{m,n} \rangle_i + \langle \sigma_i \widetilde{P}_{ik}^{m,n}, \widetilde{P}_{ik}^{m,n} \rangle_i \} \, dt$$

$$+ \int_{\Sigma_{ik}} \{ \alpha_i \frac{\beta}{\gamma} |\nu_i \wedge \widetilde{E}_{ik}^{m,n}|^2 + (1 + \alpha_i \frac{\beta}{\gamma}) |\nu_i \wedge \widetilde{P}_{ik}^{m,n}|^2 + \frac{\beta}{\gamma} \mathrm{Re}\, (\nu_i \wedge \widetilde{E}_{ik}^{m,n}) \cdot (\nu_i \wedge \widetilde{P}_{ik}^{m,n}) \} d\Sigma \Big\}.$$

By iteration in n of the fundamental recursion we obtain

$$\mathcal{E}^{m,n+1} \le \mathcal{E}^{m,1} - \sum_{p=1}^{n+1} c_p(\epsilon) \mathcal{F}^{m,p}, \quad m = 1, 2, \dots, \tag{7.5.1.9}$$

where the coefficients $c_p(\epsilon)$ are defined in (6.4.2.26). Part (i) of Theorem 7.5.1.1 is just the observation that $\mathcal{F}^{m,n} \ge 0$ under the stated hypotheses on the parameters β, γ, θ and thus, from (7.5.1.9), $\mathcal{F}^{m,n} \to 0$ as $n \to \infty$.

Part (ii) of Theorem 7.5.1.1 is proved by a calculation essentially identical to that carried out in the proof of Part (i) of Theorem 7.4.2.1.

7.5.2 Sequential Time-Space Domain Decomposition

In sequential time-space domain decomposition, we apply the time domain decomposition algorithm of Section 7.3 to each of the spatially localized problems arising in the spatial domain decomposition algorithm of Section 7.4. The result is a family of problems localized on the cylinders $\mathcal{Q}_{ik}$ given by

$$\begin{cases} \varepsilon_i (E_{ik}^{m+1,n})' - \mathrm{rot}\, H_{ik}^{m+1,n} + \sigma_i E_{ik}^{m+1,n} = F_{ik} \\ \mu_i (H_{ik}^{m+1,n})' + \mathrm{rot}\, E_{ik}^{m+1,n} = G_{ik} \qquad \text{in } \mathcal{Q}_{ik} \end{cases} \tag{7.5.2.1}$$

$$H_{ik\tau}^{m+1,n} - \alpha_i (\nu_i \wedge E_{ik}^{m+1,n}) = \nu_i \wedge P_{ik}^{m+1,n} \text{ on } \Sigma_{ik}$$

$$\begin{cases} \varepsilon_i (P_{ik}^{m+1,n})' - \mathrm{rot}\, Q_{ik}^{m+1,n} - \sigma_i P_{ik}^{m+1,n} = 0 \\ \mu_i (Q_{ik}^{m+1,n})' + \mathrm{rot}\, P_{ik}^{m+1,n} = 0 \qquad \text{in } \mathcal{Q}_{ik} \end{cases} \tag{7.5.2.2}$$

$$Q_{ik\tau}^{m+1,n} + \alpha_i (\nu_i \wedge P_{ik}^{m+1,n}) = 0 \text{ on } \Sigma_{ik}$$

$$\begin{cases} H_{ik\tau}^{m+1,n} - \beta(\nu_i \wedge E_{ik}^{m+1,n}) - \gamma(\nu_i \wedge P_{ik}^{m+1,n}) = \lambda_{ik}^{n-1} \\ Q_{ik\tau}^{m+1,n} + \beta(\nu_i \wedge P_{ik}^{m+1,n}) - \gamma(\nu_i \wedge E_{ik}^{m+1,n}) = \rho_{ik}^{n-1} \quad \text{on } \Sigma_{ik}^{\text{int}} \end{cases} \tag{7.5.2.3}$$

$$\begin{cases} E_{i0}^{m+1,n}(0) = E_{i0}, \quad H_{i0}^{m+1,n}(0) = H_{i0} \\ E_{ik}^{m+1,n}(T_k) = \dfrac{1}{\theta}(\mu_{i,k,k-1}^{m,n} - P_{ik}^{m+1,n}(T_k)) \\ H_{ik}^{m+1,n}(T_k) = \dfrac{1}{\theta}(\eta_{i,k,k-1}^{m,n} - Q_{ik}^{m+1,n}(T_k)) \quad \text{in } \Omega_i \end{cases} \tag{7.5.2.4}$$

$$\begin{cases} P_{ik}^{m+1,n}(T_{k+1}) = \theta E_{ik}^{m+1,n}(T_{k+1}) - \mu_{i,k,k+1}^{m,n} \\ Q_{ik}^{m+1,n}(T_{k+1}) = \theta H_{ik}^{m+1,n}(T_{k+1}) - \eta_{i,k,k+1}^{m,n} \\ P_{iK}^{m+1,n}(T) = \kappa(E_{iK}^{m+1,n}(T) - E_{iT}) \\ Q_{iK}^{m+1,n}(T) = \kappa(H_{iK}^{m+1,n}(T) - H_{iT}) \quad \text{in } \Omega_i. \end{cases} \tag{7.5.2.5}$$

The system (7.5.2.1)–(7.5.2.5) is, except for indexing, the same as the system (7.5.1.1)–(7.5.1.5). However, in the present context, λ_{ik}^{n-1}, ρ_{ik}^{n-1} are treated as given data while

$$\begin{aligned} \mu_{i,k,k+1}^{m,n} =& (1-\epsilon)\big(\theta E_{i,k+1}^{m,n}(T_{k+1}) - P_{i,k+1}^{m,n}(T_{k+1})\big) \\ &+ \epsilon\big(\theta E_{i,k}^{m,n}(T_{k+1}) - P_{i,k}^{m,n}(T_{k+1})\big) \\ \eta_{i,k,k+1}^{m,n} =& (1-\epsilon)\big(\theta H_{i,k+1}^{m,n}(T_{k+1}) - Q_{i,k+1}^{m,n}(T_{k+1})\big) \\ &+ \epsilon\big(\theta H_{i,k}^{m,n}(T_{k+1}) - Q_{i,k}^{m,n}(T_{k+1})\big) \\ \mu_{i,k,k-1}^{m,n} =& (1-\epsilon)\big(\theta E_{i,k-1}^{m,n}(T_k) + P_{i,k-1}^{m,n}(T_k)\big) \\ &+ \epsilon\big(\theta E_{i,k}^{m,n}(T_k) + P_{i,k}^{m,n}(T_k)\big) \\ \eta_{i,k,k-1}^{m,n} =& (1-\epsilon)\big(\theta H_{i,k-1}^{m,n}(T_k) + Q_{i,k-1}^{m,n}(T_k)\big) \\ &+ \epsilon\big(\theta H_{i,k}^{m,n}(T_k) + Q_{i,k}^{m,n}(T_k)\big), \end{aligned} \tag{7.5.2.6}$$

where $\epsilon \in [0,1)$ is a relaxation parameter.

As in the proceeding section, if the global data satisfies our standing assumptions and if λ_{ik}^{n-1}, $\rho_{ik}^{n-1} \in L^2(\Sigma_{ik}^{\text{int}})$, (7.5.2.1)–(7.5.2.5) is an optimality system concentrated on the cylinder $\mathcal{Q}_{ik}$ provided, in addition, that $(\mu_{i,k,k\pm1}^{m,n}, \eta_{i,k,k\pm1}^{m,n}) \in \mathcal{H}_i$. Since the latter requirement propagates with m, the iteration (in m) (7.5.2.1)–(7.5.2.6) is well defined if $(\mu_{i,k,k\pm1}^{0,n}, \eta_{i,k,k\pm1}^{0,n}) \in \mathcal{H}_i$. The object of this section is to verify that, as $m \to \infty$,

$$\begin{aligned} (E_{ik}^{m,n}, H_{ik}^{m,n}) &\to (E_{ik}^n, H_{ik}^n) := (E_i^n, H_i^n)\big|_{I_k} \\ (P_{ik}^{m,n}, Q_{ik}^{m,n}) &\to (P_{ik}^n, Q_{ik}^n) := (P_i^n, Q_i^n)\big|_{I_k} \end{aligned} \tag{7.5.2.7}$$

where (E_i^n, H_i^n), (P_i^n, Q_i^n) is the solution on Ω_i of the spatial domain decomposition scheme (7.4.1.5)–(7.4.1.7) at iteration level n in which

$$\lambda_{ik}^{n-1} = \{\lambda_{ij}^{n-1}\big|_{I_k} : j \in \mathcal{J}_i\}, \quad \rho_{ik}^{n-1} = \{\rho_{ij}^{n-1}\big|_{I_k} : j \in \mathcal{J}_i\}. \tag{7.5.2.8}$$

It is to be expected that (7.5.2.7) is valid since the relaxed iteration (7.5.2.1)–(7.5.2.6) has the same structure as the pure time domain decomposition algorithm of Section 7.3, except for the fact that the boundary conditions (7.5.2.3) are inhomogeneous and an additional term, $\gamma(\nu_i \wedge E_{ik}^{m+1,n})$, appears in $(7.5.2.3)_2$. The inhomogeneity causes no difficulties with regard to convergence because of (7.5.2.8); the inhomogeneity disappears in the system satisfied by the local errors

$$\widetilde{E}_{ik}^{m,n} := E_{ik}^{m,n} - E_{ik}^{n}, \quad \widetilde{H}_{ik}^{m,n} := H_{ik}^{m,n} - H_{ik}^{n}$$
$$\widetilde{P}_{ik}^{m,n} := P_{ik}^{m,n} - P_{ik}^{n}, \quad \widetilde{Q}_{ik}^{m,n} := Q_{ik}^{m,n} - Q_{ik}^{n}.$$

Further, the proof of convergence of the pure time domain decomposition algorithm carries over almost verbatim to a proof of convergence of (7.5.2.1)–(7.5.2.6) and leads to the following result.

Theorem 7.5.2.1 *Let β, γ, θ be positive constants. Then as $m \to \infty$ we have, for $i = 1, \ldots, I$ and $n = 1, 2, \ldots$:*
(i) *for any $\epsilon \in [0, 1)$*

$$\nu_i \wedge \widetilde{P}_{ik}^{m,n}\big|_{\Sigma_{ik} \cup \Sigma_{ik}^{int}} \to 0 \;\; \textit{strongly in } L^2(\Sigma_{ik} \cup \Sigma_{ik}^{int}),\; k = 0, \ldots, K$$
$$\nu_i \wedge \widetilde{E}_{ik}^{m,n}\big|_{\Sigma_{ik}^{int}} \to 0 \;\; \textit{strongly in } L^2(\Sigma_{ik}^{int}),\; k = 0, \ldots, K$$
$$\big(\widetilde{E}_{i0}^{m,n}, \widetilde{H}_{i0}^{m,n}\big) \to 0 \;\; \textit{in } C(I_0; \mathcal{H}_i), \quad \big(\widetilde{P}_{iK}^{m,n}, \widetilde{Q}_{iK}^{m,n}\big) \to 0 \;\; \textit{in } C(I_K; \mathcal{H}_i)$$
$$\big(\widetilde{E}_{iK}^{m,n}(T), \widetilde{H}_{iK}^{m,n}(T)\big) \to 0 \;\; \textit{strongly in } \mathcal{H}_i.$$

(ii) *In addition, if $\epsilon > 0$ then for $k = 0, \ldots, K$*

$$\big(\widetilde{E}_{ik}^{m,n}, \widetilde{H}_{ik}^{m,n}\big) \to 0, \quad \big(\widetilde{P}_{ik}^{m,n}, \widetilde{Q}_{ik}^{m,n}\big) \to 0 \;\; \textit{in } C(I_k; \mathcal{H}_i)$$
$$\nu_i \wedge \widetilde{E}_{ik}^{m,n}\big|_{\Sigma_{ik}} \to 0 \;\; \textit{strongly in } L^2(\Sigma_{ik}).$$

Indeed, the same arguments used in the proof of Theorem 7.3.2.1 lead to the usual recursive inequality

$$\mathcal{E}^{m+1,n} + \mathcal{F}^{m+1,n} \le \mathcal{E}^{m,n} + (1 - 2\epsilon)\mathcal{F}^{m,n} \tag{7.5.2.9}$$

where

$$\mathcal{E}^{m,n} = \sum_{i=1}^{I} \sum_{k=0}^{K-1} \big[\mathcal{E}_{ik}^{m,n}(T_{k+1}) + \mathcal{E}_{i,k+1}^{m,n}(T_{k+1})\big],$$
$$\mathcal{E}_{ik}^{m,n}(t) = \theta^2 \|(\widetilde{E}_{ik}^{m,n}(t), \widetilde{H}_{ik}^{m,n}(t))\|_{\mathcal{H}_i}^2 + \|(\widetilde{P}_{ik}^{m,n}(t), \widetilde{Q}_{ik}^{m,n}(t))\|_{\mathcal{H}_i}^2,$$

$$\mathcal{F}^{m,n} = 2\kappa\theta \sum_{i=1}^{I} \|(\widetilde{E}_{iK}^{m,n}(T), \widetilde{H}_{iK}^{m,n}(T))\|_{\mathcal{H}_i}^2 + 2\theta \sum_{i=1}^{I} \sum_{k=0}^{K} \Big\{ \int_{\Sigma_{ik}} |\nu_i \wedge \widetilde{P}_{ik}^{m,n}|^2 d\Sigma$$
$$+ \int_{\Sigma_{ik}^{\rm int}} \gamma\big(|\nu_i \wedge \widetilde{P}_{ik}^{m,n}|^2 + |\nu_i \wedge \widetilde{E}_{ik}^{m,n}|^2\big)\, d\Sigma \Big\}.$$

The inequality (7.5.2.9) is valid for all positive constants β, γ, θ and all $\epsilon \in [0,1)$. It follows that $\mathcal{F}^{m,n} \to 0$ as $m \to \infty$, from which one immediately deduces the conclusions of (i) of Theorem 7.5.2.1. The remainder of the proof of Theorem 7.5.2.1 follows exactly the proof of part (ii) of Theorem 7.3.2.1.

Chapter 8

Optimal Final Value Boundary Control of Conservative Wave Equations

8.1 Introduction

This chapter is concerned with the problem with of optimal final value control of the second-order hyperbolic system

$$\begin{gathered}\frac{\partial^2 w}{\partial t^2} - \nabla \cdot (A\nabla w) + cw = F \quad \text{in } \Omega \times (0,T) \\ w = 0 \quad \text{on } \Gamma^D \times (0,T) \\ \frac{\partial w}{\partial \nu_A} = f \text{ on } \Gamma^N \times (0,T) \\ w(0) = w_0, \quad \frac{\partial w}{\partial t}(0) = v_0 \text{ in } \Omega.\end{gathered} \tag{8.1.1.1}$$

where $f \in L^2(\Gamma^N \times (0,T))$ is the control input and F is a given distributed system input. We retain the notation of Section 6.2.1, as well as the assumptions on the coefficients A and c, the region Ω and its boundary components Γ^D, Γ^N delineated there. When $f = 0$ and $F = 0$, this system conserves energy: $E(t) \equiv E(0)$, where the energy functional $E(t)$ is given by

$$E(t) = \int_\Omega \left[\left| \frac{\partial w}{\partial t}(x,t) \right|^2 + A\nabla w(x,t) \cdot \nabla w(x,t) + c|w(x,t)|^2 \right] dx.$$

When $f = 0$ and $F \in L^2(Q)$, say, the solution of (8.1.1.1) satisfies $E(t) < \infty$ for all $t > 0$ provided $E(0) < \infty$. However, when $f \neq 0$, there is a loss of regularity: the solution of (8.1.1.1) in general does not have finite energy for any $t > 0$ regardless of the regularity of the remaining data and regardless of the geometry of Ω. Therefore, when considering the problem of optimal final value $L^2(\Sigma^N)$ boundary control, the deviation of the solution at the final time from its target must be penalized in a norm strictly weaker than the energy norm. The resulting optimality system then has a different structure than the one considered in Chapter 6, where the control entered through a dissipative boundary condition that improved the regularity of the solution. This leads us to a domain decomposition algorithm for the optimality system that is different from the one introduced in Chapter 6. In fact, in the present case we use a domain decomposition method that is based on an under relaxation of the nonoverlapping Schwarz alternating algorithm and the introduction of skew-symmetric, Robin iterative transmission conditions. The use of this type of decomposition, without relaxation, in problems of optimal control was first proposed by Benamou as discussed earlier in this book.

In this chapter we shall also develop an iterative domain decomposition algorithm for the solution of the optimality system associated with the problem of minimum norm *exact controllability* of solutions of (8.1.1.1), that is, the state constrained optimal control problem

$$\inf_{f\in L^2(\Gamma^N\times(0,T))}\frac{1}{2}\int_\Sigma |f|^2 d\Sigma \tag{8.1.1.2}$$

subject to (8.1.1.1) and

$$(w(T),\frac{\partial w}{\partial t}(T)) = (z_0, z_1). \tag{8.1.1.3}$$

Assuming that for T suitably large this problem has a solution for all (w_0, v_0) and (z_0, z_1) in an appropriate function space, we will in fact establish the validity of the commutative diagram

$$\begin{array}{ccc} (w^n(\cdot\,;\kappa), p^n(\cdot\,;\kappa)) & \xrightarrow{n\to\infty} & (w(\cdot\,;\kappa), p(\cdot\,;\kappa)) \\ {\scriptstyle\kappa\to\infty}\big\downarrow & & {\scriptstyle\kappa\to\infty}\big\downarrow \\ (w^n(\cdot\,;\infty), p^n(\cdot\,;\infty)) & \xrightarrow{n\to\infty} & (w(\cdot\,;\infty), p(\cdot\,;\infty)) \end{array} \tag{8.1.1.4}$$

in appropriate norms, where the parameter κ penalizes the deviation of the solution at time T from its target (z_0, z_1). In (8.1.1.4), p denotes the adjoint variable and $(w(\cdot\,;\kappa), p(\cdot\,;\kappa))$ is the solution of the optimality system associated with the optimal final value control of (8.1.1.1) for a fixed value of the penalty parameter κ; $(w^n(\cdot\,;\kappa), p^n(\cdot\,;\kappa))$ is the nth iterate in its domain decomposition; $(w(\cdot\,;\infty), p(\cdot\,;\infty))$ is the solution of the optimality system associated with (8.1.1.2), (8.1.1.3), and $(w^n(\cdot\,;\infty), p^n(\cdot\,;\infty))$ is the nth iterate in its domain decomposition.

8.2 Optimal Boundary Control

8.2.1 Setting the Problem

We set $H = L^2(\Omega)$ with its standard norm denoted by $\|\cdot\|$, and $V = H^1_D(\Omega)$ with norm $\|\cdot\|_V$ defined by

$$\|\phi\|_V^2 = \int_\Omega [A\nabla\phi\cdot\nabla\phi + |\phi|^2]\,dx.$$

The space H is identified with its dual space and V^* denotes the dual space of V with respect to H, so that we have $V \subset H \subset V^*$ with dense and continuous embedding. We further set $\mathcal{H} = V \times H$ with the usual product norm. If $F \in L^1(0,T;V^*)$, $(v_0, w_0) \in \mathcal{H}^*$ and $f \in L^2(\Sigma^N)$, the system (8.1.1.1) has a unique solution, which may be defined by transposition, for example. Even if the coefficients A and c are C^∞ in Ω, $F = 0$, $w_0 = v_0 = 0$ and $\Gamma \in C^\infty$, the best regularity of the solution is $(w, \partial w/\partial t) \in C([0,T]; H^s_D(\Omega) \times H^{s-1}(\Omega))$, where $s > 1/2$ but $s \leq 5/6$, the precise value of s depending on the geometry of Ω (see [97] and [62]). When the data is less regular as is the situation under consideration, the precise regularity of the solution is more difficult to determine but, in any case, one has $(w, \partial w/\partial t) \in C([0,T]; H \times V^*)$ (see the next section). Thus for any $(z_0, z_1) \in H \times V^*$ we may consider the optimal control problem

$$\inf_{f \in L^2(\Sigma^N)} \left\{ \frac{1}{2} \int_{\Sigma^N} |f|^2 d\Sigma + \frac{\kappa}{2} \left(\|(w(T) - z_0\|^2 + \|\frac{\partial w}{\partial t}(T)) - z_1\|^2_{V^*} \right) \right\} \tag{8.2.1.1}$$

subject to (8.1.1.1), where $\kappa > 0$. There is a unique optimal control f_{opt} given by (see Section 8.2.3)

$$f_{\text{opt}} = -p|_{\Sigma^N} \tag{8.2.1.2}$$

where p is the solution of the backwards running adjoint system

$$\begin{aligned} &\frac{\partial^2 p}{\partial t^2} - \nabla \cdot (A \nabla p) + cp = 0 \text{ in } Q \\ &p = 0 \text{ on } \Sigma^D, \qquad \frac{\partial p}{\partial \nu_A} = 0 \text{ on } \Sigma^N \\ p(T) = \kappa \mathcal{A}^{-1}\big(&\frac{\partial w}{\partial t}(T) - z_1\big), \quad \frac{\partial p}{\partial t}(T) = -\kappa(w(T) - z_0) \text{ in } \Omega \end{aligned} \tag{8.2.1.3}$$

and where $\mathcal{A}$ is the Riesz isomorphism of V onto V^*. Since

$$\langle \mathcal{A}\phi, \psi \rangle_V = \int_\Omega [A \nabla \phi(x) \cdot \nabla \psi(x) + \phi(x)\psi(x)]\, dx, \quad \forall \phi, \psi \in V,$$

the condition $p(T) = \kappa \mathcal{A}^{-1}\big(\frac{\partial w}{\partial t}(T) - z_1\big)$ in (8.2.1.3) is the elliptic boundary value problem

$$\begin{aligned} -\nabla \cdot (A \nabla p(T)) + p(T) &= \kappa\big(\frac{\partial w}{\partial t}(T) - z_1\big) \text{ in } \Omega \\ p(T) = 0 \text{ on } \Gamma^D, \quad &\frac{\partial p(T)}{\partial \nu_A} = 0 \text{ on } \Gamma^N. \end{aligned} \tag{8.2.1.4}$$

Thus, any spatial domain decomposition algorithm for the optimality system (8.1.1.1), (8.2.1.2), (8.2.1.3) must include a spatial domain decomposition procedure for the elliptic boundary value problem (8.2.1.4).

8.2.2 Existence and Regularity of Solutions

Consider the problem

$$\begin{aligned} &\frac{\partial^2\phi}{\partial t^2} - \nabla\cdot(A\nabla\phi) + c\phi = G \text{ in } Q \\ &\phi = 0 \text{ on } \Sigma^D, \quad \frac{\partial\phi}{\partial\nu_A} = 0 \text{ on } \Sigma^N \end{aligned} \tag{8.2.2.1}$$

$$\phi(0) = \phi_0, \quad \frac{\partial\phi}{\partial t}(0) = \phi_1 \text{ in } \Omega. \tag{8.2.2.2}$$

One may write (8.2.2.1), (8.2.2.2) as the first-order abstract initial value problem in $\mathcal{H}$

$$\frac{d\Phi}{dt} = \Lambda\Phi + \mathcal{G}, \quad \Phi(0) = \Phi_0, \tag{8.2.2.3}$$

where

$$\Phi = \begin{pmatrix} \phi \\ \partial\phi/\partial t \end{pmatrix}, \quad \Phi_0 = \begin{pmatrix} \phi_0 \\ \phi_1 \end{pmatrix}, \quad \mathcal{G} = \begin{pmatrix} 0 \\ G \end{pmatrix},$$

$$\begin{aligned} \Lambda &= \begin{pmatrix} 0 & I \\ \nabla\cdot(A(\cdot)\nabla) - c(\cdot)I & 0 \end{pmatrix} \\ &= \begin{pmatrix} 0 & I \\ -\mathcal{A} & 0 \end{pmatrix} + \begin{pmatrix} 0 & 0 \\ (1-c(\cdot))I & 0 \end{pmatrix} := \Lambda_0 + \mathcal{K}, \end{aligned}$$

where

$$D(\Lambda) = D(\Lambda_0) = \{(\phi,\psi) \in \mathcal{H} : \psi \in V, \ \mathcal{A}\phi \in H\} := D(\mathcal{A}) \times V. \tag{8.2.2.4}$$

If the coefficients A, c, and the boundary Γ are sufficiently regular, one has

$$(\phi,\psi) \in D(\Lambda) \Leftrightarrow (\phi,\psi) \in (H^2(\Omega) \cap V) \times V, \ \frac{\partial\phi}{\partial\nu_A}|_{\Gamma^N} = 0.$$

The operator Λ_0 generates a unitary group of bounded linear operators on $\mathcal{H}$, hence Λ is the infinitesimal generator of a C_0 group $\{S(t) : t \in \mathbb{R}\}$ of bounded linear operators on $\mathcal{H}$ such that

$$\|S(t)\| \le \exp(t\|\mathcal{K}\|) \le \exp\big(t(1 + \|c\|_{L^\infty(\Omega)})\big).$$

It follows that if $\mathcal{G} \in L^1(0,T;\mathcal{H})$ and $\Phi_0 \in \mathcal{H}$, (8.2.2.3) has a unique mild solution $\Phi \in C([0,T];\mathcal{H})$ that satisfies the a priori inequality

$$\|\Phi\|_{L^\infty(0,T;\mathcal{H})} \le C_T\big[\|\Phi_0\|_{\mathcal{H}} + \|\mathcal{G}\|_{L^1(0,T;\mathcal{H})}\big], \tag{8.2.2.5}$$

where

$$C_T := \sup_{0\le t\le T} \|S(t)\| \le \exp\big(T(1 + \|c\|_{L^\infty(\Omega)})\big). \tag{8.2.2.6}$$

Obviously the same conclusions are valid if the initial conditions (8.2.2.2) are replaced by the final conditions

$$\phi(T) = \phi_0, \quad \frac{\partial \phi}{\partial t}(T) = \phi_1 \text{ in } \Omega. \tag{8.2.2.7}$$

Remark 8.2.2.1 If $c(x) > 0$ on a set in Ω of positive measure, or if $\Gamma^D \neq \emptyset$, we may renorm V by setting

$$\|\phi\|_V^2 = \int_\Omega [A\nabla\phi \cdot \nabla\phi + c|\phi|^2]\, dx,$$

in which case $\Lambda = \Lambda_0$ and $\{S(t) : t \in \mathbb{R}\}$ is a unitary group on $\mathcal{H}$, hence we may take $C_T = 1$ in (8.2.2.5).

Now let ϕ be the solution of (8.2.2.1) with $G = 0$ and (8.2.2.7). Suppose that $F \in L^1(0,T;V^*)$, $f \in L^2(\Sigma^N)$ and $(v_0, w_0) \in \mathcal{H}^* = V^* \times H$. If w a solution of (8.1.1.1) and if formal integrations by parts are carried out in

$$\int_0^t \langle F, \phi\rangle_V\, ds = \int_0^t \langle (\frac{\partial^2 w}{\partial t^2} - \nabla \cdot (A\nabla w) + cw), \phi \rangle_V\, ds$$

one obtains

$$\begin{aligned}\langle (\frac{\partial w}{\partial t}(t), -w(t)), (\phi(t), \frac{\partial \phi}{\partial t}(t))\rangle_{\mathcal{H}} &= \langle (v_0, -w_0), (\phi(0), \frac{\partial \phi}{\partial t}(0))\rangle_{\mathcal{H}} \\ &+ \int_0^t \langle F, \phi\rangle_V\, ds + \int_0^t \int_{\Gamma^N} f\phi\, d\Gamma ds, \quad \forall (\phi_0, \phi_1) \in \mathcal{H}. \end{aligned} \tag{8.2.2.8}$$

From (8.2.2.5) we have, for each $t \in [0,T]$,

$$\begin{aligned}|\langle (\frac{\partial w}{\partial t}(t), -w(t)), (\phi(t), \frac{\partial \phi}{\partial t}(t))\rangle_{\mathcal{H}}| &\le C_T \|(v_0, -w_0)\|_{\mathcal{H}^*} \|(\phi(t), \frac{\partial \phi}{\partial t}(t))\|_{\mathcal{H}} \\ &+ \|F\|_{L^1(0,T;V^*)} \|\phi\|_{L^\infty(0,t;V)} + \|f\|_{L^2(\Sigma^N)} \|\phi\|_{L^2(0,t;\Gamma^N)} \\ &\le C_T [\|(v_0, -w_0)\|_{\mathcal{H}^*} + \|F\|_{L^1(0,T;V^*)} + C_\Omega \sqrt{T} \|f\|_{L^2(\Sigma^N)}] \|(\phi(t), \frac{\partial \phi}{\partial t}(t))\|_{\mathcal{H}}\end{aligned}$$

for all $(\phi_0, \phi_1) \in \mathcal{H}$, where C_Ω is a constant such that

$$\|\phi\|_{L^2(\Gamma^N)} \le C_\Omega \|\phi\|_V, \quad \forall \phi \in V.$$

Since the mapping $(\phi_0, \phi_1) \mapsto (\phi(t), \frac{\partial \phi}{\partial t}(t))$ is an isomorphism on $\mathcal{H}$, it follows that for each $t \in (0,T]$, (8.2.2.8) has a unique solution $(\frac{\partial w}{\partial t}(t), -w(t)) \in \mathcal{H}^*$ satisfying the a priori estimate

$$\|(\frac{\partial w}{\partial t}(t), -w(t))\|_{\mathcal{H}^*} \le C_T [\|(v_0, -w_0)\|_{\mathcal{H}^*} + \|F\|_{L^1(0,T;V^*)} + C_\Omega \sqrt{T} \|f\|_{L^2(\Sigma^N)}] \tag{8.2.2.9}$$

for $0 \le t \le T$. Thus $(\frac{\partial w}{\partial t}, -w) \in L^\infty(0,T;\mathcal{H}^*)$. In fact, this pair is in $C([0,T];\mathcal{H}^*)$, that is, it is uniformly continuous on $[0,T]$ into $\mathcal{H}^*$, which may be proved as in [48, Theorem 4.7]. Thus we have the following result.

Theorem 8.2.2.1 *Suppose that* $F \in L^1(0,T;V^*)$, $f \in L^2(\Sigma^N)$ *and* $(w_0, v_0) \in H \times V^*$. *There is a unique function* $w \in C([0,T];H) \cap C^1([0,T];V^*)$ *satisfying* (8.2.2.8), (8.2.2.9) *for all* $t \in [0,T]$, *where* ϕ *is the solution of* (8.2.2.1), (8.2.2.7).

By definition, the function w of Theorem 8.2.2.1 is the solution of (8.1.1.1).

Remark 8.2.2.2 If $f = 0$, $F \in L^1(0,T;H)$ and $(w_0, v_0) \in V \times H$ it follows from (8.2.2.5) that the trace $w|_{\Sigma^N} \in L^2(\Sigma^N)$ and the mapping from the data to this trace is continuous in the indicated topologies. Lasiecka and Triggiani [62] have generalized this result to arbitrary $f \in L^2(\Sigma^N)$ provided the coefficients A, c and the boundary Γ are sufficiently regular. They show that in fact

$$(w, \frac{\partial w}{\partial t}) \in C([0,T]; H^\alpha_D(\Omega) \times H^{\alpha-1}(\Omega)), \quad w|_{\Sigma^N} \in H^{2\alpha-1}(\Sigma^N)$$

where $\alpha = 3/5 - \varepsilon$, $\forall \varepsilon > 0$, and that the map from the data to $\big((w, \frac{\partial w}{\partial t}), w|_{\Sigma^N}\big)$ is continuous in the indicated topologies. This result was later strengthened by Tataru [97], who showed that one may take $\alpha = 2/3$.

8.2.3 The Global Optimality System

Consider the optimal control problem (8.2.1.1) subject to (8.1.1.1). By virtue of properties of the mapping $((w_0, v_0), f) \mapsto (w, \partial w/\partial t)$, the existence of a unique optimal control $f_{\rm opt} \in L^2(\Sigma^N)$ is standard. It is characterized by the requirement that the directional derivative of J at $f_{\rm opt}$ in every direction must vanish. Therefore

$$\int_{\Sigma^N} fh\, d\Sigma + \kappa\big[(\psi(T), w(T) - z_0) + \langle \frac{\partial \psi}{\partial t}(T), \mathcal{A}^{-1}(\frac{\partial w}{\partial t}(T) - z_1)\rangle_V\big] = 0,$$
$$\forall h \in L^2(\Sigma^N), \quad (8.2.3.1)$$

where ψ is the solution of

$$\begin{aligned} &\frac{\partial^2 \psi}{\partial t^2} - \nabla \cdot (A \nabla \psi) + c\psi = 0 \text{ in } Q \\ &\psi = 0 \text{ on } \Sigma^D, \quad \frac{\partial \psi}{\partial \nu_A} = h \text{ on } \Sigma^N \\ &\psi(0) = \frac{\partial \psi}{\partial t}(0) = 0 \text{ in } \Omega. \end{aligned} \qquad (8.2.3.2)$$

Let $p \in C([0,T];V) \cap C^1([0,T];H)$ be the solution of (8.2.1.3). By virtue of the definition of the solution of (8.2.3.2), ψ satisfies

$$0 = \left\langle \frac{\partial \psi}{\partial t}(T), \kappa \mathcal{A}^{-1}\left(\frac{\partial w}{\partial t}(T) - z_1\right)\right\rangle_V + \big(\psi(T), \kappa(w(T) - z_0)\big) - \int_{\Sigma^N} ph \, d\Sigma, \quad \forall h \in L^2(\Sigma^N). \tag{8.2.3.3}$$

It follows immediately from (8.2.3.1) and (8.2.3.3) that

$$f_{\text{opt}} = -p|_{\Sigma^N}.$$

Thus the optimality system for the optimal control problem (8.2.1.1) consists of (8.1.1.1), (8.2.1.3) and (8.2.1.2).

Remark 8.2.3.1 Suppose that $(w_0, v_0) \in V \times H$ and $F \in L^1(0,T;H)$. Note that the optimal control satisfies

$$f_{\text{opt}} \in C([0,T];H^{1/2}(\Gamma^N)).$$

It then follows from a result of Miyatake [85] that the optimal trajectory satisfies $(w, \partial w/\partial t) \in C([0,T];V \times H)$ provided the coefficients A, c are piecewise smooth and sufficiently regular near Γ^N and Γ is itself sufficiently regular. Then if $(z_0, z_1) \in V \times H$, we obtain $(p, \partial p/\partial t) \in C([0,T];D(\mathcal{A}) \times V)$ and therefore $f_{\text{opt}} \in C^1([0,T];H^{1/2}(\Gamma^N))$. If also $(w_0, v_0) \in D(\mathcal{A}) \times V$ and $F \in W^{1,\infty}(0,T;H)$, we may conclude in particular that $(w, \partial w/\partial t) \in C([0,T];D(\mathcal{A}) \times V)$.

8.3 Time Domain Decomposition

8.3.1 Description of the Algorithm

We decompose the time interval $(0,T)$ into $K+1$ subintervals as in Section 6.3.1 and impose the continuity conditions (6.3.1.2) at the break points $t = T_k$, $k = 1, \ldots, N$. However, because of regularity considerations the iterative uncoupling (6.3.1.2), (6.3.1.3) of the transmission conditions does not make sense. Rather, we iteratively uncouple (6.3.1.2) according to

$$\begin{aligned}
\frac{\partial w_k^{n+1}}{\partial t}(T_{k+1}) - \beta \mathcal{A} p_k^{n+1}(T_{k+1}) &= \mu_{k,k+1}^n \\
\frac{\partial p_k^{n+1}}{\partial t}(T_{k+1}) + \beta w_k^{n+1}(T_{k+1}) &= \eta_{k,k+1}^n \\
-\frac{\partial w_k^{n+1}}{\partial t}(T_k) - \beta \mathcal{A} p_k^{n+1}(T_k) &= \mu_{k,k-1}^n \\
-\frac{\partial p_k^{n+1}}{\partial t}(T_k) + \beta w_k^{n+1}(T_k) &= \eta_{k,k-1}^n
\end{aligned} \tag{8.3.1.1}$$

where

$$\begin{aligned}\mu^n_{k,k+1} &= \frac{\partial w^n_{k+1}}{\partial t}(T_{k+1}) - \beta\mathcal{A}p^n_{k+1}(T_{k+1})\\ \eta^n_{k,k+1} &= \frac{\partial p^n_{k+1}}{\partial t}(T_{k+1}) + \beta w^n_{k+1}(T_{k+1})\\ \mu^n_{k,k-1} &= -\frac{\partial w^n_{k-1}}{\partial t}(T_k) - \beta\mathcal{A}p^n_{k-1}(T_k)\\ \eta^n_{k,k-1} &= -\frac{\partial p^n_{k-1}}{\partial t}(T_k) + \beta w^n_{k-1}(T_k).\end{aligned} \tag{8.3.1.2}$$

On I_0 and I_K, respectively, only the first two or last two equations in (8.3.1.1) are imposed. Thus the decoupled local problems are

$$\begin{cases}\dfrac{\partial^2 w^{n+1}_k}{\partial t^2} - \nabla\cdot(A\nabla w^{n+1}_k) + cw^{n+1}_k = F_k\\ \dfrac{\partial^2 p^{n+1}_k}{\partial t^2} - \nabla\cdot(A\nabla p^{n+1}_k) + cp^{n+1}_k = 0 \ \text{ in } Q_k := \Omega\times I_k\end{cases} \tag{8.3.1.3}$$

$$w^{n+1}_k = p^{n+1}_k = 0 \ \text{ on } \Sigma^D_k := \Gamma^D \times I_k, \tag{8.3.1.4}$$

$$\frac{\partial w^{n+1}_k}{\partial \nu_A} + p^{n+1}_k = \frac{\partial p^{n+1}_k}{\partial \nu_A} = 0 \ \text{ on } \Sigma^N_k := \Gamma^N \times I_k \tag{8.3.1.5}$$

$$\begin{aligned}&w^{n+1}_0(0) = w_0, \quad \frac{\partial w^{n+1}_0}{\partial t}(0) = v_0 \ \text{ in } \Omega\\ &p^{n+1}_K(T) = \kappa\mathcal{A}^{-1}(\frac{\partial w^{n+1}_K}{\partial t}(T) - z_1) \in V\\ &\frac{\partial p^{n+1}_K}{\partial t}(T) = -\kappa(w^{n+1}_K(T) - z_0) \in H\end{aligned} \tag{8.3.1.6}$$

subject to (8.3.1.1), (8.3.1.2).

As in Section 6.3.1, one may show that the local problem on I_k, $k = 0, \ldots, K$, is itself an optimality system. For example, on an internal interval the local problem is the optimality system for the optimal control problem

$$\inf_{\substack{f_k\in L^2(\Sigma^N)\\ (g_{k,k-1},h_{k,k-1})\in\mathcal{H}^*}} J^n_k(f_k, g_{k,k-1}, g_{k,k+1}), \tag{8.3.1.7}$$

where

$$\begin{aligned}J^n_k(f_k, g_{k,k-1}, h_{k,k-1}) := \frac{1}{2}\int_{\Sigma^N_k} f^2_k d\Gamma dt + \frac{1}{2\beta}\{\|\beta w_k(T_{k+1}) - \eta^n_{k,k+1}\|^2\\ + \|\frac{\partial w_k}{\partial t}(T_{k+1}) - \mu^n_{k,k+1}\|^2_{V^*} + \|g_{k,k-1}\|^2_{V^*} + \|h_{k,k-1}\|^2\},\end{aligned} \tag{8.3.1.8}$$

subject to

$$\begin{aligned}
&\frac{\partial^2 w_k}{\partial t^2} - \nabla \cdot (A \nabla w_k) + c w_k = F_k \ \text{ in } Q_k \\
&w_k = 0 \ \text{ on } \Sigma_k^D, \qquad \frac{\partial w_k}{\partial \nu_A} = f_k \ \text{ on } \Sigma_k^N \\
&w_k(T_k) = \frac{1}{\beta}(\eta_{k,k-1}^n + h_{k,k-1}) \ \text{ in } H \\
&\frac{\partial w_k}{\partial t}(T_k) = -(\mu_{k,k-1}^n + g_{k,k-1}) \ \text{ in } V^*.
\end{aligned} \tag{8.3.1.9}$$

In this problem, $(\mu_{k,k+1}^n, \eta_{k,k+1}^n) \in \mathcal{H}^*$ and $(\mu_{k,k-1}^n, \eta_{k,k-1}^n) \in \mathcal{H}^*$ are given data. We omit the calculation since it is essentially identical to that in Section 6.3.1.

The optimal control problem on the boundary interval I_0 is given by (8.3.1.7)–(8.3.1.9) with $k = 0$ but with the initial conditions replaced by $w_0(0) = w_0$, $(\partial w_0/\partial t)(0) = v_0$ and the $g_{0,-1}$ and $h_{0,-1}$ terms dropped in (8.3.1.8). The optimal control problem on the boundary interval I_K is given by (8.3.1.7)–(8.3.1.9) with $k = K$ but with the terms

$$\frac{1}{2\beta}\{\|\beta w_K(T_{K+1}) - \eta_{K,K+1}^n\|^2 + \|\frac{\partial w_K}{\partial t}(T_{K+1}) - \mu_{K,K+1}^n\|_{V^*}^2\}$$

replaced by

$$\frac{\kappa}{2}\{\|w_K(T_{k+1}) - z_0\|^2 + \|\frac{\partial w_K}{\partial t}(T_{K+1}) z_1\|_{V^*}^2\}.$$

Since the regularity properties $(\mu_{k,k+1}^n, \eta_{k,k+1}^n) \in \mathcal{H}^*$ and $(\mu_{k,k-1}^n, \eta_{k,k-1}^n) \in \mathcal{H}^*$ propagate with n, it follows that the iterative procedure (8.3.1.1)–(8.3.1.6) is well defined provided $(\mu_{k,k+1}^0, \eta_{k,k+1}^0) \in \mathcal{H}^*$ and $(\mu_{k,k-1}^0, \eta_{k,k-1}^0) \in \mathcal{H}^*$ for $k = 0, \ldots, K-1$ and $k = 1, \ldots, K$, respectively. The local solutions have regularity

$$(w_k^{n+1}, \frac{\partial w_k^{n+1}}{\partial t}) \in C(I_k; H \times V^*), \quad (p_k^{n+1}, \frac{\partial p_k^{n+1}}{\partial t}) \in C(I_k; V \times H).$$

8.3.2 Convergence of the Iterates

The proof of the convergence of the scheme (8.3.1.1)–(8.3.1.6) closely follows the proof of Theorem 6.3.2.1 so it is only sketched here. As usual, we introduce the local errors

$$\tilde{w}_k^{n+1} = w_k^{n+1} - w_k, \quad \tilde{p}_k^{n+1} = \tilde{p}_k^{n+1} - \tilde{p}_k,$$

which satisfy

$$\begin{cases}
\dfrac{\partial^2 \tilde{w}_k^{n+1}}{\partial t^2} - \nabla \cdot (A \nabla \tilde{w}_k^{n+1}) + c\tilde{w}_k^{n+1} = 0 \\
\dfrac{\partial^2 \tilde{p}_k^{n+1}}{\partial t^2} - \nabla \cdot (A \nabla \tilde{p}_k^{n+1}) + c\tilde{p}_k^{n+1} = 0 \ \text{ in } Q_k
\end{cases} \tag{8.3.2.1}$$

$$\tilde{w}_k^{n+1} = \tilde{p}_k^{n+1} = 0 \ \text{ on } \Sigma_k^D, \tag{8.3.2.2}$$

$$\frac{\partial \tilde{w}_k^{n+1}}{\partial \nu_A} + \tilde{p}_k^{n+1} = \frac{\partial \tilde{p}_k^{n+1}}{\partial \nu_A} = 0 \quad \text{on } \Sigma_k^N \tag{8.3.2.3}$$

$$\begin{gathered} \tilde{w}_0^{n+1}(0) = \frac{\partial \tilde{w}_0^{n+1}}{\partial t}(0) = 0 \quad \text{in } \Omega \\ \tilde{p}_K^{n+1}(T) = \kappa \mathcal{A}^{-1} \frac{\partial \tilde{w}_K^{n+1}}{\partial t}(T), \quad \frac{\partial \tilde{p}_K^{n+1}}{\partial t}(T) = -\kappa \tilde{w}_K^{n+1}(T) \end{gathered} \tag{8.3.2.4}$$

subject to

$$\begin{aligned} &\frac{\partial \tilde{w}_k^{n+1}}{\partial t}(T_{k+1}) - \beta \mathcal{A} \tilde{p}_k^{n+1}(T_{k+1}) = \tilde{\mu}_{k,k+1}^n \\ &\frac{\partial \tilde{p}_k^{n+1}}{\partial t}(T_{k+1}) + \beta \tilde{w}_k^{n+1}(T_{k+1}) = \tilde{\eta}_{k,k+1}^n \\ &-\frac{\partial \tilde{w}_k^{n+1}}{\partial t}(T_k) - \beta \mathcal{A} \tilde{p}_k^{n+1}(T_k) = \tilde{\mu}_{k,k-1}^n \\ &-\frac{\partial \tilde{p}_k^{n+1}}{\partial t}(T_k) + \beta \tilde{w}_k^{n+1}(T_k) = \tilde{\eta}_{k,k-1}^n \end{aligned} \tag{8.3.2.5}$$

where, for $n \geq 1$,

$$\begin{aligned} \tilde{\mu}_{k,k+1}^n &= \frac{\partial \tilde{w}_{k+1}^n}{\partial t}(T_{k+1}) - \beta \mathcal{A} \tilde{p}_{k+1}^n(T_{k+1}) \\ \tilde{\eta}_{k,k+1}^n &= \frac{\partial \tilde{p}_{k+1}^n}{\partial t}(T_{k+1}) + \beta \tilde{w}_{k+1}^n(T_{k+1}) \\ \tilde{\mu}_{k,k-1}^n &= -\frac{\partial \tilde{w}_{k-1}^n}{\partial t}(T_k) - \beta \mathcal{A} \tilde{p}_{k-1}^n(T_k) \\ \tilde{\eta}_{k,k-1}^n &= -\frac{\partial \tilde{p}_{k-1}^n}{\partial t}(T_k) + \beta \tilde{w}_{k-1}^n(T_k), \end{aligned} \tag{8.3.2.6}$$

and

$$\begin{aligned} \tilde{\mu}_{k,k+1}^0 &= \mu_{k,k+1}^0 - \Big(\frac{\partial w_{k+1}}{\partial t}(T_{k+1}) - \beta \mathcal{A} p_{k+1}(T_{k+1})\Big) \\ \tilde{\eta}_{k,k+1}^0 &= \eta_{k,k+1}^0 - \Big(\frac{\partial p_{k+1}}{\partial t}(T_{k+1}) + \beta w_{k+1}(T_{k+1})\Big) \\ \tilde{\mu}_{k,k-1}^0 &= \mu_{k,k-1}^0 + \Big(\frac{\partial w_{k-1}}{\partial t}(T_k) + \beta \mathcal{A} p_{k-1}(T_k)\Big) \\ \tilde{\eta}_{k,k-1}^0 &= \eta_{k,k-1} - \Big(-\frac{\partial p_{k-1}}{\partial t}(T_k) + \beta w_{k-1}(T_k)\Big). \end{aligned} \tag{8.3.2.7}$$

Set

$$\mathcal{X} = (V^* \times H)^{2K} = (\mathcal{H}^*)^{2K}$$

with the standard product norm. Let

$$X = \left\{ \left(\tilde{\mu}_{k,k+1}, \eta_{k,k+1}\right)_{k=0}^{K-1}, \left(\tilde{\mu}_{k,k-1}, \eta_{k,k-1}\right)_{k=1}^{K} \right\} \in \mathcal{X}$$

and $(\tilde{w}_k, \tilde{p}_k)_{k=0}^K$ be the solution of (8.3.2.1)–(8.3.2.5) with the superscripts $n\, n+1$ removed. Define $\mathcal{T} : \mathcal{X} \mapsto \mathcal{X}$ by

$$\mathcal{T}X = \Big\{ \Big(\frac{\partial \tilde{w}_{k+1}}{\partial t}(T_{k+1}) - \beta \mathcal{A} \tilde{p}_{k+1}(T_{k+1}), \frac{\partial \tilde{p}_{k+1}}{\partial t}(T_{k+1}) + \beta \tilde{w}_{k+1}(T_{k+1}) \Big)_{k=0}^{K-1}, \\ \Big(- \frac{\partial \tilde{w}_{k-1}}{\partial t}(T_k) - \beta \mathcal{A} \tilde{p}_{k-1}(T_k), - \frac{\partial \tilde{p}_{k-1}}{\partial t}(T_k) + \beta \tilde{w}_{k-1}(T_k) \Big)_{K=1}^{K} \Big\}. \tag{8.3.2.8}$$

The mapping $\mathcal{T}$ has $X = 0$ as its sole fixed point. Further, if we set

$$X^n = \big\{ \big(\tilde{\mu}_{k,k+1}^{n-1}, \tilde{\eta}_{k,k+1}^{n-1} \big)_{k=0}^{K-1}, \big(\tilde{\mu}_{k,k-1}^{n-1}, \tilde{\eta}_{k,k-1}^{n-1} \big)_{k=1}^{K} \big\} \in \mathcal{X}$$

and let $\big(\tilde{w}_k^n, \tilde{p}_k^n\big)_{k=0}^K$ be the solutions of (8.3.2.1)–(8.3.2.5) with n replaced by $n-1$, then the iteration step (8.3.2.5), (8.3.2.6) may be expressed as the fixed point iteration

$$X^{n+1} = \mathcal{T} X^n. \tag{8.3.2.9}$$

As before, we shall consider not only the iteration (8.3.2.9) but also the under relaxed iteration

$$X^{n+1} = (1-\epsilon)\mathcal{T}X^n + \epsilon X^, \quad \epsilon \in [0,1). \tag{8.3.2.10}$$

The first step in proving convergence of the iterates is to show that $\mathcal{T}$ is nonexpansive.

Lemma 8.3.2.1 *For any $X \in \mathcal{X}$,*

$$\|\mathcal{T}X\|_{\mathcal{X}}^2 = \|X\|^2 - 2\mathcal{F} \tag{8.3.2.11}$$

where

$$\mathcal{F} = 2\beta \sum_{k=0}^{K} \int_{\Sigma_k^N} |\tilde{p}_k|^2 \, d\Sigma + 2\beta\kappa \big(\|\tilde{w}_K(T)\|^2 + \|\frac{\partial \tilde{w}_K}{\partial t}(T)\|_{V^*}^2 \big). \tag{8.3.2.12}$$

Proof. The proof follows closely the proof of Lemma 6.3.2.1. A direct calculation yields

$$\|X\|_{\mathcal{X}}^2 = \sum_{k=1}^{K-1} [\mathcal{E}_k(T_{k+1}) + \mathcal{E}_k(T_k)] + \mathcal{E}_0(T_1) + \mathcal{E}_K(T_K) \\ - 2\beta \sum_{k=0}^{K-1} \Big[\big\langle \frac{\partial \tilde{w}_k}{\partial t}(t), \tilde{p}_k(t) \big\rangle_V \Big|_{t=T_k}^{T_{k+1}} - \big(\tilde{w}_k(t), \frac{\partial \tilde{p}_k}{\partial t}(t) \big) \Big|_{t=T_k}^{T_{k+1}} \Big] \\ + 2\beta \Big[\big\langle \frac{\partial \tilde{w}_K}{\partial t}(T_K), \tilde{p}_K(T_k) \big\rangle_V - \big(\tilde{w}_K(T_K), \frac{\partial \tilde{p}_K}{\partial t}(T_K) \big) \Big], \tag{8.3.2.13}$$

where

$$\mathcal{E}_k(t) = \beta^2 \|\tilde{w}_k(t)\|^2 + \|\frac{\partial \tilde{w}_k}{\partial t}(t)\|_{V^*}^2 + \beta^2 \|\tilde{p}_k(t)\|_V^2 + \|\frac{\partial \tilde{p}_k}{\partial t}(t)\|^2.$$

By a formal integration by parts that is justified by the *definition* of the solution $\tilde{w}_k$ (see (8.2.2.8)) we have, for $k = 0, \dots, K$,

$$\langle \frac{\partial \tilde{w}_k}{\partial t}(t), \tilde{p}_k(t) \rangle_V \Big|_{t=T_k}^{T_{k+1}} - \left(\tilde{w}_k(t), \frac{\partial \tilde{p}_k}{\partial t}(t)\right)\Big|_{t=T_k}^{T_{k+1}} = -\int_{\Sigma_k^N} |\tilde{p}_k|^2 d\Sigma. \qquad (8.3.2.14)$$

In particular, when $k = K$ we obtain by $(8.3.2.4)_2$

$$\langle \frac{\partial \tilde{w}_K}{\partial t}(T_K), \tilde{p}_K(T_K) \rangle_V - \left(\tilde{w}_K(T_K), \frac{\partial \tilde{p}_K}{\partial t}(T_K)\right)$$
$$= \kappa\left(\left\|\frac{\partial \tilde{w}_K}{\partial t}(T)\right\|_{V^*}^2 + \|\tilde{w}_K(T)\|^2\right) + \int_{\Sigma_K^N} |\tilde{p}_K|^2 d\Sigma. \qquad (8.3.2.15)$$

Thus from (8.3.2.13)–(8.3.2.15) we obtain

$$\|X\|_{\mathcal{X}}^2 = \mathcal{E} + \mathcal{F} \qquad (8.3.2.16)$$

where

$$\begin{aligned} \mathcal{E} :&= \sum_{k=1}^{K-1} [\mathcal{E}_k(T_{k+1}) + \mathcal{E}_k(T_k)] + \mathcal{E}_0(T_1) + \mathcal{E}_K(T_K) \\ &= \sum_{k=0}^{K-1} [\mathcal{E}_k(T_{k+1}) + \mathcal{E}_{k+1}(T_{k+1})]. \end{aligned} \qquad (8.3.2.17)$$

In a completely similar manner we also obtain

$$\|\mathcal{T}X\|_{\mathcal{X}}^2 = \mathcal{E} - \mathcal{F}. \qquad (8.3.2.18)$$

Lemma 8.3.2.1 follows from (8.3.2.16) and (8.3.2.18).

We can now prove the following convergence result.

Theorem 8.3.2.1 *Let $\beta > 0$. Then*

(i) *for any $\epsilon \in [0, 1)$*

$$\tilde{p}_k^n|_{\Sigma_k^N} \to 0 \text{ strongly in } L^2(\Sigma_k^N),\ k = 0, \dots, K$$
$$\left(\tilde{w}_0^n, \frac{\partial \tilde{w}_0^n}{\partial t}\right) \to 0 \text{ in } C(I_0; H \times V^*)$$
$$\left(\tilde{p}_K^n, \frac{\partial \tilde{p}_K^n}{\partial t}\right) \to 0 \text{ in } C(I_K; V \times H).$$

(ii) *For any $\epsilon \in (0, 1)$ and for $k = 0, \dots, K$*

$$\left(\tilde{w}_k^n, \frac{\partial \tilde{w}_k^n}{\partial t}\right) \to 0 \quad \text{in } C(I_k; H \times V^*)$$
$$\left(\tilde{p}_k^n, \frac{\partial \tilde{p}_k^n}{\partial t}\right) \to 0 \quad \text{in } C(I_k; V \times H).$$

The proof follows from Lemma 8.3.2.1 and, especially, (8.3.2.16) and (8.3.2.18). We introduce

$$\mathcal{E}_k^n(t) = \beta^2\|\tilde{w}_k^n(t)\|^2 + \Big\|\frac{\partial \tilde{w}_k^n}{\partial t}(t)\Big\|_{V^*}^2 + \beta\|\tilde{p}_k^n(t)\|_V^2 + \Big\|\frac{\partial \tilde{p}_k^n}{\partial t}(t)\Big\|^2$$

$$\mathcal{E}^n = \sum_{k=0}^{K-1}[\mathcal{E}_k^n(T_{k+1}) + \mathcal{E}_{k+1}^n(T_{k+1})] \tag{8.3.2.19}$$

$$\mathcal{F}^n = 2\beta\sum_{k=0}^{K}\int_{\Sigma_k^N}|\tilde{p}_k^n|^2\,d\Sigma + 2\beta\kappa\Big(\|\tilde{w}_K^n(T)\|^2 + \Big\|\frac{\partial \tilde{w}_K^n}{\partial t}(T)\Big\|_{V^*}^2\Big).$$

From (8.3.2.16) and (8.3.2.18) we have

$$\|X^n\|_{\mathcal{X}}^2 = \mathcal{E}^n + \mathcal{F}^n, \quad \|\mathcal{T}X^n\|_{\mathcal{X}}^2 = \mathcal{E}^n - \mathcal{F}^n. \tag{8.3.2.20}$$

By using (8.3.2.10) and (8.3.2.20) we obtain the fundamental recursion formula

$$\mathcal{E}^{n+1} + \mathcal{F}^{n+1} \le \mathcal{E}^n - (1-2\epsilon)\mathcal{F}^n, \quad \epsilon \in [0,1),\ n = 0,1,\ldots. \tag{8.3.2.21}$$

Part (i) of Theorem 8.3.2.1 follows from (8.3.2.21) exactly as in the proof of part (i) of Theorem 6.3.2.1.

For part (ii) we again use Proposition 2.3.7.1 as in the proof of Theorem 6.3.2.1. For this one has to calculate

$$\|\mathcal{T}X^n - X^n\|_{\mathcal{X}}^2 = 2(\mathcal{E}^n - (\mathcal{T}X^n, X^n)_{\mathcal{X}})$$

Not surprisingly, it turns out that

$$\begin{aligned}\mathcal{E}^n - (\mathcal{T}X^n, X^n)_{\mathcal{X}} = \sum_{k=0}^{K-1}\Big\{&\beta^2\big(\|\tilde{w}_k^n(T_{k+1}) - \tilde{w}_{k+1}^n(T_{k+1})\|^2\\ &+ \|\tilde{p}_k^n(T_{k+1}) - \tilde{p}_{k+1}^n(T_{k+1})\|_V^2\big) + \Big\|\frac{\partial \tilde{w}_k^n}{\partial t}(T_{k+1}) - \frac{\partial \tilde{w}_{k+1}^n}{\partial t}(T_{k+1})\Big\|_{V^*}^2\\ &+ \Big\|\frac{\partial \tilde{p}_k^n}{\partial t}(T_{k+1}) - \frac{\partial \tilde{p}_{k+1}^n}{\partial t}(T_{k+1})\Big\|^2\Big\}.\end{aligned} \tag{8.3.2.22}$$

The rest of the argument follows exactly the proof of part (ii) Theorem 6.3.2.1.

8.3.3 A Posteriori Error Estimates

We define *the error* e^n *at the nth iteration* by

$$e^n = \max_{0\le k\le K}\Big\|\Big(\tilde{w}_k^n, \frac{\partial \tilde{w}_k^n}{\partial t}\Big)\Big\|_{L^\infty(I_k;H\times V^*)}^2 + \max_{0\le k\le K}\Big\|\Big(\tilde{p}_k^n, \frac{\partial \tilde{p}_k^n}{\partial t}\Big)\Big\|_{L^\infty(I_k;V\times H)}^2$$

and further set

$$\mathcal{E}^n_{k,k+1}(T_{k+1}) = \big\|\big(w^n_{k+1}(T_{k+1}), \frac{\partial w^n_{k+1}}{\partial t}(T_{k+1})\big) - \big(w^n_k(T_{k+1}), \frac{\partial w^n_k}{\partial t}(T_{k+1})\big)\big\|^2_{H\times V^*}$$
$$+ \big\|\big(p^n_k(T_{k+1}), \frac{\partial p^n_k}{\partial t}(T_{k+1})\big) - \big(p^n_{k+1}(T_{k+1}), \frac{\partial p^n_{k+1}}{\partial t}(T_{k+1})\big)\big\|^2_{V\times H}.$$

Our first a posteriori error estimate is given by the following result.

Theorem 8.3.3.1 *Let* $\epsilon \in [0,1)$. *Then*

$$e^n \leq C_{K,T}\big[2K \sum_{k=0}^{K-1} \mathcal{E}^n_{k,k+1}(T_{k+1}) + C_n\big\{\sum_{k=0}^{K-1} \mathcal{E}^n_{k,k+1}(T_{k+1})\big\}^{1/2}\big], \tag{8.3.3.1}$$

where

$$C_{K,T} = C_T^{2(K+1)}, \quad C_n = \sqrt{2}\max\big(1, \frac{1}{\beta}\big)\max\big(C_\Omega^2 T, \kappa\big)(\mathcal{E}^n + \mathcal{F}^n)^{1/2},$$

and where C_T *is given by* (8.2.2.6) *(cf. Remark* 8.2.2.1*).*

As shall be shown below, $\mathcal{E}^n + \mathcal{F}^n$ is bounded by a constant depending only on β, K and the input data to the global and local optimal control problems. Therefore we have the following corollary.

Corollary 8.3.3.1 *Let* $\epsilon \in [0,1)$. *There is an explicitly computable constant* C *depending only on* β, K *and the input data to the global and local optimal control problems such that*

$$e^n \leq C_{K,T}\big[2K \sum_{k=0}^{K-1} \mathcal{E}^n_{k,k+1}(T_{k+1}) + C\big\{\sum_{k=0}^{K-1} \mathcal{E}^n_{k,k+1}(T_{k+1})\big\}^{1/2}\big]. \tag{8.3.3.2}$$

Proof of Theorem 8.3.3.1. Set

$$\alpha_k = C_\Omega\sqrt{|I_k|}.$$

We begin by applying the estimate (8.2.2.9) to $\tilde{w}^n_k$ on the interval I_k to obtain

$$\big\|\big(\tilde{w}^n_k(t), \frac{\partial \tilde{w}^n_k}{\partial t}(t)\big)\big\|_{H\times V^*} \leq C_T\big[\big\|\big(\tilde{w}^n_k(T_k), \frac{\partial \tilde{w}^n_k}{\partial t}(T_k)\big)\big\|_{H\times V^*} + \alpha_k\|\tilde{p}^n_k\|_{L^2(I_k;U)}\big] \tag{8.3.3.3}$$

where $U := L^2(\Gamma^n)$. The first term on the right is written

$$\big\|\big(\tilde{w}^n_k(T_k), \frac{\partial \tilde{w}^n_k}{\partial t}(T_k)\big)\big\|_{H\times V^*} - \big\|\big(\tilde{w}^n_{k-1}(T_k), \frac{\partial \tilde{w}^n_{k-1}}{\partial t}(T_k)\big)\big\|_{H\times V^*}$$
$$+ \big\|\big(\tilde{w}^n_{k-1}(T_k), \frac{\partial \tilde{w}^n_{k-1}}{\partial t}(T_k)\big)\big\|_{H\times V^*}$$
$$\leq \big\|\big(\tilde{w}^n_k(T_k), \frac{\partial \tilde{w}^n_k}{\partial t}(T_k)\big)\big\|_{H\times V^*} - \big\|\big(\tilde{w}^n_{k-1}(T_k), \frac{\partial \tilde{w}^n_{k-1}}{\partial t}(T_k)\big)\big\|_{H\times V^*}$$
$$+ C_T\big[\big\|\big(\tilde{w}^n_{k-1}(T_{k-1}), \frac{\partial \tilde{w}^n_{k-1}}{\partial t}(T_{k-1})\big)\big\|_{H\times V^*} + C_{k-1}\|\tilde{p}^n_{k-1}\|_{L^2(I_{k-1};U)}\big] \tag{8.3.3.4}$$

where we have used (8.2.2.9) again. By replacing the third term on the right side of (8.3.3.4) in the same manner and iterating the argument, and using $\tilde{w}_0^n(0) - (\partial\tilde{w}_0^n/\partial t)(0) = 0$, we obtain for $t \in I_k$

$$\|(\tilde{w}_k^n(t), \frac{\partial\tilde{w}_k^n}{\partial t}(t))\|_{H\times V^*} \le \sum_{m=0}^{k-1} C_T^{k-m}(\|(\tilde{w}_{m+1}^n(T_{m+1}), \frac{\partial\tilde{w}_{m+1}^n}{\partial t}(T_{m+1}))\|_{H\times V^*} - \|(\tilde{w}_m^n(T_{m+1}), \frac{\partial\tilde{w}_m^n}{\partial t}(T_{m+1}))\|_{H\times V^*}) + \sum_{m=0}^{k} C_T^{k-m+1}\alpha_m\|\tilde{p}_m^n\|_{L^2(I_m;U)}.$$

We apply the Cauchy-Schwarz inequality to the two sums to obtain

$$\max_{0\le k\le K}\|(\tilde{w}_k^n, \frac{\partial\tilde{w}_k^n}{\partial t})\|^2_{L^\infty(I_k;H\times V^*)} \le 2C_{K,T}\Big[K\sum_{k=0}^{K-1}\|(w_{k+1}^n(T_{k+1}), \frac{\partial w_{k+1}^n}{\partial t}(T_{k+1})) - (w_k^n(T_{k+1}), \frac{\partial w_k^n}{\partial t}(T_{k+1}))\|^2_{H\times V^*} + C_\Omega^2 T\sum_{k=0}^{K}\|\tilde{p}_k^n\|^2_{L^2(I_k;U)}\Big]. \tag{8.3.3.5}$$

Similarly, for $t \in I_k$ we have from (8.2.2.5) with $\mathcal{G} = 0$

$$\|(\tilde{p}_k^n(t), \frac{\partial\tilde{p}_k^n}{\partial t}(t))\|_{V\times H} \le C_T\|(\tilde{p}_k^n(T_{k+1}), \frac{\partial\tilde{p}_k^n}{\partial t}(T_{k+1}))\|_{V\times H} \le \sum_{m=k}^{K-1} C_T^{m-k+1}(\|(\tilde{p}_m^n(T_{m+1}), \frac{\partial\tilde{p}_m^n}{\partial t}(T_{m+1}))\|_{V\times H} - \|(\tilde{p}_{m+1}^n(T_{m+1}), \frac{\partial\tilde{p}_{m+1}^n}{\partial t}(T_{m+1}))\|_{V\times H}) + C_T^{K-k+1}\|(\tilde{p}_K^n(T), \frac{\partial\tilde{p}_K^n}{\partial t}(T))\|_{V\times H}. \tag{8.3.3.6}$$

It follows from (8.3.3.6) and (8.3.2.4) that

$$\max_{0\le k\le K}\|(\tilde{p}_k^n, \frac{\partial\tilde{p}_k^n}{\partial t})\|^2_{L^\infty(I_k;V\times H)} \le 2C_{K,T}\Big[K\sum_{k=0}^{K-1}\|(p_k^n(T_{k+1}), \frac{\partial p_k^n}{\partial t}(T_{k+1})) - (p_{k+1}^n(T_{k+1}), \frac{\partial p_{k+1}^n}{\partial t}(T_{k+1}))\|^2_{V\times H} + \kappa^2\|(\tilde{w}_K^n(T), \frac{\partial\tilde{w}_K^n}{\partial t}(T))\|^2_{H\times V^*}\Big]. \tag{8.3.3.7}$$

From (8.3.3.5) and (8.3.3.7) we thus obtain

$$e^n \le 2C_{K,T}\Big[K\sum_{k=0}^{K-1}\mathcal{E}^n_{k,k+1}(T_{k+1}) + C_\Omega^2\,T\sum_{k=0}^{K}\|\tilde{p}^n_k\|^2_{L^2(I_k;U)} + \kappa^2\big\|\big(\tilde{w}^n_K(T),\frac{\partial \tilde{w}^n_K}{\partial t}(T)\big)\big\|^2_{H\times V^*}\Big]$$

$$\le C_{K,T}\Big[2K\sum_{k=0}^{K-1}\mathcal{E}^n_{k,k+1}(T_{k+1}) + \frac{1}{\beta}\max\big(C_\Omega^2\,T,\kappa\big)\mathcal{F}^n\Big]. \quad (8.3.3.8)$$

Finally, as in the proof of Lemma 6.3.3.1 we have

$$\mathcal{F}^n \le \sqrt{2}\max(1,\beta)(\mathcal{E}^n+\mathcal{F}^n)^{1/2}\Big\{\sum_{k=0}^{K-1}\mathcal{E}^n_{k,k+1}(T_{k+1})\Big\}^{1/2}. \quad (8.3.3.9)$$

Theorem 8.3.3.1 follows from (8.3.3.8) and (8.3.3.9).

Proof of Corollary 8.3.3.1. It suffices to show that $\mathcal{E}^n + \mathcal{F}^n \le C$ where C is a constant depending only on β, K and the input data to the global and local optimal control problems. One has

$$\mathcal{E}^n+\mathcal{F}^n = \|X^n\|^2_{\mathcal{X}} \le \|X^1\|^2_{\mathcal{X}} = \sum_{k=0}^{K-1}\big[\|(\tilde{\mu}^0_{k,k+1},\tilde{\eta}^0_{k,k+1})\|^2_{\mathcal{H}^*} + \|(\tilde{\mu}^0_{k+1,k},\tilde{\eta}^0_{k+1,k})\|^2_{\mathcal{H}^*}\big].$$

From (8.3.2.7) we have

$$\sum_{k=0}^{K-1}\big[\|(\tilde{\mu}^0_{k,k+1},\tilde{\eta}^0_{k,k+1})\|^2_{\mathcal{H}^*} + \|(\tilde{\mu}^0_{k+1,k},\tilde{\eta}^0_{k+1,k})\|^2_{\mathcal{H}^*}\big]$$

$$\le 2\sum_{k=0}^{K-1}\big[\|(\mu^0_{k,k+1},\eta^0_{k,k+1})\|^2_{\mathcal{H}^*} + \|(\mu^0_{k+1,k},\eta^0_{k+1,k})\|^2_{\mathcal{H}^*}\big]$$

$$+4\sum_{k=0}^{K-1}\Big[\beta^2\|w(T_{k+1})\|^2 + \Big\|\frac{\partial w}{\partial t}(T_{k+1})\Big\|^2_{V^*} + \beta^2\|p(T_{k+1})\|^2_V + \Big\|\frac{\partial p}{\partial t}(T_{k+1})\Big\|^2\Big]. \quad (8.3.3.10)$$

The first sum in (8.3.3.10) represents data input to the local optimal control problems, so it remains to estimate the second sum in terms of global data, where w,p is the solution of the global optimality system (8.1.1.1), (8.2.1.2), (8.2.1.3).

From (8.2.2.9) and (8.2.2.5) we have

$$\Big\|\big(w,\frac{\partial w}{\partial t}\big)\Big\|^2_{L^\infty(0,T;H\times V^*)} \le 3C_T^2\big[\|(w_0,v_0)\|^2_{H\times V^*} + \|F\|^2_{L^1(0,T;V^*)} + C_\Omega^2\,T\|p\|^2_{L^2(\Sigma^N)}\big] \quad (8.3.3.11)$$

$$\Big\|\big(p,\frac{\partial p}{\partial t}\big)\Big\|^2_{L^\infty(0,T;V\times H)} \le C_T^2\,\kappa^2\Big\|\big(w(T)-z_0,\frac{\partial w}{\partial t}(T)-z_1\big)\Big\|^2_{H\times V^*}.$$

Therefore by (8.3.3.10) and (8.3.3.11) we obtain

$$\begin{aligned}\mathcal{E}^n + \mathcal{F}^n \le 2 \sum_{k=0}^{K-1} \big[\|(\mu^0_{k,k+1}, \eta^0_{k,k+1})\|^2_{\mathcal{H}^*} + \|(\mu^0_{k+1,k}, \eta^0_{k+1,k})\|^2_{\mathcal{H}^*}\big] \\ + 4K \max(1, \beta^2) C_T^2 \big[3\|(w_0, v_0)\|^2_{H\times V^*} + 3\|F\|^2_{L^1(0,T;V^*)} \\ + 3C_\Omega^2 T \|p\|^2_{L^2(\Sigma^N)} + \kappa^2 \big\|\big(w(T) - z_0, \frac{\partial w}{\partial t}(T) - z_1\big)\big\|^2_{H\times V^*}\big]. \quad (8.3.3.12)\end{aligned}$$

Since $p|_{\Sigma^N}$ is the optimal control, one has

$$\begin{aligned}\|p\|^2_{L^2(\Sigma^N)} + \kappa \big\|\big(w(T) - z_0, \frac{\partial w}{\partial t}(T) - z_1\big)\big\|^2_{H\times V^*} \\ \le \kappa \big\|\big(\hat{w}(T) - z_0, \frac{\partial \hat{w}}{\partial t}(T) - z_1\big)\big\|^2_{H\times V^*}, \quad (8.3.3.13)\end{aligned}$$

where $\hat{w}$ is the solution of (8.1.1.1) corresponding to $f = 0$. Therefore

$$\big\|\big(\hat{w}(T), \frac{\partial \hat{w}}{\partial t}(T)\big)\big\|^2_{H\times V^*} \le 2C_T^2 \big[\|(w_0, v_0)\|^2_{H\times V^*} + \|F\|^2_{L^1(0,T;V^*)}\big]. \quad (8.3.3.14)$$

It follows from (8.3.3.13) and (8.3.3.14) that

$$\begin{aligned}3C_\Omega^2 T \|p\|^2_{L^2(\Sigma^N)} + \kappa^2 \big\|\big(w(T) - z_0, \frac{\partial w}{\partial t}(T) - z_1\big)\big\|^2_{H\times V^*} \\ \le \max(3C_\Omega^2 T, \kappa)\kappa \big\|\big(\hat{w}(T) - z_0, \frac{\partial \hat{w}}{\partial t}(T) - z_1\big)\big\|^2_{H\times V^*} \\ \le 2\kappa \max(3C_\Omega^2 T, \kappa)\big\{\|(z_0, z_1)\|^2_{H\times V^*} \\ + 2C_T^2 \big[\|(w_0, v_0)\|^2_{H\times V^*} + \|F\|^2_{L^1(0,T;V^*)}\big]\big\}. \quad (8.3.3.15)\end{aligned}$$

If (8.3.3.15) is inserted into (8.3.3.12) we obtain

$$\begin{aligned}\mathcal{E}^n + \mathcal{F}^n \le 2 \sum_{k=0}^{K-1} \big[\|(\mu^0_{k,k+1}, \eta^0_{k,k+1})\|^2_{\mathcal{H}^*} + \|(\mu^0_{k+1,k}, \eta^0_{k+1,k})\|^2_{\mathcal{H}^*}\big] \\ + 4K \max(1, \beta^2) C_T^2 \big[3 + 4\kappa C_T^2 \max(3C_\Omega^2 T, \kappa)\big] \\ \times \big[\|(w_0, v_0)\|^2_{H\times V^*} + \|F\|^2_{L^1(0,T;V^*)}\big] \\ + 8K\kappa C_T^2 \max(1, \beta^2) \max(3C_\Omega^2 T, \kappa) \|(z_0, z_1)\|^2_{H\times V^*}.\end{aligned}$$

The right side of this inequality is computable from local and global data, completing the proof of Corollary 8.3.3.1.

8.3.4 Extension to General Conservative Control Systems

The time domain decomposition procedure described above for the system (8.1.1.1) may be extended without difficulty to general second-order control systems of the following type. Let V and H be Hilbert spaces such that $V \hookrightarrow H$. We denote by

V^* the dual space of V with respect to the pivot space H and by $\mathcal{A}$ the Riesz isomorphism of V onto V^*. Let $\mathcal{C} \in \mathcal{L}(H)$, let U be another Hilbert space and $\mathcal{B}$ be a bounded linear operator from U into V^*. We consider systems of the form

$$\begin{aligned} \frac{d^2w}{dt^2} + \mathcal{A}w + \mathcal{C}w = \mathcal{B}f + F \ \text{ in } V^* \\ w(0) = w_0, \quad \frac{dw}{dt}(0) = v_0. \end{aligned} \tag{8.3.4.1}$$

In this system the initial data $(w_0, v_0) \in H \times V^*$, $F \in L^1(0,T;V^*)$, and the control function $f \in L^2(0,T;U)$. If $\mathcal{C}$ is self-adjoint, a condition that is not required for the present discussion, the system (8.3.4.1) is conservative in the sense that when w is the solution corresponding to $F = f = 0$ the quantity

$$\|\frac{dw}{dt}(t)\|^2 + \langle \mathcal{A}w(t), w(t)\rangle_V + (\mathcal{C}w(t), w(t))$$

is invariant with respect to t. The system (8.1.1.1) fits into this framework wherein the spaces V and H are those defined in Section 8.2.1, $\mathcal{C}w = c(\cdot)w - w$, $U = L^2(\Gamma^N)$, and the operator $\mathcal{B}$ is defined by

$$\langle \mathcal{B}f, \phi\rangle_V = \int_{\Gamma^N} f\phi \, d\Gamma, \quad \forall f \in U, \phi \in V.$$

One may also consider locally distributed controls, or boundary control of thin, elastic places, etc., within the framework of (8.3.4.1), as discussed in the examples of Section 6.3.4.

If $F \in L^2(0,T;H)$, $f = 0$, and $(w_0, v_0) \in \mathcal{H}$, by standard semigroup theory the system (8.3.4.1) has a unique solution such that

$$\begin{aligned} (w, \frac{dw}{dt}) \in C([0,T];\mathcal{H}), \\ \|(w, \frac{dw}{dt})\|_{L^\infty(0,T;\mathcal{H})} \le C_T\big[\|(w_0, v_0)\|_{\mathcal{H}} + \|F\|_{L^1(0,T;H)}\big] \end{aligned} \tag{8.3.4.2}$$

where $C_T = \sup_{0\le t\le T} \|S(t)\| \le \exp(\|\mathcal{C}\|T)$ and $S(t)$, $t \in \mathbb{R}$, is the C_0 group of bounded operators on $\mathcal{H}$ associated with the system (8.3.4.1) (cf. Section 8.2.2). In the general case, the solution of (8.3.4.1) may be defined by transposition. It is the unique pair

$$(\frac{dw}{dt}, -w) \in C([0,T];\mathcal{H}^*)$$

such that for every $t \in [0,T]$

$$\begin{aligned} \langle(\frac{dw}{dt}(t), -w(t)), (\phi(t), \frac{d\phi}{dt}(t))\rangle_{\mathcal{H}} = \langle(v_0, -w_0), (\phi(0), \frac{d\phi}{dt}(0))\rangle_{\mathcal{H}} \\ + \int_0^t [\langle F, \phi\rangle_V + (f, \mathcal{B}^*\phi)_U] \, dt, \quad \forall(\phi_0, \phi_1) \in \mathcal{H}, \end{aligned} \tag{8.3.4.3}$$

where

$$\frac{d^2\phi}{dt^2} + \mathcal{A}\phi + \mathcal{C}^*\phi = 0, \quad 0 \le t < T,$$
$$\phi(T) = \phi_0, \quad \frac{d\phi}{dt}(T) = \phi_1. \tag{8.3.4.4}$$

It follows from (8.3.4.4) and the estimate (8.3.4.2), applied to the solution ϕ, that

$$\|(w, \frac{dw}{dt})\|_{L^\infty(0,T;\mathcal{H}^*)} \le C_T^*\big[\|(v_0, -w_0)\|_{\mathcal{H}^*} + \|F\|_{L^1(0,T;V^*)} + \sqrt{T}\|\mathcal{B}^*\|\|f\|_{L^2(0,T;U)}\big] \tag{8.3.4.5}$$

where $C_T^* = \sup_{0\le t\le T}\|S^*(t)\| \le \exp(\|\mathcal{C}^*\|T)$.

We may therefore consider the optimal control problem

$$\inf_{f\in L^2(0,T;U)} \left[\int_0^T \|f\|_U^2 dt + \kappa\big[\|w(T) - z_0\|_H^2 + \|\frac{dw}{dt}(T) - z_1\|_{V^*}^2\right]$$

subject to (8.3.4.1), where $\kappa > 0$ and $(z_0, z_1) \in H \times V^*$ is the intended target state. There is a unique optimal control which, by the usual argument, is given by

$$f_{\text{opt}} = -\mathcal{B}^* p \tag{8.3.4.6}$$

where p is the solution of

$$\frac{d^2p}{dt^2} + \mathcal{A}p + \mathcal{C}^*p = 0, \quad 0 \le t < T,$$
$$p(T) = \kappa\mathcal{A}^{-1}\big(\frac{dw}{dt}(T) - z_1\big), \quad \frac{dp}{dt}(T) = -\kappa\big(w(T) - z_0\big). \tag{8.3.4.7}$$

A time domain decomposition of the optimality system, consisting of (8.3.4.1), (8.3.4.6) and (8.3.4.7), proceeds along the same lines as Section 8.3.1. It is given by

$$\frac{d^2w_k^{n+1}}{dt^2} + \mathcal{A}w_k^{n+1} + \mathcal{C}w_k^{n+1} = F_k - \mathcal{B}^*p_k^{n+1},$$
$$\frac{d^2p_k^{n+1}}{dt^2} + \mathcal{A}p_k^{n+1} + \mathcal{C}^*p_k^{n+1} = 0, \quad t \in I_k; \tag{8.3.4.8}$$

$$w_0^{n+1} = w_0, \quad \frac{dw_k^{n+1}}{dt}(0) = v_0,$$
$$p_k^{n+1}(T) = \kappa\mathcal{A}^{-1}\big(\frac{dw_k^{n+1}}{dt}(T) - z_1\big), \tag{8.3.4.9}$$
$$\frac{dp_k^{n+1}}{dt}(T) = -\kappa\big(w_k^{n+1}(T) - z_0\big).$$

The initial conditions $w_k^{n+1}(T_k)$, $k = 1, \dots, K$, and the final conditions $p_k^{n+1}(T_{k+1})$, $k = K-1, \dots, 0$, are determined by (8.3.1.1) in which the quantities $\mu_{k,k\pm1}$, $\eta_{k,k\pm1}$

are given by (8.3.1.2), just as before. The local problem on the interval I_k, $k = 0, \dots, K$, is itself the optimality system for a certain optimal control problem concentrated on I_k, the specific form of which may easily be deduced from the discussion of Section 8.3.1.

Convergence of the iterates, and a posteriori error estimates, may be developed by the same arguments employed in Sections 8.3.2 and 8.3.3, respectively. In particular, for the iterative scheme with or without relaxation there is strong convergence in $L^2(0,T;U)$ of the aggregate of effective local optimal controls, $\{-\mathcal{B}^* p_k^n\}_{k=0}^K$, to the global optimal control $-\mathcal{B}^* p$; and strong convergence in $H \times V^*$ of the final state $\big(w_K(T), \frac{dw_K^n}{dt}(T)\big)$ of the local optimal trajectory $\big(w_K^n, \frac{dw_K^n}{dt}\big)$ to the final state of the global optimal trajectory.

8.4 Decomposition of the Spatial Domain

8.4.1 The Local Optimality Systems

We decompose the region Ω into subregions $\{\Omega_i\}_{i=1}^I$ as in Section 6.4.1 and retain all of the notation of the first paragraph of that section. We set $H_i = L^2(\Omega_i)$ with the standard norm, and $W_i = H_D^1(\Omega_i)$ with norm

$$\|\phi\|_{W_i} = \left(\int_{\Omega_i} (A_i \nabla\phi \cdot \nabla\phi + |\phi|^2) dx + \alpha \int_{\Gamma_i^{\text{int}}} |\phi|^2 d\Gamma \right)^{1/2}, \quad \alpha > 0.$$

The dual space of W_i with respect to H_i is denoted by W_i^*, and $\mathcal{B}_i$ is defined as the canonical isomorphism of W_i onto W_i^*. The duality pairing between elements $\phi^* \in W_i^*$ and $\phi \in W_i$ is denoted by $\langle \phi^*, \phi \rangle_{W_i}$.

We also denote by $\mathcal{A}_i$ the canonical isomorphism of $V_i := H_D^1(\Omega_i)$ onto its dual space V_i^* with respect to H_i when $H_D^1(\Omega_i)$ is endowed with the norm

$$\|\phi\|_{V_i} = \left(\int_{\Omega_i} (A_i \nabla\phi \cdot \nabla\phi + |\phi|^2) dx \right)^{1/2}.$$

Of course W_i and V_i (resp., W_i^* and V_i^*) are the same as sets and have equivalent norms, but $\mathcal{B}_i$ and $\mathcal{A}_i$ are not the same. They are, however, connected through the bounded, invertible mapping K_i on W_i defined by $(\phi, \psi)_{V_i} = (K_i \phi, \psi)_{W_i}$, $\forall \phi, \psi \in W_i$. Note that if $\phi \in V$ and $\phi_i = \phi|_{\Omega_i}$, then

$$\|\phi\|_V^2 = \sum_{i=1}^I \|\phi_i\|_{V_i}^2 \qquad \text{and} \qquad \mathcal{A}\phi = (\mathcal{A}_1 \phi_1, \dots, \mathcal{A}_I \phi_I)$$

in the sense that

$$\langle \mathcal{A}\phi, \psi \rangle_V = \sum_{i=1}^I \langle \mathcal{A}_i \phi_i, \psi_i \rangle_{V_i}, \quad \forall \phi, \psi \in V.$$

For τ_i given in $L^2(\Gamma_i^{\text{int}})$ we define $\sigma_i \in W_i^*$ by

$$\langle \sigma_i, \phi \rangle_{W_i} = \int_{\Gamma_i^{\text{int}}} \tau_i \phi \, d\Gamma, \quad \forall \phi \in V_i.$$

Let λ_i, $\mu_i \in L^2(\Sigma_i^{\text{int}})$. We introduce the following local optimal control problems on the regions Q_i, $i = 1, \ldots, I$:

$$\inf_{(f_1, f_2) \in U_i} J_i(f_1, f_2), \qquad U_i = L^2(\Sigma_i^{\text{int}}) \times L^2(\Sigma_i^N), \tag{8.4.1.1}$$

$$\begin{aligned} J_i(f_1, f_2) =& \frac{1}{2} \int_{\Sigma_i^N} |f_2|^2 d\Sigma + \frac{1}{2\beta} \int_{\Sigma_i^{\text{int}}} (|f_1|^2 + |\beta w_i + \mu_i|^2 d\Sigma \\ &+ \frac{\kappa}{2} \Big(\|w_i(T) - z_{0i}\|_{H_i}^2 + \Big\| \frac{\partial w_i}{\partial t}(T) - z_{1i} + \kappa^{-1} \sigma_i \Big\|_{W_i^*}^2 \Big), \quad \beta > 0, \end{aligned}$$

subject to

$$\begin{aligned} \frac{\partial^2 w_i}{\partial t^2} - \nabla \cdot (A_i \nabla w_i) + c_i w_i &= F_i \text{ in } Q_i \\ w_i &= 0 \text{ on } \Sigma_i^D \\ \frac{\partial w_i}{\partial \nu_{A_i}} &= f_2 \text{ on } \Sigma_i^N \\ \frac{\partial w_i}{\partial \nu_{A_i}} &= f_1 + \lambda_i \text{ on } \Sigma_i^{\text{int}} \\ w_i(0) = w_{0i}, \quad \frac{\partial w_i}{\partial t}(0) &= v_{0i} \text{ in } \Omega_i. \end{aligned} \tag{8.4.1.2}$$

The optimal control problem (8.4.1.1), (8.4.1.2) is well posed if the following regularity condition is satisfied:

Regularity Assumption: If $w_{0i} = v_{0i} = 0$, $F_i = 0$, $f_2 \in L^2(\Sigma_i^N)$, $f_1 + \lambda_i \in L^2(\Sigma_i^{\text{int}})$, the solution of (8.4.1.2) satisfies

$$w_i|_{\Gamma_i^N \cup \Gamma_i^{\text{int}}} \in L^2(\Gamma_i^N \cup \Gamma_i^{\text{int}}) \tag{8.4.1.3}$$

and the map $(f_2, f_1 + \lambda_i) \mapsto w_i|_{\Gamma_i^N \cup \Gamma_i^{\text{int}}}$ is continuous in the indicated spaces.

If this assumption holds then, by superposition, (8.4.1.3) remains true if $(w_{0i}, v_{0i}) \in V_i \times H_i$ and $F_i \in L^2(0, T; H_i)$, and the mapping from the data to $w_i|_{\Gamma_i^N \cup \Gamma_i^{\text{int}}}$ is continuous in the indicated spaces. Condition (8.4.1.3) is known to be true if, for example, the coefficients A_i, c_i and the boundary set $\Gamma_i^N \cup \Gamma_i^{\text{int}}$ are sufficiently regular (see Remark 8.2.2.2).

Using the same argument as for the global optimal control problem, it follows that there are unique optimal controls given by

$$f_1 = -\beta p_i|_{\Sigma_i^{\text{int}}} \quad f_2 = -p_i|_{\Sigma_i^N}, \tag{8.4.1.4}$$

where p is the solution of

$$\begin{aligned}
&\frac{\partial^2 p_i}{\partial t^2} - \nabla\cdot(A_i\nabla p_i) + c_i p_i = 0 \text{ in } Q_i \\
&p_i = 0 \text{ on } \Sigma_i^D, \qquad \frac{\partial p_i}{\partial \nu_{A_i}} = 0 \text{ on } \Sigma_i^N \\
&\frac{\partial p_i}{\partial \nu_{A_i}} = \beta w_i + \mu_i \text{ on } \Sigma_i^{\text{int}} \\
&\begin{cases} p_i(T) = \kappa \mathcal{B}_i^{-1}\left(\dfrac{\partial w_i}{\partial t}(T) - z_{1i} + \kappa^{-1}\sigma_i\right) \\ \dfrac{\partial p_i}{\partial t}(T) = -\kappa(w_i(T) - z_{0i}) \text{ in } \Omega_i. \end{cases}
\end{aligned} \tag{8.4.1.5}$$

Equations (8.4.1.2)–(8.4.1.5) comprise the *local optimality systems.*

Note that $(8.4.1.5)_4$ means that $p_i(T) \in W_i$ is determined by the elliptic boundary value problem

$$\begin{aligned}
&-\nabla\cdot(A_i\nabla p_i(T)) + p_i(T) = \kappa\left(\frac{\partial w_i}{\partial t}(T) - z_{1i}\right) \text{ in } \Omega_i \\
&p_i(T) = 0 \text{ on } \Gamma_i^D, \qquad \frac{\partial p_i(T)}{\partial \nu_{A_i}} = 0 \text{ on } \Gamma_i^N \\
&\frac{\partial p_i(T)}{\partial \nu_{A_i}} + \alpha p_i(T) = \tau_i \text{ on } \Gamma_i^{\text{int}}.
\end{aligned} \tag{8.4.1.6}$$

8.4.2 The Domain Decomposition Algorithm

For the remainder of this chapter we assume that the global data satisfy

$$(w_0, v_0) \in V \times H, \quad (z_0, z_1) \in V \times H, \quad F \in L^1(0,T;H),$$

and that the regularity assumption (8.4.1.3) holds. We introduce the following iterative scheme based on the local optimality systems (8.4.1.2)–(8.4.1.5):

$$\begin{aligned}
&\frac{\partial^2 w_i^{n+1}}{\partial t^2} - \nabla\cdot(A_i\nabla w_i^{n+1}) + c_i w_i^{n+1} = F_i \text{ in } Q_i \\
&w_i^{n+1} = 0 \text{ on } \Sigma_i^D, \qquad \frac{\partial w_i^{n+1}}{\partial \nu_{A_i}} + p_i^{n+1} = 0 \text{ on } \Sigma_i^N \\
&\frac{\partial w_i^{n+1}}{\partial \nu_{A_i}} + \beta p_i^{n+1} = \lambda_{ij}^n \text{ on } \Sigma_{ij}, \ \forall j \in \mathcal{J}_i \\
&w_i^{n+1}(0) = w_{0i}, \qquad \frac{\partial w_i^{n+1}}{\partial t}(0) = v_{0i} \text{ in } \Omega_i,
\end{aligned} \tag{8.4.2.7}$$

$$
\begin{aligned}
&\frac{\partial^2 p_i^{n+1}}{\partial t^2} - \nabla \cdot (A_i \nabla p_i^{n+1}) + c_i p_i^{n+1} = 0 \text{ in } Q_i \\
&p_i^{n+1} = 0 \text{ on } \Sigma_i^D, \qquad \frac{\partial p_i^{n+1}}{\partial \nu_{A_i}} = 0 \text{ on } \Sigma_i^N \\
&\frac{\partial p_i^{n+1}}{\partial \nu_{A_i}} - \beta w_i^{n+1} = \mu_{ij}^n \text{ on } \Sigma_{ij},\ \forall j \in \mathcal{J}_i \\
&\begin{cases} p_i^{n+1}(T) = \kappa \mathcal{B}_i^{-1}\big(\frac{\partial w_i^{n+1}}{\partial t}(T) - z_{1i} + \kappa^{-1}\sigma_i^n\big) \\ \frac{\partial p_i^{n+1}}{\partial t}(T) = -\kappa(w_i^{n+1}(T) - z_{0i}) \text{ in } \Omega_i, \end{cases}
\end{aligned}
\tag{8.4.2.8}
$$

where $\lambda_{ij}^0,\ \mu_{ij}^0 \in L^2(\Sigma_{ij})$, $\tau_{ij}^0 \in L^2(\Gamma_{ij})$ are arbitrary,

$$
\begin{aligned}
\lambda_{ij}^n &= -\frac{\partial w_j^n}{\partial \nu_{A_j}} + \beta p_j^n\big|_{\Sigma_{ij}}, \quad \mu_{ij}^n = -\frac{\partial p_j^n}{\partial \nu_{A_j}} - \beta w_j^n\big|_{\Sigma_{ij}} \\
\tau_{ij}^n &= -\frac{\partial p_j^n(T)}{\partial \nu_j} + \alpha p_j^n(T)\big|_{\Gamma_{ij}}, \quad n \geq 1,
\end{aligned}
\tag{8.4.2.9}
$$

and where $\sigma_i^n \in W_i^*$ is defined by

$$
\langle \sigma_i^n, \phi \rangle_{W_i} = \sum_{j \in \mathcal{J}_i} \int_{\Gamma_{ij}} \tau_{ij}^n \phi \, d\Gamma, \quad \forall \phi \in W_i. \tag{8.4.2.10}
$$

Thus we see that $p_i^{n+1}(T)$ is determined by the elliptic boundary value problem

$$
\begin{aligned}
-\nabla \cdot (A_i \nabla p_i^{n+1}(T)) + p_i^{n+1}(T) &= \kappa\big(\frac{\partial w_i^{n+1}}{\partial t}(T) - z_{1i}\big) \text{ in } \Omega_i \\
p_i^{n+1}(T) = 0 \text{ on } \Gamma_i^D, \quad \frac{\partial p_i^{n+1}(T)}{\partial \nu_{A_i}} &= 0 \text{ on } \Gamma_i^N \\
\frac{\partial p_i^{n+1}(T)}{\partial \nu_{A_i}} + \alpha p_i^{n+1}(T) &= \tau_{ij}^n \text{ on } \Gamma_{ij},\ \forall j \in \mathcal{J}_i.
\end{aligned}
\tag{8.4.2.11}
$$

Note that, in the limit, the conditions on the interfaces Σ_{ij} and Γ_{ij} become

$$
\begin{aligned}
\frac{\partial w_i}{\partial \nu_{A_i}} + \beta p_i &= -\frac{\partial w_j}{\partial \nu_{A_j}} + \beta p_j \\
\frac{\partial p_i}{\partial \nu_{A_i}} - \beta w_i &= -\frac{\partial p_j}{\partial \nu_{A_j}} - \beta w_j
\end{aligned}
$$

and

$$
\frac{\partial p_i(T)}{\partial \nu_{A_i}} + \alpha p_i(T) = -\frac{\partial p_j(T)}{\partial \nu_{A_j}} + \alpha p_j(T)
$$

respectively. By interchanging i and j and adding the resulting equations to the original ones we recover the *transmission conditions*

$$\begin{aligned} w_i = w_j, \quad p_i = p_j, \quad \frac{\partial w_i}{\partial \nu_{A_i}} = -\frac{\partial w_j}{\partial \nu_{A_j}}, \quad \frac{\partial p_i}{\partial \nu_{A_i}} = -\frac{\partial p_j}{\partial \nu_{A_j}} \text{ on } \Sigma_{ij} \\ p_i(T) = p_j(T), \quad \frac{\partial p_i(T)}{\partial \nu_{A_i}} = -\frac{\partial p_j(T)}{\partial \nu_{A_j}} \text{ on } \Gamma_{ij}. \end{aligned} \tag{8.4.2.12}$$

Thus the algorithm (8.4.2.7)–(8.4.2.10) represents a domain decomposition of the global optimality system that is consistent with the transmission conditions (8.4.2.12).

If λ_{ij}^n, $\mu_{ij}^n \in L^2(\Sigma_{ij})$, $\tau_{ij}^n \in L^2(\Gamma_{ij})$, $\forall j \in \mathcal{J}_i$, then, as discussed in the previous subsection, (8.4.2.8), (8.4.2.9) is the optimality system for the optimal control problem

$$\inf_{(f_1,f_2)\in U_i} J_i^{n+1}(f_1,f_2), \quad U_i = L^2(\Sigma_i^{\text{int}}) \times L^2(\Sigma_i^N), \tag{8.4.2.13}$$

$$\begin{aligned} J_i^{n+1}(f_1,f_2) =& \frac{1}{2}\int_{\Sigma_i^N} |f_2|^2 d\Sigma + \frac{1}{2\beta}\sum_{j\in\mathcal{J}_i}\int_{\Sigma_{ij}} (|f_1|^2 + |\beta w_i^{n+1} + \mu_{ij}^n|^2 d\Sigma \\ &+ \frac{\kappa}{2}(\|w_i^{n+1}(T) - z_{0i}\|_{H_i}^2 + \|\frac{\partial w_i^{n+1}}{\partial t}(T) - z_{1i} + \kappa^{-1}\sigma_i^n\|_{W_i^*}^2) \end{aligned}$$

subject to

$$\begin{aligned} \frac{\partial^2 w_i^{n+1}}{\partial t^2} - \nabla\cdot(A_i \nabla w_i^{n+1}) + c_i w_i^{n+1} &= F_i \text{ in } Q_i \\ w_i^{n+1} &= 0 \text{ on } \Sigma_i^D \\ \frac{\partial w_i^{n+1}}{\partial \nu_{A_i}} &= f_2 \text{ on } \Sigma_i^N \\ \frac{\partial w_i^{n+1}}{\partial \nu_{A_i}} &= f_1 + \lambda_{ij}^n \text{ on } \Sigma_{ij} \\ w_i^{n+1}(0) = w_{0i}, \quad \frac{\partial w_i^{n+1}}{\partial t}(0) &= v_{0i} \text{ in } \Omega_i. \end{aligned} \tag{8.4.2.14}$$

Since the traces $w_i^{n+1}|_{\Sigma_{ij}}$, $p_i^{n+1}|_{\Sigma_{ij}} \in L^2(\Sigma_{ij})$ and $p_i^{n+1}(T)|_{\Gamma_{ij}} \in L^2(\Gamma_{ij})$, it follows from the boundary conditions on Σ_{ij} and Γ_{ij} that the normal derivative traces are also in $L^2(\Sigma_{ij})$ or $L^2(\Gamma_{ij})$, as appropriate, and therefore λ_{ij}^{n+1}, $\mu_{ij}^{n+1} \in L^2(\Sigma_{ij})$, $\tau_{ij}^{n+1} \in L^2(\Gamma_{ij})$. Thus the iterative scheme is well defined provided the local input data satisfy λ_{ij}^0, $\mu_{ij}^0 \in L^2(\Sigma_{ij})$, $\tau_{ij}^0 \in L^2(\Gamma_{ij})$, $\forall j \in \mathcal{J}_i$.

8.4.3 Convergence of the Iterates

We introduce the local errors

$$\tilde{w}_i^{n+1} = w_i^{n+1} - w_i, \quad \tilde{p}_i^{n+1} = p_i^{n+1} - p_i$$

where $(\tilde{w}_i^{n+1}, \tilde{p}_i^{n+1})$ is the solution of the local optimality system (8.4.2.7)–(8.4.2.9), and $w_i = w|_{Q_i}$, $p_i = p|_{Q_i}$, where (w, p) is the solution of the global optimality system (8.1.1.1), (8.2.1.3), (8.2.1.2).

The errors satisfy

$$\begin{aligned}
&\frac{\partial^2 \tilde{w}_i^{n+1}}{\partial t^2} - \nabla \cdot (A_i \nabla \tilde{w}_i^{n+1}) + c_i \tilde{w}_i^{n+1} = 0 \text{ in } Q_i \\
&\tilde{w}_i^{n+1} = 0 \text{ on } \Sigma_i^D, \quad \frac{\partial \tilde{w}_i^{n+1}}{\partial \nu_{A_i}} + \tilde{p}_i^{n+1} = 0 \text{ on } \Sigma_i^N \\
&\frac{\partial \tilde{w}_i^{n+1}}{\partial \nu_{A_i}} + \beta \tilde{p}_i^{n+1} = \tilde{\lambda}_{ij}^n \text{ on } \Sigma_{ij}, \forall j \in \mathcal{J}_i \\
&\tilde{w}_i^{n+1}(0) = \frac{\partial \tilde{w}_i^{n+1}}{\partial t}(0) = 0 \text{ in } \Omega_i,
\end{aligned} \tag{8.4.3.1}$$

$$\begin{aligned}
&\frac{\partial^2 \tilde{p}_i^{n+1}}{\partial t^2} - \nabla \cdot (A_i \nabla \tilde{p}_i^{n+1}) + c_i \tilde{p}_i^{n+1} = 0 \text{ in } Q_i \\
&\tilde{p}_i^{n+1} = 0 \text{ on } \Sigma_i^D, \quad \frac{\partial \tilde{p}_i^{n+1}}{\partial \nu_{A_i}} = 0 \text{ on } \Sigma_i^N \\
&\frac{\partial \tilde{p}_i^{n+1}}{\partial \nu_{A_i}} - \beta \tilde{w}_i^{n+1} = \tilde{\mu}_{ij}^n \text{ on } \Sigma_{ij}, \forall j \in \mathcal{J}_i \\
&\begin{cases} \tilde{p}_i^{n+1}(T) = \kappa \mathcal{B}_i^{-1} \left(\dfrac{\partial \tilde{w}_i^{n+1}}{\partial t}(T) + \kappa^{-1} \tilde{\sigma}_i^n \right) \\ \dfrac{\partial \tilde{p}_i^{n+1}}{\partial t}(T) = -\kappa \tilde{w}_i^{n+1}(T) \text{ in } \Omega_i, \end{cases}
\end{aligned} \tag{8.4.3.2}$$

where

$$\begin{aligned}
\tilde{\lambda}_{ij}^n &= \lambda_{ij}^n - \left(-\frac{\partial w_j}{\partial \nu_{A_j}} + \beta p_j\right)\Big|_{\Sigma_{ij}} \\
\tilde{\mu}_{ij}^n &= \mu_{ij}^n - \left(-\frac{\partial p_j^n}{\partial \nu_{A_j}} - \beta w_j^n\right)\Big|_{\Sigma_{ij}} \\
\tilde{\tau}_{ij}^n &= \tau_{ij}^n - \left(-\frac{\partial p_j^n(T)}{\partial \nu_j} + \alpha p_j^n(T)\right)\Big|_{\Gamma_{ij}}
\end{aligned} \tag{8.4.3.3}$$

with λ_{ij}^0, $\mu_{ij}^0 \in L^2(\Sigma_{ij})$, $\tau_{ij}^0 \in L^2(\Gamma_{ij})$ arbitrary, λ_{ij}^n, μ_{ij}^n, τ_{ij}^n given by (8.4.2.9) for $n \geq 1$, and where $\tilde{\sigma}_i^n \in W_i^*$ is defined by

$$\langle \tilde{\sigma}_i^n, \phi \rangle_{W_i} = \sum_{j \in \mathcal{J}_i} \int_{\Gamma_{ij}} \tilde{\tau}_{ij}^n \phi \, d\Gamma, \quad \forall \phi \in W_i. \tag{8.4.3.4}$$

As usual, we wish to write the iterative step $(8.4.3.1)_3$, $(8.4.3.2)_3$, $(8.4.3.2)_4$, as a fixed point iteration. To this end we define

$$\mathcal{X} := \prod_{i=1}^{I} L^2(\Sigma_i^{\text{int}}) \times L^2(\Sigma_i^{\text{int}}) \times L^2(\Gamma_i^{\text{int}})$$

$$\|\mathcal{X}\|_{\mathcal{X}}^2 = \sum_{i=1}^{I} \left\{ \frac{1}{\beta} \int_{\Sigma_i^{\text{int}}} (|\lambda_i|^2 + |\mu_i|^2)\, d\Sigma + \frac{1}{\alpha\kappa} \int_{\Gamma_i^{\text{int}}} |\tau_i|^2 d\Gamma \right\}$$

where $X = \{(\lambda_i, \mu_i, \tau_i) : i = 1, \ldots, I\}$. For $X \in \mathcal{X}$ let (w_i, p_i) be the solution of the local optimality system (8.4.1.2), (8.4.1.4), (8.4.1.5) with

$$F_i = 0, \; w_{0i} = v_{0i} = z_{0i} = z_{1i} = 0 \tag{8.4.3.5}$$

and define $\mathcal{T} \in \mathcal{L}(\mathcal{X})$ by

$$\mathcal{T}X = \{(-\frac{\partial w_j}{\partial \nu_{A_j}} + \beta p_j)\big|_{\Sigma_{ij}}, (-\frac{\partial p_j}{\partial \nu_{A_j}} - \beta w_j)\big|_{\Sigma_{ij}},$$
$$(-\frac{\partial p_j(T)}{\partial \nu_{A_j}} + \alpha p_j(T))\big|_{\Gamma_{ij}} : j \in \mathcal{J}_i;\, i = 1, \ldots, I\}. \tag{8.4.3.6}$$

We see that X is a fixed point of $\mathcal{T}$ if and only if $w := \{w_i\}_{i=1}^I$, $p := \{p_i\}_{i=1}^I$ is the solution of the global optimality system corresponding to vanishing data (8.4.3.5). Since, in this case, the optimal control is $f = 0$, it follows that the only fixed point of $\mathcal{T}$ is $X = 0$.

Now set

$$X^n = \{(\tilde{\lambda}_{ij}^{n-1}, \tilde{\mu}_{ij}^{n-1}, \tilde{\tau}_{ij}^{n-1}) : j \in \mathcal{J}_i;\, i = 1, \ldots, I\} \tag{8.4.3.7}$$

and let $\{(\tilde{w}_i^n, \tilde{p}_i^n) : i = 1, \ldots, I\}$ be the solutions of (8.4.3.1), (8.4.3.2) with n replaced by $n-1$. We then have

$$X^n = \{(\frac{\partial \tilde{w}_i^n}{\partial \nu_{A_i}} + \beta \tilde{p}_i^n)\big|_{\Sigma_{ij}}, (\frac{\partial \tilde{p}_i^n}{\partial \nu_{A_i}} - \beta \tilde{w}_i^n)\big|_{\Sigma_{ij}},$$
$$(\frac{\partial \tilde{p}_i^n(T)}{\partial \nu_{A_i}} + \alpha \tilde{p}_i^n(T))\big|_{\Gamma_{ij}} : j \in \mathcal{J}_i;\, i = 1, \ldots, I\}. \tag{8.4.3.8}$$

and the iterative step $(8.4.3.1)_3$, $(8.4.3.2)_3$, $(8.4.3.2)_4$, with $\tilde{\lambda}_{ij}^n$, $\tilde{\mu}_{ij}^n$, $\tilde{\tau}_{ij}^n$ given by (8.4.3.3), may be expressed as the fixed point iteration

$$X^{n+1} = \mathcal{T}X^n. \tag{8.4.3.9}$$

We consider in what follows the iterative step (8.4.3.9) with under relaxation, that is, we consider the fixed point iteration

$$X^{n+1} = (1-\epsilon)\mathcal{T}X^n + \epsilon X^n, \quad \epsilon \in [0,1). \tag{8.4.3.10}$$

The main convergence result of this subsection is based on the following properties of $\mathcal{T}$.

Lemma 8.4.3.1 *For any $X \in \mathcal{X}$ one may write*

$$\|X\|_{\mathcal{X}}^2 = \mathcal{E} + \mathcal{F}, \qquad \|\mathcal{T}X\|_{\mathcal{X}}^2 = \mathcal{E} - \mathcal{F} \tag{8.4.3.11}$$

where

$$\mathcal{E} = \sum_{i=1}^{I} \Big\{ \int_{\Sigma_i^{int}} \big[\frac{1}{\beta}(|\frac{\partial w_i}{\partial \nu_{A_i}}|^2 + |\frac{\partial p_i}{\partial \nu_{A_i}}|^2) + \beta(|w_i|^2 + |p_i|^2)\big]\, d\Sigma + \frac{1}{\kappa}\int_{\Gamma_i^{int}} (\frac{1}{\alpha}|\frac{\partial p_i(T)}{\partial \nu_{A_i}}|^2 + \alpha|p_i(T)|^2)\, d\Gamma \Big\}$$

$$\mathcal{F} = 2\sum_{i=1}^{I} \Big\{ \int_{\Sigma_i^N} |p_i|^2 d\Sigma + \frac{1}{\kappa}\|(p_i(T), \frac{\partial p_i}{\partial t}(T))\|_{V_i \times H_i}^2 \Big\}.$$

In particular, $\mathcal{T}$ is a nonexpansive mapping.

Proof. For $X = \{(\lambda_i, \mu_i, \tau_i) : i = 1, \dots, I\} \in \mathcal{X}$ we may write

$$\|X\|_{\mathcal{X}}^2 = \sum_{i=1}^{I} \Big\{ \int_{\Sigma_i^{\rm int}} (|\frac{\partial w_i}{\partial \nu_{A_i}} + \beta p_i|^2 + |\frac{\partial p_i}{\partial \nu_{A_i}} - \beta w_i|^2)\, d\Sigma + \frac{1}{\alpha\kappa}\int_{\Gamma_i^{\rm int}} |\frac{\partial p_i(T)}{\partial \nu_{A_i}} + \alpha p_i(T)|^2 d\Gamma$$

where (w_i, p_i) is the solution of (8.4.1.2), (8.4.1.4), (8.4.1.5) with data (8.4.3.5). Thus

$$\|X\|_{\mathcal{X}}^2 = \mathcal{E} + 2\sum_{i=1}^{I} \Big\{ \int_{\Sigma_i^{\rm int}} (p_i \frac{\partial w_i}{\partial \nu_{A_i}} - w_i \frac{\partial p_i}{\partial \nu_{A_i}})\, d\Sigma + \frac{1}{\kappa}\int_{\Gamma_i^{\rm int}} p_i(T)\frac{\partial p_i(T)}{\partial \nu_{A_i}}\, d\Gamma \Big\}. \tag{8.4.3.12}$$

From

$$0 = \int_{Q_i} p_i(\frac{\partial^2 w_i}{\partial t^2} - \nabla \cdot (A_i \nabla w_i) + c_i w_i)\, dxdt$$

we obtain, after formal integration by parts (which may be justified because of the regularity properties of w_i and p_i)

$$\langle \frac{\partial w_i}{\partial t}(T), p_i(T)\rangle_{V_i} + \frac{1}{\kappa}\|\frac{\partial p_i}{\partial t}(T)\|_{H_i}^2 + \int_{\Sigma_i^N} |p_i|^2 d\Sigma = \int_{\Sigma_i^{\rm int}} (p_i \frac{\partial w_i}{\partial \nu_{A_i}} - w_i \frac{\partial p_i}{\partial \nu_{A_i}})\, d\Sigma. \tag{8.4.3.13}$$

Similarly, from

$$\kappa\langle\frac{\partial w_i}{\partial t}(T),p_i(T)\rangle_{V_i}=\int_{\Omega_i}\big[-\nabla\cdot(A_i\nabla p_i(T))+p_i(T)\big]p_i(T)\,dx$$

we have

$$\int_{\Gamma_i^{\rm int}}p_i(T)\frac{\partial p_i(T)}{\partial\nu_{A_i}}\,d\Gamma=\|p_i(T)\|^2_{V_i}-\kappa\langle\frac{\partial w_i}{\partial t}(T),p_i(T)\rangle_{V_i}. \tag{8.4.3.14}$$

It follows from (8.4.3.12), (8.4.3.13) and (8.4.3.14) that

$$\begin{aligned}\|X\|^2_{\mathcal{X}}&=\mathcal{E}+2\sum_{i=1}^{I}\Big\{\int_{\Sigma_i^N}|p_i|^2d\Sigma+\frac{1}{\kappa}\|\big(p_i(T),\frac{\partial p_i}{\partial t}(T)\|^2_{V_i\times H_i}\Big\}\\&=\mathcal{E}+\mathcal{F}.\end{aligned}$$

In a completely analogous manner we calculate

$$\begin{aligned}\|\mathcal{T}X\|^2_{\mathcal{X}}=\sum_{i=1}^{I}\sum_{j\in\mathcal{J}_i}\Big\{\int_{\Sigma_{ij}}\Big(\Big|-\frac{\partial w_j}{\partial\nu_{A_j}}+\beta p_j\Big|^2+\Big|-\frac{\partial p_j}{\partial\nu_{A_j}}-\beta w_j\Big|^2\Big)\,d\Sigma\\+\frac{1}{\alpha\kappa}\int_{\Gamma_{ij}}\Big|-\frac{\partial p_j(T)}{\partial\nu_{A_j}}+\alpha p_j(T)\Big|^2d\Gamma\Big\}.\end{aligned}$$

Since $\sum_{i=1}^{I}\sum_{j\in\mathcal{J}_i}=\sum_{j=1}^{I}\sum_{i\in\mathcal{J}_j}$ and since the integrands are independent of i we have, by the same calculation as above,

$$\begin{aligned}\|\mathcal{T}X\|^2_{\mathcal{X}}&=\sum_{j=1}^{I}\Big\{\int_{\Sigma_j^{\rm int}}\Big(\Big|-\frac{\partial w_j}{\partial\nu_{A_j}}+\beta p_j\Big|^2+\Big|-\frac{\partial p_j}{\partial\nu_{A_j}}-\beta w_j\Big|^2\Big)\,d\Sigma\\&\qquad+\frac{1}{\alpha\kappa}\int_{\Gamma_j^{\rm int}}\Big|-\frac{\partial p_j(T)}{\partial\nu_{A_j}}+\alpha p_j(T)\Big|^2d\Gamma\Big\}\\&=\mathcal{E}-\mathcal{F}.\end{aligned}$$

The main convergence result of this subsection is as follows.

Theorem 8.4.3.1 *With the iterative transmission conditions given by* (8.4.3.10), *we have*

(i) *for any* $\epsilon\in[0,1)$ *and for* $i=1,\ldots,I$

$$\tilde{p}_i^n\big|_{\Sigma_i^N}\to 0\ \ \textit{strongly in}\ L^2(\Sigma_i^n)$$

$$\big(\tilde{p}_i^n(T),\frac{\partial\tilde{p}_i^n}{\partial t}(T)\big)\to 0\ \ \textit{strongly in}\ V_i\times H_i$$

$$\big(\tilde{w}_i^n(T),\frac{\partial\tilde{w}_i^n}{\partial t}(T)\big)\to 0\ \ \textit{strongly in}\ H_i\times H^{-1}(\Omega_i).$$

(ii) *In addition, if* $\epsilon > 0$ *we have*

$$
\begin{gathered}
(\tilde{w}_i^n, \frac{\partial \tilde{w}_i^n}{\partial t}) \to 0, \quad (\tilde{p}_i^n, \frac{\partial \tilde{p}_i^n}{\partial t}) \to 0 \ \textit{weakly* in } L^\infty(0,T; H_i \times V_i^*) \\
\tilde{w}_i^n|_{\Sigma_i^N} \to 0 \ \textit{weakly in } L^2(\Sigma_i^N) \\
\tilde{w}_i^n|_{\Sigma_i^{\text{int}}} \to 0, \quad \tilde{p}_i^n|_{\Sigma_i^{\text{int}}} \to 0 \ \textit{weakly in } L^2(\Sigma_i^{\text{int}}) \\
\frac{\partial \tilde{w}_i^n}{\partial \nu_{A_i}}|_{\Sigma_i^{\text{int}}} \to 0, \quad \frac{\partial \tilde{p}_i^n}{\partial \nu_{A_i}}|_{\Sigma_i^{\text{int}}} \to 0 \ \textit{weakly in } L^2(\Sigma_i^{\text{int}}) \\
\frac{\partial \tilde{p}_i^n(T)}{\partial \nu_{A_i}}|_{\Gamma_i^{int}} \to 0 \ \textit{weakly in } L^2(\Gamma_i^{int}) \\
\tilde{w}_i^n - \tilde{w}_j^n|_{\Sigma_{ij}} \to 0, \quad \tilde{p}_i^n - \tilde{p}_j^n|_{\Sigma_{ij}} \to 0 \ \textit{strongly in } L^2(\Sigma_{ij}) \\
\frac{\partial \tilde{w}_i^n}{\partial \nu_{A_i}} + \frac{\partial \tilde{w}_j^n}{\partial \nu_{A_j}}|_{\Sigma_{ij}} \to 0, \quad \frac{\partial \tilde{p}_i^n}{\partial \nu_{A_i}} + \frac{\partial \tilde{p}_j^n}{\partial \nu_{A_j}}|_{\Sigma_{ij}} \to 0 \ \textit{strongly in } L^2(\Sigma_{ij}) \\
\frac{\partial \tilde{p}_i^n(T)}{\partial \nu_{A_i}} + \frac{\partial \tilde{p}_j^n(T)}{\partial \nu_{A_j}}|_{\Gamma_{ij}} \to 0 \ \textit{strongly in } L^2(\Gamma_{ij}).
\end{gathered}
$$

Remark 8.4.3.1 Part (i) shows that for any $\epsilon \in [0,1)$ the effective local optimal control $-p_i^n|_{\Sigma_i^N}$ converges strongly in $L^2(\Sigma_i^N)$ to $-p|_{\Sigma_i^N}$, where $-p|_{\Sigma^N}$ is the global optimal control. Further, the final state $(w_i^n(T), (\partial w_i^n/\partial t)(T))$ of the local optimal trajectory converges strongly in $L^2(\Omega_i) \times H^{-1}(\Omega_i)$ to the final state of the global optimal trajectory, restricted to Ω_i.

Remark 8.4.3.2 If the coefficients A_i, c_i and the boundary part $\Gamma_i^N \cup \Gamma_i^{\text{int}}$ are sufficiently regular, the boundedness of $\mathcal{E}^n$ implies that

$$
\begin{gathered}
\{\tilde{w}_i^n\}, \{\tilde{p}_i^n\} \ \text{are bounded in } C([0,T]; H_D^{2/3}(\Omega)) \\
\tilde{w}_i^n|_{\Sigma_i^N} \ \text{is bounded in } H^{1/3}(\Sigma_i^N) \\
\tilde{w}_i^n|_{\Sigma_{ij}}, \tilde{p}_i^n|_{\Sigma_{ij}} \ \text{are bounded in } H^{1/3}(\Sigma_{ij})
\end{gathered}
$$

(see Remark 8.2.2.2). We may then conclude, by using a compactness result of J. Simon [96, Corollary 5], that for $\epsilon \in (0,1)$

$$
\begin{gathered}
(\tilde{w}_i^n, \frac{\partial \tilde{w}_i^n}{\partial t}) \to 0, \quad (\tilde{p}_i^n, \frac{\partial \tilde{p}_i^n}{\partial t}) \to 0 \ \textit{in } C([0,T]; H_i \times V_i^*) \\
\tilde{w}_i^n|_{\Sigma_i^N} \to 0 \ \textit{strongly in } L^2(\Sigma_i^N) \\
\tilde{w}_i^n|_{\Sigma_{ij}} \to 0, \quad \tilde{p}_i^n|_{\Sigma_{ij}} \to 0 \ \textit{strongly in } L^2(\Sigma_{ij}).
\end{gathered}
$$

Proof of Theorem 8.4.3.1. With X^n given by (8.4.3.8) we have from Lemma 8.4.3.1

$$
\|X^n\|_{\mathcal{X}}^2 = \mathcal{E}^n + \mathcal{F}^n, \quad \|\mathcal{T}X^n\|_{\mathcal{X}}^2 = \mathcal{E}^n - \mathcal{F}^n \tag{8.4.3.15}
$$

where

$$\mathcal{E}^n = \sum_{i=1}^{I}\Big\{\int_{\Sigma_i^{\rm int}}\big[\frac{1}{\beta}(|\frac{\partial \tilde{w}_i^n}{\partial \nu_{A_i}}|^2 + |\frac{\partial \tilde{p}_i^n}{\partial \nu_{A_i}}|^2) + \beta(|\tilde{w}_i^n|^2 + |\tilde{p}_i^n|^2)\big]\, d\Sigma$$

$$+ \frac{1}{\kappa}\int_{\Gamma_i^{\rm int}}(\frac{1}{\alpha}|\frac{\partial \tilde{p}_i^n(T)}{\partial \nu_{A_i}}|^2 + \alpha|\tilde{p}_i^n(T)|^2)\, d\Gamma\Big\} \tag{8.4.3.16}$$

$$\mathcal{F}^n = 2\sum_{i=1}^{I}\Big\{\int_{\Sigma_i^N}|\tilde{p}_i^n|^2 d\Sigma + \frac{1}{\kappa}\|(\tilde{p}_i^n(T), \frac{\partial \tilde{p}_i^n}{\partial t}(T))\|^2_{V_i\times H_i}\Big\}. \tag{8.4.3.17}$$

From (8.4.3.10) we calculate, with the aid of (8.4.3.15),

$$\begin{aligned}\mathcal{E}^{n+1} + \mathcal{F}^{n+1} &= ((1-\epsilon)^2+\epsilon^2)\mathcal{E}^n - (1-2\epsilon)\mathcal{F}^2 + 2\epsilon(1-\epsilon)(\mathcal{T}X^n, X^n)_{\mathcal{X}}\\ &\le ((1-\epsilon)^2+\epsilon^2)\mathcal{E}^n - (1-2\epsilon)\mathcal{F}^2 + 2\epsilon(1-\epsilon)\sqrt{(\mathcal{E}^n)^2 - (\mathcal{F}^n)^2}\\ &\le \mathcal{E}^n - (1-2\epsilon)\mathcal{F}^n,\end{aligned} \tag{8.4.3.18}$$

hence

$$\mathcal{E}^{n+1} \le \mathcal{E}^n - \mathcal{F}^{n+1} - (1-2\epsilon)\mathcal{F}^n. \tag{8.4.3.19}$$

Upon iterating (8.4.3.19) down to $n=1$ we obtain

$$\mathcal{E}^{n+1} \le \mathcal{E}^1 - \sum_{p=1}^{n+1} c_p(\epsilon)\mathcal{F}^p \tag{8.4.3.20}$$

where

$$c_p(\epsilon) = 2(1-\epsilon),\ p=2,\dots,n;\quad c_1(\epsilon) = 1-2\epsilon,\ c_{n+1}(\epsilon) = 1. \tag{8.4.3.21}$$

It follows from (8.4.3.20) that

$$\mathcal{F}^n \to 0,$$

that is to say,

$$\begin{aligned}&\tilde{p}_i^n|_{\Sigma_i^N} \to 0 \text{ strongly in } L^2(\Sigma_i^N)\\ &(\tilde{p}_i^n(T), \frac{\partial \tilde{p}_i^n(T)}{\partial t}) \to 0 \text{ strongly in } V_i\times H_i.\end{aligned} \tag{8.4.3.22}$$

From $(8.4.3.22)_2$ and $(8.4.3.2)_5$ we have

$$\tilde{w}_i^n(T) \to 0 \text{ strongly in } H_i.$$

Further, since $\langle \tilde{\sigma}_i^n, \phi\rangle_{W_i} = 0$, $\forall \phi \in H_0^1(\Omega_i)$, we obtain from $(8.4.3.2)_4$

$$\kappa|\langle \frac{\partial \tilde{w}_i^n}{\partial t}(T), \phi\rangle| = |\langle \mathcal{B}_i\tilde{p}_i^{n+1}(T), \phi\rangle_{W_i}| \le \|\tilde{p}_i^{n+1}(T)\|_{W_i}\|\phi\|_{H_0^1(\Omega_i)},\quad \forall \phi\in H_0^1(\Omega_i),$$

hence

$$\frac{\partial \tilde{w}_i^n}{\partial t}(T)\big|_{H_0^1(\Omega_i)} \to 0 \text{ strongly in } H^{-1}(\Omega_i).$$

This proves part (i) of Theorem 8.4.3.1.

Now suppose that $\epsilon > 0$. We first invoke Proposition 2.3.7.3 to deduce that

$$X^n \to 0 \text{ weakly in } \mathcal{X}$$

and, therefore,

$$(1-\epsilon)\mathcal{T}X^n = X^{n+1} - \epsilon X^n \to 0 \text{ weakly in } \mathcal{X}.$$

Thus we may conclude that

$$\pm\frac{\partial \tilde{w}_i^n}{\partial \nu_{A_i}} + \beta \tilde{p}_i^n\big|_{\Sigma_i^{\text{int}}} \to 0, \quad \pm\frac{\partial \tilde{p}_i^n}{\partial \nu_{A_i}} - \beta \tilde{w}_i^n\big|_{\Sigma_i^{\text{int}}} \to 0 \text{ weakly in } L^2(\Sigma_i^{\text{int}})$$
$$\frac{\partial \tilde{p}_i^n(T)}{\partial \nu_{A_i}}\big|_{\Gamma_i^{\text{int}}} \to 0 \text{ weakly in } L^2(\Gamma_i^{\text{int}})$$

hence

$$\tilde{w}_i^n\big|_{\Sigma_i^{\text{int}}} \to 0, \quad \tilde{p}_i^n\big|_{\Sigma_i^{\text{int}}} \to 0 \text{ weakly in } L^2(\Sigma_i^{\text{int}})$$
$$\frac{\partial \tilde{w}_i^n}{\partial \nu_{A_i}}\big|_{\Sigma_i^{\text{int}}} \to 0, \quad \frac{\partial \tilde{p}_i^n}{\partial \nu_{A_i}}\big|_{\Sigma_i^{\text{int}}} \to 0 \text{ weakly in } L^2(\Sigma_i^{\text{int}})$$
$$\frac{\partial \tilde{p}_i^n(T)}{\partial \nu_{A_i}}\big|_{\Gamma_i^{\text{int}}} \to 0 \text{ weakly in } L^2(\Gamma_i^{\text{int}}).$$

Therefore

$$\tilde{\lambda}_{ij}^n \to 0, \quad \tilde{\mu}_{ij}^n \to 0 \text{ weakly in } L^2(\Sigma_{ij})$$
$$\tilde{\tau}_{ij}^n \to 0 \text{ weakly in } L^2(\Gamma_{ij}), \quad \forall j \in \mathcal{J}_i,\ i = 1, \dots, I.$$

Since the map taking the data

$$\left(-\tilde{p}_i^{n+1}|_{\Sigma_i^N}, \{\lambda_{ij}^n\}_{j\in\mathcal{J}_i} - \beta\tilde{p}_i^{n+1}|_{\Sigma_i^{\text{int}}}\right) \mapsto \left((\tilde{w}_i^{n+1}, \partial\tilde{w}_i^{n+1}/\partial t), \tilde{w}_i^{n+1}|_{\Sigma_i^N\cup\Sigma_i^{\text{int}}}\right),$$

where $\tilde{w}_i^{n+1}$ is the solution of (8.4.3.1), is continuous from $L^2(\Sigma_i^N \cup \Sigma_i^{\text{int}})$ into $L^\infty(0,T; H_i \times V_i^*) \times L^2(\Sigma_i^N \cup \Sigma_i^{\text{int}})$, it follows that

$$\left(\tilde{w}_i^{n+1}, \frac{\partial \tilde{w}_i^{n+1}}{\partial t}\right) \to 0 \text{ weakly* in } L^\infty(0,T; H_i \times V_i^*)$$
$$\tilde{w}_i^{n+1}\big|_{\Sigma_i^N} \to 0 \text{ weakly in } L^2(\Sigma_i^N).$$

Similarly

$$\left(\tilde{p}_i^{n+1}, \frac{\partial \tilde{p}_i^{n+1}}{\partial t}\right) \to 0 \text{ weakly* in } L^\infty(0,T; H_i \times V_i^*).$$

It remains to prove that for all $j \in \mathcal{J}_i$ and $i = 1, \dots, I$,

$$\frac{\partial \tilde{w}_i^n}{\partial \nu_{A_i}} + \frac{\partial \tilde{w}_j^n}{\partial \nu_{A_j}}\Big|_{\Sigma_{ij}} \to 0, \quad \frac{\partial \tilde{p}_i^n}{\partial \nu_{A_i}} + \frac{\partial \tilde{p}_j^n}{\partial \nu_{A_j}}\Big|_{\Sigma_{ij}} \to 0 \text{ strongly in } L^2(\Sigma_{ij})$$
$$\tilde{w}_i^n - \tilde{w}_j^n\big|_{\Sigma_{ij}} \to 0, \quad \tilde{p}_i^n - \tilde{p}_j^n\big|_{\Sigma_{ij}} \to 0 \text{ strongly in } L^2(\Sigma_{ij}) \tag{8.4.3.23}$$
$$\frac{\partial \tilde{p}_i^n(T)}{\partial \nu_{A_i}} + \frac{\partial \tilde{p}_j^n(T)}{\partial \nu_{A_j}}\Big|_{\Gamma_{ij}} \to 0 \text{ strongly in } L^2(\Gamma_{ij}).$$

For this we invoke Proposition 2.3.7.1 to obtain

$$\|X^{n+1} - X^n\|_{\mathcal{X}} = (1-\epsilon)\|\mathcal{T}X^n - X^n\|_{\mathcal{X}} \to 0. \tag{8.4.3.24}$$

From (8.4.3.15) we have

$$\|\mathcal{T}X^n - X^n\|_{\mathcal{X}}^2 = 2[\mathcal{E}^n - (\mathcal{T}X^n, X^n)_{\mathcal{X}}]. \tag{8.4.3.25}$$

A direct calculation gives

$$\begin{aligned}(\mathcal{T}X^n, X^n)_{\mathcal{X}} &= \sum_{i=1}^{I}\sum_{j\in\mathcal{J}_i}\Big\{\frac{1}{\beta}\int_{\Sigma_{ij}}\Big[-\frac{\partial \tilde{w}_i^n}{\partial \nu_{A_i}}\frac{\partial \tilde{w}_j^n}{\partial \nu_{A_j}} - \frac{\partial \tilde{p}_i^n}{\partial \nu_{A_i}}\frac{\partial \tilde{p}_j^n}{\partial \nu_{A_j}} \\ &+ \beta^2\big(\tilde{w}_i^n\tilde{w}_j^n + \tilde{p}_i^n\tilde{p}_j^n\big) + \beta\big(\tilde{p}_j^n\frac{\partial \tilde{w}_i^n}{\partial \nu_{A_i}} - \tilde{p}_i^n\frac{\partial \tilde{w}_j^n}{\partial \nu_{A_j}}\big) - \beta\big(\tilde{w}_j^n\frac{\partial \tilde{p}_i^n}{\partial \nu_{A_i}} - \tilde{w}_i^n\frac{\partial \tilde{p}_j^n}{\partial \nu_{A_j}}\big)\Big]\,d\Sigma \\ &+ \frac{1}{\alpha\kappa}\int_{\Gamma_{ij}}\Big[-\frac{\partial \tilde{p}_i^n(T)}{\partial \nu_{A_i}}\frac{\partial \tilde{p}_j^n(T)}{\partial \nu_{A_j}} + \alpha^2\tilde{p}_i^n(T)\tilde{p}_j^n(T) \\ &+ \alpha\big(\tilde{p}_j^n(T)\frac{\partial \tilde{p}_i^n(T)}{\partial \nu_{A_i}} - \tilde{p}_i^n(T)\frac{\partial \tilde{p}_j^n(T)}{\partial \nu_{A_j}}\big)\Big]\,d\Gamma\Big\}.\end{aligned}$$

One has

$$\sum_{i=1}^{I}\sum_{j\in\mathcal{J}_i}\int_{\Sigma_{ij}}\big(\tilde{p}_j^n\frac{\partial \tilde{w}_i^n}{\partial \nu_{A_i}} - \tilde{p}_i^n\frac{\partial \tilde{w}_j^n}{\partial \nu_{A_j}}\big)\,d\Sigma = 0$$

and, similarly, for the other integrals with mixed terms. Since also

$$\sum_{i=1}^{I}\sum_{j\in\mathcal{J}_i}\int_{\Sigma_{ij}} f_i f_j\,d\Sigma = 2\sum_{i=1}^{I-1}\sum_{\substack{j\in\mathcal{J}_i\\ j>i}}\int_{\Sigma_{ij}} f_i f_j\,d\Sigma$$

we may write

$$\begin{aligned}&(\mathcal{T}X^n, X^n)_{\mathcal{X}} \\ &= 2\sum_{i=1}^{I-1}\sum_{\substack{j\in\mathcal{J}_i\\ j>i}}\Big\{\frac{1}{\beta}\int_{\Sigma_{ij}}\Big[-\frac{\partial \tilde{w}_i^n}{\partial \nu_{A_i}}\frac{\partial \tilde{w}_j^n}{\partial \nu_{A_j}} - \frac{\partial \tilde{p}_i^n}{\partial \nu_{A_i}}\frac{\partial \tilde{p}_j^n}{\partial \nu_{A_j}} + \beta^2\big(\tilde{w}_i^n\tilde{w}_j^n + \tilde{p}_i^n\tilde{p}_j^n\big)\Big]\,d\Sigma \\ &+ \frac{1}{\alpha\kappa}\int_{\Gamma_{ij}}\Big[-\frac{\partial \tilde{p}_i^n(T)}{\partial \nu_{A_i}}\frac{\partial \tilde{p}_j^n(T)}{\partial \nu_{A_j}} + \alpha^2\tilde{p}_i^n(T)\tilde{p}_j^n(T)\Big]\,d\Gamma\Big\}.\end{aligned} \tag{8.4.3.26}$$

Since we may write

$$\sum_{i=1}^{I}\int_{\Sigma_i^{\rm int}}|f_i|^2 d\Sigma=\sum_{i=1}^{I-1}\sum_{\substack{j\in\mathcal{J}_i\\ j>i}}\int_{\Sigma_{ij}}(|f_i|^2+|f_j|^2)\,d\Sigma$$

if follows from (8.4.3.16) and (8.4.3.26) that

$$\begin{aligned}\mathcal{E}^n-(\mathcal{T}X^n,X^n)_{\mathcal{X}}=\sum_{i=1}^{I-1}\sum_{\substack{j\in\mathcal{J}_i\\ j>i}}\Big\{\frac{1}{\beta}\int_{\Sigma_{ij}}&\Big[\Big|\frac{\partial\tilde{w}_i^n}{\partial\nu_{A_i}}+\frac{\partial\tilde{w}_j^n}{\partial\nu_{A_j}}\Big|^2+\Big|\frac{\partial\tilde{p}_i^n}{\partial\nu_{A_i}}+\frac{\partial\tilde{p}_j^n}{\partial\nu_{A_j}}\Big|^2\\ &+\beta^2\big(|\tilde{w}_i^n-\tilde{w}_j^n|^2+|\tilde{p}_i^n-\tilde{p}_j^n|^2\big)\Big]d\Sigma\\ +\frac{1}{\alpha\kappa}\int_{\Gamma_{ij}}\Big[\Big|\frac{\partial\tilde{p}_i^n(T)}{\partial\nu_{A_i}}+\frac{\partial\tilde{p}_j^n(T)}{\partial\nu_{A_j}}\Big|^2&+\alpha^2|\tilde{p}_i^n(T)-\tilde{p}_j^n(T)|^2\Big]d\Gamma\Big\}.\end{aligned}\qquad(8.4.3.27)$$

The conclusions (8.4.3.23) now follow from (8.4.3.24), (8.4.3.25) and (8.4.3.27).

8.5 Domain Decomposition in the Exact Reachability Problem

8.5.1 The Global Optimality System

In this section we consider the state constrained optimal control problem

$$\inf_{f\in L^2(\Sigma^N)} J(f),\quad J(f)=\frac{1}{2}\int_{\Sigma^N}|f|^2 d\Sigma \qquad (8.5.1.1)$$

subject to (8.1.1.1) and the state constraint

$$w(T)=z_0,\quad \frac{\partial w}{\partial t}(T)=z_1. \qquad (8.5.1.2)$$

We assume that for each T sufficiently large, say $T>T_0$, this problem has a unique solution for each $(w_0,v_0)\in\mathcal{H}:=V\times H$, $(z_0,z_1)\in\mathcal{H}$ and $F\in L^1(0,T;H)$. Because of the time reversible character of the system (8.1.1.1), we may and do assume without loss of generality that

$$w_0=v_0=0,\quad F=0. \qquad (8.5.1.3)$$

In this case the optimal control problem is called the *exact reachability problem.*

The solvability of the exact reachability problem for each $(z_0,z_1)\in\mathcal{H}$ means that

$$V\times H\subset \mathrm{Rg}\,(\mathcal{L}_T):=\big\{\big(w(T),\frac{\partial w}{\partial t}(T)\big):\, f\in L^2(\Sigma^N)\big\} \qquad (8.5.1.4)$$

where

$$\mathcal{L}_T : f \mapsto \left(w(T), \frac{\partial w}{\partial t}(T)\right), \ L^2(\Sigma^N) \mapsto H \times V^*$$

is the *control-to-state mapping*. It is well known that (8.5.1.4) is equivalent to the *continuous observability from* Σ^N of the system (8.2.2.1) (with $G = 0$), (8.2.2.2), that is, to the *observability estimate*

$$\|(\phi_0, \phi_1)\|^2_{H\times V^*} \leq C_T \int_{\Sigma^N} |\phi|^2 d\Sigma, \quad \forall (\phi_0, \phi_1) \in V \times H, \tag{8.5.1.5}$$

for some constant C_T independent of (ϕ_0, ϕ_1).

When (8.5.1.5) holds, the optimal control f_{opt} may be constructed via the *Hilbert Uniqueness Method* of J.-L. Lions [72]:

$$f_{\text{opt}} = -p\big|_{\Sigma^N} \tag{8.5.1.6}$$

where p is the solution of

$$\begin{gathered} \frac{\partial^2 p}{\partial t^2} - \nabla \cdot (A \nabla p) + cp = 0 \ \text{ in } Q \\ p = 0 \ \text{ on } \Sigma^D, \quad \frac{\partial p}{\partial \nu_A} = 0 \text{ on } \Sigma^N \\ p(T) = p_0, \quad \frac{\partial p}{\partial t}(T) = p_1 \text{ in } \Omega \end{gathered} \tag{8.5.1.7}$$

and where $(p_0, p_1) \in H \times V^*$ is the unique pair satisfying

$$\big\langle (p_0, p_1), (-z_1, z_0) \big\rangle_{\mathcal{H}} = \int_{\Sigma^N} |p|^2 d\Sigma. \tag{8.5.1.8}$$

The coupled systems (8.5.1.7) and (8.1.1.1), with (p_0, p_1) given by (8.5.1.8) and f given by (8.5.1.6), comprise the *optimality system* for the optimal control problem (8.5.1.1) subject to (8.1.1.1) and (8.5.1.2).

The optimality system (8.1.1.1), (8.5.1.6), (8.5.1.7), (8.5.1.8) may, in fact, be obtained in the limit of the optimality system (8.1.1.1), (8.2.1.2), (8.2.1.3) for the unconstrained optimal control problem (8.2.1.1) as the penalty parameter $\kappa \to \infty$, as described by the following theorem.

Theorem 8.5.1.1 *Suppose that the data satisfies* (8.5.1.3), *that* $(z_0, z_1) \in V \times H$, *and that* (8.5.1.5) *holds. Then as* $\kappa \to \infty$ *the solution* $w(\cdot\,;\kappa)$, $p(\cdot\,;\kappa)$ *of the optimality system* (8.1.1.1), (8.2.1.2), (8.2.1.3) *satisfies*

$$\begin{gathered} \left(w(\cdot\,;\kappa), \frac{\partial w}{\partial t}(\cdot\,;\kappa)\right) \to \left(w(\cdot), \frac{\partial w}{\partial t}(\cdot)\right) \ \textit{in } C([0,T]; H \times V^*) \\ \left(p(\cdot\,;\kappa), \frac{\partial p}{\partial t}(\cdot\,;\kappa)\right) \to \left(p(\cdot), \frac{\partial p}{\partial t}(\cdot)\right) \ \textit{in } C([0,T]; H \times V^*) \\ p(\cdot\,;\kappa)\big|_{\Sigma^N} \to p(\cdot)\big|_{\Sigma^N} \ \textit{strongly in } L^2(\Sigma^N), \end{gathered}$$

where $w(\cdot)$, $p(\cdot)$ *is the solution of the optimality system* (8.1.1.1), (8.5.1.6), (8.5.1.7), (8.5.1.8).

We omit the proof of Theorem 8.5.1.1 since it is outside the main theme of this monograph. Complete details may, however, be found in [56, Theorem 4.1].

In the following subsections we shall introduce a nonoverlapping spatial domain decomposition for the computation of the solution of the global optimality system (8.1.1.1), (8.5.1.6), (8.5.1.7), (8.5.1.8) similar to the one introduced in subsection 8.4.2 for the optimality system associated with the unconstrained optimal control problem (8.2.1.1). In fact, if $w_i^{n+1}(\cdot\,;\kappa)$, $p_i^{n+1}(\cdot\,;\kappa)$, $i = 1, \ldots, I$, denote the solutions of the iterative scheme (8.4.2.7), (8.4.2.8), (8.4.2.9) with data satisfying (8.5.1.3), and if λ_{ij}^0, μ_{ij}^0 and τ_{ij}^0 are chosen independent of κ, it shall be proved that as $\kappa \to \infty$

$$\begin{aligned}
\Big(w_i^n(\cdot\,;\kappa), \frac{\partial w_i^n}{\partial t}(\cdot\,;\kappa)\Big) &\to \Big(w_i^n(\cdot), \frac{\partial w_i^n}{\partial t}(\cdot)\Big) \ \text{ in } C([0,T]; H_i \times V_i^*) \\
\Big(p_i^n(\cdot\,;\kappa), \frac{\partial p_i^n}{\partial t}(\cdot\,;\kappa)\Big) &\to \Big(p_i^n(\cdot), \frac{\partial p_i^n}{\partial t}(\cdot)\Big) \ \text{ in } C([0,T]; H_i \times V_i^*) \\
p_i^n(\cdot\,;\kappa)\big|_{\Sigma_i^N} &\to p_i^n(\cdot)\big|_{\Sigma_i^N} \ \text{ strongly in } L^2(\Sigma_i^N) \\
w_i^n(\cdot\,;\kappa)\big|_{\Sigma_i^{\text{int}}} &\to w_i^n(\cdot)\big|_{\Sigma_i^{\text{int}}} \ \text{ strongly in } L^2(\Sigma_i^{\text{int}}) \\
p_i^n(\cdot\,;\kappa)\big|_{\Sigma_i^{\text{int}}} &\to p_i^n(\cdot)\big|_{\Sigma_i^{\text{int}}} \ \text{ strongly in } L^2(\Sigma_i^{\text{int}})
\end{aligned} \tag{8.5.1.9}$$

where $w_i^{n+1}(\cdot)$, $p_i^{n+1}(\cdot)$ is the solution of a certain local, state constrained optimal control problem concentrated on the cylinder Q_i. These functions are determined through an iterative scheme that represents a nonoverlapping spatial domain decomposition of the global optimality system (8.1.1.1), (8.5.1.6), (8.5.1.7), (8.5.1.8), and they converge, in appropriate spaces, to its solution. Thus we have the commutative diagram

$$\begin{array}{ccc}
(w_i^n(\cdot\,;\kappa), p_i^n(\cdot\,;\kappa)) & \xrightarrow{n\to\infty} & (w_i(\cdot\,;\kappa), p_i(\cdot\,;\kappa)) \\
\kappa\to\infty \big\downarrow & & \kappa\to\infty \big\downarrow \\
(w_i^n(\cdot), p_i^n(\cdot)) & \xrightarrow{n\to\infty} & (w_i(\cdot), p_i(\cdot))
\end{array} \tag{8.5.1.10}$$

where $w_i(\cdot\,;\kappa)$, $p_i(\cdot\,;\kappa)$ (resp., $w_i(\cdot)$, $p_i(\cdot)$) denotes the restriction to Q_i of the solution of the global optimality system for the unconstrained (resp., constrained) optimal control problem (8.2.1.1) (resp., (8.5.1.1)). The top branch of the diagram was proved in Subsection 8.4.3, and the right branch represents the content of Theorem 8.5.1.1. The left branch is the assertion (8.5.1.9). The bottom branch will be proved in the subsections that follow.

8.5.2 The Limit of the Local Optimality Systems

In this subsection we will examine the limiting behavior as $\kappa \to \infty$ of the solution $w_i^{n+1}(\cdot\,;\kappa)$, $p_i^{n+1}(\cdot\,;\kappa)$ of the local optimality system (8.4.2.7)–(8.4.2.10) with data satisfying

$$w_{0i} = v_{0i} = F_i = 0, \quad i = i, \dots, I.$$

To this end we first consider the limiting behavior as $\kappa \to \infty$ of the solution $w_i(\cdot\,;\kappa)$, $p_i(\cdot\,;\kappa)$ of the system

$$\begin{aligned} &\frac{\partial^2 w_i}{\partial t^2} - \nabla\cdot(A_i\nabla w_i) + c_i w_i = 0 \text{ in } Q_i \\ &w_i = 0 \text{ on } \Sigma_i^D, \quad \frac{\partial w_i}{\partial \nu_{A_i}} = f_2^\kappa \text{ on } \Sigma_i^N \\ &\frac{\partial w_i}{\partial \nu_{A_i}} = f_1^\kappa + \lambda_i(\cdot\,;\kappa) \text{ on } \Sigma_i^{\text{int}} \\ &w_i(0;\kappa) = \frac{\partial w_i}{\partial t}(0;\kappa) = 0, \end{aligned} \tag{8.5.2.1}$$

$$\begin{aligned} &\frac{\partial^2 p_i}{\partial t^2} - \nabla\cdot(A_i\nabla p_i) + c_i p_i = 0 \text{ in } Q_i \\ &p_i = 0 \text{ on } \Sigma_i^D, \quad \frac{\partial p_i}{\partial \nu_{A_i}} = 0 \text{ on } \Sigma_i^N \\ &\frac{\partial p_i}{\partial \nu_{A_i}} = \beta w_i + \mu_i(\cdot\,;\kappa) \quad \text{on } \Sigma_i^{\text{int}}, \\ &p_i(T;\kappa) = \kappa\mathcal{B}_i^{-1}\Big(\frac{\partial w_i}{\partial t}(T;\kappa) - z_{1i} + \kappa^{-1}\sigma_i(\kappa)\Big), \\ &\frac{\partial p_i}{\partial t}(T;\kappa) = -\kappa(w_i(T;\kappa) - z_{0i}), \end{aligned} \tag{8.5.2.2}$$

$$f_1^\kappa = -\beta p_i(\cdot\,;\kappa)|_{\Sigma_i^{\text{int}}}, \quad f_2^\kappa = -p_i(\cdot\,;\kappa)|_{\Sigma_i^N}, \tag{8.5.2.3}$$

where $\lambda_i(\cdot\,;\kappa)$, $\mu_i(\cdot\,;\kappa) \in L^2(\Sigma_i^{\text{int}})$, $\tau_i(\cdot\,;\kappa) \in L^2(\Gamma_i^{\text{int}})$ are given and $\sigma_i(\kappa) \in W_i^*$ is defined by

$$\langle\sigma_i(\kappa), \phi\rangle_{W_i} = \int_{\Gamma_i^{\text{int}}} \tau_i(\cdot\,;\kappa)\phi(\cdot)\,d\Gamma, \quad \forall\phi \in W_i.$$

The system (8.5.2.1)–(8.5.2.3) is the optimality system for the problem

$$\begin{aligned} J_i^\kappa(f_1, f_2) = \frac{1}{2}\int_{\Sigma_i^N} |f_2|^2 d\Sigma + \frac{1}{2\beta}\int_{\Sigma_i^{\text{int}}} (|f_1|^2 + |\beta w_i(\cdot\,;\kappa) + \mu_i(\cdot\,;\kappa)|^2) d\Sigma \\ + \frac{\kappa}{2}(\|w_i(T;\kappa) - z_{0i}\|_{H_i}^2 + \Big\|\frac{\partial w_i}{\partial t}(T;\kappa) - z_{1i} + \kappa^{-1}\sigma_i(\kappa)\Big\|_{W_i^*}^2) \\ \to \inf_{(f_1,f_2)\in\mathcal{U}} J_i^\kappa(f_1, f_2), \qquad \mathcal{U} = L^2(\Sigma_i^{\text{int}}) \times L^2(\Sigma_i^N), \end{aligned}$$

subject to

$$
\begin{aligned}
&\frac{\partial^2 w_i}{\partial t^2} - \nabla \cdot (A_i \nabla w_i) + c_i w_i = 0 \ \text{ in } Q_i \\
&w_i = 0 \ \text{ on } \Sigma_i^D, \quad \frac{\partial w_i}{\partial \nu_{A_i}} = f_2 \ \text{ on } \Sigma_i^N \\
&\frac{\partial w_i}{\partial \nu_{A_i}} = f_1 + \lambda_i(\cdot\,;\kappa) \ \text{ on } \Sigma_i^{\text{int}} \\
&w_i(0;\kappa) = \frac{\partial w_i}{\partial t}(0;\kappa) = 0.
\end{aligned}
\tag{8.5.2.4}
$$

Let ϕ_i be the solution of

$$
\begin{aligned}
&\frac{\partial^2 \phi_i}{\partial t^2} - \nabla \cdot (A_i \nabla \phi_i) + c_i \phi_i = 0 \ \text{ in } Q_i \\
&\phi_i = 0 \ \text{ on } \Sigma_i^D, \quad \frac{\partial \phi_i}{\partial \nu_{A_i}} = 0 \ \text{ on } \Sigma_i^N \cup \Sigma_i^{\text{int}} \\
&\phi_i(0) = \phi_i^0, \quad \frac{\partial \phi_i}{\partial t}(0) = \phi_i^1 \ \text{ in } \Omega_i.
\end{aligned}
\tag{8.5.2.5}
$$

It is assumed that (8.5.2.5) is continuously observable from $\Sigma_i^N \cup \Sigma_i^{\text{int}}$ for $T > T_0$ for some $T_0 > 0$:

$$
\|\phi_i^0\|_{H_i}^2 + \|\phi_i^1\|_{V_i^*}^2 \le C_T \left(\int_{\Sigma_i^{\text{int}}} |\phi_i|^2 d\Sigma + \int_{\Sigma_i^N} |\phi_i|^2 d\Sigma \right), \quad \forall (\phi_i^0, \phi_i^1) \in V_i \times H_i, \tag{8.5.2.6}
$$

for some constant C_T. It follows from (8.5.2.6) that if $(z_{0i}, z_{1i}) \in V_i \times H_i$ there are controls $f_1 \in L^2(\Sigma_i^{\text{int}})$, $f_2 \in L^2(\Sigma_i^N)$ such that the solution of (8.5.2.4) satisfies

$$
w_i(T;\kappa) = z_{0i}, \quad \frac{\partial w_i}{\partial t}(T;\kappa) = z_{1i}. \tag{8.5.2.7}
$$

Theorem 8.5.2.1 *Suppose that the regularity requirement* (8.4.1.3) *is satisfied, that the observability estimate* (8.5.2.6) *holds, that* $T > T_0$*, and that* $(z_{0i}, z_{1i}) \in V_i \times H_i$*. Suppose further that as* $\kappa \to \infty$

$$
\begin{aligned}
&\lambda_i(\cdot\,;\kappa) \to \lambda_i^\infty(\cdot), \quad \mu_i(\cdot\,;\kappa) \to \mu_i^\infty(\cdot) \quad \textit{strongly in } L^2(\Sigma_i^{\text{int}}), \\
&\frac{1}{\sqrt{\kappa}} \tau_i(\cdot\,;\kappa) \to 0 \quad \textit{strongly in } L^2(\Gamma_i^{int}).
\end{aligned}
\tag{8.5.2.8}
$$

Then the solution of (8.5.2.1)–(8.5.2.3) *satisfies*

$$\begin{aligned}
\Big(w_i(\cdot\,;\kappa), \frac{\partial w_i}{\partial t}(\cdot\,;\kappa)\Big) &\to \Big(w_i(\cdot), \frac{\partial w_i}{\partial t}(\cdot)\Big) \quad \textit{in } C([0,T];H_i)\\
\sqrt{\kappa}\big(w_i(T;\kappa) - z_{0i}\big) &\to 0 \quad \textit{strongly in } H_i\\
\sqrt{\kappa}\Big(\frac{\partial w_i}{\partial t}(T;\kappa) - z_{1i}\Big) &\to 0 \quad \textit{strongly in } V_i^*\\
w_i(\cdot\,;\kappa)\big|_{\Sigma_i^{\text{int}}\cup\Sigma_i^N} &\to w_i(\cdot)\big|_{\Sigma_i^{\text{int}}\cup\Sigma_i^N} \quad \textit{strongly in } L^2(\Sigma_i^{\text{int}}\cup\Sigma_i^N)\\
\Big(p_i(\cdot\,;\kappa), \frac{\partial p_i}{\partial t}(\cdot\,;\kappa)\Big) &\to \Big(p_i(\cdot), \frac{\partial p_i}{\partial t}(\cdot)\Big) \quad \textit{in } C([0,T];H_i)\\
p_i(\cdot\,;\kappa)\big|_{\Sigma_i^{\text{int}}\cup\Sigma_i^N} &\to p_i(\cdot)\big|_{\Sigma_i^{\text{int}}\cup\Sigma_i^N} \quad \textit{strongly in } L^2(\Sigma_i^{\text{int}}\cup\Sigma_i^N)
\end{aligned}$$

where w_i, p_i *is the solution of*

$$\begin{aligned}
&\frac{\partial^2 w_i}{\partial t^2} - \nabla\cdot(A_i\nabla w_i) + c_i w_i = 0 \quad \textit{in } Q_i\\
&w_i = 0 \quad \textit{on } \Sigma_i^D, \qquad \frac{\partial w_i}{\partial \nu_{A_i}} = -p_i \quad \textit{on } \Sigma_i^N\\
&\frac{\partial w_i}{\partial \nu_{A_i}} = -\beta p_i + \lambda_i^\infty \quad \textit{on } \Sigma_i^{\text{int}}\\
&w_i(0) = \frac{\partial w_i}{\partial t}(0) = 0, \qquad w_i(T) = z_{0i}, \ \frac{\partial w_i}{\partial t}(T) = z_{1i},
\end{aligned} \tag{8.5.2.9}$$

$$\begin{aligned}
&\frac{\partial^2 p_i}{\partial t^2} - \nabla\cdot(A_i\nabla p_i) + c_i p_i = 0 \quad \textit{in } Q_i\\
&p_i = 0 \quad \textit{on } \Sigma_i^D, \qquad \frac{\partial p_i}{\partial \nu_{A_i}} = 0 \quad \textit{on } \Sigma_i^N\\
&\frac{\partial p_i}{\partial \nu_{A_i}} = \beta w_i + \mu_i^\infty \quad \textit{on } \Sigma_i^{\text{int}}\\
&p_i(T) = p_{0i}, \quad \frac{\partial p_i}{\partial t}(T) = p_{1i},
\end{aligned} \tag{8.5.2.10}$$

and where $(p_{0i}, p_{1i}) \in H_i \times V_i^*$ *is the unique solution of*

$$\begin{aligned}
&\big\langle (p_{0i}, p_{1i}), (-z_{1i}, z_{0i})\big\rangle_{H_i\times V_i}\\
&\quad = \int_{\Sigma_i^{\text{int}}} [\beta(|p_i|^2 + |w_i|^2) - \lambda_i^\infty p_i + \mu_i^\infty w_i]\, d\Sigma + \int_{\Sigma_i^N} |p_i|^2 d\Sigma.
\end{aligned} \tag{8.5.2.11}$$

Remark 8.5.2.1 It will be shown in the course of the proof that the system (8.5.2.9)–(8.5.2.11) is the optimality system for the problem

$$\inf_{(f_1,f_2)\in\mathcal{U}_{\text{ad}}} J_i(f_1,f_2)$$

$$J_i(f_1,f_2) = \frac{1}{2}\int_{\Sigma_i^N} |f_2|^2 d\Sigma + \frac{1}{2\beta}\int_{\Sigma_i^{\text{int}}} \big[|f_1|^2 + |\beta w_i + \mu_i^\infty|^2\big]\, d\Sigma,$$

where $\mathcal{U}_{\text{ad}}$ consists of controls $f_1 \in L^2(\Sigma_i^{\text{int}})$, $f_2 \in L^2(\Sigma_i^N)$ such that the solution of

$$\begin{aligned}
&\frac{\partial^2 w_i}{\partial t^2} - \nabla\cdot(A_i \nabla w_i) + c_i w_i = 0 \text{ in } Q_i \\
&w_i = 0 \text{ on } \Sigma_i^D, \quad \frac{\partial w_i}{\partial \nu_{A_i}} = f_2 \text{ on } \Sigma_i^N \\
&\frac{\partial w_i}{\partial \nu_{A_i}} = f_1 + \lambda_i^\infty \text{ on } \Sigma_i^{\text{int}} \\
&w_i(0) = \frac{\partial w_i}{\partial t}(0) = 0
\end{aligned} \tag{8.5.2.12}$$

satisfies the state constraint

$$w_i(T) = z_{0i}, \quad \frac{\partial w_i}{\partial t}(T) = z_{1i}.$$

Proof of Theorem 8.5.2.1. Let $T > T_0$ and $f_1^0 \in L^2(\Sigma_i^{\text{int}})$, $f_2^0 \in L^2(\Sigma_i^N)$ be such that the solution $\hat{w}_i$ of

$$\begin{aligned}
&\frac{\partial^2 \hat{w}_i}{\partial t^2} - \nabla\cdot(A_i \nabla \hat{w}_i) + c_i \hat{w}_i = 0 \text{ in } Q_i, \\
&\hat{w}_i = 0 \text{ on } \Sigma_i^D, \quad \frac{\partial \hat{w}_i}{\partial \nu_{A_i}} = f_2^0 \text{ on } \Sigma_i^N, \\
&\frac{\partial \hat{w}_i}{\partial \nu_{A_i}} = f_1^0 \text{ on } \Sigma_i^{\text{int}}, \\
&\hat{w}_i(0) = \frac{\partial \hat{w}_i}{\partial t}(0) = 0 \quad \text{in } \Omega_i
\end{aligned}$$

satisfies

$$\hat{w}_i(T) = z_{0i}, \quad \frac{\partial \hat{w}_i}{\partial t}(T) = z_{1i}.$$

By the regularity assumption (8.4.1.3) we have

$$\|\hat{w}_i\|_{L^2(\Sigma_i^{\text{int}}\cup\Sigma_i^N)} \le C(\|f_1^0\|_{L^2(\Sigma_i^{\text{int}})} + \|f_2^0\|_{L^2(\Sigma_i^N)}).$$

Set $\hat{f}_1^\kappa(\cdot\,;\kappa) = f_1^0(\cdot) - \lambda_i(\cdot\,;\kappa)$. Then

$$\begin{aligned}
J_i^\kappa(f_1^\kappa, f_2^\kappa) \le J_i^\kappa(\hat{f}_1^\kappa, f_2^0) &= \frac{1}{2}\|f_2^0\|^2_{L^2(\Sigma_i^N)} + \frac{1}{2\beta}\|\hat{f}_1^\kappa\|_{L^2(\Sigma_i^{\text{int}})} \\
&+ \frac{1}{2\beta}\int_{\Sigma_i^{\text{int}}} |\beta\hat{w}_i(\cdot) + \mu_i(\cdot\,;\kappa)|^2 d\Sigma + \frac{1}{2\kappa}\|\sigma_i(\kappa)\|^2_{W_i^*}.
\end{aligned}$$

It follows that

$$\begin{aligned}
&f_1^\kappa \quad \text{is bounded in } L^2(\Sigma_i^{\text{int}}) \\
&f_2^\kappa \quad \text{is bounded in } L^2(\Sigma_i^N) \\
&\sqrt{\kappa}(w_i(T;\kappa) - z_{0i}) \quad \text{is bounded in } H_i \\
&\sqrt{\kappa}(\frac{\partial w_i}{\partial t}(T;\kappa) - z_{1i} + \kappa^{-1}\sigma_i(\kappa)) \quad \text{is bounded in } W_i^*.
\end{aligned} \tag{8.5.2.13}$$

and, therefore,

$$\begin{aligned}
\Big(w_i(\cdot\,;\kappa), \frac{\partial w_i}{\partial t}(\cdot\,;\kappa)\Big) &\quad \text{is bounded in } L^\infty(0,T;H_i\times V_i^*)\\
w_i(\cdot\,;\kappa)|_{\Sigma_i^{\text{int}}\cup\Sigma_i^N} &\quad \text{is bounded in } L^2(\Sigma_i^{\text{int}}\cup\Sigma_i^N)\\
w_i(T;\kappa)\to z_{0i} &\quad \text{strongly in } H_i,\\
\frac{\partial w_i}{\partial t}(T;\kappa)\to z_{1i} &\quad \text{strongly in } V_i^*.
\end{aligned}\tag{8.5.2.14}$$

We may conclude that on a sequence $\kappa=\kappa_m\to\infty$,

$$\begin{aligned}
f_1^\kappa\to f_1^\infty &\quad \text{weakly in } L^2(\Sigma_i^{\text{int}})\\
f_2^\kappa\to f_2^\infty &\quad \text{weakly in } L^2(\Sigma_i^N)\\
\Big(w_i(\cdot\,;\kappa), \frac{\partial w_i}{\partial t}(\cdot\,;\kappa)\Big)\to\Big(w_i(\cdot), \frac{\partial w_i}{\partial t}(\cdot)\Big) &\quad \text{weakly}^* \text{ in } L^\infty(0,T;H_i\times V_i^*)\\
w_i(\cdot\,;\kappa)|_{\Sigma_i^{\text{int}}\cup\Sigma_i^N}\to w_i(\cdot)|_{\Sigma_i^{\text{int}}\cup\Sigma_i^N} &\quad \text{weakly in } L^2(\Sigma_i^{\text{int}}\cup\Sigma_i^N)
\end{aligned}\tag{8.5.2.15}$$

where w_i satisfies

$$\begin{aligned}
&\frac{\partial^2 w_i}{\partial t^2}-\nabla\cdot(A_i\nabla w_i)+c_iw_i=0 \quad \text{in } Q_i,\\
&\frac{\partial w_i}{\partial\nu_{A_i}}=f_1^\infty+\lambda_i^\infty \quad \text{on } \Sigma_i^{\text{int}},\\
&w_i=0 \text{ on } \Sigma_i^D,\quad \frac{\partial w_i}{\partial\nu_{A_i}}=f_2^\infty \quad \text{on } \Sigma_i^N,\\
&w_i(0)=\frac{\partial w_i}{\partial t}(0)=0,\qquad w_i(T)=z_{0i},\quad \frac{\partial w_i}{\partial t}(T)=z_{1i}.
\end{aligned}\tag{8.5.2.16}$$

Further

$$\begin{aligned}
&\|f_2^\infty\|^2_{L^2(\Sigma_i^N)}+\frac{1}{\beta}\|f_1^\infty\|^2_{L^2(\Sigma_i^{\text{int}})}+\frac{1}{\beta}\|\beta w_i(\cdot)+\mu_i^\infty(\cdot)\|^2_{L^2(\Sigma_i^{\text{int}})}\\
&\quad\le\liminf_{\kappa_m\to\infty}\Big(\|f_2^\kappa\|^2_{L^2(\Sigma_i^N)}+\frac{1}{\beta}\|f_1^\kappa\|^2_{L^2(\Sigma_i^{\text{int}})}+\frac{1}{\beta}\|\beta w_i(\cdot\,;\kappa)+\mu_i(\cdot\,;\kappa)\|^2_{L^2(\Sigma_i^{\text{int}})}\Big)
\end{aligned}\tag{8.5.2.17}$$

Set

$$\mathcal{U}_{\text{ad}}^\kappa=\{f_1\in L^2(\Sigma_i^{\text{int}}),\ f_2\in L^2(\Sigma_i^N): \text{the solution } z_i(\cdot\,;\kappa) \text{ of (8.5.2.4) satisfies } z_i(T;\kappa)=z_{0i},\ \frac{\partial z_i}{\partial t}(T;\kappa)=z_{1i}\},$$

$$\mathcal{U}_{\text{ad}}=\{f_1\in L^2(\Sigma_i^{\text{int}}),\ f_2\in L^2(\Sigma_i^N): \text{the solution } z_i(\cdot) \text{ of (8.5.2.12) satisfies } z_i(T)=z_{0i},\ \frac{\partial z_i}{\partial t}(T)=z_{1i}\}.$$

In addition, set

$$J_i(f_1, f_2) = \frac{1}{2}\int_{\Sigma_i^N} |f_2|^2 d\Sigma + \frac{1}{2\beta}\int_{\Sigma_i^{\text{int}}} \left[|f_1|^2 + |\beta z_i + \mu_i^\infty|^2\right] d\Sigma, \quad \forall (f_1, f_2) \in \mathcal{U}_{\text{ad}}.$$

We have

$$\begin{aligned} J_i^\kappa(f_1^\kappa, f_2^\kappa) &\le J_i^\kappa(f_1, f_2) \\ &= \frac{1}{2}\int_{\Sigma_i^N} |f_2|^2 d\Sigma + \frac{1}{2\beta}\int_{\Sigma_i^{\text{int}}} \left[|f_1|^2 + |\beta z_i(\cdot\,;\kappa) + \mu_i(\cdot\,;\kappa)|^2\right] d\Sigma \\ &\qquad + \frac{1}{2\sqrt{\kappa}}\sigma_i(\kappa), \quad \forall (f_1, f_2) \in \mathcal{U}_{\text{ad}}^\kappa. \end{aligned} \tag{8.5.2.18}$$

If $(f_1, f_2) \in \mathcal{U}_{\text{ad}}$ and if we define

$$\hat{f}_1^\kappa(\cdot\,;\kappa) = f_1(\cdot) - \lambda_i(\cdot\,;\kappa) + \lambda_i^\infty(\cdot), \quad \hat{f}_2^\kappa(\cdot\,;\kappa) = f_2(\cdot),$$

then $(\hat{f}_1^\kappa, \hat{f}_2^\kappa) \in \mathcal{U}_{\text{ad}}^\kappa$ and $z_i(\cdot\,;\kappa) \equiv z_i(\cdot)$, where $z_i(\cdot\,;\kappa)$ is the solution (8.5.2.4) corresponding to $\hat{f}_1^\kappa$, $\hat{f}_2^\kappa$. From (8.5.2.18) and hypothesis (8.5.2.8) we then obtain

$$\limsup_{\kappa\to\infty} J_i^\kappa(f_1^\kappa, f_2^\kappa) \le J_i(f_1, f_2), \quad \forall (f_1, f_2) \in \mathcal{U}_{\text{ad}}. \tag{8.5.2.19}$$

In particular, we have

$$\limsup_{\kappa\to\infty} J_i^\kappa(f_1^\kappa, f_2^\kappa) \le J_i(f_1^\infty, f_2^\infty). \tag{8.5.2.20}$$

On the other hand, from (8.5.2.17) and (8.5.2.19) we also have

$$J_i(f_1^\infty, f_2^\infty) \le \liminf_{\kappa_m\to\infty} J_i^\kappa(f_1^\kappa, f_2^\kappa) \le J_i(f_1, f_2), \quad \forall (f_1, f_2) \in \mathcal{U}_{\text{ad}}. \tag{8.5.2.21}$$

Therefore f_1^∞, f_2^∞ is the unique solution of the state constrained optimal control problem

$$\inf_{(f_1, f_2)\in\mathcal{U}_{\text{ad}}} J_i(f_1, f_2) \tag{8.5.2.22}$$

and

$$J_i^\kappa(f_1^\kappa, f_2^\kappa) \to J_i(f_1^\infty, f_2^\infty) \tag{8.5.2.23}$$

through all real $\kappa \to \infty$, from which we may conclude that

$$\begin{aligned} &(f_1^\kappa, f_2^\kappa) \to (f_1^\infty, f_2^\infty) \quad \text{strongly in } L^2(\Sigma_i^{\text{int}}) \times L^2(\Sigma_i^N) \\ &\sqrt{\kappa}\big(w_i(T;\kappa) - z_{0i}\big) \to 0 \quad \text{strongly in } H_i \\ &\sqrt{\kappa}\Big(\frac{\partial w_i}{\partial t}(T;\kappa) - z_{1i}\Big) \to 0 \quad \text{strongly in } V_i^* \\ &\Big(w_i(\cdot\,;\kappa), \frac{\partial w_i}{\partial t}(\cdot\,;\kappa)\Big) \to \Big(w_i(\cdot), \frac{\partial w_i}{\partial t}(\cdot)\Big) \quad \text{in } C([0,T]; H_i \times V_i^*) \\ &w_i(\cdot\,;\kappa)\big|_{\Sigma_i^{\text{int}}\cup\Sigma_i^N} \to w_i(\cdot)\big|_{\Sigma_i^{\text{int}}\cup\Sigma_i^N} \quad \text{strongly in } L^2(\Sigma_i^{\text{int}} \cup \Sigma_i^N). \end{aligned} \tag{8.5.2.24}$$

The solution $p_i(\cdot\,;\kappa)$ of (8.5.2.2) may be expressed as $p_i = q_i + r_i$, where r_i satisfies (8.5.2.2) except that

$$r_i(T;\kappa) = r_{i,t}(T;\kappa) = 0,$$

and q_i satisfies (8.5.2.2) except that

$$\frac{\partial q_i}{\partial \nu_{A_i}} = 0 \quad \text{on } \Sigma_i^{\text{int}}.$$

By virtue of the regularity assumption (8.4.1.3) we have

$$\|r_i(\cdot\,;\kappa)\|^2_{L^2(\Sigma_i^N)} + \|r_i(\cdot\,;\kappa)\|^2_{L^2(\Sigma_i^{\text{int}})} \le C_T \int_{\Sigma_i^{\text{int}}} |\beta w_i(\cdot\,;\kappa) + \mu_i(\cdot\,;\kappa)|^2 d\Sigma \quad (8.5.2.25)$$

and, by virtue of the observability assumption (8.5.2.6), we have, for $T > T_0$,

$$\|p_i(T;\kappa)\|^2_{H_i} + \|\frac{\partial p_i}{\partial t}(T;\kappa)\|^2_{V_i^*} \le C_T \left(\int_{\Sigma_i^{\text{int}}} |q_i|^2 d\Sigma + \int_{\Sigma_i^N} |q_i|^2 d\Sigma \right). \quad (8.5.2.26)$$

By replacing q_i by $p_i - r_i$ in (8.5.2.26) and then using (8.5.2.25) we obtain

$$\begin{aligned} &\|p_i(T;\kappa)\|^2_{H_i} + \|\frac{\partial p_i}{\partial t}(T;\kappa)\|^2_{V_i^*} \\ &\le C_T \left(\int_{\Sigma_i^{\text{int}}} (|p_i(\cdot\,;\kappa)|^2 + |\beta w_i(\cdot\,;\kappa) + \mu_i(\cdot\,;\kappa)|^2) d\Sigma + \int_{\Sigma_i^N} |p_i(\cdot\,;\kappa)|^2 d\Sigma \right) \\ &= C_T \left(\int_{\Sigma_i^{\text{int}}} (|f_1^\kappa|^2 + |\beta w_i(\cdot\,;\kappa) + \mu_i(\cdot\,;\kappa)|^2) d\Sigma + \int_{\Sigma_i^N} |f_2^\kappa|^2 d\Sigma \right). \quad (8.5.2.27) \end{aligned}$$

Next, we note that $w_i(\cdot;\kappa_1) - w_i(\cdot;\kappa_2)$, $p_i(\cdot;\kappa_1) - p_i(\cdot;\kappa_2)$ satisfy (8.5.2.1) and $(8.5.2.2)_{1,2,3}$ with f_1^κ, f_2^κ, $\lambda_i(\cdot\,;\kappa)$, $\mu_i(\cdot\,;\kappa)$ replaced by $f_1^{\kappa_1} - f_1^{\kappa_2}$, $f_2^{\kappa_1} - f_2^{\kappa_2}$, and so forth. The inequality (8.5.2.27) therefore holds for the difference $p_i(T;\kappa_1) - p_i(T;\kappa_2)$ by replacing the terms on the right side by the corresponding differences $f_1^{\kappa_1} - f_1^{\kappa_2}$, $f_2^{\kappa_1} - f_2^{\kappa_2}$, and so forth. It follows from (8.5.2.24) that

$$\begin{aligned} &p_i(T;\kappa) \to p_{0i} \quad \text{strongly in } H_i \\ &\frac{\partial p_i}{\partial t}(T;\kappa) \to p_{1i} \quad \text{strongly in } V_i^* \end{aligned} \quad (8.5.2.28)$$

for some $(p_{0i}, p_{1i}) \in H_i \times V_i^*$. This, together with (8.5.2.24), allows us to conclude that

$$\begin{aligned} &(p_i(\cdot\,;\kappa), \frac{\partial p_i}{\partial t}(\cdot\,;\kappa)) \to (p_i(\cdot), \frac{\partial p_i}{\partial t}(\cdot)) \quad \text{in } C([0,T]; H_i \times V_i^*) \\ &p_i(\cdot\,;\kappa)|_{\Sigma_i^{\text{int}} \cup \Sigma_i^N} \to p_i(\cdot)|_{\Sigma_i^{\text{int}} \cup \Sigma_i^N} \quad \text{strongly in } L^2(\Sigma_i^{\text{int}} \cup \Sigma_i^N) \end{aligned}$$

where p_i is the solution of

$$\begin{aligned}
&\frac{\partial^2 p_i}{\partial t^2} - \nabla\cdot(A_i\nabla p_i) + c_i p_i = 0 \quad \text{in } Q_i,\\
&\frac{\partial p_i}{\partial \nu_{A_i}} = \beta w_i + \mu_i^\infty \quad \text{on } \Sigma_i^{\text{int}},\\
&p_i = 0 \ \text{ on } \Sigma_i^D, \quad \frac{\partial p_i}{\partial \nu_{A_i}} = 0 \ \text{ on } \Sigma_i^N\\
&p_i(T) = p_{0i}, \quad \frac{\partial p_i}{\partial t}(T) = p_{1i}.
\end{aligned} \tag{8.5.2.29}$$

A formal integration by parts (which may be justified by approximating f_1^∞, f_2^∞, μ_i^∞, λ_i^∞ by smoother data) gives

$$0 = (z_{1i}, p_{0i})_{H_i} - \langle p_{1i}, z_{0i}\rangle_{V_i} - \int_{\Sigma_i^{\text{int}}}\left(p_i\frac{\partial w_i}{\partial \nu_{A_i}} - w_i\frac{\partial p_i}{\partial \nu_{A_i}}\right)d\Sigma - \int_{\Sigma_i^N} p_i\frac{\partial w_i}{\partial \nu_{A_i}}\,d\Sigma,$$

that is,

$$\begin{aligned}
&\big\langle (p_{0i}, p_{1i}), (-z_{1i}, z_{0i})\big\rangle_{H_i\times V_i}\\
&\quad = \int_{\Sigma_i^{\text{int}}} [\beta(|p_i|^2 + |w_i|^2) - \lambda_i^\infty p_i + \mu_i^\infty w_i]d\Sigma + \int_{\Sigma_i^N} |p_i|^2 d\Sigma.
\end{aligned} \tag{8.5.2.30}$$

This completes the proof of Theorem 8.5.2.1.

Remark 8.5.2.2 Let us comment further on equation (8.5.2.30). The pair (p_{0i}, p_{1i}) may be calculated in a manner reminiscent of the Hilbert Uniqueness Method. In fact, this pair is chosen so that the solution of

$$\begin{aligned}
&\frac{\partial^2 w_i}{\partial t^2} - \nabla\cdot(A_i\nabla w_i) + c_i w_i = 0 \quad \text{in } Q_i\\
&\frac{\partial w_i}{\partial \nu_{A_i}} = -\beta p_i + \lambda_i^\infty \quad \text{on } \Sigma_i^{\text{int}}\\
&w_i = 0 \ \text{ on } \Sigma_i^D, \quad \frac{\partial w_i}{\partial \nu_{A_i}} = -p_i \ \text{ on } \Sigma_i^N\\
&w_i(0) = \frac{\partial w_i}{\partial t}(0) = 0,
\end{aligned} \tag{8.5.2.31}$$

$$\begin{aligned}
&\frac{\partial^2 p_i}{\partial t^2} - \nabla\cdot(A_i\nabla p_i) + c_i p_i = 0 \quad \text{in } Q_i\\
&\frac{\partial p_i}{\partial \nu_{A_i}} = \beta w_i + \mu_i^\infty \quad \text{on } \Sigma_i^{\text{int}}\\
&p_i = 0 \ \text{ on } \Sigma_i^D, \quad \frac{\partial p_i}{\partial \nu_{A_i}} = 0 \ \text{ on } \Sigma_i^N\\
&p_i(T) = p_{0i}, \quad \frac{\partial p_i}{\partial t}(T) = p_{1i},
\end{aligned} \tag{8.5.2.32}$$

satisfies

$$w_i(T) = z_{0i}, \qquad \frac{\partial w_i}{\partial t}(T) = z_{1i}. \tag{8.5.2.33}$$

For arbitrary $(p_{0i}, p_{1i}) \in V_i \times H_i$ the system (8.5.2.31), (8.5.2.32) has a unique solution since it is the optimality system for the problem

$$\inf_{f\in L^2(\Sigma_i^{\text{int}})} \left(\frac{1}{2\beta} \int_{\Sigma_i^{\text{int}}} (|f|^2 + |\beta p_i - \lambda_i^\infty|^2)\, d\Sigma + \frac{1}{2} \int_{\Sigma_i^N} |p_i|^2 d\Sigma \right)$$

subject to

$$\frac{\partial^2 p_i}{\partial t^2} - \nabla \cdot (A_i \nabla p_i) + c_i p_i = 0 \quad \text{in } Q_i,$$

$$\frac{\partial p_i}{\partial \nu_{A_i}} = f + \mu_i^\infty \quad \text{on } \Sigma_i^{\text{int}},$$

$$p_i = 0 \ \text{ on } \Sigma_i^D, \quad \frac{\partial p_i}{\partial \nu_{A_i}} = 0 \ \text{ on } \Sigma_i^N,$$

$$p_i(T) = p_{0i}, \quad \frac{\partial p_i}{\partial t}(T) = p_{1i}.$$

The solution of (8.5.2.31), (8.5.2.32) may be written $w_i = w_i^1 + w_i^2$, $p_i = p_i^1 + p_i^2$, where

(w_i^1, p_i^1) is the solution corresponding to $\lambda_i^\infty = \mu_i^\infty = 0$,
(w_i^2, p_i^2) is the solution corresponding to $p_{0i} = p_{1i} = 0$.

By Holmgren's theorem, we may define a norm $\|(p_{0i}, p_{1i})\|_{F_i}$ by setting

$$\|(p_{0i}, p_{1i})\|_{F_i}^2 := \int_{\Sigma_i^{\text{int}}} \beta(|p_i^1|^2 + |w_i^1|^2) d\Sigma + \int_{\Sigma_i^N} |p_i^1|^2 d\Sigma,$$

and a corresponding Hilbert space

$$F_i = \text{completion of } V_i \times H_i \text{ in } \| \cdot \|_{F_i}.$$

For $T > T_0$ one has (cf. (8.5.2.25)–(8.5.2.27))

$$\|(p_{0i}, p_{1i})\|_{H_i \times V_i^*} \leq C_T \|(p_{0i}, p_{1i})\|_{F_i},$$

hence $F_i \subset H_i \times V_i^*$, $H_i \times V_i \subset F_i'$. In addition, by Green's formula we have

$$\left\langle (p_{0i}, p_{1i}), \left(-\frac{\partial w_i^1}{\partial t}(T), w_i^1(T)\right) \right\rangle_{H_i \times V_i} = \|(p_{0i}, p_{1i})\|_{F_i}^2,$$

where $\langle \cdot, \cdot \rangle_{H_i \times V_i}$ denotes the duality pairing in the $H_i \times V_i$, $H_i \times V_i^*$ duality. Therefore if $(-\hat{z}_{1i}, \hat{z}_{0i}) \in F_i'$ and if we set

$$(p_{0i}, p_{1i}) = \Lambda_i^{-1}(-\hat{z}_{1i}, \hat{z}_{0i}),$$

where Λ_i is the canonical isomorphism $F_i \mapsto F_i'$, then we will have

$$w_i^1(T) = \hat{z}_{0i}, \qquad \frac{\partial w_i^1}{\partial t}(T) = \hat{z}_{1i}.$$

It follows that (8.5.2.33) will be satisfied by choosing

$$(p_{0i}, p_{1i}) = \Lambda_i^{-1}\big((-z_{1i}, z_{0i}) - (-\frac{\partial w_i^2}{\partial t}(T), w_i^2(T))\big). \tag{8.5.2.34}$$

It may be checked that (8.5.2.34) is exactly the same as (8.5.2.30).

8.5.3 Application of Theorem 8.5.2.1 to Domain Decomposition

We now apply Theorem 8.5.2.1 to the solution $w_i^{n+1}(\cdot\,;\kappa)$, $p_i^{n+1}(\cdot\,;\kappa)$ of the local optimality system (8.4.2.7), (8.4.2.8) with under relaxation in the iteration step, that is with $\lambda_{ij}^n(\cdot\,;\kappa)$, $\mu_{ij}^n(\cdot\,;\kappa)$, $\tau_{ij}^n(\cdot\,;\kappa)$ given by

$$\begin{aligned}
\lambda_{ij}^n &= (1-\epsilon)\big(-\frac{\partial w_j^n}{\partial \nu_{A_j}} + \beta p_j^n\big) + \epsilon\big(\frac{\partial w_i^n}{\partial \nu_{A_i}} + \beta p_i^n\big)\big|_{\Sigma_{ij}}\\
\mu_{ij}^n &= (1-\epsilon)\big(-\frac{\partial p_j^n}{\partial \nu_{A_j}} - \beta w_j^n\big) + \epsilon\big(\frac{\partial p_i^n}{\partial \nu_{A_i}} - \beta w_i^n\big)\big|_{\Sigma_{ij}}\\
\tau_{ij}^n &= (1-\epsilon)\big(-\frac{\partial p_j^n(T)}{\partial \nu_{A_j}} + \alpha p_j^n(T)\big) + \epsilon\big(\frac{\partial p_i^n(T)}{\partial \nu_i} + \alpha p_i^n(T)\big)\big|_{\Gamma_{ij}}
\end{aligned} \tag{8.5.3.35}$$

for $n \geq 1$, and where $\sigma_i^n(\kappa)$ is defined by (8.4.2.10). As in the last section we assume vanishing data $w_{0i} = v_{0i} = 0$ and $F_i = 0$. We further assume that the starting data $\lambda_{ij}^0(\cdot\,;\kappa)$, $\mu_{ij}^0(\cdot\,;\kappa)$ and $\tau_{ij}^0(\cdot\,;\kappa)$ satisfy hypothesis (8.5.2.8), that is

$$\begin{aligned}
&\lambda_{ij}^0(\cdot\,;\kappa) \to \lambda_{ij}^0(\cdot),\ \mu_{ij}^0(\cdot\,;\kappa) \to \mu_{ij}^0(\cdot) \quad \text{strongly in } L^2(\Sigma_{ij})\\
&\frac{1}{\sqrt{\kappa}}\tau_{ij}^0(\cdot\,;\kappa) \to 0 \quad \text{strongly in } L^2(\Gamma_{ij}),\ \forall j \in \mathcal{J}_i
\end{aligned} \tag{8.5.3.36}$$

for some $\lambda_{ij}^0, \mu_{ij}^0 \in L^2(\Sigma_{ij})$. Hypothesis (8.5.3.36) will hold if, for example, $\lambda_{ij}^0(\cdot\,;\kappa)$, $\mu_{ij}^0(\cdot\,;\kappa)$ and $\tau_{ij}^0(\cdot\,;\kappa)$ are chosen so that they are independent of κ.

It then follows from Theorem 8.5.2.1 that, as $\kappa \to \infty$, the solution $w_i^1(\cdot\,;\kappa)$, $p_i^1(\cdot\,;\kappa)$ converges in the manner described in that theorem to the solution $w_i^1(\cdot)$, $p_i^1(\cdot)$ of the system

$$\begin{gathered}
\frac{\partial^2 w_i^1}{\partial t^2} - \nabla\cdot(A_i\nabla w_i) + c_i w_i^1 = 0 \quad \text{in } Q_i \\
\frac{\partial^2 p_i^1}{\partial t^2} - \nabla\cdot(A_i\nabla p_i) + c_i p_i^1 = 0 \quad \text{in } Q_i \\
\frac{\partial w_i^1}{\partial \nu_{A_i}} + \beta p_i^1 = \lambda_{ij}^0, \quad \frac{\partial p_i^1}{\partial \nu_{A_i}} - \beta w_i^1 = \mu_{ij}^0 \quad \text{on } \Sigma_{ij}, \\
w_i^1 = p_i^1 = 0 \quad \text{on } \Sigma_i^D \\
\frac{\partial w_i^1}{\partial \nu_{A_i}} + p_i^1 = 0, \quad \frac{\partial p_i^1}{\partial \nu_{A_i}} = 0 \quad \text{on } \Sigma_i^N \\
w_i^1(0) = \frac{\partial w_i^1}{\partial t}(0) = 0, \quad p_i^1(T) = p_{0i}^1, \quad \frac{\partial p_i^1}{\partial t}(T) = p_{1i}^1, \\
w_i^1(T) = z_{0i}, \quad \frac{\partial w_i^1}{\partial t}(T) = z_{1i},
\end{gathered}$$

where $(p_{0i}^1, p_{1i}^1) \in H_i \times V_i^*$ is the solution of

$$\begin{aligned}
&\big\langle (p_{0i}^1, p_{1i}^1), (-z_{1i}, z_{0i}) \big\rangle_{H_i\times V_i} \\
&\quad = \sum_{j\in\mathcal{J}_i} \int_{\Sigma_{ij}} [\beta(|p_i^1|^2 + |w_i^1|^2) d\Sigma - \lambda_{ij}^0 p_i^1 + \mu_{ij}^0 w_i^1] d\Sigma + \int_{\Sigma_i^N} |p_i^1|^2 d\Sigma.
\end{aligned}$$

According to Theorem 8.5.2.1,

$$w_i^1(\cdot\,;\kappa)|_{\Sigma_i^{\text{int}}} \to w_i^1(\cdot)|_{\Sigma_i^{\text{int}}}, \quad p_i^1(\cdot\,;\kappa)|_{\Sigma_i^{\text{int}}} \to p_i^1(\cdot)|_{\Sigma_i^{\text{int}}} \quad \text{strongly in } L^2(\Sigma_i^{\text{int}})$$

and therefore

$$\begin{gathered}
\frac{\partial w_i^1(\cdot\,;\kappa)}{\partial \nu_{A_i}} \to \lambda_{ij}^0(\cdot) - \beta p_i^1(\cdot) = \frac{\partial w_i^1(\cdot)}{\partial \nu_{A_i}}, \\
\frac{\partial p_i^1(\cdot\,;\kappa)}{\partial \nu_{A_i}} \to \mu_{ij}^0(\cdot) + \beta w_i^1(\cdot) = \frac{\partial p_i^1(\cdot)}{\partial \nu_{A_i}} \quad \text{strongly in } L^2(\Sigma_{ij}).
\end{gathered}$$

As a result

$$\begin{aligned}
\lambda_{ij}^1(\cdot\,;\kappa) &\to (1-\epsilon)(-\frac{\partial w_j^1(\cdot)}{\partial \nu_{A_j}} + \beta p_j^1(\cdot))|_{\Sigma_{ij}} + \epsilon(\frac{\partial w_i^1(\cdot)}{\partial \nu_{A_i}} + \beta p_i^1(\cdot))|_{\Sigma_{ij}} \\
&:= \lambda_{ij}^1(\cdot) \quad \text{strongly in } L^2(\Sigma_{ij}), \\
\mu_{ij}^1(\cdot\,;\kappa) &\to (1-\epsilon)(-\frac{\partial p_j^1(\cdot)}{\partial \nu_{A_j}} - \beta w_j^1(\cdot))|_{\Sigma_{ij}} + \epsilon(\frac{\partial p_i^1(\cdot)}{\partial \nu_{A_i}} - \beta w_i^1(\cdot))|_{\Sigma_{ij}} \\
&:= \mu_{ij}^1(\cdot) \quad \text{strongly in } L^2(\Sigma_{ij}).
\end{aligned} \tag{8.5.3.37}$$

The proof of Theorem 8.5.2.1 showed that

$$\sqrt{\kappa}\Big(\frac{\partial w_i^1}{\partial t}(T;\kappa) - z_{1i} + \kappa^{-1}\sigma_i^0(\kappa)\Big) \to 0 \quad \text{strongly in } W_i^*$$

and therefore

$$\frac{1}{\sqrt{\kappa}} p_i^1(T;\kappa) = \mathcal{B}_i^{-1}\Big[\sqrt{\kappa}\Big(\frac{\partial w_i^1}{\partial t}(T;\kappa) - z_{1i} + \kappa^{-1}\sigma_i^0(\kappa)\Big)\Big] \to 0 \quad \text{strongly in } W_i.$$

In particular,

$$\frac{1}{\sqrt{\kappa}} p_i^1(T;\kappa)|_{\Gamma_i^{\rm int}} \to 0 \quad \text{strongly in } L^2(\Gamma_i^{\rm int}).$$

Since

$$\frac{\partial p_i^1(T;\kappa)}{\partial \nu_{A_i}} + \alpha p_i^1(T;\kappa) = \tau_{ij}^0(\cdot) \quad \text{on } \Gamma_{ij}$$

it follows that

$$\frac{1}{\sqrt{\kappa}} \frac{\partial p_i^1(T;\kappa)}{\partial \nu_{A_i}}\Bigg|_{\Gamma_{ij}} \to 0 \quad \text{strongly in } L^2(\Gamma_{ij}),$$

hence

$$\begin{aligned} \frac{1}{\sqrt{\kappa}} \tau_{ij}^1(\cdot\,;\kappa) &= \frac{1-\epsilon}{\sqrt{\kappa}} \left(-\frac{\partial p_j^1(T;\kappa)}{\partial \nu_{A_j}} + \alpha p_j^1(T;\kappa) \right)\Bigg|_{\Gamma_{ij}} \\ &+ \frac{\epsilon}{\sqrt{\kappa}} \left(\frac{\partial p_i^1(T;\kappa)}{\partial \nu_{A_i}} + \alpha p_i^1(T;\kappa) \right)\Bigg|_{\Gamma_{ij}} \to 0 \quad \text{strongly in } L^2(\Gamma_{ij}). \end{aligned}$$

We may now apply Theorem 8.5.2.1 to conclude that as $\kappa \to \infty$ the solution $w_i^2(\cdot\,;\kappa)$, $p_i^2(\cdot\,;\kappa)$ of (8.4.2.7), (8.4.2.8) with $n+1 = 2$ converges in the manner described in that Theorem to the solution $w_i^2(\cdot)$, $p_i^2(\cdot)$ of

$$\begin{gathered} \frac{\partial^2 w_i^2}{\partial t^2} - \nabla\cdot(A_i \nabla w_i) + c_i w_i^2 = 0 \quad \text{in } Q_i \\ \frac{\partial^2 p_i^2}{\partial t^2} - \nabla\cdot(A_i \nabla p_i) + c_i p_i^2 = 0 \quad \text{in } Q_i \\ \frac{\partial w_i^2}{\partial \nu_{A_i}} + \beta p_i^2 = \lambda_{ij}^1, \quad \frac{\partial p_i^2}{\partial \nu_{A_i}} - \beta w_i^2 = \mu_{ij}^1 \quad \text{on } \Sigma_{ij} \\ w_i^2 = p_i^2 = 0 \quad \text{on } \Sigma_i^D \\ \frac{\partial w_i^2}{\partial \nu_{A_i}} + p_i^2 = 0, \quad \frac{\partial p_i^2}{\partial \nu_{A_i}} = 0 \quad \text{on } \Sigma_i^N \\ w_i^2(0) = \frac{\partial w_i^2}{\partial t}(0) = 0, \quad p_i^2(T) = p_{0i}^2, \quad \frac{\partial p_i^2}{\partial t}(T) = p_{1i}^2 \\ w_i^2(T) = z_{0i}, \quad \frac{\partial w_i^2}{\partial t}(T) = z_{1i}, \end{gathered}$$

where λ^1_{ij}, μ^1_{ij} are given by (8.5.3.37) and where (p^2_{0i}, p^2_{1i}) is the solution of

$$\langle (p^2_{0i}, p^2_{1i}), (-z_{1i}, z_{0i}) \rangle_{H_i \times V_i} = \sum_{j \in \mathcal{J}_i} \int_{\Sigma_{ij}} [\beta(|p^2_i|^2 + |w^2_i|^2) d\Sigma - \lambda^1_{ij} p^2_i + \mu^1_{ij} w^2_i] d\Sigma + \int_{\Sigma^N_i} |p^2_i|^2 d\Sigma.$$

One may now proceed inductively to obtain the following result.

Theorem 8.5.3.1 *Suppose that the regularity requirement* (8.4.1.3) *is satisfied, that the observability estimate* (8.5.2.6) *holds, that* $T > T_0$*, and that* $(z_{0i}, z_{1i}) \in V_i \times H_i$*. Assume also that the starting data* $\lambda^0_{ij}(\cdot\,;\kappa)$, $\mu^0_{ij}(\cdot\,;\kappa)$, $\tau^0_{ij}(\cdot\,;\kappa)$ *satisfy* (8.5.3.36) *(which will be the case if they are chosen independent of* κ*, for example). Then, as* $\kappa \to \infty$*, for every* $n = 0, 1, \dots$ *the solution* $w^{n+1}_i(\cdot\,;\kappa)$, $p^{n+1}_i(\cdot\,;\kappa)$ *of* (8.4.2.7), (8.4.2.8), (8.5.3.35) *converges in the sense of Theorem* 8.5.2.1 *to the solution* $w^{n+1}_i(\cdot)$, $p^{n+1}_i(\cdot)$ *of*

$$\begin{gathered} \frac{\partial^2 w^{n+1}_i}{\partial t^2} - \nabla \cdot (A_i \nabla w^{n+1}_i) + c_i w^{n+1}_i = 0 \quad \text{in } Q_i \\ \frac{\partial w^{n+1}_i}{\partial \nu_{A_i}} + \beta p^{n+1}_i = \lambda^n_{ij} \quad \text{on } \Sigma_{ij} \\ w^{n+1}_i = 0 \quad \text{on } \Sigma^D_i, \quad \frac{\partial w^{n+1}_i}{\partial \nu_{A_i}} + p^{n+1}_i = 0 \quad \text{on } \Sigma^N_i \\ w^{n+1}_i(0) = \frac{\partial w^{n+1}_i}{\partial t}(0) = 0 \end{gathered} \tag{8.5.3.38}$$

$$\begin{gathered} \frac{\partial^2 p^{n+1}_i}{\partial t^2} - \nabla \cdot (A_i \nabla p^{n+1}_i) + c_i p^{n+1}_i = 0 \quad \text{in } Q_i \\ \frac{\partial p^{n+1}_i}{\partial \nu_{A_i}} - \beta w^{n+1}_i = \mu^n_{ij} \quad \text{on } \Sigma_{ij} \\ p^{n+1}_i = 0 \quad \text{on } \Sigma^D_i, \quad \frac{\partial p^{n+1}_i}{\partial \nu_{A_i}} = 0 \quad \text{on } \Sigma^N_i \\ p^{n+1}_i(T) = p^{n+1}_{0i}, \quad \frac{\partial p^{n+1}_i}{\partial t}(T) = p^{n+1}_{1i} \end{gathered} \tag{8.5.3.39}$$

$$w^{n+1}_i(T) = z_{0i}, \quad \frac{\partial w^{n+1}_i}{\partial t}(T) = z_{1i} \tag{8.5.3.40}$$

where, for $n \ge 1$,

$$\begin{aligned} \lambda^n_{ij} &= (1-\epsilon)\Big(-\frac{\partial w^n_j}{\partial \nu_{A_j}} + p^n_j\Big) + \epsilon\Big(\frac{\partial w^n_i}{\partial \nu_{A_i}} + p^n_i\Big)\Big|_{\Sigma_{ij}} \\ \mu^n_{ij} &= (1-\epsilon)\Big(-\frac{\partial p^n_j}{\partial \nu_{A_j}} - w^n_j\Big) + \epsilon\Big(\frac{\partial p^n_i}{\partial \nu_{A_i}} - w^n_i\Big)\Big|_{\Sigma_{ij}} \end{aligned} \tag{8.5.3.41}$$

and where, for $n \geq 0$, $(p_{0i}^{n+1}, p_{1i}^{n+1}) \in H_i \times V_i^*$ *is the solution of*

$$\langle (p_{0i}^{n+1}, p_{1i}^{n+1}), (-z_{1i}, z_{0i}) \rangle_{H_i \times V_i} = \int_{\Sigma_i^N} |p_i^{n+1}|^2 d\Sigma$$
$$+ \sum_{j \in \mathcal{J}_i} \int_{\Sigma_{ij}} [\beta(|p_i^{n+1}|^2 + |w_i^{n+1}|^2) d\Sigma - \lambda_{ij}^n p_i^{n+1} + \mu_{ij}^n w_i^{n+1}] d\Sigma. \quad (8.5.3.42)$$

Remark 8.5.3.1 Analogous to Remark 8.5.2.1, the system (8.5.3.39)–(8.5.3.42) is the optimality system for the optimal control problem

$$\inf_{(f_1, f_2) \in \mathcal{U}_{\text{ad}}^n} J_i^{n+1}(f_1, f_2)$$

$$J_i^{n+1}(f_1, f_2) = \frac{1}{2} \int_{\Sigma_i^N} |f_2|^2 d\Sigma + \frac{1}{2\beta} \sum_{j \in \mathcal{J}_i} \int_{\Sigma_{ij}} \left[|f_1|^2 + |\beta w_i^{n+1} + \mu_{ij}^n|^2 \right] d\Sigma$$

where $\mathcal{U}_{\text{ad}}^n$ consists of controls $f_1 \in L^2(\Sigma_i^{\text{int}})$, $f_2 \in L^2(\Sigma_i^N)$, such that the solution of

$$\frac{\partial^2 w_i^{n+1}}{\partial t^2} - \nabla \cdot (A_i \nabla w_i^{n+1}) + c_i w_i^{n+1} = 0 \text{ in } Q_i$$
$$w_i^{n+1} = 0 \text{ on } \Sigma_i^D, \quad \frac{\partial w_i^{n+1}}{\partial \nu_{A_i}} = f_2 \text{ on } \Sigma_i^N$$
$$\frac{\partial w_i^{n+1}}{\partial \nu_{A_i}} = f_1 + \lambda_{ij}^n \text{ on } \Sigma_{ij}, \ j \in \mathcal{J}_i$$
$$w_i^{n+1}(0) = \frac{\partial w_i^{n+1}}{\partial t}(0) = 0$$

satisfies the state constraint

$$w_i^{n+1}(T) = z_{0i}, \quad \frac{\partial w_i^{n+1}}{\partial t}(T) = z_{1i}.$$

8.5.4 Convergence to the Global Optimality System

In this section is it proved that the solutions (w_i^{n+1}, p_i^{n+1}), $i = 1, \ldots, I$, of the local optimality systems (8.5.3.38), (8.5.3.39), (8.5.3.41), (8.5.3.42), in aggregate converge to the solution (w, p) of the global optimality system (8.1.1.1), (8.5.1.6)–(8.5.1.8). Let (w_i, p_i) denote the restriction of (w, p) to Q_i.

Theorem 8.5.4.1 *Assume that* $(z_0, z_1) \in V \times H$, *that* $T > T_0$, *that the local and global observability estimates* (8.5.1.6) *and* (8.5.2.6) *hold, and that the regularity requirement* (8.4.1.3) *is satisfied. Then*

(i) *for any* $\epsilon \in [0, 1)$

$$p_i^n\big|_{\Sigma_i^N} \to 0 \quad \textit{strongly in } L^2(\Sigma_i^N).$$

(ii) *In addition, if* $\epsilon \in (0,1)$, *then*

$$(w_i^n, \frac{\partial w_i^n}{\partial t}) \to (w_i, \frac{\partial w_i^n}{\partial t}) \ \textit{weakly}^* \textit{ in } L^\infty(0,T; H_i \times V_i^*)$$
$$(p_i^n, \frac{\partial p_i^n}{\partial t}) \to (p_i, \frac{\partial p_i^n}{\partial t}) \ \textit{weakly}^* \textit{ in } L^\infty(0,T; H_i \times V_i^*)$$
$$w_i^n\big|_{\Sigma_i^N} \to w_i\big|_{\Sigma_i^N} \ \textit{weakly in } L^2(\Sigma_i^N)$$
$$w_i^n\big|_{\Sigma_i^{\text{int}}} \to w_i\big|_{\Sigma_i^{\text{int}}}, \quad \tilde{p}_i^n\big|_{\Sigma_i^{\text{int}}} \to p_i\big|_{\Sigma_i^{\text{int}}} \ \textit{weakly in } L^2(\Sigma_i^{\text{int}})$$
$$\frac{\partial w_i^n}{\partial \nu_{A_i}}\Big|_{\Sigma_i^{\text{int}}} \to \frac{\partial w_i}{\partial \nu_{A_i}}\Big|_{\Sigma_i^{\text{int}}}, \quad \frac{\partial p_i^n}{\partial \nu_{A_i}}\Big|_{\Sigma_i^{\text{int}}} \to \frac{\partial p_i}{\partial \nu_{A_i}}\Big|_{\Sigma_i^{\text{int}}} \ \textit{weakly in } L^2(\Sigma_i^{\text{int}})$$
$$w_i^n - w_j^n\big|_{\Sigma_{ij}} \to 0, \quad p_i^n - p_j^n\big|_{\Sigma_{ij}} \to 0 \ \textit{strongly in } L^2(\Sigma_{ij})$$
$$\frac{\partial w_i^n}{\partial \nu_{A_i}} + \frac{\partial w_j^n}{\partial \nu_{A_j}}\Big|_{\Sigma_{ij}} \to 0, \quad \frac{\partial p_i^n}{\partial \nu_{A_i}} + \frac{\partial p_j^n}{\partial \nu_{A_j}}\Big|_{\Sigma_{ij}} \to 0 \ \textit{strongly in } L^2(\Sigma_{ij})$$

Proof. Since the proof is similar to that of Theorem 8.4.3.1 it is only sketched here. We introduce the space

$$\mathcal{X} = \prod_{i=1}^{I} L^2(\Sigma_i^{\text{int}}) \times L^2(\Sigma_i^{\text{int}})$$

with norm defined by

$$\|X\|_{\mathcal{X}}^2 = \sum_{i=1}^{I} \frac{1}{\beta} \int_{\Sigma_i^{\text{int}}} (|\lambda_i|^2 + |\mu_i|^2)\, d\Sigma, \quad X = \{(\lambda_i, \mu_i) : i = 1, \ldots, I\}.$$

For $X \in \mathcal{X}$ let $(\tilde{w}_i, \tilde{p}_i)$ be the solution of

$$\begin{aligned}
&\frac{\partial^2 \tilde{w}_i}{\partial t^2} - \nabla \cdot (A_i \nabla \tilde{w}_i) + c_i \tilde{w}_i = 0 \ \text{ in } Q_i \\
&\tilde{w}_i = 0 \ \text{ on } \Sigma_i^D, \quad \frac{\partial \tilde{w}_i}{\partial \nu_{A_i}} = -\tilde{p}_i \ \text{ on } \Sigma_i^N \\
&\frac{\partial \tilde{w}_i}{\partial \nu_{A_i}} = -\beta \tilde{p}_i + \lambda_i \ \text{ on } \Sigma_i^{\text{int}} \\
&\tilde{w}_i(0) = \frac{\partial \tilde{w}_i}{\partial t}(0) = 0,
\end{aligned} \tag{8.5.4.1}$$

$$\begin{aligned}
&\frac{\partial^2 \tilde{p}_i}{\partial t^2} - \nabla \cdot (A_i \nabla \tilde{p}_i) + c_i \tilde{p}_i = 0 \ \text{ in } Q_i \\
&\tilde{p}_i = 0 \ \text{ on } \Sigma_i^D, \quad \frac{\partial \tilde{p}_i}{\partial \nu_{A_i}} = 0 \ \text{ on } \Sigma_i^N \\
&\frac{\partial \tilde{p}_i}{\partial \nu_{A_i}} = \beta \tilde{w}_i + \mu_i \ \text{ on } \Sigma_i^{\text{int}} \\
&\tilde{p}_i(T) = \tilde{p}_{0i}, \quad \frac{\partial \tilde{p}_i}{\partial t}(T) = \tilde{p}_{1i},
\end{aligned} \tag{8.5.4.2}$$

and where $(\tilde{p}_{0i}, \tilde{p}_{1i}) \in H_i \times V_i^*$ is chosen so that

$$\tilde{w}_i(T) = \frac{\partial \tilde{w}_i}{\partial t}(T) = 0 \tag{8.5.4.3}$$

(see Remark 8.5.2.2, especially (8.5.2.34)). The system (8.5.4.1), (8.5.4.2) is then an optimality system; see Remark 8.5.2.1. Define the bounded linear map $\mathcal{T} : \mathcal{X} \mapsto \mathcal{X}$ by

$$\mathcal{T}X = \{(-\frac{\partial \tilde{w}_j}{\partial \nu_{A_j}} + \beta \tilde{p}_j\big|_{\Sigma_{ij}}, -\frac{\partial \tilde{p}_j}{\partial \nu_{A_j}} - \beta \tilde{w}_j\big|_{\Sigma_{ij}}) : j \in \mathcal{J}_i, i = 1, \dots, I\}.$$

Then $\mathcal{T}X = X$ if and only if the global transmission conditions hold on Σ_{ij}. Further, $X = 0$ is the only fixed point of $\mathcal{T}$.

By a calculation similar to that in Lemma 8.4.3.1, for any $X \in \mathcal{X}$ we may write

$$\|X\|_{\mathcal{X}}^2 = \mathcal{E} + \mathcal{F}, \quad \|\mathcal{T}X\|_{\mathcal{X}}^2 = \mathcal{E} - \mathcal{F} \tag{8.5.4.4}$$

where

$$\mathcal{E} = \sum_{i=1}^{I} \int_{\Sigma_i^{\text{int}}} \left[\frac{1}{\beta} \left(\left| \frac{\partial \tilde{w}_i}{\partial \nu_{A_i}} \right|^2 + \left| \frac{\partial \tilde{p}_i}{\partial \nu_{A_i}} \right|^2 \right) + \beta \left(|\tilde{w}_i|^2 + |\tilde{p}_i|^2 \right) \right] d\Sigma$$

$$\mathcal{F} = 2 \sum_{i=1}^{I} \int_{\Sigma_i^N} |\tilde{p}_i|^2 d\Sigma.$$

For λ_{ij}^n, μ_{ij}^n given in $L^2(\Sigma_{ij})$, let w_i^{n+1}, p_i^{n+1} be the solution of the local optimality system (8.5.3.38), (8.5.3.39), with $(p_{0i}^{n+1}, p_{1i}^{n+1})$ given according to (8.5.3.42). Let w_i, p_i be the restriction to Q_i of the solution of the global optimality system (8.1.1.1), (8.5.1.7) with (p_0, p_1) given by (8.5.1.8) and f by (8.5.1.6). Then

$$w_i^{n+1}(T) = w_i(T) = z_{0i}, \quad \frac{\partial w_i^{n+1}}{\partial t}(T) = \frac{\partial w_i}{\partial t}(T) = z_{1i}.$$

Introduce the local errors

$$\tilde{w}_i^{n+1} = w_i^{n+1} - w_i, \quad \tilde{p}_i^{n+1} = p_i^{n+1} - p_i.$$

These satisfy

$$\begin{aligned} &\frac{\partial^2 \tilde{w}_i^{n+1}}{\partial t^2} - \nabla \cdot (A_i \nabla \tilde{w}_i^{n+1}) + c_i \tilde{w}_i^{n+1} = 0 \quad \text{in } Q_i \\ &\frac{\partial \tilde{w}_i^{n+1}}{\partial \nu_{A_i}} + \beta \tilde{p}_i^{n+1} = \tilde{\lambda}_{ij}^n \quad \text{on } \Sigma_{ij} \\ &\tilde{w}_i^{n+1} = 0 \ \text{ on } \Sigma_i^D, \quad \frac{\partial \tilde{w}_i^{n+1}}{\partial \nu_{A_i}} + \tilde{p}_i^{n+1} = 0 \ \text{ on } \Sigma_i^N \\ &\tilde{w}_i^{n+1}(0) = \frac{\partial \tilde{w}_i^{n+1}}{\partial t}(0) = 0 \end{aligned} \tag{8.5.4.5}$$

$$\begin{aligned}
&\frac{\partial^2 \tilde{p}_i^{n+1}}{\partial t^2} - \nabla\cdot(A_i \nabla \tilde{p}_i^{n+1}) + c_i \tilde{p}_i^{n+1} = 0 \quad \text{in } Q_i\\
&\frac{\partial \tilde{p}_i^{n+1}}{\partial \nu_{A_i}} - \beta \tilde{w}_i^{n+1} = \tilde{\mu}_{ij}^n \quad \text{on } \Sigma_{ij}\\
&\tilde{p}_i^{n+1} = 0 \ \text{ on } \Sigma_i^D, \quad \frac{\partial \tilde{p}_i^{n+1}}{\partial \nu_{A_i}} = 0 \ \text{ on } \Sigma_i^N\\
&\tilde{p}_i^{n+1}(T) = \tilde{p}_{0i}^{n+1}, \quad \frac{\partial \tilde{p}_i^{n+1}}{\partial t}(T) = \tilde{p}_{1i}^{n+1}
\end{aligned} \tag{8.5.4.6}$$

where

$$\begin{aligned}
\tilde{\lambda}_{ij}^n &= \lambda_{ij}^n - \Big(\frac{\partial w_i}{\partial \nu_{A_i}} + \beta p_i\Big)\\
\tilde{\mu}_{ij}^n &= \mu_{ij}^n - \Big(\frac{\partial p_i}{\partial \nu_{A_i}} - \beta w_i\Big)
\end{aligned} \tag{8.5.4.7}$$

and

$$\tilde{p}_{0i}^{n+1} = p_{0i}^{n+1} - p_{0i}, \quad \tilde{p}_{01}^{n+1} = p_{1i}^{n+1} - p_{1i}.$$

Further

$$\tilde{w}_i^{n+1}(T) = \frac{\partial \tilde{w}_i^{n+1}}{\partial t}(T) = 0.$$

By setting

$$X^n = \{(\tilde{\lambda}_{ij}^{n-1}, \tilde{\mu}_{ij}^{n-1}) : j \in \mathcal{J}_i,\ i = 1,\dots,I\}$$

and solving (8.5.4.5), (8.5.4.6) with n replaced by $n-1$, the iterative step $(8.5.4.5)_2$, $(8.5.4.6)_2$, with $\tilde{\lambda}_{ij}^n$, $\tilde{\mu}_{ij}^n$ given by (8.5.4.7) and λ_{ij}^n, μ_{ij}^n given by (8.5.3.41), may be expressed as the relaxed fixed point iteration

$$X^{n+1} = (1-\epsilon)\mathcal{T}X^n + \epsilon X^n. \tag{8.5.4.8}$$

From (8.5.4.4) we have

$$\|X^n\|_{\mathcal{X}}^2 = \mathcal{E}^n + \mathcal{F}^n, \quad \|\mathcal{T}X^n\|_{\mathcal{X}}^2 = \mathcal{E}^n - \mathcal{F}^n \tag{8.5.4.9}$$

with

$$\mathcal{E}^n = \sum_{i=1}^I \int_{\Sigma_i^{\text{int}}} \Big[\frac{1}{\beta}\Big(\Big|\frac{\partial \tilde{w}_i^{n+1}}{\partial \nu_{A_i}}\Big|^2 + \Big|\frac{\partial \tilde{p}_i^{n+1}}{\partial \nu_{A_i}}\Big|^2\Big) + \beta\big(|\tilde{w}_i^{n+1}|^2 + |\tilde{p}_i^{n+1}|^2\big)\Big] d\Sigma$$

$$\mathcal{F}^n = 2\sum_{i=1}^I \int_{\Sigma_i^N} |\tilde{p}_i^{n+1}|^2 d\Sigma.$$

By taking the square norm of each side of (8.5.4.8) and using (8.5.4.9), we obtain the fundamental recursion inequality

$$\mathcal{E}^{n+1} + \mathcal{F}^{n+1} \le \mathcal{E}^n - (1-2\epsilon)\mathcal{F}^n$$

and then, by iterating down to $n = 1$,

$$\mathcal{E}^{n+1} \le \mathcal{E}^1 - \sum_{p=1}^{n+1} c_p(\epsilon)\mathcal{F}^p$$

where $\mathbf{c}_p(\epsilon)$ is given by (8.4.3.21). It follows, in particular, that $\mathcal{F}^n \to 0$ for any $\epsilon \in [0,1)$, that is

$$\tilde{p}_i^n\big|_{\Sigma_i^N} \to 0 \quad \text{strongly in } L^2(\Sigma_i^N). \tag{8.5.4.10}$$

Now suppose that $\epsilon \in (0,1)$ Then, according to Proposition 2.3.7.3, $X^x \to X$ weakly in $\mathcal{X}$, where X is a fixed point of $\mathcal{T}$. Thus $X^n \to 0$ and $\mathcal{T}X^n \to 0$ weakly in $\mathcal{X}$, from which follows that

$$\begin{aligned} &\tilde{w}_i^n\big|_{\Sigma_i^{\text{int}}} \to 0, \quad \tilde{p}_i^n\big|_{\Sigma_i^{\text{int}}} \to 0 \text{ weakly in } L^2(\Sigma_i^{\text{int}}) \\ &\frac{\partial \tilde{w}_i^{n+1}}{\partial \nu_{A_i}} \to 0, \quad \frac{\partial \tilde{p}_i^{n+1}}{\partial \nu_{A_i}} \to 0 \text{ weakly in } L^2(\Sigma_i^{\text{int}}). \end{aligned} \tag{8.5.4.11}$$

Therefore

$$\tilde{\lambda}_{ij}^n \to 0, \quad \tilde{\mu}_{ij}^n \to 0 \text{ weakly in } L^2(\Sigma_{ij}),\ j \in \mathcal{J}_i,\ i = 1,\dots,I. \tag{8.5.4.12}$$

It follows from (8.5.4.5) and (8.5.4.10)–(8.5.4.12) that

$$\Big(\tilde{w}_i^n, \frac{\partial \tilde{w}_i^n}{\partial t}\Big) \to 0 \quad \text{weakly}^* \text{ in } L^\infty(0,T; H_i \times V_i^*).$$

From the observability estimate assumption (8.5.2.6) we have (cf. (8.5.2.27))

$$\begin{aligned} &\|\tilde{p}_{0i}^{n+1}\|_{H_i}^2 + \|\tilde{p}_{1i}^{n+1}\|_{V_i^*}^2 \\ &\quad \le C_T\Big(\sum_{j\in\mathcal{J}_i}\int_{\Sigma_{ij}} (\beta^2|\tilde{p}_i^{n+1}|^2 + |\beta\tilde{w}_i^{n+1} + \tilde{\mu}_{ij}^n|^2)\,d\Sigma + \int_{\Sigma_i^N} |\tilde{p}_i^{n+1}|^2 d\Sigma\Big). \end{aligned}$$

Therefore, on a sequence $n = n_k \to \infty$,

$$(\tilde{p}_{0i}^{n+1}, \tilde{p}_{1i}^{n+1}) \to (\tilde{p}_{0i}, \tilde{p}_{1i}) \quad \text{weakly}^* \text{ in } H_i \times V_i^* \tag{8.5.4.13}$$

for some $(\tilde{p}_{0i}, \tilde{p}_{1i}) \in H_i \times V_i^*$. It then follows from (8.5.4.6) and (8.5.4.11)–(8.5.4.13) that, on $n = n_k$,

$$\Big(\tilde{p}_i^{n+1}, \frac{\partial \tilde{p}_i^{n+1}}{\partial t}\Big) \to \Big(\tilde{p}_i, \frac{\partial \tilde{p}_i}{\partial t}\Big) \quad \text{weakly}^* \text{ in } L^\infty(0,T; H_i \times V_i^*),$$

where $\tilde{p}_i$ is the solution of

$$\begin{aligned} &\frac{\partial^2 \tilde{p}_i}{\partial t^2} - \nabla\cdot(A_i\nabla\tilde{p}_i) + c_i\tilde{p}_i = 0 \text{ in } Q_i \\ &\tilde{p}_i\big|_{\Sigma_i^D} = 0, \quad \frac{\partial \tilde{p}_i}{\partial \nu_{A_i}}\Big|_{\Sigma_i^N \cup \Sigma_i^{\text{int}}} = 0 \\ &\tilde{p}_i(T) = \tilde{p}_{0i}, \quad \frac{\partial \tilde{p}_i}{\partial t}(T) = \tilde{p}_{1i}. \end{aligned}$$

In addition, from (8.5.4.1) and (8.5.4.11) we have

$$\tilde{p}_i\big|_{\Sigma_i^N \cup \Sigma_i^{\text{int}}} = 0.$$

It then follows from the observability estimate that $\tilde{p}_{0i} = \tilde{p}_{1i} = 0$, hence

$$(\tilde{p}_i^n, \frac{\partial \tilde{p}_i^n}{\partial t}) \to 0 \quad \text{weakly}^* \text{ in } L^\infty(0,T; H_i \times V_i^*).$$

To complete the proof of Theorem 8.5.4.1, we invoke Proposition 2.3.7.1 to conclude that

$$\|X^{n+1} - X^n\|_{\mathcal{X}} = (1-\epsilon)\|\mathcal{T}X^n - X^n\|_{\mathcal{X}} \to 0.$$

From (8.5.4.9) we have

$$\|\mathcal{T}X^n - X^n\|_{\mathcal{X}}^2 = 2\big(\mathcal{E}^n - (\mathcal{T}X^n, X^n)_{\mathcal{X}}\big).$$

A calculation completely analogous to one carried out at the end of the proof of Theorem 8.4.3.1 gives

$$\mathcal{E}^n - (\mathcal{T}X^n, X^n)_{\mathcal{X}} = \sum_{i=1}^{I-1} \sum_{\substack{j \in \mathcal{J}_i \\ j>i}} \Big\{ \frac{1}{\beta} \int_{\Sigma_{ij}} \Big[\Big| \frac{\partial \tilde{w}_i^n}{\partial \nu_{A_i}} + \frac{\partial \tilde{w}_j^n}{\partial \nu_{A_j}} \Big|^2 + \Big| \frac{\partial \tilde{p}_i^n}{\partial \nu_{A_i}} + \frac{\partial \tilde{p}_j^n}{\partial \nu_{A_j}} \Big|^2$$

$$+ \beta^2 \big(|\tilde{w}_i^n - \tilde{w}_j^n|^2 + |\tilde{p}_i^n - \tilde{p}_j^n|^2 \big) \Big] \, d\Sigma.$$

In this expression we may replace $\tilde{w}_i^n$, $\tilde{w}_j^n$ and $\tilde{p}_i^n$, $\tilde{p}_j^n$ by w_i^n, w_j^n and p_i^n, p_j^n, respectively, since

$$\tilde{w}_i^n - \tilde{w}_j^n = w_i^n - w_j^n, \quad \frac{\partial \tilde{w}_i^n}{\partial \nu_{A_i}} + \frac{\partial \tilde{w}_j^n}{\partial \nu_{A_j}} = \frac{\partial w_i^n}{\partial \nu_{A_i}} + \frac{\partial w_j^n}{\partial \nu_{A_j}}$$

$$\tilde{p}_i^n - \tilde{p}_j^n = p_i^n - p_j^n, \quad \frac{\partial \tilde{p}_i^n}{\partial \nu_{A_i}} + \frac{\partial \tilde{p}_j^n}{\partial \nu_{A_j}} = \frac{\partial p_i^n}{\partial \nu_{A_i}} + \frac{\partial p_j^n}{\partial \nu_{A_j}}.$$

The proof of Theorem 8.5.4.1 is now complete.

Chapter 9

Domain Decomposition for Distributed Parameter Systems on 2-D Networks

9.1 Elliptic Systems on 2-D Networks

This chapter is mainly concerned with domain decomposition in optimal final value control of hyperbolic systems on two-dimensional polygonal networks $\mathcal{P}$ in $\mathbb{R}^N$. However, in order to fix the ideas in the context of a somewhat simpler problem we shall begin by considering a domain decomposition procedure for the calculation of the solutions of elliptic systems on such networks.

A *two-dimensional polygonal network* $\mathcal{P}$ in $\mathbb{R}^N$ is a finite union of nonempty subsets $\mathcal{P}_i$, $i \in \mathcal{I}$, such that

(i) each $\mathcal{P}_i$ is a simply connected open polygonal subset of a plane Π_i in $\mathbb{R}^N$;
(ii) $\bigcup_{i\in\mathcal{I}} \overline{\mathcal{P}}_i$ is connected;
(iii) for all $i, j \in \mathcal{I}$, $\overline{\mathcal{P}}_i \bigcap \overline{\mathcal{P}}_j$ is either empty, a common vertex, or a whole common side.

The reader is referred to S. Nicaise [86], whose notation we adopt, for more details about such 2-d networks.

For each $i \in \mathcal{I}$ we fix once and for all a system of coordinates in Π_i. Thus

$$\Pi_i = \{p_{i0} + x_1\eta_{i1} + x_2\eta_{i2}\}, \quad x_1, x_2 \in \mathbb{R},$$

where p_{i0} is the origin of coordinates in Π_i and η_{i1}, η_{i2} are orthonormal vectors that span Π_i. We assume that the boundary $\partial\mathcal{P}_i$ of $\mathcal{P}_i$ is the union of a finite number of linear segments $\overline{\Gamma}_{ij}$, $j = 1, \dots, N_i$. It is convenient to assume that Γ_{ij} is open in $\partial\mathcal{P}_i$. The collection of all Γ_{ij} are the *edges* of $\mathcal{P}$ and will be denoted by $\mathcal{E}$. An edge Γ_{ij} corresponding to an $e \in \mathcal{E}$ will be denoted by Γ_{ie} and the *index set* $\mathcal{I}_e$ of e is $\mathcal{I}_e = \{i |\, e = \Gamma_{ie}\}$. The *degree* of an edge is the cardinality of $\mathcal{I}_e$ and is denoted by $d(e)$. For each $i \in \mathcal{I}_e$ we will denote by ν_{ie} the unit outer normal to $\mathcal{P}_i$ along Γ_{ie}. The coordinates of ν_{ie} in the given coordinate system of $\mathcal{P}_i$ are denoted by (ν_{ie}^1, ν_{ie}^2).

We partition the edges of $\mathcal{E}$ into two disjoint subsets $\mathcal{D}$ and $\mathcal{N}$, corresponding respectively to edges along which Dirichlet conditions hold and along which

Neumann or transmission conditions hold. The Dirichlet edges are assumed to be *exterior edges*, that is, edges for which $d(e) = 1$. The Neumann edges consist of exterior edges $\mathcal{N}^{\text{ext}}$ and *interior edges* $\mathcal{N}^{\text{int}} := \mathcal{N} \backslash \mathcal{N}^{\text{ext}}$.

Let $m \geq 1$ be a given integer. For a function $W : \mathcal{P} \mapsto \mathbb{R}^m$, W_i will denote the restriction of W to $\mathcal{P}_i$, that is

$$W_i : \mathcal{P}_i \mapsto \mathbb{R}^m : x \mapsto W(x).$$

We introduce real $m \times m$ matrices

$$A_i^{\alpha\beta},\ B_i^{\beta},\ C_i, \quad i \in \mathcal{I},\ \alpha, \beta = 1, 2,$$

where

$$A_i^{\alpha\beta} = (A_i^{\beta\alpha})^*, \quad C_i = C_i^*$$

and where the * superscript denotes transpose. For sufficiently regular $W, \Phi : \mathcal{P} \mapsto \mathbb{R}^m$ we define the symmetric bilinear form

$$a(W, \Phi) = \sum_{i \in \mathcal{I}} \int_{\mathcal{P}_i} [A_i^{\alpha\beta}(W_{i,\beta} + B_i^{\beta} W_i) \cdot (\Phi_{i,\alpha} + B_i^{\alpha} \Phi_i) + C_i W_i \cdot \Phi_i]\, dx, \quad (9.1.1.1)$$

where repeated lower case Greek indices are summed over 1,2. A subscript following a comma indicates differentiation with respect to the corresponding variable, e.g., $W_{i,\beta} = \partial W_i / \partial x_\beta$. The matrices $A_i^{\alpha\beta}$, B_i^{β}, C_i may depend on $(x_1, x_2) \in \mathcal{P}_i$ and $a(W, \Phi)$ is required to be $\mathcal{V}$-elliptic for an appropriate function space $\mathcal{V}$ specified below.

We shall consider the variational problem

$$a(W, \Phi) = \langle F, \Phi \rangle_{\mathcal{V}'}, \quad \forall \Phi \in \mathcal{V},\ 0 < t < T, \quad (9.1.1.2)$$

where $\mathcal{V}$ is a certain space of test functions and F is a given, sufficiently regular function. The variational equation (9.1.1.2) obviously implies, in particular, that the W_i, $i \in \mathcal{I}$, formally satisfy the system of equations

$$-\frac{\partial}{\partial x_\alpha}[A_i^{\alpha\beta}(W_{i,\beta} + B_i^{\beta} W_i)] + (B_i^{\alpha})^* A_i^{\alpha\beta}(W_{i,\beta} + B_i^{\beta} W_i)$$
$$+ C_i W_i = F_i \ \text{ in } \mathcal{P}_i,\ i \in \mathcal{I}. \quad (9.1.1.3)$$

To determine the space $\mathcal{V}$, we need to specify the conditions satisfied by W along the edges of $\mathcal{P}$. These conditions are of two types: *geometric edge conditions*, and *mechanical edge conditions*. As usual, the space $\mathcal{V}$ is then defined in terms of the geometric edge conditions. At a Dirichlet edge we set

$$W_i = 0 \ \text{ on } e \text{ when } \Gamma_{ie} \in \mathcal{D}. \quad (9.1.1.4)$$

Further, along each $e \in \mathcal{N}^{\text{int}}$ we impose the condition

$$Q_{ie} W_i = Q_{je} W_j \ \text{ on } e \text{ when } \Gamma_{ie} = \Gamma_{je},\ e \in \mathcal{N}^{\text{int}}, \quad (9.1.1.5)$$

where, for each $i \in \mathcal{I}_e$, Q_{ie} is a real, nontrivial $p_e \times m$ matrix of rank $p_e \leq m$ with p_e independent of $i \in \mathcal{I}_e$. If $p_e < m$ additional conditions *may* be imposed, such as

$$\Pi_{ie} W_i = 0 \text{ on } e, \forall i \in \mathcal{I}_e, \, e \in \mathcal{N}^{\text{int}}, \tag{9.1.1.6}$$

where Π_{ie} is the orthogonal projection onto the kernel of Q_{ie}. (Note that (9.1.1.5) is a condition on only the components $\Pi_{ie}^{\perp} W_i$, $i \in \mathcal{I}_e$, where $\Pi_{ie}^{\perp}$ is the orthogonal projection onto the orthogonal complement in $\mathbb{R}^m$ of the kernel of Q_{ie}.) For definiteness we always assume that (9.1.1.6) is imposed and leave to the reader the minor modifications that occur in the opposite case. Thus the geometric edge conditions are taken to be (9.1.1.4)–(9.1.1.6), and the space $\mathcal{V}$ of test functions consists of sufficiently regular functions $\Phi : \mathcal{P} \mapsto \mathbb{R}^m$ that satisfy the geometric edge conditions.

Formal integration by parts in (9.1.1.2) shows that, in addition to (9.1.1.3), W_i must satisfy the variational boundary condition

$$\sum_{e \in \mathcal{N}} \sum_{i \in \mathcal{I}_e} \int_{\Gamma_{ie}} \nu_{ie}^{\alpha} A_i^{\alpha\beta} (W_{i,\beta} + B_i^{\beta} W_i) \cdot \Phi_i \, d\Gamma = 0, \quad \forall \Phi \in \mathcal{V}.$$

It follows in particular that

$$\nu_{ie}^{\alpha} A_i^{\alpha\beta} (W_{i,\beta} + B_i^{\beta} W_i) = 0 \text{ on } e \text{ when } \Gamma_{ie} \in \mathcal{N}^{\text{ext}} \tag{9.1.1.7}$$

and that

$$\sum_{e \in \mathcal{N}^{\text{int}}} \sum_{i \in \mathcal{I}_e} \int_{\Gamma_{ie}} \nu_{ie}^{\alpha} A_i^{\alpha\beta} (W_{i,\beta} + B_i^{\beta} W_i) \cdot \Phi_i \, d\Gamma = 0, \quad \forall \Phi \in \mathcal{V}. \tag{9.1.1.8}$$

For each $\Gamma_{ie} \in \mathcal{N}^{\text{int}}$ write $\Phi_i = \Pi_{ie} \Phi_i + \Pi_{ie}^{\perp} \Phi_i$, and let Q_{ie}^{+} denote the generalized inverse of Q_{ie}, that is Q_{ie}^{+} is a $m \times p_e$ matrix such that

$$Q_{ie} Q_{ie}^{+} = I_{p_e}, \quad Q_{ie}^{+} Q_{ie} = \Pi_{ie}^{\perp}.$$

Then (9.1.1.8) may be written

$$\sum_{e \in \mathcal{N}^{\text{int}}} \sum_{i \in \mathcal{I}_e} \int_{\Gamma_{ie}} (Q_{ie}^{+})^* \nu_{ie}^{\alpha} A_i^{\alpha\beta} (W_{i,\beta} + B_i^{\beta} W_i) \cdot Q_{je} \Phi_j \, d\Gamma = 0, \quad \forall \Phi \in \mathcal{V},$$

where j is any index selected from $\mathcal{I}_e$. We deduce that

$$\sum_{i \in \mathcal{I}_e} (Q_{ie}^{+})^* \nu_{ie}^{\alpha} A_i^{\alpha\beta} (W_{i,\beta} + B_i^{\beta} W_i) = 0 \text{ on } e \text{ if } e \in \mathcal{N}^{\text{int}}. \tag{9.1.1.9}$$

Conditions (9.1.1.7) and (9.1.1.9) are called the *mechanical edge conditions.* We also refer to (9.1.1.5) and (9.1.1.9) as the geometric and mechanical *transmission conditions*, respectively..

To summarize, the edge conditions are comprised of the geometric edge conditions (9.1.1.4)–(9.1.1.6), and the mechanical edge conditions (9.1.1.7), (9.1.1.9). The geometric transmission conditions are (9.1.1.5) and (9.1.1.6), while the mechanical transmission conditions are given by (9.1.1.9).

9.1.2 Examples

Example 9.1.2.1 Suppose that $m = 1$. In this case the matrices $A_i^{\alpha\beta}$, B_i^{α}, C_i, Q_{ie} reduce to scalars $a_i^{\alpha\beta}$, b_i^{α}, c_i, q_{ie}, where $a_i^{\alpha\beta} = a_i^{\beta\alpha}$. Set

$$A_i = \left(a_i^{\alpha\beta}\right), \quad b_i = \mathrm{col}\left(b_i^{\alpha}\right).$$

The system (9.1.1.3) takes the form

$$-\nabla\cdot(A_i\nabla W_i) + [-\nabla\cdot(A_i b_i) + b_i^* A_i b_i + c_i]W_i = F_i. \tag{9.1.2.1}$$

Suppose that all $q_{ie} = 1$. The geometric edge conditions (9.1.1.4), (9.1.1.5) are then

$$W_i = 0 \text{ on } e \text{ when } e \in \mathcal{D}$$
$$W_i = W_j \text{ on } e \text{ when } e \in \mathcal{N}^{\text{int}}$$

while the mechanical edge conditions are

$$\sum_{i\in\mathcal{I}_e}[\nu_{ie}\cdot(A_i\nabla W_i) + (\nu_{ie}\cdot A_i b_i)W_i] = 0 \text{ on } e \text{ when } e \in \mathcal{N}.$$

Example 9.1.2.2 *(Membrane networks in $\mathbb{R}^3$.)* In this case $m = N = 3$. For each $i \in \mathcal{I}$ set $\eta_{i3} = \eta_{i1}\wedge\eta_{i2}$, where η_{i1}, η_{i2} are the unit coordinate vectors in Π_i. Suppose that $B_i = C_i = 0$, $Q_{ie} = I_3$, where I_3 denotes the identity matrix with respect to the $\{\eta_{ik}\}_{k=1}^3$ basis. With respect to this basis the matrices $A_i^{\alpha\beta}$ are given by

$$A_i^{11} = \begin{pmatrix} 2\mu_i+\lambda_i & 0 & 0\\ 0 & \mu_i & 0\\ 0 & 0 & \mu_i\end{pmatrix}, \quad A_i^{22} = \begin{pmatrix} \mu_i & 0 & 0\\ 0 & 2\mu_i+\lambda_i & 0\\ 0 & 0 & \mu_i\end{pmatrix},$$
$$A_i^{12} = \begin{pmatrix} 0 & \lambda_i & 0\\ \mu_i & 0 & 0\\ 0 & 0 & 0\end{pmatrix}, \quad A_i^{21} = \begin{pmatrix} 0 & \mu_i & 0\\ \lambda_i & 0 & 0\\ 0 & 0 & 0\end{pmatrix}$$

Write

$$W_i = \sum_{k=1}^{3} W_{ik}\eta_{ik}, \quad w_i = W_{i\alpha}\eta_{i\alpha},$$
$$\varepsilon_{\alpha\beta}(w_i) = \frac{1}{2}(W_{i\alpha,\beta} + W_{i\beta,\alpha}),$$
$$\sigma_i^{\alpha\beta}(w_i) = 2\mu_i\varepsilon_{\alpha\beta}(w_i) + \lambda_i\varepsilon_{\gamma\gamma}(w_i)\delta^{\alpha\beta}.$$

The bilinear form (9.1.1.1) may be written

$$a(W,\Phi) = \sum_{i\in\mathcal{I}}\int_{\mathcal{P}_i}[\sigma_i^{\alpha\beta}(w_i)\varepsilon_{\alpha\beta}(\phi_i) + \mu_i W_{i3,\alpha}\Phi_{i3,\alpha}]\,dx$$

where $\Phi_i = \sum_{k=1}^3 \Phi_{ik}\eta_{ik} := \phi_i + \Phi_{i3}\eta_{i3}$. The corresponding system (9.1.1.3) is

$$-\sigma_{i,\alpha}^{\alpha\beta}(w_i) = F_{i\beta},\ \beta = 1, 2, \quad -\mu_i W_{i3,\alpha\alpha} = F_{i3} \quad \text{in } \mathcal{P}_i,\ i \in \mathcal{I} \tag{9.1.2.2}$$

where $F_i = \sum_{k=1}^3 F_{ik}\eta_{ik}$. The geometric edge conditions (9.1.1.4), (9.1.1.5) are

$$\begin{aligned} W_i &= 0 \quad \text{on } e \text{ when } \Gamma_{ie} \in \mathcal{D}, \\ W_i &= W_j \quad \text{on } e \text{ when } \Gamma_{ie} = \Gamma_{je}. \end{aligned} \tag{9.1.2.3}$$

The mechanical edge conditions (9.1.1.9) are

$$\sum_{i \in \mathcal{I}_e} [\sigma_i^{\alpha\beta}(w_i)\eta_{i\beta} + \mu_i W_{i3,\alpha}\eta_{i3}]\nu_{ie}^{\alpha} = 0 \text{ on } e \text{ when } e \in \mathcal{N}. \tag{9.1.2.4}$$

The system (9.1.2.2)–(9.1.2.4) models the small, static deformation of a network of homogeneous isotropic membranes $\{\mathcal{P}_i\}_{i\in\mathcal{I}}$ in $\mathbb{R}^3$ of uniform density one and Lamé parameters λ_i and μ_i under distributed loads F_i, $i \in \mathcal{I}$; $W_i(x_1, x_2)$ represents the displacement of the material particle situated at $(x_1, x_2) \in \mathcal{P}_i$ in the reference configuration; and the transmission conditions have the interpretation of continuity of displacements and balance of forces at a junction. The reader is referred to [61] where this model is introduced and analyzed.

Example 9.1.2.3 *(Interface problems for serial Reissner-Mindlin plates.)* We take $N = m = 3$ and assume that the regions $\mathcal{P}_i$ are coplanar. The region $\cup_{i\in\mathcal{I}}\overline{\mathcal{P}}_i$ represents the middle surface of a thin plate of uniform thickness h. Within the region $\mathcal{P}_i \times (-h/2, h/2)$ the plate is assumed homogeneous and elastically isotropic with Lamé parameters λ_i, μ_i and shear modulus K_i. Let η_1, η_2 be unit coordinate vectors in the plane Π containing the regions $\mathcal{P}_i$, and set $\xi_1 = \eta_2$, $\xi_2 = -\eta_1$ and $\xi_3 = \xi_1 \wedge \xi_2$. The matrix $C_i = 0$ and, with respect to the $\{\xi_k\}_{k=1}^3$ basis,

$$\begin{aligned} A_i^{11} &= h \operatorname{diag}\left(\frac{h^2}{12}(2\mu_i + \sigma_i), \frac{h^2}{12}\mu_i, K_i\right), \\ A_i^{22} &= h \operatorname{diag}\left(\frac{h^2}{12}\mu_i, \frac{h^2}{12}(2\mu_i + \sigma_i), K_i\right) \end{aligned}$$

where $\sigma_i = 2\mu_i\lambda_i/(2\mu_i + \lambda_i)$,

$$A_i^{12} = \frac{h^3}{12}\begin{pmatrix} 0 & \sigma_i & 0 \\ \mu_i & 0 & 0 \\ 0 & 0 & 0 \end{pmatrix}, \quad A_i^{21} = \frac{h^3}{12}\begin{pmatrix} 0 & \mu_i & 0 \\ \sigma_i & 0 & 0 \\ 0 & 0 & 0 \end{pmatrix}$$

$$B_i^1 = \begin{pmatrix} 0 & 0 & 0 \\ 0 & 0 & 0 \\ 1 & 0 & 0 \end{pmatrix}, \quad B_i^2 = \begin{pmatrix} 0 & 0 & 0 \\ 0 & 0 & 0 \\ 0 & 1 & 0 \end{pmatrix}$$

Write $W_i = \sum_{k=1}^{3} W_{ik}\xi_k := w_i + W_{i3}\xi_3$, and set

$$\sigma_i^{\alpha\beta}(w_i) = \frac{h^3}{12}(2\mu_i\varepsilon_{\alpha\beta}(w_i) + \sigma_i\varepsilon_{\gamma\gamma}(w_i)\delta^{\alpha\beta}). \tag{9.1.2.5}$$

With this notation the bilinear form (9.1.1.1) is

$$a(W,\Phi) = \sum_{i\in\mathcal{I}} \int_{\mathcal{P}_i} [\sigma_i^{\alpha\beta}(w_i)\varepsilon_{\alpha\beta}(\phi_i) + K_i h(W_{i3,\alpha} + W_{i\alpha})(\Phi_{i3,\alpha} + \Phi_{i\alpha})]\, dx$$

where $\Phi_i = \sum_{k=1}^{2} \Phi_{ik}\xi_k := \phi_i + \Phi_{i3}\xi_3$. The system (9.1.1.3) takes the form

$$\begin{aligned} -\sigma_{i,\alpha}^{\alpha\beta}(w_i) + hK_i(W_{i3,\beta} + W_{i\beta}) &= M_{i\beta},\ \beta = 1,2, \\ -hK_i(W_{i3,\beta} + W_{i\beta})_{,\beta} &= F_{i3} \quad \text{in } \mathcal{P}_i,\ i \in \mathcal{I}. \end{aligned} \tag{9.1.2.6}$$

For fixed $i \in \mathcal{I}$ the system (9.1.2.6) is the Reissner-Mindlin model for the static deformation of a thin elastic plate under distributed loads

$$G_i(x_1,x_2,x_3) = \sum_{k=1}^{3} G_{ik}(x_1,x_2,x_3)\xi_k, \quad (x_1,x_2) \in \mathcal{P}_i,\ -h/2 \le x_3 \le h/2$$

(see, for example, [93] or [61]). In (9.1.2.6)

$$F_{i3}(x_1,x_2) = \int_{-h/2}^{h/2} G_{i3}(x_1,x_2,x_3)\, dx_3,$$

$$M_{i\beta}(x_1,x_2,t) = \int_{-h/2}^{h/2} x_3 G_{i\beta}(x_1,x_2,x_3)\, dx_3.$$

In particular, if G_i does not vary through the thickness of the plate, then $F_{i3} = hG_{i3}$ and $M_{i\beta} = 0$, $\beta = 1,2$. The function $W_{i3}(x_1,x_2)$ represents the transverse displacement of the material particle situated at $(x_1,x_2) \in \mathcal{P}_i$ in the reference configuration, and W_{i1}, W_{i2} measure transverse shearing. With the choice $Q_{ie} = I_3$ the geometric boundary and transmission conditions are

$$\begin{aligned} W_i &= 0 \quad \text{on } e \text{ when } \Gamma_{ie} \in \mathcal{D}, \\ W_{ik} &= W_{jk} \quad \text{on } e \text{ when } \Gamma_{ie} = \Gamma_{je},\ k = 1,2,3, \end{aligned}$$

(the degree $d_e = 2$ for each interior edge) while the mechanical edge conditions are

$$\sum_{i\in\mathcal{I}_e} \nu_{ie}^{\alpha}\sigma_i^{\alpha\beta}(w_i) = 0, \quad \beta = 1,2,$$

$$\sum_{i\in\mathcal{I}_e} K_i\nu_{ie}^{\alpha}(W_{i3,\alpha} + W_{i\alpha}) = 0 \text{ on } e \text{ when } e \in \mathcal{N}.$$

These have the interpretations of balance of angular and linear momenta, respectively.

Example 9.1.2.4 *(Interface problems for non coplanar thin plates.)* We again take $m = N = 3$. Let $\mathbf{i}, \mathbf{j}, \mathbf{k}$ denote the standard basis for $\mathbb{R}^3$ and (y_1, y_2, y_3) denote the Cartesian coordinates of a point in $\mathbb{R}^3$ with respect to that basis. We consider a 2-d network in $\mathbb{R}^3$ of consisting of planar regions $\mathcal{P}_i$, $i = 1, \dots, n$, such that

(i) $\mathcal{P}_i$ is a unit square;

(ii) $\overline{\mathcal{P}}_i \cap \overline{\mathcal{P}}_j = \{(0, 0, y_3) | \, |y_3| \le 1/2\} := e$, for all $i \neq j$.

If $n = 2$ it is further assumed that $\mathcal{P}_1$ and $\mathcal{P}_2$ are not coplanar. Let ν_{ie} denote the exterior unit normal to $\partial\mathcal{P}_i$ along e and choose $\eta_{i1} = -\nu_{ie}$, $\eta_{i2} = \mathbf{k}$ as unit coordinate vectors in $\mathcal{P}_i$. One may represent $\mathcal{P}_i$ as

$$\mathcal{P}_i = x_1\eta_{i1} + (x_2 - 1/2)\eta_{i2}, \quad 0 < x_1, x_2 < 1.$$

The network $\{\mathcal{P}_i\}_{i\in\mathcal{I}}$ has e as its only interior edge and the degree of this edge is n. The coordinates of e with respect to η_{i1}, η_{i2} are $(0, x_2)$, $0 < x_2 < 1$.

Set $\xi_{i1} = \eta_{i2}$, $\xi_{i2} = -\eta_{i1}$, $\xi_{i3} = \xi_{i1} \wedge \xi_{i2}$ and $W_i = \sum_{k=1}^{3} W_{ik}\xi_{ik} := w_i + W_{i3}\xi_{i3}$. Let $\omega = [0,1] \times [0,1]$. We consider the following problem on the network (see [51] and Remark 9.1.2.1 below):

$$\begin{aligned} -\sigma_{i,\alpha}^{\alpha\beta}(w_i) + hK_i(W_{i3,\beta} + W_{i\beta}) &= M_{i\beta}, \ \beta = 1, 2, \\ -hK_i(W_{i3,\beta} + W_{i\beta})_{,\beta} &= F_{i3}, \quad (x_1, x_2) \in \omega, \end{aligned} \tag{9.1.2.7}$$

where $\sigma_i^{\alpha\beta}(w_i)$ is given by (9.1.2.5),

$$\begin{aligned} \nu_{ie}^{\alpha}\sigma_i^{\alpha\beta}(w_i) &= 0, \quad \beta = 1, 2, \\ K_i\nu_{ie}^{\alpha}(W_{i3,\alpha} + W_{i\alpha}) &= 0, \quad (x_1, x_2) \in \partial\omega\backslash e. \end{aligned} \tag{9.1.2.8}$$

As in the Example 9.1.2.3, equations (9.1.2.7) and (9.1.2.8) represent the small static deformations of a network of Reissner-Mindlin plates whose outer edges are free (with respect to transverse displacement and shear), with $\mathcal{P}_i$ representing the mid-surface of the ith plate. In addition, there are junction conditions that must hold on the interior edge e. The geometric junction conditions are taken to be (see Remark 9.1.2.1 below)

$$\begin{aligned} W_{i1}(0, x_2) = W_{j1}(0, x_2), \quad & \forall i, j \in \mathcal{I}, \ 0 < x_2 < 1 \\ W_{i2}(0, x_2) = -k_i'\vartheta, \quad & W_{i3}(0, x_2) = k_i'[(x_2 - 1/2)\vartheta + b], \end{aligned} \tag{9.1.2.9}$$

where k_i' is a constant depending on the angle between $\mathcal{P}_1$ and $\mathcal{P}_i$ and ϑ, b are constants that are among the unknowns of the problem. The mechanical junction conditions on e are

$$\begin{aligned}
&\sum_{i=1}^{n} \frac{\mu_i(\mu_i+\lambda_i)}{2\mu_i+\lambda_i} W_{i1,1}(0,x_2) = 0,\\
&\sum_{i=1}^{n} k_i' K_i h \int_e (W_{i1}+W_{i3,1})\,dx_2 + \sum_{i=2}^{n} k_i \int_\omega F_{i1}\,d\omega = 0,\\
&\sum_{i=1}^{n} k_i' h \int_e \Big[-\frac{\mu_i h^2}{12}(W_{i1,2}+W_{i2,1}) + K_i(W_{i1}+W_{i3,1})(x_2-1/2)\Big] dx_2\\
&\qquad + \sum_{i=2}^{n} k_i \int_\omega [F_{i1}(x_2-1/2) - F_{i2}x_1]\,d\omega = 0.
\end{aligned} \tag{9.1.2.10}$$

where k_i is a constant that also depends on the angle between $\mathcal{P}_1$ and $\mathcal{P}_i$ and

$$F_{i\beta}(x_1,x_2) = \int_{-h/2}^{h/2} G_{i\beta}(x_1,x_2,x_3)\,dx_3.$$

Let us remark that solutions of (9.1.2.7)–(9.1.2.10) are unique only up to rigid motions of the form

$$W_{i1} = -a, \quad W_{i2} = -k_i'\theta, \quad W_{i3} = ax_1 + k_i'[(x_2-1/2)\theta + c],$$

where a, c, θ are arbitrary constants. Further, the system has a solution if and only if the applied load satisfies

$$\begin{aligned}
&\sum_{i=1}^{n} \int_\omega (x_1 F_{i3} - M_{i1})\,d\omega = 0,\\
&\sum_{i=1}^{n} k_i' \int_\omega F_{i3}\,d\omega + \sum_{i=2}^{n} k_i \int_\omega F_{i1}\,d\omega = 0,\\
&\sum_{i=1}^{n} k_i' \int_\omega [(x_2-1/2)F_{i3} - M_{i2}]\,d\omega\\
&\qquad + \sum_{i=2}^{n} k_i \int_\omega [(x_2-1/2)F_{i1} - x_1 F_{i2}]\,d\omega = 0.
\end{aligned} \tag{9.1.2.11}$$

Suppose that $\{W_i\}_{i=1}^n, \vartheta, b$ is a solution of (9.1.2.7)–(9.1.2.10) and let us set

$$\begin{aligned}
U_{i1} &= W_{i1}, \quad U_{i2} = W_{i2} + k_i'\vartheta,\\
U_{i3} &= W_{i3} - k_i'[(x_2-1/2)\vartheta + b].
\end{aligned}$$

Then it is seen that $\{U_i\}_{i=1}^n$ is a solution of

$$\begin{aligned}
&-\sigma_{i,\alpha}^{\alpha\beta}(u_i) + hK_i(U_{i3,\beta}+U_{i\beta}) = M_{i\beta}, \ \beta = 1,2,\\
&-hK_i(U_{i3,\beta}+U_{i\beta})_{,\beta} = F_{i3}, \quad (x_1,x_2)\in\omega,
\end{aligned} \tag{9.1.2.12}$$

where $u_i = U_{i\alpha}\xi_{i\alpha}$,

$$\begin{aligned} \nu_{ie}^{\alpha}\sigma_i^{\alpha\beta}(u_i) &= 0, \quad \beta = 1,2, \\ K_i\nu_{ie}^{\alpha}(U_{i3,\alpha} + U_{i\alpha}) &= 0, \quad (x_1,x_2) \in \partial\omega\backslash e, \end{aligned} \tag{9.1.2.13}$$

$$\begin{aligned} &U_{i1}(0,x_2) = U_{j1}(0,x_2), \quad i,j = 1,\dots,n, \\ &U_{i2}(0,x_2) = U_{i3}(0,x_2) = 0, \quad i = 1,\dots,n, \\ &\sum_{i=1}^{n} \frac{\mu_i(\mu_i+\lambda_i)}{2\mu_i+\lambda_i} U_{i1,1}(0,x_2) = 0, \ \ 0 < x_2 < 1. \end{aligned} \tag{9.1.2.14}$$

Solutions of (9.1.2.12)–(9.1.2.14) are unique up to rigid motions of the form

$$U_{i1} = -a, \quad U_{i2} = 0, \quad U_{i3} = ax_1,$$

where a is arbitrary. Further, this system has a solution if and only if the first equation in (9.1.2.11) is satisfied. If $\{U_i\}_{i=1}^n$ is a solution of (9.1.2.12)–(9.1.2.14) and if we set

$$\begin{aligned} W_{i1} = U_{i1}, \quad W_{i2} &= U_{i2} - k_i'\vartheta, \\ W_{i3} = U_{i3} &+ k_i'[(x_2 - 1/2)\vartheta + b], \end{aligned}$$

where ϑ, b are arbitrary constants, it may be proved that $\{W_i\}_{i=1}^n$, ϑ, b is a solution of (9.1.2.7)–(9.1.2.10) provided (9.1.2.11) holds.

The system (9.1.2.12)–(9.1.2.14) fits into the general framework of the previous section. Indeed, let $A_i^{\alpha\beta}$, B_i^{α}, C_i be the matrices in Example 9.1.2.3 with respect to the $\{\xi_{ik}\}_{k=1}^3$ basis, $\mathcal{D} = \emptyset$, and let

$$Q_{ie} = \begin{pmatrix} 1 & 0 & 0 \end{pmatrix}, \quad i = 1,\dots,n,$$

also with respect to the $\{\xi_{ik}\}_{k=1}^3$ basis. The orthogonal projection Π_{ie} onto the kernel of Q_{ie} and the pseudo-inverse Q_{ie}^+ of Q_{ie} are then given by

$$\Pi_{ie} = \begin{pmatrix} 0 & 0 & 0 \\ 0 & 1 & 0 \\ 0 & 0 & 1 \end{pmatrix}, \quad Q_{ie}^+ = \begin{pmatrix} 1 \\ 0 \\ 0 \end{pmatrix}.$$

Therefore the geometric transmission conditions (9.1.1.5), (9.1.1.6) reduce to the first two conditions in (9.1.2.14), while the mechanical edge conditions (9.1.1.7) and (9.1.1.9) give (9.1.2.13) and the last of the conditions in (9.1.2.14), respectively.

Remark 9.1.2.1 A dynamic version of (9.1.2.7)–(9.1.2.10) was derived in [51] as a model for the deformation of a network of Reissner-Mindlin plates connected at a single rigid joint. There it was assumed that (a) the in-plane displacements of each plate are rigid motions; (b) the displacement vectors of the various plates coincide along the junction e; (c) the "rotation vectors" of the various plates coincide along the junction (roughly speaking, this means that the angles between plates are preserved under the deformation); (d) the outer edge of $\mathcal{P}_1$ opposite e is fully clamped, the other outside edges of the network being free. These assumptions lead to the system (9.1.2.7)–(9.1.2.10) except that the boundary conditions (9.1.2.8) hold only for $i = 2, \dots, n$ and on the two edges of $\mathcal{P}_1$ immediately adjacent to e, while on the remaining edge of $\mathcal{P}_1$ we have $W_1(1, \mathbf{x}_2) = 0$. In fact, the derivation given in [51] leads exactly to the system (9.1.2.7)–(9.1.2.10) if condition (d) is replaced by (d'): the in-plane displacements in $\mathcal{P}_1$ vanish. According to condition (a), this is the same as saying that the in-plane displacements vanish at the edge of $\mathcal{P}_1$ opposite e. When $n = 2$ and $\mathcal{P}_1$ is orthogonal to $\mathcal{P}_2$, the model in [51] is analogous to one derived by Le Dret [17] in the spirit of the Ciarlet-Destuynder method of asymptotic expansions. More precisely, the Le Dret model is obtained in the limit as the shear moduli $K_i \to \infty$.

9.1.3 Existence and Uniqueness of Solutions

In this section existence and uniqueness of solutions of the variational equation (9.1.1.2) are considered. It is assumed that the elements of the matrices $A_i^{\alpha\beta}$, B_i^{β}, C_i are all in $L^\infty(\mathcal{P}_i)$. For a function $\Phi : \mathcal{P} \mapsto \mathbb{R}^m$ we denote by Φ_i the restriction of Φ to $\mathcal{P}_i$ and we set

$$\mathcal{H}^s(\mathcal{P}) = \{\Phi : \Phi_i \in \mathcal{H}^s(\mathcal{P}_i), \forall i \in \mathcal{I}\}$$

$$\|\Phi\|_{\mathcal{H}^s(\mathcal{P})} = \left(\sum_{i \in \mathcal{I}} \|\Phi_i\|^2_{\mathcal{H}^s(\mathcal{P}_i)}\right)^{1/2}$$

where $\mathcal{H}^s(\mathcal{P}_i)$ denotes the usual (vector) Sobolev space of order s on $\mathcal{P}_i$. Set $\mathcal{H} = \mathcal{H}^0(\mathcal{P})$ and define a closed subspace $\mathcal{V}$ of $\mathcal{H}^1(\mathcal{P})$ by

$$\begin{aligned}\mathcal{V} = \{\Phi \in \mathcal{H}^1(\mathcal{P}) |\ & \Phi_i = 0 \text{ on } e \text{ when } \Gamma_{ie} \in \mathcal{D}, \\ & Q_{ie}\Phi_i = Q_{je}\Phi_j \text{ on } e \text{ when } \Gamma_{ie} = \Gamma_{je}, \\ & \Pi_{ie}\Phi_i = 0 \text{ on } e \text{ when } e \in \mathcal{N}^{\text{int}}, i \in \mathcal{I}\}.\end{aligned}$$

The space $\mathcal{V}$ is densely and compactly embedded in $\mathcal{H}$. It is assumed that $a(\Phi, \Phi)$ is elliptic on $\mathcal{V}$: there are constants $k \geq 0$, $K > 0$, such that

$$a(\Phi, \Phi) + k\|\Phi\|^2_{\mathcal{H}} \geq K\|\Phi\|^2_{\mathcal{H}^1(\mathcal{P})}, \quad \forall \Phi \in \mathcal{V}. \tag{9.1.3.1}$$

Let $F = \{F_i\}_{i \in \mathcal{I}} \in \mathcal{H}$. It follows from standard variational theory and the Fredholm alternative that the variational equation (9.1.1.2) has a solution if and only if F is orthogonal in $\mathcal{H}$ to all solutions $W \in \mathcal{V}$ of $a(W, \Phi) = 0$, $\forall \Phi \in \mathcal{V}$. If it is known that $a(\Phi, \Phi) \geq 0$ for each $\Phi \in \mathcal{V}$, the last equation has only the trivial solution if, and only if, (9.1.3.1) holds with $k = 0$.

Let us apply this to the four examples of the last section.

Example 9.1.3.1 Referring to Example 9.1.2.1, it is clear that (9.1.3.1) is satisfied for k large enough if each 2×2 matrix A_i is uniformly positive definite on $\mathcal{P}_i$, $\forall i \in \mathcal{I}$. Further, (9.1.3.1) will hold with $k = 0$ if, for each $i \in \mathcal{I}$, we have $c_i \geq 0$ and the conditions $\Phi \in \mathcal{V}$, $\Phi_{i,\alpha} + b_i^\alpha \Phi_i = 0$, $\alpha = 1, 2$, $c_i \Phi_i = 0$ in $\mathcal{P}_i$ implies $\Phi_i = 0$ in $\mathcal{P}_i$. This will obviously be the case if, for example, $c_i > 0$ on a set of positive measure in $\mathcal{P}_i$, or if $b_i^1 = b_i^2 = 0$ and $\mathcal{D}_i \neq \emptyset$.

Example 9.1.3.2 Referring to Examples 9.1.2.2 and 9.1.2.3, it is a consequence of Korn's Lemma for 3-d linear elasticity that (9.1.3.1) holds for some constants $k > 0$ and $K > 0$. If $\mathcal{D} \neq \emptyset$ then (9.1.3.1) holds for $k = 0$ (see [61, Chapter VI, Section 2.4].

Example 9.1.3.3 Consider the system (9.1.2.12)–(9.1.2.14) in Example 9.1.2.4. This system corresponds to the variational equation

$$a(U, \Phi) = \sum_{i=1}^{n} \int_\omega (M_{i\alpha} U_{i\alpha} + F_{i3} U_{i3})\, dx, \quad \forall \Phi \in \mathcal{V}, \tag{9.1.3.2}$$

where

$$a(U, \Phi) = \sum_{i=1}^{n} \int_{\mathcal{P}_i} [\sigma_i^{\alpha\beta}(u_i) \varepsilon_{\alpha\beta}(\phi_i) + K_i h (U_{i3,\alpha} + U_{i\alpha})(\Phi_{i3,\alpha} + \Phi_{i\alpha})]\, dx$$

and

$$\mathcal{V} = \{\Phi |\, \Phi_{ik} \in H^1(\omega),\ k = 1, 2, 3,\ \Phi_{i1}(0, x_2) = \Phi_{j1}(0, x_2),$$
$$\Phi_{i2}(0, x_2) = \Phi_{i3}(0, x_2) = 0;\ i, j = 1, \ldots, n;\ 0 < x_2 < 1\}.$$

Here $\Phi_i = \sum_{k=1}^3 \Phi_{ik} \xi_{ik} := \phi_i + \Phi_{i3} \xi_{i3}$. The inequality (9.1.3.1) holds for some $k > 0$ but not for $k = 0$ since the equation of $a(U, \Phi) = 0$, $\forall \Phi \in \mathcal{V}$, has nontrivial solutions $U \in \mathcal{V}$. These solutions are those rigid motions for which $\varepsilon_{\alpha\beta}(u_i) = 0$, $U_{i3,\alpha} + U_{i\alpha} = 0$, $I = 1, \ldots, n$, $\alpha, \beta = 1, 2$. They are given by

$$U_{i1} = -a, \quad U_{i2} = 0, \quad U_{i3} = a x_1$$

where a is an arbitrary constant. It follows that (9.1.3.2) has a solution in $\mathcal{V}$ if and only if

$$\sum_{i=1}^{n} \int_\omega (x_1 F_{i3} - M_{i1})\, d\omega = 0.$$

Remark 9.1.3.1 In the general case, (9.1.3.1) will be satisfied if the matrices $A_i^{\alpha\beta}$ satisfy, for each $i \in \mathcal{I}$, the inequality

$$(A_i^{\alpha\beta} \xi^\alpha) \cdot \xi^\beta \geq c(|\xi^1|^2 + |\xi^2|^2), \quad \forall \xi^1, \xi^2 \in \mathbb{R}^m \tag{9.1.3.3}$$

for some $c > 0$. Further, if the matrices C_i are all nonnegative definite, (9.1.3.1) will hold with $k = 0$ when, for each $i \in \mathcal{I}$, the conditions $\Phi \in \mathcal{V}$, $\Phi_{i,\alpha} + B_i^\alpha \Phi_i = 0$, $\alpha = 1, 2$, $C_i \Phi_i = 0$ in $\mathcal{P}_i$ implies $\Phi_i = 0$ in $\mathcal{P}_i$.

9.1.4 Domain Decomposition

In the present context the goal of domain decomposition is to replace the above global problem on the network $\mathcal{P}$ by a sequence of local problems, each of which lives on a single element $\mathcal{P}_i$, and whose solutions converge to a solution of the global problem in an appropriate sense. In the decomposition, the coupling conditions on the interfaces between adjacent regions in the global network are replaced by local iterative Robin-type boundary conditions in which the effect of adjacent regions is modeled by certain inhomogeneities in the boundary conditions. These iterative conditions are analogous to ones introduced earlier in the study of 1-d networks.

For each index $i \in \mathcal{I}$ we denote by $\mathcal{D}_i$ the Dirichlet edges in $\partial\mathcal{P}_i$ and set $\mathcal{N}_i = \partial\mathcal{P}_i \backslash \mathcal{D}_i$, which is partitioned into $\mathcal{N}_i^{\text{ext}} \cup \mathcal{N}_i^{\text{int}}$ with obvious notation. We shall write $\mathcal{H}_i^s$ for $\mathcal{H}^s(\mathcal{P}_i)$ and set $\mathcal{H}_i := \mathcal{H}_i^0$. We define

$$\mathcal{V}_i = \{\Phi_i | \, \Phi_i \in \mathcal{H}_i^1, \ \Phi_i = 0 \text{ on } e \text{ when } \Gamma_{ie} \in \mathcal{D}_i, \\ \Pi_{ie}\Phi_i = 0 \text{ on } e \text{ when } \Gamma_{ie} \in \mathcal{N}_i^{\text{int}}\}$$

A continuous, symmetric bilinear form a_i is defined on $\mathcal{V}_i$ by

$$a_i(W_i, V_i) = \int_{\mathcal{P}_i} [A_i^{\alpha\beta}(W_{i,\beta} + B_i^\beta W_i) \cdot (\Phi_{i,\alpha} + B_i^\alpha \Phi_i) + C_i W_i \cdot \Phi_i] \, dx.$$

For each $i \in \mathcal{I}$ we consider the following sequence of local problems for functions $W_i^{n+1} \in \mathcal{V}_i$:

$$a_i(W_i^{n+1}, \Phi_i) + \sum_{e \in \mathcal{N}_i^{\text{int}}} k_e \int_{\Gamma_{ie}} Q_{ie} W_i^{n+1} \cdot Q_{ie}\Phi_i \, d\Gamma \\ = \int_{\mathcal{P}_i} F_i \cdot \Phi_i \, dx + \sum_{e \in \mathcal{N}_i^{\text{int}}} \int_{\Gamma_{ie}} \lambda_{ie}(W^n) \cdot Q_{ie}\Phi_i \, d\Gamma, \quad \forall \Phi_i \in \mathcal{V}_i, \quad (9.1.4.1)$$

where $n = 0, 1, \ldots$, $k_e > 0$ is constant and

$$\lambda_{ie}(W^n) = -\frac{2}{d_e} \sum_{j \in \mathcal{I}_e} (Q_{je}^+)^* \nu_{je}^\alpha A_j^{\alpha\beta}(W_{j,\beta}^n + B_j^\beta W_j^n) \\ + (Q_{ie}^+)^* \nu_{ie}^\alpha A_i^{\alpha\beta}(W_{i,\beta}^n + B_i^\beta W_i^n) + \frac{2}{d_e} \sum_{j \in \mathcal{I}_e} k_e Q_{je} W_j^n - k_e Q_{ie} W_i^n.$$

The inhomogeneity $\lambda_{ie}(W^n)$ reflects the influence of the regions adjacent to Γ_{ie}, $e \in \mathcal{N}_i^{\text{int}}$, on the solution in $\mathcal{P}_i$ at level $n+1$. The choice of $\{W_i^0\}_{i \in \mathcal{I}}$ is arbitrary save for a regularity requirement on $\lambda_{ie}(W^0)$

We note that (9.1.4.1) is the variational formulation of the boundary value problem

$$-\frac{\partial}{\partial x_\alpha}\left[A_i^{\alpha\beta}(W_{i,\beta}^{n+1}+B_i^\beta W_i^{n+1})\right]+(B_i^\alpha)^*A_i^{\alpha\beta}(W_{i,\beta}^{n+1}+B_i^\beta W_i^{n+1}) + C_iW_i^{n+1}=F_i \quad \text{in } \mathcal{P}_i, \qquad (9.1.4.2)$$

$$\begin{aligned} W_i^{n+1}&=0 \text{ on } e \text{ when } \Gamma_{ie}\in\mathcal{D}_i,\\ \Pi_{ie}W_i^{n+1}&=0 \text{ on } e \text{ when } \Gamma_{ie}\in\mathcal{N}_i^{\rm int}, \end{aligned} \qquad (9.1.4.3)$$

$$\begin{aligned} D_{\nu_{ie}}W_i^{n+1}&=0, \quad \Gamma_{ie}\in\mathcal{N}_i^{\rm ext},\\ (Q_{ie}^+)^*D_{\nu_{ie}}W_i^{n+1}+k_eQ_{ie}W_i^{n+1}&=\lambda_{ie}(W^n), \quad \Gamma_{ie}\in\mathcal{N}_i^{\rm int}, \end{aligned} \qquad (9.1.4.4)$$

where, for convenience, we have set

$$D_{\nu_{ie}}W_i=\nu_{ie}^\alpha A_i^{\alpha\beta}(W_{i,\beta}+B_i^\beta W_i).$$

With this notation, $\lambda_{ie}(W^n)$ is written

$$\lambda_{ie}(W^n)=-\frac{2}{d_e}\sum_{j\in\mathcal{I}_e}(Q_{je}^+)^*D_{\nu_{je}}W_j^n+(Q_{ie}^+)^*D_{\nu_{ie}}W_i^n + \frac{2}{d_e}\sum_{j\in\mathcal{I}_e}k_eQ_{je}W_j^n-k_eQ_{ie}W_i^n. \qquad (9.1.4.5)$$

Suppose that

$$F_i\in L^2(\mathcal{P}_i), \quad \lambda_{ie}(W^n)\in L^2(\Gamma_{ie}), \; \forall e\in\mathcal{N}_i^{\rm int},$$

and that a_i satisfies, for some $C>0$,

$$a_i(\Phi_i,\Phi_i)+\sum_{e\in\mathcal{N}_i^{\rm int}}\int_{\Gamma_{ie}}|Q_{ie}\Phi_i|^2\,d\Gamma\ge C\|\Phi_i\|^2_{\mathcal{H}_i^1}, \quad \forall\Phi_i\in\mathcal{V}_i. \qquad (9.1.4.6)$$

Then (9.1.4.1) has a *unique* solution $W_i^{n+1}\in\mathcal{V}_i$. Thus $Q_{ie}W_i^{n+1}|_{\Gamma_{ie}}\in L^2(\Gamma_{ie})$ so that, from (9.1.4.4), $(Q_{ie}^+)^*D_{\nu_{ie}}W_i^{n+1}\in L^2(\Gamma_{ie})$. It then follows that $\lambda_{ie}(W^{n+1})\in L^2(\Gamma_{ie})$. As a consequence, if W_i^0 is chosen so that $\lambda_{ie}(W^0)\in L^2(\Gamma_{ie})$, $\forall e\in\mathcal{N}_i^{\rm int}$, the variational equation (9.1.4.1) has a unique solution $W_i^{n+1}\in\mathcal{V}_i$, $n=0,1,\ldots$.

Let us note that if $U_i\in\mathcal{V}_i$ has the trace regularity $D_{\nu_{ie}}U_i\in L^2(\Gamma_{ie})$ for $e\in\mathcal{N}_i$, then the following Green's formula is valid:

$$\int_{\mathcal{P}_i}\mathcal{A}_iU_i\cdot\Phi_i\,dx=a_i(U_i,\Phi_i)-\sum_{e\in\mathcal{N}_i^{\rm ext}}\int_{\Gamma_{ie}}(D_{\nu_{ie}}U_i)\cdot\Phi_i\,d\Gamma - \sum_{e\in\mathcal{N}_i^{\rm int}}\int_{\Gamma_{ie}}((Q_{ie}^+)^*D_{\nu_{ie}}U_i)\cdot(Q_{ie}\Phi_i)\,d\Gamma, \quad \forall\Phi_i\in\mathcal{V}_i, \qquad (9.1.4.7)$$

where

$$\mathcal{A}_i U_i = -\frac{\partial}{\partial x_\alpha}\left[A_i^{\alpha\beta}(U_{i,\beta} + B_i^\beta U_i)\right] + (B_i^\alpha)^* A_i^{\alpha\beta}(U_{i,\beta} + B_i^\beta U_i) + C_i U_i$$

and the integral on the left side of (9.1.4.7) is interpreted in the sense of $\mathcal{V}_i^*$-$\mathcal{V}_i$ duality, where we have identified $\mathcal{H}_i$ with its dual space and where $\mathcal{V}_i^*$ denotes the dual space of $\mathcal{V}_i$ with respect to $\mathcal{H}_i$, so that $\mathcal{V}_i \subset \mathcal{H}_i \subset \mathcal{V}_i^*$ with dense and compact embedding.

Let us comment on condition (9.1.4.6). It is a consequence of Korn's lemma that (9.1.4.6) is satisfied by the forms a_i in Examples 9.1.2.2–9.1.2.4 above. It also holds in the case of Example 9.1.2.1 if, for instance, A_i is uniformly positive definite on $\mathcal{P}_i$ and $c_i > 0$ on a set of positive measure in $\mathcal{P}_i$. In the general case, (9.1.4.6) will hold if $A_i^{\alpha\beta}$ satisfies (9.1.3.3), C_i is nonnegative definite, and

$$\Phi_i \in \mathcal{V}_i, \quad a_i(\Phi_i, \Phi_i) = 0, \quad \Phi_i = 0 \text{ on } e \text{ when } e \in \mathcal{N}_i^{\text{int}} \text{ implies } \Phi_i = 0.$$

Indeed, under the stated hypothesis we have

$$a_i(\Phi_i, \Phi_i) + k_i\|\Phi_i\|^2_{\mathcal{H}_i} \ge K_i\|\Phi_i\|^2_{\mathcal{H}_i^1}, \quad \forall \Phi_i \in \mathcal{V}_i \tag{9.1.4.8}$$

for some constants $k_i \ge 0$, $K_i > 0$. If (9.1.4.6) is false, there is a sequence $\{\Phi_i^k\} \subset \mathcal{V}_i$ such that

$$\begin{gathered} \|\Phi_i^k\|_{\mathcal{H}_i^1} = 1, \quad a_i(\Phi_i^k, \Phi_i^k) \to 0, \\ Q_{ie}\Phi_i^k|_{\Gamma_{ie}} \to 0 \text{ strongly in } L^2(\Gamma_{ie}). \end{gathered} \tag{9.1.4.9}$$

It follows that, on a subsequence, $\Phi_i^k \to \Phi_i$ weakly in $\mathcal{H}_i^1$ and, by compactness, $\Phi_i^k \to \Phi_i$ strongly in $\mathcal{H}_i$. It then follows from (9.1.4.8) and (9.1.4.9) that $\Phi_i^k \to \Phi_i \in \mathcal{V}_i$ strongly in $\mathcal{H}_i^1$, where

$$\|\Phi_i\|_{\mathcal{H}_i^1} = 1, \quad a_i(\Phi_i, \Phi_i) = 0, \quad Q_{ie}\Phi_i = 0 \text{ on } e, \forall e \in \mathcal{N}_i. \tag{9.1.4.10}$$

Since $\Phi_i \in \mathcal{V}_i$ we also have $\Pi_{ie}\Phi_i = 0$ on e, $\forall e \in \mathcal{N}_i^{\text{int}}$, hence $\Phi_i = 0$ on e, $\forall e \in \mathcal{N}_i^{\text{int}}$. It follows that $\Phi_i = 0$ on Γ_{ie} when $e \in \mathcal{N}_i^{\text{int}}$ and therefore $\Phi_i = 0$, contradicting (9.1.4.10).

9.1.5 Convergence of the Algorithm

In order to obtain a convergence result we shall consider a relaxation of the basic iteration step $(9.1.4.4)_2$. Thus we introduce a relaxation parameter $\epsilon \in [0, 1)$ and replace $(9.1.4.4)_2$ by

$$\begin{aligned} &(Q_{ie}^+)^* D_{\nu_{ie}} W_i^{n+1} + k_e Q_{ie} W_i^{n+1} \\ &\quad = (1-\epsilon)\lambda_{ie}(W^n) + \epsilon\left[(Q_{ie}^+)^* D_{\nu_{ie}} W_i^n + k_e Q_{ie} W_i^n\right] := \lambda_{ie}^\epsilon(W^n). \end{aligned} \tag{9.1.5.1}$$

Thus the $(n+1)$st local iterate is calculated as the solution of the variational equation

$$a_i(W_i^{n+1}, \Phi_i) + \sum_{e\in\mathcal{N}_i^{\text{int}}} k_e \int_{\Gamma_{ie}} Q_{ie}W_i^{n+1} \cdot Q_{ie}\Phi_i \, d\Gamma$$
$$= \int_{\mathcal{P}_i} F_i \cdot \Phi_i \, dx + \sum_{e\in\mathcal{N}_i^{\text{int}}} \int_{\Gamma_{ie}} \lambda_{ie}^{\epsilon}(W^n) \cdot Q_{ie}\Phi_i \, d\Gamma, \quad \forall \Phi_i \in \mathcal{V}_i, \quad (9.1.5.2)$$

where the starting data $\lambda_{ie}^{\epsilon}(W^0) \in L^2(\Gamma_{ie})$.

We next note that a function $W : \mathcal{P} \mapsto \mathbb{R}^m$ satisfies the global transmission conditions (9.1.1.5) and (9.1.1.9) if and only if $W_i^n \equiv W_i$, $i \in \mathcal{I}$, $n = 0, 1, \ldots$, satisfies $(9.1.4.4)_2$ (and, consequently, (9.1.5.1)). In fact, from (9.1.1.5) and (9.1.1.9) we have

$$(Q_{ie}^{+})^* D_{\nu_{ie}} W_i + k_e Q_{ie} W_i = \lambda_{ie}(W) \text{ on } e \in \mathcal{N}_i^{\text{int}}. \quad (9.1.5.3)$$

Conversely, if (9.1.5.3) is satisfied we obtain

$$\sum_{i\in\mathcal{I}_e} k_e Q_{ie} W_i = -2 \sum_{j\in\mathcal{I}_e} (Q_{je}^{+})^* D_{\nu_{je}} W_j + 2 \sum_{j\in\mathcal{I}_e} k_e Q_{je} W_j - \sum_{i\in\mathcal{I}_e} k_e Q_{ie} W_i,$$

hence

$$\sum_{j\in\mathcal{I}_e} (Q_{je}^{+})^* D_{\nu_{je}} W_j = 0$$

and then, from (9.1.5.3),

$$Q_{ie} W_i = \frac{1}{d_e} \sum_{j\in\mathcal{I}_e} Q_{je} W_j \text{ on } e \in \mathcal{N}_i^{\text{int}}, \; i \in \mathcal{I}_e,$$

that is, $Q_{ie} W_i|_{\Gamma_{ie}}$ is independent of the index $i \in \mathcal{I}_e$.

It follows from the above discussion that if W_i^{n+1} is the solution of (9.1.5.2) and W is a solution of (9.1.1.2) (recall that solutions of (9.1.1.2) are not necessarily unique) then the *errors*

$$U_i^n := W_i^n - W_i, \quad i \in \mathcal{I}, \; n = 1, 2, \ldots,$$

formally satisfy (9.1.4.2)–(9.1.4.4) with $F_i = 0$, that is

$$a_i(U_i^{n+1}, \Phi_i) + \sum_{e\in\mathcal{N}_i^{\text{int}}} k_e \int_{\Gamma_{ie}} Q_{ie}U_i^{n+1} \cdot Q_{ie}\Phi_i$$
$$= \sum_{e\in\mathcal{N}_i^{\text{int}}} \int_{\Gamma_{ie}} \lambda_{ie}^{\epsilon}(U^n) \cdot Q_{ie}\Phi_i, \quad \forall \Phi_i \in \mathcal{V}_i. \quad (9.1.5.4)$$

As discussed above, if $\lambda_{ie}^{\epsilon}(U^0) \in L^2(\Gamma_{ie})$, the problems (9.1.5.4) for $n = 0, 1, \ldots$ have unique solutions with regularity

$$U_i^{n+1} \in \mathcal{V}_i, \quad (Q_{ie}^{+})^* D_{\nu_{ie}} U_i^{n+1} \in L^2(\Gamma_{ie}) \text{ on } e \in \mathcal{N}_i^{\text{int}}.$$

Since for the system (9.1.4.2)–(9.1.4.4) the W_i^0 are chosen so that $\lambda_{ie}^{\epsilon}(W^0) \in L^2(\Gamma_{ie})$, in order that $\lambda_{ie}^{\epsilon}(U^0) \in L^2(\Gamma_{ie})$ it is necessary that $\lambda_{ie}^{\epsilon}(W) \in L^2(\Gamma_{ie})$, which is equivalent to

$$(Q_{ie}^{+})^* D_{\nu_{ie}} W_i \in L^2(\Gamma_{ie}) \text{ on } e \in \mathcal{N}_i^{\text{int}},\ i \in \mathcal{I}. \tag{9.1.5.5}$$

The regularity *assumption* (9.1.5.5) is problematical in as much as precise knowledge of the regularity of solutions of (9.1.1.2) is generally unavailable. The regularity of W will depend on the F_i, on the set of coefficients $A_i^{\alpha\beta}$, B_i^{α}, C_i and the particular configuration and geometry of the 2-d network. For the case of interface problems for general elliptic equations on polygonal domains, i.e., when all $\mathcal{P}_i$ are coplanar, much is known and, in fact, (9.1.5.5) does not hold for all possible choices of the coefficients and the regions $\mathcal{P}_i$. The reader may consult the work of Nicaise and Sändig [87], [88], Nicaise [86], and references therein. For a study of the Laplace and biharmonic operators on general 2-d networks, see [86].

We shall prove the following convergence result.

Theorem 9.1.5.1 *Assume that the coercivity conditions* (9.1.3.1) *and* (9.1.4.6) *are satisfied and that the global problem* (9.1.1.2) *has at least one solution having the regularity* (9.1.5.5). *Let* $W = \{W_i\}_{i\in\mathcal{I}}$ *be such a solution. Suppose that* $k_e \equiv k$ *is independent of* $e \in \mathcal{N}^{int}$ *and let* $W_i^{n+1} \in \mathcal{V}_i$ *be the solution of* (9.1.5.2) *for* $n = 0, 1, \ldots$. *Then the following hold for the errors* $U_i^n = W_i^n - W_i$:

(i) *For each* $\epsilon \in (0, 1)$

$$U_i^n \to U_i \ \textit{strongly in } H^1(\mathcal{P}_i), \ \forall i \in \mathcal{I},$$
$$\textit{where } U = \{U_i\}_{i\in\mathcal{I}} \in \mathcal{V} \ \textit{ and } a(U, \Phi) = 0, \ \forall \Phi \in \mathcal{V}.$$

(ii) *If* $\epsilon = 0$ *we have*

$$a_i(U_i^n, U_i^n) \to 0, \quad \forall i \in \mathcal{I}.$$

Remark 9.1.5.1 The conclusion of part (i) of Theorem 9.1.5.1 is equivalent to saying that $U^n = \{U_i^n\}_{i\in\mathcal{I}}$ converges to zero in the quotient space $\mathcal{W}\backslash\mathcal{R}$, where $\mathcal{W} = \prod_{i\in\mathcal{I}} \mathcal{V}_i$ and $\mathcal{R}$ denotes the null space of the bilinear form $a(W, \Phi)$, that is

$$\mathcal{R} = \{W \in \mathcal{V} : a(W, \Phi) = 0, \ \forall \Phi \in \mathcal{V}\}.$$

Remark 9.1.5.2 In Examples 9.1.2.2 and 9.1.2.3 above, $a(U, \Phi) = 0$, $\forall \Phi \in \mathcal{V}$, implies that $U = 0$ and, further, $a_i(U_i^n, U_i^n) \to 0$ implies that $U_i^n \to 0$ strongly in $H^1(\mathcal{P}_i)$. The same is true in Example 9.1.2.1 under certain conditions on the coefficients as discussed above. However, for the system (9.1.2.12)–(9.1.2.14) in

Example 9.1.2.4, solutions in $\mathcal{V}$ of $a(U,\Phi)=0$, $\forall\Phi\in\mathcal{V}$, consist of rigid motions of the form $U_i=-a\xi_{i1}+ax_1\xi_{i3}$, $a\in\mathbb{R}$, $I=1,\dots,n$. Thus the conclusion in part (i) means that

$$W_{i2}^n - W_{i2} \to 0 \text{ in } H^1(\omega),$$
$$W_{i1,\alpha}^n - W_{i1,\alpha} \to 0 \text{ in } L^2(\omega),\ \alpha=1,2,$$
$$W_{i3,2}^n - W_{i3,2} \to 0 \text{ in } L^2(\omega),$$

and

$$\int_\omega \{|W_{i1}^n - W_{i1} - a|^2 + |W_{i3}^n - W_{i3} + ax_1|^2 + |W_{i3,1}^n - W_{i3,1} + a|^2\}\, d\omega \to 0$$

for some $a\in\mathbb{R}$.

Proof of Theorem 9.1.5.1. Set

$$\mathcal{X} = \prod_{i\in\mathcal{I}} \prod_{e\in\mathcal{N}_i^{\text{int}}} L^2(\Gamma_{ie})$$

with the standard product norm. Let

$$X = \{\lambda_{ie} : i\in\mathcal{I},\ e\in\mathcal{N}_i^{\text{int}}\}\in\mathcal{X}$$

and, for each $i\in\mathcal{I}$, let U_i be the solution of

$$a_i(U_i,\Phi_i) + \sum_{e\in\mathcal{N}_i^{\text{int}}} k_e \int_{\Gamma_{ie}} Q_{ie}U_i\cdot Q_{ie}\Phi_i$$
$$= \sum_{e\in\mathcal{N}_i^{\text{int}}} \int_{\Gamma_{ie}} \lambda_{ie}\cdot Q_{ie}\Phi_i, \quad \forall\Phi_i\in\mathcal{V}_i. \quad (9.1.5.6)$$

We define the bounded linear operator $\mathcal{T}:\mathcal{X}\mapsto\mathcal{X}$ by

$$(\mathcal{T}X)_{ie} = -\frac{2}{d_e}\sum_{j\in\mathcal{I}_e} (Q_{je}^+)^* D_{\nu_{je}}U_j + (Q_{ie}^+)^* D_{\nu_{ie}}U_i$$
$$+\frac{2}{d_e}\sum_{j\in\mathcal{I}_e} k_eQ_{je}U_j - k_eQ_{ie}U_i, \quad (9.1.5.7)$$

$$\mathcal{T}X = \{(\mathcal{T}X)_{ie} : i\in\mathcal{I},\ e\in\mathcal{N}_i^{\text{int}}\}.$$

With this notation, X is a fixed point of $\mathcal{T}$ if and only if the corresponding solution $U=\{U_i\}_{i\in\mathcal{I}}$ satisfies the global transmission conditions (9.1.1.5) and (9.1.1.9); that is, if and only if U is a solution of the global problem $a(U,\Phi)=0$, $\forall\Phi\in\mathcal{V}$. Furthermore, if we set

$$X^n = \{\lambda_{ie}^{n-1} : i\in\mathcal{I},\ e\in\mathcal{N}_i^{\text{int}}\},\quad n=1,2,\dots$$

and let U_i^n be the solution of (9.1.5.6) corresponding to λ_{ie}^{n-1}, then

$$X^n = \{(Q_{ie}^+)^* D_{\nu_{ie}} U_i^n + k_e Q_{ie} U_i^n : i \in \mathcal{I},\, e \in \mathcal{N}_i^{\text{int}}\},$$

$$\mathcal{T} X^n = \{\lambda_{ie}(U^n) : i \in \mathcal{I},\, e \in \mathcal{N}_i^{\text{int}}\},$$

so that the iteration step $(9.1.4.4)_2$ may be written as the fixed point iteration

$$X^{n+1} = \mathcal{T} X^n,$$

while the relaxed iteration step (9.1.5.1) is the same as the relaxed fixed point iteration

$$X^{n+1} = (1-\epsilon)\mathcal{T} X^n + \epsilon X^n.$$

We shall need the following properties of $\mathcal{T}$.

Lemma 9.1.5.1 *Suppose that $k_e \equiv k$ is independent of $e \in \mathcal{N}^{int}$. Then for any $X = \{\lambda_{ie} : i \in \mathcal{I},\, e \in \mathcal{N}_i^{int}\} \in \mathcal{X}$ we may write*

$$\|X\|_{\mathcal{X}}^2 = \mathcal{E} + \mathcal{F}, \quad \|\mathcal{T} X\|_{\mathcal{X}}^2 = \mathcal{E} - \mathcal{F} \tag{9.1.5.8}$$

where

$$\mathcal{E} = \sum_{e \in \mathcal{N}^{int}} \sum_{i \in \mathcal{I}_e} \int_{\Gamma_{ie}} \left[|(Q_{ie}^+)^* D_{\nu_{ie}} U_i|^2 + k_e^2 |Q_{ie} U_i|^2 \right] d\Gamma, \tag{9.1.5.9}$$

$$\mathcal{F} = 2k \sum_{i \in \mathcal{I}} a_i(U_i, U_i), \tag{9.1.5.10}$$

and where U_i is the solution of (9.1.5.6). *In particular, $\mathcal{T}$ is a nonexpansive map.*

Proof of Lemma 9.1.5.1. Since $\Sigma_{I \in \mathcal{I}} \sum_{e \in \mathcal{N}_i^{\text{int}}} = \sum_{e \in \mathcal{N}^{\text{int}}} \sum_{i \in \mathcal{I}_e}$ we have

$$\|X\|_{\mathcal{X}}^2 = \sum_{e \in \mathcal{N}^{\text{int}}} \sum_{i \in \mathcal{I}_e} \int_{\Gamma_{ie}} \Big[|(Q_{ie}^+)^* D_{\nu_{ie}} U_i|^2 + k_e^2 |Q_{ie} U_i|^2 + 2k_e((Q_{ie}^+)^* D_{\nu_{ie}} U_i) \cdot (Q_{ie} U_i) \Big] d\Gamma. \tag{9.1.5.11}$$

$$\begin{aligned} \|(\mathcal{T} X)_{ie}\|^2 = {} & \frac{4k_e^2}{d_e^2} \Big| \sum_{j \in \mathcal{I}_e} Q_{je} U_j \Big|^2 + k_e^2 |Q_{ie} U_i|^2 - \frac{4k_e^2}{d_e} Q_{ie} U_i \cdot \sum_{j \in \mathcal{I}_e} Q_{je} U_j \\ & + \frac{4}{d_e^2} \Big| \sum_{j \in \mathcal{I}_e} (Q_{je}^+)^* D_{\nu_{je}} U_j \Big|^2 + |(Q_{ie}^+)^* D_{\nu_{ie}} U_i|^2 \\ & - \frac{4}{d_e} (Q_{ie}^+)^* D_{\nu_{ie}} U_i \cdot \sum_{j \in \mathcal{I}_e} (Q_{je}^+)^* D_{\nu_{je}} U_j \\ & - 2k_e \Big(\frac{2}{d_e} \sum_{j \in \mathcal{I}_e} Q_{je} U_j - Q_{ie} U_i \Big) \cdot \Big(\frac{2}{d_e} \sum_{j \in \mathcal{I}_e} (Q_{je}^+)^* D_{\nu_{je}} U_j - (Q_{ie}^+)^* D_{\nu_{ie}} U_i \Big). \end{aligned}$$

After some algebra we obtain

$$\sum_{i\in\mathcal{I}_e}\|(\mathcal{T}X)_{ie}\|^2 = k_e^2\sum_{i\in\mathcal{I}_e}|Q_{ie}U_i|^2+\sum_{i\in\mathcal{I}_e}|(Q_{ie}^+)^*D_{\nu_{ie}}U_i|^2 \\ -2k_e\sum_{i\in\mathcal{I}_e}(Q_{ie}U_i)\cdot((Q_{ie}^+)^*D_{\nu_{ie}}U_i),\quad e\in\mathcal{N}. \tag{9.1.5.12}$$

It follows from (9.1.5.11) and (9.1.5.12) that

$$\begin{aligned}\|X\|_{\mathcal{X}}^2 &= \mathcal{E}+2\sum_{i\in\mathcal{I}}\sum_{e\in\mathcal{N}_i^{\text{int}}}\int_{\Gamma_{ie}} k_e((Q_{ie}^+)^*D_{\nu_{ie}}U_i)\cdot(Q_{ie}U_i)\,d\Gamma\\ \|\mathcal{T}X\|_{\mathcal{X}}^2 &= \mathcal{E}-2\sum_{i\in\mathcal{I}}\sum_{e\in\mathcal{N}_i^{\text{int}}}\int_{\Gamma_{ie}} k_e((Q_{ie}^+)^*D_{\nu_{ie}}U_i)\cdot(Q_{ie}U_i)\,d\Gamma.\end{aligned} \tag{9.1.5.13}$$

From Green's formula (9.1.4.7) we have

$$\sum_{e\in\mathcal{N}_i^{\text{int}}}\int_{\Gamma_{ie}}((Q_{ie}^+)^*D_{\nu_{ie}}U_i)\cdot(Q_{ie}U_i)\,d\Gamma = a_i(U_i,U_i). \tag{9.1.5.14}$$

Since $k_e\equiv k$ is independent of $e\in\mathcal{N}^{\text{int}}$, the conclusions of the Lemma follow from (9.1.5.13) and (9.1.5.14).

Continuing with the proof of Theorem 9.1.5.1, we consider the relaxed fixed point iteration

$$X^{n+1}=(1-\epsilon)\mathcal{T}X^n+\epsilon X^n$$

and define

$$\mathcal{E}^n=\sum_{e\in\mathcal{N}^{\text{int}}}\sum_{i\in\mathcal{I}_e}\int_{\Gamma_{ie}}\left[|(Q_{ie}^+)^*D_{\nu_{ie}}U_i^n|^2+k_e^2|Q_{ie}U_i^n|^2\right]d\Gamma, \tag{9.1.5.15}$$

$$\mathcal{F}^n=2k\sum_{i\in\mathcal{I}}a_i(U_i^n,U_i^n), \tag{9.1.5.16}$$

from Lemma 9.1.5.1 we have

$$\begin{aligned}\|X^{n+1}\|_{\mathcal{X}}^2 &= \mathcal{E}^{n+1}+\mathcal{F}^{n+1}\\ &=(1-\epsilon)^2\|\mathcal{T}X^n\|_{\mathcal{X}}^2+\epsilon^2\|X^n\|_{\mathcal{X}}^2+2\epsilon(1-\epsilon)(X^n,\mathcal{T}X^n)_{\mathcal{X}}\\ &=\big((1-\epsilon)^2+\epsilon^2\big)\mathcal{E}^n-(1-2\epsilon)\mathcal{F}^n+2\epsilon(1-\epsilon)(X^n,\mathcal{T}X^n)_{\mathcal{X}}\end{aligned}$$

as well as

$$\big|(X^n,\mathcal{T}X^n)_{\mathcal{X}}\big|\le\sqrt{(\mathcal{E}^n)^2-(\mathcal{F}^n)^2}\le\mathcal{E}^n.$$

We thus obtain the recursion formula

$$\mathcal{E}^{n+1}\le\mathcal{E}^n-\mathcal{F}^{n+1}-(1-2\epsilon)\mathcal{F}^n.$$

and therefore, upon iterating this formula, we have

$$\mathcal{E}^{n+1} \leq \mathcal{E}^1 - \sum_{p=1}^{n+1} c_p(\epsilon)\mathcal{F}^p \tag{9.1.5.17}$$

where

$$c_1(\epsilon) = 1 - 2\epsilon, \quad c_{n+1} = 1, \quad c_p(\epsilon) = 2(1-\epsilon),\ p = 2, \ldots, n.$$

It follows from (9.1.5.17) that, for any $\epsilon \in [0,1)$,

$$\sum_{p=1}^{\infty} \sum_{i\in\mathcal{I}} a_i(U_i^p, U_i^p) < \infty,$$

$$\{Q_{ie}U_i^n|_{\Gamma_{ie}}\} \text{ is bounded in } L^2(\Gamma_{ie}),\ e \in \mathcal{N}^{\text{int}},\ i \in \mathcal{I}_e,$$

$$\{(Q_{ie}^+)^* D_{\nu_{ie}} U_i^n|_{\Gamma_{ie}}\} \text{ is bounded in } L^2(\Gamma_{ie}),\ e \in \mathcal{N}^{\text{int}},\ i \in \mathcal{I}_e.$$

In particular, for each $i \in \mathcal{I}$ we have

$$a_i(U_i^n, U_i^n) \to 0 \tag{9.1.5.18}$$

and, in view of (9.1.4.6), that $\{U_i^n\}$ is bounded in $\mathcal{H}^1(\mathcal{P}_i)$, hence $\{Q_{ie}U_i^n|_{\Gamma_{ie}}\}$ is bounded in $H^{1/2}(\Gamma_{ie})$. Therefore on a subsequence $n = n_k$ of the positive integers we have

$$\begin{aligned} U_i^n &\to \widehat{U}_i \text{ weakly in } \mathcal{H}^1(\mathcal{P}_i), \\ Q_{ie}U_i^n &\to Q_{ie}\widehat{U}_i \text{ strongly in } L^2(\Gamma_{ie}),\ \forall e \in \mathcal{N}^{\text{int}}, \\ (Q_{ie}^+)^* D_{\nu_{ie}} U_i^n &\to \widehat{U}_{ie} \text{ weakly in } L^2(\Gamma_{ie}),\ \forall e \in \mathcal{N}^{\text{int}} \end{aligned} \tag{9.1.5.19}$$

for some $\widehat{U}_i \in \mathcal{V}_i$ and $\widehat{U}_{ie} \in L^2(\Gamma_{ie})$. Conclusions (9.1.5.18) and (9.1.5.19) are valid for any $\epsilon \in [0,1)$.

Now let $\epsilon \in (0,1)$. According to Proposition 2.3.7.3 we have

$$X^n \to X, \quad \mathcal{T}X^n \to X \ \text{ weakly in } \mathcal{X} \tag{9.1.5.20}$$

where X is a fixed point of $\mathcal{T}$. If we set $X = \{\lambda_{ie} : i \in I,\ e \in \mathcal{N}_i^{\text{int}}\}$ and let U_i be the solution of (9.1.5.6) corresponding to λ_{ie}, $e \in \mathcal{N}_i^{\text{int}}$, then $U = \{U_i\}_{i\in\mathcal{I}}$ is a solution of $a(U, \Phi) = 0$, $\forall \Phi \in \mathcal{V}$.

Since

$$\{\lambda_{ie}^\epsilon(U^n) : i \in \mathcal{I},\ e \in \mathcal{N}^{\text{int}}\} = (1-\epsilon)\mathcal{T}X^n + \epsilon X^n$$

we have $\lambda_{ie}^\epsilon(U^n) \to \lambda_{ie}$ weakly in $L^2(\Gamma_{ie})$. If we pass to the limit in (9.1.5.4) through the subsequence $n+1 = n_k$ we find that $\widehat{U}_i$ satisfies (9.1.5.6), that is, $\widehat{U}_i = U_i$, $\forall i \in \mathcal{I}$. Thus the convergence in (9.1.5.19) is through the entire sequence of positive integers and, furthermore, $\widehat{U}_{ie} = (Q_{ie}^+)^* D_{\nu_{ie}} U_i$.

It remains to show that $U_i^n \to U_i$ strongly in $\mathcal{H}^1(\mathcal{P}_i)$. Since $U_i^n \to U_i$ weakly in $\mathcal{V}_i$ we have $a_i(U_i^n, \Phi_i) \to a_i(U_i, \Phi_i)$, $\forall \Phi_i \in \mathcal{V}_i$. From (9.1.4.6) we have the estimate

$$C\|U_i^n - U_i\|^2_{\mathcal{H}_i^1} \leq a_i(U_i^n, U_i^n) + a_i(U_i, U_i) - 2a_i(U_i^n, U_i) + \sum_{e \in \mathcal{N}_i^{\text{int}}} \int_{\Gamma_{ie}} |Q_{ie}(U_i^n - U_i)|^2 d\Gamma.$$

Thus

$$C \sum_{i \in \mathcal{I}} \limsup_{n \to \infty} \|U_i^n - U_i\|^2_{\mathcal{H}_i^1} \leq -a(U, U) = 0$$

since $U \in \mathcal{V}$. This completes the proof of Theorem 9.1.5.1.

9.2 Optimal Control of Hyperbolic Systems on 2-D Networks

9.2.1 Optimal Final Value Control

We retain the notation of Section 9.1. For each $i \in \mathcal{I}$ and each edge $e = \Gamma_{ie} \in \mathcal{N}^C$, let α_{ie} be a positive scalar function defined on Γ_{ie}. We introduce continuous, symmetric, bilinear functional b on $\mathcal{V}$ by

$$b(W, \Phi) = \sum_{i \in \mathcal{I}} \sum_{e \in \mathcal{N}_i^{\text{ext}}} \int_{\Gamma_{ie}} \alpha_{ie} W_i \cdot \Phi_i \, d\Gamma. \tag{9.2.1.1}$$

Fix an interval $\{t : 0 < t < T\}$ and let $F : \mathcal{P} \times (0, T) \mapsto \mathbb{R}^m$ and, for each $i \in \mathcal{I}$ and $e \in \mathcal{N}_i^{\text{ext}}$, $f_{ie} : \Gamma_{ie} \times (0, T) \mapsto \mathbb{R}^m$ be given functions. We shall consider the evolutionary variational problem

$$\langle \frac{\partial^2 W}{\partial t^2}, \Phi \rangle_{\mathcal{V}} + a(W, \Phi) + b(\frac{\partial W}{\partial t}, \Phi) = (F, \Phi)_{\mathcal{H}} + \sum_{i \in \mathcal{I}} \sum_{e \in \mathcal{N}_i^{\text{ext}}} \int_{\Gamma_{ie}} f_{ie} \cdot \Phi_i \, d\Gamma, \quad \forall \Phi \in \mathcal{V},\ 0 < t < T, \tag{9.2.1.2}$$

$$W(0) = W_0, \quad \frac{\partial W}{\partial t}(0) = V_0. \tag{9.2.1.3}$$

A calculation similar to that of Section 9.1 shows that the W_i formally satisfy the system of equations

$$\frac{\partial^2 W_i}{\partial t^2} - \frac{\partial}{\partial x_\alpha}[A_i^{\alpha\beta}(W_{i,\beta} + B_i^\beta W_i)] + (B_i^\alpha)^* A_i^{\alpha\beta}(W_{i,\beta} + B_i^\beta W_i) + C_i W_i = F_i \ \text{ in } \mathcal{Q}_i := \mathcal{P}_i \times (0, T), \tag{9.2.1.4}$$

the exterior edge boundary conditions

$$\begin{aligned} &W_i = 0 \text{ on } \Sigma_{ie} := \Gamma_{ie} \times (0,T) \text{ when } \Gamma_{ie} \in \mathcal{D} \\ &\nu_{ie}^\alpha A_i^{\alpha\beta}(W_{i,\beta} + B_i^\beta W_i) + \alpha_{ie}\frac{\partial W_i}{\partial t} = f_{ie} \text{ on } \Sigma_{ie} \text{ when } \Gamma_{ie} \in \mathcal{N}^{\text{ext}} \end{aligned} \tag{9.2.1.5}$$

together with the interior edge geometric and mechanical conditions

$$\begin{aligned} &\Pi_{ie} W_i = 0 \text{ on } \Sigma_{ie}, \forall i \in \mathcal{I}_e,\ e \in \mathcal{N}^{\text{int}} \\ &Q_{ie}W_i = Q_{je}W_j \text{ on } e \times (0,T) \text{ when } \Gamma_{ie} = \Gamma_{je},\ e \in \mathcal{N}^{\text{int}} \\ &\sum_{i \in \mathcal{I}_e} (Q_{ie}^+)^* \nu_{ie}^\alpha A_i^{\alpha\beta}(W_{i,\beta} + B_i^\beta W_i) = 0 \text{ on } e \times (0,T) \text{ if } e \in \mathcal{N}^{\text{int}}. \end{aligned} \tag{9.2.1.6}$$

Let $L^2(\Gamma_{ie})$ denote the space of square summable $\mathbb{R}^m$-valued functions defined on Γ_{ie} with the standard norm and set $L^2(\Sigma_{ie}) = L^2(0,T;L^2(\Gamma_{ie}))$. It shall be shown below that under suitable conditions on the coefficient matrices and assuming that

$$(W_0, V_0) \in \mathcal{V} \times \mathcal{H}, \quad F \in L^1(0,T;\mathcal{H}), \quad f_{ie} \in L^2(\Sigma_{ie}) \tag{9.2.1.7}$$

the variational problem (9.2.1.2), (9.2.1.3) has a unique solution with regularity

$$\begin{aligned} &W \in C([0,T];\mathcal{V}) \cap C^1([0,T];\mathcal{H}) \cap C^2([0,T];\mathcal{V}^*) \\ &\left.\frac{\partial W_i}{\partial t}\right|_{\Sigma_{ie}} \in L^2(\Sigma_{ie}), \quad \forall e \in \mathcal{N}_i^{\text{ext}},\ i \in \mathcal{I}. \end{aligned} \tag{9.2.1.8}$$

Thus we may consider the optimal final value control problem

$$\inf_{f \in L^2(0,T;\mathcal{U})} J(f) \text{ subject to (9.2.1.2), (9.2.1.3)} \tag{9.2.1.9}$$

where

$$\begin{aligned} &\mathcal{U} = \prod_{i \in \mathcal{I}} \prod_{e \in \mathcal{N}_i^{\text{ext}}} L^2(\Gamma_{ie}) \\ &J(f) = \frac{1}{2}\int_0^T \|f(t)\|_{\mathcal{U}}^2\, dt + \frac{\kappa}{2}\|(W(T), \frac{\partial W}{\partial t}(T)) - (Z_0, Z_1)\|_{\mathcal{V}\times\mathcal{H}}^2 \end{aligned} \tag{9.2.1.10}$$

and where the target state $(Z_0, Z_1) \in \mathcal{V} \times \mathcal{H}$ is given and the penalty parameter $\kappa > 0$. As is noted in the following section, it follows from the material developed in Section 6.2.2 that the problem (9.2.1.9), (9.2.1.10) has a unique optimal control given by

$$f = \{f_{ie}\}, \text{ where } f_{ie} = -P_i|_{\Sigma_{ie}}, \quad i \in \mathcal{I},\ e \in \mathcal{N}_i^{\text{ext}}. \tag{9.2.1.11}$$

In (9.2.1.11), $P_i = P|_{\mathcal{Q}_i}$ where P is the solution of the backwards running adjoint system

$$\begin{aligned} \frac{\partial^2 P_i}{\partial t^2} - \frac{\partial}{\partial x_\alpha}[A_i^{\alpha\beta}(P_{i,\beta} + B_i^\beta P_i)] + (B_i^\alpha)^* A_i^{\alpha\beta}(P_{i,\beta} + B_i^\beta P_i)& \\ + C_i P_i = 0 \text{ in } \mathcal{Q}_i& \end{aligned} \tag{9.2.1.12}$$

$$\begin{aligned} &P_i = 0 \text{ on } \Sigma_{ie} \text{ when } \Gamma_{ie} \in \mathcal{D} \\ &\nu_{ie}^{\alpha} A_i^{\alpha\beta}(P_{i,\beta} + B_i^{\beta} P_i) - \alpha_{ie} d_t P_i = 0 \text{ on } \Sigma_{ie} \text{ when } \Gamma_{ie} \in \mathcal{N}^{\text{ext}} \end{aligned} \tag{9.2.1.13}$$

$$\begin{aligned} &\Pi_{ie} P_i = 0 \text{ on } \Sigma_{ie},\ \forall i \in \mathcal{I}_e,\ e \in \mathcal{N}^{\text{int}} \\ &Q_{ie} P_i = Q_{je} P_j \text{ on } e \times (0,T) \text{ when } \Gamma_{ie} = \Gamma_{je},\ e \in \mathcal{N}^{\text{int}} \\ &\sum_{i \in \mathcal{I}_e} (Q_{ie}^{+})^* \nu_{ie}^{\alpha} A_i^{\alpha\beta}(P_{i,\beta} + B_i^{\beta} P_i) = 0 \text{ on } e \times (0,T) \text{ if } e \in \mathcal{N}^{\text{int}}. \end{aligned} \tag{9.2.1.14}$$

$$P(T) = \kappa\Big(\frac{\partial W}{\partial t}(T) - Z_1\Big), \quad \frac{\partial P}{\partial t}(T) = -\kappa \mathcal{A}\big(W(T) - Z_0\big), \tag{9.2.1.15}$$

In (9.2.1.15), $\mathcal{A}$ denotes the Riesz isomorphism of $\mathcal{V}$ onto its dual space $\mathcal{V}^*$ while, in (9.2.1.13), d_t is the bounded linear operator

$$d_t : L^2(0,T; L^2(\Gamma_{ie})) \mapsto \big(H^1(0,T; L^2(\Gamma_{ie}))\big)^*$$

given by (6.2.2.27). Obviously the form of d_t is independent of the particular edge Γ_{ie}. The problem (9.2.1.12)–(9.2.1.15) has a unique solution, appropriately defined, with regularity

$$P \in C([0,T]; \mathcal{H}) \cap C^1([0,T]; \mathcal{V}^*), \quad P_i\big|_{\Sigma_{ie}} \in L^2(\Sigma_{ie}), \ i \in \mathcal{I},\ e \in \mathcal{N}_i^{\text{ext}}. \tag{9.2.1.16}$$

In fact, if $\mathcal{V}$ is appropriately normed the solution of (9.2.1.12)–(9.2.1.15) may be expressed in terms of a more regular quantity R as

$$P = \frac{\partial R}{\partial t}, \quad \frac{\partial P}{\partial t} = -\mathcal{A}R \tag{9.2.1.17}$$

where R is the solution of

$$\langle \frac{\partial^2 R}{\partial t^2}, \Phi \rangle_{\mathcal{V}} + a(R, \Phi) - b(\frac{\partial R}{\partial t}, \Phi) = 0, \quad \forall \Phi \in \mathcal{V},\ 0 < t < T, \tag{9.2.1.18}$$

$$R(T) = \kappa\big(W(T) - Z_0\big), \quad \frac{\partial R}{\partial t}(T) = \kappa\Big(\frac{\partial W}{\partial t}(T) - Z_1\Big). \tag{9.2.1.19}$$

9.2.2 Existence and Regularity of Solutions

It is assumed that the matrices $A_i^{\alpha\beta}$ B_i^{α} and C_i have entries that are in $L^\infty(\mathcal{P}_i)$, and that α_{ie} and $1/\alpha_{ie}$ are in $L^\infty(\Gamma_{ie})$ for all $i \in \mathcal{I}$ and $e \in \mathcal{N}_i^{\text{ext}}$. We further assume that the form $a(\Phi, \Phi)$ is strictly coercive on $\mathcal{V}$, that is

$$a(\Phi, \Phi) \geq K \|\Phi\|^2_{\mathcal{H}^1(\mathcal{P})}, \qquad \forall \Phi \in \mathcal{V}. \tag{9.2.2.1}$$

Then $\|\Phi\|_{\mathcal{V}} := \sqrt{a(\Phi, \Phi)}$ defines a norm on $\mathcal{V}$ that is equivalent to the induced $\mathcal{H}^1(\mathcal{P})$ norm. We shall always assume that $\mathcal{V}$ is normed according to $\|\Phi\|_{\mathcal{V}}$. Then the canonical isomorphism $\mathcal{A}$ of $\mathcal{V}$ onto its dual space $\mathcal{V}^*$ satisfies $\langle \mathcal{A}\Phi, \Psi \rangle_{\mathcal{V}} =$

$a(\Phi, \Psi)$ for all Φ, $\Psi \in \mathcal{V}$, where $\langle \Phi^*, \Phi \rangle_{\mathcal{V}}$ denotes the duality pairing between the elements $\Phi^* \in \mathcal{V}^*$ and $\Phi \in \mathcal{V}$.

We define the bounded linear operator $\mathcal{B} : \mathcal{U} \mapsto \mathcal{V}^*$ by

$$\langle \mathcal{B}g, \Phi \rangle_{\mathcal{V}} = \sum_{i \in \mathcal{I}} \sum_{e \in \mathcal{N}_i^{\text{ext}}} \int_{\Gamma_{ie}} \sqrt{\alpha_{ie}}\, g_{ie} \cdot \Phi_i \, d\Gamma, \quad \forall g = \{g_{ie}\} \in \mathcal{U},\ \Phi \in \mathcal{V}. \tag{9.2.2.2}$$

If we identify $\mathcal{U}$ with its dual space, the dual operator $\mathcal{B}^* : \mathcal{V} \mapsto \mathcal{U}$ of $\mathcal{B}$ is then given by

$$\mathcal{B}^* \Phi = \{ \sqrt{\alpha_{ie}}\, \Phi_i|_{\Gamma_{ie}} : i \in \mathcal{I},\ e \in \mathcal{N}_i^{\text{ext}} \}. \tag{9.2.2.3}$$

With this notation, the variational problem (9.2.1.2), (9.2.1.3) may be expressed in the abstract form

$$\begin{gathered} \frac{d^2 W}{dt^2} + \mathcal{B}\mathcal{B}^* \frac{dW}{dt} + \mathcal{A}W = F + \mathcal{B} f_\alpha \ \text{ in } \mathcal{V}^* \\ W(0) = W_0, \quad \frac{dW}{dt}(0) = V_0, \end{gathered} \tag{9.2.2.4}$$

where

$$f_\alpha = \{ f_{ie}/\sqrt{\alpha_{ie}} : i \in \mathcal{I},\ e \in \mathcal{N}_i^{\text{ext}} \}.$$

The problem (9.2.2.4) has the same structure as the problem (6.2.2.3) considered earlier, and theory developed in Section 6.2.2 applies to the present problem. Thus if the data satisfies (9.2.1.7), then (9.2.2.4) has a unique solution with regularity (9.2.1.8). Further, for any $\delta \in (0,1)$ we have the a priori estimate

$$\begin{aligned} (1-\delta)\|(W, \frac{\partial W}{\partial t})\|^2_{L^\infty(0,T;\mathcal{V}\times\mathcal{H})} + \sum_{i \in \mathcal{I}} \sum_{e \in \mathcal{N}_i^{\text{ext}}} \int_{\Sigma_{ie}} \left[\frac{1}{\alpha_{ie}} |D_{\nu_{ie}} W_i|^2 + \alpha_{ie} |\frac{\partial W_i}{\partial t}|^2 \right] d\Sigma \\ \le \|(W_0, V_0)\|^2_{\mathcal{V}\times\mathcal{H}} + \frac{1}{\delta}\|F\|^2_{L^1(0,T;\mathcal{H})} + \sum_{i \in \mathcal{I}} \sum_{e \in \mathcal{N}_i^{\text{ext}}} \int_{\Sigma_{ie}} \frac{1}{\alpha_{ie}} |f_{ie}|^2 \, d\Sigma \end{aligned} \tag{9.2.2.5}$$

where

$$D_{\nu_{ie}} W_i = \nu_{ie}^\alpha A_i^{\alpha\beta} (W_{i,\beta} + B_i^\beta W_i). \tag{9.2.2.6}$$

If $F = 0$ one has the identity in T

$$\begin{aligned} \|(W(T), \frac{\partial W}{\partial t}(T))\|^2_{\mathcal{V}\times\mathcal{H}} + \sum_{i \in \mathcal{I}} \sum_{e \in \mathcal{N}_i^{\text{ext}}} \int_{\Sigma_{ie}} \left[\frac{1}{\alpha_{ie}} |D_{\nu_{ie}} W_i|^2 + \alpha_{ie} |\frac{\partial W_i}{\partial t}|^2 \right] d\Sigma \\ = \|(W_0, V_0)\|^2_{\mathcal{V}\times\mathcal{H}} + \sum_{i \in \mathcal{I}} \sum_{e \in \mathcal{N}_i^{\text{ext}}} \int_{\Sigma_{ie}} \frac{1}{\alpha_{ie}} |f_{ie}|^2 \, d\Sigma. \end{aligned} \tag{9.2.2.7}$$

It also follows from the material of Section 6.2.2 that for any $f \in L^2(0,T;\mathcal{U})$ and $(P_0, R_0) \in \mathcal{H} \times \mathcal{V}^*$ the following problem has a unique solution with regularity (9.2.1.16):

$$\frac{\partial^2 P_i}{\partial t^2} - \frac{\partial}{\partial x_\alpha}[A_i^{\alpha\beta}(P_{i,\beta} + B_i^\beta P_i)] + (B_i^\alpha)^* A_i^{\alpha\beta}(P_{i,\beta} + B_i^\beta P_i) + C_i P_i = 0 \text{ in } Q_i \quad (9.2.2.8)$$

$$\begin{aligned} &P_i = 0 \text{ on } \Sigma_{ie} \text{ when } \Gamma_{ie} \in \mathcal{D} \\ &\nu_{ie}^\alpha A_i^{\alpha\beta}(P_{i,\beta} + B_i^\beta P_i) - \alpha_{ie} d_t P_i = d_t f_{ie} \text{ on } \Sigma_{ie} \text{ when } \Gamma_{ie} \in \mathcal{N}^{\text{ext}} \end{aligned} \quad (9.2.2.9)$$

$$\begin{aligned} &\Pi_{ie} P_i = 0 \text{ on } \Sigma_{ie},\ \forall i \in \mathcal{I}_e,\ e \in \mathcal{N}^{\text{int}} \\ &Q_{ie} P_i = Q_{je} P_j \text{ on } e \times (0,T) \text{ when } \Gamma_{ie} = \Gamma_{je},\ e \in \mathcal{N}^{\text{int}} \\ &\sum_{i \in \mathcal{I}_e} (Q_{ie}^+)^* \nu_{ie}^\alpha A_i^{\alpha\beta}(P_{i,\beta} + B_i^\beta P_i) = 0 \text{ on } e \times (0,T) \text{ if } e \in \mathcal{N}^{\text{int}}. \end{aligned} \quad (9.2.2.10)$$

$$P(T) = P_0, \quad \frac{\partial P}{\partial t}(T) = -R_0 \quad (9.2.2.11)$$

This solution may be defined by transposition and satisfies

$$\begin{aligned} &\langle (-\frac{\partial P}{\partial t}(0), P(0)), (V_0, V_1) \rangle_{\mathcal{V}\times\mathcal{H}} + \int_0^T (P(t), G(t))_{\mathcal{H}}\, dt + \int_0^T (\mathcal{B}^* P(t), g_\alpha(t))_{\mathcal{U}}\, dt \\ &= \langle (R_0, P_0), (V(T), \frac{\partial V}{\partial t}(T)) \rangle_{\mathcal{V}\times\mathcal{H}} - \int_0^T \big(f_\alpha(t), \mathcal{B}^* \frac{\partial V}{\partial t}(t)\big)_{\mathcal{U}}\, d\Sigma, \\ &\forall (V_0, V_1) \in \mathcal{V} \times \mathcal{H},\ G \in L^1(0,T;\mathcal{H}),\ g \in L^2(0,T;U), \end{aligned} \quad (9.2.2.12)$$

where V is the solution of

$$\begin{aligned} &\frac{d^2 V}{dt^2} + \mathcal{B}\mathcal{B}^* \frac{dV}{dt} + \mathcal{A} V = G + \mathcal{B} g_\alpha \\ &V(0) = V_0, \quad \frac{dV}{dt}(0) = V_1. \end{aligned}$$

In addition, this solution may be expressed as (9.2.1.17), where R is the solution of

$$\langle \frac{\partial^2 R}{\partial t^2}, \Phi \rangle_{\mathcal{V}} + a(R, \Phi) - b(\frac{\partial R}{\partial t}, \Phi) = \langle \mathcal{B} f_\alpha, \Phi \rangle_{\mathcal{V}}, \quad \forall \Phi \in \mathcal{V},\ 0 < t < T, \quad (9.2.2.13)$$

and

$$R(T) = -\mathcal{A}^{-1} R_0, \quad \frac{\partial R}{\partial t}(T) = P_0. \quad (9.2.2.14)$$

It is now easy to show that the optimality system for the optimal control problem (9.2.1.9), (9.2.1.10) is indeed given by the forward running problem (9.2.1.2),

(9.2.1.3), the backwards running problem (9.2.1.12)–(9.2.1.15), with the optimal control given by (9.2.1.11). Indeed, the optimal control $f = f_{\rm opt}$ is the unique solution of the variational equation

$$\int_0^T (f(t), g(t))_{\mathcal{U}} dt + \kappa\big[\langle \mathcal{A}(w(T) - Z_0), U(T)\rangle_{\mathcal{V}} + \big(\frac{\partial W}{\partial t}(T) - Z_1, \frac{\partial U}{\partial t}(T)\big)_{\mathcal{H}}\big] = 0, \quad \forall g \in \mathcal{U}, \quad (9.2.2.15)$$

where U is the solution of

$$\frac{d^2U}{dt^2} + \mathcal{B}\mathcal{B}^* \frac{dU}{dt} + \mathcal{A}U = \mathcal{B}g_\alpha, \quad U(0) = \frac{dU}{dt}(0) = 0.$$

Let P be the solution of (9.2.1.12)–(9.2.1.15). Then P satisfies (9.2.2.12) with

$$P_0 = \kappa\big(\frac{\partial W}{\partial t}(T) - Z_1\big), \quad R_0 = \kappa\mathcal{A}(W(T) - Z_0).$$

If we insert U in place of V in (9.2.2.12) we obtain

$$\int_0^T (\mathcal{B}^* P(t), g_\alpha(t))_{\mathcal{U}}\, dt$$
$$= \kappa\big[\langle \mathcal{A}(w(T) - Z_0), U(T)\rangle_{\mathcal{V}} + \big(\frac{\partial W}{\partial t}(T) - Z_1, \frac{\partial U}{\partial t}(T)\big)_{\mathcal{H}}\big], \quad \forall g \in \mathcal{U}. \quad (9.2.2.16)$$

It follows from (9.2.2.15) and (9.2.2.16) that $f_{\rm opt}$ is given by (9.2.1.11).

9.3 Decomposition of the Spatial Domain

As in the case of one-dimensional networks, the object of spatial domain decomposition on two-dimensional networks is to uncouple the global optimality system transmission conditions on the interior edges $\mathcal{N}^{\rm int}$ through an iteration such that the original global system on the network $\mathcal{P}$ is replaced by a collection of independent subsystems, each of which resides on a single polygon $\mathcal{P}_i$. As usual, the recursive conditions imposed on the interior edges must be such that communication among the polygons is restored in the limit, that is, the transmission conditions must be recovered in the limit.

Some preliminary material is required in order to properly formulate the subsystems in the decomposition. Let $\mathcal{H}_i$, $\mathcal{V}_i$, and $a_i(W_i, V_i)$ be the spaces and bilinear form defined in Section 9.1.4. It is assumed that

$$a_i(\Phi_i, \Phi_i) \geq K_i \|\Phi_i\|^2_{\mathcal{H}^1(\mathcal{P}_i)}, \quad \forall \Phi_i \in \mathcal{V}_i, \quad (9.3.1.1)$$

for some constant $K_i > 0$. We may then define a norm on $\mathcal{V}_i$ equivalent to the induced $\mathcal{H}^1(\mathcal{P}_i)$ by setting

$$\|\Phi_i\|_{\mathcal{V}_i} = \sqrt{a_i(\Phi_i, \Phi_i)}.$$

We identify the dual of $\mathcal{H}_i$ with $\mathcal{H}_i$ and denote by $\mathcal{V}_i^*$ the dual space of $\mathcal{V}_i$ with respect to $\mathcal{H}_i$. We define the continuous bilinear functional b_i on $\mathcal{V}_i$ by

$$b_i(W_i, \Phi_i) = \sum_{e \in \mathcal{N}_i^{\text{ext}}} \int_{\Gamma_{ie}} \alpha_{ie} W_i \cdot \Phi_i \, d\Gamma + \sum_{e \in \mathcal{N}_i^{\text{int}}} \int_{\Gamma_{ie}} \beta_e Q_{ie} W_i \cdot Q_{ie} \Phi_i \, d\Gamma.$$

where β_e is a positive constant independent of $i \in \mathcal{I}_e$. For each $i \in \mathcal{I}$ we consider the following local problems for functions W_i defined on $\mathcal{Q}_i$:

$$\begin{aligned} &\langle \frac{\partial^2 W_i}{\partial t^2}, \Phi_i \rangle_{\mathcal{V}_i} + a_i(W_i, \Phi_i) + b_i(\frac{\partial W_i}{\partial t}, \Phi_i) \\ &\quad = \big(F_i, \Phi_i\big)_{\mathcal{H}_i} + \sum_{e \in \mathcal{N}_i^{\text{ext}}} \int_{\Gamma_{ie}} f_{ie} \cdot \Phi_i \, d\Gamma + \sum_{e \in \mathcal{N}_i^{\text{int}}} \int_{\Gamma_{ie}} g_{ie} \cdot Q_{ie} \Phi_i \, d\Gamma, \\ &\qquad \forall \Phi \in \mathcal{V}_i, \ 0 < t < T, \end{aligned} \tag{9.3.1.2}$$

$$W_i(0) = W_{0i}, \quad \frac{\partial W_i}{\partial t}(0) = V_{0i} \tag{9.3.1.3}$$

where $(W_{0i}, V_{0i}) = (W_0, V_0)\big|_{\mathcal{P}_i}$. The boundary value problem corresponding to (9.3.1.2) is

$$\frac{\partial^2 W_i}{\partial t^2} - \frac{\partial}{\partial x_\alpha}\left[A_i^{\alpha\beta}(W_{i,\beta} + B_i^\beta W_i)\right] + (B_i^\alpha)^* A_i^{\alpha\beta}(W_{i,\beta} + B_i^\beta W_i) + C_i W_i = F_i \ \text{ in } \mathcal{Q}_i, \tag{9.3.1.4}$$

$$\begin{aligned} W_i &= 0 \ \text{ on } \Sigma_{ie}, \ e \in \mathcal{D}_i \\ D_{\nu_{ie}} W_i + \alpha_{ie} \frac{\partial W_i}{\partial t} &= f_{ie} \ \text{ on } \Sigma_{ie}, \ e \in \mathcal{N}_i^{\text{ext}} \\ \Pi_{ie} W_i &= 0 \ \text{ on } \Sigma_{ie}, \ e \in \mathcal{N}_i^{\text{int}} \\ (Q_{ie}^+)^* D_{\nu_{ie}} W_i + \beta_e Q_{ie} \frac{\partial W_i}{\partial t} &= g_{ie} \ \text{ on } \Sigma_{ie}, \ e \in \mathcal{N}_i^{\text{int}}. \end{aligned} \tag{9.3.1.5}$$

where the operator $D_{\nu_{ie}}$ is defined in (9.2.2.6).

Set

$$\mathcal{U}_i = \prod_{e \in \mathcal{N}_i^{\text{ext}}} L^2(\Gamma_{ie}) \times \prod_{e \in \mathcal{N}_i^{\text{int}}} L^2(\Gamma_{ie})$$

and define the bounded linear mapping $\mathcal{B}_i : \mathcal{U}_i \mapsto \mathcal{V}_i^*$ by

$$\langle \mathcal{B}_i(g_i, h_i), \Phi_i \rangle_{\mathcal{V}_i} = \sum_{e \in \mathcal{N}_i^{\text{ext}}} \int_{\Gamma_{ie}} \sqrt{\alpha_{ie}}\, g_{ie} \cdot \Phi_i \, d\Gamma + \sum_{e \in \mathcal{N}_i^{\text{int}}} \int_{\Gamma_{ie}} \sqrt{\beta_e}\, h_{ie} \cdot Q_{ie} \Phi_i \, d\Gamma.$$

Then the dual operator $\mathcal{B}_i^* : \mathcal{V}_i \mapsto \mathcal{U}_i$ (identifying $\mathcal{U}_i$ with its dual space) is given by

$$\mathcal{B}_i^* \Phi_i = \big(\{\sqrt{\alpha_{ie}}\, \Phi_i\big|_{\Gamma_{ie}}\}_{e \in \mathcal{N}^{\text{ext}}}, \{\sqrt{\beta_e}\, Q_{ie} \Phi_i\big|_{\Gamma_{ie}}\}_{e \in \mathcal{N}^{\text{int}}}\big).$$

With this notation the variational equation may be written as the abstract equation

$$\frac{d^2W_i}{dt^2} + \mathcal{B}_i\mathcal{B}_i^*\frac{dW_i}{dt} + \mathcal{A}_iW_i = F_i + \mathcal{B}_i(f_{i\alpha}, g_{i\beta}) \text{ in } \mathcal{V}_i^*, \tag{9.3.1.6}$$

where

$$f_{i\alpha} = \{\frac{1}{\sqrt{\alpha_{ie}}}f_{ie}\}_{e\in\mathcal{N}_i^{\text{ext}}}, \quad g_{i\beta} = \{\frac{1}{\sqrt{\beta_e}}g_{ie}\}_{e\in\mathcal{N}_i^{\text{int}}}.$$

From the previously developed existence theory, assuming that the global data satisfy (9.2.1.7), and that $g_{ie} \in L^2(\Sigma_{ie})$, $\forall e \in \mathcal{N}_i^{\text{int}}$, it follows that (9.3.1.2), (9.3.1.3) has a unique solution such that

$$\begin{aligned}
&(W_i, \frac{\partial W_i}{\partial t}) \in C([0,T]; \mathcal{V}_i \times \mathcal{H}_i)\\
&\frac{\partial W_i}{\partial t}\Big|_{\Sigma_{ie}} \in L^2(\Sigma_{ie}), \quad \forall e \in \mathcal{N}_i^{\text{ext}},\\
&Q_{ie}\frac{\partial W_i}{\partial t}\Big|_{\Sigma_{ie}} \in L^2(\Sigma_{ie}), \quad \forall e \in \mathcal{N}_i^{\text{int}}.
\end{aligned}$$

Further, one may derive an a priori estimate and an identity for this solution analogous to (9.2.2.5) and (9.2.2.6) above. For example, when $F = 0$ we have the identity in T

$$\begin{aligned}
&\|(W_i(T), \frac{\partial W_i}{\partial t}(T))\|^2_{\mathcal{V}_i\times\mathcal{H}_i} + \sum_{e\in\mathcal{N}_i^{\text{ext}}}\int_{\Sigma_{ie}}\Big[\frac{1}{\alpha_{ie}}|D_{\nu_{ie}}W_i|^2 + \alpha_{ie}|\frac{\partial W_i}{\partial t}|^2\Big]\,d\Sigma\\
&\quad + \sum_{e\in\mathcal{N}_i^{\text{int}}}\int_{\Sigma_{ie}}\Big[\frac{1}{\beta_e}|(Q_{ie}^+)^*D_{\nu_{ie}}W_i|^2 + \beta_e|Q_{ie}\frac{\partial W_i}{\partial t}|^2\Big]\,d\Sigma\\
&= \|(W_{0i}, V_{0i})\|^2_{\mathcal{V}_i\times\mathcal{H}_i} + \sum_{e\in\mathcal{N}_i^{\text{ext}}}\int_{\Sigma_{ie}}\frac{1}{\alpha_{ie}}|f_{ie}|^2\,d\Sigma + \sum_{e\in\mathcal{N}_i^{\text{int}}}\int_{\Sigma_{ie}}\frac{1}{\beta_e}|g_{ie}|^2\,d\Sigma.
\end{aligned} \tag{9.3.1.7}$$

Now consider the following problem

$$\frac{\partial^2 P_i}{\partial t^2} - \frac{\partial}{\partial x_{i,\alpha}}[A_i^{\alpha\beta}(P_{i,\beta} + B_i^\beta P_i)] + (B_i^\alpha)^*A_i^{\alpha\beta}(P_{i,\beta} + B_i^\beta P_i) + C_iP_i = 0 \text{ in } \mathcal{Q}_i, \tag{9.3.1.8}$$

$$\begin{aligned}
&P_i = 0 \text{ on } \Sigma_{ie},\ e \in \mathcal{D}_i\\
&D_{\nu_{ie}}P_i - \alpha_{ie}d_tP_i = d_th_{ie} \text{ on } \Sigma_{ie},\ e \in \mathcal{N}_i^{\text{ext}}\\
&\Pi_{ie}P_i = 0 \text{ on } \Sigma_{ie},\ e \in \mathcal{N}_i^{\text{int}}\\
&(Q_{ie}^+)^*D_{\nu_{ie}}P_i - \beta_ed_tQ_{ie}P_i = d_t\tilde{h}_{ie} \text{ on } \Sigma_{ie},\ e \in \mathcal{N}_i^{\text{int}},
\end{aligned} \tag{9.3.1.9}$$

$$P_i(T) = P_{0i}, \quad \frac{\partial P_i}{\partial t}(T) = -R_{0i}. \tag{9.3.1.10}$$

If

$$(P_{0i}, R_{0i}) \in \mathcal{H}_i \times \mathcal{V}_i^*, \quad (h, \tilde{h}) = \left(\{h_{ie}\}_{e\in\mathcal{N}_i^{\text{ext}}}, \{\tilde{h}_{ie}\}_{e\in\mathcal{N}_i^{\text{int}}}\right) \in \mathcal{U}_i$$

(9.3.1.9), (9.3.1.10) has a unique solution which may be defined by transposition. This solution satisfies (cf. (9.2.2.12))

$$\begin{aligned}
&\langle(-\frac{\partial P_i}{\partial t}(0), P_i(0)), (V_{0i}, V_{1i})\rangle_{\mathcal{V}_i\times\mathcal{H}_i} \\
&\qquad + \int_0^T (P_i(t), G_i(t))_{\mathcal{H}_i}\, dt + \int_0^T (\mathcal{B}_i^* P_i(t), (f_{i\alpha}(t), g_{i\beta}(t))_{\mathcal{U}_i}\, dt \\
&= \langle(R_{0i}, P_{0i}), (V_i(T), \frac{\partial V_i}{\partial t}(T))\rangle_{\mathcal{V}_i\times\mathcal{H}_i} - \int_0^T \left((h_{i\alpha}(t), \tilde{h}_{i\beta}(t)), \mathcal{B}_i^* \frac{\partial V_i}{\partial t}(t)\right)_{\mathcal{U}_i} d\Sigma, \\
&\quad \forall\, (V_{0i}, V_{1i}) \in \mathcal{V}_i \times \mathcal{H}_i,\ G_i \in L^1(0,T;\mathcal{H}_i),\ (f_i, g_i) \in L^2(0,T;U_i),
\end{aligned} \tag{9.3.1.11}$$

where V_i is the solution of

$$\begin{gathered}
\frac{d^2 V_i}{dt^2} + \mathcal{B}_i\mathcal{B}_i^* \frac{dV_i}{dt} + \mathcal{A}_i V_i = G_i + \mathcal{B}_i(f_{i\alpha}, g_{i\beta}) \\
V_i(0) = V_{0i}, \quad \frac{dV_i}{dt}(0) = V_{1i}.
\end{gathered}$$

The solution of (9.3.1.11) has regularity

$$\begin{gathered}
(P_i, \frac{\partial P_i}{\partial t}) \in C([0,T]; \mathcal{H}_i \times \mathcal{V}_i^*) \\
P_i\big|_{\Sigma_{ie}} \in L^2(\Sigma_{ie}), \quad \forall e \in \mathcal{N}_i^{\text{ext}} \\
Q_{ie} P_i\big|_{\Sigma_{ie}} \in L^2(\Sigma_{ie}), \quad \forall e \in \mathcal{N}_i^{\text{int}}.
\end{gathered} \tag{9.3.1.12}$$

This is due to the fact that P_i may be expressed as

$$P_i = \frac{\partial R_i}{\partial t}, \quad \frac{\partial P_i}{\partial t} = -\mathcal{A}_i R_i \tag{9.3.1.13}$$

where R_i is the solution of the more regular problem

$$\frac{d^2 R_i}{dt^2} - \mathcal{B}_i\mathcal{B}_i^* \frac{dR_i}{dt} + \mathcal{A}_i R_i = \mathcal{B}_i(h_{i\alpha}, \tilde{h}_{i\beta}) \tag{9.3.1.14}$$

and

$$R_i(T) = -\mathcal{A}_i^{-1} R_{0i}, \quad \frac{\partial R_i}{\partial t}(T) = P_{0i}. \tag{9.3.1.15}$$

9.3.2 The Decomposition Algorithm

In place of the transmission conditions

$$\begin{gathered}
Q_{ie}W_i = Q_{je}W_j, \quad Q_{ie}W_i = Q_{je}W_j \ \text{ on } e \times (0,T) \text{ if } \Gamma_{ie} = \Gamma_{je},\ e \in \mathcal{N}^{\text{int}} \\
\sum_{i\in\mathcal{I}_e} (Q_{ie}^+)^* D_{\nu_{ie}} W_i = \sum_{i\in\mathcal{I}_e} (Q_{ie}^+)^* D_{\nu_{ie}} P_i = 0 \ \text{ on } e \times (0,T) \text{ if } e \in \mathcal{N}^{\text{int}}
\end{gathered} \tag{9.3.2.1}$$

on the global network $\mathcal{P}$ consider the following iteration on an interior edge $e \in \mathcal{N}_i^{\text{int}}$ of the polygon $\mathcal{P}_i$:

$$\begin{aligned}(Q_{ie}^{+})^* D_{\nu_{ie}} W_i^{n+1} + \beta_e Q_{ie} \frac{\partial W_i^{n+1}}{\partial t} + \gamma_e Q_{ie} P_i^{n+1} &= \lambda_{ie}^n \\ (Q_{ie}^{+})^* D_{\nu_{ie}} P_i^{n+1} - \beta_e d_t Q_{ie} P_i^{n+1} + \gamma_e d_t Q_{ie} \frac{\partial W_i^{n+1}}{\partial t} &= \mu_{ie}^n\end{aligned} \tag{9.3.2.2}$$

where β_e, γ_e are positive constants independent of $i \in \mathcal{I}_e$, and where, for $n \geq 1$,

$$\begin{aligned}\lambda_{ie}^n = \lambda_{ie}(W^n, P^n) := &-\frac{2}{d_e} \sum_{j \in \mathcal{I}_e} (Q_{je}^{+})^* D_{\nu_{je}} W_j^n + (Q_{ie}^{+})^* D_{\nu_{ie}} W_i^n \\ &+ \frac{2}{d_e} \sum_{j \in \mathcal{I}_e} \beta_e Q_{je} \frac{\partial W_j^n}{\partial t} - \beta_e Q_{ie} \frac{\partial W_i^n}{\partial t} \\ &+ \frac{2}{d_e} \sum_{j \in \mathcal{I}_e} \gamma_e Q_{je} P_j^n - \gamma_e Q_{ie} P_i^n\end{aligned} \tag{9.3.2.3}$$

$$\begin{aligned}\mu_{ie}^n = \mu_{ie}(W^n, P^n) := &-\frac{2}{d_e} \sum_{j \in \mathcal{I}_e} (Q_{je}^{+})^* D_{\nu_{je}} P_j^n + (Q_{ie}^{+})^* D_{\nu_{ie}} P_i^n \\ &+ \frac{2}{d_e} \sum_{j \in \mathcal{I}_e} \gamma_e d_t Q_{je} \frac{\partial W_j^n}{\partial t} - \gamma_e d_t Q_{ie} \frac{\partial W_i^n}{\partial t} \\ &- \frac{2}{d_e} \sum_{j \in \mathcal{I}_e} \beta_e d_t Q_{je} P_j^n + \beta_e d_t Q_{ie} P_i^n.\end{aligned} \tag{9.3.2.4}$$

The data λ_{ie}^0, μ_{ie}^0 are arbitrary save for a regularity requirement.

The iteration (9.3.2.2)–(9.3.2.4) is consistent with the transmission conditions (9.3.2.1). Indeed, if we formally pass to the limit in (9.3.2.2)–(9.3.2.4) and then sum each equation in (9.3.2.2) over the index i we immediately obtain $(9.3.2.1)_2$, and therefore (9.3.2.2)–(9.3.2.4) reduce to

$$\begin{aligned}\beta_e Q_{ie} \frac{\partial W_i}{\partial t} + \gamma_e Q_{ie} P_i &= \frac{1}{d_e} \sum_{j \in \mathcal{I}_e} \beta_e Q_{je} \frac{\partial W_j}{\partial t} + \frac{1}{d_e} \sum_{j \in \mathcal{I}_e} \gamma_e Q_{je} P_j^n \\ -\beta_e d_t Q_{ie} P_i + \gamma_e d_t Q_{ie} \frac{\partial W_i}{\partial t} &= -\frac{1}{d_e} \sum_{j \in \mathcal{I}_e} \beta_e d_t Q_{je} P_j + \frac{1}{d_e} \sum_{j \in \mathcal{I}_e} \gamma_e d_t Q_{je} \frac{\partial W_j}{\partial t}.\end{aligned} \tag{9.3.2.5}$$

Since $\mathcal{N}(d_t) = \{0\}$ we may drop the operator d_t in (9.3.2.5). If we solve the resulting linear system for $Q_{ie} P_i$ and $Q_{ie} \partial W_i / \partial t$ we obtain

$$Q_{ie} \frac{\partial W_i}{\partial t}\bigg|_{\Sigma_{ie}} = \frac{1}{d_e} \sum_{j \in \mathcal{I}_e} Q_{je} \frac{\partial W_j}{\partial t}\bigg|_{\Sigma_{je}}, \qquad Q_{ie} P_i|_{\Sigma_{ie}} = \frac{1}{d_e} \sum_{j \in \mathcal{I}_e} Q_{je} P_j|_{\Sigma_{je}}$$

and therefore

$$Q_{ie}P_i = Q_{je}P_j, \qquad Q_{ie}\frac{\partial W_i}{\partial t} = Q_{je}\frac{\partial W_j}{\partial t} \quad \text{on } e \times (0,T) \text{ when } \Gamma_{ie} = \Gamma_{je}.$$

Since $W(0) \in \mathcal{V}$ we have $Q_{ie}W_i(0)\big|_{\Gamma_{ie}} = Q_{je}W_j(0)\big|_{\Gamma_{je}}$, hence

$$Q_{ie}W_i = Q_{je}W_j \quad \text{on } e \times (0,T) \text{ when } \Gamma_{ie} = \Gamma_{je}.$$

For reference we write down the iteratively decoupled optimality system.

$$\frac{\partial^2 W_i^{n+1}}{\partial t^2} - \frac{\partial}{\partial x_\alpha}\left[A_i^{\alpha\beta}(W_{i,\beta}^{n+1} + B_i^\beta W_i^{n+1})\right] + (B_i^\alpha)^* A_i^{\alpha\beta}\big(W_{i,\beta}^{n+1} + B_i^\beta W_i^{n+1}\big) + C_i W_i^{n+1} = F_i \ \text{ in } \mathcal{Q}_i, \tag{9.3.2.6}$$

$$\frac{\partial^2 P_i^{n+1}}{\partial t^2} - \frac{\partial}{\partial x_{i,\alpha}}\left[A_i^{\alpha\beta}(P_{i,\beta}^{n+1} + B_i^\beta P_i^{n+1})\right] + (B_i^\alpha)^* A_i^{\alpha\beta}\big(P_{i,\beta}^{n+1} + B_i^\beta P_i^{n+1}\big) + C_i P_i^{n+1} = 0 \ \text{ in } \mathcal{Q}_i; \tag{9.3.2.7}$$

$$\begin{aligned} W_i^{n+1} &= P_i^{n+1} = 0 \ \text{ on } \Sigma_{ie},\ e \in \mathcal{D}_i,\\ \Pi_{ie}W_i^{n+1} &= \Pi_{ie}P_i^{n+1} = 0 \ \text{ on } \Sigma_{ie},\ e \in \mathcal{N}_i^{\text{int}}, \end{aligned} \tag{9.3.2.8}$$

$$\begin{cases} D_{\nu_{ie}}W_i^{n+1} + \alpha_{ie}\dfrac{\partial W_i^{n+1}}{\partial t} + P_i^{n+1} = 0,\\ D_{\nu_{ie}}P_i^{n+1} - \alpha_{ie}d_t P_i^{n+1} = 0 \qquad \text{on } \Sigma_{ie},\ e \in \mathcal{N}_i^{\text{ext}}; \end{cases} \tag{9.3.2.9}$$

$$\begin{cases} (Q_{ie}^+)^* D_{\nu_{ie}}W_i^{n+1} + \beta_e Q_{ie}\dfrac{\partial W_i^{n+1}}{\partial t} + \gamma_e Q_{ie}P_i^{n+1} = \lambda_{ie}^n,\\ (Q_{ie}^+)^* D_{\nu_{ie}}P_i^{n+1} - \beta_e d_t Q_{ie}P_i^{n+1} + \gamma_e d_t Q_{ie}\dfrac{\partial W_i^{n+1}}{\partial t} = \mu_{ie}^n \\ \qquad\qquad \text{on } \Sigma_{ie},\ e \in \mathcal{N}_i^{\text{int}}, \end{cases} \tag{9.3.2.10}$$

where λ_{ie}^n, μ_{ie}^n are given by (9.3.2.3), (9.3.2.4);

$$\begin{aligned} W_i^{n+1}(0) &= W_{0i}, \quad \frac{\partial W_i^{n+1}}{\partial t}(0) = V_{0i}\\ P_i^{n+1}(T) &= \kappa\Big(\frac{\partial W_i^{n+1}}{\partial t}(T) - Z_{1i}\Big),\\ \frac{\partial P_i^{n+1}}{\partial t}(T) &= -\kappa\mathcal{A}_i\big(W_i^{n+1}(T) - Z_{0i}\big). \end{aligned} \tag{9.3.2.11}$$

Of course, it has to be demonstrated that the iterative system (9.3.2.6)–(9.3.2.11) is well posed for $n = 0, 1, \ldots$, provided the global data satisfy (9.2.1.7) and the local data λ_{ie}^0, μ_{ie}^0 are suitably regular. As usual, this is done by showing that (9.3.2.6)–(9.3.2.11) is itself an optimality system concentrated on $\mathcal{Q}_i$. Suppose that for a fixed n, λ_{ie}^n and μ_{ie}^n have regularity

$$\lambda_{ie}^n \in L^2(\Sigma_{ie}), \quad \mu_{ie}^n = d_t\rho_{ie}^n \text{ where } \rho_{ie}^n \in L^2(\Sigma_{ie}). \tag{9.3.2.12}$$

Consider the LQR problem

$$\inf_{(f_i,g_i)\in\mathcal{U}_i} J_i^n(f,g),$$

where

$$\begin{aligned} J_i^n(f,g) = \frac{1}{2}\sum_{e\in\mathcal{N}_i^{\text{ext}}} \int_{\Sigma_{ie}} |f_{ie}|^2\, d\Sigma \\ + \frac{1}{2}\sum_{e\in\mathcal{N}_i^{\text{int}}} \frac{1}{\gamma_e}\int_{\Sigma_{ie}} \left(|g_{ie}|^2 + \left|\gamma_e Q_{ie}\frac{\partial W_i}{\partial t} - \rho_{ie}^n\right|^2\right) d\Sigma \\ + \frac{\kappa}{2}\left\|\left(W_i(T), \frac{\partial W_i}{\partial t}(T)\right) - (Z_{0i}, Z_{1i})\right\|^2_{\mathcal{V}_i\times\mathcal{H}_i} \end{aligned}$$

subject to

$$\frac{\partial^2 W_i}{\partial t^2} - \frac{\partial}{\partial x_\alpha}\left[A_i^{\alpha\beta}(W_{i,\beta} + B_i^\beta W_i)\right] + (B_i^\alpha)^* A_i^{\alpha\beta}(W_{i,\beta} + B_i^\beta W_i) + C_i W_i = F_i \text{ in } \mathcal{Q}_i \tag{9.3.2.13}$$

$$\begin{aligned} W_i &= 0 \text{ on } \Sigma_{ie},\ e\in\mathcal{D}_i \\ D_{\nu_{ie}} W_i + \alpha_{ie}\frac{\partial W_i}{\partial t} &= f_{ie} \text{ on } \Sigma_{ie},\ e\in\mathcal{N}_i^{\text{ext}} \\ \Pi_{ie} W_i &= 0 \text{ on } \Sigma_{ie},\ e\in\mathcal{N}_i^{\text{int}} \\ (Q_{ie}^+)^* D_{\nu_{ie}} W_i + \beta_e Q_{ie}\frac{\partial W_i}{\partial t} &= \lambda_{ie}^n + g_{ie} \text{ on } \Sigma_{ie},\ e\in\mathcal{N}_i^{\text{int}} \end{aligned} \tag{9.3.2.14}$$

$$W_i(0) = W_{0i}, \quad \frac{\partial W_i}{\partial t}(0) = V_{0i}. \tag{9.3.2.15}$$

This optimal problem is well posed. The unique optimal controls (f,g) are the solutions of the variational equation

$$\begin{aligned} \sum_{e\in\mathcal{N}_i^{\text{ext}}} \int_{\Sigma_{ie}} f_{ie}\cdot\hat{f}_{ie}\, d\Sigma + \sum_{e\in\mathcal{N}_i^{\text{int}}} \frac{1}{\gamma_e}\int_{\Sigma_{ie}} g_{ie}\cdot\hat{g}_{ie}\, d\Sigma \\ + \sum_{e\in\mathcal{N}_i^{\text{int}}} \int_{\Sigma_{ie}} \left(\gamma_e Q_{ie}\frac{\partial W_i}{\partial t} - \rho_{ie}^n\right)\cdot\left(Q_{ie}\frac{\partial \widehat{W}_i}{\partial t}\right) d\Sigma \\ + \kappa\left[\left\langle \mathcal{A}_i(W_i(T) - Z_{0i}), \widehat{W}_i(T)\right\rangle_{\mathcal{V}_i} + \left(\frac{\partial W_i}{\partial t}(T) - Z_{1i}, \frac{\partial \widehat{W}_i}{\partial t}(T)\right)_{\mathcal{H}_i}\right] = 0, \\ \forall(\hat{h}_i, \hat{g}_i)\in\mathcal{U}_i, \end{aligned} \tag{9.3.2.16}$$

where $\widehat{W}_i$ is the solution of

$$\frac{\partial^2 \widehat{W}_i}{\partial t^2} - \frac{\partial}{\partial x_\alpha}[A_i^{\alpha\beta}(\widehat{W}_{i,\beta}+B_i^\beta \widehat{W}_i)] + (B_i^\alpha)^* A_i^{\alpha\beta}(\widehat{W}_{i,\beta}+B_i^\beta \widehat{W}_i) + C_i \widehat{W}_i = 0 \text{ in } Q_i$$

$$\widehat{W}_i = 0 \text{ on } \Sigma_{ie},\ e \in \mathcal{D}_i$$

$$D_{\nu_{ie}} \widehat{W}_i + \alpha_{ie} \frac{\partial \widehat{W}_i}{\partial t} = \hat{f}_{ie} \text{ on } \Sigma_{ie},\ e \in \mathcal{N}_i^{\text{ext}}$$

$$\Pi_{ie} \widehat{W}_i = 0 \text{ on } \Sigma_{ie},\ e \in \mathcal{N}_i^{\text{int}}$$

$$(Q_{ie}^+)^* D_{\nu_{ie}} \widehat{W}_i + \beta_e Q_{ie} \frac{\partial \widehat{W}_i}{\partial t} = \hat{g}_{ie} \text{ on } \Sigma_{ie},\ e \in \mathcal{N}_i^{\text{int}}$$

$$\widehat{W}_i(0) = \frac{\partial \widehat{W}_i}{\partial t}(0) = 0.$$

Let P_i be the solution of

$$\frac{\partial^2 P_i}{\partial t^2} - \frac{\partial}{\partial x_{i,\alpha}}[A_i^{\alpha\beta}(P_{i,\beta} + B_i^\beta P_i)] + (B_i^\alpha)^* A_i^{\alpha\beta}(P_{i,\beta} + B_i^\beta P_i) + C_i P_i = 0 \text{ in } Q_i, \quad (9.3.2.17)$$

$$\begin{aligned} P_i &= 0 \text{ on } \Sigma_{ie},\ e \in \mathcal{D}_i \\ D_{\nu_{ie}} P_i - \alpha_{ie} d_t P_i &= 0 \text{ on } \Sigma_{ie},\ e \in \mathcal{N}_i^{\text{ext}} \\ \Pi_{ie} P_i &= 0 \text{ on } \Sigma_{ie},\ e \in \mathcal{N}_i^{\text{int}} \end{aligned} \quad (9.3.2.18)$$

$$(Q_{ie}^+)^* D_{\nu_{ie}} P_i - \beta_e d_t Q_{ie} P_i = -\gamma_e d_t Q_{ie} \frac{\partial W_i}{\partial t} + \mu_{ie}^n \text{ on } \Sigma_{ie},\ e \in \mathcal{N}_i^{\text{int}},$$

$$P_i(T) = \kappa\Big(\frac{\partial W_i}{\partial t}(T) - Z_{1i}\Big), \quad \frac{\partial P_i}{\partial t}(T) = -\kappa \mathcal{A}_i\big(W_i(T) - Z_{0i}\big). \quad (9.3.2.19)$$

According to the *definition* of the solution of this problem, P_i satisfies (9.3.1.11) with

$$P_0 = \kappa\Big(\frac{\partial W_i}{\partial t}(T) - Z_{1i}\Big), \quad R_0 = \kappa \mathcal{A}_i\big(W_i(T) - Z_{0i}\big)$$

$$h_{ie} = 0, \quad \tilde{h}_{ie} = -\gamma_e Q_{ie} \frac{\partial W_i}{\partial t} + \rho_{ie}^n$$

If in (9.3.1.11) we replace V_i by $\widehat{W}_i$ we obtain

$$\sum_{e \in \mathcal{N}_i^{\text{ext}}} \int_{\sigma_{ie}} P_i \cdot \hat{f}_{ie}\, d\Sigma + \sum_{e \in \mathcal{N}_i^{\text{int}}} \int_{\sigma_{ie}} Q_{ie} P_i \cdot \hat{g}_{ie}\, d\Sigma$$

$$= \kappa\Big[\big\langle \mathcal{A}_i(W_i(T) - Z_{0i}), \widehat{W}_i(T)\big\rangle_{\mathcal{V}_i} + \Big(\frac{\partial W_i}{\partial t}(T) - Z_{1i}, \frac{\partial \widehat{W}_i}{\partial t}(T)\Big)_{\mathcal{H}_i}\Big]$$

$$- \sum_{e \in \mathcal{N}_i^{\text{int}}} \int_{\sigma_{ie}} \Big(-\gamma_e Q_{ie} \frac{\partial W_i}{\partial t} + \rho_{ie}^n\Big) \cdot \Big(Q_{ie} \frac{\partial \widehat{W}_i}{\partial t}\Big)\, d\Sigma, \quad \forall (\hat{h}_i, \hat{g}_i) \in \mathcal{U}_i. \quad (9.3.2.20)$$

It follows from (9.3.2.16) and (9.3.2.20) that the optimal controls are given by

$$f_{ie} = -P_i\big|_{\Sigma_{ie}}, \quad e \in \mathcal{N}_i^{\text{ext}},$$
$$g_{ie} = -\gamma_e Q_{ie} P_i\big|_{\Sigma_{ie}}, \quad e \in \mathcal{N}_i^{\text{int}}. \tag{9.3.2.21}$$

Upon inserting the optimal controls into the system (9.3.2.13)–(9.3.2.15), it is seen that the *optimality system* consisting of (9.3.2.13)–(9.3.2.15), (9.3.2.17)–(9.3.2.19), and (9.3.2.21), coincides with the system (9.3.2.6)–(9.3.2.11). Thus the latter is well posed if (9.3.2.12) is satisfied.

Let us once again observe that the adjoint variable P_i^{n+1} in (9.3.2.6)–(9.3.2.11) may be expressed as

$$P_i^{n+1} = \frac{\partial R_i^{n+1}}{\partial t}, \quad \frac{\partial P_i^{n+1}}{\partial t} = -\mathcal{A}_i R_i^{n+1} \tag{9.3.2.22}$$

where R_i is the solution of

$$\begin{aligned}
&\frac{\partial^2 R_i^{n+1}}{\partial t^2} - \frac{\partial}{\partial x_{i,\alpha}}\big[A_i^{\alpha\beta}(R_{i,\beta}^{n+1} + B_i^\beta R_i^{n+1})\big] + (B_i^\alpha)^* A_i^{\alpha\beta}(R_{i,\beta}^{n+1} + B_i^\beta R_i^{n+1})\\
&\qquad\qquad\qquad\qquad\qquad\qquad\qquad\qquad + C_i R_i^{n+1} = 0 \text{ in } Q_i,\\
&R_i^{n+1} = 0 \text{ on } \Sigma_{ie},\ e \in \mathcal{D}_i\\
&D_{\nu_{ie}} R_i^{n+1} - \alpha_{ie}\frac{\partial R_i^{n+1}}{\partial t} = 0 \text{ on } \Sigma_{ie},\ e \in \mathcal{N}_i^{\text{ext}}\\
&\Pi_{ie} R_i^{n+1} = 0 \text{ on } \Sigma_{ie},\ e \in \mathcal{N}_i^{\text{int}}\\
&(Q_{ie}^+)^* D_{\nu_{ie}} R_i^{n+1} - \beta_e Q_{ie}\frac{\partial R_i^{n+1}}{\partial t} + \gamma_e Q_{ie}\frac{\partial W_i^{n+1}}{\partial t} = \rho_{ie}^n \text{ on } \Sigma_{ie},\ e \in \mathcal{N}_i^{\text{int}},\\
&R_i^{n+1}(T) = \kappa\big(W_i^{n+1}(T) - Z_{0i}\big), \quad \frac{\partial R_i^{n+1}}{\partial t}(T) = \kappa\big(\frac{\partial W_i^{n+1}}{\partial t}(T) - Z_{1i}\big).
\end{aligned} \tag{9.3.2.23}$$

From (9.3.2.22) and $(9.3.2.23)_5$ we have, for each $j \in \mathcal{I}_e$,

$$\begin{aligned}
\langle d_t(Q_{je}^+)^* D_{\nu_{je}} R_j^{n+1}, \Phi\rangle_{\Sigma_{je}} &= -\int_{\Sigma_{je}} \big(\beta_e Q_{je}\frac{\partial R_j^{n+1}}{\partial t} - \gamma_e Q_{je}\frac{\partial W_j^{n+1}}{\partial t} + \rho_{je}^n\big)\cdot\frac{\partial \Phi}{\partial t}\, d\Sigma\\
&= \int_{\Sigma_{je}} \big(\beta_e d_t Q_{je} P_j^{n+1} - \gamma_e d_t Q_{je}\frac{\partial W_j^{n+1}}{\partial t} + \mu_{je}^n\big)\cdot\Phi\, d\Sigma\\
&= \langle (Q_{je}^+)^* D_{\nu_{je}} P_j^{n+1}, \Phi\rangle_{\Sigma_{ie}}, \quad \forall \Phi \in H^1(0,T; L^2(\Gamma_{ie})).
\end{aligned}$$

Therefore

$$\mu_{ie}^{n+1} = d_t \rho_{ie}^{n+1}, \qquad \forall i \in \mathcal{I}_e,$$

where μ_{ie}^{n+1} is given by (9.3.2.4) and where

$$\begin{aligned}\rho_{ie}^{n+1} = \rho_{ie}(W^{n+1}, R^{n+1}) &:= -\frac{2}{d_e}\sum_{j\in\mathcal{I}_e}(Q_{je}^{+})^* D_{\nu_{je}} R_j^{n+1} + (Q_{ie}^{+})^* D_{\nu_{ie}} R_i^{n+1} \\ &+ \frac{2}{d_e}\sum_{j\in\mathcal{I}_e}\gamma_e Q_{je}\frac{\partial W_j^{n+1}}{\partial t} - \gamma_e Q_{ie}\frac{\partial W_i^{n+1}}{\partial t} \\ &- \frac{2}{d_e}\sum_{j\in\mathcal{I}_e}\beta_e Q_{je}\frac{\partial R_j^{n+1}}{\partial t} + \beta_e Q_{ie}\frac{\partial R_i^{n+1}}{\partial t} \in L^2(\Sigma_{ie}). \quad (9.3.2.24)\end{aligned}$$

It follows that the iterative scheme (9.3.2.6)–(9.3.2.11) is well posed for $n = 0, 1, \ldots$ provided that the global data satisfy (9.2.1.7) and the local data λ_{ie}^0, μ_{ie}^0 satisfy (9.3.2.12) for $n = 0$.

Let us summarize the above discussion in the following theorem.

Theorem 9.3.2.1 *Let β_e, γ_e be positive constants for each $e \in \mathcal{N}^{int}$. Assume that the global data satisfy (9.2.1.7) and that the local data satisfy $\lambda_{ie}^0 \in L^2(\Sigma_{ie})$, $\mu_{ie}^0 = d_t\rho_{ie}^0$ for some $\rho_{ie}^0 \in L^2(\Sigma_{ie})$ and each $i \in \mathcal{I}$ and $e \in \mathcal{N}_i^{int}$. Then the iterative scheme (9.3.2.6)–(9.3.2.11) is well posed for $n = 0, 1, \ldots$. Its solution has regularity*

$$\begin{aligned}&(W_i^{n+1}, \frac{\partial W_i^{n+1}}{\partial t}) \in C([0,T]; \mathcal{V}_i \times \mathcal{H}_i), \quad (P_i^{n+1}, \frac{\partial P_i^{n+1}}{\partial t}) \in C([0,T]; \mathcal{H}_i \times \mathcal{V}_i^*) \\ &\left.\frac{\partial W_i^{n+1}}{\partial t}\right|_{\Sigma_{ie}} \in L^2(\Sigma_{ie}), \quad P_i^{n+1}|_{\Sigma_{ie}} \in L^2(\Sigma_{ie}), \quad \forall e \in \mathcal{N}_i^{ext} \cup \mathcal{N}_i^{int}.\end{aligned}$$

The adjoint variable P_i^{n+1} may be expressed in terms of a more regular variable R_i^{n+1} according to (9.3.2.22), where R_i^{n+1} is the solution of (9.3.2.23) and ρ_{ie}^n is given by (9.3.2.24) for $n \geq 1$.

9.3.3 Convergence of the Algorithm

As usual we first relax the iteration step (9.3.2.2) by introducing the relaxation parameter $\epsilon \in [0, 1)$ and replacing (9.3.2.2) by

$$\begin{aligned}(Q_{ie}^{+})^* D_{\nu_{ie}} W_i^{n+1} + \beta_e Q_{ie}\frac{\partial W_i^{n+1}}{\partial t} + \gamma_e Q_{ie} P_i^{n+1} &= (1-\epsilon)\lambda_{ie}^n \\ +\epsilon\big((Q_{ie}^{+})^* D_{\nu_{ie}} W_i^n + \beta_e Q_{ie}\frac{\partial W_i^n}{\partial t} &+ \gamma_e Q_{ie} P_i^n\big)\end{aligned} \quad (9.3.3.1)$$

$$\begin{aligned}(Q_{ie}^{+})^* D_{\nu_{ie}} P_i^{n+1} - \beta_e d_t Q_{ie} P_i^{n+1} + \gamma_e d_t Q_{ie}\frac{\partial W_i^{n+1}}{\partial t} &= (1-\epsilon)\mu_{ie}^n \\ +\epsilon\big((Q_{ie}^{+})^* D_{\nu_{ie}} P_i^n - \beta_e d_t Q_{ie} P_i^n + \gamma_e d_t Q_{ie}&\frac{\partial W_i^n}{\partial t}\big)\end{aligned} \quad (9.3.3.2)$$

where λ_{ie}^n, μ_{ie}^n are given by (9.3.2.3), (9.3.2.4). We then introduce the *local errors*

$$\widetilde{W}_i^{n+1} = W_i^{n+1} - W_i, \qquad \widetilde{P}_i^{n+1} = P_i^{n+1} - P_i,$$

where $\{(W_i, P_i)\}_{i\in\mathcal{I}}$ is the solution of the global optimality system (9.2.1.2), (9.2.1.3), (9.2.1.11)–(9.2.1.15). The local errors satisfy the system (9.3.2.6)–(9.3.2.11) with homogeneous data

$$F_i = 0, \quad W_{0i} = V_{0i} = Z_{0i} = Z_{1i} = 0, \quad \forall i \in \mathcal{I} \tag{9.3.3.3}$$

and with λ_{ie}^n, μ_{ie}^n replaced by

$$\tilde{\lambda}_{ie}^n = (1-\epsilon)\lambda_{ie}(\widetilde{W}^n, \widetilde{P}^n) + \epsilon\big((Q_{ie}^+)^* D_{\nu_{ie}} \widetilde{W}_i^n + \beta_e Q_{ie} \frac{\partial \widetilde{W}_i^n}{\partial t} + \gamma_e Q_{ie} \widetilde{P}_i^n\big) \tag{9.3.3.4}$$

$$\tilde{\mu}_{ie}^n = (1-\epsilon)\mu_{ie}(\widetilde{W}^n, \widetilde{P}^n) + \epsilon\big((Q_{ie}^+)^* D_{\nu_{ie}} \widetilde{P}_i^n - \beta_e d_t Q_{ie} \widetilde{P}_i^n + \gamma_e d_t Q_{ie} \frac{\partial \widetilde{W}_i^n}{\partial t}\big). \tag{9.3.3.5}$$

The above are valid for $n \geq 1$, while

$$\tilde{\lambda}_{ie}^0 = \lambda_{ie}^0 - \lambda_{ie}(W, P), \quad \tilde{\mu}_{ie}^0 = \mu_{ie}^0 - \mu_{ie}(W, P). \tag{9.3.3.6}$$

As above, we may express $\widetilde{P}_i^{n+1}$ in terms of a more regular variable $\widetilde{R}_i^{n+1}$ according to (9.3.2.22), where $\widetilde{R}_i^{n+1}$ satisfies the system (9.3.2.23) with homogeneous data $Z_{0i} = Z_{1i} = 0$, and with ρ_{ie}^n replaced by

$$\tilde{\rho}_{ie}^n = (1-\epsilon)\rho_{ie}(\widetilde{W}^n, \widetilde{R}^n) + \epsilon\big((Q_{ie}^+)^* D_{\nu_{ie}} \widetilde{R}_i^n - \beta_e Q_{ie} \frac{\partial \widetilde{R}_i^n}{\partial t} + \gamma_e Q_{ie} \frac{\partial \widetilde{W}_i^n}{\partial t}\big) \tag{9.3.3.7}$$

for $n \geq 1$, while

$$\tilde{\rho}_{ie}^0 = \rho_{ie}^0 - \rho_{ie}(W, R).$$

For reference, we write down the complete system satisfied by $\widetilde{W}_i^{n+1}$, $\widetilde{R}_i^{n+1}$.

$$\frac{\partial^2 \widetilde{W}_i^{n+1}}{\partial t^2} - \frac{\partial}{\partial x_\alpha}[A_i^{\alpha\beta}(\widetilde{W}_{i,\beta}^{n+1} + B_i^\beta \widetilde{W}_i^{n+1})] + (B_i^\alpha)^* A_i^{\alpha\beta}(\widetilde{W}_{i,\beta}^{n+1} + B_i^\beta \widetilde{W}_i^{n+1}) + C_i \widetilde{W}_i^{n+1} = 0 \text{ in } \mathcal{Q}_i, \tag{9.3.3.8}$$

$$\frac{\partial^2 \widetilde{R}_i^{n+1}}{\partial t^2} - \frac{\partial}{\partial x_{i,\alpha}}[A_i^{\alpha\beta}(\widetilde{R}_{i,\beta}^{n+1} + B_i^\beta \widetilde{R}_i^{n+1})] + (B_i^\alpha)^* A_i^{\alpha\beta}(\widetilde{R}_{i,\beta}^{n+1} + B_i^\beta \widetilde{R}_i^{n+1}) + C_i \widetilde{R}_i^{n+1} = 0 \text{ in } \mathcal{Q}_i; \tag{9.3.3.9}$$

$$\begin{aligned} \widetilde{W}_i^{n+1} &= \widetilde{R}_i^{n+1} = 0 \ \text{ on } \Sigma_{ie},\ e \in \mathcal{D}_i, \\ \Pi_{ie}\widetilde{W}_i^{n+1} &= \Pi_{ie}\widetilde{R}_i^{n+1} = 0 \ \text{ on } \Sigma_{ie},\ e \in \mathcal{N}_i^{\text{int}}, \end{aligned} \tag{9.3.3.10}$$

$$\begin{cases} D_{\nu_{ie}}\widetilde{W}_i^{n+1} + \alpha_{ie}\dfrac{\partial \widetilde{W}_i^{n+1}}{\partial t} + \dfrac{\partial \widetilde{R}_i^{n+1}}{\partial t} = 0, \\ D_{\nu_{ie}}\widetilde{R}_i^{n+1} - \alpha_{ie}\dfrac{\partial \widetilde{R}_i^{n+1}}{\partial t} = 0 \quad \text{on } \Sigma_{ie},\ e \in \mathcal{N}_i^{\text{ext}}; \end{cases} \tag{9.3.3.11}$$

$$\begin{cases} (Q_{ie}^+)^* D_{\nu_{ie}}\widetilde{W}_i^{n+1} + \beta_e Q_{ie}\dfrac{\partial \widetilde{W}_i^{n+1}}{\partial t} + \gamma_e Q_{ie}\dfrac{\partial \widetilde{R}_i^{n+1}}{\partial t} = \tilde{\lambda}_{ie}^n, \\ (Q_{ie}^+)^* D_{\nu_{ie}}\widetilde{R}_i^{n+1} - \beta_e Q_{ie}\dfrac{\partial \widetilde{R}_i^{n+1}}{\partial t} + \gamma_e Q_{ie}\dfrac{\partial \widetilde{W}_i^{n+1}}{\partial t} = \tilde{\rho}_{ie}^n \\ \qquad\qquad\qquad\qquad \text{on } \Sigma_{ie},\ e \in \mathcal{N}_i^{\text{int}}; \end{cases} \tag{9.3.3.12}$$

$$\begin{cases} \widetilde{W}_i^{n+1}(0) = \dfrac{\partial \widetilde{W}_i^{n+1}}{\partial t}(0) = 0, \\ \widetilde{R}_i^{n+1}(T) = \kappa \widetilde{W}_i^{n+1}(T), \quad \dfrac{\partial \widetilde{R}_i^{n+1}}{\partial t}(T) = \kappa \dfrac{\partial \widetilde{W}_i^{n+1}}{\partial t}(T). \end{cases} \tag{9.3.3.13}$$

We have the following convergence result.

Theorem 9.3.3.1 *Assume that β_e, $\gamma_e \equiv \beta$, γ are independent of $e \in \mathcal{N}^{int}$. Assume further that*

$$\begin{aligned} \alpha_0 := \inf\{\alpha_{ie} : i \in \mathcal{I},\ e \in \mathcal{N}_i^{ext}\} &> \frac{1}{2}\left(-\frac{\gamma}{\beta} + \sqrt{\frac{\gamma^2}{\beta^2} + 1}\right), \\ \frac{\gamma}{\beta} &> \frac{\kappa^2 - 1}{2\kappa}. \end{aligned} \tag{9.3.3.14}$$

(i) *For any $\epsilon \in [0, 1)$ we have*

$$\begin{gathered} \|(\widetilde{W}_i^n(T), \frac{\partial \widetilde{W}_i^n}{\partial t}(T))\|_{\mathcal{V}_i \times \mathcal{H}_i} \to 0, \quad \|\frac{\partial \widetilde{W}_i^n}{\partial t}\|_{L^2(\Sigma_{ie})} \to 0, \\ \|(\widetilde{R}_i^n(0), \frac{\partial \widetilde{R}_i^n}{\partial t}(0))\|_{\mathcal{V}_i \times \mathcal{H}_i} \to 0, \quad \|(\widetilde{R}_i^n(T), \frac{\partial \widetilde{R}_i^n}{\partial t}(T))\|_{\mathcal{V}_i \times \mathcal{H}_i} \to 0, \\ \|\frac{\partial \widetilde{R}_i^n}{\partial t}\|_{L^2(\Sigma_{ie})} \to 0, \quad \forall i \in \mathcal{I},\ e \in \mathcal{N}_i^{ext}. \end{gathered}$$

(ii) *In addition, if* $\epsilon > 0$, *we have*

$$(\widetilde{W}_i^n, \frac{\partial \widetilde{W}_i^n}{\partial t}) \to 0, \quad (\widetilde{R}_i^n, \frac{\partial \widetilde{R}_i^n}{\partial t}) \to 0 \ \text{ in } C([0,T]; \mathcal{V}_i \times \mathcal{H}_i), \ \forall i \in \mathcal{I},$$

$$(Q_{ie}^+)^* D_{\nu_{ie}} \widetilde{W}_i^n\big|_{\Sigma_{ie}} \to 0, \quad (Q_{ie}^+)^* D_{\nu_{ie}} \widetilde{R}_i^n\big|_{\Sigma_{ie}} \to 0,$$

$$Q_{ie} \frac{\partial \widetilde{W}_i^n}{\partial t}\bigg|_{\Sigma_{ie}} \to 0, \quad Q_{ie} \frac{\partial \widetilde{R}_i^n}{\partial t}\bigg|_{\Sigma_{ie}} \to 0 \ \text{ weakly in } L^2(\Sigma_{ie}), \ \forall i \in \mathcal{I}_e, \ e \in \mathcal{N}^{int},$$

$$\|\sum_{i \in \mathcal{I}_e} (Q_{ie}^+)^* D_{\nu_{ie}} W_i^n \|_{L^2(\Sigma_e)} \to 0, \quad \|\sum_{i \in \mathcal{I}_e} (Q_{ie}^+)^* D_{\nu_{ie}} R_i^n \|_{L^2(\Sigma_e)} \to 0,$$

$$\|Q_{ie} \frac{\partial W_i^n}{\partial t} - Q_{je} \frac{\partial W_j^n}{\partial t}\|_{L^2(\Sigma_e)} \to 0, \quad \|Q_{ie} \frac{\partial R_i^n}{\partial t} - Q_{je} \frac{\partial R_j^n}{\partial t}\|_{L^2(\Sigma_e)} \to 0,$$
$$\forall i, j \in \mathcal{I}_e, \ \forall e \in \mathcal{N}^{int},$$

where $\Sigma_e := \Sigma_{ie}$ *for any* $i \in \mathcal{I}_e$ *(recall that* $\Sigma_{ie} = \Sigma_{je}$ *for all* $i, j \in \mathcal{I}_e$*).*

Remark 9.3.3.1 The conclusions of Theorem 9.3.3.1 may be stated in terms of the dual variables $\widetilde{W}_i^n$, $\widetilde{P}_i^n$, and W_i^n, P_i^n by recalling that $\widetilde{P}_i^n = \partial \widetilde{R}_i^n / \partial t$, $\partial \widetilde{P}_i^n / \partial t = -\mathcal{A}_i \widetilde{R}_i^n$, and $P_i^n = \partial R_i^n / \partial t$, $\partial P_i^n / \partial t = -\mathcal{A}_i R_i^n$.

Proof. The first step is to write the iteration step (9.3.3.12), where $\tilde{\lambda}_{ie}^n$ and $\tilde{\rho}_{ie}^n$ are given by (9.3.3.4) and (9.3.3.7) respectively, as the relaxed fixed point iteration of a mapping $\mathcal{T}$ that is defined in the following way. Set

$$\mathcal{X} = \prod_{i \in \mathcal{I}} \prod_{e \in \mathcal{N}_i^{\text{int}}} L^2(\Sigma_{ie}) \times L^2(\Sigma_{ie}).$$

For $X = \{(\lambda_{ie}, \rho_{ie}) : i \in \mathcal{I}, \ e \in \mathcal{N}_i^{\text{int}}\} \in \mathcal{X}$ set

$$\|X\|_{\mathcal{X}}^2 = \sum_{i \in \mathcal{I}} \sum_{e \in \mathcal{N}_i^{\text{int}}} \frac{1}{2\gamma_e} \int_{\Sigma_{ie}} (|\lambda_{ie}|^2 + |\rho_{ie}|^2) \, d\Sigma.$$

Let W_i, R_i be the solution of

$$\frac{\partial^2 W_i}{\partial t^2} - \frac{\partial}{\partial x_\alpha} [A_i^{\alpha\beta} (W_{i,\beta} + B_i^\beta W_i)] + (B_i^\alpha)^* A_i^{\alpha\beta} (W_{i,\beta} + B_i^\beta W_i) + C_i W_i = 0 \ \text{ in } \mathcal{Q}_i, \quad (9.3.3.15)$$

$$\frac{\partial^2 R_i}{\partial t^2} - \frac{\partial}{\partial x_{i,\alpha}} [A_i^{\alpha\beta} (R_{i,\beta} + B_i^\beta R_i)] + (B_i^\alpha)^* A_i^{\alpha\beta} (R_{i,\beta} + B_i^\beta R_i) + C_i R_i = 0 \ \text{ in } \mathcal{Q}_i; \quad (9.3.3.16)$$

$$\begin{aligned} &W_i = R_i = 0 \ \text{ on } \Sigma_{ie},\ e \in \mathcal{D}_i, \\ &\Pi_{ie} W_i = \Pi_{ie} R_i = 0 \ \text{ on } \Sigma_{ie},\ e \in \mathcal{N}_i^{\text{int}}, \end{aligned} \tag{9.3.3.17}$$

$$\begin{cases} D_{\nu_{ie}} W_i + \alpha_{ie} \dfrac{\partial W_i}{\partial t} + \dfrac{\partial R_i}{\partial t} = 0, \\ D_{\nu_{ie}} R_i - \alpha_{ie} \dfrac{\partial R_i}{\partial t} = 0 \quad \text{on } \Sigma_{ie},\ e \in \mathcal{N}_i^{\text{ext}}; \end{cases} \tag{9.3.3.18}$$

$$\begin{cases} (Q_{ie}^{+})^* D_{\nu_{ie}} W_i + \beta_e Q_{ie} \dfrac{\partial W_i}{\partial t} + \gamma_e Q_{ie} \dfrac{\partial R_i}{\partial t} = \lambda_{ie}, \\ (Q_{ie}^{+})^* D_{\nu_{ie}} R_i - \beta_e Q_{ie} \dfrac{\partial R_i}{\partial t} + \gamma_e Q_{ie} \dfrac{\partial W_i}{\partial t} = \rho_{ie} \\ \qquad\qquad\qquad\qquad \text{on } \Sigma_{ie},\ e \in \mathcal{N}_i^{\text{int}}; \end{cases} \tag{9.3.3.19}$$

$$\begin{cases} W_i(0) = \dfrac{\partial W_i}{\partial t}(0) = 0, \\ R_i(T) = \kappa W_i(T), \quad \dfrac{\partial R_i}{\partial t}(T) = \kappa \dfrac{\partial W_i}{\partial t}(T). \end{cases} \tag{9.3.3.20}$$

We then define $\mathcal{T} : \mathcal{X} \mapsto \mathcal{X}$ by setting

$$\mathcal{T} X = \{(\hat{\lambda}_{ie}, \hat{\rho}_{ie}) : i \in \mathcal{I},\ e \in \mathcal{N}_i^{\text{int}}\} \tag{9.3.3.21}$$

where (cf. (9.3.2.3) and (9.3.2.24))

$$\begin{aligned} \hat{\lambda}_{ie} = \lambda_{ie}\Big(W, \frac{\partial R}{\partial t}\Big) = &-\frac{2}{d_e} \sum_{j \in \mathcal{I}_e} (Q_{je}^{+})^* D_{\nu_{je}} W_j + (Q_{ie}^{+})^* D_{\nu_{ie}} W_i \\ &+ \frac{2}{d_e} \sum_{j \in \mathcal{I}_e} \beta_e Q_{je} \frac{\partial W_j}{\partial t} - \beta_e Q_{ie} \frac{\partial W_i}{\partial t} \\ &\qquad + \frac{2}{d_e} \sum_{j \in \mathcal{I}_e} \gamma_e Q_{je} \frac{\partial R_j}{\partial t} - \gamma_e Q_{ie} \frac{\partial R_i}{\partial t} \Big|_{\Sigma_{ie}}, \end{aligned} \tag{9.3.3.22}$$

$$\begin{aligned} \hat{\rho}_{ie} = \rho_{ie}(W, R) = &-\frac{2}{d_e} \sum_{j \in \mathcal{I}_e} (Q_{je}^{+})^* D_{\nu_{je}} R_j + (Q_{ie}^{+})^* D_{\nu_{ie}} R_i \\ &+ \frac{2}{d_e} \sum_{j \in \mathcal{I}_e} \gamma_e Q_{je} \frac{\partial W_j}{\partial t} - \gamma_e Q_{ie} \frac{\partial W_i}{\partial t} \\ &\qquad - \frac{2}{d_e} \sum_{j \in \mathcal{I}_e} \beta_e Q_{je} \frac{\partial R_j}{\partial t} + \beta_e Q_{ie} \frac{\partial R_i}{\partial t} \Big|_{\Sigma_{ie}}. \end{aligned} \tag{9.3.3.23}$$

With these definitions, we see that X is a fixed point of $\mathcal{T}$ if and only if $\{(W_i, P_i) : i \in \mathcal{I}\}$, where $P_i = \partial R_i/\partial t$, is the solution of the global optimality system corresponding to the homogeneous data (9.3.3.3). It follows that the only fixed point of $\mathcal{T}$ is $X = 0$.

For $n \geq 1$ let

$$X^n := \{(\tilde{\lambda}_{ie}^{n-1}, \tilde{\rho}_{ie}^{n-1}) : i \in \mathcal{I},\, e \in \mathcal{N}_i^{\text{int}}\} \in \mathcal{X}$$

and let $\widetilde{W}_i^n$, $\widetilde{R}_i^n$ be the solution of (9.3.3.8)–(9.3.3.13) with n replaced by $n-1$. Then we may write

$$\begin{aligned} X^n = \Big\{\Big((Q_{ie}^+)^* D_{\nu_{ie}} \widetilde{W}_i^n + \beta_e Q_{ie} \frac{\partial \widetilde{W}_i^n}{\partial t} + \gamma_e Q_{ie} \frac{\partial \widetilde{R}_i^n}{\partial t}, \\ (Q_{ie}^+)^* D_{\nu_{ie}} \widetilde{R}_i^{n+1} - \beta_e Q_{ie} \frac{\partial \widetilde{R}_i^{n+1}}{\partial t} + \gamma_e Q_{ie} \frac{\partial \widetilde{W}_i^{n+1}}{\partial t}\Big)\Big|_{\Sigma_{ie}} \\ : i \in \mathcal{I},\, e \in \mathcal{N}_i^{\text{int}}\Big\} \end{aligned} \quad (9.3.3.24)$$

so that the iteration step defined by (9.3.3.12), (9.3.3.4) and (9.3.3.7) is expressible as the relaxed fixed point iteration

$$X^{n+1} = (1-\epsilon)\mathcal{T} X^n + \epsilon X^n. \quad (9.3.3.25)$$

We shall need the following Lemma.

Lemma 9.3.3.1 *Assume that* β_e, $\gamma_e \equiv \beta, \gamma$ *are independent of* $e \in \mathcal{N}^{int}$. *For* $X = \{(\lambda_{ie}, \rho_{ie}) : i \in \mathcal{I},\, e \in \mathcal{N}_i^{int}\} \in \mathcal{X}$ *we may write*

$$\|X\|_{\mathcal{X}}^2 = \mathcal{E} + \mathcal{F}, \quad \|\mathcal{T}X\|_{\mathcal{X}}^2 = \mathcal{E} - \mathcal{F} \quad (9.3.3.26)$$

where

$$\begin{aligned} \mathcal{E} = \sum_{i \in \mathcal{I}} \sum_{e \in \mathcal{N}_i^{int}} \int_{\Sigma_{ie}} \Big\{ \frac{1}{2\gamma}\big(|(Q_{ie}^+)^* D_{\nu_{ie}} W_i|^2 + |(Q_{ie}^+)^* D_{\nu_{ie}} R_i|^2\big) \\ + \frac{\beta^2+\gamma^2}{2\gamma}\Big(\Big|\frac{\partial W_i}{\partial t}\Big|^2 + \Big|\frac{\partial R_i}{\partial t}\Big|^2\Big)\Big\} d\Sigma, \end{aligned} \quad (9.3.3.27)$$

$$\begin{aligned} \mathcal{F} = \sum_{i \in \mathcal{I}} \sum_{e \in \mathcal{N}_i^{int}} \Big\{ \Big(\kappa + \frac{\beta}{2\gamma}(1-\kappa^2)\Big) \Big\|\Big(W_i(T), \frac{\partial W_i}{\partial t}(T)\Big)\Big\|_{\mathcal{V}_i \times \mathcal{H}_i}^2 \\ + \frac{\beta}{2\gamma}\Big\|\Big(R_i(0), \frac{\partial R_i}{\partial t}(0)\Big)\Big\|_{\mathcal{V}_i \times \mathcal{H}_i}^2 \\ + \int_{\Sigma_{ie}} \Big[\Big(1 + \frac{\alpha_{ie}\beta}{\gamma}\Big)\Big|\frac{\partial R_i}{\partial t}\Big|^2 + \frac{\beta}{\gamma}\frac{\partial W_i}{\partial t} \cdot \frac{\partial R_i}{\partial t} + \frac{\alpha_{ie}\beta}{\gamma}\Big|\frac{\partial W_i}{\partial t}\Big|^2\Big]\, d\Sigma\Big\}. \end{aligned} \quad (9.3.3.28)$$

In particular, $\mathcal{T}$ *is a nonexpansive mapping.*

Proof of Lemma 9.3.3.1. We have

$$\begin{aligned}
\|X\|_{\mathcal{X}}^2 = &\sum_{i\in\mathcal{I}}\sum_{e\in\mathcal{N}_i^{\rm int}}\int_{\Sigma_{ie}}\{\big|(Q_{ie}^{+})^* D_{\nu_{ie}} W_i + \beta_e Q_{ie}\frac{\partial W_i}{\partial t} + \gamma_e Q_{ie}\frac{\partial R_i}{\partial t}\big|^2 \\
&+ \big|(Q_{ie}^{+})^* D_{\nu_{ie}} R_i - \beta_e Q_{ie}\frac{\partial R_i}{\partial t} + \gamma_e Q_{ie}\frac{\partial W_i}{\partial t}\big|^2\} d\Sigma \\
= &\sum_{i\in\mathcal{I}}\sum_{e\in\mathcal{N}_i^{\rm int}}\int_{\Sigma_{ie}}\{\frac{1}{2\gamma_e}\big(\big|(Q_{ie}^{+})^* D_{\nu_{ie}} W_i\big|^2 + \big|(Q_{ie}^{+})^* D_{\nu_{ie}} R_i\big|^2\big) \\
&+ \frac{\beta_e^2+\gamma_e^2}{2\gamma_e}\big(\big|Q_{ie}\frac{\partial W_i}{\partial t}\big|^2 + \big|Q_{ie}\frac{\partial R_i}{\partial t}\big|^2\big) \\
&+ \frac{\beta_e}{\gamma_e}\big((Q_{ie}^{+})^* D_{\nu_{ie}} W_i\cdot Q_{ie}\frac{\partial W_i}{\partial t} - (Q_{ie}^{+})^* D_{\nu_{ie}} R_i\cdot Q_{ie}\frac{\partial R_i}{\partial t}\big) \qquad (9.3.3.29)\\
&+ \big((Q_{ie}^{+})^* D_{\nu_{ie}} W_i\cdot Q_{ie}\frac{\partial R_i}{\partial t} + (Q_{ie}^{+})^* D_{\nu_{ie}} R_i\cdot Q_{ie}\frac{\partial W_i}{\partial t}\big)\} d\Sigma \\
= &\,\mathcal{E} + \sum_{i\in\mathcal{I}}\sum_{e\in\mathcal{N}_i^{\rm int}}\int_{\Sigma_{ie}}\{\frac{\beta_e}{\gamma_e}\big((Q_{ie}^{+})^* D_{\nu_{ie}} W_i\cdot Q_{ie}\frac{\partial W_i}{\partial t} \\
&- (Q_{ie}^{+})^* D_{\nu_{ie}} R_i\cdot Q_{ie}\frac{\partial R_i}{\partial t}\big) \\
&+ \big((Q_{ie}^{+})^* D_{\nu_{ie}} W_i\cdot Q_{ie}\frac{\partial R_i}{\partial t} + (Q_{ie}^{+})^* D_{\nu_{ie}} R_i\cdot Q_{ie}\frac{\partial W_i}{\partial t}\big)\} d\Sigma.
\end{aligned}$$

The function W_i satisfies the variational problem (9.3.1.2), (9.3.1.3) with $F_i = 0$, $W_{0i} = V_{0i} = 0$, and

$$f_{ie} = -\frac{\partial R_i}{\partial t}\Big|_{\Sigma_{ie}},$$

$$g_{ie} = \lambda_{ie} - \gamma_e Q_{ie}\frac{\partial R_i}{\partial t}\Big|_{\Sigma_{ie}} = (Q_{ie}^{+})^* D_{\nu_{ie}} W_i + \beta_e Q_{ie}\frac{\partial W_i}{\partial t}\Big|_{\Sigma_{ie}}.$$

We formally set $\Phi_i = \partial W_i/\partial t$ in (9.3.1.2) and integrate over $(0,T)$. (This may be justified by approximating $\partial W_i/\partial t$ by functions $\Phi_i^n \in C([0,T];\mathcal{V}_i)$ such that $\Phi_i^n \to \partial W_i/\partial t$ in $C([0,T];\mathcal{H}_i)$ and $\Phi_i^n|_{\Sigma_{ie}} \to \partial W_i/\partial t|_{\Sigma_{ie}}$ in $L^2(\Sigma_{ie})$, $\forall e \in \mathcal{N}_i^{\rm int}\cup\mathcal{N}_i^{\rm ext}$.) We obtain

$$\begin{aligned}
\sum_{e\in\mathcal{N}_i^{\rm int}}\int_{\Sigma_{ie}} (Q_{ie}^{+})^* D_{\nu_{ie}} W_i\cdot Q_{ie}\frac{\partial W_i}{\partial t}\, d\Sigma = &\frac{1}{2}\big\|\big(W_i(T),\frac{\partial W_i}{\partial t}(T)\big)\big\|^2_{\mathcal{V}_i\times\mathcal{H}_i} \\
&+ \sum_{e\in\mathcal{N}_i^{\rm ext}}\int_{\Sigma_{ie}}\big(\alpha_{ie}\big|\frac{\partial W_i}{\partial t}\big|^2 + \frac{\partial W_i}{\partial t}\cdot\frac{\partial R_i}{\partial t}\big)\, d\Sigma. \qquad (9.3.3.30)
\end{aligned}$$

Similarly, R_i satisfies the variational problem

$$\langle \frac{\partial^2 R_i}{\partial t^2}, \Phi_i \rangle_{\mathcal{V}_i} + a_i(R_i, \Phi_i) - b_i\left(\frac{\partial R_i}{\partial t}, \Phi_i\right) = \sum_{e \in \mathcal{N}_i^{\text{int}}} \int_{\Gamma_{ie}} \left((Q_{ie}^+)^* D_{\nu_{ie}} R_i - \beta_e Q_{ie} \frac{\partial R_i}{\partial t}\right) \cdot Q_{ie} \Phi_i \, d\Sigma,$$

$$\forall \Phi \in \mathcal{V}_i, \; 0 < t < T, \quad (9.3.3.31)$$

$$R_i(T) = \kappa W_i(T), \quad \frac{\partial R_i}{\partial t}(T) = \kappa \frac{\partial W_i}{\partial t}(T). \quad (9.3.3.32)$$

If we formally replace Φ_i in (9.3.3.31) by $\partial R_i/\partial t$ and integrate over $(0,T)$ we obtain

$$\sum_{e \in \mathcal{N}_i^{\text{int}}} \int_{\Sigma_{ie}} (Q_{ie}^+)^* D_{\nu_{ie}} R_i \cdot Q_{ie} \frac{\partial R_i}{\partial t} \, d\Sigma = \frac{\kappa^2}{2} \left\| \left(W_i(T), \frac{\partial W_i}{\partial t}(T)\right) \right\|^2_{\mathcal{V}_i \times \mathcal{H}_i} - \frac{1}{2} \left\| \left(R_i(0), \frac{\partial R_i}{\partial t}(0)\right) \right\|^2_{\mathcal{V}_i \times \mathcal{H}_i} - \sum_{e \in \mathcal{N}_i^{\text{ext}}} \int_{\Sigma_{ie}} \alpha_{ie} \left| \frac{\partial R_i}{\partial t} \right|^2 d\Sigma. \quad (9.3.3.33)$$

Since β_e, $\gamma_e \equiv \beta$, γ are independent of e, it follows from (9.3.3.30) and (9.3.3.33) that

$$\sum_{e \in \mathcal{N}_i^{\text{int}}} \int_{\Sigma_{ie}} \frac{\beta_e}{\gamma_e} \left((Q_{ie}^+)^* D_{\nu_{ie}} W_i \cdot Q_{ie} \frac{\partial W_i}{\partial t} - (Q_{ie}^+)^* D_{\nu_{ie}} R_i \cdot Q_{ie} \frac{\partial R_i}{\partial t}\right) d\Sigma$$
$$= \frac{\beta}{2\gamma}(1-\kappa^2) \left\| \left(W_i(T), \frac{\partial W_i}{\partial t}(T)\right) \right\|^2_{\mathcal{V}_i \times \mathcal{H}_i} + \frac{\beta}{2\gamma} \left\| \left(R_i(0), \frac{\partial R_i}{\partial t}(0)\right) \right\|^2_{\mathcal{V}_i \times \mathcal{H}_i} + \sum_{e \in \mathcal{N}_i^{\text{ext}}} \frac{\beta}{\gamma} \int_{\Sigma_{ie}} \left[\alpha_{ie} \left(\left| \frac{\partial R_i}{\partial t} \right|^2 + \left| \frac{\partial W_i}{\partial t} \right|^2 \right) + \frac{\partial W_i}{\partial t} \cdot \frac{\partial R_i}{\partial t}\right] d\Sigma. \quad (9.3.3.34)$$

We next utilize the variational equation (9.3.1.2) with $\Phi_i = \partial R_i/\partial t$ and integrate over $(0,T)$. (Again, this may be justified by working with suitable smooth approximations of $\partial R_i/\partial t$ and $\partial W_i/\partial t$.) The formal calculation yields

$$\left(\frac{\partial W_i}{\partial t}(T), \frac{\partial R_i}{\partial t}(T)\right)_{\mathcal{H}_i} + (W_i(T), R_i(T))_{\mathcal{V}_i} - \langle \frac{\partial^2 R_i}{\partial t}, \frac{\partial W_i}{\partial t} \rangle_{\mathcal{V}_i} - a_i\left(R_i, \frac{\partial R_i}{\partial t}\right) + b_i\left(\frac{\partial W_i}{\partial t}, \frac{\partial R_i}{\partial t}\right)$$
$$= -\sum_{e \in \mathcal{N}_i^{\text{ext}}} \int_{\Sigma_{ie}} \left| \frac{\partial R_i}{\partial t} \right|^2 d\Sigma + \sum_{e \in \mathcal{N}_i^{\text{int}}} \int_{\Sigma_{ie}} \left((Q_{ie}^+)^* D_{\nu_{ie}} W_i + \beta_e Q_{ie} \frac{\partial W_i}{\partial t}\right) \cdot Q_{ie} \frac{\partial R_i}{\partial t} \, d\Sigma$$

and, therefore,

$$\sum_{e\in\mathcal{N}_i^{\text{int}}}\int_{\Sigma_{ie}}(Q_{ie}^{+})^{*}D_{\nu_{ie}}W_i\cdot Q_{ie}\frac{\partial R_i}{\partial t}\,d\Sigma=\kappa\big\|\big(W_i(T),\frac{\partial W_i}{\partial t}(T)\big)\big\|^2_{\mathcal{V}_i\times\mathcal{H}_i}$$
$$+\sum_{e\in\mathcal{N}_i^{\text{ext}}}\int_{\Sigma_{ie}}\big|\frac{\partial R_i}{\partial t}\big|^2 d\Sigma-\sum_{e\in\mathcal{N}_i^{\text{int}}}\int_{\Sigma_{ie}}\beta_e Q_{ie}\frac{\partial W_i}{\partial t}\cdot Q_{ie}\frac{\partial R_i}{\partial t}\,d\Sigma$$
$$-\big\langle\frac{\partial^2 R_i}{\partial t},\frac{\partial W_i}{\partial t}\big\rangle_{\mathcal{V}_i}-a_i\big(R_i,\frac{\partial R_i}{\partial t}\big)+b_i\big(\frac{\partial W_i}{\partial t},\frac{\partial R_i}{\partial t}\big).\quad(9.3.3.35)$$

By utilizing (9.3.3.31) with $\Phi_i=\partial W_i/\partial t$ it is seen that the last four terms on the right side of (9.3.3.35) combine to equal

$$-\sum_{e\in\mathcal{N}_i^{\text{int}}}\int_{\Sigma_{ie}}(Q_{ie}^{+})^{*}D_{\nu_{ie}}R_i\cdot Q_{ie}\frac{\partial W_i}{\partial t}\,d\Sigma.$$

It then follows from (9.3.3.35) that

$$\sum_{e\in\mathcal{N}_i^{\text{int}}}\int_{\Sigma_{ie}}\big[(Q_{ie}^{+})^{*}D_{\nu_{ie}}W_i\cdot Q_{ie}\frac{\partial R_i}{\partial t}+(Q_{ie}^{+})^{*}D_{\nu_{ie}}R_i\cdot Q_{ie}\frac{\partial W_i}{\partial t}\big]\,d\Sigma$$
$$=\kappa\big\|\big(W_i(T),\frac{\partial W_i}{\partial t}(T)\big)\big\|^2_{\mathcal{V}_i\times\mathcal{H}_i}+\sum_{e\in\mathcal{N}_i^{\text{ext}}}\int_{\Sigma_{ie}}\big|\frac{\partial R_i}{\partial t}\big|^2 d\Sigma.\quad(9.3.3.36)$$

From (9.3.3.34) and (9.3.3.36) we thus obtain

$$\sum_{i\in\mathcal{I}}\sum_{e\in\mathcal{N}_i^{\text{int}}}\int_{\Sigma_{ie}}\Big\{\frac{\beta_e}{\gamma_e}\big((Q_{ie}^{+})^{*}D_{\nu_{ie}}W_i\cdot Q_{ie}\frac{\partial W_i}{\partial t}$$
$$-(Q_{ie}^{+})^{*}D_{\nu_{ie}}R_i\cdot Q_{ie}\frac{\partial R_i}{\partial t}\big)$$
$$+\big((Q_{ie}^{+})^{*}D_{\nu_{ie}}W_i\cdot Q_{ie}\frac{\partial R_i}{\partial t}+(Q_{ie}^{+})^{*}D_{\nu_{ie}}R_i\cdot Q_{ie}\frac{\partial W_i}{\partial t}\big)\Big\}d\Sigma=\mathcal{F}\quad(9.3.3.37)$$

and then, from (9.3.3.29),

$$\|X\|^2_{\mathcal{X}}=\mathcal{E}+\mathcal{F}$$

with $\mathcal{E}$, $\mathcal{F}$ defined by (9.3.3.27) and (9.3.3.28), respectively.

The calculation of $\|\mathcal{T}X\|^2_{\mathcal{X}}$ is lengthy but elementary and we will not show all of the details. One has

$$\|\mathcal{T}X\|^2_{\mathcal{X}}=\sum_{i\in\mathcal{I}}\sum_{e\in\mathcal{N}_i^{\text{int}}}\frac{1}{2\gamma_e}\int_{\Sigma_{ie}}\big(|\hat\lambda_{ie}|^2+|\hat\rho_{ie}|^2\big)\,d\Sigma,$$

where $\hat\lambda_{ie}$, $\hat\rho_{ie}$ are given by (9.3.3.22) and (9.3.3.23), respectively. Since

$$\sum_{i\in\mathcal{I}}\sum_{e\in\mathcal{N}_i^{\text{int}}}=\sum_{e\in\mathcal{N}^{\text{int}}}\sum_{i\in\mathcal{I}_e}\qquad\text{and}\qquad\Sigma_{ie}=\Sigma_{je}:=\Sigma_e$$

for all $i, j \in \mathcal{I}_e$, we may write

$$\|\mathcal{T}X\|_{\mathcal{X}}^2 = \sum_{e\in\mathcal{N}^{\text{int}}} \frac{1}{2\gamma_e}\int_{\Sigma_e}\sum_{i\in\mathcal{I}_e}\{|\hat{\lambda}_{ie}|^2 + |\hat{\rho}_{ie}|^2\}\, d\Sigma. \tag{9.3.3.38}$$

One has

$$\begin{aligned}
|\hat{\lambda}_{ie}|^2 =& \frac{4}{d_e^2}\Big|\sum_{j\in\mathcal{I}_e}(Q_{je}^+)^* D_{\nu_{je}}W_j\Big|^2 + |(Q_{ie}^+)^* D_{\nu_{ie}}W_i|^2 \\
&\qquad - \frac{4}{d_e}(Q_{ie}^+)^* D_{\nu_{ie}}W_i \cdot \sum_{j\in\mathcal{I}_e}(Q_{je}^+)^* D_{\nu_{je}}W_j \\
&+ \frac{4}{d_e^2}\Big|\sum_{j\in\mathcal{I}_e}\beta_e Q_{je}\frac{\partial W_j}{\partial t}\Big|^2 + \beta_e^2\Big|Q_{ie}\frac{\partial W_i}{\partial t}\Big|^2 - \frac{4\beta_e}{d_e}Q_{ie}\frac{\partial W_i}{\partial t}\cdot\sum_{j\in\mathcal{I}_e}\beta_e Q_{je}\frac{\partial W_j}{\partial t} \\
&+ \frac{4}{d_e^2}\Big|\sum_{j\in\mathcal{I}_e}\gamma_e Q_{je}\frac{\partial R_j}{\partial t}\Big|^2 + \gamma_e^2\Big|Q_{ie}\frac{\partial R_i}{\partial t}\Big|^2 - \frac{4\gamma_e}{d_e}Q_{ie}\frac{\partial R_i}{\partial t}\cdot\sum_{j\in\mathcal{I}_e}\gamma_e Q_{je}\frac{\partial R_j}{\partial t} \\
&+ 2\Big(-\frac{2}{d_e}\sum_{j\in\mathcal{I}_e}(Q_{je}^+)^* D_{\nu_{je}}W_j + (Q_{ie}^+)^* D_{\nu_{ie}}W_i\Big) \\
&\qquad \cdot\Big(\frac{2}{d_e}\sum_{j\in\mathcal{I}_e}\beta_e Q_{je}\frac{\partial W_j}{\partial t} - \beta_e Q_{ie}\frac{\partial W_i}{\partial t}\Big) \\
&+ 2\Big(-\frac{2}{d_e}\sum_{j\in\mathcal{I}_e}(Q_{je}^+)^* D_{\nu_{je}}W_j + (Q_{ie}^+)^* D_{\nu_{ie}}W_i\Big) \\
&\qquad \cdot\Big(\frac{2}{d_e}\sum_{j\in\mathcal{I}_e}\gamma_e Q_{je}\frac{\partial R_j}{\partial t} - \gamma_e Q_{ie}\frac{\partial R_i}{\partial t}\Big) \\
&+ 2\Big(\frac{2}{d_e}\sum_{j\in\mathcal{I}_e}\beta_e Q_{je}\frac{\partial W_j}{\partial t} - \beta_e Q_{ie}\frac{\partial W_i}{\partial t}\Big) \\
&\qquad \cdot\Big(\frac{2}{d_e}\sum_{j\in\mathcal{I}_e}\gamma_e Q_{je}\frac{\partial R_j}{\partial t} - \gamma_e Q_{ie}\frac{\partial R_i}{\partial t}\Big).
\end{aligned}$$

After some algebra, one finds that

$$\begin{aligned}
\sum_{i\in\mathcal{I}_e}|\hat{\lambda}_{ie}|^2 =& \sum_{i\in\mathcal{I}_e}\Big\{|(Q_{ie}^+)^* D_{\nu_{ie}}W_i|^2 + \beta_e^2\Big|Q_{ie}\frac{\partial W_i}{\partial t}\Big|^2 + \gamma_e^2\Big|Q_{ie}\frac{\partial R_i}{\partial t}\Big|^2\Big\} \\
&- 2\sum_{i\in\mathcal{I}_e}\Big\{(Q_{ie}^+)^* D_{\nu_{ie}}W_i\cdot\beta_e Q_{ie}\frac{\partial W_i}{\partial t} + (Q_{ie}^+)^* D_{\nu_{ie}}W_i\cdot\gamma_e Q_{ie}\frac{\partial R_i}{\partial t} \\
&\qquad - \beta_e Q_{ie}\frac{\partial W_i}{\partial t}\cdot\gamma_e Q_{ie}\frac{\partial R_i}{\partial t}\Big\}.
\end{aligned}$$

Similarly

$$\sum_{i\in\mathcal{I}_e} |\hat{\rho}_{ie}|^2 = \sum_{i\in\mathcal{I}_e} \{|(Q_{ie}^+)^* D_{\nu_{ie}} R_i|^2 + \beta_e^2 |Q_{ie} \frac{\partial R_i}{\partial t}|^2 + \gamma_e^2 |Q_{ie} \frac{\partial W_i}{\partial t}|^2\}$$
$$- 2\sum_{i\in\mathcal{I}_e} \{(Q_{ie}^+)^* D_{\nu_{ie}} R_i \cdot \gamma_e Q_{ie} \frac{\partial W_i}{\partial t} - (Q_{ie}^+)^* D_{\nu_{ie}} R_i \cdot \beta_e Q_{ie} \frac{\partial R_i}{\partial t}$$
$$+ \gamma_e Q_{ie} \frac{\partial W_i}{\partial t} \cdot \beta_e Q_{ie} \frac{\partial R_i}{\partial t}\}.$$

It follows that

$$\sum_{i\in\mathcal{I}_e} \frac{1}{2\gamma_e} \{|\hat{\lambda}_{ie}|^2 + |\hat{\rho}_{ie}|^2\}$$
$$= \sum_{i\in\mathcal{I}_e} \{\frac{1}{2\gamma_e} [|(Q_{ie}^+)^* D_{\nu_{ie}} W_i|^2 + |(Q_{ie}^+)^* D_{\nu_{ie}} R_i|^2]$$
$$+ \frac{\beta_e^2+\gamma_e^2}{2\gamma_e} [|Q_{ie} \frac{\partial R_i}{\partial t}|^2 + \gamma_e^2 |Q_{ie} \frac{\partial W_i}{\partial t}|^2]\}$$
$$- \sum_{i\in\mathcal{I}_e} \{\frac{\beta_e}{\gamma_e} ((Q_{ie}^+)^* D_{\nu_{ie}} W_i \cdot Q_{ie} \frac{\partial W_i}{\partial t} - (Q_{ie}^+)^* D_{\nu_{ie}} R_i \cdot Q_{ie} \frac{\partial R_i}{\partial t})$$
$$+ ((Q_{ie}^+)^* D_{\nu_{ie}} W_i \cdot Q_{ie} \frac{\partial R_i}{\partial t} + (Q_{ie}^+)^* D_{\nu_{ie}} R_i \cdot Q_{ie} \frac{\partial W_i}{\partial t})\}. \tag{9.3.3.39}$$

From (9.3.3.34), (9.3.3.36), (9.3.3.38) and (9.3.3.39) we obtain

$$\|\mathcal{T}X\|_{\mathcal{X}}^2 = \mathcal{E} - \mathcal{F}. \tag{9.3.3.40}$$

This concludes the proof of Lemma 9.3.3.1.

Continuing with the proof of Theorem 9.3.3.1, we set

$$\mathcal{E}^n = \sum_{i\in\mathcal{I}} \sum_{e\in\mathcal{N}_i^{\rm int}} \int_{\Sigma_{ie}} \{\frac{1}{2\gamma} (|(Q_{ie}^+)^* D_{\nu_{ie}} \widetilde{W}_i^n|^2 + |(Q_{ie}^+)^* D_{\nu_{ie}} \widetilde{R}_i^n|^2)$$
$$+ \frac{\beta^2+\gamma^2}{2\gamma} (|\frac{\partial \widetilde{W}_i^n}{\partial t}|^2 + |\frac{\partial \widetilde{R}_i^n}{\partial t}|^2)\} d\Sigma, \tag{9.3.3.41}$$

$$\mathcal{F}^n = \sum_{i\in\mathcal{I}} \sum_{e\in\mathcal{N}_i^{\rm int}} \{(\kappa + \frac{\beta}{2\gamma}(1-\kappa^2)) \|(\widetilde{W}_i^n(T), \frac{\partial \widetilde{W}_i^n}{\partial t}(T)\|_{\mathcal{V}_i\times\mathcal{H}_i}^2$$
$$+ \frac{\beta}{2\gamma} \|(\widetilde{R}_i^n(0), \frac{\partial \widetilde{R}_i^n}{\partial t}(0)\|_{\mathcal{V}_i\times\mathcal{H}_i}^2$$
$$+ \int_{\Sigma_{ie}} [(1+\frac{\alpha_{ie}\beta}{\gamma}) |\frac{\partial \widetilde{R}_i^n}{\partial t}|^2 + \frac{\beta}{\gamma} \frac{\partial \widetilde{W}_i^n}{\partial t} \cdot \frac{\partial \widetilde{R}_i^n}{\partial t} + \frac{\alpha_{ie}\beta}{\gamma} |\frac{\partial \widetilde{W}_i^n}{\partial t}|^2]\, d\Sigma\}. \tag{9.3.3.42}$$

From Lemma 9.3.3.1 we have

$$\|X^n\|^2_{\mathcal{X}} = \mathcal{E}^n + \mathcal{F}^n, \quad \|\mathcal{T}X^n\|^2_{\mathcal{X}} = \mathcal{E}^n - \mathcal{F}^n \tag{9.3.3.43}$$

where X^n is given by (9.3.3.24). From (9.3.3.25) and (9.3.3.43) we obtain in the usual way the fundamental recursive inequality

$$\mathcal{E}^{n+1} + \mathcal{F}^{n+1} \leq \mathcal{E}^n - (1-2\epsilon)\mathcal{F}^n$$

which, by iteration, leads to

$$\mathcal{E}^{n+1} \leq \mathcal{E}^1 - \sum_{p=1}^{n+1} c_p(\epsilon)\mathcal{F}^p \tag{9.3.3.44}$$

where

$$c_1(\epsilon) = 1-2\epsilon,\ c_{n+1}(\epsilon) = 1,\ c_p(\epsilon) = 2(1-\epsilon), \quad p = 2,\dots,n.$$

The assumption (9.3.3.14) guarantees that the quadratic integrand in (9.3.3.42) is positive definite and that $\kappa + \beta(1-\kappa^2)/2\gamma > 0$. It then follows from (9.3.3.44) that for any $\epsilon \in [0,1)$

$$\sum_{p=1}^{\infty} c_p(\epsilon)\mathcal{F}^p \text{ converges and } \{\mathcal{E}^n\}_{n=1}^{\infty} \text{ is a bounded sequence.}$$

In particular

$$\mathcal{F}^n \to 0 \text{ as } n \to \infty,\ \forall \epsilon \in [0,1)$$

and, therefore

$$\|(\widetilde{W}_i^n(T), \frac{\partial \widetilde{W}_i^n}{\partial t}(T))\|_{\mathcal{V}_i\times\mathcal{H}_i} \to 0, \quad \|(\widetilde{R}_i^n(0), \frac{\partial \widetilde{R}_i^n}{\partial t}(0))\|_{\mathcal{V}_i\times\mathcal{H}_i} \to 0, \tag{9.3.3.45}$$

$$\|\frac{\partial \widetilde{W}_i^n}{\partial t}\|_{L^2(\Sigma_{ie})} \to 0, \quad \|\frac{\partial \widetilde{R}_i^n}{\partial t}\|_{L^2(\Sigma_{ie})} \to 0, \quad \forall i \in \mathcal{I},\ e \in \mathcal{N}_i^{\text{ext}}. \tag{9.3.3.46}$$

From (9.3.3.45) and $(9.3.3.13)_2$ we obtain

$$\|(\widetilde{R}_i^n(T), \frac{\partial \widetilde{R}_i^n}{\partial t}(T))\|_{\mathcal{V}_i\times\mathcal{H}_i} \to 0, \quad \forall i \in \mathcal{I}.$$

This proves part (i) of Theorem 9.3.3.1.

Now suppose that $\epsilon > 0$. Since $X = 0$ is the sole fixed point of $\mathcal{T}$, by Proposition 2.3.7.3 we may conclude that

$$X^n \to 0, \quad \mathcal{T}X^n \to 0 \text{ weakly in } \mathcal{X}.$$

Thus

$$(Q_{ie}^{+})^{*} D_{\nu_{ie}} \widetilde{W}_i^n + \beta Q_{ie} \frac{\partial \widetilde{W}_i^n}{\partial t} + \gamma \frac{\partial \widetilde{R}_i^n}{\partial t} \to 0, \tag{9.3.3.47}$$

$$(Q_{ie}^{+})^{*} D_{\nu_{ie}} \widetilde{R}_i^n - \beta Q_{ie} \frac{\partial \widetilde{R}_i^n}{\partial t} + \gamma \frac{\partial \widetilde{W}_i^n}{\partial t} \to 0, \tag{9.3.3.48}$$

$$\begin{aligned} -\frac{2}{d_e} \sum_{j \in \mathcal{I}_e} (Q_{je}^{+})^{*} D_{\nu_{je}} \widetilde{W}_j^n + (Q_{ie}^{+})^{*} D_{\nu_{ie}} \widetilde{W}_i^n + \frac{2\beta}{d_e} \sum_{j \in \mathcal{I}_e} Q_{je} \frac{\partial \widetilde{W}_j^n}{\partial t} \\ - \beta Q_{ie} \frac{\partial \widetilde{W}_i^n}{\partial t} + \frac{2\gamma}{d_e} \sum_{j \in \mathcal{I}_e} Q_{je} \frac{\partial \widetilde{R}_j^n}{\partial t} - \gamma Q_{ie} \frac{\partial \widetilde{R}_i^n}{\partial t} \to 0, \end{aligned} \tag{9.3.3.49}$$

$$\begin{aligned} -\frac{2}{d_e} \sum_{j \in \mathcal{I}_e} (Q_{je}^{+})^{*} D_{\nu_{je}} \widetilde{R}_j^n + (Q_{ie}^{+})^{*} D_{\nu_{ie}} \widetilde{R}_i^n + \frac{2\gamma}{d_e} \sum_{j \in \mathcal{I}_e} Q_{je} \frac{\partial \widetilde{W}_j^n}{\partial t} \\ - \gamma Q_{ie} \frac{\partial \widetilde{W}_i^n}{\partial t} - \frac{2\beta}{d_e} \sum_{j \in \mathcal{I}_e} Q_{je} \frac{\partial \widetilde{R}_j^n}{\partial t} + \beta Q_{ie} \frac{\partial \widetilde{R}_i^n}{\partial t} \to 0 \end{aligned} \tag{9.3.3.50}$$

weakly in $L^2(\Sigma_{ie})$, $\forall i \in \mathcal{I}_e$, $e \in \mathcal{N}^{\text{int}}$. We sum (9.3.3.49) and (9.3.3.50) over $i \in \mathcal{I}_e$ and obtain

$$-\sum_{i \in \mathcal{I}_e} (Q_{ie}^{+})^{*} D_{\nu_{ie}} \widetilde{W}_i^n + \beta \sum_{i \in \mathcal{I}_e} Q_{ie} \frac{\partial \widetilde{W}_i^n}{\partial t} + \gamma \sum_{i \in \mathcal{I}_e} Q_{ie} \frac{\partial \widetilde{R}_i^n}{\partial t} \to 0, \tag{9.3.3.51}$$

$$-\sum_{i \in \mathcal{I}_e} (Q_{ie}^{+})^{*} D_{\nu_{ie}} \widetilde{R}_i^n + \gamma \sum_{i \in \mathcal{I}_e} Q_{ie} \frac{\partial \widetilde{W}_i^n}{\partial t} - \beta \sum_{i \in \mathcal{I}_e} Q_{ie} \frac{\partial \widetilde{R}_i^n}{\partial t} \to 0 \tag{9.3.3.52}$$

weakly in $L^2(\Sigma_e)$, $\forall e \in \mathcal{N}^{\text{int}}$. If we now sum (9.3.3.47) and (9.3.3.48) over $i \in \mathcal{I}_e$ and add the results to (9.3.3.51) and (9.3.3.52), respectively, we obtain

$$\beta \sum_{i \in \mathcal{I}_e} Q_{ie} \frac{\partial \widetilde{W}_i^n}{\partial t} + \gamma \sum_{i \in \mathcal{I}_e} Q_{ie} \frac{\partial \widetilde{R}_i^n}{\partial t} \to 0,$$

$$\gamma \sum_{i \in \mathcal{I}_e} Q_{ie} \frac{\partial \widetilde{W}_i^n}{\partial t} - \beta \sum_{i \in \mathcal{I}_e} Q_{ie} \frac{\partial \widetilde{R}_i^n}{\partial t} \to 0,$$

hence

$$\sum_{i \in \mathcal{I}_e} Q_{ie} \frac{\partial \widetilde{W}_i^n}{\partial t} \to 0, \quad \sum_{i \in \mathcal{I}_e} Q_{ie} \frac{\partial \widetilde{R}_i^n}{\partial t} \to 0 \text{ weakly in } L^2(\Sigma_e). \tag{9.3.3.53}$$

From (9.3.3.51), (9.3.3.52) we then conclude that

$$\sum_{i\in\mathcal{I}_e}(Q_{ie}^{+})^* D_{\nu_{ie}}\widetilde{W}_i^n \to 0, \quad \sum_{i\in\mathcal{I}_e}(Q_{ie}^{+})^* D_{\nu_{ie}}\widetilde{R}_i^n \to 0 \text{ weakly in } L^2(\Sigma_e). \quad (9.3.3.54)$$

It now follows from (9.3.3.49), (9.3.3.50), (9.3.3.53) and (9.3.3.54) that

$$(Q_{ie}^{+})^* D_{\nu_{ie}}\widetilde{W}_i^n - \beta Q_{ie}\frac{\partial \widetilde{W}_i^n}{\partial t} - \gamma\frac{\partial \widetilde{R}_i^n}{\partial t} \to 0, \quad (9.3.3.55)$$

$$(Q_{ie}^{+})^* D_{\nu_{ie}}\widetilde{R}_i^n + \beta Q_{ie}\frac{\partial \widetilde{R}_i^n}{\partial t} - \gamma\frac{\partial \widetilde{W}_i^n}{\partial t} \to 0 \quad (9.3.3.56)$$

weakly in $L^2(\Sigma_{ie})$. From (9.3.3.47), (9.3.3.48), (9.3.3.55) and (9.3.3.56) we may conclude that

$$(Q_{ie}^{+})^* D_{\nu_{ie}}\widetilde{W}_i^n \to 0, \quad (Q_{ie}^{+})^* D_{\nu_{ie}}\widetilde{R}_i^n \to 0, \quad Q_{ie}\frac{\partial \widetilde{W}_i^n}{\partial t} \to 0, \quad Q_{ie}\frac{\partial \widetilde{R}_i^n}{\partial t} \to 0$$

weakly in $L^2(\Sigma_{ie})$, $\forall i \in \mathcal{I}_e$, $e \in \mathcal{N}^{\text{int}}$.

We next show that

$$\|\sum_{i\in\mathcal{I}_e}(Q_{ie}^{+})^* D_{\nu_{ie}}W_i^n\|_{L^2(\Sigma_e)} \to 0, \quad \|\sum_{i\in\mathcal{I}_e}(Q_{ie}^{+})^* D_{\nu_{ie}}R_i^n\|_{L^2(\Sigma_e)} \to 0,$$
$$\|Q_{ie}\frac{\partial W_i^n}{\partial t} - Q_{je}\frac{\partial W_j^n}{\partial t}\|_{L^2(\Sigma_e)} \to 0, \quad \|Q_{ie}\frac{\partial R_i^n}{\partial t} - Q_{je}\frac{\partial R_j^n}{\partial t}\|_{L^2(\Sigma_e)} \to 0,$$
$$\forall i,j \in \mathcal{I}_e, \ \forall e \in \mathcal{N}^{\text{int}}. \quad (9.3.3.57)$$

The conclusions in (9.3.3.57) are a consequence of Proposition 2.3.7.1, which tells us that

$$\|\mathcal{T}_\epsilon^n X^1 - \mathcal{T}_\epsilon^{n-1}X^1\|_{\mathcal{X}} = (1-\epsilon)\|\mathcal{T}X^n - X^n\|_{\mathcal{X}} \to 0.$$

From (9.3.3.43) we have

$$\|\mathcal{T}X^n - X^n\|_{\mathcal{X}} = 2\big(\mathcal{E}^n - (\mathcal{T}X^n, X^n)_{\mathcal{X}}\big).$$

To obtain the value of the right-hand side we use the following result.

Lemma 9.3.3.2 *For any $X = \{(\lambda_{ie},\rho_{ie}) : e \in \mathcal{N}^{int}, i \in \mathcal{I}_e\} \in \mathcal{X}$ we may write*

$$\mathcal{E} - (\mathcal{T}X,X)_{\mathcal{X}} = \sum_{e\in\mathcal{N}^{int}} \frac{1}{\gamma d_e}\int_{\Sigma_e}\{|\sum_{i\in\mathcal{I}_e}(Q_{ie}^{+})^* D_{\nu_{ie}}W_i^n|^2 + |\sum_{i\in\mathcal{I}_e}(Q_{ie}^{+})^* D_{\nu_{ie}}R_i^n|^2$$
$$+\frac{\beta^2+\gamma^2}{2}\sum_{i,j\in\mathcal{I}_e}[|Q_{ie}\frac{\partial W_i^n}{\partial t} - Q_{je}\frac{\partial W_j^n}{\partial t}|^2 + |Q_{ie}\frac{\partial R_i^n}{\partial t} - Q_{je}\frac{\partial R_j^n}{\partial t}|^2]\}d\Sigma,$$

where W_i, R_i is the solution of (9.3.3.15)–(9.3.3.20) *and $\mathcal{E}$ is given by* (9.3.3.27).

Lemma 9.3.3.2 is proved below.

We apply Lemma 9.3.3.2 with $X = X^n$ given by (9.3.3.24) and conclude that

$$\|\sum_{i\in\mathcal{I}_e}(Q_{ie}^+)^* D_{\nu_{ie}}\widetilde{W}_i^n\|_{L^2(\Sigma_e)} \to 0, \quad \|\sum_{i\in\mathcal{I}_e}(Q_{ie}^+)^* D_{\nu_{ie}}\widetilde{R}_i^n\|_{L^2(\Sigma_e)} \to 0,$$

$$\|Q_{ie}\frac{\partial \widetilde{W}_i^n}{\partial t} - Q_{je}\frac{\partial \widetilde{W}_j^n}{\partial t}\|_{L^2(\Sigma_e)} \to 0, \quad \|Q_{ie}\frac{\partial \widetilde{R}_i^n}{\partial t} - Q_{je}\frac{\partial \widetilde{R}_j^n}{\partial t}\|_{L^2(\Sigma_e)} \to 0,$$
$$\forall i,j \in \mathcal{I}_e,\ \forall e \in \mathcal{N}^{\rm int}.$$

The conclusions in (9.3.3.57) follow immediately since

$$\sum_{i\in\mathcal{I}_e}(Q_{ie}^+)^* D_{\nu_{ie}} W_i = \sum_{i\in\mathcal{I}_e}(Q_{ie}^+)^* D_{\nu_{ie}} R_i = 0,$$
$$Q_{ie}\frac{\partial W_i}{\partial t} = Q_{je}\frac{\partial W_j}{\partial t}, \quad Q_{ie}\frac{\partial R_i}{\partial t} = Q_{je}\frac{\partial R_j}{\partial t} \quad \text{on } \Sigma_e$$

where $\{W_i, P_i\}_{i\in\mathcal{I}}$ is the solution of the global optimality system (9.2.1.2), (9.2.1.3), (9.2.1.11)–(9.2.1.15), and where $P_i = \partial R_i/\partial t$, $\partial P_i/\partial t = -\mathcal{A}_i R_i$.

To complete the proof of Theorem 9.3.3.1 it remains to show that

$$(\widetilde{W}_i^n, \frac{\partial \widetilde{W}_i^n}{\partial t}) \to 0, \quad (\widetilde{R}_i^n, \frac{\partial \widetilde{R}_i^n}{\partial t}) \to 0 \ \text{ in } C([0,T]; \mathcal{V}_i \times \mathcal{H}_i).$$

To do so we employ the following result, analogous to Proposition 6.4.2.1.

Proposition 9.3.3.1 *For $i \in \mathcal{I}$ let Φ_i^n satisfy*

$$\frac{\partial^2 \Phi_i^n}{\partial t^2} - \frac{\partial}{\partial x_\alpha}[A_i^{\alpha\beta}(\Phi_{i,\beta}^n + B_i^\beta \Phi_i^n)]$$
$$+ (B_i^\alpha)^* A_i^{\alpha\beta}(\Phi_{i,\beta}^n + B_i^\beta\Phi_i^n) + C_i\Phi_i^n = 0 \ \ in\ \mathcal{Q}_i;$$

$$\Phi_i^n = 0 \ \ on\ \Sigma_{ie},\ e \in \mathcal{D}_i, \quad \Pi_{ie}\Phi_i^n = 0 \ \ on\ \Sigma_{ie},\ e \in \mathcal{N}_i^{int};$$

$$D_{\nu_{ie}}\Phi_i^n + \alpha_{ie}\frac{\partial \Phi_i^n}{\partial t} = f_{ie}^n \quad on\ \Sigma_{ie},\ e \in \mathcal{N}_i^{ext};$$

$$\Phi_i^n(0) = \Phi_{0i}^n, \quad \frac{\partial \Phi_i^n}{\partial t}(0) = \Phi_{1i}^n \quad in\ \Omega_i,$$

where

$$(\Phi_{0i}^n, \Phi_{1i}^n) \in \mathcal{V}_i \times \mathcal{H}_i, \qquad f_{ie}^n \in L^2(\Sigma_{ie}).$$

Assume that

$$(Q_{ie}^+)^* D_{\nu_{ie}}\Phi_i^n\big|_{\Sigma_{ie}}, \quad Q_{ie}\frac{\partial \Phi_i^n}{\partial t}\Big|_{\Sigma_{ie}} \quad \textit{are bounded in } L^2(\Sigma_{ie}),\ \forall e \in \mathcal{N}_i^{int}. \quad (9.3.3.58)$$

If

$$\|(\Phi_{0i}^n, \Phi_{1i}^n)\|_{\mathcal{V}_i\times\mathcal{H}_i} \to 0, \quad \|f_{ie}^n\|_{L^2(\Sigma_{ie})} \to 0,$$
$$\|\sum_{i\in\mathcal{I}_e}(Q_{ie}^+)^* D_{\nu_{ie}}\Phi_i^n\|_{L^2(\Sigma_e)} \to 0, \quad \|Q_{ie}\frac{\partial\Phi_i^n}{\partial t} - Q_{je}\frac{\partial\Phi_j^n}{\partial t}\|_{L^2(\Sigma_e)} \to 0 \tag{9.3.3.59}$$

for all $i,j\in\mathcal{I}_e$ *and* $e\in\mathcal{N}^{int}$, *then*

$$(\Phi_i^n, \frac{\partial\Phi_i^n}{\partial t}) \to 0 \ \ \textit{strongly in } L^\infty(0,T;\mathcal{V}_i\times\mathcal{H}_i),\ \forall i\in\mathcal{I}.$$

The proof of this proposition is given below.

We apply Proposition 9.3.3.1 with $\Phi_i^n = \widetilde{W}_i^n$. In this case,

$$\Phi_{0i} = \Phi_{1i} = 0, \quad f_{ie}^n = -\frac{\partial R_i^n}{\partial t}\Big|_{\Sigma_{ie}}.$$

Condition (9.3.3.58) is satisfied by virtue of the boundedness of $\mathcal{E}^n$, and (9.3.3.59) has already been established. Thus we may conclude that

$$(\widetilde{W}_i^n, \frac{\partial\widetilde{W}_i^n}{\partial t}) \to 0 \ \ \text{strongly in } L^\infty(0,T;\mathcal{V}_i\times\mathcal{H}_i),\ \forall i\in\mathcal{I}.$$

Similarly, after the change of variable $t\mapsto T-t$, we may apply Proposition 9.3.3.1 to $\widetilde{R}_i^n$. In this case

$$\Phi_{0i} = \widetilde{R}_i^n(T), \quad \Phi_{1i} = \frac{\partial\widetilde{R}_i^n}{\partial t}(T), \quad f_{ie}^n = 0.$$

Again, condition (9.3.3.58) is satisfied by virtue of the boundedness of $\mathcal{E}^n$, and (9.3.3.59) is satisfied by virtue of what has already been proved. This completes the proof of Theorem 9.3.3.1.

Proof of Lemma 9.3.3.2. We have

$$\begin{aligned}(\mathcal{T}X,X)_{\mathcal{X}} &= \sum_{i\in\mathcal{I}}\sum_{e\in\mathcal{N}_i^{\rm int}}\frac{1}{2\gamma}\int_{\Sigma_{ie}}(\lambda_{ie}\cdot\hat{\lambda}_{ie} + \rho_{ie}\cdot\hat{\rho}_{ie})\,d\Sigma\\ &= \sum_{e\in\mathcal{N}^{\rm int}}\frac{1}{2\gamma}\int_{\Sigma_e}\sum_{i\in\mathcal{I}_e}(\lambda_{ie}\cdot\hat{\lambda}_{ie} + \rho_{ie}\cdot\hat{\rho}_{ie})\,d\Sigma\end{aligned} \tag{9.3.3.60}$$

where $\hat{\lambda}_{ie}$ and $\hat{\rho}_{ie}$ are given by (9.3.3.22) and (9.3.3.23), respectively. We may write

$$\lambda_{ie} = (Q_{ie}^+)^* D_{\nu_{ie}} W_i + \beta Q_{ie}\frac{\partial W_i}{\partial t} + \gamma Q_{ie}\frac{\partial R_i}{\partial t}\Big|_{\Sigma_{ie}}$$
$$\rho_{ie} = (Q_{ie}^+)^* D_{\nu_{ie}} W_i - \beta Q_{ie}\frac{\partial R_i}{\partial t} + \gamma Q_{ie}\frac{\partial W_i}{\partial t}\Big|_{\Sigma_{ie}}.$$

The calculation of (9.3.3.60) is elementary but tedious. One finds after some computation that

$$\sum_{i\in\mathcal{I}_e}\lambda_{ie}\cdot\hat{\lambda}_{ie}=-\frac{2}{d_e}\Big|\sum_{i\in\mathcal{I}_e}(Q_{ie}^{+})^{*}D_{\nu_{ie}}W_i\Big|^2+\sum_{i\in\mathcal{I}_e}\big|(Q_{ie}^{+})^{*}D_{\nu_{ie}}W_i\big|^2$$
$$+\frac{2\beta^2}{d_e}\Big|\sum_{i\in\mathcal{I}_e}Q_{ie}\frac{\partial W_i}{\partial t}\Big|^2-\beta^2\sum_{i\in\mathcal{I}_e}\Big|Q_{ie}\frac{\partial W_i}{\partial t}\Big|^2$$
$$+\frac{2\gamma^2}{d_e}\Big|\sum_{i\in\mathcal{I}_e}Q_{ie}\frac{\partial R_i}{\partial t}\Big|^2-\gamma^2\sum_{i\in\mathcal{I}_e}\Big|Q_{ie}\frac{\partial R_i}{\partial t}\Big|^2$$
$$+\frac{4\beta\gamma}{d_e}\sum_{i\in\mathcal{I}_e}Q_{ie}\frac{\partial W_i}{\partial t}\cdot\sum_{i\in\mathcal{I}_e}Q_{ie}\frac{\partial R_i}{\partial t}-2\beta\gamma\sum_{i\in\mathcal{I}_e}Q_{ie}\frac{\partial W_i}{\partial t}\cdot Q_{ie}\frac{\partial R_i}{\partial t}.\qquad(9.3.3.61)$$

Similarly,

$$\sum_{i\in\mathcal{I}_e}\rho_{ie}\cdot\hat{\rho}_{ie}=-\frac{2}{d_e}\Big|\sum_{i\in\mathcal{I}_e}(Q_{ie}^{+})^{*}D_{\nu_{ie}}R_i\Big|^2+\sum_{i\in\mathcal{I}_e}\big|(Q_{ie}^{+})^{*}D_{\nu_{ie}}R_i\big|^2$$
$$+\frac{2\beta^2}{d_e}\Big|\sum_{i\in\mathcal{I}_e}Q_{ie}\frac{\partial R_i}{\partial t}\Big|^2-\beta^2\sum_{i\in\mathcal{I}_e}\Big|Q_{ie}\frac{\partial R_i}{\partial t}\Big|^2$$
$$+\frac{2\gamma^2}{d_e}\Big|\sum_{i\in\mathcal{I}_e}Q_{ie}\frac{\partial W_i}{\partial t}\Big|^2-\gamma^2\sum_{i\in\mathcal{I}_e}\Big|Q_{ie}\frac{\partial W_i}{\partial t}\Big|^2$$
$$-\frac{4\beta\gamma}{d_e}\sum_{i\in\mathcal{I}_e}Q_{ie}\frac{\partial W_i}{\partial t}\cdot\sum_{i\in\mathcal{I}_e}Q_{ie}\frac{\partial R_i}{\partial t}+2\beta\gamma\sum_{i\in\mathcal{I}_e}Q_{ie}\frac{\partial W_i}{\partial t}\cdot Q_{ie}\frac{\partial R_i}{\partial t}.\qquad(9.3.3.62)$$

It follows from (9.3.3.61), (9.3.3.62) and the definition (9.3.3.27) of $\mathcal{E}$ that

$$\mathcal{E}-(\mathcal{T}X,X)_{\mathcal{X}}=\sum_{e\in\mathcal{N}^{\text{int}}}\frac{1}{\gamma}\int_{\Sigma_e}\Big\{\frac{1}{d_e}\Big[\Big|\sum_{i\in\mathcal{I}_e}(Q_{ie}^{+})^{*}D_{\nu_{ie}}W_i\Big|^2+\Big|\sum_{i\in\mathcal{I}_e}(Q_{ie}^{+})^{*}D_{\nu_{ie}}R_i\Big|^2\Big]$$
$$+(\beta^2+\gamma^2)\Big[\sum_{i\in\mathcal{I}_e}\Big|Q_{ie}\frac{\partial W_i}{\partial t}\Big|^2-\frac{1}{d_e}\Big|\sum_{i\in\mathcal{I}_e}Q_{ie}\frac{\partial W_i}{\partial t}\Big|^2$$
$$+\sum_{i\in\mathcal{I}_e}\Big|Q_{ie}\frac{\partial R_i}{\partial t}\Big|^2-\frac{1}{d_e}\Big|\sum_{i\in\mathcal{I}_e}Q_{ie}\frac{\partial R_i}{\partial t}\Big|^2\Big]\Big\}\,d\Sigma.$$

We obtain the conclusion of Lemma 9.3.3.2 by noting that

$$\frac{1}{2d_e}\sum_{i,j\in\mathcal{I}_e}|a_i-a_j|^2=\sum_{i\in\mathcal{I}_e}|a_i|^2-\frac{1}{d_e}\Big|\sum_{i\in\mathcal{I}_e}a_i\Big|^2.$$

Proof of Proposition 9.3.3.1. Set

$$g_{ie}^n=(Q_{ie}^{+})^{*}D_{\nu_{ie}}\Phi_i^n+\beta_e Q_{ie}\frac{\partial\Phi_i^n}{\partial t}\Big|_{\Sigma_{ie}},\qquad e\in\mathcal{N}_i^{\text{int}},$$

where $\beta_e > 0$ is constant. Then Φ_i^n satisfies the abstract equation (cf. (9.3.1.6))

$$\frac{d^2\Phi_i^n}{dt^2} + \mathcal{B}_i\mathcal{B}_i^*\frac{d\Phi_i^n}{dt} + \mathcal{A}_i\Phi_i^n = \mathcal{B}_i(f_{i\alpha}^n, g_{i\beta}^n) \text{ in } \mathcal{V}_i^*.$$

By using a standard multiplier method, one easily obtains the a priori identity (cf. (9.3.1.7))

$$\begin{aligned}\|(\Phi_i^n(t), \frac{\partial\Phi_i^n}{\partial t}(t))\|^2_{\mathcal{V}_i\times\mathcal{H}_i} &+ \sum_{e\in\mathcal{N}_i^{\text{ext}}}\int_0^t\int_{\Gamma_{ie}}\left[\frac{1}{\alpha_{ie}}|D_{\nu_{ie}}\Phi_i^n|^2 + \alpha_{ie}|\frac{\partial\Phi_i^n}{\partial t}|^2\right]d\Gamma dt\\
&= \|(\Phi_{0i}^n, \Phi_{1i}^n)\|^2_{\mathcal{V}_i\times\mathcal{H}_i} + \sum_{e\in\mathcal{N}_i^{\text{ext}}}\int_0^t\int_{\Gamma_{ie}}\frac{1}{\alpha_{ie}}|f_{ie}^n|^2\, d\Gamma dt\\
&+ \sum_{e\in\mathcal{N}_i^{\text{int}}} 2\beta_e\int_0^t\int_{\Gamma_{ie}}(Q_{ie}^+)^* D_{\nu_{ie}}\Phi_i^n\cdot Q_{ie}\frac{\partial\Phi_i^n}{\partial t}\, d\Gamma dt, \quad 0\le t\le T. \qquad (9.3.3.63)\end{aligned}$$

We sum this identity over $i\in\mathcal{I}$. After summation, the last term on the right may be written

$$\sum_{e\in\mathcal{N}^{\text{int}}} 2\beta_e\int_0^t\int_{\Gamma_e}\sum_{i\in\mathcal{I}_e}(Q_{ie}^+)^* D_{\nu_{ie}}\Phi_i^n\cdot Q_{ie}\frac{\partial\Phi_i^n}{\partial t}\, d\Gamma dt.$$

We have

$$\begin{aligned}\sum_{i\in\mathcal{I}_e}(Q_{ie}^+)^* D_{\nu_{ie}}\Phi_i^n\cdot Q_{ie}\frac{\partial\Phi_i^n}{\partial t} &= \frac{1}{d_e}\sum_{i,j\in\mathcal{I}_e}(Q_{ie}^+)^* D_{\nu_{ie}}\Phi_i^n\cdot(Q_{ie}\frac{\partial\Phi_i^n}{\partial t} - Q_{je}\frac{\partial\Phi_j^n}{\partial t})\\
&+ \frac{1}{d_e}\sum_{i\in\mathcal{I}_e}(Q_{ie}^+)^* D_{\nu_{ie}}\Phi_i^n\cdot\sum_{i\in\mathcal{I}_e}Q_{ie}\frac{\partial\Phi_i^n}{\partial t}. \qquad (9.3.3.64)\end{aligned}$$

It follows from assumptions (9.3.3.58) and (9.3.3.59) that the right side of (9.3.3.64) converges to zero in $L^1(\Sigma_e)$ as $n\to\infty$. We may then conclude from (9.3.3.63) and (9.3.3.59) that

$$\sup_{0\le t\le T}\sum_{i\in\mathcal{I}}\|(\Phi_i^n(t), \frac{\partial\Phi_i^n}{\partial t}(t))\|_{\mathcal{V}_i\times\mathcal{H}_i}\to 0.$$

This completes the proof.

We close this chapter with some numerical experiment on optimal null-controllability by the way of so-called instantaneous controls explained at the end of Chapter 5. Here we concentrate on a very simple example of three square shaped domains coupled in order to form an L-shaped domain. For the sake of simplicity we consider a scalar out-of-the-plane wave equation. The boundary conditions are given by uncontrolled (Figures 9.1 to 9.3) and controlled Neumann conditions Figures 9.4 to 9.6) along the L-shaped boundary on the 'right side' (the one opposite to the re-entrant corner). Along the remaining boundary (the one including the edges forming the re-entrant corner and the two corresponding adjacent edges) we impose Dirichlet conditions. This setup represents an exactly controllable situation. The numerical simulation is based on the standard relaxed domain decomposition technique discussed in this chapter and throughout this monograph. The time discretization is performed using the (conservative) Newmark method. The first set of figures (Fig. 9.1, 9.2, 9.3) show the free evolution of the waves. The total energy is conserved. The second set of figures (Fig. 9.4, 9.5, 9.6) show the effect of the boundary controls. It is evident from the pictures, and can be demonstrated also by error plots, that the controls completely absorb the energy and steer the system to rest. As mentioned above, the technique employed here is a suboptimal one based on the concept of instantaneous (or receding horizon and, in particular, rolling horizon) controls. Such a control strategy is real-time capable. See Hundhammer and Leugering [47] and Hundhammer [46], and the references therein, for further information. We note that we have also performed numerical simulations of scalar wave equations on 2-D networks in three-space. The results are, however, more difficult to visualize in the format of this monograph. The result displayed below, and many more, can be seen in [47] and [46].

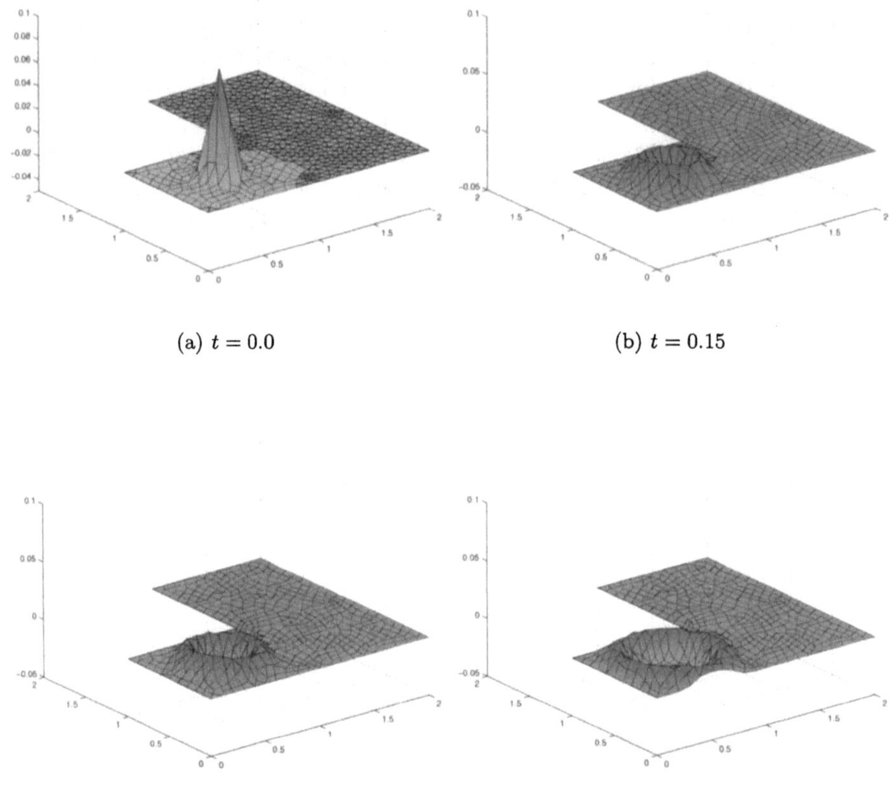

(a) $t = 0.0$ (b) $t = 0.15$

(c) $t = 0.2$ (d) $t = 0.3$

Figure 9.1: Simulation of a vibrating membrane on an L-shaped domain using domain decomposition: $t = 0.0 \ldots t = 0.3$

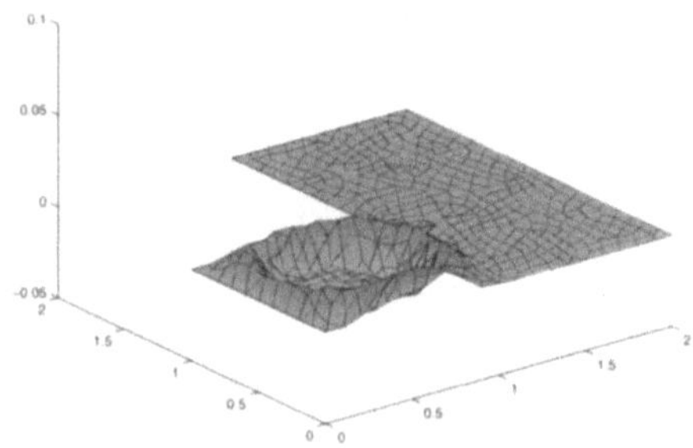

(a) $t = 0.35$

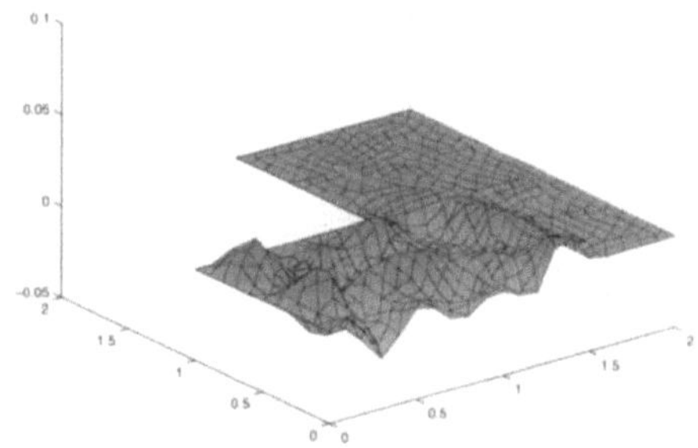

(b) $t = 0.75$

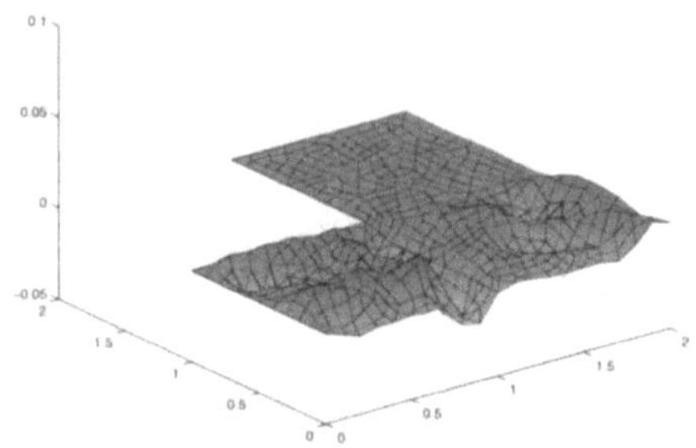

(c) $t = 1.15$

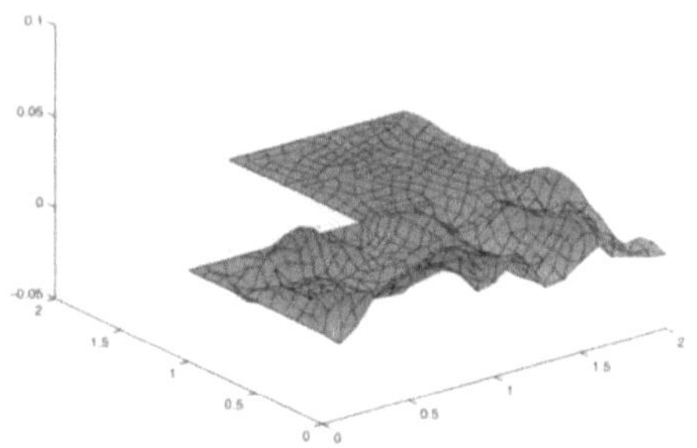

(d) $t = 1.500$

Figure 9.2: Simulation of a vibrating membrane on an L-shaped domain using domain decomposition: $t = 0.75 \ldots\ t = 1.5$

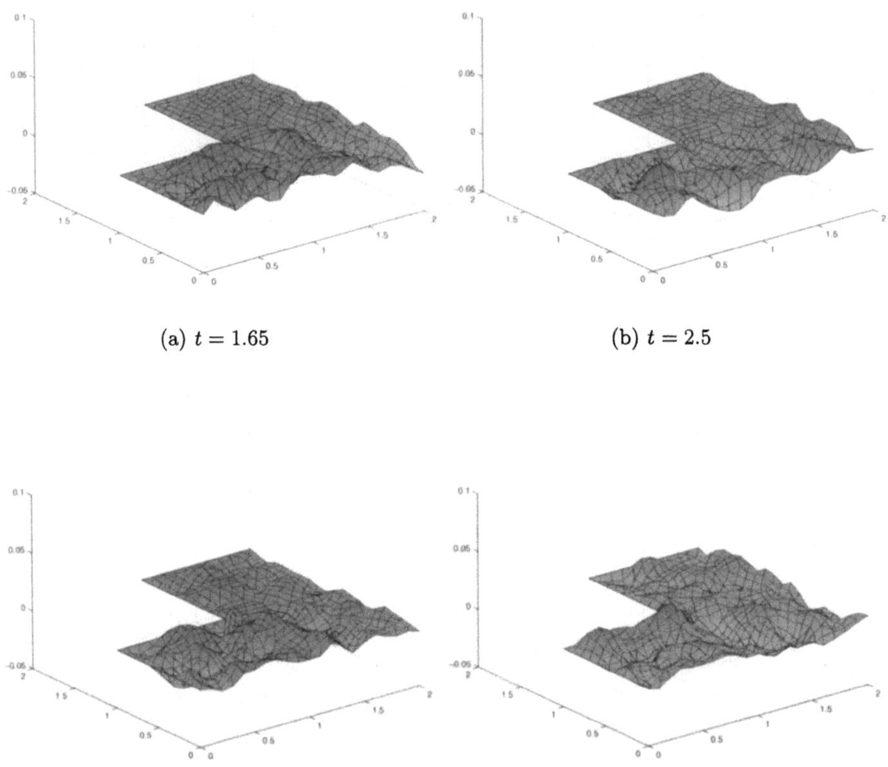

(a) $t = 1.65$

(b) $t = 2.5$

(c) $t = 4.0$

(d) $t = 12.0$

Figure 9.3: Simulation of a vibrating membrane on an L-shaped domain using domain decomposition: $t = 1.65 \ldots t = 12.0$

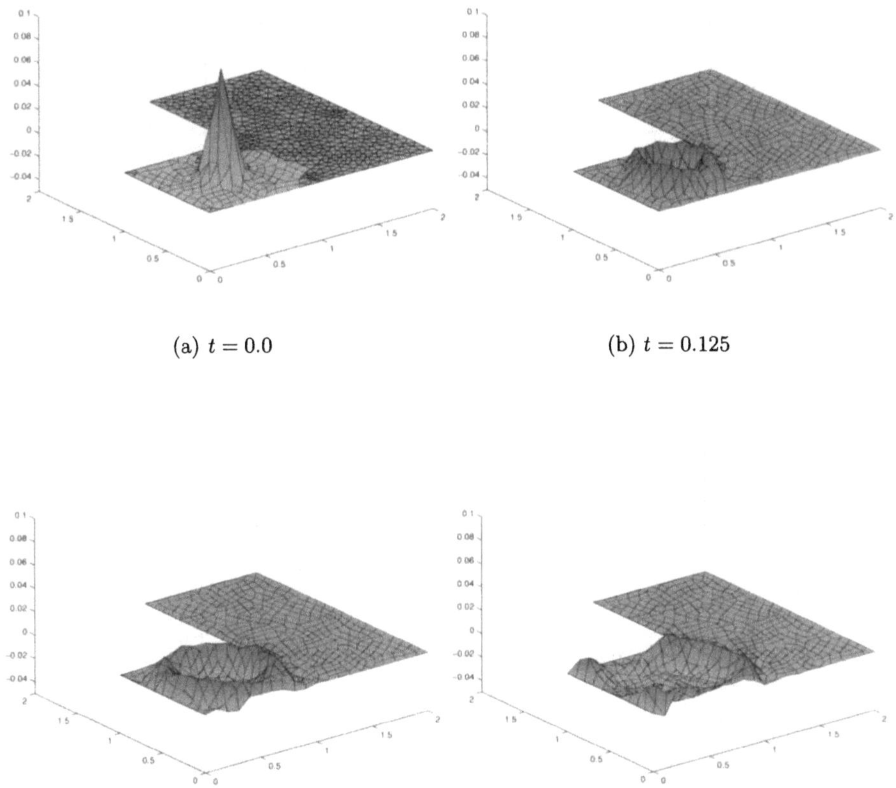

(a) $t = 0.0$ (b) $t = 0.125$

(c) $t = 0.2$ (d) $t = 0.275$

Figure 9.4: Instantaneous control with domain decomposition: $t = 0.0 \ldots\ t = 0.275$

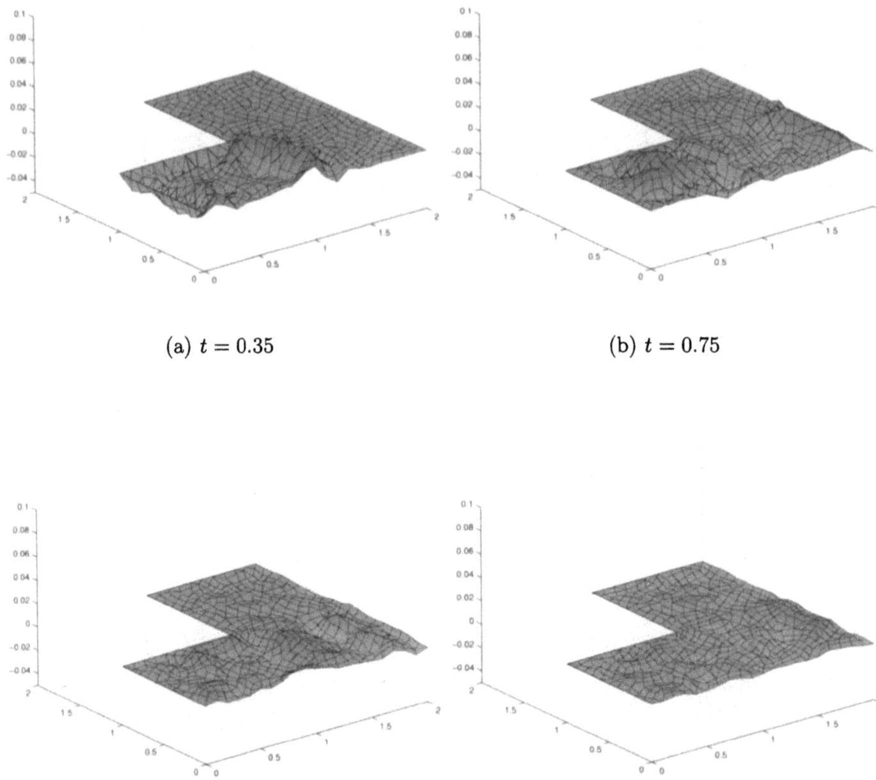

(a) $t = 0.35$

(b) $t = 0.75$

(c) $t = 1.15$

(d) $t = 1.475$

Figure 9.5: Instantaneous control with domain decomposition: $t = 0.35 \ldots t = 1.475$

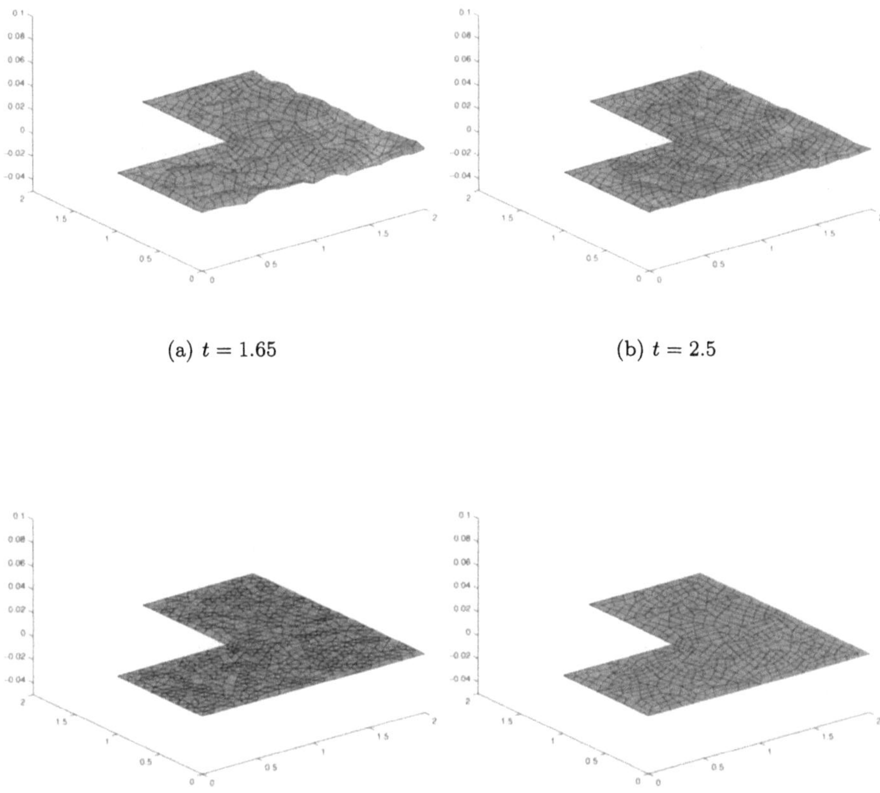

(a) $t = 1.65$

(b) $t = 2.5$

(c) $t = 4.0$

(d) $t = 12.0$

Figure 9.6: Instantaneous control with domain decomposition: $t = 1.65 \dots t = 12.0$

Bibliography

[1] J.-L. Lions. Virtual and effective control for distributed systems and decomposition of everything. *J. d'Anal. Math.*, 80:257–297, 2000.

[2] F. Ali Mehmeti, S. Nicaise, and J. von Below. *Partial Differential equations on multistrucutres.* Marcel Dekker, 2001.

[3] U.M. Ascher, R.M.M. Mattheij, and R.D. Russell. *Numerical solution of boundary value problems for ordinary differential equations.* Classics in Applied Mathematics. 13. Philadelphia, PA: SIAM, Society for Industrial and Applied Mathematics. xxv, 595 p. $ 39.50 , 1995.

[4] A. Auge, G. Lube, and F.C. Otto. A non-overlapping domain decomposition method with adaptive interface conditions for elliptic problems. In *Hackbusch, Wolfgang (ed.) et al., Numerical treatment of multi-scale problems. Proceedings of the 13th GAMM-seminar, Kiel, Germany, January 24–26, 1997. Braunschweig: Vieweg. Notes Numer. Fluid Mech. 70, 12–23* . 1999.

[5] F.B. Belgacem, C. Bernardi, M. Costabel, and M. Dauge. Un résultat de densité pour les équations de Maxwell. *C. R. Acad. Sci. Paris*, 324, Ser. I:731–736, 1997.

[6] J.-D. Benamou. A domain decomposition method for the polar factorization of vector-valued mappings. *SIAM J. Numer. Anal.*, 32(6):1808–1838, 1995.

[7] J.-D. Benamou. A domain decomposition method with coupled transmission conditions for the optimal control of systems governed by elliptic partial differential equations. *SIAM J. Numer. Anal.*, 33(6):2401–2416, 1996.

[8] J.-D. Benamou. Décomposition de domaine pour le contrôle de systèmes gouvernés par des équations d'évolution. (Domain decomposition for the control of systems governed by evolution equations). 1997.

[9] J.-D. Benamou. Domain decomposition, optimal control of systems governed by partial differential equations, and synthesis of feedback laws. *J. Optimization Theory Appl.*, 102(1):15–36, 1999.

[10] J.-D. Benamou and Bruno Després. A domain decomposition method for the Helmholtz equation and related optimal control problems. *J. Comput. Phys.*, 136(1):68–82, 1997.

[11] H. Bialy. Iterative Behandlung linearer Funktionalgleichungen. *Arch. Ration. Mech. Anal.*, 4:166–176, 1959.

[12] Haecheon Choi, Michael Hinze, and Karl Kunisch. Instantaneous control of backward-facing step flows. *Appl. Numer. Math.*, 31(2):133–158, 1999.

[13] J. de Halleux, C. Prieur, J.-M. Coron, B. d'Andrea Novel, and G. Bastin. Boundary feedback control in networks of open channels. 2003. preprint, to appear.

[14] S. Delpino, J.-L. Lions, and O. Pironneau. A priori domain decomposition of PDE systems and applications. In *Babuska, Ivo (ed.) et al., Mathematical modeling and numerical simulation in continuum mechanics. Proceedings of the international symposium, Yamaguchi, Japan, September 29–October 3, 2000. Berlin: Springer. Lect. Notes Comput. Sci. Eng. 19, 125–135* . 2002.

[15] Q. Deng. An analysis for a nonoverlapping domain decomposition iterative procedure. *SIAM J. Sci. Comput.*, 18(5):1517–1525, 1997.

[16] B. Després. *Méthodes de décomposition de domaine pour les problèmes de propagation d'ondes en régime harmonique. Le théorème de Borg pour l'équation de Hill vectorielle. (Domain decomposition methods for harmonic wave-propagation problems. The Borg theorem for vectorial Hill equation).* PhD thesis, Rocquencourt: Institut National de Recherche en Informatique et en Automatique (INRIA). vi, 233 p. (Univ. de Paris IX, 1991) , 1991.

[17] H. Le Dret. Vibrations of a folded plate. *RAIRO Modél. Math. Anal. Numér.*, 24:501–521, 1990.

[18] M. Dryja and W. Hackbusch. On the nonlinear domain decomposition method. *BIT*, 37(2):296–311, 1997.

[19] Q. Du and M.D. Gunzburger. A gradient method approach to optimization-based multidisciplinary simulations and nonoverlapping domain decomposition algorithms. *SIAM J. Numer. Anal.*, 37(5):1513–1541, 2000.

[20] M. Eller. Unique continuation for solutions to Maxwell's system with non-analytic anisotropic coefficients. *J. Math. Anal. and Appl.*, to appear.

[21] M. Eller. Uniqueness of continuation theorems. In R.P. Gilbert, J. Kajiwara, and Y.S. Xu, editors, *Direct and Inverse Problems of Mathematical Physics.* Kluver, 1999.

[22] T. Fischer and G. Leugering. On instantaneous control of singularly perturbed hyperbolic problems on graphs. In *Ali Mehmeti et al. eds.: Partial differential equations on mutlistructures*, pages 69–97. Marcel Dekker, New York, 2001.

[23] M. Fortin and R. Glowinski. *Augmented Lagrangian Methods: Applications to the Numerical Solution of Boundary-Value Problems.* North-Holland, 1983.

[24] M.J. Gander. A waveform relaxation algorithm with overlapping splitting for reaction diffusion equations. *Numer. Linear Algebra Appl.*, 6(2):125–145, 1999.

[25] M.J. Gander and L. Halpern. Méthodes de décomposition de domaines pour l'équation des ondes en dimension 1. (Optimal Schwarz waveform relaxation for the one-dimensional wave equation). 2001.

[26] M.J. Gander and L. Halpern. Un algorithm discret de décomposition de domaines pour l'équation des ondes en dimension 1. (A discrete domain decomposition algorithm for the one-dimensional wave equation). 2001.

[27] M.J. Gander, L. Halpern, and F. Nataf. Domain decomposition methods for wave propagation. In *Bermédez, Alfredo (ed.) et al., Mathematical and numerical aspects of wave propagation. 5th international conference, Santiago de Compostela, Spain, July 10–14, 2000. Philadelphia, PA: SIAM, Society for Industrial and Applied Mathematics. 807–811* . 2000.

[28] R. Glowinski and P. Le Tallec. *Augmented Lagrangian and operator-splitting methods in nonlinear mechanics.* SIAM Studies in Applied Mathematics, 9. Philadelphia, PA: Society for Industrial and Applied Mathematics, 1989.

[29] R. Glowinski and P. Le Tallec. Augmented Lagrangian interpretation of the nonoverlapping Schwarz alternating method. In *Domain decomposition methods for partial differential equations, Proc. 3rd Int. Symp. Houston/TX (USA) 1989, 224–231* . 1990.

[30] P. Grisvard. *Elliptic Problems in Nonsmooth Domains.* Pitman, Boston, 1985.

[31] M. Gugat. Problems of optimal boundary control in flood management. 2003. submitted.

[32] M. Gugat, M. Herty, A. Klar, and G. Leugering. Optimal control for traffic flow networks. *J. Optimization, Theory and Applications*, 2003.

[33] M. Gugat and G. Leugering. Global boundary controllability of the de St. Venant equations between steady states. *Annales de l'Institut Henri Poincaré, Nonlinear Analysis*, 20:1–11, 2003.

[34] M. Gugat, G. Leugering, and E.J.P.G. Schmidt. Global controllability between steady supercritical flows in channel networks. *Math. Meth. Appl. Sci.*, 26, 2004. to appear.

[35] M.D. Gunzburger, M. Heinkenschloss, and H.K. Lee. Solution of elliptic partial differential equations by an optimization-based domain decomposition method. *Appl. Math. Comput.*, 113(2-3):111–139, 2000.

[36] M.D. Gunzburger and H.K. Lee. An optimization-based domain decomposition method for the Navier-Stokes equations. *SIAM J. Numer. Anal.*, 37(5):1455–1480, 2000.

[37] M.D. Gunzburger and Jeehyun Lee. A domain decomposition method for optimization problems for partial differential equations. *Comput. Math. Appl.*, 40(2-3):177–192, 2000.

[38] M.D. Gunzburger, J.S. Peterson, and H. Kwon. An optimization based domain decomposition method for partial differential equations. *Comput. Math. Appl.*, 37(10):77–93, 1999.

[39] W. Hachbusch. *Iterative Solutions of Large Sparse Systems of Equations.* Springer-Verlag, 1994.

[40] W. Hackbusch. Direct domain decomposition using the hierarchical matrix technique., 2002. MPI-Leipzig, Preprint.

[41] L. Halpern. Absorbing boundary conditions for the discretization schemes of the one-dimensional wave equation. *Math. Comput.*, 38:415–429, 1982.

[42] F. Hecht, J.-L. Lions, and O. Pironneau. Domain decomposition algorithm for computer aided design. In *Sequeira, Adélia (ed.) et al., Applied nonlinear analysis. In honor of the 70th birthday of Professor Jindrich Necas. New York, NY: Kluwer Academic/Plenum Publishers. 185–198* . 1999.

[43] E. Hendrickson and I. Lasiecka. Finite-dimensional approximations of boundary control problems arising in partially observed hyperbolic systems. *Dynam. Cont. Discrete Impuls. Systems*, 1:101–142, 1995.

[44] Michael Hinze and Karl Kunisch. On suboptimal control strategies for the Navier-Stokes equations. *ESAIM, Proc.*, 4:181–198, 1998.

[45] L. Hörmander. A uniqueness theorem for second-order hyperbolic differential equations. *Commun. PDEs*, 17:699–714, 1992.

[46] R. Hundhammer. *Simulation und Echtzeitsteuerung schwingender Saitennetzwerke.* PhD thesis, 2002.

[47] R. Hundhammer and G. Leugering. Instantaneous control of vibrating string networks. In *Grötschel, Martin (ed.) et al., Online optimization of large scale systems*, pages 229–249. Springer-Verlag, Berlin, 2001.

[48] V. Komornik. *Exact Controllability and Stabilization: The Multiplier Method.* Masson, Paris, 1994.

[49] J.E. Lagnese. *Boundary Stabilization of Thin Plates.* Studies in Applied Mathematics. SIAM, Philadelphia, PA, 1989.

[50] J.E. Lagnese. The Hilbert Uniqueness Method: A retrospective. In K.-H. Hoffmann and W. Krabs, editors, *Lecture Notes in Control and Information Science*, volume 149, pages 158–181, Berlin, 1991. Springer-Verlag.

[51] J.E. Lagnese. A comparison of models of a folded plate. *Proc. Eur. Ser. Appl. Indus. Math.*, 2:113–131 (electronic), 1997.

[52] J.E. Lagnese. Domain decomposition in optimal boundary control problems of second-order hyperbolic systems on 2-d networks. In *Proceedings of the 37th IEEE CDC*, pages 2964–2969, Tampa, FL, December 1998.

[53] J.E. Lagnese. Domain decomposition in exact controllability of second-order hyperbolic systems on 1-d networks. *Control and Cybernetics*, 28:531–556, 1999.

[54] J.E. Lagnese. A nonoverlapping domain decomposition for optimal boundary control of the dynamic Maxwell system. In G. Chen, I. Lasiecka, and J. Zhou, editors, *Control of Nonlinear Distributed Parameter Systems*, pages 157–176, New York, 2000. Marcel Dekker.

[55] J.E. Lagnese. Domain decomposition in optimal boundary control of hyperbolic equations with boundary damping. In *International Series in Numerical Mathematics*, volume 139, pages 157–169, Basel, 2001. Birkhäuser Verlag.

[56] J.E. Lagnese and G. Leugering. Dynamic domain decomposition in approximate and exact boundary control in problems of transmission for wave equations. *SIAM J. Contrl. and Optim.*, 38:503–537, 2000.

[57] J.E. Lagnese and G. Leugering. Domain decomposition in optimal control problems for distributed parameter systems. *Jber. dt. Math. Ver.*, 104(3):91–124, 2002.

[58] J.E. Lagnese and G. Leugering. A posteriori error estimates in time-domain decomposition of final value optimal control of the acoustic wave equation. *Applied Math. and Opt.*, 46:263–290, 2002.

[59] J.E. Lagnese and G. Leugering. Time domain decomposition in final value optimal control of the Maxwell system. *ESAIM: Control, Optimisation and Calculus of Variations*, 8:775–799, 2002.

[60] J.E. Lagnese and G. Leugering. Time-domain decomposition of optimal control problems for the wave equation. *System and Control Letters*, 48:229–242, 2003.

[61] J.E. Lagnese, G. Leugering, and E.J.P.G. Schmidt. *Modeling, analysis and control of dynamic elastic multi-link structures.* Systems and Control: Foundations and Applications. Birkhäuser, Boston, 1994.

[62] I. Lasiecka and R. Triggiani. Regularity theory of hyperbolic equations with nonhomogeneous Neumann boundary conditions, Part II: General boundary data. *J. Diff. Eqs.*, 94:112–164, 1991.

[63] I. Lasiecka and R. Triggiani. *Control Theory for Partial Differential Equations: Continuous and Approximation Theories II*. Cambridge University Press, Cambridge UK, 2000.

[64] G. Leugering. Domain decomposition of optimal control problems for multi-link flexible structures. In *Bock, Hans Georg (ed.) et al., ENUMATH 97. Proceedings of the 2nd European conference on numerical mathematics and advanced applications held in Heidelberg, Germany, September 28-October 3, 1997. World Scientific. 38-53.* 1998.

[65] G. Leugering. Dynamic domain decomposition of optimal control problems for networks of Euler-Bernoulli beams. *ESAIM, Proc.*, 4:223–233, 1998.

[66] G. Leugering. On dynamic domain decomposition of controlled networks of elastic strings and joint-masses. In *Desch, W. (ed.) et al., Control and estimation of distributed parameter systems.*, ISNM 126, pages 191–205. Birkhäuser-Verlag, 1998.

[67] G. Leugering. Dynamic domain decomposition of optimal control problems for networks of strings and Timoshenko beams. *SIAM J. Control Optimization*, 37(6):1649–1675, 1999.

[68] G. Leugering. Domain decomposition of optimal control problems for dynamic networks of elastic strings. *Comput. Optim. Appl.*, 16(1):5–27, 2000.

[69] G. Leugering. On the semi-discretization of optimal control problems for networks of elastic strings: Global optimality systems and domain decomposition. *J. Comput. Appl. Math.*, 120(1-2):133–157, 2000.

[70] G. Leugering and W. Rathmann. On modeling, analyis and simualtion of optimal control problems for dynamic networks of networks of euler-bernoulli and rayleigh beams. In *G. Chen et al. eds.: Control of nonlinear distributed parameter systems*, pages 199–232. Marcel Dekker, New York, 2001.

[71] G. Leugering and E.J.P.G. Schmidt. On the modelling and Stabilization of flows in networks of open channels. *SIAM J. Control and Optimization*, 41(1):164–180, 2002.

[72] J.-L. Lions. *Contrôlabilité Exacte, Perturbations et Stabilisation de Systèmes Distribués: Tome I, Contrôlabilité Exacte*, volume 8 of *Coll. RMA*. Masson, Paris, 1988.

[73] J.-L. Lions. Virtual and effective control for distributed systems and decomposition of everything. *J. d'Anal. Math.*, 80:257–297, 2000.

[74] J.-L. Lions. Decomposition of energy space and virtual control for parabolic equations. In *Domain decomposition methods in sciences and engineering (Chiba, 1999)*, pages 41–53 (electronic). DDM.org, Augsburg, 2001.

[75] J.-L. Lions, Y. Maday, and G. Turinici. Résolution d'EDP par un schéma en temps "pararéel". (A "parareal" in time discretization of PDE's). 2001.

[76] J.-L. Lions and O. Pironneau. Algorithmes parallèles pour la solution de problèmes aux limites. *C.R. Acad. Sci. Paris Ser. I*, 327:947–952, 1998.

[77] J.-L. Lions and O. Pironneau. Sur le contrôle parallèle des systèmes distribués. *C.R. Acad. Sci. Paris*, 327, Ser. I:993–998, 1998.

[78] J.-L. Lions and O. Pironneau. Domain decomposition methods for CAD. *C.R. Acad. Sci. Paris, Séries I*, 328:73–80, 1999.

[79] J.-L. Lions and O. Pironneau. A domain decomposition algorithm. In *Neittaanmäki, Pekka (ed.) et al., ENUMATH 99. Numerical mathematics and advanced applica-*

tions. Proceedings of the 3rd European conference, Jyväskylä, Finland, July 26-30, 1999. Singapore: World Scientific. 42-57 . 2000.

[80] J.-L. Lions and O. Pironneau. Non-overlapping domain decomposition for evolution operators. *C.R. Acad. Sci. Paris Ser. I*, 330:943–950, 2000.

[81] J.-L. Lions and O. Pironneau. Overlapping domain decomposition of evolution operators. *C.R. Acad. Sci. Paris Ser. I*, 330:937–942, 2000.

[82] P.-L. Lions. On the Schwarz alternating method. III: A variant for nonoverlapping subdomains. In *Domain decomposition methods for partial differential equations, Proc. 3rd Int. Symp. Houston/TX (USA) 1989, 202-223* . 1990.

[83] G. Lube, L. Müller, and F. C. Otto. A nonoverlapping domain decomposition method for stabilized finite element approximations of the Oseen equations. *J. Comput. Appl. Math.*, 132(2):211–236, 2001.

[84] Y. Maday and G. Turinici. A parareal in time procedure for the control of partial differential equations. *C. R., Math., Acad. Sci. Paris*, 335(4):387–392, 2002.

[85] S. Miyatake. Mixed problems for hyperbolic equations of second order. *J. Math. Kyoto Univ.*, 13:435–487, 1973.

[86] S. Nicaise. *Polygonal Interface Problems*, volume 39 of *Methoden und Verfahren der Mathematischen Physik*. Peter Lang, Frankfurt, 1993.

[87] S. Nicaise and A.-M. Sändig. General interface problems I. *Math. Meth. Appl. Sci.*, 17:395–429, 1994.

[88] S. Nicaise and A.-M. Sändig. General interface problems II. *Math. Meth. Appl. Sci.*, 17:431–450, 1994.

[89] Z. Opial. Weak convergence of the sequence of successive approximations for nonexpansive mappings. *Bull. Am. Math. Soc.*, 73:591–597, 1967.

[90] F. C. Otto and G. Lube. A posteriori estimates for a non-overlapping domain decomposition method. *Computing*, 62(1):27–43, 1999.

[91] A. Quarteroni and A. Valli. *Domain decomposition methods for partial differential equations*. Numerical Mathematics and Scientific Computation. Oxford: Clarendon Press. xv, 360 p. , 1999.

[92] P.-A. Raviart and J. M. Thomas. *Introduction à l'analyse numérique des équations aux dérivées partielles*. Collection Mathématiques Appliquées pour la Maîtrise. Paris etc.: Masson. 224 p. , 1983.

[93] E. Reissner. On bending of elastic plates. *Quart. Appl. Math.*, V:55–68, 1947.

[94] H. Schaefer. Über die Methode sukzessiver Approximationen. *Jber. Deutsch. Math.-Verein*, 59:131–140, 1957.

[95] H. Schwarz. Über einige Abbildungsaufgaben. 1869.

[96] J. Simon. Compact sets in $L^p(0,T;B)$. *Ann. Mat. Pura et Appl.*, CXLVI:65–96, 1987.

[97] D. Tataru. On the regularity of boundary traces for the wave equation. *Ann. Scuola Norm. Sup. Pisa, Cl. Sci. (4)*, 26:185–206, 1998.

[98] J. von Below. Sturm-Liouville eigenvalue problems on networks. *Math. Methods Appl. Sci.*, 10(4):383–395, 1988.

Index

Free Boundary Problems

MIX
Papier aus verantwortungsvollen Quellen
Paper from responsible sources
FSC® C105338

If you have any concerns about our products,
you can contact us on
ProductSafety@springernature.com

In case Publisher is established outside the EU,
the EU authorized representative is:
Springer Nature Customer Service Center GmbH
Europaplatz 3, 69115 Heidelberg, Germany

Printed by Libri Plureos GmbH
in Hamburg, Germany